Spon's Civil Engineering and Highway Works Price Book

2013

Spon's Civil Engineering and Highway Works Price Book

Edited by

Davis Langdon, An AECOM Company

2013

Twenty-seventh edition

Spon Press
an imprint of Taylor & Francis
LONDON AND NEW YORK

First edition 1984
Twenty-seventh edition published 2013
by Spon Press
2 Park Square, Milton Park, Abingdon, Oxon, OX14 4RN

Simultaneously published in the USA and Canada
by Spon Press
711 Third Avenue, New York, NY 10017

Spon Press is an imprint of the Taylor & Francis Group, an informa business

© 2013 Spon Press

The right of Davis Langdon to be identified as the Author of this Work has been asserted by them in accordance with the Copyright, Designs and Patents Act 1988

All rights reserved. No part of this book may be reprinted or reproduced or utilized in any form or by any electronic, mechanical, or other means, now known or hereafter invented, including photocopying and recording, or in any information storage or retrieval system, without permission in writing from the publishers.

The publisher makes no representation, express or implied, with regard to the accuracy of the information contained in this book and cannot accept legal responsibility or liability for any errors or omissions that may be made.

British Library Cataloguing in Publication Data
A catalogue record for this book is available from the British Library

ISBN 13: 978-0-415-69078-2 (hardback)
ISBN 13: 978-0-203-07707-8 (ebook)

ISSN: 0957-171X

Typeset in Arial by Taylor & Francis Books

Printed and bound in Great Britain by
TJ International Ltd, Padstow, Cornwall

Contents

Preface to the twenty-seventh edition	ix
Abbreviations	xi
Acknowledgements	xiii

PART 1: GENERAL

Purpose and content of the book	3
Outline of the tendering and estimating process	5
The aggregates levy	13
Capital allowances	19
Value added tax	27
Landfill tax	35
Land remediation	37

PART 2: ON COSTS AND PROFIT

Risk/opportunity	49
Head office overheads	49
Profit	49
Finance costs – assessment of contract investment	49
Value added tax	51
Tender summary	51
Preliminaries and general items	53

PART 3: COST AND TENDER PRICES INDICES

Introduction	71
Price adjustment formula indices	71
A constructed cost index based on the price adjustment formula indices	73
The road construction tender price index	74

PART 4: RESOURCES

Basis of this section	77
Labour costs	80
Plant costs	85

PART 5: APPROXIMATE ESTIMATING

Building prices per square metre	93	Underpasses	100
Approximate estimating rates	95	Landscaping	100
Earth retention and stabilization	95	Temporary works	101
Bridgework	96	Approximate estimating rates – General building	
Tunnelling field costs	97	works	104
Highway works	97	Substructure	104

Frame and upper floors	108	Space heating and air conditioning and ventilation	130
Roof	110	Electrical installations	130
Stairs and ramps	116	Lift and conveyor installations	131
External walls	117	Fire and lighning protection	131
Windows and external doors	120	Building work in connection with services	132
Internal walls, partitions and doors	121	Roads, Paths and surfaces	132
Internal doors	123	Soft landscaping, planting and irrigation systems	133
Wall finishes	124	Fencing, railings and walls	133
Floor finishes	125	External fixtures	134
Ceiling finishes	128	External drainage	134
Fittings and furnishings	128	External services	135
Sanitary installations	129		

PART 6: UNIT COSTS – CIVIL ENGINEERING WORKS

Introductory notes	137	Class M: Structural metalwork	242
Guidance notes	139	Class N: Miscellaneous metalwork	244
Class A: General items	141	Class O: Timber	249
Class B: Ground investigation	144	Class P: Piling	256
Class C: Geotechnical and other specialist processes	148	Class Q: Piling ancillaries	269
Class D: Demolition and site clearance	155	Class R: Roads and pavings	276
Class E: Excavation	157	Class S: Rail track	293
Class F: In situ concrete	179	Class T: Tunnels	300
Class G: Concrete ancillaries	185	Class U: Brickwork, blockwork and masonry	304
Class H: Precast concrete	198	Class V: Painting	318
Class I: Pipework - pipes	200	Class W: Waterproofing	326
Class J: Pipework – fittings and valves	222	Class X: Miscellaneous work	331
Class K: Pipework – manholes and pipework	223	Class Y: Sewer and water main renovation and ancillary works	336
Class L: Pipework accessories ancillaries supports and pipework	237	Class Z: Simple building works incidental to civil engineering works	338

PART 7: UNIT COSTS – HIGHWAY WORKS

Introductory notes	355	Series 1600: Piling and embedded retaining walls	450
Guidance notes	357	Series 1700: Structural concrete	468
Series 100: Preliminaries	361	Series 1800: Steelwork for structures	477
Series 200: Site clearance	363	Series 1900: Protection of steelwork against corrosion	478
Series 300: Fencing	366	Series 2000: Waterproofing for structures	480
Series 400: Road restraint systems (vehicle and pedestrian)	370	Series 2100: Bridge bearings	482
Series 500: Drainage and service ducts	377	Series 2300: Bridge expansion joints and sealing of gaps	484
Series 600: Earthworks	401	Series 2400: Brickwork, blockwork and stonework	486
Series 700: Pavements	421	Series 2500: Special structures	493
Series 1100: Kerbs, footways & paved areas	428	Series 2700: Accommodation works, works for statutory undertakers, provisional sums and prime cost items	494
Series 1200: Traffic signs and road markings	436	Series 3000: Landscape and ecology	495
Series 1300: Road lighting columns and brackets, CCTV masts and cantilever masts	442	Series 5000: Maintenance painting of steelwork	500
Series 1400: Electrical work for road lighting and traffic signs	445		
Series 1500: Motorway communications	448		

PART 8: DAYWORK

Introduction	503
Schedules of dayworks carried out incidental to contract work	503
Application of dayworks	506

PART 9: PROFESSIONAL FEES

Consulting engineers' fees	513
Quantity surveyors' fees	518

PART 10: OUTPUTS

Disposal of excavated materials	523
Breaking out pavements, brickwork, concrete and masonry by hand and pneumatic breaker	524
Loading loose materials and items by hand	524
Placing ready mixed concrete in the works	525
Fixing bar reinforcement	526
Erecting formwork to beams and walls	527
Erecting formwork to slabs	527
Multipliers for formwork	527
Laying and jointing flexible-jointed clayware pipes	528
Precast concrete manholes in sections	528

PART 11: USEFUL ADDRESSES FOR FURTHER INFORMATION — 529

PART 12: TABLES AND MEMORANDA — 537

INDEX — 631

SPON'S PRICEBOOKS 2013

Spon's Architects' and Builders' Price Book 2013
DAVIS LANGDON

The most detailed, professionally relevant source of UK construction price information currently available anywhere.

What's new in this year's Spon's A&B Price Book? There's a new cost model for an office fit out. And new measured works: Concrete canvas, Dreadnought Handformed roof tiles, UPVC external doors, Thermalite paint grade concrete blocks, EPS/PIR hybrid roof insulation, Omnidec composite flooring systems, sheep wool insulation, Kingspan TEK cladding panels, 750 & 900 dia CFA piling, Lignacite architectural blocks. The book also provides more on cork flooring. Plus lots else has been changed throughout.

Hbk & Spon's Online
800pp approx.: 978-0-415-69077-5: **£150**

ebook & Spon's Online
978-0-203-07708-5: **£150**
(inc. sales tax where appropriate)

Spon's Mechanical and Electrical Services Price Book 2013
DAVIS LANGDON ENGINEERING SERVICES

Our M&E price book continues to be the most comprehensive and best annual services engineering price book currently available, providing detailed pricing information across the full range of mechanical and electrical services, together with higher-level costs for a diverse range of systems and different building applications. This year sees very slight increases in material and labour rates. The book keeps its focus on CO2 control and renewables. Feed-In Tariffs have been brought up to date with the current rates and processes.

Hbk & Spon's Online
864pp approx.: 978-0-415-69080-5: **£150**

ebook & Spon's Online
978-0-203-07705-4: **£150**
(inc. sales tax where appropriate)

Spon's External Works and Landscape Price Book 2013
DAVIS LANGDON

Now in its 32nd edition, the External Works and Landscape Price Book provides a special focus on water related issues. It remains the indispensable source of UK cost information for detailed external works and landscaping. This year brings rainwater harvesting tanks; Cellweb load support systems and erosion control; outdoor rooms – with several design options of home-offices and garden studios; green roofs – updated systems of green roof treatments and installations; and permeable pavings – permeable macadams, bricks and more.

Hbk & Spon's Online
720pp approx.: 978-0-415-69079-9: **£125**

ebook & Spon's Online
978-0-203-07706-1: **£125**
(inc. sales tax where appropriate)

Spon's Civil Engineering and Highway Works Price Book 2013
DAVIS LANGDON

With pressure through the National Infrastructure Plan to deliver "value for money", accurate pricing counts. This new edition of the Civils and Highways Price Book reflects the moderate overall price rises of the last year, and the higher than average price increases of anything affected by crude oil costs. This is more than just a price book. It provides a comprehensive work manual for the UK's civil engineering, surveying and construction business.

Hbk & Spon's Online
752pp approx.: 978-0-415-69078-2: **£160**

ebook & Spon's Online
978-0-203-07707-8: **£160**
(inc. sales tax where appropriate)

Receive *Spon's Online* free when you order any hard copy or ebook Spon 2013 Price Book. Visit **www.pricebooks.co.uk**

To Order: Tel: +44 (0) 1235 400524 **Fax:** +44 (0) 1235 400525
or Post: Taylor and Francis Customer Services, Bookpoint Ltd, Unit T1, 200 Milton Park, Abingdon, Oxon, OX14 4SB, UK
Email: book.orders@tandf.co.uk
A complete listing of all our books is on: **www.sponpress.com**

Spon Press
an imprint of Taylor & Francis

Preface to the twenty-seventh edition

At the time of writing the construction industry continues to experience a marked downturn in both the Private and Public Sectors. Raw materials, supplier and manufacturer prices continue to harden yet tender prices continue at the best to flat-line. This plus the effect of the continuing current tight credit market price levels for the coming year again continue to be extremely difficult to forecast. Infrastructure works associated with the London 2012 Olympics tail off and have yet to be replaced in volume by other significant infrastructure projects such as London Crossrail. Out of London infrastructure work levels continue at a very low level. There is much publicity about new infrastructure programmes from government departments but these are still in the early planning stages and will not become physical work for many years yet.

The drop in demand for steel products in 2009/10 has flattened in 2010/2011 and there is evidence of hardening of prices. Oil prices have softened since the second quarter of 2011. Petroleum and diesel prices have dropped slightly to reflect this but are not yet back to their pre 2010 levels and are dependent on demand from the Near and Far East. Oil prices coupled with low worldwide demand continue to impact on manufacturing, energy and transportation costs and the effect is to make it extremely difficult to predict price levels for 2012 / 2013. The continuing economic turmoil in the Euro Zone is an added uncertainty to future construction activity in the region and elsewhere.

Under these market conditions the prices in the various Parts of this publication can only be taken as a guide to actual costs. Any sustained upturn may give rise to marked increases in selling prices as manufacturers and specialists seek to recover margins lost in the downturn.

For the 2013 edition, we have carried out a general review of all prices up to April/May 2012 in consultation with leading manufacturers, suppliers and specialist contractors and included revisions as necessary. Our efforts have been directed at reviewing, revising and consolidating the scope, range and detail of information to help the reader to compare or adjust any unit costs with reference to allocated resources or outputs.

The rates, prices and outputs included in the Resources and Unit Cost calculations, including allowances for wastage, normal productivity and efficiency, are based on medium sized Civil Engineering schemes of about £8–£10 million in value, with no acute access or ground condition problems. However, they are equally applicable, with little or no adjustment, to a wide range of construction projects from £2 million to £50 million. Where suitable, tables of multipliers have been given to enable easy adjustment to outputs or costs for varying work conditions.

As with all attempts to provide price guidance on a general basis, this must be loaded with caveats. In applying the rates to any specific project, the user must take into account the general nature of the project, i.e. matters such as scale, site difficulties, locale, tender climate etc. This book aims at providing as much information as possible about the nature of the rate so as to assist the user to adapt it if necessary.

This edition continues to provide the reader with cost guidance at a number of levels, varying from the more general functional costs shown in Part 5, through the detailed unit costs in Parts 6 and 7 which relate respectively to the CESMM3 and the Highways Method of Measurement bills of quantities formats, down to the detailed resource costings given in Part 4 supplemented by the further advice on output factors in Part 10.

The outputs used in this work have been compiled in detail from the editors' wide ranging experience and are based almost exclusively on studies and records derived from a large number of recent Civil Engineering schemes. This information is constantly being re-appraised to ensure consistency with current practice. A number of prices and outputs are based upon detailed specialist advice and acknowledgements to the main contributors are included within the book.

The current market is unlikely to change very rapidly, however to monitor this and maintain accuracy levels readers should use the free price book update service, which advises of any significant changes to the published information over the period covered by this edition. The Update is posted free every three months on the publishers' website, until the publication of the next annual Price Book, to those readers who have registered with the publisher. Details of how to register can be found at the end of the book.

Whilst all efforts are made to ensure the accuracy of the data and information used in the book, neither the editors nor the publishers can in any way accept liability for loss of any kind resulting from the use made by any person of such information.

Davis Langdon, an AECOM Company
MidCity Place
71 High Holborn
London
WC1V 6QS

Abbreviations

BS	British Standard	μu	micron (10^{-3} millimetre)
BSS	British Standard Specification	mm	millimetre
DERV	diesel engine road vehicle	mm²	square millimetre
DfT	Department for Transport	N	newton
Ha	hectare	ne	not exceeding
Hp	horsepower	nr	number
Hr	hour	pa	per annum
In	inch	PC	Prime Cost
kg	kilogramme	sq	square
km	kilometre	t	tonne
kVA	kilovolt ampere	TSO	The Stationery Office
kW	kilowatt	wk	week
m	metre	yd	yard
m²	square metre	yd³	cubic yard
m³	cubic metre		

ESSENTIAL READING FROM TAYLOR AND FRANCIS

Spon's Asia-Pacific Construction Costs Handbook

4th Edition

Edited by **Davis Langdon & Seah**

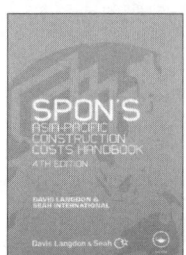

Spon's Asia Pacific Construction Costs Handbook includes construction cost data for 20 countries. This new edition has been extended to include Pakistan and Cambodia. Australia, UK and America are also included, to facilitate comparison with construction costs elsewhere. Information is presented for each country in the same way, as follows:

- key data on the main economic and construction indicators.

- an outline of the national construction industry, covering structure, tendering and contract procedures, materials cost data, regulations and standards

- labour and materials cost data

- Measured rates for a range of standard construction work items

- Approximate estimating costs per unit area for a range of building types

- price index data and exchange rate movements against £ sterling, $US and Japanese Yen

The book also includes a Comparative Data section to facilitate country-to-country comparisons. Figures from the national sections are grouped in tables according to national indicators, construction output, input costs and costs per square metre for factories, offices, warehouses, hospitals, schools, theatres, sports halls, hotels and housing.

This unique handbook will be an essential reference for all construction professionals involved in work outside their own country and for all developers or multinational companies assessing comparative development costs.

April 2010: 234x156: 480pp
Hb: 978-0-415-46565-6: **£120.00**

To Order: Tel: +44 (0) 1235 400524 **Fax:** +44 (0) 1235 400525
or Post: Taylor and Francis Customer Services,
Bookpoint Ltd, Unit T1, 200 Milton Park, Abingdon, Oxon, OX14 4TA UK
Email: book.orders@tandf.co.uk

**For a complete listing of all our titles visit:
www.tandf.co.uk**

Acknowledgements

The Editors wish to record their appreciation of the assistance given by many individuals and organizations in the compilation of this edition.

Materials suppliers and subcontractors who have contributed this year include:

Abacus Lighting Ltd
Oddicroft Lane
Sutton-in-Ashfield
Notts NG17 5FT
Lighting and street furniture
Tel: 01623 511111
Fax: 01623 552133
Website: www.abacuslighting.com
Email: sales@abacuslighting.com

ACO Technologies Plc
Hitchin Road
Shefford
Beds SG17 5TE
Drainage systems
Tel: 01462 816666
Fax: 01462 815895
Website: www.aco.co.uk

Aggregate Industries
Hulland Ward
Ashbourne
Derby DE6 3ET
Kerbs, edgings and pavings
Tel: 01335 372222
Fax: 01335 370074
Website: www.aggregate.com

Anaco Trading Ltd
P.O. Box 39
Oundle
Peterborough PE8 4JT
Reinforcing Bar
Tel: 01832 272109
Fax: 01832 275759
Website: www.anacotrading.com
Email: anacotrading@btconnect.com

Arcelor Commercial RPS UK Ltd
Arcelor House
4 Princes Way
Solihull
West Midlands B91 3 AL
Steel sheet piling
Tel: 0870 770 8057
Fax: 0870 770 8058
Website: www.sheet-piling.arcelor.com
Email: sheet-piling@arcelor.com

Arthur Fischer (UK) Ltd
Hithercroft Road
Wallingford
Oxon OX10 9 AT
Fixings
Tel: 01491 833000
Fax: 01491 827953
Website: www.fischer.co.uk
Email: sales@fischer.co.uk

Asset International Ltd
Stephenson Street
Newport
South Wales NP19 0XH
Steel culverts
Tel: 01633 271906
Fax: 01633 290519
Website: www.assetint.co.uk

Avon Manufacturing (Warwick) Ltd
Avon House
Kineton Road
Southam
Warwickshire CV47 2DG
Fixings
Tel: 01926 817292
Fax: 01926 814156
Website: www.avonmanufacturing.co.uk/

Bachy Soletanche Ltd
Foundation Court
Godalming Business Centre
Catteshall Lane
Godalming
Surrey GU7 1XW
Diaphragm walling
Tel: 01483 427311
Fax: 01483 417021
Website: www.bacsol.co.uk

BRC Special Products
Astonfield Industrial Estate
Stafford ST16 3BP
Masonry reinforcement and accessories
Tel: 01785 222288
Fax: 01785 240029
Website: www.brc-special-products.co.uk

Capital Demolition
Capital House
Woodham Park Road
Addlestone
Surrey KT15 3TG
Demolition and clearance
Tel: 01932 346222
Fax: 01932 340244
Website: www.capitaldemolition.co.uk

Caradon Parker Ltd
Units 1-2 Longwall Avenue
Queen's Drive Industrial Estate
Nottingham NG2 1NA
Geotextiles
Tel: 0115 986 2121

Castle Cement
Park Square
3160 Solihull Parkway
Birmingham Business Park
Birmingham B37 7YN
Cement products
Tel: 0121 606 4000
Fax: 0121 606 1412
Website: www.castlecement.co.uk
Email: information@castlecement.co.uk

CCL Stressing Systems
Unit 4
Park 2000
Millennium Way
Westland Road
Leeds LS11 5 AL
Bridge bearings
Tel: 0113 270 1221
Fax: 0113 271 4184
Website: www.cclstressing.com
Email: sales@cclstressing.com

Celsa Steel (UK) Ltd
Building 58
East Moors Road
Cardiff CF24 5NN
Bar reinforcement
Tel: 0292 035 1800
Fax: 0292 035 1801
Website: www.celsauk.com

Cementation Foundations Skanska Ltd
Maple Cross House
Denham Way
Rickmansworth
Herts WD3 9 AS
Piling
Tel: 01923 423 100
Fax: 01923 777 834
Website: www.cementationfoundationsskanska.co.uk
Email: cementation.foundations@skanska.co.uk

Civils and Lintels
Stock Lane
Chadderton
Oldham OL9 9EY
Drainage Materials
Tel: 01616 270888
Fax: 01616271888
Website: www.civilsandlintels.co.uk
Email: oldham@civilsandlintels.co.uk

Cogne UK Ltd
Uniformity Steel Works
19 Don Road
Newhall
Sheffield S9 2UD
Steel works
Tel: 01142212020
Fax: 01142213030

CPM Group
Mells Road
Nr Frome, Somerset BA11 3PD
Drainage materials
Tel: 01179 812791
Fax: 01179 814511
Website: www.cpm-group.com
Email: sales@cpm-group.com

Craigton Industries
Craigton Works
Milngavie
Glasgow G62 7HF
Precast concrete manholes etc.
Tel: 0141 956 6585
Fax: 0141 956 3757

Don & Low Ltd
Newford Park House
Glamis Road
Forfar
Angus DD8 1FR
Geotextiles
Tel: 01307 452200
Fax: 01307 452422
Website: www.donlow.co.uk

Dorsey Construction Materials Ltd
Unit 11 Nimrod Industrial Estate
Elgar Road
Reading
Berks RG2 0EB
Water stops
Tel: 0118 975 3377
Fax: 0118 975 3393

Ecochoice Ltd
External Solutions in Certified Timber
Trinity House
Cowley Road
Cambridge CB4 0WZ
Timber piles
Tel: 0845 638 1340
Fax: 0870 642 1340
Website: www.ecochoice.co.uk
Email: info@ecochoice.co.uk

Euromix Concrete Ltd
Chelmsford
Unit 1 Boreham Industrial Estate
Waltham Road
Chelmsford, CM3 3 AW
Ready mix concrete
Tel: 01245 464545
Fax: 01245 451507

Expamet Building Products
Longhill Industrial Estate (North)
Greatham Street
Hartlepool TS25 1PU
General building materials
Tel: 01429 866688
Fax: 014129 866633
Website: www.expamet.co.uk
Email: sales@expamet.net

Fosroc Ltd
Coleshill Road
Tamworth
Staffordshire B78 3TL
Waterproofing and expansion joints
Tel: 01827 262222
Fax: 01827 262444
Website: www.fosrocuk.com
Email: sales@fosrocuk.com

Freyssinet Ltd
Hollinswood Court
Stafford Park
Telford
Shropshire TF3 3DE
Reinforced earth systems
Tel: 01952 201901
Fax: 01952 201753
Website: www.freyssinet.co.uk

Grace Construction Products Ltd
Ajax Avenue
Slough
Berks SL1 4BH
Waterproofing/expansion joints
Tel: 01753 692929
Fax: 01753 691623
Website: www.graceconstruction.com

Greenham Trading Ltd
Greenham House
671 London Road
Isleworth
Middx TW7 4EX
Contractors site equipment
Tel: 020 8560 1244
Fax: 020 8568 8423
Website: www.greenham.com
Email: isleworth.sales@greenham.com

Griffiths Sign & Equipment
Unit 2, Wern Industrial Estate
Rogerstone
Newport
South Wales NP10 9FQ
Traffic signs
Tel: 01633 895566
Fax: 01633 895577
Website: www.griffiths-signs.co.uk

Hanson Building Products
Stewartby
Bedford
Beds MK43 9LZ
Bricks
Tel: 08705 258258
Fax: 01234 762040
Email: info@hansonbrick.com

Hanson Concrete Products
PO Box 14
Appleford Road
Sutton Courtney
Abingdon
Oxfordshire OX14 4UB
Blockwork and precast concrete products
Tel: 01235 848808
Fax: 01235 846613
Website: www.hanson-europe.com

Hepworth Building Products Ltd
Hazlehead
Crow Edge
Sheffield S36 4HG
Clayware pipes
Tel: 01226 763561
Fax: 01226 764827
Website: www.hepworthdrainage.co.uk
Email: info@hepworthdrainage.co.uk

Hughes Concrete Ltd
Barnfields
Leek
Staffs ST13 5QG
Precast concrete manhole units
Tel: 01538 380500
Fax: 01538 380510
Website: www.hughesconcrete.co.uk

Iko Roofing Specification Division
Permanite Asphalt
IKO Plc
Appley Lane North
Appley Bridge
Wigan
Lancashire WN6 9 AB
Asphalt products
Tel: 0844 412 7224
Fax: 0844 412 7205
Website: www.kogroup.co.uk

Inform UK Ltd
Industrial Park
Ely Road
Waterbeach
Cambridge CB5 9PG
Formwork and accessories
Tel: 01223 862230
Fax: 01223 440246
Website: www.inform.co.uk
Email: general@inform.co.uk

Jacksons Fencing Ltd
Stowting Common
Ashford
Kent TN25 6BN
Fencing
Tel: 01233 750393
Fax: 01233 750403
Website: www.jackson-fencing.co.uk
Email: sales@jacksons-fencing.co.uk

James Latham Plc
Unit 3
Finway Road
Hemel Hempstead HP2 7QU
Timber merchants
Tel: 01442 849100
Fax: 01442 267241
Website: www.lathamtimber.co.uk
Email: plc@lathams.co.uk

K B Rebar Ltd
Unit 5, Dobson Park Industrial Estate
Dobson
Park Way, Ince
Wigan WN2 2DY
Steel reinforcement
Tel: 01617 908635
Website: www.kbrebar.co.uk

Maccaferri Ltd
The Quorum
Oxford Business Park
Nr Garsington Road
Oxford OX4 2JZ
River and sea gabions
Tel: 01865 770555
Fax: 01865 774550
Website: www.maccaferri.co.uk
Email: oxford@maccaferri.co.uk

Marley Plumbing & Drainage Ltd
Dickley Lane
Lenham
Maidstone
Kent ME17 2DE
uPVC drainage and rainwater systems
Tel: 01622 858888
Fax: 01622 858725
Website: www.marley.co.uk

Marshalls Mono Ltd
Southowram
Halifax
West Yorkshire HX3 9SY
Kerbs, edgings and pavings
Tel: 01422 312000
Fax: 01422 330185
Website: www.marshalls.co.uk

M & T Pipe Ltd
Unit 6, Queenslie Point,
Glagow G33 3NQ
Pipe products
Tel: 0141 774 8300
Fax: 0141 774 7300
Website: www.mtpipe.co.uk
Email: info@mtpipr.co.uk

Milton Pipes Limited
Milton Regis
Sittingbourne
Kent ME10 2QF
Precast cconcrete solutions
Tel: 01795 425191
Fax: 01795 478232
Website: www.miltonpipes.co.uk

Morelock Signs
Morelock House
Strawberry Lane
Willenhall
Wolverhampton WV13 3RS
Traffic signs
Tel: 01902 637575
Fax: 01902 637576
Website: morelock.co.uk

Naylor Industries Ltd
Clough Green
Cawthorne
Barnsley
South Yorkshire S75 4 AD
Clayware pipes and fittings
Tel: 01226 790591
Fax: 01226 790531
Website: www.naylor.co.uk
Email: sales@naylor.co.uk

PHI Group Ltd
Harcourt House, Royal Crescent
Cheltenham
Glos GL50 3DA
Retaining walls, gabions
Tel: 0870 3334120
Fax: 0870 3334121
Website: www.phigroup.co.uk
Email: info@phigroup.co.uk

Platipus Anchors Ltd
Kingsfield Business Centre
Philanthropic Road
Redhill
Surrey RH1 4DP
Earth anchors
Tel: 01737 762300
Fax: 01737 773395
Website: www.platipus-anchors.com
Email: info@platipus-anchors.com

Redman Fisher
Halesfield 18
Telford
Shropshire TF7 4JS
Steel products
Tel: 01902 880880
Fax: 01902 880446
Website: www.redmanfisher.co.uk
Email: IanFrench@redmanfisher.co.uk

Roger Bullivant Ltd
Innovation House
7 Kingdom Avenue
North acre Industrial Park
Westbury
Wiltshire BA13 4FG
Pilling and foundations
Tel: 01373 865012
Fax: 0 1373 865559
Website: www.roger-bullivant.co.uk

ROM Ltd
710 Rightside Lane
Sheffield
South Yorkshire S9 2BR
Formwork and accessories
Tel: 0114 231 7900
Fax: 0114 231 7905
Website: www.rom.co.uk
E-mail: sales@rom.co.uk

ROM (NE) Ltd
Unit 10
Blaydon Industrial Estate
Chainbridge Road
Blaydon-on-Tyne
Northumberland NE21 5 AB
Reinforcement
Tel: 0191 414 9600
Fax: 0191 414 9650
Website: www.rom.co.uk

Saint-Gobain Pipelines Plc
Lows Lane
Stanton-By-Dale
Derbyshire DE7 4QU
**Cast iron pipes and fittings,
polymer concrete channels,
manhole covers and street furniture**
Tel: 0115 9305000
Fax: 0115 9329513
Website. www.saint-gobain-pipelines.co.uk

Stent
Pavilion C2
Ashwood Park
Ashwood Way
Basingstoke RG23 8BG
Piling and foundations
Tel: 01256 763161
Fax: 01256 768614
Website: www.stent.co.uk

Stocksigns Ltd
Ormside Way
Redhill
Surrey RH1 2LG
Road signs and posts
Tel: 01737 764764
Fax: 01737 763763
Website: www.stocksigns.co.uk

Structural Soil Ltd
The Old School
Fell Bank
Bristol BS3 4EB.
Ground Investigation
Tel: 0117-947-1000
Fax: 0117-947-1004,
Email: admin@soils.co.uk
Website: www.soils.co.uk

Tarmac Precast Concrete Ltd
Tallington
Barhom Road
Stamford
Lincs PE9 4RL
Prestressed concrete beams
Tel: 01778 381000
Fax: 01778 348041
Website: www.tarmacprecast.com
Email: tall@tarmac.co.uk

Tarmac Topblock Ltd
Airfield Industrial Estate
Ford
Arundel
West Sussex BN18 0HY
Blockwork
Tel: 01903 723333
Fax: 01903 711043
Website: www.topblock.co.uk

Tata Steel Ltd
P.O. Box 101
Weldon Road
Corby
Northants NN17 5UA
Steel products
Tel: 01536 402121
Website: www.tatasteeleurpe.com
Email: customer-service@tatasteel.com

Tensar International
Cunningham Court
Shadsworth Business Park
Blackburn BB1 2QX
Soil reinforcement
Tel: 01254 262431
Fax: 01254 26686
www.tensar.co.uk

Terry's Timber Ltd
35 Regent Road
Liverpool L5 9SR
Timber
Tel: 0151 207 4444
Fax: 0151 298 1443
Website: www.terrystimber.co.uk

Travis Perkins Trading Co. Ltd
Baltic Wharf
Boyn Valley Road
Maidenhead
Berkshire SL6 4EE
Builders merchants
Tel: 01628 770577
Fax: 01628 625919
Website: www.travisperkins.co.uk

Varley and Gulliver Ltd
Alfred Street
Sparkbrook
Birmingham B12 8JR
Bridge parapets
Tel: 0121 773 2441
Fax: 0121 766 6875
Website: www.v-and-g.co.uk
Email: sales@v-and-g.co.uk

Vibro Projects Ltd
Unit D3
Red Scar Business Park
Longridge Road
Preston PR2 5NQ
Ground consolidation – vibroflotation
Tel: 01772 703113
Fax: 01772 703114
Website: www.vibro.co.uk
Email: sales@vibro.co.uk

Wavin Plastics Ltd
Parsonage Way
Chippenham
Wilts SN15 5PN
uPVC drain pipes and fittings
Tel: 01249 654121
Fax: 01249 443286
Website: www.wavin.co.uk

Wells Spiral Tubes Ltd
Prospect Works
Airedale Road
Keighley
West Yorkshire BD21 4LW
Steel culverts
Tel: 01535 664231
Fax: 01535 664235
Website: www.wells-spiral.co.uk
Email: sales@wells-spiral.co.uk

Winn & Coales (Denso) Ltd
Denso House
33–35 Chapel Road
West Norwood
London SE27 0TR
Anti-corrosion and sealing products
Tel: 0208 670 7511
Fax: 0208 761 2456
Website: www.denso.net
Email: mail@denso.net

Davis Langdon,
An AECOM Company

Program, Cost, Consultancy
www.davislangdon.com
www.aecom.com

A leading light in our industry

SPON'S PRICEBOOKS 2013

Spon's Architects' and Builders' Price Book 2013
DAVIS LANGDON

The most detailed, professionally relevant source of UK construction price information currently available anywhere.

What's new in this year's Spon's A&B Price Book? There's a new cost model for an office fit out. And new measured works: Concrete canvas, Dreadnought Handformed roof tiles, UPVC external doors, Thermalite paint grade concrete blocks, EPS/PIR hybrid roof insulation, Omnidec composite flooring systems, sheep wool insulation, Kingspan TEK cladding panels, 750 & 900 dia CFA piling, Lignacite architectural blocks. The book also provides more on cork flooring. Plus lots else has been changed throughout.

Hbk & Spon's Online
800pp approx.: 978-0-415-69077-5: **£150**

ebook & Spon's Online
978-0-203-07708-5: **£150**
(inc. sales tax where appropriate)

Spon's Mechanical and Electrical Services Price Book 2013
DAVIS LANGDON ENGINEERING SERVICES

Our M&E price book continues to be the most comprehensive and best annual services engineering price book currently available, providing detailed pricing information across the full range of mechanical and electrical services, together with higher-level costs for a diverse range of systems and different building applications. This year sees very slight increases in material and labour rates. The book keeps its focus on CO2 control and renewables. Feed-In Tariffs have been brought up to date with the current rates and processes.

Hbk & Spon's Online
864pp approx.: 978-0-415-69080-5: **£150**

ebook & Spon's Online
978-0-203-07705-4: **£150**
(inc. sales tax where appropriate)

Spon's External Works and Landscape Price Book 2013
DAVIS LANGDON

Now in its 32nd edition, the External Works and Landscape Price Book provides a special focus on water related issues. It remains the indispensable source of UK cost information for detailed external works and landscaping. This year brings rainwater harvesting tanks; Cellweb load support systems and erosion control; outdoor rooms – with several design options of home-offices and garden studios; green roofs – updated systems of green roof treatments and installations; and permeable pavings – permeable macadams, bricks and more.

Hbk & Spon's Online
720pp approx.: 978-0-415-69079-9: **£125**

ebook & Spon's Online
978-0-203-07706-1: **£125**
(inc. sales tax where appropriate)

Spon's Civil Engineering and Highway Works Price Book 2013
DAVIS LANGDON

With pressure through the National Infrastructure Plan to deliver "value for money", accurate pricing counts. This new edition of the Civils and Highways Price Book reflects the moderate overall price rises of the last year, and the higher than average price increases of anything affected by crude oil costs. This is more than just a price book. It provides a comprehensive work manual for the UK's civil engineering, surveying and construction business.

Hbk & Spon's Online
752pp approx.: 978-0-415-69078-2: **£160**

ebook & Spon's Online
978-0-203-07707-8: **£160**
(inc. sales tax where appropriate)

Receive *Spon's Online* free when you order any hard copy or ebook Spon 2013 Price Book.
Visit **www.pricebooks.co.uk**

To Order: Tel: +44 (0) 1235 400524 **Fax:** +44 (0) 1235 400525
or Post: Taylor and Francis Customer Services, Bookpoint Ltd, Unit T1, 200 Milton Park, Abingdon, Oxon, OX14 4SB, UK
Email: book.orders@tandf.co.uk
A complete listing of all our books is on: **www.sponpress.com**

Spon Press
an imprint of Taylor & Francis

PART 1

General

	page
Purpose and content of the book	3
Outline of the tendering and estimating process	5
The Aggregates Levy	13
Capital Allowances	19
Value Added Tax	27
The Landfill Tax	35
Land Remediation	37

ESSENTIAL READING FROM TAYLOR AND FRANCIS

Spon's Asia-Pacific Construction Costs Handbook

4th Edition

Edited by **Davis Langdon & Seah**

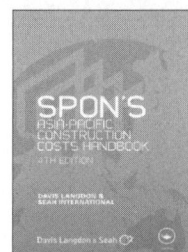

Spon's Asia Pacific Construction Costs Handbook includes construction cost data for 20 countries. This new edition has been extended to include Pakistan and Cambodia. Australia, UK and America are also included, to facilitate comparison with construction costs elsewhere. Information is presented for each country in the same way, as follows:

• key data on the main economic and construction indicators.

• an outline of the national construction industry, covering structure, tendering and contract procedures, materials cost data, regulations and standards

• labour and materials cost data

• Measured rates for a range of standard construction work items

• Approximate estimating costs per unit area for a range of building types

• price index data and exchange rate movements against £ sterling, $US and Japanese Yen

The book also includes a Comparative Data section to facilitate country-to-country comparisons. Figures from the national sections are grouped in tables according to national indicators, construction output, input costs and costs per square metre for factories, offices, warehouses, hospitals, schools, theatres, sports halls, hotels and housing.

This unique handbook will be an essential reference for all construction professionals involved in work outside their own country and for all developers or multinational companies assessing comparative development costs.

April 2010: 234x156: 480pp
Hb: 978-0-415-46565-6: **£120.00**

To Order: Tel: +44 (0) 1235 400524 **Fax:** +44 (0) 1235 400525
or Post: Taylor and Francis Customer Services,
Bookpoint Ltd, Unit T1, 200 Milton Park, Abingdon, Oxon, OX14 4TA UK
Email: book.orders@tandf.co.uk

For a complete listing of all our titles visit:
www.tandf.co.uk

PURPOSE AND CONTENT OF THE BOOK

For many years the Editors have compiled a price book for use in the building industry with, more recently, companion volumes for use in connection with engineering services contracts and landscaping work. All of these price books take their reliability from the established practice within these sectors of the construction industry of pricing work by the application of unit rates to quantities measured from the designer's drawings. This practice is valid because most building work can be carried out under similar circumstances regardless of site location; a comparatively low proportion of contract value is subject to the risks that attend upon work below ground level; and once the building envelope is complete most trades can proceed without serious disruption from the weather.

This is not, however, the general method of pricing Civil Engineering work: the volume of work below ground, increased exposure to weather and the tremendous variety of projects, in terms of type, complexity and scale, makes the straightforward use of unit rates less reliable. So, whilst even in building work similar or identical measured items attract a fairly broad range of prices, the range is much greater in Civil Engineering Bills. This uncertainty is compounded by the lower number of bill items generated under Civil Engineering Methods of Measurement, so that the precise nature of the work is less apparent from the bill descriptions and the statistical effect of 'swings and roundabouts' has less scope to average out extremes of pricing.

To prepare a price for a Civil Engineering project, then, it is necessary to have regard to the method to be adopted in executing the work, draw up a detailed programme and then cost out the resources necessary to prosecute the chosen method. Because the first part of this process is the province of the Contractor's planner, there has been a tendency to postpone detailed estimating until the tendering stage itself, with the employer relying up to that point upon an estimate prepared on a 'broad brush' basis.

The result has been a growing pressure on the part of project sponsors for an improvement in budgetary advice, so that a decision to commit expenditure to a particular project is taken on firmer grounds. The absence of a detailed pricing method during the pre-contract phase also inhibits the accurate costing of alternative designs and regular cost checking to ensure that the design is being developed within the employer's budget.

This book therefore seeks to draw together the information appropriate to two methods of pricing: the cost of resources for use where an operational plan has been outlined, and unit rates for use where quantities can be taken from available drawings.

To take some note of the range of unit rates that might apply to an item, the rates themselves are in some cases related to working method – for example by identifying the different types of plant that would suit varying circumstances. Nonetheless, it would be folly to propose that all types of Civil Engineering work could be covered by a price book such as this. The Editors have therefore had in mind the type and scale of work commissioned by a local authority, a public corporation or a large private company.

This does embrace the great majority of work undertaken by the industry each year. Although almost all projects will have individual features that require careful attention in pricing, there will be some projects that are so specialist that they will not conform to standard pricing information at all.

But for most projects, within the range of work covered, this book should provide a firm foundation of cost information upon which a job-specific estimate can be built.

The contents of the book are therefore set out in a form that permits the user to follow the estimating process through in a structured way, as follows:

PURPOSE AND CONTENT OF THE BOOK

Part 1: General

The balance of this section describes in narrative form the work stages normally followed in a Contractor's office from receipt of the tender documents through to the submission of the tender.

The Land Remediation section reviews the general background of ground contamination and discusses the impact of the introduction of the Landfill Directive in July 2004 and recent changes thereto.

Part 2: Oncosts and Profit

Having produced an estimate for the predicted cost of the work, being the sum of the preliminaries and the measured work, the estimate must be converted to a tender by the application of any adjustment made by management (which follows the Management Appraisal described later in this part of the book) and by additions for financing charges, head office overheads and profit. These additions are discussed in this section and also included is a worked example of a tender summary.

Part 3: Costs and Tender Price Indices

The costs and tender price indices included in this part of the book provide a basis for updating historical cost or price information, by presenting changes in the indices since 1988. Caution must be taken when applying these indices as individual price fluctuations outside the general trend may have a significant effect on contract cost.

Part 4: Resources

This deals with the basic cost of resources, so that a resource-based system of estimating can be adopted where it is possible to develop an outline programme and method statement. Reference to this section will also assist the user to make adjustments to unit rates where different labour or material costs are thought to apply and to calculate analogous rates for work based on the hypothetical examples given. It is stressed that all of the costs given in this section are exclusive of the items costed with the preliminaries and of financing charges, head office overheads and profit. The materials and plant costs as shown are gross, with no deduction of discount.

Part 5: Approximate Estimating

The prices in this section have been assembled from a number of sources, including the relevant items in the unit costs section and recovered data from recent projects. They are intended to give broad price guides or to assist in comparison exercises.

Unit Costs (Ancillary Building Work)

This section is to be utilized in conjunction with Parts 4 and 5 to enable the user to incorporate within the estimate items more normally associated with Building Work rather than Civil Engineering and which do not fall readily under recognized methods of measurement for Civil Engineering Work. Due to the diversity of items that fall under such a definition, because of specification differences, the format for this section is structured to incorporate a range of items to allow the production of the estimate for such items prior to detailed design information being available. Additionally this section includes, in the same format, items covering simple Building Works which occur in connection with a Civil Engineering Contract but which do not form a significant proportion of the overall value and therefore do not need to be estimated in great detail using Parts 4 and 5 unit rates.

Preliminaries and General Items

Containing a checklist of items to be priced with Preliminaries and General Items (or Method Related Charges) and a worked example containing specific cost information.

PURPOSE AND CONTENT OF THE BOOK

Parts 6 and 7: Unit Costs

These sections are structured around methods of measurement for Civil Engineering Work and Highway Works and give 'trade by trade' unit rates for those circumstances where the application of unit rates to measured quantities is possible and practical. Again, it is stressed that the rates are exclusive of the items costed with preliminaries and of financing charges, head office overheads and profit. Both materials and plant costs are adjusted to allow a normal level of discount, with allowances for materials wastage and plant usage factors.

Part 8: Daywork

Including details of the CECA dayworks schedule and advice on costing excluded items.

Part 9: Professional Fees

These contain reference to standard documentation relating to professional fees for Consulting Engineers and Quantity Surveyors.

Part 10: Outputs

Scheduled here are various types of operations and the outputs expected of them. Also listed are man hours for various trades found in Civil Engineering.

Part 11: Useful Addresses for Further Information

This section provides useful contact addresses in the UK construction industry.

Part 12: Tables and Memoranda

These include conversion tables, formulae, and a series of reference tables structured around trade headings. It also includes a review of current Capital Allowances, Value Added Tax and The Aggregates Levy.

OUTLINE OF THE TENDERING AND ESTIMATING PROCESS

This section of the book outlines the nature and purpose of Civil Engineering estimating and provides background information for users. It comprises an outline of the estimating and tendering process with supporting notes and commentaries on particular aspects. Some worked examples on tender preparation referred to in this part are included at the end of Part 2.

It must be emphasized that the main purpose of this book is to aid the estimating process. Thus it is concerned more with the predicted cost of Civil Engineering work than with the prices in a bill of quantities. To ensure the correct interpretation of the information provided it is important to distinguish clearly between estimating and tendering; the following definitions are followed throughout.

The estimate is the prediction of the cost of a project to the Contractor. The Tender is the price submitted by the Contractor to the Employer.

The tender is based on the estimate with adjustments being made after review by management; these include allowances for risk, overheads, profit and finance charges. As discussed later in this section, prices inserted against individual items in a bill of quantities may not necessarily reflect the true cost of the work so described due to the view taken by the Contractor on the risks and financial aspects involved in executing the work.

Whilst projects are now constructed using many different forms of contract the core estimating process falls into two main divisions namely 'Design & Construct' and 'Construct only'. The following list summarizes the activities involved in the preparation of a tender for a typical construct only Civil Engineering project where the client issues full drawings, specifications and Bills describing the extent of the works to be priced. The diagram that follows illustrates the relationships between the activities. After the diagram, notes on factors affecting each stage of the process are given.

PURPOSE AND CONTENT OF THE BOOK

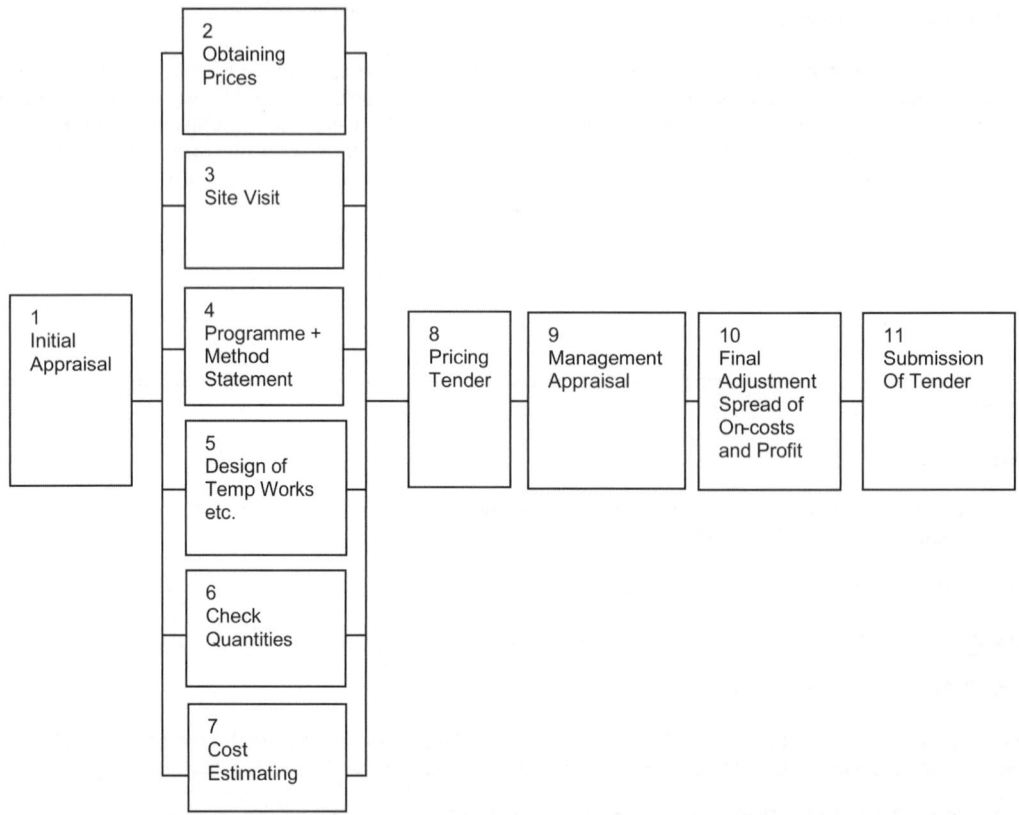

1. An overall appraisal of the project is made including any variations to the standard contract form, insurance provisions and any other unusual or onerous requirement.
2. Material requirements are abstracted from the tender documents and prices are obtained from suppliers. Details of work normally subcontracted are abstracted, together with relevant extracts from the tender documents and prices are obtained from subcontractors.
3. The site of the works and the surrounding area is visited and studied. Local information is obtained on factors affecting the execution of the contract.
4. A programme and detailed method of working for the execution of the contract is prepared, to include details of plant requirements, temporary works, unmeasured work, appropriate supervisory staff, etc.
5. Designs are made for temporary works and other features left to be designed by the Contractor, and quantities are taken off for costing.
6. Major quantities given in the tender documents are checked.
7. The cost estimate for the project is prepared by costing out all of the resources identified by stages 2 to 6. A more detailed report is made on the conditions of contract, financial requirements, etc. and an assessment of risk/opportunity is prepared.
   The tender documents are priced.
8. Management reviews the estimate, evaluates the risks and makes allowances for overheads, profit and finance.
9. The tender is completed and submitted.

PURPOSE AND CONTENT OF THE BOOK

1. INITIAL APPRAISAL

The purpose of the initial overall appraisal is to highlight any high value areas or any particular problems which may require specialist attention; it can also identify possible alternative methods of construction or temporary works.

PURPOSE AND CONTENT OF THE BOOK

Other points to be considered:

- the location and type of project and its suitability to the tenderer's particular expertise
- the size of project, its financing requirement, the proportion of annual turnover it would represent and the availability of resources
- the size of the tender list and the nature of the competition
- the identity of the employer and his professional consultants
- the adequacy of the tender documents
- an initial appraisal of risk and opportunity (see Section 8)

Corporate governance requires that the directors of companies are aware of all the liabilities inherent in the contract being sought. Few contracts are offered on totally unamended standard forms and specifications and it is the estimator's duty to ensure that a report is prepared to advise on the precise terms and conditions being offered and their potential implications.

It is essential for the estimator to study the contract documents issued with the enquiry or made available for inspection and to note those parts which will affect pricing or involve the Contractor in liabilities which must be evaluated and drawn to the attention of management. The following comments are indicative only:

Conditions of Contract

For Civil Engineering work these are normally, but not exclusively, based on either the ICE. or ECC (NEC) standard forms. However these forms are rarely offered without addition and amendment and it is imperative that the full implications are understood and directors informed. Any required bonds, guarantees and warranties must be identified, reported and included in subcontract enquiries where appropriate. Insurance requirements and excesses must be checked against company policies.

Bill of Quantities

Where a Bill is provided it serves three purposes: first and foremost it must be prepared with the objective of providing the estimator with as accurate a picture of the project as possible, so as to provide a proper basis for pricing. Second, it should enable the employer to compare tenders on an equal basis and third it will be used to evaluate the work executed for payment purposes. Individual items in the Bill do not necessarily describe in detail the work to be carried out under a particular item; reference must be made to the specification and the drawings to ascertain the full scope of the work involved.

The method of preparing the Bill may be based on the 'Civil Engineering Standard Method of Measurement' issued by the Institution of Civil Engineers or the 'Method of Measurement for Highway Works' issued by the Highways Agency, but some employing authorities have evolved their own methods and it is important for the estimator to study the Bill and its preambles to ensure that his rates and prices are comprehensive.

In all cases the quantities given in the bill are not a guarantee and the drawings usually have precedence. The estimator must understand whether he is pricing a remeasurable or fixed price contract and make due allowance.

Specification

This gives a detailed description of the workmanship, finish and materials to be used in the construction of the work. It may also give completion periods for sections and/or the whole of the work together with details of the requirements for the employer and/or the Consulting Engineer in connection with their supervision on site.

Water Authority and Highway Works in particular are based around a standard specification. However even standard specifications will have contract-specific appendices and tables. The estimator must take due note of these requirements and ensure that this information is issued and taken into account by all potential subcontractors and material suppliers.

PURPOSE AND CONTENT OF THE BOOK

Drawings

These give details of the work to be carried out and must be read in conjunction with the specification. It is important for the estimator to study the notes and descriptions given on the drawings as these amplify the specification.

Should the estimator discover any conflict between the various documents, it is important to have such discrepancies clarified by the Employer or Engineer prior to submission of any offer.

2. OBTAINING PRICES

(a) Materials

When pricing materials, the following points must be noted:

- checks must be made to ensure that the quality of materials to be supplied meets with the requirements of the specification. If necessary, samples should be obtained and tested for compliance
- checks must be made to ensure that the rates of delivery and fabrication periods can meet the demands of the programme. It is sometimes necessary to use more than one supplier, with differing prices, to ensure a sufficient flow of materials
- tests should be carried out to ascertain allowances to be made for operations such as compaction of soils and aggregates. Records of past contracts using similar materials can give this information, providing such records are accurate and reliable

(b) Subcontractors

It is common practice among Civil Engineering Contractors to subcontract a significant proportion of their work.

Subcontracted work can represent the bulk of the value of measured work.

When utilizing subcontractors' prices it is extremely important to ensure that the rates given cover the full extent of the work described in the main contract, and that the subcontractors quotation allows for meeting the main Tender programme and methods of working.

Unless an exclusive relationship has been entered into prior to the tendering process it is likely that the same subcontractor will submit prices to a number of competing tenderers. It is important for the estimator to ensure that the price offered represents the method intended and is not a generic sum which is subject to variation if successful.

3. SITE VISIT

Factors to check during the site visit include:

- access
- limitations of working space
- existing overhead and underground services
- nearby public transport routes
- availability of services – water, gas, electricity, telephone, etc.
- availability of labour and subcontractors
- availability of materials – particularly aggregates and fill materials and location of nearest tipping facility
- nature of the ground, including inspection of trial bores/pits if dug ground water level
- presence of other Contractors working on or adjacent to the site

PURPOSE AND CONTENT OF THE BOOK

4. METHOD STATEMENT AND PROGRAMME

As previously stated whilst an estimate is being prepared it is necessary that a detailed method of working and a programme for the execution of the works is drawn up; the latter can take the form of a bar chart or, for large and more complex projects, may be prepared on more sophisticated computerised platforms. Compliance with the employer's target completion dates is, of course, essential. The method of working will depend on this programme in so far as the type and size of plant and the gang sizes to be used must be capable of achieving the output necessary to meet the programmed times. Allowance must be made for delays due to adverse weather, other hazards particular to the site and the requirements of the specification, particularly with regard to periods to be allowed for service diversions and other employer's requirements.

A method statement is prepared in conjunction with the programme, setting out the resources required, outputs to be achieved and the requirements in respect of temporary works, etc.

At the same time, separate bar charts may be produced giving:

- plant requirements
- staff and site supervision requirements
- labour requirements

These programmes and method statements will form the basis of the actual contract programme should the tender be accepted. They will also enable the Contractor to assure himself that he has available or can gain access to the necessary resources in plant, labour, materials and supervision to carry out the work should he be awarded the contract.

5. DESIGN OF TEMPORARY WORKS, ETC.

Normally the period of time allowed for tendering is relatively short and therefore it is important that those aspects requiring design work are recognized as early as possible in the tender period. Design can be carried out either by the Contractor using his own engineers or by utilizing the services of a Consulting Engineer. There are three aspects of design to be considered:

(a) Temporary works to the contractor's requirements to enable the works to be constructed

Design of temporary works covers the design of those parts of the work for which the contractor accepts full responsibility in respect of stability and safety during construction. Such parts include support structures, cofferdams, temporary bridges, jetties and river diversions, special shuttering, scaffolding, haul roads and hardstandings, compounds, traffic management etc. Design must be in sufficient detail to enable materials, quantities and work content to be assessed and priced by the estimator. In designing such work, it is important that adequate attention is given to working platforms and access for labour and plant and also to ease of dismantling and re-erection for further uses without damage.

It should be noted that many specialist subcontractors will provide a design service when submitting quotations. For example, scaffolding contractors will design suitable support work for soffit shuttering, etc.

(b) Specific items of the permanent works to meet a performance specification set out by the client or the Consulting Engineer

It is common practice for certain parts of the work to be specified by means of a performance specification. For example, concrete is specified by strength only, piles by load carrying capacity, etc. It is then left to the contractor to use those materials, workmanship and design which he feels are most suited to the particular site and conditions. In many cases such design will be carried out by specialist suppliers or subcontractors.

PURPOSE AND CONTENT OF THE BOOK

(c) Alternative designs for sections of the permanent works where the contractor's experience leads him to consider that a more economical design could be used

It is possible for a Contractor to use his expertise and experience to design and submit alternative proposals and prices for complete sections of the permanent work without, of course, altering the basic requirements of the original design. Examples would be foundations in difficult ground conditions, bridge superstructures, use of precast in place of in situ concrete, etc. Such designs may be carried out by the contractor's own staff or in conjunction with Consulting Engineers or specialist contractors.

Obviously this is only done when the Contractor can offer considerable savings in cost and/or a reduced construction period. It is necessary to include a price for the original design in the tender, but the decision to submit a keen tender may be underpinned by the hope of sharing such savings with the Employer.

In all cases, the Contractor's designs and calculations must be checked and approved by the Employer's Consulting Engineer, but such checks and approvals do not relieve the Contractor of his responsibilities for the safety and stability of the work.

Where any design work is undertaken it is important to ensure that adequate design insurance is maintained.

6. QUANTITIES CHECK

Working within obvious time constraints, the estimating team will endeavour to complete a quantities check on at least the major and high price quantities of the Bill, as this could affect pricing strategy, for example, in pricing provisional items. Any major discrepancies noted should be referred to the management appraisal agenda.

7. COST ESTIMATE

At this stage the estimator draws together the information and prepares a cost estimate made up of:

- preliminaries and general items)
- temporary works
- labour, costed by reference to the method statement, with appropriate allowances for labour oncosts (non-productive overtime, travelling time/fares, subsistence, guaranteed time, bonus, Employer's Liability Insurance, training, statutory payments, etc.)
- material costs taken from price lists or suppliers' quotations, with appropriate allowances for waste
- plant, whether owned or hired, with appropriate allowances for transport to/from site, erection/dismantling, maintenance, fuel and operation. Heavy static plant (batching plant, tower cranes, etc.) will normally be priced with general items; the remainder will normally be allocated to the unit rates)
- sublet work, as quoted by specialist subcontractors, with appropriate allowances for all attendances required to be provided by the main contractor

At the same time, a preliminary assessment of risk/opportunity will be made for consideration with the Management Appraisal (see Section 8 below). On major tenders a formal quantified risk assessment (QRA) will be undertaken – the estimator will be expected to evaluate the effects of the risks for inclusion in the calculations. This will include a look at:

- weather conditions – costs not recoverable by claims under the Conditions of Contract, ground becoming unsuitable for working due to the effect of weather, etc.
- flooding – liability of site to flooding and the consequent costs
- suitability of materials – particular risk can arise if prices are based on the use of borrow pits or quarries and inadequate investigation has been carried out or development is refused by the local authorities
- reliability of subcontractors – failure of a subcontractor to perform can result in higher costs through delays to other operations and employing an alternative subcontractor at higher cost
- non-recoverable costs – such as excesses on insurance claims
- estimator's ability – e.g. outputs allowed. This can only be gauged from experience

PURPOSE AND CONTENT OF THE BOOK

- cost increase allowances for fixed price contract
- terms and conditions contained in the contract documents
- ability to meet specification requirements for the prices allowed
- availability of adequate and suitable labour

Risk is, of course, balanced by opportunity and consideration needs to be given to areas for which particular expertise possessed by the Contractor will lead to a price advantage against other tenderers.

8. PRICING THE TENDER

Once the cost estimate is complete, the estimator prices the items in the Bill. The rate to be entered against the items at this stage should be the correct rate for doing the job, whatever the quantity shown.

Where the overall operation covers a number of differing Bill items, the estimator will allocate the cost to the various items in reasonable proportions; the majority of the work is priced in this manner. The remaining items are normally priced by unit rate calculation.

Resources schedules, based on the programme and giving details of plant, labour and staff, perform an important role in enabling the estimator to check that he has included the total cost of resources for the period they are required on site. It is not unusual for an item of plant to be used intermittently for more than one operation. A reconciliation of the total time for which the cost has been included in the estimate against the total time the item is needed on site as shown on the programme, gives a period of non-productive time. The cost of this is normally included in site oncosts and preliminaries. A similar situation can arise in the cases of skilled labour, craftsmen and plant drivers.

Having priced the Bill at cost, there will remain a sum to be spread over the Bill items. The way in which this is done depends on the view taken by the Contractor of the project; for example:

- sums can be put against the listed general items in the Preliminaries Bill
- a fixed 'Adjustment Item' can be included in the Bill for the convenience of the estimator. This can be used for the adjustment made following the Management Appraisal, and for taking advantage of any late but favourable quotations received from subcontractors or suppliers
- the balance can be spread over the rates by equal percentage on all items, or by unequal percentages to assist in financing the contract or to take advantage of possible contract variations, or expected quantity changes (see notes against 7)

The Contractor will normally assess the financial advantages to be gained from submitting his bid in this manner and possibly enabling him to submit a more competitive offer.

After completing the pricing of all aspects of the tender, the total costs are summarized and profit, risk, etc., added to arrive at the total value of the Tender. A suggested form of this summary is set out in Part 2: Oncosts and Profits.

Finally, reasonable forecasts of cash flow and finance requirements are essential for the successful result of the project. Preliminary assessments may have been made for the information of management, but contract cash flow and the amount of investment required to finance the work can now be estimated by plotting income against expenditure using the programme of work and the priced Bill of Quantities. Payment in arrears and retentions, both from the Employer and to the suppliers and subcontractors, must be taken into account.

It is unlikely that sufficient time will be available during the tender period to produce such information accurately, but an approximate figure, for use as a guide for finance requirements, can be assessed.

A worked example is set out in Oncosts and Profits.

The Aggregates Levy

The Aggregates Levy came into operation on 1 April 2002 in the UK, except for Northern Ireland where it has been phased in over five years from 2003.

It was introduced to ensure that the external costs associated with the exploitation of aggregates are reflected in the price of aggregate, and to encourage the use of recycled aggregate. There continues to be strong evidence that the levy is achieving its environmental objectives, with sales of primary aggregate down and production of recycled aggregate up. The Government expects that the rates of the levy will at least keep pace with inflation over time, although it accepts that the levy is still bedding in.

The rate of the levy increased to £2.10 per tonne from 1 April 2012 and is levied on anyone considered to be responsible for commercially exploiting virgin aggregates in the UK and should naturally be passed by price increase to the ultimate user.

All materials falling within the definition of Aggregates are subject to the levy unless specifically exempted.

It does not apply to clay, soil, vegetable or other organic matter.

The intention is that it will:

- Encourage the use of alternative materials that would otherwise be disposed of to landfill sites
- Promote development of new recycling processes, such as using waste tyres and glass
- Promote greater efficiency in the use of virgin aggregates
- Reduce noise and vibration, dust and other emissions to air, visual intrusion, loss of amenity and damage to wildlife habitats

Definitions

'Aggregates' means any rock, gravel or sand which is extracted or dredged in the UK for aggregates use. It includes whatever substances are for the time being incorporated in it or naturally occur mixed with it.

'Exploitation' is defined as involving any one or a combination of any of the following:

- Being removed from its original site
- Becoming subject to a contract or other agreement to supply to any person
- Being used for construction purposes
- Being mixed with any material or substance other than water, except in permitted circumstances

Incidence

It is a tax on primary aggregates production – i.e. virgin aggregates won from a source and used in a location within the UK territorial boundaries (land or sea). The tax is not levied on aggregates which are exported or on aggregates imported from outside the UK territorial boundaries.

It is levied at the point of sale.

THE AGGREGATES LEVY

Exemption from tax

An aggregate is exempt from the levy if it is:

- Material which has previously been used for construction purposes
- Aggregate that has already been subject to a charge to the Aggregates Levy
- Aggregate which was previously removed from its originating site before the start date of the levy
- Aggregate which is being returned to the land from which it was won
- Aggregate won from a farm land or forest where used on that farm or forest
- Rock which has not been subjected to an industrial crushing process
- Aggregate won by being removed from the ground on the site of any building or proposed building in the course of excavations carried out in connection with the modification or erection of the building and exclusively for the purpose of laying foundations or of laying any pipe or cable
- Aggregate won by being removed from the bed of any river, canal or watercourse or channel in or approach to any port or harbour (natural or artificial), in the course of carrying out any dredging exclusively for the purpose of creating, restoring, improving or maintaining that body of water
- Aggregate won by being removed from the ground along the line of any highway or proposed highway in the course of excavations for improving, maintaining or constructing the highway otherwise than purely to extract the aggregate
- Drill cuttings from petroleum operations on land and on the seabed
- Aggregate resulting from works carried out in exercise of powers under the New Road and Street Works Act 1991, the Roads (Northern Ireland) Order 1993 or the Street Works (Northern Ireland) Order 1995
- Aggregate removed for the purpose of cutting of rock to produce dimension stone, or the production of lime or cement from limestone.
- Aggregate arising as a waste material during the processing of the following industrial minerals:
    - ball clay
    - barytes
    - calcite
    - china clay
    - coal, lignite, slate or shale
    - feldspar
    - flint
    - fluorspar
    - fuller's earth
    - gems and semi-precious stones
    - gypsum
    - any metal or the ore of any metal
    - muscovite
    - perlite
    - potash
    - pumice
    - rock phosphates
    - sodium chloride
    - talc
    - vermiculite

However, the levy is still chargeable on any aggregates arising as the spoil or waste from or the by-products of the above exempt processes. This includes quarry overburden.

Anything that consists 'wholly or mainly' of the following is exempt from the levy (note that 'wholly' is defined as 100% but 'mainly' as more than 50%, thus exempting any contained aggregates amounting to less than 50% of the original volumes):

- clay, soil, vegetable or other organic matter
- coal, slate or shale
- china clay waste and ball clay waste

THE AGGREGATES LEVY

Relief from the levy either in the form of credit or repayment is obtainable where:

- it is subsequently exported from the UK in the form of aggregate
- it is used in an exempt process
- where it is used in a prescribed industrial or agricultural process
- it is waste aggregate disposed of by dumping or otherwise, e.g. sent to landfill or returned to the originating site

The Aggregates Levy Credit Scheme (ALCS) for Northern Ireland was suspended with effect from 1 December 2010 following a ruling by the European General Court.

A new exemption for aggregate obtained as a by-product of railway, tramway and monorail improvement, maintenance and construction was introduced in 2007.

Discounts

From 1 July 2005 the standard added water percentage discounts listed below can be used. Alternatively a more exact percentage can be agreed and this must be done for dust dampening of aggregates.

- washed sand = 7%
- washed gravel = 3.5%
- washed rock/aggregate = 4%

Impact

The British Aggregates Association suggests that the additional cost imposed by quarries is more likely to be in the order of £3.40 per tonne on mainstream products, applying an above average rate on these in order that by-products and low grade waste products can be held at competitive rates, as well as making some allowance for administration and increased finance charges.

With many gravel aggregates costing in the region of £16.00 to £18.00 per tonne, there is a significant impact on construction costs.

Avoidance

An alternative to using new aggregates in filling operations is to crush and screen rubble which may become available during the process of demolition and site clearance as well as removal of obstacles during the excavation processes.

THE AGGREGATES LEVY

Example:

Assuming that the material would be suitable for fill material under buildings or roads, a simple cost comparison would be as follows (note that for the purpose of the exercise, the material is taken to be 1.80 tonne per m^3 and the total quantity involved less than 1,000 m^3):

	£/m^3	£/tonne
Importing fill material:		
Cost of 'new' aggregates delivered to site	31.23	17.35
Addition for Aggregates Tax	3.78	2.10
Total cost of importing fill materials	35.01	19.45
Disposing of site material:		
Cost of removing materials from site materials	21.52	11.95
Crushing site materials:		
Transportation of material from excavations or demolition to stockpiles	3.00	1.67
Transportation of material from temporary stockpiles to the crushing plant	4.00	2.22
Establishing plant and equipment on site; removing on completion	2.00	1.11
Maintain and operate plant	9.00	5.00
Crushing hard materials on site	13.00	7.22
Screening material on site	2.00	1.11
Total cost of crushing site materials	33.00	18.33

From the above it can be seen that potentially there is a great benefit in crushing site materials for filling rather than importing fill materials.

Setting the cost of crushing against the import price would produce a saving of £2.10 per m^3. If the site materials were otherwise intended to be removed from the site, then the cost benefit increases by the saved disposal cost to £23.53 per m^3.

Even if there is no call for any or all of the crushed material on site, it ought to be regarded as a useful asset and either sold on in crushed form or else sold with the prospects of crushing elsewhere.

THE AGGREGATES LEVY

Specimen Unit rates	Unit³	£
Establishing plant and equipment on site; removing on completion		
crushing plant	trip	1,200.00
screening plant	trip	600.00
Maintain and operate plant		
crushing plant	week	7,200.00
screening plant	week	1,800.00
Transportation of material from excavations or demolition places to temporary stockpiles	m³	3.00
Transportation of material from temporary stockpiles to the crushing plant	m³	2.40
Breaking up material on site using impact breakers		
mass concrete	m³	14.00
reinforced concrete	m³	16.00
brickwork	m³	6.00
Crushing material on site		
mass concrete not exceeding 1000 m³	m³	13.00
mass concrete 1000–5000 m³	m³	12.00
mass concrete over 5000 m³	m³	11.00
reinforced concrete not exceeding 1000 m³	m³	15.00
reinforced concrete 1000–5000 m³	m³	14.00
reinforced concrete over 5000 m³	m³	13.00
brickwork not exceeding 1000 m³	m³	12.00
brickwork 1000–5000 m³	m³	11.00
brickwork over 5000 m³	m³	10.00
Screening material on site	m³	2.00

More detailed information can be found on the HMRC website (www.hmrc.gov.uk) in Notice AGL 1 Aggregates Levy published in May 2009.

ESSENTIAL READING FROM TAYLOR AND FRANCIS

Manual for Detailing Reinforced Concrete Structures to EC2

J. Calavera

Detailing 210 structural elements, each with a double page spread, this book is an indispensible guide to designing for reinforced concrete. Drawing on Jose Calavera's many years of experience, each entry provides commentary and recommendations for the elementconcerned, as 0077ell as summarising the appropriate Eurocode legislation and referring to further standards and literature.

Includes a CD ROM with AutoCad files of all of the models, which you can adapt to suit your own purposes.

Section 1. General Rules for Bending, Placing and Lapping of Reinforcement
Section 2. Constructive Details

1. Foundations 2. Walls 3. Columns and Joints 4. Walls Subjected to Axial Loads. Shear Walls and Cores 5. Beams and Lintels 6. Slabs 7. Flat Slabs 8. Stairs 9. Supports
10. Brackets 11. Pavements and Galleries 12. Cylindrical Chimneys, Towers and Hollow Piles
13. Rectangular Silos, Caissons and Hollow Piles 14. Reservoirs, Tanks and Swimming Pools
15. Special Details for Seismic Areas

October 2011: 297 x210: 528pp
Hb: 978-0-415-66348-9: **£80.00**

To Order: Tel: +44 (0) 1235 400524 **Fax:** +44 (0) 1235 400525
or Post: Taylor and Francis Customer Services,
Bookpoint Ltd, Unit T1, 200 Milton Park, Abingdon, Oxon, OX14 4TA UK
Email: book.orders@tandf.co.uk

For a complete listing of all our titles visit:
www.tandf.co.uk

Taylor & Francis
Taylor & Francis Group

Capital Allowances

Introduction

Capital Allowances provide tax relief by prescribing a statutory rate of depreciation for tax purposes in place of that used for accounting purposes. They are utilized by government to provide an incentive to invest in capital equipment, including assets within commercial property, by allowing the majority of taxpayers a deduction from taxable profits for certain types of capital expenditure, thereby reducing or deferring tax liabilities.

The capital allowances most commonly applicable to real estate are those given for capital expenditure on existing commercial buildings in disadvantaged areas, and plant and machinery in all buildings other than residential dwellings. Relief for certain expenditure on industrial buildings and hotels was withdrawn from April 2011, although the ability to claim plant and machinery remains.

Enterprise Zone Allowances are also available for capital expenditure within designated areas only where there is a focus on high value manufacturing. Enhanced rates of allowances are available on certain types of energy and water saving plant and machinery assets, whilst reduced rates apply to 'integral features' and items with an expected economic life of more than 25 years.

The Act

The primary legislation is contained in the Capital Allowances Act 2001. Major changes to the system were announced by the Government in 2007 and there have been further changes in subsequent Finance Acts.

The Act is arranged in 12 Parts (plus two addenda) and was published with an accompanying set of Explanatory Notes.

Plant and Machinery

The Finance Act 1994 introduced major changes to the availability of Capital Allowances on real estate. A definition was introduced which precludes expenditure on the provision of a building from qualifying for plant and machinery, with prescribed exceptions.

List A in Section 21 of the 2001 Act sets out those assets treated as parts of buildings:

- Walls, floors, ceilings, doors, gates, shutters, windows and stairs.
- Mains services, and systems, for water, electricity and gas.
- Waste disposal systems.
- Sewerage and drainage systems.
- Shafts or other structures in which lifts, hoists, escalators and moving walkways are installed.
- Fire safety systems.

Similarly, List B in Section 22 identifies excluded structures and other assets.

Both sections are, however, subject to Section 23. This section sets out expenditure, which although being part of a building, may still be expenditure on the provision of Plant and Machinery.

List C in Section 23 is reproduced below:

CAPITAL ALLOWANCES

Sections 21 and 22 do not affect the question whether expenditure on any item in List C is expenditure on the provision of Plant or Machinery

1) Machinery (including devices for providing motive power) not within any other item in this list.
2) Gas and sewerage systems provided mainly –
   a) to meet the particular requirements of the qualifying activity, or
   b) to serve particular plant or machinery used for the purposes of the qualifying activity.
3) Omitted.
4) Manufacturing or processing equipment; storage equipment (including cold rooms); display equipment; and counters, checkouts and similar equipment.
5) Cookers, washing machines, dishwashers, refrigerators and similar equipment; washbasins, sinks, baths, showers, sanitary ware and similar equipment; and furniture and furnishings.
6) Hoists.
7) Sound insulation provided mainly to meet the particular requirements of the qualifying activity.
8) Computer, telecommunication and surveillance systems (including their wiring or other links).
9) Refrigeration or cooling equipment.
10) Fire alarm systems; sprinkler and other equipment for extinguishing or containing fires.
11) Burglar alarm systems.
12) Strong rooms in bank or building society premises; safes.
13) Partition walls, where moveable and intended to be moved in the course of the qualifying activity.
14) Decorative assets provided for the enjoyment of the public in hotel, restaurant or similar trades.
15) Advertising hoardings; signs, displays and similar assets.
16) Swimming pools (including diving boards, slides & structures on which such boards or slides are mounted).
17) Any glasshouse constructed so that the required environment (namely, air, heat, light, irrigation and temperature) for the growing of plants is provided automatically by means of devices forming an integral part of its structure.
18) Cold stores.
19) Caravans provided mainly for holiday lettings.
20) Buildings provided for testing aircraft engines run within the buildings.
21) Moveable buildings intended to be moved in the course of the qualifying activity.
22) The alteration of land for the purpose only of installing Plant or Machinery.
23) The provision of dry docks.
24) The provision of any jetty or similar structure provided mainly to carry Plant or Machinery.
25) The provision of pipelines or underground ducts or tunnels with a primary purpose of carrying utility conduits.
26) The provision of towers to support floodlights.
27) The provision of –
   a) any reservoir incorporated into a water treatment works, or
   b) any service reservoir of treated water for supply within any housing estate or other particular locality.
28) The provision of –
   a) silos provided for temporary storage, or
   b) storage tanks.
29) The provision of slurry pits or silage clamps.
30) The provision of fish tanks or fish ponds.
31) The provision of rails, sleepers and ballast for a railway or tramway.
32) The provision of structures and other assets for providing the setting for any ride at an amusement park or exhibition.
33) The provision of fixed zoo cages.

Capital Allowances on plant and machinery are given in the form of writing down allowances at the rate of 18% per annum on a reducing balance basis. For every £100 of qualifying expenditure £18 is claimable in year 1, £14.76 in year 2 and so on until either the all the allowances have been claimed or the asset is sold.

CAPITAL ALLOWANCES

Integral Features

The category of qualifying expenditure on integral features was introduced with effect from April 2008. The following items are integral features:

- An electrical system (including a lighting system)
- A cold water system
- A space or water heating system, a powered system of ventilation, air cooling or air purification, and any floor or ceiling comprised in such a system
- A lift, an escalator or a moving walkway
- External solar shading

The draft legislation also included active facades but these were subsequently omitted, the explanation being that allowances are already given on the additional inner skin because it is part of the air conditioning system.

A reduced writing down allowance of 8% per annum is available on integral features.

The legislation also includes a rule that prevents a revenue deduction being obtained where expenditure is incurred that is more than 50% of the cost of replacing the integral feature.

Thermal Insulation

For many years the addition of thermal insulation to an existing industrial building has been treated as qualifying for plant and machinery allowances. From April 2008 this has been extended to include all commercial buildings but not residential buildings.

A reduced writing down allowance of 8% per annum is available on thermal insulation.

Long-Life Assets

A reduced writing down allowance of 8% per annum is available on long-life assets. Allowances were given at the rate of 6% before April 2008.

A long-life asset is defined as plant and machinery that can reasonably be expected to have a useful economic life of at least 25 years. The useful economic life is taken as the period from first use until it is likely to cease to be used as a fixed asset of any business. It is important to note that this likely to be a shorter period than an item's physical life.

Plant and machinery provided for use in a building used wholly or mainly as dwelling house, showroom, hotel, office or retail shop or similar premises, or for purposes ancillary to such use, cannot be long-life assets.

In contrast plant and machinery assets in buildings such as factories, cinemas, hospitals and so on are all potentially long-life assets.

Case Law

The fact that an item appears in List C does not automatically mean that it will qualify for capital allowances. It only means that it may potentially qualify.

Guidance about the meaning of plant has to be found in case law. The cases go back a long way, beginning in 1887. The current state of the law on the meaning of plant derives from the decision in the case of *Wimpy International Ltd and Associated Restaurants Ltd v Warland* in the late 1980s.

CAPITAL ALLOWANCES

The Judge in that case said that there were three tests to be applied when considering whether or not an item is plant:

1. Is the item stock in trade? If the answer yes, then the item is not plant.
2. Is the item used for carrying on the business? In order to pass the business use test the item must be employed in carrying on the business; it is not enough for the asset to be simply used in the business. For example, product display lighting in a retail store may be plant but general lighting in a warehouse would fail the test.
3. Is the item the business premises or part of the business premises? An item cannot be plant if it fails the premises test, i.e. if the business use is as the premises (or part of the premises) or place on which the business is conducted. The meaning of part of the premises in this context should not be confused with the law of real property. The Inland Revenue's internal manuals suggest there are four general factors to be considered, each of which is a question of fact and degree:
   - does the item appear visually to retain a separate identity
   - with what degree of permanence has it been attached to the building
   - to what extent is the structure complete without it
   - to what extent is it intended to be permanent or alternatively is it likely to be replaced within a short period.

There is obviously a core list of items that will usually qualify in the majority of cases. However, many others still need to be looked at on a case-by-case basis. For example, decorative assets in a hotel restaurant may be plant but similar assets in an office reception area would almost certainly not be.

One of the benefits of the integral features rules, apart from simplification, is that items that did not qualify by applying these rules, such as general lighting in an office building, will now qualify albeit at a reduced rate.

Refurbishment Schemes

Building refurbishment projects will typically be a mixture of capital costs and revenue expenses, unless the works are so extensive that they are more appropriately classified a redevelopment. A straightforward repair or a 'like for like' replacement of part of an asset would be a revenue expense, meaning that the entire amount can be deducted from taxable profits in the same year.

Where capital expenditure is incurred that is incidental to the installation of plant or machinery then Section 25 of the 2001 Act allows it to be treated as part of the expenditure on the qualifying item. Incidental expenditure will often include parts of the building that would be otherwise disallowed, as shown in the Lists reproduced above. For example, the cost of forming a lift shaft inside an existing building would be deemed to be part of the expenditure on the provision of the lift.

The extent of the application of section 25 was reviewed for the first time by the Special Commissioners in December 2007 and by the First Tier Tribunal (Tax Chamber) in December 2009, in the case of JD Wetherspoon. The key areas of expenditure considered were overheads and preliminaries where it was held that such costs could be allocated on a pro-rata basis; decorative timber panelling which was found to be part of the premises and so ineligible for allowances; toilet lighting which was considered to provide an attractive ambience and qualified for allowances; and incidental building alterations of which enclosing walls to toilets and kitchens and floor finishes did not qualify but tiled splash backs, toilet cubicles and drainage did qualify along with the related sanitary fittings and kitchen equipment.

Annual Investment Allowance

The annual investment allowance is available to all businesses of any size and allows a deduction for the whole of the first £25,000 (£100,000 before April 2012) of qualifying expenditure on plant and machinery, including integral features and long-life assets.

CAPITAL ALLOWANCES

The Enhanced Capital Allowances Scheme

The scheme is one of a series of measures introduced to ensure that the UK meets its target for reducing greenhouse gases under the Kyoto Protocol. 100% first year allowances are available on products included on the Energy Technology List published on the website at www.eca.gov.uk and other technologies supported by the scheme. All businesses will be able to claim the enhanced allowances, but only investments in new and unused machinery and plant can qualify.

There are currently 14 technologies and 53 sub-technologies currently covered by the scheme:

1. Air-to-air energy recovery
2. Automatic monitoring and targeting (AMT)
3. Boiler equipment
4. Combined heat and power (CHP)
5. Compact heat exchangers
6. Compressed air equipment
7. Heat pumps for space heating
8. Heating ventilation and air conditioning zone controls
9. Lighting
10. Motors and drives
11. Pipework insulation
12. Refrigeration equipment
13. Solar thermal systems
14. Warm air and radiant heaters

The Finance Act 2003 introduced a new category of environmentally beneficial plant and machinery qualifying for 100% first year allowances. The Water Technology List includes 14 technologies:

- Cleaning in place equipment
- Efficient showers
- Efficient taps
- Efficient toilets
- Efficient washing machines
- Flow controllers
- Leakage detection equipment
- Meters and monitoring equipment
- Rainwater harvesting equipment
- Small-scale slurry and sludge dewatering equipment
- Vehicle wash water reclaim units
- Water efficient industrial cleaning equipment
- Water management equipment for mechanical seals
- Efficient membrane systems

Buildings and structures and long life assets as defined above cannot qualify under the scheme. However, following the introduction of the integral features rules lighting in any non-residential building may potentially qualify for enhanced capital allowances if it meets the relevant criteria.

A limited payable ECA tax credit equal to 19% of the loss surrendered was also introduced for UK companies in April 2008.

From April 2012 expenditure on plant and machinery for which tariff payments are received under the renewable energy schemes introduced by the Department of Energy and Climate Change (Feed-in Tariffs or Renewable Heat Incentives) will not be entitled to enhanced capital allowances.

CAPITAL ALLOWANCES

Enterprise Zones

The creation of 21 new Enterprise Zones was announced in the 2011 Budget. An additional zone in Humber was added in February 2012, with a further five confirmed in Budget 2012. Originally introduced in the early 1980s as a stimulus to commercial development and investment, they had virtually faded from the real estate psyche.

The original zones benefited from a 100% first year allowance on capital expenditure incurred on the construction (or the purchase within two years of first use) of any commercial building within a designated enterprise zone, within 10 years of the site being so designated. Like other allowances given under the industrial buildings code the building has a life of 25 years for tax purposes.

The majority of these enterprise zones had reached the end of their 10-year life by 1993. However, in certain very limited circumstances it may still be possible to claim these allowances up to 20 years after the site was first designated.

Enterprise zones benefit from a number of reliefs, including a 100% first year allowance for new and unused non-leased plant and machinery assets, where there is a focus on high-value manufacturing.

Flat Conversion Allowances

Tax relief is available on capital expenditure incurred on or after 11 May 2001 on the renovation or conversion of vacant or underused space above shops and other commercial premises to provide flats for rent.

In order to qualify the property must have been built before 1980 and the expenditure incurred on, or in connection with:

- Converting part of a qualifying building into a qualifying flat.
- Renovating an existing flat in a qualifying building if the flat is, or will be a qualifying flat.
- Repairs incidental to conversion or renovation of a qualifying flat, and
- The cost of providing access to the flat(s).

The property must not have more than four storeys above the ground floor and it must appear that, when the property was constructed, the floors above the ground floor were primarily for residential use. The ground floor must be authorized for business use at the time of the conversion work and for the period during which the flat is held for letting. Each new flat must be a self-contained dwelling, with external access separate from the ground-floor premises. It must have no more than four rooms, excluding kitchen and bathroom. None of the flats can be 'high value' flats, as defined in the legislation. The new flats must be available for letting as a dwelling for a period of not more than five years.

An initial allowance of 100% is available or, alternatively, a lower amount may be claimed, in which case the balance may be claimed at a rate of 25% per annum in subsequent a years. The allowances may be recovered if the flat is sold or ceases to be let within seven years.

Business Premises Renovation Allowance

The Business Premises Renovation Allowance (BPRA) was first announced in December 2003. The idea behind the scheme is to bring long-term vacant properties back into productive use by providing 100% capital allowances for the cost of renovating and converting unused premises in disadvantaged areas. The legislation was included in the Finance Act 2005 and was finally implemented on 11 April 2007 following EU state aid approval.

The legislation is identical in many respects to that for flat conversion allowances. The scheme will apply to properties within one of the areas specified in the Assisted Areas Order 2007 and Northern Ireland.

CAPITAL ALLOWANCES

BPRA will be available to both individuals and companies who own or lease business property that has been unused for 12 months or more. Allowances will be available to a person who incurs qualifying capital expenditure on the renovation of business premises.

An announcement to extend the scheme by a further five years to 2017 was made within the 2011 Budget.

Other Capital Allowances

Other types of allowances include those available for capital expenditure on Mineral Extraction, Research and Development, Know-How, Patents, Dredging and Assured Tenancy.

ESSENTIAL READING FROM TAYLOR AND FRANCIS

Construction Contracts Questions and Answers

2nd Edition

David Chappell

What they said about the first edition: "A fascinating concept, full of knowledgeable gems put in the most frank of styles... A book to sample when the time is right and to come back to when another time is right, maybe again and again."
– David A Simmonds, Building Engineer magazine

- Is there a difference between inspecting and supervizing?
- What does 'time-barred' mean?
- Is the contractor entitled to take possession of a section of the work even though it is the contractor's fault that possession is not practicable?

Construction law can be a minefield. Professionals need answers which are pithy and straightforward, as well as legally rigorous. The two hundred questions in the book are real questions, picked from the thousands of telephone enquiries David Chappell has received as a Specialist Adviser to the Royal Institute of British Architects. Although the enquiries were originally from architects, the answers to most of them are of interest to project managers, contractors, QSs, employers and others involved in construction.

The material is considerably updated from the first edition – weeded, extended and almost doubled in coverage. The questions range in content from extensions of time, liquidated damages and loss and/or expense to issues of warranties, bonds, novation, practical completion, defects, valuation, certificates and payment, architects' instructions, adjudication and fees. Brief footnotes and a table of cases will be retained for those who may wish to investigate further.

August 2010: 216x138: 344pp
Pb: 978-0-415-56650-6: **£34.99**

To Order: Tel: +44 (0) 1235 400524 **Fax:** +44 (0) 1235 400525
or Post: Taylor and Francis Customer Services,
Bookpoint Ltd, Unit T1, 200 Milton Park, Abingdon, Oxon, OX14 4TA UK
Email: book.orders@tandf.co.uk

**For a complete listing of all our titles visit:
www.tandf.co.uk**

Taylor & Francis
Taylor & Francis Group

Value Added Tax

Introduction

Value Added Tax (VAT) is a tax on the consumption of goods and services. The UK adopted VAT when it joined the European Community in 1973. The principal source of European law in relation to VAT is Council Directive 2006/112/EC, a recast of Directive 77/388/EEC which is currently restated and consolidated in the UK through the VAT Act 1994 and various Statutory Instruments, as amended by subsequent Finance Acts.

VAT Notice 708: Buildings and construction (June 2007) gives an interpretation of the law in connection with construction works from the point of view of HM Revenue & Customs. VAT tribunals and court decisions since the date of this publication will affect the application of the law in certain instances. The Notice is available on HM Revenue & Customs website at www.hmrc.gov.uk.

The scope of VAT

VAT is payable on:

- Supplies of goods and services made in the UK
- By a taxable person
- In the course or furtherance of business; and
- Which are not specifically exempted or zero-rated.

Rates of VAT

There are three rates of VAT:

- A standard rate, currently 20% since January 2011
- A reduced rate, currently 5%; and
- A zero rate.

Additionally some supplies are exempt from VAT and others are outside the scope of VAT.

Recovery of VAT

When a taxpayer makes taxable supplies he must account for VAT at the appropriate rate of either 20% or 5%. This VAT then has to be passed to HM Revenue & Customs and will normally be charged to the taxpayer's customers.

As a VAT registered person, the taxpayer can reclaim from HM Revenue & Customs as much of the VAT incurred on their purchases as relates to the standard-rated, reduced-rated and zero-rated onward supplies they make. A person cannot, however, reclaim VAT that relates to any non-business activities (but see below) or to any exempt supplies they make.

At predetermined intervals the taxpayer will pay to HM Revenue & Customs the excess of VAT collected over the VAT they can reclaim. However, if the VAT reclaimed is more than the VAT collected, the taxpayer can reclaim the difference from HM Revenue & Customs.

VALUE ADDED TAX

Example

X Ltd constructs a block of flats. It sells long leases to buyers for a premium. X Ltd has constructed a new building designed as a dwelling and will have granted a long lease. This sale of a long lease is VAT zero-rated. This means any VAT incurred in connection with the development which X Ltd will have paid (e.g. payments for consultants and certain preliminary services) will be reclaimable. For reasons detailed below the contractor employed by X Ltd will not have charged VAT on his construction services.

Use for Business and Non-Business Activities

Where VAT relates partly to business use and partly to non-business use then the basic rule is that it must be apportioned so that only the business element is potentially recoverable. In some cases VAT on land, buildings and certain construction services purchased for both business and non-business use could be recovered in full by applying what is known as 'Lennartz' accounting to reclaim VAT relating to the non-business use and account for VAT on the non-business use over a maximum period of 10 years. HM Revenue & Customs revised their policy in January 2010 following an ECJ case restricting the scope and its application to immovable property will be removed completely from January 2011 when the UK law is amended to comply with EU Council Directive 2009/162/EU.

Taxable Persons

A taxable person is an individual, firm, company etc. who is required to be registered for VAT. A person who makes taxable supplies above certain value limits is required to be registered. The current registration limit is £77,000 for 2012–13. The threshold is exceeded if at the end of any month the value of taxable supplies in the period of one year then ending is over the limit, or at any time, if there are reasonable grounds for believing that the value of the taxable supplies in the period of 30 days then beginning will exceed £77,000.

A person who makes taxable supplies below these limits is entitled to be registered on a voluntary basis if they wish, for example in order to recover VAT incurred in relation to those taxable supplies.

In addition, a person who is not registered for VAT in the UK but acquires goods from another EC member state, or make distance sales in the UK, above certain value limits may be required to register for VAT in the UK.

VAT Exempt Supplies

If a supply is exempt from VAT this means that no tax is payable – but equally the person making the exempt supply cannot normally recover any of the VAT on their own costs relating to that supply.

Generally property transactions such as leasing of land and buildings are exempt unless a landlord chooses to standard-rate its supplies by a process known as opting to tax. This means that VAT is added to rental income and also that VAT incurred on, say, an expensive refurbishment, is recoverable.

Supplies outside the scope of VAT

Supplies are outside the scope of VAT if they are:

- Made by someone who is not a taxable person;
- Made outside the UK; or
- Not made in the course or furtherance of business.

VALUE ADDED TAX

In course or furtherance of business

VAT must be accounted for on all taxable supplies made in the course or furtherance of business with the corresponding recovery of VAT on expenditure incurred.

If a taxpayer also carries out non-business activities then VAT incurred in relation to such supplies is generally not recoverable.

In VAT terms, business means any activity continuously performed which is mainly concerned with making supplies for a consideration. This includes:

- Any one carrying on a trade, vocation or profession;
- The provision of membership benefits by clubs, associations and similar bodies in return for a subscription or other consideration; and
- Admission to premises for a charge.

It may also include the activities of other bodies including charities and non-profit-making organizations. Examples of non-business activities are:

- Providing free services or information;
- Maintaining some museums or particular historic sites;
- Publishing religious or political views.

Construction Services

In general the provision of construction services by a contractor will be VAT standard rated at 20%, however, there are a number of exceptions for construction services provided in relation to certain residential and charitable use buildings.

The supply of building materials is VAT standard rated at 20%, however, where these materials are supplied as part of the construction services the VAT liability of those materials follows that of the construction services supplied.

Zero-rated construction services

The following construction services are VAT zero-rated including the supply of related building materials.

The construction of new dwellings

The supply of services in the course of the construction of a building designed for use as a dwelling or number of dwellings is zero-rated other than the services of an architect, surveyor or any other person acting as a consultant or in a supervisory capacity.

The following conditions must be satisfied in order for the works to qualify for zero-rating:

1. The work must not amount to the conversion, reconstruction or alteration of an existing building;
2. The work must not be an enlargement of, or extension to, an existing building except to the extent that the enlargement or extension creates an additional dwelling or dwellings;
3. The building must be designed as a dwelling or number of dwellings. Each dwelling must consist of self-contained living accommodation with no provision for direct internal access from the dwelling to any other dwelling or part of a dwelling;
4. Statutory planning consent must have been granted for the construction of the dwelling, and construction carried out in accordance with that consent;
5. Separate use or disposal of the dwelling must not be prohibited by the terms of any covenant, statutory planning consent or similar provision.

VALUE ADDED TAX

The construction of a garage at the same time as the dwelling can also be zero-rated as can the demolition of any existing building on the site of the new dwelling.

A building only ceases to be an existing building (see points 1. and 2. above) when it is:

1. Demolished completely to ground level; or when
2. The part remaining above ground level consists of no more than a single facade (or a double facade on a corner site) the retention of which is a condition or requirement of statutory planning consent or similar permission.

The construction of a new building for 'relevant residential or charitable' use

The supply of services in the course of the construction of a building designed for use as a relevant residential or charitable building is zero-rated other than the services of an architect, surveyor or any other person acting as a consultant or in a supervisory capacity.

A relevant residential use building means:

1. A home or other institution providing residential accommodation for children;
2. A home or other institution providing residential accommodation with personal care for persons in need of personal care by reason of old age, disablement, past or present dependence on alcohol or drugs or past or present mental disorder;
3. A hospice;
4. Residential accommodation for students or school pupils;
5. Residential accommodation for members of any of the armed forces;
6. A monastery, nunnery, or similar establishment; or
7. An institution which is the sole or main residence of at least 90% of its residents.

A relevant residential purpose building does not include use as a hospital, a prison or similar institution or as a hotel, inn or similar establishment.

A relevant charitable use means use by a charity:

1. Otherwise than in the course or furtherance of a business; or
2. As a village hall or similarly in providing social or recreational facilities for a local community.

Non-qualifying use which is not expected to exceed 10% of the time the building is normally available for use can be ignored. The calculation of business use can be based on time, floor area or head count subject to approval being acquired from HM Revenue & Customs.

The construction services can only be zero-rated if a certificate is given by the end user to the contractor carrying out the works confirming that the building is to be used for a qualifying purpose, i.e. for a 'relevant residential or charitable' purpose. It follows that such services can only be zero-rated when supplied to the end user and, unlike supplies relating to dwellings, supplies by subcontractors cannot be zero-rated.

The construction of an annexe used for a 'relevant charitable' purpose

Construction services provided in the course of construction of an annexe for use entirely or partly for a 'relevant charitable' purpose can be zero-rated.

In order to qualify the annexe must:

1. Be capable of functioning independently from the existing building;
2. Have its own main entrance; and
3. Be covered by a qualifying use certificate.

VALUE ADDED TAX

The conversion of a non-residential building into dwellings or the conversion of a building from non-residential use to 'relevant residential' use where the supply is to a 'relevant' housing association

The supply to a 'relevant' housing association in the course of conversion of a non-residential building or non-residential part of a building into:

1. A building or part of a building designed as a dwelling or number of dwellings; or
2. A building or part of a building for use solely for a relevant residential purpose, of any services related to the conversion other than the services of an architect, surveyor or any person acting as a consultant or in a supervisory capacity are zero-rated.

A relevant housing association is defined as:

1. A private registered provider of social housing;
2. A registered social landlord within the meaning of Part I of the Housing Act 1996 (Welsh registered social landlords);
3. A registered social landlord within the meaning of the Housing (Scotland) Act 2001 (Scottish registered social landlords); or
4. A registered housing association within the meaning of Part II of the Housing (Northern Ireland) Order 1992 (Northern Irish registered housing associations).

If the building is to be used for a 'relevant residential' purpose the housing association should issue a qualifying use certificate to the contractor completing the works.

The construction of a permanent park for residential caravans

The supply in the course of the construction of any civil engineering work 'necessary for' the development of a permanent park for residential caravans of any services related to the construction can be VAT zero-rated. This includes access roads, paths, drainage, sewerage and the installation of mains water, power and gas supplies.

Certain building alterations for disabled persons

Certain goods and services supplied to a disabled person, or a charity making these items and services available to 'disabled' persons can be zero-rated. The recipient of these goods or services needs to give the supplier an appropriate written declaration that they are entitled to benefit from zero rating.

The following services (amongst others) are zero-rated:

1. the installation of specialist lifts and hoists and their repair and maintenance;
2. the construction of ramps, widening doorways or passageways including any preparatory work and making good work;
3. the provision, extension and adaptation of a bathroom, washroom or lavatory; and
4. emergency alarm call systems.

Approved alterations to protected buildings

A supply in the course of an approved alteration to a protected building of any services other than the services of an architect, surveyor or any person acting as consultant or in a supervisory capacity can be zero-rated.

A protected building is defined as a building that is:

- designed to remain as or become a dwelling or number of dwellings after the alterations; or
- is intended for use for a 'relevant residential or charitable purpose' after the alterations; and which is;
- a listed building or scheduled ancient monument.

VALUE ADDED TAX

A listed building does not include buildings that are in conservation areas, but not on the statutory list, or buildings included in non-statutory local lists.

An approved alteration is an alteration to a 'protected building' that requires and has obtained listed building consent or scheduled monument consent. This consent is necessary for any works that affect the character of a building of special architectural or historic interest.

It is important to note that approved alterations do not include any works of repair or maintenance or any incidental alteration to the fabric of a building that results from the carrying out of repairs or maintenance work.

A protected building that is intended for use for a relevant residential or charitable purpose will require the production of a qualifying use certificate by the end user to the contractor providing the alteration services.

Listed Churches are relevant charitable use buildings and where approved alterations are being carried out zero-rate VAT can be applied. Additionally since April 1 2004, listed places of worship can apply for a grant for repair and maintenance works equal to the full amount of VAT paid on eligible works carried out on or after 1 April 2004. Information relating to the scheme can be obtained from the website at www.lpwscheme.org.uk.

Budget 2012 announced plans to consult on the abolition of the zero percent rate for approved alterations to protected buildings from 1 October 2012. Transitional arrangements will be put in place to protect contracts entered into before 21 March 2012 for expenditure incurred up to 20 March 2013.

Sale of Reconstructed Buildings

Budget 2012 also announced a change in the rules that currently enable developers of substantially reconstructed listed residential properties to recover the VAT they incur on costs relating to the redevelopment through zero-rating. From 1 October 2012 a protected building shall not be regarded as substantially reconstructed unless, when the reconstruction is completed, the reconstructed building incorporates no more of the original building than the external walls, together with other external features of architectural or historical interest. Transitional arrangements protect contracts entered into before 21 March 2012 for the first grant of a major interest in the protected building made on or before 20 March 2013.

DIY Builders and Converters

Private individuals who decide to construct their own home are able to reclaim VAT they pay on goods they use to construct their home by use of a special refund mechanism made by way of an application to HM Revenue & Customs. This also applies to services provided in the conversion of an existing non-residential building to form a new dwelling.

The scheme is meant to ensure that private individuals do not suffer the burden of VAT if they decide to construct their own home.

Charities may also qualify for a refund on the purchase of materials incorporated into a building used for non-business purposes where they provide their own free labour for the construction of a 'relevant charitable' use building.

Reduced-rated construction services

The following construction services are subject to the reduced rate of VAT of 5%, including the supply of related building materials.

VALUE ADDED TAX

A changed number of dwellings conversion

In order to qualify for the 5% rate there must be a different number of 'single household dwellings' within a building than there were before commencement of the conversion works. A 'single household dwelling' is defined as a dwelling that is designed for occupation by a single household.

These conversions can be from relevant residential purpose buildings, non-residential buildings and houses in multiple occupations.

A house in multiple occupation conversion

This relates to construction services provided in the course of converting a 'single household dwelling', a number of single household dwellings, a non-residential building or a relevant residential purpose building into a house for multiple occupation such as a bed sit accommodation.

A special residential conversion

A special residential conversion involves the conversion of a single household dwelling, a house in multiple occupation or a non-residential building into a 'relevant residential' purpose building such as student accommodation or a care home.

Renovation of derelict dwellings

The provision of renovation services in connection with a dwelling or 'relevant residential' purpose building that has been empty for two or more years prior to the date of commencement of construction works can be carried out at a reduced rate of VAT of 5%.

Installation of energy saving materials

The supply and installation of certain energy saving materials including insulation, draught stripping, central heating and hot water controls and solar panels in a residential building or a building used for a relevant charitable purpose.

Budget 2012 announced that buildings that are used by charities for non-business purposes, and/or as village halls, will be removed from the scope of the reduced rate for the supply of energy saving materials under legislation to be introduced in Finance Bill 2013.

Grant-funded heating equipment or connection of a gas supply

The grant-funded supply and installation of heating appliances, connection of a mains gas supply, supply, installation, maintenance and repair of central heating systems, and supply and installation of renewable source heating systems, to qualifying persons. A qualifying person is someone aged 60 or over or is in receipt of various specified benefits.

Installation of security goods

The grant-funded supply and installation of security goods to a qualifying person.

Housing alterations for the elderly

Certain home adaptations that support the needs of elderly people were reduced rated with effect from 1 July 2007.

VALUE ADDED TAX

Building Contracts

Design and build contracts

If a contractor provides a design and build service relating to works to which the reduced or zero rate of VAT is applicable then any design costs incurred by the contractor will follow the VAT liability of the principal supply of construction services.

Management contracts

A management contractor acts as a main contractor for VAT purposes and the VAT liability of his services will follow that of the construction services provided. If the management contractor only provides advice without engaging trade contractors his services will be VAT standard rated.

Construction Management and Project Management

The project manager or construction manager is appointed by the client to plan, manage and coordinate a construction project. This will involve establishing competitive bids for all the elements of the work and the appointment of trade contractors. The trade contractors are engaged directly by the client for their services.

The VAT liability of the trade contractors will be determined by the nature of the construction services they provide and the building being constructed.

The fees of the construction manager or project manager will be VAT standard rated. If the construction manager also provides some construction services these works may be zero or reduced rated if the works qualify.

Liquidated and Ascertained Damages

Liquidated damages are outside of the scope of VAT as compensation. The employer should not reduce the VAT amount due on a payment under a building contract on account of a deduction of damages. In contrast, an agreed reduction in the contract price will reduce the VAT amount.

Similarly, in certain circumstances HM Revenue & Customs may agree that a claim by a contractor under a JCT or other form of contract is also compensation payment and outside the scope of VAT.

The Landfill Tax

The Tax

The Landfill Tax came into operation on 1 October 1996. It is levied on operators of licensed landfill sites at the following rates with effect from 1 April 2012:

• Inactive or inert wastes	£2.50 per tonne	Included are soil, stones, brick, plain and reinforced concrete, plaster and glass
• All other taxable wastes	£64 per tonne	Included are timber, paint and other organic wastes generally found in demolition work and builders skips

The rate for all other taxable wastes will be increased by £8 per tonne each year at least until 2014 when the rate will be £80 per tonne. There will also be a floor under the standard rate, so that the rate will not fall below £80 per tonne from 2014/15 to 2019/20. The lower rate for inactive or inert wastes will be frozen at £2.50 per tonne to 31 March 2013.

The Landfill Tax (Qualifying Material) Order 2011 came into force on 1 April 2011. This has amended the qualifying criteria that are eligible for the lower rate of landfill tax. The revisions introduced arose primarily from the need to reflect changes in wider environmental policy and legislation since 1996, such as the implementation of the European Landfill Directive.

A waste will be lower rated for landfill tax from 1 April 2011 only if it is listed as a qualifying material in the Landfill Tax (Qualifying Material) Order 2011.

The principal changes to qualifying material are:

- Rocks and subsoils that are currently lower rated will remain so.
- Topsoil and peat will be removed from the lower rate, as these are natural resources that can always be recycled/reused.
- Used foundry sand, which has in practice been lower rated by extra-statutory concession since the tax's introduction in 1996, will now be included in the lower rate Order.
- Definitions of qualifying ash arising from the burning of coal and petroleum coke (including when burnt with biomass) will be clarified.
- The residue from titanium dioxide manufacture will qualify, rather than titanium dioxide itself, reflecting industry views.
- Minor changes will be made to the wording of the calcium sulphate group of wastes to reflect the implementation of the Landfill Directive since 2001.
- Water will be removed from the lower rate – water is now banned from landfill so its inclusion in the list of lower rated wastes is unnecessary; where water is used as a waste carrier the water is not waste and therefore not taxable.

Calculating the Weight of Waste

There are two options:

- If licensed sites have a weighbridge, tax will be levied on the actual weight of waste.
- If licensed sites do not have a weighbridge, tax will be levied on the permitted weight of the lorry based on an alternative method of calculation based on volume to weight factors for various categories of waste.

THE LANDFILL TAX

Effect on Prices

The tax is paid by Landfill site operators only. Tipping charges reflect this additional cost.

As an example, Spon's A & B rates for mechanical disposal will be affected as follows:

• Inactive waste	Spon's A & B 2013 net rate	£12.51 per m³
	Tax, 1.9 tonne per m³ (un-bulked) @ £2.50	£4.75 per m³
	Spon's rate including tax	£17.26 per m³
• Active waste	Active waste will normally be disposed of by skip and will probably be mixed with inactive waste. The tax levied will depend on the weight of materials in the skip which can vary significantly.	

Exemptions

The following disposals are exempt from Landfill Tax subject to meeting certain conditions:

- dredgings which arise from the maintenance of inland waterways and harbours
- naturally occurring materials arising from mining or quarrying operations
- reclamation of contaminated land
- inert waste used to restore landfill sites and to fill working and old quarries where a planning condition or obligation is in existence
- waste from visiting NATO forces

The exemption for waste from contaminated land has been phased out completely by 1 April 2012 and no new applications for landfill tax exemption are now accepted.

For further information contact the National Advisory Service, Telephone: 0845 010 9000.

Land Remediation

The purpose of this section is to review the general background of ground contamination, the cost implications of current legislation and to consider the various remedial measures and to present helpful guidance on the cost of Land Remediation.

It must be emphasized that the cost advice given is an average and that costs can vary considerably from contract to contract depending on individual Contractors, site conditions, type and extent of contamination, methods of working and various other factors as diverse as difficulty of site access and distance from approved tips.

We have structured this Unit Cost section to cover as many aspects of Land Remediation works as possible.

The introduction of the Landfill Directive in July 2004 has had a considerable impact on the cost of Remediation works in general and particularly on the practice of Dig and Dump. The number of Landfill sites licensed to accept Hazardous Waste has drastically reduced and inevitably this has led to increased costs.

Market forces will determine future increases in cost resulting from the introduction of the Landfill Directive and the cost guidance given within this section will require review in light of these factors.

Statutory framework

In July 1999 new contaminated land provisions, contained in Part IIA of the Environmental Protection Act 1990 were introduced. Primary objectives of the measures included a legal definition of Contaminated Land and a framework for identifying liability, underpinned by a 'polluter pays' principle meaning that remediation should be paid for by the party (or parties) responsible for the contamination. A secondary, and indirect, objective of Part IIA is to provide the legislative context for remediation carried out as part of development activity which is controlled through the planning system. This is the domain where other related objectives, such as encouraging the recycling of brownfield land, are relevant.

Under the Act action to remediate land is required only where there are unacceptable actual or 'significant possibility of significant harm' to health, controlled waters or the environment. Only Local Authorities have the power to determine a site as Contaminated Land and enforce remediation. Sites that have been polluted from previous land use may not need remediating until the land use is changed; this is referred to as 'land affected by contamination'. This is a risk-based assessment on the site specifics in the context of future end uses. As part of planning controls, the aim is to ensure that a site is incapable of meeting the legal definition of Contaminated Land post-development activity. In addition, it may be necessary to take action only where there are appropriate, cost-effective remediation processes that take the use of the site into account.

The Environment Act 1995 amended the Environment Protection Act 1990 by introducing a new regime designed to deal with the remediation of sites which have been seriously contaminated by historic activities. The regime became operational on 1 April 2000. Local authorities and/or the Environment Agency regulate seriously contaminated sites which are known as 'special sites'. The risks involved in the purchase of potentially contaminated sites are high, particularly considering that a transaction can result in the transfer of liability for historic contamination from the vendor to the purchaser.

LAND REMEDIATION

The contaminated land provisions of the Environmental Protection Act 1990 are only one element of a series of statutory measures dealing with pollution and land remediation that have been and are to be introduced. Others include:

- Groundwater regulations, including pollution prevention measures
- An integrated prevention and control regime for pollution
- Sections of the Water Resources Act 1991, which deals with works notices for site controls, restoration and clean up

April 2012 saw the first revision of the accompanying Part IIA Statutory Guidance. This has introduced a new categorization scheme for assessing sites under Part IIA. Category 1 is land which definitely is Contaminated Land and Category 4 is for land which definitely is not Contaminated Land. This is intended to assist prioritisation of sites which pose the greatest risk. Still included in the statutory guidance are matters of inspection, definition, remediation, apportionment of liabilities and recovery of costs of remediation. The measures are to be applied in accordance with the following criteria:

- the planning system
- the standard of remediation should relate to the present use
- the costs of remediation should be reasonable in relation to the seriousness of the potential harm
- the proposals should be practical in relation to the availability of remediation technology, impact of site constraints and the effectiveness of the proposed clean-up method

Liability for the costs of remediation rests with either the party that 'caused or knowingly permitted' contamination, or with the current owners or occupiers of the land.

Apportionment of liability, where shared, is determined by the local authority. Although owners or occupiers become liable only if the polluter cannot be identified, the liability for contamination is commonly passed on when land is sold.

If neither the polluter nor owner can be found, the clean up is funded from public resources.

The ability to forecast the extent and cost of remedial measures is essential for both parties, so that they can be accurately reflected in the price of the land.

At the end of March 2012, the National Planning Policy Framework replaced relevant planning guidance relating to remediation, most significantly PPS 23 Planning and Pollution Control. This has been replaced by a need to investigate and assess land contamination, which must be carried out by a competent person.

The EU Landfill Directive

The Landfill (England and Wales) Regulations 2002 came into force on 15 June 2002 followed by Amendments in 2004 and 2005. These new regulations implement the Landfill Directive (Council Directive 1999/31/EC), which aims to prevent, or to reduce as far as possible, the negative environmental effects of landfill. These regulations have had a major impact on waste regulation and the waste management industry in the UK.

The Scottish Executive and the Northern Ireland Assembly will be bringing forward separate legislation to implement the Directive within their regions.

In summary, the Directive requires that:

- Sites are to be classified into one of three categories: hazardous, non-hazardous or inert, according to the type of waste they will receive
- Higher engineering and operating standards will be followed
- Biodegradable waste will be progressively diverted away from landfills
- Certain hazardous and other wastes, including liquids, explosive waste and tyres will be prohibited from landfills
- Pretreatment of wastes prior to landfilling will become a requirement

LAND REMEDIATION

On 15 July 2004 the co-disposal of hazardous and non-hazardous waste in the same landfill site ended and in July 2005 new waste acceptance criteria (WAC) were introduced which also prevents the disposal of materials contaminated by coal tar.

The effect of this Directive has been to dramatically reduce the hazardous disposal capacity post July 2004, resulting in a **SIGNIFICANT** increase in remediating costs. There are now approximately 20 commercial landfills licensed to accept hazardous waste as a direct result of the implementation of the Directive! There is one site in Scotland, none in Wales and only limited capacity in the South of England. This has significantly increased travelling distance and cost for disposal to landfill. The increase in operating expenses incurred by the landfill operators has also resulted in higher tipping costs.

However, there is now a growing number of opportunities to dispose of hazardous waste to other facilities such as soil treatment centres, often associated with registered landfills potentially eliminating landfill tax. Equally, improvements in on-site treatment technologies is helping to reduce the costs of disposal by reducing the hazardous properties of materials going off site.

All hazardous materials designated for disposal off site are subject to WAC tests. Samples of these materials are taken from site to laboratories in order to classify the nature of the contaminants. These tests, which cost approximately £200 each, have resulted in increased costs for site investigations and as the results may take up to 3 weeks this can have a detrimental effect on programme.

As from 1 July 2008 the WAC derogations which have allowed oil contaminated wastes to be disposed in landfills with other inert substances were withdrawn. As a result the cost of disposing of oil contaminated solids has increased.

There has been a marked slowdown in brownfield development in the UK with higher remediation costs, longer clean-up programmes and a lack of viable treatment options for some wastes.

The UK Government established the Hazardous Waste Forum in December 2002 to bring together key stakeholders to advise on the way forward on the management of hazardous waste.

Effect on Disposal Costs

Although most landfills are reluctant to commit to future tipping prices, tipping costs have generally stabilized. However, there are significant geographical variances, with landfill tip costs in the North of England typically being less than their counterparts in the Southern regions.

For most projects to remain viable there is an increasing need to treat soil in situ by bioremediation, soil washing or other alternative long-term remediation measures. Waste untreatable on site such as coal tar remains a problem. Development costs and programmes need to reflect this change in methodology.

Types of hazardous waste

- Sludges, acids and contaminated wastes from the oil and gas industry
- Acids and toxic chemicals from chemical and electronics industries
- Pesticides from the agrochemical industry
- Solvents, dyes and sludges from leather and textile industries
- Hazardous compounds from metal industries
- Oil, oil filters and brake fluids from vehicles and machines
- Mercury-contaminated waste from crematoria
- Explosives from old ammunition, fireworks and airbags
- Lead, nickel, cadmium and mercury from batteries
- Asbestos from the building industry
- Amalgam from dentists
- Veterinary medicines

[Source: Sepa]

LAND REMEDIATION

Foam insulation materials containing ODP (Ozone Depletant Potential) are also considered as hazardous waste under the EC Regulation 2037/2000.

LAND REMEDIATION TECHNIQUES

There are two principal approaches to remediation – dealing with the contamination in situ or ex situ. The selection of the approach will be influenced by factors such as: initial and long-term cost, timeframe for remediation, types of contamination present, depth and distribution of contamination, the existing and planned topography, adjacent land uses, patterns of surface drainage, the location of existing on-site services, depth of excavation necessary for foundations and below-ground services, environmental impact and safety, interaction with geotechnical performance, prospects for future changes in land use and long-term monitoring and maintenance of in situ treatment.

On most sites, contamination can be restricted to the top couple of metres, although gasholder foundations, for example, can go down 8 to 10 metres. Underground structures can interfere with the normal water regime and trap water pockets.

There could be a problem if contaminants get into fissures in bedrock.

In situ techniques

A range of in situ techniques is available for dealing with contaminants, including:

- Clean cover – a layer of clean soil is used to segregate contamination from receptor. This technique is best suited to sites with widely dispersed contamination. Costs will vary according to the need for barrier layers to prevent migration of the contaminant.
- On-site encapsulation – the physical containment of contaminants using barriers such as slurry trench cut-off walls. The cost of on-site encapsulation varies in relation to the type and extent of barriers required, the costs of which range from £50/m² to more than £175/m².

There are also in situ techniques for treating more specific contaminants, including:

- Bioremediation – for removal of oily, organic contaminants through natural digestion by microorganisms. Most bioremediation is ex situ, i.e. it is dug out and then treated on site in bio-piles. The process can be slow, taking up to three years depending upon the scale of the problem, but is particularly effective for the long-term improvement of a site, prior to a change of use.
- Phytoremediation – the use of plants that mitigate the environmental problem without the need to excavate the contaminant material and dispose of it elsewhere. Phytoremediation consists of mitigating pollutant concentrations in contaminated soils, water or air, with plants able to contain, degrade or eliminate metals, pesticides, solvents, explosives, crude oil and its derivatives and various other contaminants from the media that contain them.
- Vacuum extraction – involving the extraction of volatile organic compounds (e.g. benzene) from soil and groundwater by vacuum.
- Thermal treatment – the incineration of contaminated soils on site. Thermal processes use heat to increase the volatility to burn, decompose, destroy or melt the contaminants. Cleaning soil with thermal methods may take only a few months to several years.
- Stabilization – cement or lime, is used to physically or chemically bind oily or metal contaminants to prevent leaching or migration. Stabilization can be used in both in situ and ex situ conditions.
- Aeration – if the ground contamination is highly volatile, e.g. fuel oils, then the ground can be ploughed and rotovated to allow the substance to vaporize.
- Air sparging – the injection of contaminant-free air into the subsurface enabling a phase transfer of hydrocarbons from a dissolved state to a vapour phase.
- Chemical oxidization – the injection of reactive chemical oxidants directly into the soil for the rapid destruction of contaminants.
- Pumping – to remove liquid contaminants from boreholes or excavations. Contaminated water can be pumped into holding tanks and allowed to settle; testing may well prove it to be suitable for discharging into the foul sewer subject to payment of a discharge fee to the local authority. It may be necessary to process the water through an approved water treatment system to render it suitable for discharge.

LAND REMEDIATION

Ex situ techniques

Removal for landfill disposal has, historically, been the most common and cost-effective approach to remediation in the UK, providing a broad spectrum solution by dealing with all contaminants. As mentioned above, the implementation of the Landfill Directive has resulted in other techniques becoming more competitive for the disposal of hazardous waste.

If used in combination with material-handling techniques such as soil washing, the volume of material disposed at landfill sites can be significantly reduced. The disadvantages of these techniques include the fact that the contamination is not destroyed, there are risks of pollution during excavation and transfer; road haulage may also cause a local nuisance. Ex situ techniques include:

- Soil washing – involving the separation of a contaminated soil fraction or oily residue through a washing process. This also involves the excavation of the material for washing ex situ. The dewatered contaminant still requires disposal to landfill. In order to be cost effective, 70–90% of soil mass needs to be recovered. It will involve constructing a hard area for the washing, intercepting the now-contaminated water and taking it away in tankers.
- Thermal treatment – the incineration of contaminated soils ex situ. The uncontaminated soil residue can be recycled. By-products of incineration can create air pollution and exhaust air treatment may be necessary.

Soil treatment centres are now beginning to be established. These use a combination of treatment technologies to maximize the potential recovery of soils and aggregates and render them suitable for disposal to the landfill. The technologies include:

- Physicochemical treatment – a method which uses the difference in grain size and density of the materials to separate the different fractions by means of screens, hydrocyclones and upstream classification.
- Bioremediation – the aerobic biodegradation of contaminants by naturally occurring microorganisms placed into stockpiles/windrows.
- Stabilization/solidification – a cement or chemical stabilization unit capable of immobilizing persistent leachable components.

COST CONSIDERATIONS

Cost drivers

Cost drivers relate to the selected remediation technique, site conditions and the size and location of a project.

The wide variation of indicative costs of land remediation techniques shown below is largely because of differing site conditions.

Indicative costs of land remediation techniques for 2012 (excluding general items, testing, landfill tax and backfilling)		
Remediation technique	Unit	Rate (£/unit)
Removal – non-hazardous	disposed material (m^3)	30–50
Removal – hazardous	disposed material (m^3)	75–200
Note: excluding any pretreatment of material		
Clean cover	surface area of site (m^2)	20–45
On-site encapsulation	encapsulated material (m^3)	30–95
Bioremediation (in situ)	treated material (m^3)	15–45
Bioremediation (ex situ)	treated material (m^3)	20–50
Chemical oxidation	treated material (m^3)	30–80
Stabilization/solidification	treated material (m^3)	20–65
Vacuum extraction	treated material (m^3)	25–65
Soil washing	treated material (m^3)	35–80
Thermal treatment	treated material (m^3)	100–400

LAND REMEDIATION

Many other on-site techniques deal with the removal of the contaminant from the soil particles and not the wholesale treatment of bulk volumes. Costs for these alternative techniques are very much Engineer designed and site specific.

Factors that need to be considered include:

- waste classification of the material
- underground obstructions, pockets of contamination and live services
- ground water flows and the requirement for barriers to prevent the migration of contaminants
- health and safety requirements and environmental protection measures
- location, ownership and land use of adjoining sites
- distance from landfill tips, capacity of the tip to accept contaminated materials, and transport restrictions
- the cost of diesel fuel, currently approximately £1.47 per litre (at April 2012 prices)

Other project-related variables include size, access to disposal sites and tipping charges; the interaction of these factors can have a substantial impact on overall unit rates.

The tables below set out the costs of remediation using dig-and-dump methods for different sizes of project, differentiated by the disposal of non-hazardous and hazardous material. Variation in site establishment and disposal cost accounts for 60–70% of the range in cost.

Variation in the costs of land remediation by removal: Non-hazardous Waste			
Item	Disposal volume (less than 3000 m^3) (£/m^3)	Disposal volume (3000–10 000 m^3) (£/m^3)	Disposal volume (more than 10 000 m^3) (£/m^3)
General items and site organization costs	55–90	25–40	7–20
Site investigation and testing	5–12	2–7	2–6
Excavation and backfill	18–35	12–25	10–20
Disposal costs (including tipping charges but not landfill tax)	20–35	20–35	20–35
Haulage	15–35	15–35	15–35
Total (£/m^3)	113–207	74–142	54–116
Allowance for site abnormals	0–10 +	0–15 +	0–10 +

Variation in the costs of land remediation by removal: Hazardous Waste			
Item	Disposal volume (less than 3000 m^3) (£/m^3)	Disposal volume (3000–10 000 m^3) (£/m^3)	Disposal volume (more than 10 000 m^3) (£/m^3)
General items and site organization costs	55–90	25–40	7–20
Site investigation and testing	10–18	5–12	5–12
Excavation and backfill	18–35	12–25	10–20
Disposal costs (including tipping charges but not landfill tax)	80–170	80–170	80–170
Haulage	25–120	25–120	25–120
Total (£/m^3)	188–433	147–367	127–342
Allowance for site abnormals	0–10 +	0–15 +	0–10 +

The strict health and safety requirements of remediation can push up the overall costs of site organization to as much as 50% of the overall project cost (see the above tables). A high proportion of these costs are fixed and, as a result, the unit costs of site organization increase disproportionally on smaller projects.

LAND REMEDIATION

Haulage costs are largely determined by the distances to a licensed tip. Current average haulage rates, based on a return journey, range from £1.95 to £4.50 per mile. Short journeys to tips, which involve proportionally longer standing times, typically incur higher mileage rates, up to £9.00 per mile.

A further source of cost variation relates to tipping charges. The table below summarizes typical tipping charges for 2012, exclusive of landfill tax:

Typical 2012 tipping charges (excluding landfill tax)	
Waste classification	Charges (£/tonne)
Non-hazardous wastes	8–15
Hazardous wastes	25–85
Contaminated liquid	40–75
Contaminated sludge	55–200

Tipping charges fluctuate in relation to the grades of material a tip can accept at any point in time. This fluctuation is a further source of cost risk. Furthermore, tipping charges in the North of England are generally less than other areas of the country.

Prices at licensed tips can vary by as much as 50%. In addition, landfill tips generally charge a tip administration fee of approximately £25 per load, equivalent to £1.25 per tonne. This charge does not apply to non-hazardous wastes.

Landfill Tax, increased on 1 April 2012 to £64 a tonne for active waste, is also payable. Exemptions are no longer available for the disposal of historically contaminated material (refer also to 'Landfill Tax' section).

Tax Relief for Remediation of Contaminated Land

The Finance Act 2001 included provisions that allow companies (but not individuals or partnerships) to claim tax relief on capital and revenue expenditure on the 'remediation of contaminated land' in the United Kingdom. The relief is available for expenditure incurred on or after 11 May 2001.

From 1 April 2009 there was an increase in the scope of costs that qualify for Land Remediation Relief where they are incurred on long-term derelict land. The list includes costs that the Treasury believe to be primarily responsible for causing dereliction, such as additional costs for removing building foundations and machine bases. However, while there is provision for the list to be extended, the additional condition for the site to have remained derelict since 1998 is likely to render this relief redundant in all but a handful of cases. The other positive change is the fact that Japanese Knotweed removal and treatment (on site only) will now qualify for the relief under the existing legislation, thereby allowing companies to make retrospective claims for any costs incurred since May 2001 – provided all other entitlement conditions are met.

A company is able to claim an additional 50% deduction for 'qualifying land remediation expenditure' allowed as a deduction in computing taxable profits, and may elect for the same treatment to be applied to qualifying capital expenditure.

With Landfill Tax Exemption (LTE) now phased out, Land Remediation Relief (LRR) for contaminated and derelict land is the Government's primary tool to create incentives for brownfield development. LRR is available to companies engaged in land remediation that are not responsible for the original contamination.

Over 7 million tonnes of waste each year were being exempted from Landfill Tax in England alone, so this change could have a major impact on the remediation industry. The modified LRR scheme, which provides Corporation Tax relief on any costs incurred on qualifying land remediation expenditure, is in the long run designed to yield benefits roughly equal to those lost through the withdrawal of LTE, although there is some doubt about this stated equity in reality.

LAND REMEDIATION

However, with much remediation undertaken by polluters or public authorities, who cannot benefit from tax relief benefits, the change could result in a net withdrawal of Treasury support to a vital sector. Lobbying and consultation continues to ensure the Treasury maintains its support for remediation.

While there are no financial penalties for not carrying out remediation, a steep escalator now affects the rate of Landfill Tax for waste material other than inert or inactive wastes, which will rise at the rate of £8/tonne per year until 2014–15. By then, the rate will be £80/tonne. This means that for schemes where there is no alternative to dig and dump and no pre-existing LTE, the cost of remediation could rise to prohibitive levels.

Looking forward, tax-relief benefits under LRR could provide a significant cash contribution to remediation. Careful planning is the key to ensure that maximum benefits are realized, with actions taken at the points of purchase, formation of JV arrangements, procurement of the works and formulation of the Final Account (including apportionment of risk premium) all influencing the final value of the claim agreed with HM Revenue & Customs.

The Relief

Qualifying expenditure may be deducted at 150% of the actual amount expended in computing profits for the year in which it is incurred.

For example, a property trading company may buy contaminated land for redevelopment and incurs £250,000 on qualifying land remediation expenditure that is an allowable for tax purposes. It can claim an additional deduction of £125,000, making a total deduction of £375,000. Similarly, a company incurring qualifying capital expenditure on a fixed asset of the business is able to claim the same deduction provided it makes the relevant election within 2 years.

What is Remediation?

Land remediation is defined as the doing of works including preparatory activities such as condition surveys, to the land in question, any controlled waters affected by the land, or adjoining or adjacent land for the purpose of:

- Preventing or minimizing, or remedying or mitigating the effects of, any relevant harm, or any pollution of controlled waters, by reason of which the land is in a contaminated state.

Definitions

Contaminated land is defined as land that, because of substances on or under it, is in such a condition that relevant harm is or has the significant possibility of relevant harm being caused to:

- The health of living organisms
- Ecological systems
- Quality of controlled waters
- Property

Relevant harm is defined as meaning:

- death of living organisms or significant injury or damage to living organisms
- significant pollution of controlled waters
- a significant adverse impact on the ecosystem, or
- structural or other significant damage to buildings or other structures or interference with buildings or other structures that significantly compromises their use

Land includes buildings on the land, and expenditure on asbestos removal is expected to qualify for this tax relief. It should be noted that the definition is not the same as that used in the Environmental Protection Act Part IIA.

Sites with a nuclear license are specifically excluded.

LAND REMEDIATION

Conditions

To be entitled to claim LRR, the general conditions for all sites, which must all be met, are:

- Must be a company
- Must be land in the United Kingdom
- Must acquire an interest in the land
- Must not be the polluter or have a relevant connection to the polluter
- Must not be in receipt of a subsidy
- Must not also qualify for Capital Allowances (particular to capital expenditure only)

Additional conditions introduced since 1 April 2009:

- The interest in land must be major – freehold or leasehold longer than seven years
- Must not be obligated to carry out remediation under a statutory notice

Additional conditions particular to derelict land:

- Must not be in or have been in productive use at any time since at least 1 April 1998
- Must not be able to be in productive use without the removal of buildings or other structures

In order for expenditure to become qualifying, it must relate to substances present at the point of acquisition.

Furthermore, it must be demonstrated that the expenditure would not have been incurred had those substances not been present.

ESSENTIAL READING FROM TAYLOR AND FRANCIS

A Guide to Field Instrumentation in Geotechnics

R. Bassett

Geotechnical instrumentation is used for installation, monitoring and assessment on any sizeable project, particularly in urban areas, and is used for recording, controlled remedial work, and safety.

This unique and up-to-date book deals with the conceptual philosophy behind the use of instruments, and then systematically covers their practical use. It is divided into displacement dominated systems and stress recording systems. The limitations are discussed and the theoretical background for data assessment and presentation are covered in some detail, with some relevant background material in theoretical soil mechanics. Relevant advanced electronic techniques such as laser scanning in surveying and fibre-optics are also included and communication and data recovery systems are discussed.

It is written for senior designers, consulting engineers and major contractors who need a major introduction to the general purpose, availability and analysis of field instruments before details of their own project can be progressed, and serves as a text book to any specialist geotechnical MSc or professional seminar course in which instrumentation forms a major part.

November 2011: 232pp
Hb: 978-0-415-67537-6: **£90.00**

To Order: Tel: +44 (0) 1235 400524 **Fax:** +44 (0) 1235 400525
or Post: Taylor and Francis Customer Services,
Bookpoint Ltd, Unit T1, 200 Milton Park, Abingdon, Oxon, OX14 4TA UK
Email: book.orders@tandf.co.uk

**For a complete listing of all our titles visit:
www.tandf.co.uk**

Taylor & Francis
Taylor & Francis Group

PART 2

Oncosts and Profits

	page
Risk/opportunity	49
Head office overheads	49
Profit	49
Finance costs – assessment of contract investment	49
Value added tax	51
Tender summary	51
Preliminaries and general items	53

ESSENTIAL READING FROM TAYLOR AND FRANCIS

Failures in Concrete Structures: Case Studies in Reinforced and Prestressed Concrete

R. Whittle

Some lessons are only learned from mistakes. But, it's much cheaper to learn from someone else's mistakes than to have to do so from your own. Drawing on over 50 years of working with concrete structures, Robin Whittle examines the problems which he has seen occur and shows how they could have been avoided.

The first and largest part of the book tells the stories of a number of cases where things have gone wrong with concrete structures. Each case is analysed to identify its cause and how it might have been prevented. It then looks at how failures in structural modelling can lead to big problems if they are not identified before construction is undertaken. Beyond this it examines how contract arrangements can encourage or prevent problems in the designing and building processes. It concludes with an examination of the role research and development in preventing failures.

By identifying the differences between shoddy economisations and genuine efficiency savings, this book offers savings in the short term which won't be at the expense of a structure's long-term performance. Invaluable reading if you're designing or building concrete structures and want to avoid problems which could be expensive or embarrassing further down the line.

July 2012:: 234x156: 154pp
Hb: 978-0-415-56701-5: **£ 60.00**

To Order: Tel: +44 (0) 1235 400524 **Fax:** +44 (0) 1235 400525
or Post: Taylor and Francis Customer Services,
Bookpoint Ltd, Unit T1, 200 Milton Park, Abingdon, Oxon, OX14 4TA UK
Email: book.orders@tandf.co.uk

For a complete listing of all our titles visit:
www.tandf.co.uk

Taylor & Francis
Taylor & Francis Group

Oncosts and Profit

In Part 1 of this book, it is stressed that the cost information given in Parts 4 to 7 leads to a cost estimate that requires further adjustment before it is submitted as a tender. This part deals with those adjustments and includes a worked example of a Tender Summary.

RISK / OPPORTUNITY

The factors to be taken into account when gauging the possibility of the Estimator's prediction of cost being inadequate or excessive are given in Part 2. Clearly it is considered in parallel with profit and it is not possible to give any indicative guidance on the level of adjustment that might result. For the purpose of a preliminary estimate, it is suggested that no adjustment is made to the costs generated by the other parts of this book.

At the same time as making a general appraisal of risk/opportunity, management will look at major quantities and may suggest amendments to the unit rates proposed.

HEAD OFFICE OVERHEADS

An addition needs to be made to the net estimate to cover all costs incurred in operating the central services provided by head office. Apart from general management and accountancy, this will normally include the departments dealing with:

- estimating
- planning and design
- purchasing
- surveying
- insurance
- wages and bonus
- site safety

The appropriate addition varies with the extent of services provided centrally, rather than on site, and with size of organization, but a range of 4% to 8% on turnover would cover most circumstances.

Some organizations would include finance costs with head office overheads, as a general charge to the company, but for the purposes of this book finance costs are treated separately (see below).

PROFIT

Obviously, the level of profit is governed by the degree of competition applicable to the job – which is in turn a function of the industry's current workload. Again, the appropriate addition is highly variable, but for the purposes of a preliminary estimate an addition of 2% to 5% onto nett turnover is suggested.

FINANCE COSTS – ASSESSMENT OF CONTRACT INVESTMENT

The following procedure may be followed to give an indication of the average amount of capital investment required to finance the contract. It must be emphasized that this method will not give an accurate investment as this can only be done by preparing a detailed cash flow related to the programme for the contract. The example is based on the same theoretical contract used for the worked example in Part 2 and should be read in conjunction with the Tender Summary that follows.

The average monthly income must first be assessed. This is done by deducting from the Tender total the contingency items and those items for which immediate payment is necessary.

	£	£
Tender total (excluding finance charges)		10,396,313
Deduct		
Subcontractors	2,000,000	
Prime cost sums	100,000	
Employer's contingencies	245,500	2,345,500
Amount to be financed	£	8,050,813

The average monthly income is this sum (£8,050,813) divided by the contract period (12 months), that is £670,901.

The average contract investment may now be calculated as follows:

	£	£
Plant and equipment to be purchased		90,000
Non time related		
Contractor	240,000	
Employer	8,000	
Other services, charges and fees	NIL	
Subtotal £	248,000	
Take 50% as an average [1]		124,000
Stores and unfixed materials on site		20,000
Work done but not paid for		
2½ months at £670,901 (see table above) [2]	1,677,253	
Less retention at 5% [3]	−83,863	1,593,390
Retention (5% with limit of 3%)		
Average retention [4] 3% of £ 8,050,813 (see table above)		241,524
Subtotal £		2,068,914
Deduct		
Advance payment by client	NIL	
Bill loading [5]	180,000	
Creditors (suppliers)	500,000	
Average contract investment	£	1,388,914

The interest charges that must be added to the Tender price (or absorbed from profit if capital needs to be borrowed) are therefore:

    £ 1,388,914 × say 3.5% [6] × 1 year = say – £ 48,612

Notes

1. These non time related oncosts and services are incurred as lump sums during the contract and, therefore, only 50% of such costs are taken for investment purposes.
2. This period depends on the terms of payment set out in the contract.
3. Retention is deducted as full retention is taken into account later.
4. Average retention will depend on the retention condition set out in the contract, taking into account any partial completion dates.
5. The contractor assesses here any financial advantage he may obtain by varying his items.
6. Assumed Rate – current market varies greatly.

VALUE ADDED TAX

All of the figures quoted in this book exclude value added tax, which the conditions of contract normally make the subject of a separate invoicing procedure between the contractor and the employer.

Value Added Tax will be chargeable at the standard rate, currently 20%, on supplies of services in the course of:

7. The construction of a non-domestic building
8. The construction or demolition of a civil engineering work
9. The demolition of any building, and
10. The approved alteration of a non-domestic protected building

TENDER SUMMARY

This summary sets out a suggested method of collecting together the various costs and other items and sums which, together, make up the total Tender sum for the example contract.

	£	£
Preliminaries and General Items (from Part 2)		
Contractor's site oncosts – time related	909,935	
Contractor's site oncosts – non time related	238,502	
Employer's requirements – time related	145,300	
Employer's requirements – non time related	7,700	
Other services, charges and fees	NIL	
Temporary works not included in unit costs	123,920	
Plant not included in unit costs	210,693	1,636,050
Estimated net cost of measured work, priced at unit costs		5,280,000
	£	6,916,050
Allowance for fixed price contract		
6% on labour (assumed to be £1,400,000)	84,000	
4% on materials (assumed to be £2,500 000)	100,000	
4% on plant (assumed to be £400,000)	16,000	
5% on staff, overheads etc. (assumed to be £600,000)	30,000	230,000
Subcontractors (net)		2,000,000
Prime cost sums		100,000
Adjustments made at Management Appraisal		
price adjustments, add say	75,000	
risk evaluation, add say	50,000	
	£	9,246,050
Head office overheads and profit at 6%		554,763
Finance costs (from previous page)		48,612
Provisional Sums		175,000
Dayworks Bill		175,000
	£	10,199,425
Employer's contingencies		245,500
TENDER TOTAL	£	**10,444,925**

ESSENTIAL READING FROM TAYLOR AND FRANCIS

Performance Based Fire Engineering of Structures

Yong Wang et al

This authoritative book summarizes the latest knowledge on performance-based approaches to the fire safety engineering of structures. Going beyond codes of practice, it addresses issues that have become more important in recent years, including the integrity of fire-resistant compartmentation, the progressive collapse of structures in fire and provisions for connection robustness, and accurate information on material properties at high temperatures. Written by core members of the COST C26 working group on fire resistance, it explains how to use advanced fire engineering design to improve structural safety and reduce construction costs.

June 2012: 234x156: 394pp
Hb: 978-0-415-55733-7: **£80.00**

To Order: Tel: +44 (0) 1235 400524 **Fax:** +44 (0) 1235 400525
or Post: Taylor and Francis Customer Services,
Bookpoint Ltd, Unit T1, 200 Milton Park, Abingdon, Oxon, OX14 4TA UK
Email: book.orders@tandf.co.uk

For a complete listing of all our titles visit:
www.tandf.co.uk

Taylor & Francis
Taylor & Francis Group

Preliminaries and General Items

This part deals with that portion of Civil Engineering costs not, or only indirectly, related to the actual quantity of work being carried out. It comprises a definition of Method Related Charges, a checklist of items to be accounted for on a typical Civil Engineering contract and a worked example illustrating how the various items on the checklist can be dealt with.

GENERAL REQUIREMENTS AND METHOD RELATED CHARGES

Although the more familiar terminology of Preliminaries and General Items is used in this book the principle of Method Related Charges – separating non quantity related charges from quantity related charges – is adopted. Generally the former are dealt with in this part, while the latter are dealt with in Part 5: Unit Costs. In this part non quantity related charges are further subdivided into those that are time related and those that are non time related.

The concept of METHOD RELATED CHARGES can be summarized as follows:

In commissioning Civil Engineering work the Employer buys the materials left behind, but only hires from the Contractor the men and machines which manipulate them, and the management skills to manipulate them effectively. It is logical to assess their values in the same terms as the origin of their costs. It is illogical not to do so if the Employer is to retain the right at any time to vary what is left behind and if the financial uncertainties affecting Employer and Contractors are to be minimized.

Tenderers have the option to define a group of bill items and insert charges against them to cover those expected costs which are not proportional to the quantities of Permanent Works. To distinguish these items they are called Method Related Charges. They are themselves divided into charges for recurrent or time related cost elements, such as maintaining site facilities or operating major plant, and charges for elements which are neither recurrent nor directly related to quantities, such as setting up, bringing plant to site and Temporary Works.

Another hope expressed with the introduction of Method Related Charges was that they should accurately reflect the work described in the item and that they should not, as had become the practice with some of the vague general items frequently included in Civil Engineering Bills, be used as a home for lump sum tender adjustments quite unrelated to the item. Where cost information is given in the worked example presented at the end of this part of the book, therefore, it must be stressed that only direct and relevant costs are quoted.

Where no detailed information is available, it is suggested that when preparing a preliminary estimate an addition of between 12.5% and 25% of net contract value is made to cover Contractor's Site Oncosts, both time and non time related. The higher percentages' should apply to specialist works such as railways, marine and tunnelling.

CHECKLIST OF ITEMS

The following checklist is representative but not exhaustive. It lists and describes the major preliminary and general items which are included, implicitly or explicitly, in a typical Civil Engineering contract and, where appropriate, gives an indication of how they might be costed. Generally contract documents give detailed requirements for the facilities and equipment to be provided for the Employer and for the Engineer's representative and Bills of Quantities produced in conformity with CESMM3 Class A provide items against which these may be priced; no such items are provided for Contractor's site oncosts or, usually, for temporary works and general purpose plant. For completeness a checklist of both types of item is given here under the following main headings:

PRELIMINARIES AND GENERAL ITEMS

Contractor's site oncosts	time related
	non time related
Employer's and consultants' site requirements	time related
	non time related
Site requirements	time related
	non time related
Other services, charges and fees	
Temporary works (other than those included in unit costs)	
General purpose plant (other than that included in unit costs)	

CONTRACTOR'S SITE ONCOSTS – TIME RELATED

Site staff salaries

All non-productive supervisory staff on site including: agent, sub-agent, engineers, general foremen, non-productive section foremen, clerks, typists, timekeepers, checkers, quantity surveyors, cost engineers, security guards, etc. Cost includes salaries, subsistence allowance, National Health Insurance and Pension Scheme contributions, etc. Average cost approximately 3% to 5% of contract value.

Site staff expenses

Travelling, hotel and other incidental expenses incurred by staff. Average cost approximately 1% of staff salaries.

Attendant labour

Chainmen, storemen, drivers for staff vehicles, watchmen, cleaners, etc.

General yard labour

Labour employed on loading and offloading stores, general site cleaning, removal of rubbish, etc.

Plant maintenance

Fitters, electricians, and assistants engaged on general plant maintenance on site. This excludes drivers and banksmen who are provided for specifically in the Unit Costs sections.

Site transport for staff and general use

Vehicles provided for use of staff and others including running costs, licence and insurance and maintenance if not carried out by site fitters.

Transport for labour to and from site

Buses or coaches provided for transporting employees to and from site including cost of drivers and running costs, etc., or charges by coach hire company for providing this service.

Contractor's office rental

This includes:

- rental charges for provision of offices for Contractor's staff
- main office
- section offices
- timekeepers, checkers and security
- laboratory, etc.
- an allowance of approximately 8 m² per staff member should be made

PRELIMINARIES AND GENERAL ITEMS

Contractor's site huts

Rental charges for stores and other general-use site huts

Canteen and welfare huts

Rental charges for canteen and huts for other welfare facilities required under Rule XVI of the Working Rule Agreement

Rates

Chargeable by local authorities on any site, temporary buildings or quarry

General office expenditure

Provision of postage, stationery and other consumables for general office use

Telecommunications

Rental charges and charges for calls

Furniture and equipment

Rental charges for office furniture and equipment including photocopiers, calculators, personal computers and laser printers, etc.

Surveying equipment

Rental

Canteen and welfare equipment

Rental charges for canteen and other welfare equipment

Radio communication equipment

Rental

Testing and laboratory equipment

Rental

Lighting and heating for offices and huts

Electricity, gas or other charges in connection with lighting and heating site offices and hutting

Site lighting electrical consumption

Electricity charges in connection with general external site lighting

Water consumption

Water rates and charges

Canteen operation

Labour, consumables and subsidy costs in operating site canteens

Carpenter's shop equipment

Rental of building and mechanical equipment

PRELIMINARIES AND GENERAL ITEMS

Fitter's shop equipment

Rental of building and equipment

Small tools

Provision of small tools and equipment for general use on site. Average cost 5% of total labour cost

Personal protective equipment

Provision of protective clothing for labour including boots, safety helmets, etc. Average cost 2% of total labour cost

Traffic control

Hire and operation of traffic lights

Road lighting

Hire and operation of road lighting and traffic warning lights

Cleaning vehicles

Equipment and labour cleaning vehicles before entering public roads

Cleaning roads and footpaths

Equipment and labour cleaning public roads and footpaths

Progress photographs for Contractor's records

Cost of taking and processing photographs to demonstrate progress

CONTRACTOR'S SITE ONCOSTS – NON TIME RELATED

Rent of additional land

For Contractor's use for erection of huts, storage of soil and other materials, etc.

Staff removal expenses

Costs of staff moving house to new location. Generally only applies on longer-term contracts.

Erection of offices including drainage, paths, etc.

Construction of foundations, drainage, footpaths and parking areas, erection of huts, installation of electric wiring, in situ fittings and decorating, etc.

Dismantle offices and restore site on completion

Dismantling and taking away huts and furniture, disconnecting and removing services, removing temporary foundations etc. and reinstating ground surface to condition prevailing before construction.

Erection of general site huts

Construction of foundations, drainage, footpaths and parking areas, erection of huts, installation of electric wiring, in situ fittings and decorating, etc.

Dismantle general site huts

Dismantling and taking away huts and furniture, disconnecting and removing services, removing temporary foundations etc. and re-instating ground surface to condition prevailing before construction

PRELIMINARIES AND GENERAL ITEMS

Erection of canteen and welfare huts

Construction of foundations, drainage, footpaths and parking areas, erection of huts, installation of electric wiring, in situ fittings and decorating, etc.

Dismantle canteen and welfare huts

Dismantling and taking away huts and furniture, disconnecting and removing services, removing temporary foundations etc. and reinstating ground surface to condition prevailing before construction

Caravan site construction and clearance

Construction of site for employees' caravans including provision for water, electricity and drainage, and subsequently clear away and restore site on completion, allow credit for any charges to be levied

Telecommunications

Charges for initial installation and removal

Furniture and equipment

Purchase costs of furniture and equipment, allow for residual sale value

Survey equipment

Purchase costs of survey equipment including pegs, profiles, paint, etc., for setting out

Canteen and welfare equipment

Purchase costs of equipment

Testing and laboratory equipment

Purchase cost of equipment

Radio communication

Installation costs

Electrical connection and installation

Initial charges for connections to mains supply

Electrical connection site plant

Connection to site mains supply and final disconnection and removal

Electrical connection site lighting

Connection to site mains supply and final disconnection and removal

Water supply

Installation on site and connection charges

Haulage plant

Cost of transport of plant and equipment to and from site

PRELIMINARIES AND GENERAL ITEMS

Progress photographs

Depot loading and unloading charges

Carpenter's shop

Erection of building, installation of equipment including electrical installation, etc. Dismantle and clear away on completion

Fitter's shop

Erection of building, installation of equipment including electrical installation, etc. Dismantle and clear away on completion

Stores compound

Erect and dismantle stores compound

Notice boards and signs

Supply, erect and remove Contractor's signboards, traffic control signs, etc.

Insurances

Payment of premiums for all Contractor's insurance obligations (see separate section on insurances and bond below)

Bond

Charges for provision of bond (see separate section on insurances and bond below)

Plant erection

Cost of erection of Contractor's plant on site including foundations, hardstandings, drainage, etc.

Plant dismantling

Cost of removal of Contractor's plant on site including foundations, hardstandings, drainage, etc.

Clear site on completion including removal of rubbish and reinstatement

EMPLOYER'S AND CONSULTANTS' SITE REQUIREMENTS – TIME RELATED

Office and other huts

Rental of office accommodation, sub offices, laboratory, etc.

Office and site attendant labour

Office cleaning, chainmen, laboratory assistants, etc.

Site transport

Rental for vehicles for use of client and engineer

Telecommunications

Rental and cost of calls (if to be borne by Contractor)

PRELIMINARIES AND GENERAL ITEMS

Furniture and equipment

Rental of office furniture and equipment

Survey equipment
Rental of surveying equipment

Testing and laboratory equipment
Rental of testing and laboratory equipment

Radio communication equipment
Rental and maintenance

Office lighting and heating
Cost of heating and lighting all offices and huts

Office consumables
Cost of office consumables to be provided by the Contractor

EMPLOYER'S AND CONSULTANTS' SITE REQUIREMENTS – NON TIME RELATED

Erection of huts and offices
Client's and engineer's offices and other huts including foundations, pathways, parking area, electrical installation and drainage, etc.

Dismantling huts and offices
Restoration of site on completion

Telecommunications installation charges

Furniture and equipment purchase
Purchase cost for furniture and equipment

Survey equipment purchase

Testing and laboratory equipment purchase

Radio communication equipment installation

Progress photographs
Cost of professional photographer and supplying prints as required

OTHER SERVICES, CHARGES AND FEES

Design fees for alternative designs for permanent works
Design and drawing office costs and charges for preparing alternative designs and specifications and bill of quantities for alternative designs for permanent works

Design and design office charges for temporary works
Design and drawing office costs and charges for preparing designs and drawings for temporary works

PRELIMINARIES AND GENERAL ITEMS

Preparation of bending schedules
Drawing office charges for preparation of bending schedules

Fees to local authorities

Legal advice and fees
Fees and charges from legal adviser

Fencing

Traffic diversions

Lighting

Traffic signs

Traffic control

Footpath diversions

Stream or river diversion

Cofferdam installation

Cofferdam removal

Support works

Jetties

Bridges

Dewatering

General pumping
Including construction of collecting sumps, etc.

Site access roads and maintenance

Scaffolding

GENERAL PURPOSE PLANT – OTHER THAN THAT INCLUDED IN UNIT COSTS

Lorries and dumpers for general transport around site

Tractors and trailers

Craneage for general use

Compressed air plant

Pumps

Bowsers for fuelling plant

PRELIMINARIES AND GENERAL ITEMS

Bowsers for water supply

Non productive time for plant on site

Obtained by comparing plant requirements as on programme with the plant time included in the build up of bill rates

Note: For all items of plant listed above the cost of drivers and other attendants must be allowed for together with consumables and other operating costs

INSURANCES AND BOND

Contractors are legally required to insure against liability which may be incurred when employees are injured at work and when individuals are injured by owners' vehicles. There is also a statutory requirement for certain types of machinery to be inspected at regular intervals. In addition to legal requirements, companies insure against possible loss due to fire, explosion, fraud, liability incurred as a result of damage to the property of others and through serious injury to individuals.

Certain risks are excluded from insurers' policies; these include war, revolution, etc., contamination by radioactivity and risks which arise from bad management.

Generally insurance companies take into account the claims record of a Contractor when assessing premiums payable on a particular contract or policy. Premiums are related to the risks involved and, on large Civil Engineering contracts, the insurers will require full details of the work, the methods of construction, plant used and risks involved due to flood, ground conditions, etc. Insurance companies or brokers should be consulted before submitting a tender for major Civil Engineering work.

The following gives an outline of the items to be allowed for in a tender.

Employer's liability insurance

This provides indemnity to the Employer against legal liability for death of or injury to employees sustained in the course of their employment. The cost is normally allowed for in the build-up of the 'all-in' labour rate as a percentage addition to the gross cost. This will vary, depending on the Contractor's record. An allowance of 2% has been made in this book.

Vehicle insurance

This can cover individual vehicles or fleets. The cost is normally covered in the rate charged to the contract for the use of the vehicles.

All risks insurance

This provides for loss of or damage to permanent and/or temporary works being executed on the contract. It also covers plant, materials, etc. The cost is allowed for in the tender as a percentage of the total contract value. This will vary depending on the Contractor's record and type of contract undertaken and also on the value of excesses included in the policy.

Public liability insurance

This provides indemnity against legal liability which arises out of business activities resulting in bodily injury to any person (other than employees), loss of or damage to property (not owned or under the control of the company), obstruction, trespass or the like.

Such insurance can be extended to include labour-only subcontractors and self-employed persons if required. The cost is generally included with head office overheads.

PRELIMINARIES AND GENERAL ITEMS

Professional indemnity insurance

This provides against liability arising out of claims made against the conduct and execution of the business. This covers such items as design liability, etc. The cost may be with head office overheads where such insurance is considered desirable.

Loss of money insurance

This covers loss of money and other negotiable items and loss or damage to safes and strong rooms as a result of theft. It is necessary to cover cash in transit for wages, etc. The cost is included with head office overheads.

Fidelity guarantee insurance

This covers loss by reason of any act of fraud or dishonesty committed by employees. The cost is included with head office overheads.

Other insurances

Other insurances which may be carried by a Contractor include fire insurance on his permanent premises and contents, consequential loss insurance in relation to his permanent premises, personal accident insurance on a 24 hour per day basis for employees.

Contract bond

Where the contract calls for a bond to be provided, this is normally given by either banks or insurance companies. The total value of these guarantees available to any company is limited, depending on the goodwill and assets of the company. It will also affect the borrowing facilities available to the company and therefore, to some extent, can restrict his trading. An allowance should be made at a rate of 1.5% per annum on the amount of the bond for the construction period plus a rate of 0.5% per annum for the maintenance period.

Oncosts and Profit

PRELIMINARIES AND GENERAL ITEMS

Worked Example – Programme of Activities, Staffing and Plant

Heading and item	Total weeks	4	8	12	16	20	24	28	32	36	40	44	48	52	56	60
ACTIVITY																
Establish site																
Site clearance																
Setting out																
Fencing																
Excavate and fill																
Drainage																
Culverts																
Structure 1 (subway)																
Structure 2 (footbridge)																
Subbase																
Pavement reinforcement																
Concrete and joints																
Security fencing																
Lighting and signs																
Accommodation works																
Clear site																
STAFFING																
Agent	54															
Engineer 1	54															
Engineer 2	32															
Engineer 3	44															
General foreman	52															
Fitter	42															
Office manager	52															
Timekeeper	44															
Stores checker	52															
Typist / telephonist	52															
Bonus / cost clerk	52															
Quantity surveyor	54															
Watchman	52															
Chainman / tea boy 1	52															
Chainman / tea boy 2	44															
Chainman / tea boy 3	32															
PLANT																
Agent's car	54															
4WD utility vehicle	52															
Engineer 1 car	54															
Engineer 2 car	32															
Engineer 3 car	44															
Tractor and trailer	52															
Tractair	28															
30 tonne as crane 1	14															
70 tonne as crane 2	28															
15 tonne mobile as crane 3	14															
Sawbench	34															
Water bowser	32															
Scaffolding																
Towed sweeper	52															
Hymac 580 excavator	30															
JCB 3CX	38															
Concrete vibrators	28															
Concrete pump																
Dozer D6	14															
Towed vibrating roller	14															
Loading shovel tractor	43															
160 m³ compressor and tools	22															
Vibrating trench roller	28															
3 inch centrifugal pumps																

Contract Period

PRELIMINARIES AND GENERAL ITEMS

WORKED EXAMPLE

The example is of a contract for the construction of an Airport extension. The contract includes concrete surfaced aprons/runways, surface water drainage, construction of two concrete structures, minor accommodation works and two culverts. The Conditions of contract are ICE Standard conditions; the contract period is 12 months; and the approximate value is £11 million.

It is assumed that the main Contractor will sublet bulk earthworks and landscape, fencing, concrete surfacing, signs and lighting as well as waterproofing to structures. It is also taken that all materials are obtained off site (including concrete).

The worked example demonstrates a method of assessing preliminary costs and is based on the programme below together with an assessment of general purpose plant (a plant reconciliation is also given).

CONTRACTOR'S SITE ONCOSTS – TIME RELATED

Site staff salaries (see programme)				£	£
Agent	54	wks at	£1,840	99,360	
Senior Engineer	54	wks at	£1,630	88,020	
Engineers	76	wks at	£970	73,720	
General foreman	52	wks at	£1,020	53,040	
Office manager/cost clerk	52	wks at	£870	45,240	
Timekeeper/Storeman/Checker	44	wks at	£610	26,840	
Admin support	52	wks at	£4600	23,920	
Security guard	52	wks at	£510	26,520	
Quantity Surveyor	52	wks at	£1,020	53,040	
Fitter	42	wks at	£850	35,700	525,400
Site staff expenses (1% staff salaries)					5,254
Attendant labour					
Chainman	128	wks at	£580	74,240	
Driver	52	wks at	£620	32,240	
Office cleaner (part-time)	52	wks at	£170	8,840	115,320
General yard labour					
(Part-time involvement in loading and offloading, clearing site rubbish etc.)					
1 ganger	10	wks at	£620	6,200	
4 labourers	40	wks at	£590	23,600	29,800
Plant maintenance (Contractor's own plant)					
(Fitter included in Site Staff Salaries above)					
Fitter's mate	32	wks at	£580	18,560	18,560
Carried forward					£694,334

PRELIMINARIES AND GENERAL ITEMS

Site staff salaries (see programme)				£	£
Brought forward					£694,334
Site transport for staff and general use					
QS/Agent's cars	54	wks at	£160	8,640	
Engineers' cars (contribution)	130	wks at	£80	10,400	
Land Rover or similar SWB	54	wks at	£390	21,060	40,100
Site transport – Labour	48	wks at	£610	29,280	29,280
Transport for labour to and from site					NIL
Contractor's office rental					
Mobile offices (10 staff × 8 m²) = 80 m²	52	wks at	£205	10,660	
Section offices (2 nr at 10 m²) = 20 m²	52	wks at	£50	2,600	13,260
Contractor's site huts					
Stores hut, 22 m²	52	wks at	£56	2,912	2,912
Canteen and welfare huts					
Canteen 70 m² (assume 70 men)	52	wks at	£180	9,360	
Washroom 30 m²	52	wks at	£70	3,640	
Staff toilets	52	wks at	£180	9,360	
Site toilets	52	wks at	£240	12,480	34,840
Rates					0
General office expenditure etc.					
Postage, stationery and other consumables	52	wks at	£130	6,760	
Telephone/fax calls/e-mail and rental	52	wks at	£140	7,280	
Furniture and equipment rental	52	wks at	£70	3,640	
Personal computers, laser printers, scanners rental	52	wks at	£140	7,280	
Surveying equipment rental	52	wks at	£60	3,120	
Canteen and welfare equipment rental	52	wks at	£80	4,160	
Photocopier rental	52	wks at	£120	6,240	
Testing equipment rental	52	wks at	£50	2,600	
Lighting and heating offices and huts (200 m²)	52	wks at	£125	6,500	47,580
Water consumption					
2,250,000 litres at £2.75 per 5,000 litres	450	units at	£2.75	1,238	1,238
Small tools					
1% on labour costs of say £1,400,000					14,000
Protective clothing					
0.5% on labour costs of say £1,400,000					7,000
Cleaning vehicles					
Cleaning roads					
Towed sweeper (tractors elsewhere)	52	wks at	£150	7,800	
Brushes		say,		1,000	
Labour (skill rate 4)	52	wks at	£638	33,176	41,976
Progress photographs		say,			1,000
Total Contractor's site oncosts – Time related					**£927,520**

PRELIMINARIES AND GENERAL ITEMS

CONTRACTOR'S SITE ONCOSTS – NON TIME RELATED

				£	£
Erect and dismantle offices					
Mobile	108 m²	at	£9.00	972	
Site works			say,	1000	
Toilets			say,	300	
Wiring, water, etc.			say,	<u>750</u>	3,022
Erect and dismantle other buildings					
Stores and welfare	130 m²	at	£16.00	2,080	
Site works			say,	1000	
Toilets			say,	<u>700</u>	3,780
Telephone installation					500
Survey equipment and setting out					
Purchase cost including pegs, profiles, paint ranging rods, etc.					1,200
Canteen and welfare equipment					
Purchase cost less residual value					1,500
Electrical installation					2,500
Water supply					
Connection charges				1,000	
Site installation				<u>1,000</u>	2000
Transport of plant and equipment					3,500
Stores compound and huts					1,000
Sign boards and traffic signs					1,000
Insurances (dependent on Contractor's policy and record)					
Contractor's all risks 2.5% on £7,800,000				195,000	
Allow for excesses				<u>20,000</u>	215,000
General site clearance					<u>3,500</u>
Total Contractor's site oncosts – Non time related					**£238,502**

PRELIMINARIES AND GENERAL ITEMS

EMPLOYER'S AND CONSULTANTS' SITE REQUIREMENTS – TIME RELATED
(details of requirements will be defined in the contract documents)

Offices (50 m²)	52 wks at	£110		5,720
Site attendant labour (man weeks)	150 wks at	£580		87,000
Site transport (2 Land Rovers or similar)	52 wks at	£780		40,560
Telephones and calls	52 wks at	£120		6,240
Furniture and equipment	52 wks at	£50		2,600
Survey equipment	52 wks at	£60		3,120
Office heating and lighting (50 m²)	52 wks at	£45		2,340
Office consumables (provided by Contractor)	52 wks at	£50		2,600
Total employer's and consultants' requirements – time related				**£150,180**

EMPLOYER'S AND CONSULTANTS' SITE REQUIREMENTS – NON TIME RELATED
(details of requirements will be defined in the contract documents)

Erection and dismantling of huts and offices (50 m²)	50 m² at	£9.00		450
Site works, toilets, etc.				1,000
Telephone installation				500
Electrical installation				1,000
Furniture and equipment Purchase cost less residual value				750
Progress photographs	100 sets at	£40		4,000
Total employer's and consultants' requirements – non time related				**£7,700**

OTHER SERVICES, CHARGES AND FEES

Not applicable to this cost model.

TEMPORARY WORKS – OTHER THAN THOSE INCLUDED IN UNIT COSTS

Temporary Fencing					
1200 mm chestnut fencing					
Materials	5,000 m	at	£6.94	34,700	
Labour	500 hrs	at	£12.70	6,350	41,050
Traffic diversions					
Structure No. 1				8,000	
Structure No. 2				6,500	14,500
Footpath diversion					5,000
Stream diversion					15,500
Site access roads	400 m	at	£120.00		48,000
Total temporary works					**£124,050**

PRELIMINARIES AND GENERAL ITEMS

GENERAL PURPOSE PLANT – OTHER THAN THAT INCLUDED IN UNIT COSTS

Description				Labour	Plant	Fuel etc.	Total
Wheeled tractor							
Hire charge	52	wks at	£580		30,160		30,160
driver (skill rate 4)	52	wks at	£580	30,160			30,160
consumables	52	wks at	£80			4,160	4,160
Trailer							
Hire charge	52	wks at	£25		1,300		1,300
10 t Crawler Crane							
Hire charge	40	wks at	£1,000		40,000		40,000
driver (skill rate 3)	40	wks at	£713	28,520			28,520
consumables	40	wks at	£45			1,800	1,800
14.5 t hydraulic backacter							
Hire charge	6	wks at	£1,010		6,060		6,060
Driver (skill rate 3)	6	wks at	£713	4,278			4,278
Banksman (skill rate 4)	6	wks at	£639	3,834			3,834
consumables	6	wks at	£90			540	540
Concrete vibrators (two)							
Hire charge in total	24	wks at	£100		2,400		2,400
D6 Dozer or similar							
Hire charge	4	wks at	£1,550		6,200		6,200
driver (skill rate 2)	4	wks at	£796	3,184			3,184
consumables	4	wks at	£195			780	780
Towed roller BW6 or similar							
Hire charge	6	wks at	£350		2,100		2,100
consumables	6	wks at	£60			360	360
Loading shovel Cat 939 or similar							
Hire charge	16	wks at	£1,100		17,600		17,600
driver (skill rate 2)	16	wks at	£796	12,736			12,736
consumables	16	wks at	£95			1,520	1,520
Compressor 22.1 m³/min (silenced)							
Hire charge	12	wks at	£492		5,904		5,904
consumables	12	wks at	£410			4,920	4,920
Plate compactor (180 kg)							
Hire charge	12	wks at	£50		600		600
consumables	12	wks at	£15			180	180
75 mm 750 l/min pumps							
Hire charge	25	wks at	£79		1,975		1,975
consumables	25	wks at	£15			375	375
Total costs			£	82,712	114,299	14,635	211,646

SUMMARY OF PRELIMINARIES AND GENERAL ITEMS

Contractor's site oncosts – Time related	927,520
Contractor's site oncosts – Non time related	238,502
Employer's and consultants' requirements on site – Time related	150,180
Employer's and consultants' requirements on site – Non Time related	7,700
Other services, charges and fees	
Temporary works not included in unit costs	124,050
General purpose plant and plant not included in unit costs	211,646
Total of Preliminaries and General Items	**£1,659,598**

PART 3

Costs and Tender Price Indices

	page
Introduction	71
Price adjustment formula indices	71
A constructed cost index based on price adjustment formula indices	73
The road construction tender price index	74

ESSENTIAL READING FROM TAYLOR AND FRANCIS

Damage to Concrete Structures

De Schutter

Unlike the more specialist books which deal with diagnosis techniques or non-destructive testing, with repair and strengthening of concrete structures or with specific degradation mechanisms such as sulfate attack or alkali-silica reaction, this book describes and explains the different types of damage to concrete structures comprehensively. It is written as a textbook for undergraduate and masters students, and is also helpful to practitioners such as design engineers, architects or consultants.

November 2012: 234x156: 204pp
Pb: 978-0-415-60388-1: **£35.00**

To Order: Tel: +44 (0) 1235 400524 **Fax:** +44 (0) 1235 400525
or Post: Taylor and Francis Customer Services,
Bookpoint Ltd, Unit T1, 200 Milton Park, Abingdon, Oxon, OX14 4TA UK
Email: book.orders@tandf.co.uk

For a complete listing of all our titles visit:
www.tandf.co.uk

Taylor & Francis
Taylor & Francis Group

Costs and Tender Price Indices

The purpose of this part is to present historic changes in Civil Engineering costs and tender prices. It gives published and constructed indices and diagrammatic comparisons between building and Civil Engineering costs and tender prices and the retail price index, and will provide a basis for updating historical cost or tender price information.

INTRODUCTION

It is important to distinguish between costs and tender prices; Civil Engineering costs are the costs incurred by a Contractor in the course of his business; Civil Engineering tender prices are the prices for which a Contractor undertakes work. Tender prices will be based on Contractor's costs but will also take into account market considerations such as the availability of labour and materials and the prevailing workload for Civil Engineering Contractors. This can mean that in a period when work is scarce tender prices may fall as costs are rising while when there is plenty of work prices will tend to increase at a faster rate than costs. This section comprises published Civil Engineering cost and tender indices, a constructed Civil Engineering cost index and comparisons of these with building cost and tender indices and with the retail price index.

PRICE ADJUSTMENT FORMULA INDICES

The indices were originally introduced in the 1970s to reduce the time and effort in allowing for fluctuations in labour and materials in contracts with a Variation of Price clause.

Used in conjunction with the formulae price adjustment method of adjusting building and civil and specialist engineering contracts to allow for changes in the cost of labour, plant and materials. They are familiarly known as NEDO Indices, Baxter Indices or Osborne Indices and are widely used primarily on variation of price contracts.

BCIS calls them the Price Adjustment Formulae Indices (PAFI) but will recognize them by any of their previous aliases.

Index	1970 Series Title	Index	1990 Series Title
1	Labour and supervision in Civil Engineering	1	Labour and supervision
2	Plant and road vehicles: provision and maintenance	2	Plant and road vehicles
		3	Aggregates
3	Aggregates	4	Bricks and clay products
4	Bricks and clay products	5	Cements
5	Cements	6	Ready mixed concrete
6	Cast iron products	7	Cast and spun iron products
7	Coated roadstone for road pavement & bituminous products	8	Plastic products
		9	Coated macadam and bituminous products
8	DERV fuel	10	DERV fuel
9	Gas oil fuel	11	Gas oil fuel
10	Timber	12	Timber
11 A	Steel for reinforcement	13	Steel for reinforcement
11B	Metal sections	14	Metal sections
12	Fabricated structural steel	15	Sheet steel piling
13	Labour and supervision in fabricating and erecting steelwork		

Price Adjustment Formulae for Construction Contracts
Monthly Bulletin of Indices are published by BCIS

Quarterly values of price adjustment formula 1970 Series. Base: − 1970 = 100 (except for index 11 A which has a base date of July 1976).

Year	Q	1	2	3	4	5	6	7	9	10	11a
2005	1	1507	1057	1657	1873	1171	1826	3022	3291	977	508
	2	1523	1063	1705	1919	1238	1815	3079	3573	986	459
	3	1639	1105	1661	1889	1234	1820	3081	4179	988	446
	4	1639	1106	1619	1909	1243	1839	3140	3868	1020	461
2006	1	1639	1112	1659	1962	1329	1854	3292	4044	1009	474
	2	1648	1114	1689	1970	1328	1853	3346	4465	1014	521
	3	1696	1128	1656	2000	1326	1916	3527	4332	1056	592
	4	1696	1130	1614	2013	1350	1947	3500	3437	1076	594
2007	1	1696	1131	1709	2034	1395	1967	3623	3982	1100	605
	2	1710	1137	1667	2028	1404	1890	3610	4359	1156	635
	3	1768	1158	1651	2084	1416	1919	3546	4638	1194	629
	4	1768	1159	1656	2013	1408	1944	3646	5440	1190	593
2008	1	1768	1174	1735	2099	1552	1938	4051	6305	1168	647
	2	1770	1184	1890	2066	1549	2021	4280	7799	1133	830
	3	1868	1218	1832	1987	1591	2165	4318	6830	1131	879
	4	1868	1230	1780	2032	1589	2265	4369	4576	1187	703
2009	1	1869	1249	1883	2195	1733	2290	4401	4288	1158	617
	2	1867	1249	1902	2157	1686	2379	4537	4983	1127	588
	3	1867	1251	1951	2166	1657	2386	4539	5048	1253	581
	4	1867	1253	1863	2159	1638	2389	4941	5390	1318	572
2010	1	1869	1268	1863	2200	1651	2384	4774	6060	1374	632
	2	1869	1275	1887	2223	1650	2485	4966	6128	1462	727
	3	1869	1274	1977	2231	1644	2485	4857	6027	1541	709
	4	1869	1273	1962	2208	1640	2485	4939	6657	1537	695
2011	1	1869	1282	1923	2204	1665	2634	5164	8033	1499	785
	2	1876	1292	1952	2242	1673	2634	5277	7766	1504	808
	3	1913	1303	1994	2278	1672	2634	5362	7856	1514	828
	4	1916	1303	2007	2255	1679	2596	5404	7892	1476	797
2012*	1	1916	1307	2035	2338	1751	2532	5736	8447	1461	802

* Provisional
Note: The indices shown are for the third month of each quarter

A CONSTRUCTED COST INDEX BASED ON THE PRICE ADJUSTMENT FORMULA INDICES

Although the above indices are prepared and published in order to provide a common basis for calculating reimbursement of increased costs during the course of a contract, they also present time series of cost indices for the main components of Civil Engineering work. They can therefore be used as the basis of an index for Civil Engineering work. The method used here is to construct a composite index by allocating weightings to each of the 10 indices, the weightings being established from an analysis of actual projects. The composite index is calculated by applying these weightings to the appropriate price adjustment formula indices and totalling the results; this index is again presented with a base date of 1970.

Constructed Civil Engineering Cost Index base: 1970 = 100

Year	First quarter	Second quarter	Third quarter	Fourth quarter	Annual average
2005	1611	1652	1722	1718	1676
2006	1757	1785	1842	1813	1799
2007	1841	1862	1895	1939	1884
2008	2039	2168	2215	2128	2138
2009	2099	2129	2167	2225	2155
2010	2226	2284	2285	2309	2276
2011	2396	2443	2467	2490	2449
2012*	2552				

Note: * Provisional

Source: Davis Langdon based on BIS figures.

The chart illustrates the movement of the Constructed Civil Engineering Index, the Davis Langdon Building Cost Index and the Retail Price Index.

THE ROAD CONSTRUCTION TENDER PRICE INDEX

Civil Engineering work generally does not lend itself easily to the preparation of tender price indices in the same way as building work. There is, however, a published tender index for road construction and this is reproduced below with the permission of BIS. The index is intended to indicate the movement in tender prices for road construction contracts. It is based on priced rates contained in accepted tenders for Road Construction, Motorway Widening and Major Maintenance Schemes. It is published with a base of 1995 = 100.

Tender Price Index of Road Construction – Base: 1995 = 100

Year	First quarter	Second quarter	Third quarter	Fourth quarter	Annual average
2005	129	133	137	142	135
2006	144	150	153	152	150
2007	159	153	151	149	153
2008	145	144	146	147	145
2009	137	145	151	152	146
2010	157	163	168	170	165
2011*	175	169	175	179	175

Note: * Provisional

Source: BIS

The chart illustrates the movement of the BIS Road Construction Price Index; the Davis Langdon Tender Price Index and the Retail Price Index.

PART 4

Resources

	page
Basis of this section	77
Labour	80
Plant costs	85

This part comprises sections on labour, materials and plant for civil engineering work. These resources form the basis of the unit costs in Parts 6 and 7 and are given here so that users of the book may:

Calculate rates for work similar to, but differing in detail from, the unit costs given
Compare the costs given here with those used in their own organization
Calculate the effects of changes in wage rates, material prices, etc.

Adjustments should be made to the rates shown to allow for time, location, local conditions, site constraints and any other factors likely to affect the cost of the specific scheme.

ESSENTIAL READING FROM TAYLOR AND FRANCIS

Reinforced and Prestressed Concrete Design to EC2

2nd Edition

Eugene O'Brien et al

Fully revised to Eurocode 2, this second edition retains the original's emphasis on qualitative understanding of the overall behaviour of concrete structures. With a new chapter dedicated to case studies, worked examples, and exercise examples, it is an even more comprehensive guide to conceptual design, analysis, and detailed design of concrete structures, providing civil and structural engineering students with complete coverage of the analysis and design of reinforced and prestressed concrete structures.

January 2012: 246x189: 236pp
Hb: 978-0-415-57194-4: **£99.00**
Pb: 978-0-415-57195-1: **£34.99**

To Order: Tel: +44 (0) 1235 400524 **Fax:** +44 (0) 1235 400525
or Post: Taylor and Francis Customer Services,
Bookpoint Ltd, Unit T1, 200 Milton Park, Abingdon, Oxon, OX14 4TA UK
Email: book.orders@tandf.co.uk

For a complete listing of all our titles visit:
www.tandf.co.uk

Taylor & Francis
Taylor & Francis Group

BASIS OF THIS SECTION

The following are brief details of the Construction Industry Joint Council agreement on pay and conditions for the Building and Civil Engineering Industry, effective from July 2010.
Copies of the Working Rule Agreement may be obtained from:

> Construction Industry Joint Council,
> 55 Tufton Street,
> London,
> SW1P 3QL

Due to the current, based on 5th September 2011 settlement, market situation it is anticipated that wage rates will not increase significantly in the near future and as consequence only minor changes have been made to the previous 2011 rates.

Rates of Pay (Rule WR.1)

The basic and additional hourly rates of pay are:

Labourer / General Operative	£7.87
Skill Rate 4	£8.48
Skill Rate 3	£8.98
Skill Rate 2	£9.60
Skill Rate 1	£9.97
Craft Operative	£10.46

Additional Payment (Rule WR.1.4)

The WRA recognizes an entitlement to additional payments to operatives employing intermittent skill, responsibility or working in adverse conditions, for each hour so engaged as defined in Schedule 2 of the Agreement. Many Employers have differing rates agreed within their organizations which are different to the WRI agreement. The skill rates employed in the example are set out below:

Extra £/hr

A	£0.17	Tunnels	Operatives (other than Tunnel Machine Operators, Tunnel Miners and Tunnel Miners' Mates) wholly or mainly engaged in work of actual construction including the removal and dumping of mined materials but excluding operatives whose employment in the tunnel is occasional and temporary.
B	£0.29	Sewer Work	Operatives working within a totally enclosed active surface water sewer of any nature or condition.
C	£0.37	Sewer Work	Operatives working outside existing sewers excavating or removing foul materials emanating from existing sewers.
D	£0.38	Working at Height	Operatives (including drivers of tower cranes, but excluding the drivers of power driven derricks on high stages and linesmen-erectors and their mates and scaffolders) employed on detached work.
E	£0.43	Sewer Work	Operatives working within a totally enclosed active foul sewer of any nature or condition.
F	£0.65	Stone Cleaning	Operatives other than craft operatives dry-cleaning stonework by mechanical process for the removal of protective material and/or discolouration.

Bonus (Rule WR.2)

The Working Rule leaves it open to employers and employees to agree a bonus scheme based on measured output and productivity for any operation or operations on a job.

Working Hours (Rule WR.3)

Normal working hours are unchanged at 39 hour week – Monday to Thursday at eight hours per day and Friday at seven hours. The working hours for operatives working shifts are eight hours per weekday, 40 hours per week.

Rest/Meal Breaks (Rule WR.3.1)

Times fixed by the employer, not to exceed one hour per day in aggregate, including a meal break of not less than half an hour.

Overtime Rates (Rule WR.4)

During the period of Monday to Friday, the first four hours after normal working day is paid at time and a half, after four hours double time shall be paid. Saturday is paid at time and a half until completion of the first four hours. Remainder of Saturday and all Sunday paid at double time.

Daily Fare and Travelling Allowances (Rule WR.5)

This applies only to distances from home to the site of between 15 and 75 km. Again Employers have individual agreements with typical ranges of £0.99 to £8.00 for Travelling Allowance (taxed) and £3.60 to £13.00 for Fare Allowance (not taxed). For the one way distance of 15 km used in the following example we have assumed a rate of £3.96 with effect from 30th June 2011.

Rotary Shift Working (Rule WR.6)

This relates to the situation where more than one shift of minimum eight hours is worked on a job in a 24 hour period, and the operative rotates between the shifts either in the same or different pay weeks.

The basic rate shall be the operative's normal hourly rate plus 14%. Overtime beyond the 8 hour shift shall be at time and a half for the first four hours at normal rate plus 14%, thereafter double normal rate.

Night Work (Rule WR.7)

Providing night work is carried out by a separate gang from those working during daytime, an addition of 25% shall be paid on top of the normal hourly rate.

Overtime payments: during the period of Monday to Friday, the first four hours after normal working day is paid at time and a half plus the 25% addition of normal hourly rate, after four hours double time shall be paid. All hours worked on Saturday and all Sunday paid at double time.

Tide Work (Rule WR.9)

Where the operative is also employed on other work during the day, additional time beyond the normal working day shall be paid in accordance with the rules for overtime payments.

Where the operative is solely involved in work governed by tidal conditions, he shall be paid a minimum of six hours at normal rates for each tide. Payment for hours worked in excess of eight over two tides shall be calculated proportionately.

$$\frac{(\text{Total Hours Worked} - 8 \text{ hours}) \times \text{Total Hours Worked}}{8 \text{ hours}}$$

Work done after 4pm Saturday and during Sunday shall be at double time. Operatives are guaranteed eight hours at ordinary rate for any time worked between 4pm and midnight on Saturdays and 16 hours for two tides worked on a Sunday.

Tunnel Work (Rule WR.10)

The first part of a shift equivalent to the length of the normal working day shall be paid at the appropriate normal rate, the first four hours thereafter at time and a half and thereafter at double time. In the case of shifts on a Saturday, the first four hours are at time and a half, thereafter at double time. All shifts on a Sunday at double time.

Saturday shifts extending into Sunday are at double time. Sunday shifts extending into Monday, time after midnight: first four hours at time and a half, thereafter at double time.

Subsistence Allowance (Rule WR.15)

The 2011 allowance of £32.10 per night has been incorporated into the example.

Annual Holidays Allowance (Rules WR.18 and 21)

Annual and Public Holiday Pay is included in accordance with the Building & Civil Engineering Benefit Schemes (B&CE) Template Scheme. Allowances are calculated on total weekly earnings inclusive of overtime and bonus payments and the labour cost calculation assumes 21 days (4.2 weeks) annual and eight days (1.6 weeks) public holidays.

Easy Build Pension Contributions (Rule WR 21.3

The minimum employer contribution is £3.00 per week. Where the operative contributes between £3.01 and £10.00 per week the employer will increase the minimum contribution to match that of the operative up to a maximum of £10.00 per week. The calculations below assume 1 in 10 employees make contributions of £10.00 per week.

LABOUR COSTS

CALCULATION OF LABOUR COSTS

This section sets out a method used in calculating all-in labour rates used within this book. The calculations are based on the wage rates, plus rates and other conditions of the Working Rule Agreement (WRA); important points are discussed below. The calculations can be used as a model to enable the user to make adjustments to suit specific job conditions in respect of plus rate, bonus, subsistence and overhead allowances together with working hours etc. to produce alternative all-in rates which can be substituted for those printed.

All-in labour costs are calculated on the pages following for six categories of labour reflecting the different classifications set out in the WRA.

AVERAGE WORKING WEEK has been calculated to an equal balance between winter and summer (46.20 working weeks per year).

SUBSISTENCE PAYMENTS are included for key men only: for a stated percentage of the workforce.

TRAVELLING ALLOWANCES are based on rate payable for a journey of 15 km at £3.96 per day. These allowances are adjusted to cater for operatives receiving Subsistence Payments.

BONUS PAYMENTS reflect current average payments in the industry.

WORKING HOURS & NON-PRODUCTIVE OVERTIME are calculated thus:

SUMMER (hours worked 8.00 am – 6.00pm, half hour for lunch)

	Mon	Tue	Wed	Thu	Fri	Sat	Total Paid Hours	Deductions Lost & Wet time	Paid Breaks	Effective Total Hours
Total hrs	9.50	9.50	9.50	9.50	9.50	4.00	51.50	0.50	1.00	**50.00**
Overtime	0.75	0.75	0.75	0.75	1.25	2.00	6.25			
TOTAL PAID:							57.75	EFFECTIVE		**50.00**

WINTER (hours worked 7.30 am – 4.30pm, half hour for lunch)

	Mon	Tue	Wed	Thu	Fri	Sat	Total Paid Hours	Deductions: Lost & Wet time	Paid Breaks	Effective Total Hours
Total hrs	8.50	8.50	8.50	8.50	7.50	4.00	45.50	1.50	–	**44.00**
Overtime	0.25	0.25	0.25	0.25	0.25	2.00	3.25	–	–	–
TOTAL PAID							48.75	EFFECTIVE		**44.00**

AVERAGE TOTAL HOURS PAID	(57.75 + 48.75)/2 = 53.25 hours per week
AVERAGE EFFECTIVE HOURS WORKED	(50.00 + 44.00)/2 = 47.00 hours per week

		\multicolumn{6}{c	}{Categories of Labour}
Payment details (based on September 2011 settlement)		General	Skill rate 4	Skill rate 3	Skill rate 2	Skill rate 1	Craft rate
		£	£	£	£	£	£
Basic rate per hour		7.87	8.48	8.98	9.60	9.97	10.46
53.25 Hours Paid @ Total Rate		419.08	451.56	478.19	511.20	530.90	557.00
Weekly Bonus Allowance	*	30.84	30.84	61.54	51.84	92.85	123.71
TOTAL WEEKLY EARNINGS	£	**449.92**	**482.40**	**539.73**	**563.04**	**623.76**	**680.70**
Travelling Allowances (say, 15 km per day, i.e. £3.96 × 6 days)							
general, skill rate 4, skill rate 3 = 100%		23.76	23.76	23.76	–	–	–
skill rate 2, skill rate 1 = 80%		–	–	–	19.01	19.01	–
craft rate = 90%		–	–	–	–	–	21.38
Subsistence allowance (Average 6.66 nights × £29.16, + 7.00% to cover periodic travel)							
general, skill rate 4, skill rate 3 = 0%		–	–	–	–	–	–
skill rate 2, skill rate 1 = 20%		–	–	–	38.06	38.06	–
craft rate = 10%		–	–	–	–	–	19.03
TOTAL WAGES	£	**473.68**	**506.16**	**563.49**	**620.10**	**680.82**	**721.12**
National Insurance contributions – (13.8% above Earnings Threshold except Fares)		39.77	44.25	52.16	55.38	63.76	71.61
Annual Holiday Allowance – 4.2 weeks @ Total Weekly Earnings over 46 weeks		43.25	46.21	51.45	56.62	62.16	65.84
Public Holidays with Pay – 1.8 weeks @ Total Weekly Earnings over 46.2 weeks		18.54	19.81	22.05	24.26	26.64	28.22
Easy Build pension contribution		5	5	5	5	5	5
CITB Levy (0.50% Wage Bill)		2.68	2.86	3.18	3.50	3.85	4.08
	£	582.91	624.29	697.33	764.86	842.23	895.86
Allowance for Employer's liability & third party insurances, safety officer's time QA policy/inspection and all other costs & overheads:		14.57	15.61	17.43	19.12	21.06	22.40
TOTAL WEEKLY COST (47 HOURS)	£	597.48	639.90	714.76	783.99	863.28	918.26
COST PER HOUR	£	**12.71**	**13.61**	**15.21**	**16.68**	**18.37**	**19.54**
Plant Operators rates: Addition to Cost per Hour for Rule WR.11 – plant servicing time (6 hrs)	£	1.18	1.27	1.34	1.44	1.49	1.57
COST PER HOUR	£	**13.89**	**14.88**	**16.55**	**18.12**	**19.86**	**21.10**
Additional Payments under WR 1.4 (see above)		A	B	C	D	E	F
Extra rate per hour	£	0.17	0.29	0.37	0.38	0.43	0.65

* The bonus levels have been assessed to reflect the general position as at May/June 2011 regarding bonus payments limited to key personnel, as well as being site induced.

LABOUR CATEGORIES

Schedule 1 to the WRA lists specified work establishing entitlement to the Skilled Operative Pay Rate 4, 3, 2, 1 or the Craft Rate as follows:

General Operative

Unskilled general labour

Skilled Operative Rate 4

supervision	Gangers + trade chargehands
plant	Contractors plant mechanics mate; Greaser
transport	Dumper <7 t; Agricultural Tractor (towing use); Road going motor vehicle <10 t; Loco driver
scaffolding	Trainee scaffolder
drilling	Attendee on drilling
explosives	Attendee on shot firer
piling	General skilled piling operative
tunnels	Tunnel Miner's mates; Operatives driving headings over 2 m in length from the entrance (inwith drain, cable & mains laying)
excavation	Banksmen for crane/hoist/derrick; Attendee at loading/tipping; Drag shovel; Trenching machine (multi bucket) <30 hp; Power roller <4 t; Timberman's attendee
coal	Opencast coal washeries and screening plants
tools	Compressor/generator operator; Power driven tools (breakers, tamping machines etc.); Power-driven pumps; air compressors 10 KW+; Power driven winches
concrete	Concrete leveller/vibrator operator/screeder + surface finisher, concrete placer; Mixer < 21/14 or 400 litres; Pumps/booms operator
linesmen	Linesmen-erector's mate
timber	Carpentry 1st year trainee
pipes	Pipe layers preparing beds and laying pipes <300 mm diameter; Pipe jointers, stoneware or concrete pipes; pipe jointers flexible or lead joints <300 mm diameter
paving	Rolled asphalt, tar and/or bitumen surfacing: Mixing Platform Chargehand, Chipper or Tamperman; Paviors; Rammerman, kerb & paving jointer
dry lining	Trainee dry liners
cranes	Forklift truck <3 t; Crane < 5 t; Tower crane <2 t; Mobile crane/hoists/fork-lifts <5 t; Power driven hoist/crane

Skilled Operative Rate 3

transport	Road going motor vehicle 10 t+
drilling	Drilling operator
explosives	Explosives/shot firer
piling	Piling ganger/chargehand; Pile frame winch driver
tunnels	Tunnel Miner (working at face/operating drifter type machine)
excavation	Tractors (wheeled/tracked with/without equipment) <100 hp; Excavator <0.6 m³ bucket; Trenching machine (multi bucket) 30–70 hp; Dumper 7-16 t; Power roller 4 t+; Timberman
concrete	Mixer 400–1500 litres; Mobile concrete pump with/without concrete placing boom; Hydraulic jacks & other tensioning devices in post-tensioning and/or prestressing
formwork	Formwork carpenter 2nd year trainee
masonry	Face pitching or dry walling
linesmen	Linesmen-erectors 2nd grade
steelwork	Steelwork fixing simple work; Plate layer
pipes	Pipe jointers flexible or lead joints (300–535 mm diameter)
paving	Rolled asphalt, tar and/or bitumen surfacing: Raker, Power roller 4 t+, mechanical spreader operator/leveller
dry lining	Certified dry liners
cranes	Cranes with grabs fitted; Crane 3–10 t; Mobile crane/hoist/fork-lift 4–10 t; Overhead/gantry crane <10 t; Power-driven derrick <20 t; Tower crane 2-10 t; Forklift 3 t+

Skilled Operative Rate 2

plant	Maintenance mechanic; Tyre fitter on heavy earthmover
transport	Dumper 16–60 t
scaffolding	Scaffolder < 2 years scaffolding experience and < 1 year as Basic Scaffolder
piling	Rotary or specialist mobile piling rig driver
tunnels	Face tunnelling machine
excavation	Tractors (wheeled/tracked with/without equipment) 100–400 hp; Excavator 0.6–3.85 m³ bucket; Trenching machine (multi bucket) 70 hp +; Motorized scraper; Motor grader
concrete	Mixer 1500 litres+; Mixer, mobile self-loading and batching <2500 litres
linesmen	Linesmen-erectors 1st grade
welding	Gas or electric arc welder up to normal standards
pipes	Pipe jointers flexible or lead joints over 535 mm diameter
cranes	Mobile cranes/hoists/fork-lifts 5–10 t; Power-driven derrick 20 t+, with grab <20 t; Tower crane 10–20 t

Skilled Operative Rate 1

plant	Contractors plant mechanic
transport	LGV driver; Lorry driver Class C+E licence; Dumper 60–125 t
excavation	Excavator 3.85–7.65 m³; Tractors (wheeled/tracked with/without equipment) 400–650 hp
concrete	
linesmen	
steelwork	Steelwork assembly, erection and fixing of steel framed construction
welding	Electric arc welder up to highest standards for structural fabrication & simple pressure vessels (air receivers including CO_2 processes)
drilling	Drilling rig operator
cranes	Power-driven derrick with grab 20 t +; Tower crane 20 t+

Craft Operative

transport	Dumper 125 t+
scaffolding	Scaffolder a least two years scaffolding experience and at least one year as Basic and Advanced Scaffolder
excavation	Excavator 7.65 t + (see WR 1.2.2 above); Tractors (wheeled/tracked with/without equipment) 650 hp+
concrete	Reinforcement bender & fixer
formwork	Formwork carpenter
welding	Electric arc welder capable of all welding processes on all weldable materials including working on own initiative from drawings
cranes	Crane 10 t+; Mobile cranes/hoists/fork-lifts 10 t +

ESSENTIAL READING FROM TAYLOR AND FRANCIS

Self-Consolidating Concrete
Applying What we Know

Daczko

Self-consolidating, or self-compacting, concrete (SCC) uses mixture proportioning techniques and admixture technology to create a concrete mixture with increased fluidity, thereby avoiding the need for mechanical consolidation/vibration. SCC mixtures with various performance levels can be produced in the laboratory, but the know-how is sometimes lacking to do this on site. This book presents a foundation for consistently producing and applying SCC in a daily production environment. It links science with practical application and will be useful for both new practitioners and those already using SCC. The author uses stories and case studies throughout as examples of specific types of applications.

February 2012: 234x156: 304pp
Hb: 978-0-415-59064-8: **£50.00**

To Order: Tel: +44 (0) 1235 400524 **Fax:** +44 (0) 1235 400525
or Post: Taylor and Francis Customer Services,
Bookpoint Ltd, Unit T1, 200 Milton Park, Abingdon, Oxon, OX14 4TA UK
Email: book.orders@tandf.co.uk

For a complete listing of all our titles visit:
www.tandf.co.uk

Taylor & Francis
Taylor & Francis Group

PLANT COSTS

INTRODUCTION

The information, rates and prices included in this section are calculated examples of actual owning and operating costs of a range of construction plant and equipment. To an extent they will serve as a guide to prevailing commercial plant hire rates, but be aware that many factors will influence actual hire rates. For example, rates could be lower because of long-term hire, use of older or second hand machines, low market demands/loss leaders, easy tasks or sites near to the depot. Rates could be higher due to short-term hire, high production demands, low utilization factors, specialist operations, restricted working and restricted accessibility problems, high profit or overhead demands and finance fluctuations. The use factors will mean a hired machine may not be used productively for each hour on site, it is therefore up to the estimator to allow for reasonable outputs in his unit rate calculations.

These and other considerations must be borne in mind if using these costs as a comparison to hire rates, especially operated plant.

All rates quoted EXCLUDE the following:

- Cost of drivers/operators
- VAT

CONTRACTOR OWNED PLANT

Plant owned by a Contractor generally falls under two headings:

- Small plant and tools which are the subject of a direct charge to contracts and for estimating purposes are normally allowed for as a percentage of the labour cost in site oncosts (see Preliminaries and General Items), although many items are shown in this section for information.
- Power driven plant and major non-mechanical plant such as steel trestling, scaffolding, gantries etc. Such plant is normally charged to the contract on a rental basis except in the case of purpose-made or special plant bought specifically for a particular operation; this latter is normally charged in full to contracts and allowance made for disposal on completion (often at scrap value).

A very wide range of plant is readily available from plant hire companies; it is not usually economical for Contractors to own plant unless they can ensure at least 75 to 80% utilization factor based on the Contractor's normal working hours. Where a Contractor does own plant, however, it is essential that he maintains reasonably accurate records of the working hours and detailed costs of maintenance and repairs in order that he may estimate the charges to be made for each item of plant. For this reason, where a Contractor owns a large quantity of plant, it is normal for a separate plant hire department or company to be formed.

This department or company should be financially self supporting and may hire plant to other Contractors when it is not needed. Maintenance of contractor owned plant can be carried out on site; it is not necessary to maintain a centrally based repair workshop. It is, however, desirable to have some storage facilities available for plant when not in use. The cost of owning plant and hence the rental charges to be made to contracts must take into account:

- Capital cost
- Depreciation charges
- Maintenance and repairs
- Cost of finance
- Insurances and licences
- Administration, head office, depot and other overhead charges

Calculated examples of hourly owning costs of a range of plant and equipment.

	A UTILIZATION FACTOR	B P.A.	C PLANT	D LIFE	E PURCHASE PRICE*	F RESALE	G RESALE PRICE	H LOSS IN VALUE	I DEPREC.	J MAINTEN.	K FINANCE @ 5%	L INSUR.	M ADMIN.	N TOTAL COST/HR
	%	Hours	Years	Hours	£	%	£	£	£	£	£	£	£	£
ACCESS PLATFORMS														
Boom Lift, 450 kg load 18.6 m height	65	1,365	10	13,650	37,500	20.0	7,500	30,000	2.20	2.75	0.55	0.55	1.21	7.26
Boom Lift, 230 kg load 15.8 m height	65	1,365	10	13,650	59,000	20.0	11,800	47,200	3.46	4.32	0.86	0.86	1.90	11.40
ASPHALT PAVERS														
Wheeled paver, 6.4 m wide, 79.2 m/min	65	1,365	7	9,555	80,500	27.5	22,138	58,362	6.11	5.90	1.07	1.18	2.85	17.11
Wheeled paver, 7.9 m wide, 90.5 m/min	65	1,365	7	9,555	129,000	27.5	35,475	93,525	9.79	9.45	1.71	1.89	4.57	27.41
COMPRESSORS (PORTABLE)														
17 m³/min, potable	85	1,785	9	16,065	50,000	20.0	10,000	40,000	2.49	2.80	0.56	0.56	1.28	7.69
7.3 m³/min, electric motor	85	1,785	9	16,065	12,500	20.0	2,500	10,000	0.62	0.70	0.14	0.14	0.32	1.92
CRANES														
16 t mobile	60	1,260	12	15,120	193,500	25.0	48,375	145,125	9.60	15.36	2.88	3.07	6.18	37.09
35 t mobile	60	1,260	15	18,900	215,000	27.5	59,125	155,875	8.25	17.06	3.09	3.41	6.36	38.17
GENERATORS														
10 kVA, water cooled	85	1,785	5	8,925	6,500	15.0	975	5,525	0.62	0.36	0.08	0.07	0.23	1.36
10 kVA, air cooled	85	1,785	5	8,925	5,500	15.0	825	4,675	0.52	0.31	0.07	0.06	0.19	1.15
EXCAVATORS														
Tracked backhoe 0.7 m³ max bucket	80	1,680	9	15,120	86,000	27.5	23,650	62,350	4.12	5.12	0.93	1.02	2.24	13.43
Tracked backhoe 4.5 m³ max bucket	75	1,575	9	14,175	405,000	27.5	111,375	293,625	20.71	25.71	4.66	5.14	11.24	67.46
JCB 3CX	85	1,785	6	10,710	51,500	27.5	14,163	37,337	3.49	2.89	0.52	0.58	1.50	8.98
ROLLERS														
Double Drum 757 kg. 650 mm wide	85	1,785	6	10,710	16,250	27.5	4,469	11,781	1.10	0.91	0.17	0.18	0.47	2.83
Double Drum 1040 kg. 750 mm wide	85	1,785	6	10,710	21,500	27.5	5,913	15,587	1.46	1.20	0.22	0.24	0.62	3.74
Aticulated tandem 1.6 t. 800 mm wide	85	1,785	6	10,710	32,500	27.5	8,938	23,562	2.20	1.82	0.33	0.36	0.94	5.65
CRAWLER LOADERS														
20 tonne weight, 2.0 m³ bucket	80	1,680	9	15,120	150,000	27.5	41,250	108,750	7.19	8.93	1.62	1.79	3.91	23.44
TIPPERS														
10 t capacity	90	1,890	6	11,340	48,500	22.0	10,670	37,830	3.34	2.57	0.50	0.51	1.38	8.30
20 t capacity	90	1,890	6	11,340	75,250	22.0	16,555	58,695	5.18	3.98	0.78	0.80	2.15	12.89
DUMP TRUCKS														
Articulated dump 39 t, 24 m³ capacity	75	1,575	10	15,750	315,000	30.0	94,500	220,500	14.00	20.00	3.50	4.00	8.30	49.80

NOTES:
*The purchase price is illustrative only and will require a quotation from the relevant manufacturer
No allowance is included for Road Fund Licence on road-going vehicles.

PLANT COSTS

Notes – example plant ownership hourly costs

The above costs would be updated annually or bi-annually, the rental rate being revised to ensure complete recovery of the costs associated with the item of plant and return of capital to enable the machine to be replaced at the end of its life with the Contractor. The purchase price must also be adjusted to ensure recovery of the replacement cost and not the original cost of purchase.

Driver's wages and costs should be charged direct to site wages.

Column A: The utilization factors used in conjunction with the period of ownership is the percentage of time (per annum) that an item of plant can be expected to be used productively on a site or job and therefore is a very important influence of hourly costs

(Note: Utilization factors are not the same as site utilization rates – see notes regarding Fuel Consumption)

Economical owning periods/use factors:	Life (years)	Usage (%)
Hydraulic Excavators, large	9	75
Hydraulic Excavators, medium	7	76
Hydraulic Excavators, mini	6	77
Dozers/Scrapers	10	78
Loaders/Shovels	9	79
Mobile Cranes	12	80
Crawler Cranes	15	81
Dump Trucks	8	82
Dumpers	6	83
Rollers/Compaction	6	84
Compressors/Generators	7–9	85
Diesel engine road vehicles	6–7	86
Petrol engine road vehicles	5–6	87

- Column B: The result of applying the utilization factor to 2,100 hours per annum.
- Columns C and D: The period over which the Contractor owns the plant and during which time the items of plant can be maintained at reasonable efficiency will vary according to the Contractor's experience, the type of plant under consideration and the work on which is it employed. The owning period of an item of plant can vary considerably on this basis. The data we have used in our calculations varies considerably (see the tables shown above and on the next page).
- Column E: The list price being quoted by dealers, against which should be allowed a level of discount suited to the availability of the plant and the bargaining power between the purchaser and the agent.
- Column J: Maintenance costs cover major overhauls and replacement costs for wear items only, excluding insurable damage and will vary with each type of plant. This does not include day to day general servicing on site.
- Column K: Finance interest charges are taken on the average base rate over the previous 12 months plus 3%.
- Column L: Insurance premiums have been taken as 2% on the replacement value of the plant (purchase price)
- Column M: Administration, head office, depot and other overhead charges have been taken at +20%.
- Consumables should be charged direct to site costs. The costs included with this section are based upon manufacturers' data and are used in good faith. (For outline estimates fuel consumption can be taken as approximately 0.15–0.20 litres per kW per working hour; lube oils, filters, grease etc., can be taken as 2% to 10% of the fuel cost, depending upon the working conditions of the machine and its attachments). See below for details.

Please note that the costs in this section have been reviewed for recent editions, based on the latest available advice and data from manufacturers and Plant Hire firms, the basis of calculation, including retention period of the plant, has been changed in a number of cases, providing a realistic overall owning and operating cost of the machine.

PLANT COSTS

These examples are for machine costs only and do not include for operator costs or profit element or for transport costs to and from site.

Fuel consumption is based on the following site utilization rates. (Site utilization being the percentage of time that the machine is operating at its average fuel consumption during a working day).

Typical utilization factors:	Utilization factor	Typical utilization factors:	Utilization factor
Excavators	0.75	Compressors, mixers, generators	
Dozers/Scrapers	0.80	Tractors	0.75
Loaders/Shovels/Graders	0.75	Hoists, etc.	0.65
Mobile cranes	0.50	Drills and Saws	0.90
Crawler cranes	0.25	Rollers/Compaction plant	0.75
Dump Trucks (not Tippers)	0.75	Piling/Asphalt equipment	0.65
Dumpers	0.80	All other items	1.00

Approximate fuel consumption in litres per hour and percentage addition for consumables (examples only) based on the above site utilization rates, assuming that reasonably new and well maintained plant would be used.

		Fuel l/hr	consumables % of fuel			Fuel l/hr	consumables % of fuel
Tractor	116 hp	9.0	2.0%	Rollers	BW 90 AD	2.0	7.5%
Paver	Bitelli BB650	16.0	5.0%		BW 120 AD	4.4	7.5%
Compactor plate	188 kg	0.9	3.0%		BW6 (towed)	7.6	7.5%
Compressor	3.5 m³/min	5.9	3.0%	Dozers	CAT D6N	21.0	6.0%
	10 m³/min	18.0	3.0%		CAT D8T	38.0	3.0%
	Tractor mounted	9.0	5.0%	Loaders	CAT 953C	18.0	4.0%
Mixer	5/3.5	1.2	2.0%	Scrapers	CAT 621G	41.5	3.0%
Mobile crane	15 t	5.1	10.0%		CAT 657E	125.0	6.0%
Mobile crane	40 t	6.3	10.0%	Skidsteer	Bobcat 553	3.4	5.0%
Generator	4 kVA	1.6	3.0%	Graders	CAT 140H	19.0	5.0%
	25 kVA	4.9	5.0%	Dumpers	2 tonne	3.0	3.0%
Backacter	11.5 t	9.5	7.5%		Volvo A25D	17.0	3.0%
	19 t	14.0	7.5%		36 tonne	25.0	5.0%
	30 t	24.0	7.5%	Pumps	75 mm 65 m³/hr	1.1	2.0%
Backhoes	JCB 3CX	7.5	6.0%		150 mm 360 m³/hr	4.2	3.0%

Note that the above consumption figures represent 'medium' plant operation, i.e. the plant would not be operating at full throttle and working conditions such as the soil and grades are average. A study of manufacturers' figures has indicated that 'high' or 'low' usage can affect consumption on average by ± 25%; 'high' usage would involve full loads, continuous working at full throttle with difficult ground and adverse grades, whereas 'low' usage would involve more intermittent working with perhaps considerable idling periods, more easily worked ground and easier grades.

PLANT COSTS

FUEL COSTS

Fuel costs used in build up of rates in this edition

Petrol Unleaded	140.0	Pence per litre
Petrol Unleaded Super	142.5	Pence per litre
Fuel Oil (Taxed for use in licensed vehicles) DERV	144.0	Pence per litre
Fuel Oil (Lower tax for use on site) Gas Oil	70.0	Pence per litre
Lubrication Oil (15/40)	70.0	Pence per litre
Mains Electric Power	12.5	Pence per kW hr
Propane	120.5	Pence per kg

Also included here are allowances for Transmission Oils, Hydraulic Oil, Filters, Grease, etc.

ESSENTIAL READING FROM TAYLOR AND FRANCIS

Spon's Asia-Pacific Construction Costs Handbook

4th Edition

Edited by **Davis Langdon & Seah**

Spon's Asia Pacific Construction Costs Handbook includes construction cost data for 20 countries. This new edition has been extended to include Pakistan and Cambodia. Australia, UK and America are also included, to facilitate comparison with construction costs elsewhere. Information is presented for each country in the same way, as follows:

- key data on the main economic and construction indicators.

- an outline of the national construction industry, covering structure, tendering and contract procedures, materials cost data, regulations and standards

- labour and materials cost data

- Measured rates for a range of standard construction work items

- Approximate estimating costs per unit area for a range of building types

- price index data and exchange rate movements against £ sterling, $US and Japanese Yen

The book also includes a Comparative Data section to facilitate country-to-country comparisons. Figures from the national sections are grouped in tables according to national indicators, construction output, input costs and costs per square metre for factories, offices, warehouses, hospitals, schools, theatres, sports halls, hotels and housing.

This unique handbook will be an essential reference for all construction professionals involved in work outside their own country and for all developers or multinational companies assessing comparative development costs.

April 2010: 234x156: 480pp
Hb: 978-0-415-46565-6: **£120.00**

To Order: Tel: +44 (0) 1235 400524 **Fax:** +44 (0) 1235 400525
or Post: Taylor and Francis Customer Services,
Bookpoint Ltd, Unit T1, 200 Milton Park, Abingdon, Oxon, OX14 4TA UK
Email: book.orders@tandf.co.uk

For a complete listing of all our titles visit:
www.tandf.co.uk

Taylor & Francis
Taylor & Francis Group

PART 5

Approximate Estimating

	page
Building prices per square metre	93
Approximating estimating rates – General civil engineering works	95
Approximating estimating rates – General building works	104

Estimating by means of priced approximate quantities is always more accurate than by using overall prices per square metre. Prices given in this section, which is arranged in elemental order, also include for all the incidental items and labours which are normally measured separately in Bills of Quantities. They include overheads and profit but do not include for preliminaries.

Whilst every effort is made to ensure the accuracy of these figures, they have been prepared for approximate estimating purposes only and on no account should they be used for the preparation of tenders.

Unless otherwise described units denoted as m² refer to appropriate unit areas (rather than gross floor areas).

As elsewhere in this edition prices do not include Value Added Tax, which should be applied at the current rate.

ESSENTIAL READING FROM TAYLOR AND FRANCIS

Manual for Detailing Reinforced Concrete Structures to EC2

J. Calavera

Detailing 210 structural elements, each with a double page spread, this book is an indispensible guide to designing for reinforced concrete. Drawing on Jose Calavera's many years of experience, each entry provides commentary and recommendations for the elementconcerned, as 0077ell as summarising the appropriate Eurocode legislation and referring to further standards and literature.

Includes a CD ROM with AutoCad files of all of the models, which you can adapt to suit your own purposes.

Section 1. General Rules for Bending, Placing and Lapping of Reinforcement
Section 2. Constructive Details

1. Foundations 2. Walls 3. Columns and Joints 4. Walls Subjected to Axial Loads. Shear Walls and Cores 5. Beams and Lintels 6. Slabs 7. Flat Slabs 8. Stairs 9. Supports 10. Brackets 11. Pavements and Galleries 12. Cylindrical Chimneys, Towers and Hollow Piles 13. Rectangular Silos, Caissons and Hollow Piles 14. Reservoirs, Tanks and Swimming Pools 15. Special Details for Seismic Areas

October 2011: 297 x210: 528pp
Hb: 978-0-415-66348-9: **£80.00**

To Order: Tel: +44 (0) 1235 400524 **Fax:** +44 (0) 1235 400525
or Post: Taylor and Francis Customer Services,
Bookpoint Ltd, Unit T1, 200 Milton Park, Abingdon, Oxon, OX14 4TA UK
Email: book.orders@tandf.co.uk

For a complete listing of all our titles visit:
www.tandf.co.uk

Taylor & Francis
Taylor & Francis Group

Building Prices per Square Metre

Prices given under this heading are average prices on a 'fluctuating basis' for typical buildings based on the provisional first quarter 2012 Building Works tender price level index of 446 (1976 = 100). Unless otherwise stated, prices INCLUDE overheads and profit but EXCLUDE preliminaries, loose or special equipment and fees for professional services.

Prices are based upon the total floor area measured in accordance with the 6th edition of the RICS Code of Measuring Practice, September 2007.

As in previous editions it is emphasized that the prices must be treated with reserve in that they represent the average of prices from our records and cannot provide more than a rough guide to the probable cost of a building or structure.

In many instances normal commercial pressures together with a limited range of available specifications ensure that a single rate is sufficient to indicate the prevailing average price. However, where such restrictions do not apply a range has been given; this is not to suggest that figures outside this range will not be encountered, but simply that the calibre of such a type of building can itself vary significantly.

As elsewhere in this edition, prices do not include Value Added Tax, which should be applied at the current rate where applicable.

	Unit	Cost
Surface car parking	m²	54 to 82
Multi-storey car parks		
split level	m²	230 to 310
split level with brick facades	m²	320 to 440
flat slab	m²	320 to 460
warped	m²	350 to 410
Underground car parks		
partially underground under buildings	m²	430 to 510
completely underground under buildings	m²	640 to 880
completely underground with landscaped roof	m²	920 to 1090
Railway stations	m²	1720 to 2890
Bus and coach stations	m²	660 to 1420
Bus garages	m²	770 to 850
Garage showrooms	m²	790 to 1130
Garages, domestic	m²	430 to 860
Airport facilities (excluding aprons)		
airport terminals	m²	1450 to 3170
airport piers/satellites	m²	1770 to 3950
apron/runway – varying infrastructure content	m²	100 to 190
Airport campus facilities		
cargo handling bases	m²	520 to 870
distribution centres	m²	270 to 520
hangars (type C and D aircraft)	m²	1120 to 1320
hangars (type E aircraft)	m²	1550 to 3900
TV, radio and video studios	m²	1000 to 1980
Telephone exchanges	m²	810 to 1240

BUILDING PRICES PER SQUARE METRE

	Unit	Cost
Telephone engineering centres	m²	670 to 760
Branch Post Offices	m²	850 to 1140
Postal delivery offices/Sorting offices	m²	860 to 1170
Mortuaries	m²	1530 to 2120
Substations	m²	1140 to 1720
INDUSTRIAL FACILITIES		
Agricultural storage buildings	m²	410 to 690
Factories		
for letting (incoming services only)	m²	290 to 410
for letting (including lighting, power and heating)	m²	390 to 510
nursery units (including lighting, power and heating)	m²	450 to 700
workshops	m²	520 to 910
maintenance/motor transport workshops	m²	540 to 690
owner occupation – for light industrial use	m²	550 to 700
owner occupation – for heavy industrial use	m²	900 to 1040
Factory/office buildings – high technology production		
for letting (shell and core only)	m²	500 to 690
for letting (ground floor shell, first floor offices)	m²	990 to 1290
for owner occupation (controlled environment, fully finished)	m²	1090 to 1460
B1 Light industrial/office building		
economical shell and core with heating only	m²	470 to 810
medium shell with heating and ventilation	m²	710 to 1060
high quality shell and core with air conditioning	m²	950 to 1720
Warehouse & Distribution centres		
high bay (10–15 m high) for owner occupation (no heating) up to 10,000 m²	m²	260 to 330
high bay (10–15 m high) for owner occupation (no heating) 10,000 m² to 20,000 m²	m²	190 to 260
high bay (16–24 m high) for owner occupation (no heating) 10,000 m² to 20,000 m²	m²	280 to 370
high bay (16–24 m high) for owner occupation (no heating) over 20,000 m²	m²	210 to 330
Fit-out cold stores, refrigerated stores inside warehouse	m²	400 to 760
WATER AND TREATMENT FACILITIES		
Reinforced Concrete tanks		
including all excavation; fill; structural work; valves; penstocks; pipeworks per m³ of concrete in structure. treatment tanks	m³	560 to 630
fire ponds and lagoons	m³	490 to 570
reservoirs	m³	500 to 580
Reinforced Concrete river weirs, quay and wave walls including temporary dams; caissons overpumping, anchorages		
all structural works	m²	380 to 590
Reinforced Concrete dams (up to 12 m high)		
arch dam, including excavation anchorages and structural work only (m³ of structure)	m³	1270 to 1850
flat slab dam including excavation anchorages and structural work only; (m³ of structure)	m³	960 to 1500
Earth dams		
rock fill with concrete core m³ of completed structure	m³	250 to 880
hydraulic fill embankment dam m³ of completed structure	m³	140 to 390

APPROXIMATE ESTIMATING RATES – GENERAL CIVIL ENGINEERING WORKS

Item	Unit	Range £
EARTH RETENTION AND STABILIZATION		
The prices following have been assembled from the relevant items in the unit costs section. They are intended to give a broad overall price or help in comparisons between a number of basic construction procedures.		
These approximate estimates are for construction only. They do not include for preliminaries, design/supervisioncosts, land purchase or OH&P etc.		
Prices in this section are based on the same information and outputs as used in the unit costs section. Costs per m² or m³ of completed structure.		
Reinforced in situ concrete retaining wall		
(including excavation; reinforcement; formwork; expansion joints; granular backfill and 100 mm land drain; profiled formwork finish to one side typical retaining wall, allowing for profiling finishes)		
1.0 m high	m²	480.00 to 550.00
3.0 m high	m²	420.00 to 485.00
6.0 m high	m²	510.00 to 580.00
9.0 m high	m²	570.00 to 650.00
Precast concrete block, earth retaining wall		
(including granular fill; earth anchors and proprietary units or concrete panels, strips, fixings and accessories (Reinforced Earth))		
1.5 m high	m²	360.00 to 410.00
3.0 m high	m²	330.00 to 380.00
6.0 m high	m²	360.00 to 410.00
Precast, reinforced concrete unit retaining wall (in situ foundation)		
(including excavation and fill; reinforced concrete foundation; precast concrete units, joints)		
1.0 m high	m²	440.00 to 500.00
2.0 m high	m²	445.00 to 510.00
3.0 m high	m²	460.00 to 530.00
Precast concrete crib wall (including excavation; stabilization and foundation work)		
up to 1.0 m high	m²	290.00 to 330.00
up to 1.5 m high	m²	280.00 to 320.00
up to 2.5 m high	m²	280.00 to 320.00
up to 4.0 m high	m²	260.00 to 300.00
Timber crib walling		
up to 1.5 m high	m²	140.00 to 160.00
up to 3.7 m high	m²	175.00 to 200.00
up to 5.9 m high	m²	220.00 to 250.00
up to 7.4 m high	m²	250.00 to 290.00
Rock gabions		
including preparation; excluding anchoring		
1.0 m thick	m²	80.00 to 90.00

APPROXIMATE ESTIMATING RATES – GENERAL CIVIL ENGINEERING WORKS

Item	Unit	Range £
BRIDGEWORK The following prices are based on recovered data and information from approximately 50 separate structures completed as part of actual projects. Prices include for the works described to the bridge decks and abutments, but exclude any approach works. **Road bridges** per m² of deck maximum span between piers or abutments Reinforced in situ concrete viaduct (including excavation; reinforcement; formwork; concrete; bearings; expansion joints; deck waterproofing; deck finishings; P1 parapet)		
span: 15 m	m²	1975.00 to 4100.00
span: 20 m	m²	1875.00 to 3900.00
span: 25+ m	m²	1825.00 to 3800.00
Reinforced concrete bridge with precast beams (including excavation; reinforcement; formwork; concrete; bearings; expansion joints; deck waterproofing; deck finishings; P1 parapet)		
span: 12 m	m²	2375.00 to 3550.00
span: 17 m	m²	2225.00 to 3300.00
span: 22 m	m²	2125.00 to 3200.00
span: 30+ m	m²	2025.00 to 3000.00
Reinforced concrete bridge with pre-fabricated steel beams (including excavation; reinforcement; formwork; concrete; bearings expansion joints; deck waterproofing; deck finishings; P1 parapet)		
span: 20 m	m²	2050.00 to 4800.00
span: 30 m	m²	1925.00 to 4500.00
span: 40 m	m²	1875.00 to 4400.00
Footbridges Reinforced in situ concrete with precast beams widths up to 6 m, per m² of deck maximum span between piers or abutments (including excavation; reinforcement; formwork; concrete; bearings expansion joints; deck waterproofing; deck finishings; P6 parapet)		
span: 5 m	m²	2200.00 to 5100.00
span: 10 m	m²	2125.00 to 5000.00
span: 20+ m	m²	2075.00 to 4900.00
Structural steel bridge with concrete foundations width up to 4 m wide per m² of deck		
span: 10 m	m²	2175.00 to 2850.00
span: 12 m	m²	2075.00 to 2800.00
span: 16 m	m²	2075.00 to 2750.00
span: 20 m	m²	2175.00 to 2850.00
Timber footbridge (stress graded with concrete piers) per m² of deck		
span: 12 m	m²	1140.00 to 1240.00
span: 18 m	m²	1120.00 to 1280.00

APPROXIMATE ESTIMATING RATES – GENERAL CIVIL ENGINEERING WORKS

Item	Unit	Range £
TUNNELLING FIELD COSTS		
The following have been extracted from costs recorded on historical projects and are indicative of the costs expected in constructing larger diameter tunnels. All up as constructed cost based on 60 m² to 90 m² tunnels.		
The following exclude the cost of land, wayleaves, portals, vent shafts, etc., and are an indication of the order of magnitude expected but actual costs are dependent on ground conditions, location and ease of access, etc.		
Rock tunnels; 75 m² face – typical range of rates		
maximum	m³	350.00
median	m³	290.00
lowest	m³	260.00
maximum	km	26,000,000.00
median	km	22,000,000.00
lowest	km	17,000,000.00
Soft ground bored tunnel – typical range of rates		
maximum	m³	935.00
median	m³	875.00
lowest	m³	810.00
maximum	km	13,000,000.00
median	km	12,000,000.00
lowest	km	11,000,000.00
HIGHWAY WORKS		
The following prices are the approximate costs per metre run of roadway, and are based on information from a number of sources including engineers estimates, tenders, final account values etc. on a large number of highways contracts.		
Motorway and All Purpose Road prices include for earthworks, structures, drainage, pavements, line markings, reflective studs, footways signs, lighting, motorway communications, fencing and barrier works as well as allowance for accommodation works, statutory undertakings and landscaping as appropriate to the type and location of the carriageway. The earthworks elements can be adjusted by reference to factors detailed at the end of this subsection.		
Motorway and All Purpose Road prices do NOT include for the cost of associated features such as side roads, interchanges, underbridges, overbridges, culverts, subways, gantries and retaining walls.		
Motorways		
The following costs are based on an 850 mm construction comprising 40 mm wearing course, 60 mm base course, 250 mm road base, 200 mm subbase and 350 mm capping layer; central reserve incorporating two 0.7 m wide hardstrips; no footpaths or cycle paths included.		

APPROXIMATE ESTIMATING RATES – GENERAL CIVIL ENGINEERING WORKS

Item	Unit	Range £
HIGHWAY WORKS – cont		
Rural motorways		
Grassed central reserve and outer berms; no kerbs or edgings; assumption that 30% of excavated material is unsuitable for filling; costs allow for forming embankments for 50% of highway length average 4.70 m high and 50% of length in cuttings average 3.90 m deep; accommodation fencing each side; allowance of 25% of length having crash barriers and 20% of length having lighting		
dual two lane (D2M_R); 25.20 m overall width; each carriageway 7.30 m with 3.30 m hard shoulder; 4.00 m central reserve	m	3200.00 to 3950.00
dual three lane (D3M_R); 32.60 m overall width; each carriageway 11.00 m with 3.30 m hard shoulder; 4.00 m central reserve	m	4050.00 to 5000.00
dual four lane (D4M_R); 39.80 m overall width; each carriageway 14.60 m with 3.30 m hard shoulder; 4.00 m central reserve	m	4850.00 to 6000.00
Urban motorways		
Hard paved central reserve and outer berms; precast concrete kerbs; assumption that 30% of excavated material is unsuitable for filling; costs allow for forming embankments for 50% of highway length average 3.20 m high and 50% of length in cuttings average 1.60 m accommodation fencing each side; allowance of 25% of length having crash barriers and 20% of length having lighting		
dual two lane (D2M_U); 23.10 m overall width; each carriageway 7.30 m with 2.75 m hard shoulder; 3.00 m central reserve	m	3000.00 to 3850.00
dual three lane (D3M_U); 30.50 m overall width; each carriageway 11.00 m with 2.75 m hard shoulder; 3.00 m central reserve	m	3800.00 to 4800.00
dual four lane (D4M_U); 37.70 m overall width; each carriageway 14.60 m with 2.75 m hard shoulder; 3.00 m central reserve	m	4500.00 to 5800.00
All purpose roads		
The following costs are based on a 800 mm construction comprising 40 mm wearing course, 60 mm base course, 200 mm road base, 150 mm subbase and 350 mm capping layer; no footpaths or cycle paths included.		
Rural all purpose road		
Grassed central reserve; no kerbs or edgings; assumption that 30% of excavated material is unsuitable for filling; costs allow for forming embankments for 50% of highway length average 4.00 m high and 50% of length in cuttings average 3.75 m deep; allowance of 25% of length having crash barriers and 20% of length having lighting		
dual two lane (D2 AP_R); 18.60 m overall width; each carriageway 7.30 m; 4.00 m central reserve	m	2275.00 to 2800.00
dual three lane (D3 AP_R); 26.00 m overall width; each carriageway 11.00 m ; 4.00 m central reserve	m	2650.00 to 3600.00

APPROXIMATE ESTIMATING RATES – GENERAL CIVIL ENGINEERING WORKS

Item	Unit	Range £		
Urban all purpose roads Hard paved central reserve; precast concrete kerbs; assumption that 30% of excavated material is unsuitable for filling; costs allow for forming embankments for 50% of highway length in average 2.2 m high and 50% of length in cuttings average 1.62 m deep; allowance of 25% of length having crash barriers and 20% of length having lighting				
dual two lane (D2 AP_U); 23.10 m overall width; each carriageway 7.30 m with 2.75 m hard shoulder; 3.00 m central reserve	m	2900.00	to	3400.00
dual three lane (D3 AP_U); 30.50 m overall width; each carriageway 11.00 m C456with 2.75 m hard shoulder; 3.00 m central reserve	m	3800.00	to	4350.00
Other roads				
Rural all purpose roads				
single carriageway all-purpose road (carriageway is 7.3 m wide)	m	1075.00	to	1225.00
wide single carriageway all-purpose road (carriageway is 10.0 m wide)	m	1325.00	to	1525.00
Rural link roads				
two lane link road (carriageway is 7.3 m wide)	m	960.00	to	1175.00
single lane link road (carriageway is 3.7 m wide)	m	510.00	to	770.00
Rural motorway or dual carriageway slip roads				
single carriageway slip road (carriageway is 5.0 m wide)	m	620.00	to	930.00
Urban motorway or dual carriageway slip roads				
single carriageway slip road (carriageway is 6.0 m wide)	m	800.00	to	1125.00
Wide single carriageway all-purpose road with footway each side				
carriageway is 10.0 m wide, each footway is 3.0 m wide	m	1250.00	to	1700.00
nominal 3.0 m cycle track to all-purpose roads (one side only)	m	160.00	to	220.00
Urban link roads				
two lane link road (carriageway is 7.3 m wide)	m	1050.00	to	1275.00
single lane link road (carriageway is 3.7 m wide)	m	570.00	to	830.00
Type A Construction = medium duty carriageway consisting of 100 mm surfacing, roadbase up to 115 mm, subbase of 150 mm				
Type B Construction = heavy duty carriageway consisting of reinforced concrete slab 225 mm thick subbase 130 mm thick capping layer 280 mm thick				
Turning or passing bay = 35 m² overall area; suitable for cars vans				
type A construction	nr	4500.00	to	5200.00
type B construction	nr	6500.00	to	7500.00
Turning or passing bay = 100 m² overall area; suitable for semi trailer				
type A construction	nr	11000.00	to	13000.00
type B construction	nr	16500.00	to	19000.00
Bus lay-by = 40 m² overall area				
type B construction	nr	6500.00	to	7400.00
Parking lay-by = 200 m² overall area				
type A construction	nr	22000.00	to	25000.00
type B construction	nr	32000.00	to	36500.00
Footpaths and crossings				
vehicle crossing verge/footway/central reserve	m²	150.00	to	170.00
footway construction (bit-mac plus edgings)	m²	50.00	to	58.00

APPROXIMATE ESTIMATING RATES – GENERAL CIVIL ENGINEERING WORKS

Item	Unit	Range £		
HIGHWAY WORKS – cont				
Road crossings				
Major roads				
costs include road markings, beacons, lights, signs, advance danger signs etc.				
4 way traffic signal installation	nr	66000.00	to	76000.00
zebra crossing	nr	49000.00	to	56000.00
pelican crossing	nr	22000.00	to	25500.00
pedestrian guard railing	nr	6000.00	to	6900.00
Estate roads				
costs include road markings, beacons, lights, signs, advance danger signs etc.				
4 way traffic signal installation	nr	39000.00	to	45000.00
zebra crossing	nr	45000.00	to	52000.00
pelican crossing	nr	17500.00	to	20000.00
pedestrian guard railing	nr	4650.00	to	5400.00
UNDERPASSES				
Major roads				
Provision of underpasses to new major roads constructed as part of a road building programme				
pedestrian underpass 3 m wide × 2.5 m high	m	4600.00	to	5300.00
vehicle underpass 7 m wide × 5 m high	m	20500.00	to	23500.00
vehicle underpass 14 m wide × 5 m high	m	47000.00	to	55000.00
Estate roads				
Provision of underpasses to new estate roads constructed as part of a road building programme				
pedestrian underpass 3 m wide × 2.5 m high	m	3800.00	to	4400.00
vehicle underpass 7 m wide × 5 m high	m	15500.00	to	18000.00
vehicle underpass 14 m wide × 5 m high	m	37500.00	to	43000.00
LANDSCAPING				
Sports pitches				
The provision of sports facilities will involve different techniques of earth shifting and cultivation and usually will be carried out by specialist landscaping contractors. Costs include for cultivating ground, bringing to appropriate levels for the specified game, applying fertilizer, weedkiller, seeding and rolling and white line marking with nets, posts etc. as required				
football pitch (114 × 72 m)	nr	18500.00	to	21000.00
artificial football pitch including subbase bitumen macadam open textured base and heavy duty AstroTurf carpet	nr	435000.00	to	500000.00
cricket outfield (160 × 142 m)	nr	51000.00	to	58000.00

APPROXIMATE ESTIMATING RATES – GENERAL CIVIL ENGINEERING WORKS

Item	Unit	Range £
cricket square (20 × 20 m) including imported marl or clay loam, bringing to accurate levels, seeding with cricket square type grass	nr	14000.00 to 16500.00
bowling green (38 × 38 m) rink including french drain and gravel path on four sides	nr	52000.00 to 60000.00
grass tennis courts 1 court (35 × 17 m) including bringing to accurate levels, chain link perimeter fence and gate tennis posts and net	nr	22000.00 to 25000.00
two grass tennis courts (35 × 32 m) ditto	nr	37000.00 to 42500.00
artificial surface tennis courts (35 × 17 m) including chain link fencing, gate, tennis posts and net	nr	41000.00 to 47000.00
two courts (45 × 32 m) including chain link fencing, gate, tennis posts and net	nr	100000.00 to 115000.00
golf putting green	hole	1825.00 to 2100.00
pitch and putt course	hole	6100.00 to 9200.00
full length golf course, full specifications inc watering system	hole	22000.00 to 40500.00
championship, up to	hole	50000.00 to 55000.00

Parklands

As with all sports pitches parklands will involve different techniques of earth shifting and cultivation. The following rates include for normal surface excavation.

Item	Unit	Range £
Parklands including cultivating ground, applying fertilizer etc. and seeding with parks type grass	ha	16500.00 to 19000.00
general sports field	ha	19500.00 to 22500.00
Lakes including excavation up to 10 m deep, laying 1.5 mm thick butyl rubber sheet and spreading top soil evenly on top to depth of 300 mm		
under 1 ha in area	ha	360000.00 to 415000.00
between 1 and 5 ha in area	ha	340000.00 to 390000.00
extra for planting aquatic plants in lake top soil	ha	60.00 to 70.00
Playground equipment		
modern swings with flat rubber safety seats four seats, two bays	nr	3000.00 to 3500.00
stainless steel slide 3.40 m long	nr	2650.00 to 3050.00
climbing frame – Lappset Playhouse	nr	3200.00 to 3700.00
seesaw comprising timber plank on sealed ball bearings 3960 × 230 × 70 mm thick – no-bump	nr	2150.00 to 2450.00
Wicksteed safety tiles surfacing around play equipment	m^2	56.00 to 64.00
Playbark particles type safety surfacing 150 mm thick on hardcore bed	m^2	51.00 to 59.00

TEMPORARY WORKS

Access Scaffolding

The following are guideline costs for the hire and erection of proprietary scaffold systems (tube and coupling). Costs are very general and are based on a minimum area of 360 m² at 1.80 m deep

Item	Unit	Range £
approximate hire (supply and fix) of patent scaffold per 4 week (cost dependent upon length of hire, quality of system, number of toe boards, handrails etc.)	m^2	5.10 to 6.95
based on this a typical cost of a 60 × 10 m (600 m²) area for 8 weeks would be	nr	6200.00 to 8400.00
approximate hire (supply and fix of mobile access towers per 4 week hire (refer also to plant hire in Section 3)	m^2	17.50 to 26.00

APPROXIMATE ESTIMATING RATES – GENERAL CIVIL ENGINEERING WORKS

Item	Unit	Range £		
TEMPORARY WORKS – cont				
Access Scaffolding – cont				
additional costs of Pole ladder access per 4.0 m high per 4 week	nr	6.50	to	7.90
additional cost of stair towers extra over the cost of scaffold system per 2 m rise per 4 week	m^2	68.00	to	84.00
additional cost of hoarding around base perimeter (using multi-use ply sheeting) per 4 week period	m^2	1.10	to	1.80
additional cost of polythene debris netting (no reuse)	m^2	0.20	to	0.30
additional cost of Monaflex 'T' plus weather-proof sheeting including anchors and straps (based on 3 uses)	m^2	1.15	to	1.40
Erection of scaffolding system is based on 3 men erecting a 16 m^2 bay in about 1 hour and dismantling the same in 20 minutes. Note: Although scaffolding is essentially plant hire, allowance must be made for inevitable loss and damage to fittings for consumables used and for maintenance during hire periods.				
Bailey bridges				
Installation of temporary Bailey bridges (including temporary concrete abutments; erection maintenance; dismantling)				
span up to 10 m	nr	27000.00	to	30000.00
span 15 m	nr	31000.00	to	34000.00
span 20 m	nr	39500.00	to	43500.00
span 25 m	nr	42000.00	to	47000.00
Cofferdams				
Installation of cofferdams (based on driven steel sections with recovery value) (including all plant for installation and dismantling; loss of materials; pumping and maintenance excluding excavation and disposal of material – backfilling on completion)				
Cost range based on 12 weeks installation on soft-medium ground conditions				
Depth of drive up to 5 m diameter or side length				
up to 2 m	nr	3900.00	to	4300.00
up to 10 m	nr	19500.00	to	21500.00
up to 20 m	nr	34000.00	to	43000.00
Depth of drive 5 to 10 m diameter or side length				
up to 2 m	nr	6400.00	to	7500.00
up to 10 m	nr	28500.00	to	37000.00
up to 20 m	nr	30500.00	to	43000.00
Depth of drive 10 to 15 m diameter or side length				
up to 2 m	nr	9100.00	to	10000.00
up to 10 m	nr	43500.00	to	51000.00
up to 20 m	nr	72000.00	to	110000.00
Depth of drive 15 to 20 m diameter or side length				
up to 2 m	nr	13500.00	to	17000.00
up to 10 m	nr	39000.00	to	69000.00
up to 20 m	nr	96000.00	to	140000.00
extra for				
medium-hard ground	%	20.00	to	25.00
hard, but not rock	%	33.00	to	40.00

APPROXIMATE ESTIMATING RATES – GENERAL CIVIL ENGINEERING WORKS

Item	Unit	Range £
Earthwork support		
The following are comparative costs for earthwork and trench support based on the hire of modular systems (trench box hydraulic) with an allowance for consumable materials (maximum 5 day hire allowance). For larger excavations requiring Earthwork support refer to cofferdam estimates; or sheet piling within unit cost sections		
earthwork support not exceeding 1 m deep, distance between opposing faces not exceeding 2 m	m^2	5.50 to 6.35
earthwork support not exceeding 1 m deep, distance between opposing faces 2–4 m	m^2	9.50 to 11.00
earthwork support not exceeding 2 m deep, distance between opposing faces not exceeding 2 m	m^2	10.70 to 12.30
earthwork support not exceeding 2 m deep, distance between opposing faces 2–4 m	m^2	18.90 to 22.00
earthwork support not exceeding 4 m deep, distance between opposing faces not exceeding 2 m	m^2	27.00 to 31.00
earthwork support not exceeding 4 m deep, distance between opposing face 2–4 m	m^2	39.00 to 45.00
The following are approximate weekly hire costs for a range of basic support equipment used on site.		
Steel sheet piling		
AU or AZ series section	t/wk	37.00 to 37.00
Trench sheeting		
standard overlapping sheets	m^2/wk	1.60 to 1.60
interlocking type	m^2/wk	3.70 to 3.70
heavy duty overlapping sheets		
6 mm thick	m^2/wk	5.40 to 5.40
8 mm thick	m^2/wk	5.20 to 5.20
driving cap	nr/wk	14.10 to 14.10
extraction clamp	nr/wk	28.00 to 28.00
trench struts		
No 0; 0.3 to 0.40 m	nr/wk	0.75 to 0.75
No 1; 0.5 to 0.75 m	nr/wk	0.75 to 0.75
No 2; 0.7 to 1.14 m	nr/wk	0.75 to 0.75
No 3; 1.0 to 1.75 m	nr/wk	0.75 to 0.75

APPROXIMATE ESTIMATING RATES – GENERAL BUILDING WORKS

Item	Unit	Range £		
1 SUBSTRUCTURE				
Trench fill foundations				
Machine excavation, disposal, plain in situ concrete 21 N/mm² – 20 mm aggregate (1:2:4) trench fill, 300 mm high brickwork in cement mortar (1:3), pitch polymer damp-proof course with common bricks:				
width and depth of concrete 600 mm × 1200 mm				
half brick wall	m	86.00	to	110.00
one brick wall	m	99.00	to	125.00
cavity wall	m	100.00	to	130.00
width and depth of concrete 750 mm × 1200 mm				
half brick wall	m	110.00	to	140.00
one brick wall	m	120.00	to	150.00
cavity wall	m	120.00	to	150.00
Strip foundations				
Excavate trench, partial backfill, partial disposal, earthwork support (risk item), compact base of trench, plain in situ concrete 21 N/mm² – 20 mm aggregate (1:2:4) 250 mm thick, formwork, common brickwork in cement mortar (1:3), pitch polymer damp-proof course; one brick thick, 600 mm foundation, and English bond				
hand excavation, depth of wall				
600 mm	m	87.00	to	110.00
900 mm	m	105.00	to	135.00
1200 mm	m	140.00	to	175.00
1500 mm	m	160.00	to	210.00
machine excavation, depth of wall				
600 mm	m	57.00	to	73.00
900 mm	m	76.00	to	97.00
1200 mm	m	87.00	to	110.00
1500 mm	m	140.00	to	180.00
cavity, 750 mm foundation and stretcher bond				
Hand excavation, depth of wall				
600 mm	m	85.00	to	110.00
900 mm	m	130.00	to	160.00
1200 mm	m	160.00	to	210.00
1500 mm	m	190.00	to	245.00
machine excavation, depth of wall				
600 mm	m	87.00	to	110.00
900 mm	m	110.00	to	140.00
1200 mm	m	140.00	to	180.00
1500 mm	m	170.00	to	210.00
extra over for three courses of facing bricks	m	6.10	to	7.75

APPROXIMATE ESTIMATING RATES – GENERAL BUILDING WORKS

Item	Unit	Range £		
Column bases Excavate pit in firm ground, partial backfill, partial disposal, earth work support, compact base of pit, plain in situ concrete 21 N/mm² – 20 mm aggregate (1:2:4), formwork				
depth of pit – 1200 mm				
hand excavation, base size				
600 × 600 × 300 mm	nr	66.00	to	84.00
900 × 900 × 450 mm	nr	135.00	to	170.00
1200 × 1200 × 450 mm	nr	205.00	to	265.00
1500 × 1500 × 600 mm	nr	370.00	to	475.00
machine excavation, base size				
600 × 600 × 300 mm	nr	52.00	to	66.00
900 × 900 × 450 mm	nr	110.00	to	145.00
1200 × 1200 × 450 mm	nr	165.00	to	210.00
1500 × 1500 × 600 mm	nr	310.00	to	395.00
depth of pit – 1800 mm				
hand excavation, base size				
600 × 600 × 300 mm	nr	84.00	to	105.00
900 × 900 × 450 mm	nr	155.00	to	200.00
1200 × 1200 × 450 mm	nr	260.00	to	330.00
1500 × 1500 × 600 mm	nr	435.00	to	560.00
machine excavation, base size				
600 × 600 × 300 mm	nr	65.00	to	83.00
900 × 900 × 450 mm	nr	120.00	to	160.00
1200 × 1200 × 450 mm	nr	190.00	to	240.00
1500 × 1500 × 600 mm	nr	340.00	to	430.00
Excavate pit in firm ground by machine, partial backfill, partial disposal, earth work support, compact base of pit, plain in situ concrete 21 N/mm² – 20 mm aggregate (1:2:4), formwork				
depth of pit – 1200 mm				
reinforcement at 50 kg/m³ concrete, base size				
1750 × 1750 × 500 mm	nr	360.00	to	460.00
2000 × 2000 × 500 mm	nr	445.00	to	570.00
2200 × 2200 × 600 mm	nr	660.00	to	840.00
2400 × 2400 × 600 mm	nr	760.00	to	980.00
reinforcement at 75 kg/m³ concrete, base size				
1750 × 1750 × 500 mm	nr	380.00	to	490.00
2000 × 2000 × 500 mm	nr	480.00	to	610.00
2200 × 2200 × 600 mm	nr	690.00	to	880.00
2400 × 2400 × 600 mm	nr	810.00	to	1025.00
depth of pit – 1800 mm				
reinforcement at 50 kg/m³ concrete, base size				
1750 × 1750 × 500 mm	nr	400.00	to	520.00
2000 × 2000 × 500 mm	nr	500.00	to	640.00
2200 × 2200 × 600 mm	nr	720.00	to	920.00
2400 × 2400 × 600 mm	nr	830.00	to	1050.00
reinforcement at 75 kg/m³ concrete, base size				
1750 × 1750 × 500 mm	nr	420.00	to	540.00
2000 × 2000 × 500 mm	nr	530.00	to	680.00
2200 × 2200 × 600 mm	nr	740.00	to	950.00
2400 × 2400 × 600 mm	nr	870.00	to	1100.00

APPROXIMATE ESTIMATING RATES – GENERAL BUILDING WORKS

Item	Unit	Range £		

1 SUBSTRUCTURE – cont

Pile caps
Excavate pit in firm ground by machine, partial backfill, partial disposal, earthwork support, compact base of pit, cut off top of pile and prepare reinforcement, reinforced in situ concrete 26 N/mm² – 20 mm aggregate (1:2:4), formwork
depth of pit – 1500 mm
reinforcement at 50 kg/m³ concrete, cap size

Item	Unit	From		To
900 × 900 × 1400 mm, one pile	nr	340.00	to	430.00
2700 × 900 × 1400 mm, two piles	nr	850.00	to	1075.00
*2700 × 2475 × 1400 mm, three piles	nr	1550.00	to	1975.00
2700 × 2700 × 1400 mm, four piles	nr	1975.00	to	2500.00
3700 × 2700 × 1400 mm, six piles	nr	2650.00	to	3400.00

reinforcement at 75 kg/m³ concrete, cap size

900 × 900 × 1400 mm, one pile	nr	340.00	to	430.00
2700 × 900 × 1400 mm, two piles	nr	850.00	to	1075.00
*2700 × 2475 × 1400 mm, three piles	nr	1550.00	to	1975.00
2700 × 2700 × 1400 mm, four piles	nr	2125.00	to	2700.00
3700 × 2700 × 1400 mm, six piles	nr	2800.00	to	3600.00

depth of pit – 2100 mm
reinforcement at 50 kg/m³ concrete, cap size

900 × 900 × 1400 mm, one pile	nr	360.00	to	460.00
2700 × 900 × 1400 mm, two piles	nr	870.00	to	1100.00
*2700 × 2475 × 1400 mm, three piles	nr	1575.00	to	2025.00
2700 × 2700 × 1400 mm, four piles	nr	2025.00	to	2600.00
3700 × 2700 × 1400 mm, six piles	nr	2800.00	to	3600.00

reinforcement at 75 kg/m³ concrete, cap size

900 × 900 × 1400 mm, one pile	nr	360.00	to	460.00
2700 × 900 × 1400 mm, two piles	nr	900.00	to	1150.00
*2700 × 2475 × 1400 mm, three piles	nr	1675.00	to	2175.00
2700 × 2700 × 1400 mm, four piles	nr	2150.00	to	2750.00
3700 × 2700 × 1400 mm, six piles	nr	3000.00	to	3800.00

* = triangular on plan, overall size given
additional cost of alternative strength concrete

30 N/mm²	m³	1.50	to	1.90
40 N/mm²	m³	3.60	to	4.60

Strip or base foundations
Foundations in good ground; reinforced concrete bed; for one-storey development

shallow foundations per m² ground floor plan area	m²	74.00	to	95.00
deep foundations per m² ground floor plan area	m²	120.00	to	155.00

Foundations in good ground; reinforced concrete bed; for two-storey development

shallow foundations per m² ground floor plan area	m²	82.00	to	130.00
deep foundations per m² ground floor plan area	m²	130.00	to	200.00
Extra for each additional storey	m²	28.00	to	33.00

APPROXIMATE ESTIMATING RATES – GENERAL BUILDING WORKS

Item	Unit	Range £		
Raft foundations				
Raft on poor ground for development				
one storey per m² ground floor plan area	m²	85.00	to	110.00
two storey per m² ground floor plan area	m²	140.00	to	175.00
Extra for each additional storey	m²	140.00	to	180.00
Piled foundations				
Foundations in poor ground; reinforced concrete slab; for one storey commercial development per m² ground floor plan area				
short bore piles to columns only	m²	130.00	to	170.00
short bore piles	m²	160.00	to	205.00
fully piled	m²	220.00	to	280.00
Alternative concrete mixes in lieu of 21.00 N/m² – 20 mm aggregate (1:2:4); per m² ground floor plan area				
25 N/mm²	m²	0.65	to	0.80
30 N/mm²	m²	0.90	to	1.15
40 N/mm²	m²	1.60	to	2.05
Ground slabs				
(per m² ground floor area)				
Mechanical excavation to reduce levels, disposal, level and compact, hardcore bed blinded with sand, 1200 gauge polythene damp-proof membrane, concrete 21.00 N/mm² – 20 mm aggregate (1:2:4) ground slab, tamped bed				
200 mm thick concrete slab; thickness of hardcore bed				
150 mm	m²	43.00	to	55.00
175 mm	m²	45.00	to	57.00
200 mm	m²	45.00	to	57.00
300 mm thick concrete slab; thickness of hardcore bed				
150 mm	m²	51.00	to	65.00
175 mm	m²	53.00	to	68.00
200 mm	m²	53.00	to	68.00
450 mm thick concrete slab; thickness of hardcore bed				
150 mm	m²	66.00	to	85.00
175 mm	m²	68.00	to	87.00
200 mm	m²	68.00	to	87.00
Add to the foregoing prices for fabric reinforcement BS 4483, lapped; per m² ground floor plan area				
A142 (2.22 kg/m²); 1 layer	m²	4.30	to	5.50
A142 (2.22 kg/m²); 2 layers	m²	6.30	to	8.05
A193 (3.02 kg/m²); 1 layer	m²	4.30	to	5.50
A193 (3.02 kg/m²); 2 layers	m²	7.35	to	9.40
A252 (3.95 kg/m²); 1 layer	m²	5.35	to	6.85
A252 (3.95 kg/m²); 2 layers	m²	8.60	to	11.00
A393 (6.16 kg/m²); 1 layer	m²	6.30	to	8.05
A393 (6.16 kg/m²); 2 layers	m²	12.60	to	16.10
High yield bent bar reinforcement BS 4449 ; per m² ground floor plan area at a rate of				
25 kg/m²	m²	5.35	to	6.85
50 kg/m²	m²	9.65	to	12.30
75 kg/m²	m²	14.90	to	19.00
100 kg/m²	m²	19.20	to	24.50

APPROXIMATE ESTIMATING RATES – GENERAL BUILDING WORKS

Item	Unit	Range £		
1 SUBSTRUCTURE – cont				
Other foundations/alternative slabs/extras				
Reinforced concrete bed including excavation and hardcore; per m^2 ground floor plan area				
150 mm thick	m^2	42.00	to	54.00
200 mm thick	m^2	56.00	to	72.00
300 mm thick	m^2	69.00	to	98.00
Extra per m^2 ground floor plan area for				
sound reducing quil in screed	m^2	4.40	to	7.80
50 mm insultaion under slab and at edge	m^2	7.20	to	10.60
75 mm insultaion under slab and at edge	m^2	8.95	to	13.40
suspended precast concrete slabs in lieu of in situ slab	m^2	18.90	to	24.00
2.1 FRAME AND 2.2 UPPER FLOORS				
Reinforced in situ concrete columns at 30 N/mm^2, bar reinforcement at 200 kg/m^3; basic formwork (assumed four uses) ; column size				
225 × 225 mm	m	49.00	to	62.00
300 × 300 mm	m	69.00	to	88.00
300 × 450 mm	m	95.00	to	120.00
300 × 600 mm	m	115.00	to	150.00
450 × 450 mm	m	125.00	to	160.00
In situ concrete casing to steel column at 30 N/mm^2, basic framework (assumed four uses), column size				
225 × 225 mm	m	51.00	to	51.00
300 × 300 mm	m	72.00	to	72.00
300 × 450 mm	m	95.00	to	95.00
300 × 600 mm	m	110.00	to	110.00
450 × 450 mm	m	115.00	to	115.00
450 × 600 mm	m	140.00	to	140.00
450 × 900 mm	m	195.00	to	195.00
Reinforced in situ concrete isloated beams at 30 N/mm^2; bar reinforcement at 200 kg/m^3 basic formwork (assumed four uses); beam size				
225 × 450 mm	m	86.00	to	110.00
225 × 600 mm	m	105.00	to	135.00
300 × 600 mm	m	120.00	to	160.00
300 × 900 mm	m	180.00	to	230.00
300 × 1200 mm	m	220.00	to	280.00
450 × 600 mm	m	160.00	to	200.00
450 × 900 mm	m	220.00	to	280.00
450 × 1200 mm	m	280.00	to	355.00
600 × 600 mm	m	195.00	to	250.00
600 × 900 mm	m	260.00	to	335.00
600 × 1200 mm	m	335.00	to	430.00
In situ concrete casing to steel attached beams at 30 N/mm^3; basic formwork (assumed four uses); beam size				
225 × 450 mm	m	77.00	to	99.00
225 × 600 mm	m	99.00	to	125.00
300 × 600 mm	m	110.00	to	140.00

APPROXIMATE ESTIMATING RATES – GENERAL BUILDING WORKS

Item	Unit	Range £		
300 × 900 mm	m	155.00	to	200.00
300 × 1200 mm	m	195.00	to	250.00
450 × 600 mm	m	135.00	to	170.00
450 × 900 mm	m	190.00	to	240.00
450 × 1200 mm	m	240.00	to	300.00
600 × 600 mm	m	175.00	to	225.00
600 × 900 mm	m	220.00	to	285.00
600 × 1200 mm	m	280.00	to	355.00
Extra smooth finish formwork; all categories	m	6.90	to	14.90
Softwood joisted floor; no frame				
Joisted floor; no frame; 22 mm chipboard t & g flooring; herring bone strutting; no coverings or finishes; per m² of upper floor area				
150 × 50 mm joists	m²	30.00	to	38.00
175 × 50 mm joists	m²	31.50	to	40.00
200 × 50 mm joists	m²	33.00	to	42.50
225 × 50 mm joists	m²	37.00	to	47.00
250 × 50 mm joists	m²	39.50	to	51.00
275 × 50 mm joists	m²	43.00	to	55.00
Softwood joisted floor; average depth; platerboard; skim; emulsion; vinyl flooring and painted softwood skirtings; per m² of upper floor area	m²	68.00	to	87.00
Reinforced concrete floors; no frame				
Suspended slab; no coverings or finishes; per m² of upper floor area				
2.75 m span; 8.00 kN/m² loading	m²	57.00	to	72.00
3.35 m span; 8.00 kN/m² loading	m²	65.00	to	83.00
4.25 m span; 8.00 kN/m² loading	m²	81.00	to	100.00
Suspended slab; no coverings or finishes; per m² of upper floor area				
150 mm thick	m²	77.00	to	98.00
225 mm Thick	m²	120.00	to	150.00
Reinforced concrete floors and frame				
Suspended slab; average depth; no coverings or finishes; per m² of upper floor area				
up to six storeys	m²	145.00	to	190.00
Wide span suspended slab with frame; per m² of upper floor area				
up to six storeys	m²	165.00	to	210.00

APPROXIMATE ESTIMATING RATES – GENERAL BUILDING WORKS

Item	Unit	Range £		
2.1 FRAME AND 2.2 UPPER FLOORS – cont				
Reinforced concrete floors; steel frame				
Suspended slab; average depth; Holorib permanent steel shuttering; protected steel frame; no coverings or finishes; per m² of upper floor area				
up to six storeys	m²	195.00	to	250.00
Extra for spans 7.5 to 15 m	m²	22.50	to	28.50
Suspended slab; average depth; protected steel frame; no coverings or finishes; per m² of upper floor area				
up to six storeys	m²	190.00	to	240.00
Suspended slab; 75 mm screed; no coverings or finishes; per m² of upper floor area				
3 m span; 8.50 kN/m² loading	m²	61.00	to	78.00
6 m span; 8.50 kN/m² loading	m²	65.00	to	83.00
7.5 m span; 8.50 kN/m² loading	m²	68.00	to	86.00
3 m span; 12.50 kN/m² loading	m²	73.00	to	93.00
6 m span; 12.50 kN/m² loading	m²	59.00	to	76.00
Precast concrete floors; reinforced concrete frame				
Suspended slab; average depth; no coverings or finishes; per m² of upper floor area	m²	96.00	to	120.00
Precast concrete floors; steel frame				
Suspended slabs; average depth; protected steel frame; no coverings or finishes; per m² of upper floor area				
up to six storeys	m²	175.00	to	220.00
Extra per m² of upper floor area for				
wrought formwork	m²	4.10	to	5.20
sound reducing quilt in screed	m²	4.30	to	5.50
insulation to avoid cold bridging	m²	7.70	to	9.90
2.3 ROOF				
Softwood flat roofs				
Roof joists; average depth; 25 mm softwood boarding; PVC rainwater goods; plasterboard; skim and emulsion; per m² of roof plan area				
three layer felt and chippings	m²	105.00	to	130.00
two coat asphalt and chippings	m²	110.00	to	135.00
Softwood trussed pitched roofs				
Structure only comprising 100 × 38 mm Fink trusses @ 600 mm centres (measured on plan); per m² of roof plan area				
30° pitch	m²	24.00	to	29.50
35° pitch	m²	24.50	to	30.00
40° pitch	m²	27.50	to	33.50
Fink roof trusses; narrow span; 100 mm insulation; PVC rainwater goods; plasterboard; skim and emulsion per m² or roof plan area				
concrete interlocking tile coverings	m²	95.00	to	115.00
clay pantile coverings	m²	100.00	to	125.00
composition slate coverings	m²	110.00	to	130.00

APPROXIMATE ESTIMATING RATES – GENERAL BUILDING WORKS

Item	Unit	Range £		
plain clay tile coverings	m²	125.00	to	155.00
natural slate coverings	m²	135.00	to	165.00
reconstructed stone coverings	m²	110.00	to	135.00
Monopitch roof trusses; 100 mm insulation; PVC rainwater goods; plasterboard; skim and emulsion; per m² of roof plan area				
concrete interlocking tile coverings	m²	110.00	to	140.00
clay pantile coverings	m²	110.00	to	140.00
composition slate coverings	m²	115.00	to	140.00
plain clay tile coverings	m²	135.00	to	165.00
natural slate coverings	m²	140.00	to	170.00
reconstructed stone coverings	m²	115.00	to	140.00
Steel trussed pitched roofs				
Steel roof trusses and beams; thermal and accoustic insulations; per m² of roof plan area				
aluminium profiled composite cladding	m²	240.00	to	300.00
Steel roof and glulam beams; thermal and accoustic insulation; per m² of roof plan area				
aluminium profiled composite cladding	m²	240.00	to	300.00
Concrete flat roofs				
Structure only comprising reinforced concrete suspended slab; no coverings or finishes; per m² of roof plan area				
3.65 m span; 8.00 kN/m² loading	m²	60.00	to	74.00
4.25 m span; 8.00 kN/m² loading	m²	72.00	to	88.00
Precast concrete suspended slab; average depth; 100 mm insulation; PVC rainwater goods; per m² of roof plan area				
two coat asphalt coverings and chippings	m²	110.00	to	135.00
Reinforced concrete or waffle suspended slabs; average depth; 100 mm insulation; PVC rainwater goods; per m² of roof plan area				
two coat asphalt coverings and chippings	m²	120.00	to	145.00
Reinforced concrete suspended slabs; on Holorib permanent steel shuttering; average depth; 100 mm insulation; PVC rainwater goods; per m² of roof area				
two coat asphalt coverings and chippings	m²	110.00	to	135.00
Flat roof decking and finishes				
Woodwool roof decking; per m² of roof plan area				
50 mm thick; two coat asphalt coverings to BS 6925 and chippings	m²	50.00	to	61.00
Galvanized steel roof decking; 100 mm insulation; three layer felt roofing and chippings; per m² of roof plan area				
0.7 mm thick; 3.74 m span	m²	55.00	to	67.00
0.7 mm thick; 5.13 m span	m²	56.00	to	69.00
Aluminium roof decking; 100 mm insulation; three layer felt roofing and chippings; per m² of roof plan area				
0.9 mm thick; 2.34 m span	m²	64.00	to	78.00
Metal decking; 100 mm insulation; on wood/steel open lattice beams; per m² of roof plan area				
three layer felt roofing and chippings	m²	95.00	to	115.00
two layer higher performance felt roofing and chippings	m²	100.00	to	120.00

APPROXIMATE ESTIMATING RATES – GENERAL BUILDING WORKS

Item	Unit	Range £		
2.3 ROOF – cont				
Roof claddings				
Non-asbestos profiled cladding; per roof plan area				
profile 3; natural	m²	16.50	to	20.00
profile 6; natural	m²	18.30	to	22.00
profile 6; natural; insulated; inner lining panel	m²	20.00	to	25.00
Extra for colours	m²	1.80	to	2.20
Non-asbestos profiled cladding on steel purlins; per m² of roof plan area				
Insulated	m²	32.00	to	39.50
Insulated; with 10% translucent sheets	m²	36.00	to	44.50
Insulated; plasterboard inner lining on metal tees	m²	53.00	to	64.00
PVF2 coated galvanized steel profiled cladding on steel purlins; per m² of roof plan area				
cladding only; 0.72 mm thick	m²	22.00	to	27.00
insulated	m²	39.00	to	48.00
insulated; plasterboard inner lining on metal tees	m²	59.00	to	72.00
insulated; plasterboard inner lining on metal tees; with 1% fire vents	m²	69.00	to	84.00
insulated; plasterboard inner lining on metal tees; with 2.5% fire vents	m²	86.00	to	105.00
insulated; coloured inner lining panel	m²	61.00	to	74.00
insulated; coloured inner lining panel; with 1% fire vents	m²	69.00	to	84.00
insulated; coloured inner lining panel; with 2.5% fire vents	m²	86.00	to	105.00
insulated; sandwich panel	m²	130.00	to	160.00
Rooflights/patent glazing and glazed roofs				
Rooflights				
standard PVC	m²	140.00	to	170.00
feature/ventilating	m²	255.00	to	310.00
Patent glazing; including flashings				
Standard aluminium georgian wired; single glazed	m²	180.00	to	220.00
Standard aluminium georgian wired; double glazed	m²	21.50	to	26.50
Comparitive over/underlays				
Roofing felt; unreinforced				
sloping (measured on face)	m²	1.60	to	2.00
Roofing felt; reinforced				
sloping (measured on face)	m²	2.00	to	2.50
sloping (measured on plan) 20° pitch	m²	2.30	to	2.85
sloping (measured on plan) 30° pitch	m²	2.55	to	3.10
sloping (measured on plan) 40° pitch	m²	3.30	to	4.00
Building paper; per m² of roof plan area	m²	1.40	to	1.70
Vapour barrier; per m² of roof plan area	m²	2.10	to	2.60
Insulation quilt; laid over ceiling joists; per m² of roof plan area				
100 mm thick	m²	4.80	to	5.80
150 mm thick	m²	6.55	to	8.00
200 mm thick	m²	8.75	to	10.70
Wood fibre insulation boards; impregnated; density 220–350 kg/m³; per m² of roof plan area				
12.7 mm thick	m²	8.60	to	10.50
Limestone ballast; per m² of roof plan area	m²	6.40	to	7.80

APPROXIMATE ESTIMATING RATES – GENERAL BUILDING WORKS

Item	Unit	Range £		
Polyurethane insulation boards; density 32 kg/m³; per m² of roof plan area				
30 mm thick	m²	9.15	to	11.20
50 mm thick	m²	11.00	to	13.40
Glass fibre insulation boards; density 120–130 kg/m²; per m² of roof plan area				
60 mm thick	m²	16.90	to	20.50
Extruded polystyrene foam boards; per m² of roof plan area				
50 mm thick	m²	16.00	to	19.50
50 mm thick; with cement topping	m²	26.00	to	31.50
75 mm thick	m²	21.00	to	25.50
Screeds to receive roof coverings; per m² of roof plan area				
50 mm cement and sand screed	m²	9.90	to	12.10
60 mm (av.) Isocrete K screed; density 500 kg/m³	m²	11.40	to	13.90
75 mm lightweight bituminous screed and vapour barrier	m²	17.80	to	22.00
100 mm lightweight bituminous screed and vapour barrier	m²	22.00	to	27.00
50 mm woodwool slabs; unreinforced				
sloping (measured on face)	m²	9.15	to	11.20
sloping (measured on plan); 20° pitch	m²	10.10	to	12.30
sloping (measured on plan); 30° pitch	m²	11.10	to	13.60
sloping (measured on plan); 40° pitch	m²	13.70	to	16.80
50 mm woodwool slabs; unreinforced; on and including steel purlins at 600 mm centres	m²	11.40	to	13.90
25 mm tanalized softwood boarding				
sloping (measured on face)	m²	15.60	to	19.00
sloping (measured on plan); 20° pitch	m²	16.90	to	20.50
sloping (measured on plan); 30° pitch	m²	20.00	to	25.00
sloping (measured on plan); 40° pitch	m²	20.00	to	24.50
18 mm external quality plywood boarding				
sloping (measured on face)	m²	19.70	to	24.00
sloping (measured on plan); 20° pitch	m²	22.00	to	27.00
sloping (measured on plan); 30° pitch	m²	26.50	to	32.00
sloping (measured on plan); 40° pitch	m²	31.00	to	38.00
Comparative tiling and slating finishes/perimeter treatments (including underfelt, battening, eaves courses and ridges)				
Concrete troughed interlocking tiles; 413 × 300 mm; 75 mm lap				
sloping (measured on face)	m²	18.80	to	23.00
sloping (measured on plan); 30° pitch	m²	24.00	to	30.00
sloping (measured on plan); 40° pitch	m²	29.00	to	36.00
Concrete troughed interlocking slates; 430 × 330 mm; 75 mm lap				
sloping (measured on face)	m²	19.70	to	24.00
sloping (measured on plan); 30° pitch	m²	21.50	to	26.00
sloping (measured on plan); 40° pitch	m²	29.00	to	36.00
Concrete bold roll interlocking tiles; 418 × 332 mm; 75 mm lap				
sloping (measured on face)	m²	18.80	to	23.00
sloping (measured on plan); 30° pitch	m²	25.00	to	30.00
sloping (measured on plan); 40° pitch	m²	28.00	to	34.00

APPROXIMATE ESTIMATING RATES – GENERAL BUILDING WORKS

Item	Unit	Range £		

2.3 ROOF – cont

Comparative tiling and slating finishes/perimeter treatments (including underfelt, battening, eaves courses and ridges) – cont

Item	Unit	Range £		
Natural red pantiles; 337 × 241 mm; 76 mm head and 38 mm side laps				
sloping (measured on face)	m²	31.00	to	38.00
sloping (measured on plan); 30° pitch	m²	41.00	to	50.00
sloping (measured on plan); 40° pitch	m²	46.00	to	57.00
Blue composition (non asbestos) slates; 600 × 300 mm; 75 mm lap				
sloping (measured on face)	m²	32.00	to	39.50
sloping to mansard (measured on face)	m²	46.00	to	56.00
sloping (measured on plan); 30° pitch	m²	43.00	to	52.00
sloping (measured on plan); 40° pitch	m²	48.00	to	59.00
Concrete plain tiles; 267 × 165 mm; 64 mm lap				
sloping (measured on face)	m²	37.50	to	46.00
sloping (measured on plan); 30° pitch	m²	46.50	to	57.00
sloping (measured on plan); 40° pitch	m²	56.00	to	69.00
Machine-made clay plain tiles; 267 × 165 mm; 64 mm lap				
sloping (measured on face)	m²	49.00	to	60.00
sloping (measured on plan); 30° pitch	m²	61.00	to	75.00
sloping (measured on plan); 40° pitch	m²	74.00	to	91.00
Welsh natural slates; 510 × 255 mm; 76 mm lap				
sloping (measured on face)	m²	55.00	to	67.00
sloping (measured on plan); 30° pitch	m²	71.00	to	87.00
sloping (measured on plan); 40° pitch	m²	84.00	to	100.00
Reconstructed stone slates; random lengths; 80 mm lap				
sloping (measured on face)	m²	34.00	to	42.00
sloping (measured on plan); 30° pitch	m²	43.00	to	52.00
sloping (measured on plan); 40° pitch	m²	51.00	to	62.00
Verges to sloping roofs; 200 × 25 mm painted softwood bargeboard; measured perimeter length				
6 mm Masterboard soffit lining 150 mm wide	m	17.80	to	22.00
19 × 150 mm painted softwood soffit	m	21.50	to	26.00
Eaves to sloping roofs; 200 × 25 mm painted softwood fascia; 6 mm Masterboard soffit lining 225 mm wide; measured perimeter length				
100 mm PVC gutter	m	24.00	to	30.00
150 mm PVC gutter	m	31.00	to	38.00
100 mm cast iron gutter; decorated	m	39.00	to	47.50
150 mm cast iron gutter; decorated	m	47.00	to	58.00
Eaves to sloping roofs; 200 × 25 mm painted softwood fascia; 19 × 225 mm painted softwood soffit; measured perimeter length				
100 mm PVC gutter	m	29.00	to	36.00
150 mm PVC gutter	m	36.00	to	44.50
100 mm cast iron gutter; decorated	m	44.00	to	54.00
150 mm cast iron gutter; decorated	m	51.00	to	63.00
Rainwater pipes; fixed to backgrounds; including offsets and shoe measured length of pipes				
68 mm PVC	m	8.50	to	10.40
110 mm PVC	m	12.70	to	15.50
75 mm cast iron; decorated	m	31.00	to	38.00
100 mm cast iron; decorated	m	36.00	to	44.50

APPROXIMATE ESTIMATING RATES – GENERAL BUILDING WORKS

Item	Unit	Range £		
Ridges measured length of ridge				
concrete half round tiles	m	17.30	to	21.00
machine-made clay half round tiles	m	21.50	to	26.00
Hips; including mitring roof tiles measured length of hip				
concrete half round tiles	m	22.00	to	27.00
machine-made clay half round tiles	m	32.00	to	39.00
Comparative cladding finishes (including underfelt, labours etc.)				
0.91 mm Aluminium roofing; commercial grade				
flat	m²	50.00	to	61.00
0.91 mm Aluminium roofing; commercial grade; fixed to boarding (included)				
sloping (measured on face)	m²	52.00	to	64.00
sloping (measured on plan); 20° pitch	m²	59.00	to	72.00
sloping (measured on plan); 30° pitch	m²	71.00	to	86.00
sloping (measured on plan); 40° pitch	m²	84.00	to	100.00
Comparative waterproof finishes/perimeter treatments				
Liquid applied coatings				
solar reflected paint	m²	1.80	to	2.20
solar applied bitumen	m²	7.30	to	8.90
solar applied co-polymer	m²	8.60	to	10.50
solar applied polyurethane	m²	14.30	to	17.40
20 mm Two coat asphalt roofing; laid flat; on felt underlay				
to BS 6925	m²	15.20	to	18.50
to BS 6577	m²	21.50	to	26.00
Extra for				
solar reflective paint	m²	2.50	to	3.00
limestone chipping finish	m²	3.15	to	3.85
grp tiles in hot bitumen	m²	34.00	to	42.00
20 mm two coat reinforced asphaltic compound; laid flat; on felt underlay to BS 6577				
three layer glass fibre roofing	m²	20.00	to	25.00
three layer asbestos based roofing	m²	25.00	to	31.00
Extra for granite chipping finish	m²	3.15	to	3.80
Built-up self-finished asbestos based bitumen felt roofing; laid sloping				
two layer roofing (measured on face)	m²	27.00	to	33.50
two layer roofing (measured on plan); 40° pitch	m²	42.00	to	51.00
three layer roofing (measured on face)	m²	36.00	to	44.50
three layer roofing (measured on plan); 20° pitch	m²	55.00	to	67.00
three layer roofing (measured on plan); 30° pitch	m²	57.00	to	69.00
Elastomeric single ply roofing; laid flat				
EDPM membrane; laid loose	m²	24.00	to	29.50
Butyl rubber membrane; laid loose	m²	24.00	to	29.50
Extra for ballast	m²	7.30	to	8.90
Thermoplastic single ply roofing; laid flat				
laid loose	m²	23.00	to	28.00
mechanically fixed	m²	29.00	to	36.00
fully adhered	m²	32.00	to	39.50

APPROXIMATE ESTIMATING RATES – GENERAL BUILDING WORKS

| Item | Unit | Range £ | | |
|---|---|---:|---|---:|//
2.3 ROOF – cont				
Comparative waterproof finishes/perimeter treatments – cont				
Thermoplastic single ply roofing – cont				
CPE membrane; laid loose	m²	26.50	to	32.00
CSPG membrane; fully adhered	m²	26.50	to	32.00
PIB membrane; laid loose	m²	30.00	to	37.00
Extra for ballast	m²	7.30	to	8.90
High performance built-up felt roofing; laid flat				
three layer Ruberglas 120 GP felt roofing; granite chipping finish	m²	32.00	to	39.50
Andersons three layer self-finish polyester-based bitumen felt roofing	m²	32.00	to	39.50
High performance built-up felt roofing; laid flat				
three layer polyester-based modified bitumen felt roofing	m²	33.00	to	41.00
three layer Ruberfort HP 350 felt roofing; granite chipping finish	m²	41.00	to	50.00
three layer Hyload 150E elastomeric roofing; granite chipping finish	m²	41.00	to	50.00
three layer Polybit 350 elastomeric roofing; granite chipping finish	m²	43.00	to	52.00
Torch on roofing; laid flat				
three layer polyester-based modified bitumen roofing	m²	32.00	to	39.00
two layer polymeric isotropic roofing	m²	29.00	to	36.00
Extra granite chipping finish	m²	2.70	to	3.35
Edges to flat felt roofs; softwood splayed fillet; 280 × 25 mm painted softwood fascia; no gutter				
aluminium edge trim	m	34.00	to	42.00
Edges to flat roofs; code 4 lead drip dresser into gutter; 230 × 25 mm painted softwood fascia				
100 mm PVC gutter	m	32.00	to	39.50
150 mm PVC gutter	m	41.00	to	50.00
100 mm cast iron gutter; decorated	m	49.00	to	60.00
150 mm cast iron gutter; decorated	m	61.00	to	74.00
2.4 STAIRS AND RAMPS				
Timber construction				
Softwood staircase; softwood balustrades and hardwood handrail; plasterboard; skim and emulsion to soffit				
2.6 m rise; standard; straight flight	nr	730.00	to	1000.00
2.6 m rise; standard; top three treads winding	nr	890.00	to	1250.00
2.6 m rise; standard; dogleg	nr	1025.00	to	1425.00
Reinforced concrete construction				
Escape staircase; granolithic finish; mild steel balustrades and handrails				
3 m rise; dogleg	nr	3800.00	to	5300.00
Plus or minus for each 300 mm variation in storey height	nr	370.00	to	510.00
Staircase; terrazzo finish; mild steel balustrades and handrails; plastered soffit; balustrades and staircase soffit decorated				
3 m rise; dogleg	nr	7600.00	to	10500.00
Plus or minus for each 300 mm variation in storey height	nr	770.00	to	1075.00

APPROXIMATE ESTIMATING RATES – GENERAL BUILDING WORKS

Item	Unit	Range £		
Metal construction				
Steel access/fire ladder				
3 m high	nr	580.00	to	800.00
4 m high; expoxide finished	nr	810.00	to	1125.00
Light duty metal staircase; glavanized finish; perforated treads; no risers; balustrades and handrails; decorated				
3 m rise; straight; 760 mm wide	nr	2800.00	to	3900.00
Plus or minus for each 300 mm variation in storey height	nr	285.00	to	395.00
Heavy duty cast iron staircase; perforated treads; no risers; balustrades and handrails; decorated				
3 m rise; straight; 760 mm wide	nr	4450.00	to	6200.00
Plus or minus for each 300 mm variation in storey height	nr	445.00	to	620.00
Galvanized steel catwalk; nylon coated balustrading				
450 mm wide	nr	310.00	to	435.00
Finishes to treads and risers				
PVC floor tiles including screeds	storey	770.00	to	1075.00
granolithic	storey	1175.00	to	1625.00
heavy duty carpet	storey	1625.00	to	2250.00
terrazzo	storey	3250.00	to	4500.00
Comparative finishes/balustrading				
Wall handrails				
PVC mild steel rail on brackets	storey	305.00	to	420.00
hardwood handrail on brackets	storey	830.00	to	1150.00
stainless stell handrail on brackets	storey	2650.00	to	3700.00
Balustrading and handrails				
mild steel balustrading and PVC covered handrails	storey	950.00	to	1325.00
mild steel balustrading and hardwood handrails	storey	1725.00	to	2375.00
stainless steel balustrading and handrails	storey	7300.00	to	10000.00
stainless steel and glass balustrading	storey	6400.00	to	8900.00
2.5 EXTERNAL WALLS				
Brick/block walling				
Dense aggregate block walls				
100 mm thick	m^2	22.50	to	27.00
140 mm thick	m^2	32.50	to	39.00
Common brick solid walls				
half brick thick	m^2	35.00	to	42.00
one brick thick	m^2	63.00	to	76.00
one and a half brick thick	m^2	91.00	to	110.00
Add or deduct for each variation of £10.00/1000 in Pc value				
half brick thick	m^2	1.00	to	1.20
one brick thick	m^2	2.00	to	2.45
one and a half brick thick	m^2	3.05	to	3.65
Extra for fair face one side	m^2	2.00	to	2.45
Engineering brick walls; class B				
half brick thick	m^2	42.50	to	51.00
one brick thick	m^2	81.00	to	97.00

APPROXIMATE ESTIMATING RATES – GENERAL BUILDING WORKS

Item	Unit	Range £		
2.5 EXTERNAL WALLS – cont				
Brick/block walling – cont				
Facing brick walls; machine-made facings				
half brick thick; pointed one side	m²	63.00	to	76.00
half brick thick; built against concrete	m²	67.00	to	81.00
one brick thick; pointed both sides	m²	120.00	to	150.00
Add or deduct for each variation of £10.00/1000 in Pc value				
half brick thick	m²	1.00	to	1.20
one brick thick	m²	2.00	to	2.45
Composite solid walls; facing brick on outside; and common brick on outside				
one brick thick; pointed one side	m²	99.00	to	120.00
Extra for weather pointing as a separate operation	m²	5.40	to	6.45
one and a half brick thick; pointed one side	m²	120.00	to	150.00
Composite cavity wall; block outer skin; 50 mm insulation; lightwieght block inner skin				
outer block rendered	m²	70.00	to	84.00
Extra for				
heavyweight block inner skin	m²	2.75	to	3.30
fair face one side	m²	13.10	to	15.70
75 mm cavity insulation	m²	70.00	to	84.00
100 mm cavity insulation	m²	76.00	to	91.00
plaster and emulsion	m²	86.00	to	100.00
outer block rendered; no insulation; inner skin insulating	m²	1.45	to	1.70
outer block roughcast	m²	0.70	to	0.90
coloured masonry outer block	m²	1.45	to	1.70
Composite cavity wall; facing brick outer skin; 50 mm insulation; common brick inner skin; fair face on inside				
machine-made facings	m²	105.00	to	130.00
Composite cavity wall; facing brick outer skin; 50 mm insulation; common brick inner skin; plaster and emulsion				
machine-made facings	m²	100.00	to	120.00
Composite cavity wall; coloured masonry block; outer and inner skins; fair faced both sides	m²	1.00	to	1.20
Reinforced concrete walling				
In situ reinforced concrete 25.5 N/m³; 13 kg/m² reinforcement; formwork both sides				
150 mm thick	m²	110.00	to	130.00
225 mm thick	m²	135.00	to	160.00
Wall claddings				
Non-asbestos profiled cladding				
Profile 3; natural	m²	24.00	to	28.50
Profile 3; coloured	m²	26.00	to	31.00
Profile 6; natural	m²	26.00	to	31.00
Profile 6; coloured	m²	27.00	to	32.50
insulated; inner lining of plasterboard	m²	46.50	to	56.00
Profile 6; natural; insulated; inner lining panel	m²	46.50	to	56.00
insulated; with 2.8 m high inner skin; emulsion	m²	40.50	to	48.50

APPROXIMATE ESTIMATING RATES – GENERAL BUILDING WORKS

Item	Unit	Range £		
insulated; with 2.8 m high inner skin; plasterboard lining on metal tees; emulsion	m²	55.00	to	66.00
PVF2 coated galvanized steel profiled cladding				
0.60 mm thick; Profile 20B; corrugations verticle	m²	32.50	to	39.00
0.60 mm thick; Profile 30; corrugations verticle	m²	32.50	to	39.00
0.60 mm thick; Profile TOP 40; corrugations verticle	m²	30.00	to	36.50
0.60 mm thick; Profile 60B; corrugations verticle	m²	38.00	to	45.50
0.60 mm thick; Profile 30; corrugations horizontal	m²	34.00	to	40.50
0.60 mm thick; Profile 60B; corrugations horizontal	m²	38.00	to	45.50
Extra for				
80 mm insulation and 0.4 mm thick coated inner lining sheet	m²	18.30	to	22.00
PVF2 coated galvanized steel profiled cladding on steel rails				
2.8 m high insulating block inner skin; plasterboard lining on metal tees; emulsion	m²	67.00	to	80.00
2.8 m high insulated block inner skin; plasterboard lining on metal tees; emulsion	m²	76.00	to	91.00
insulated; coloured inner lining panel	m²	76.00	to	91.00
insulated; full-height insulating block inner skin; plaster and emulsion	m²	94.00	to	110.00
insulated; metal sandwich panel system	m²	190.00	to	230.00
Other cladding systems				
vitreous enamelled insulated steel sandwich panel system; with non-asbestos fibre insulating board on inner face	m²	180.00	to	220.00
Formalux sandwich panel system; with coloured lining tray; on steel cladding rails	m²	215.00	to	260.00
aluminium over-cladding on system rain screen	m²	270.00	to	325.00
natural stone cladding on full-height insulating block inner skin; plaster and emulsion	m²	520.00	to	620.00
Curtain/glazed walling				
Single glazed polyester powder coated aluminium curtain walling				
economical; including part-height block back-up wall; plaster and emulsion	m²	405.00	to	490.00
Extra over single 6 mm float glass for				
double glazing unit with two 6 mm float glass skins	m²	43.50	to	52.00
double glazing unit with one 6 mm Antisun skin and one 6 mm float glass skin	m²	84.00	to	100.00
look-a-like non-vision spandrel panels	m²	55.00	to	66.00
10% opening lights	m²	13.10	to	15.70
economical; including infill panels	m²	390.00	to	470.00
Extra for				
50 mm insulation	m²	23.00	to	28.00
anodized finish in lieu of polyester powder coating	m²	27.50	to	33.00
bronze anodizing in lieu of polyester powder coating	m²	55.00	to	66.00
good quality	m²	640.00	to	760.00
good quality; 35% opening lights	m²	800.00	to	960.00
Extra over single 6 mm float glass for double glazing unit with low E and tinted glass	m²	83.00	to	99.00
High quality structural glazing to entrance elevation	m²	870.00	to	1050.00

APPROXIMATE ESTIMATING RATES – GENERAL BUILDING WORKS

Item	Unit	Range £		
2.5 EXTERNAL WALLS – cont				
Curtain/glazed walling – cont				
Patent glazing systems; excluding opening lights and flashings				
7 mm georgian wired cast glass, aluminium alloy bars spanning up to 3 mat 600–625 mm spacing	m²	130.00	to	155.00
6.4 mm laminated safety glass polyester powder coated aluminium capped glazing bars spanning up to 3 m at 600–625 mm spacing	m²	405.00	to	490.00
Comparative external finishes				
Comparative concrete wall finishes				
wrought formwork one side including rubbing down	m²	3.80	to	4.55
shotblasting to expose aggregate	m²	4.80	to	5.75
bush hammering to expose aggregate	m²	15.40	to	18.50
two coats Sandex Matt cement paint	m²	9.00	to	10.80
cement and sand plain face rendering	m²	15.80	to	19.00
three-coat Tyrolean rendering; including backing	m²	38.50	to	46.00
Mineralite' decorative rendering; including backing	m²	76.00	to	92.00
2.6 WINDOWS AND EXTERNAL DOORS				
Softwood windows and external doors				
Standard windows; painted				
single glazed	m²	255.00	to	310.00
double glazed	m²	330.00	to	405.00
Standard external softwood doors and hardwood frames; doors painted; including ironmongery				
solid flush door	nr	740.00	to	910.00
heavy duty solid flush door; single leaf	nr	860.00	to	1050.00
heavy duty solid flush door; double leaf	nr	1450.00	to	1775.00
Extra for				
emergency fire exit door	nr	325.00	to	400.00
Steel windows and external doors				
Standard windows				
single glazed; galvanized; painted	m²	265.00	to	320.00
single glazed; powder coated	m²	270.00	to	330.00
double glazed; galvanized; painted	m²	335.00	to	410.00
double glazed; powder coated	m²	335.00	to	410.00
Steel roller shutters				
Shutters; galvanized				
manual	m²	265.00	to	325.00
electric	m²	345.00	to	420.00
manual; insulated	m²	425.00	to	520.00
electric; insulated	m²	520.00	to	630.00
electric; insulated; fire-resistant	m²	1175.00	to	1450.00

APPROXIMATE ESTIMATING RATES – GENERAL BUILDING WORKS

Item	Unit	Range £		
Hardwood windows and external doors				
Standard windows; stained or uPVC coated				
single glazed	m²	415.00	to	510.00
double glazed	m²	450.00	to	550.00
PVC-U windows and external doors				
Purpose-made windows				
double glazed	m²	710.00	to	870.00
Aluminium windows and external doors				
Standard windows; anodized finish				
single glazed; horizontal sliding sash	m²	355.00	to	435.00
single glazed; vertical sliding sash	m²	500.00	to	610.00
single glazed; casement; in hardwood subframe	m²	400.00	to	490.00
double glazed; vertical sliding sash	m²	550.00	to	670.00
double glazed; casement; in hardwood subframe	m²	490.00	to	600.00
Purpose-made entrance screens and doors				
double glazed	m²	1025.00	to	1275.00
Stainless steel entrance screens and doors				
Purpose-made screen; double glazed				
with manual doors	m²	1625.00	to	1975.00
with automatic doors	m²	1975.00	to	2425.00
Shop fronts, shutters and grilles				
Shutters and grilles per metre of plan length				
Grilles or shutters	m	990.00	to	1200.00
Fire shutters; power-operated	m	1400.00	to	1725.00
2.7 INTERNAL WALLS, PARTITIONS AND DOORS				
Timber or metal stud partitions and doors				
Softwood stud and plasterboard partitions				
100 mm partition; skim and emulsioned both sides	m²	55.00	to	68.00
150 mm partition as party wall; skim and emulsioned both sides	m²	67.00	to	82.00
Metal stud and plasterboard partitions				
170 mm partition; one hour; taped joints; emulsioned both sides	m²	58.00	to	70.00
200 mm partition; two hour; taped joints; emulsioned both sides	m²	77.00	to	94.00
Metal stud and plasterboard partitions; emulsioned both sides; softwood doors and frames; painted				
170 mm partition	m²	81.00	to	99.00
200 mm partition; insulated	m²	105.00	to	130.00
Stud or plasterboard partitions; softwood doors and frames; painted partition; plastered and emulsioned both sides	m²	89.00	to	110.00
Stud or plasterboard partitions; hardwood doors and frames; painted partition; plastered and emulsioned both sides	m²	125.00	to	155.00

APPROXIMATE ESTIMATING RATES – GENERAL BUILDING WORKS

Item	Unit	Range £		
2.7 INTERNAL WALLS, PARTITIONS AND DOORS – cont				
Brick/block partitions and doors				
Autoclaved aerated/lightweight block partitions				
75 mm thick	m²	17.80	to	22.00
100 mm thick	m²	24.50	to	30.00
130 mm thick; insulating	m²	28.00	to	34.00
150 mm thick	m²	30.00	to	37.00
190 mm thick	m²	39.50	to	48.00
Extra for				
fair face both sides	m²	4.40	to	5.40
curved work	m²	3.60	to	4.35
average thickness; fair face both sides	m²	33.00	to	40.00
average thickness; fair face and emulsioned both sides	m²	38.50	to	47.00
average thickness; plastered and emulsioned both sides	m²	58.00	to	70.00
Dense aggregate block partitions				
average thickness; fair face both sides	m²	38.00	to	47.00
average thickness; fair face and emulsioned both sides	m²	44.00	to	53.00
average thickness; plastered and emulsioned both sides	m²	63.00	to	77.00
Common brick partitions				
half brick thick	m²	33.00	to	40.00
half brick thick; fair face both sides	m²	38.00	to	46.50
half brick thick; fair face and emulsioned both sides	m²	44.00	to	53.00
half brick thick; plastered and emulsioned both sides	m²	65.00	to	80.00
one brick thick	m²	63.00	to	77.00
one brick thick; fair face both sides	m²	68.00	to	83.00
one brick thick; fair face and emulsioned both sides	m²	74.00	to	91.00
one brick thick; plastered and emulsioned both sides	m²	95.00	to	115.00
Block partitions; softwood doors and frames; painted				
partition	m²	57.00	to	70.00
partition; fair face both sides	m²	60.00	to	73.00
partition; fair face and emulsioned both sides	m²	67.00	to	82.00
partition; plastered and emulsioned both sides	m²	87.00	to	105.00
Block partitions; hardwood doors and frames				
partition	m²	83.00	to	100.00
partition; platered and emulsioned both sides	m²	110.00	to	140.00
Reinforced concrete walls				
Walls				
150 mm thick	m²	110.00	to	140.00
150 mm thick; plastered and emulsioned both sides	m²	150.00	to	185.00
Special partitioning and doors				
Demountable partitioning; hardwood doors				
medium quality; vinyl-faced	m²	140.00	to	170.00
high quality; vinyl-faced	m²	180.00	to	220.00
Aluminium internal patent glazing				
single glazed	m²	130.00	to	160.00
double glazed	m²	210.00	to	260.00

APPROXIMATE ESTIMATING RATES – GENERAL BUILDING WORKS

Item	Unit	Range £		
Demountable steel partitioning and doors				
medium quality	m²	230.00	to	280.00
high quality	m²	285.00	to	350.00
Demountable fire partitions				
enamelled steel; half hour	m²	880.00	to	1075.00
stainless steel; half hour	m²	1175.00	to	1450.00
Soundproof partitions; hardwood doors				
luxury veneered	m²	410.00	to	500.00
WC/Changing cubicles				
W/C cubicles cost per cubicle	nr	520.00	to	640.00
INTERNAL DOORS				
Comparative doors/door linings/frames				
Standard softwood doors and frames; including ironmongery and painting				
flush; hollow core	nr	270.00	to	335.00
flush; hollow core; hardwood faced	nr	270.00	to	335.00
flush; solid core				
single leaf	nr	325.00	to	395.00
double leaf	nr	485.00	to	590.00
flush; solid core; hardwood faced	nr	360.00	to	440.00
four panel door	nr	460.00	to	570.00
Purpose-made softwood doors and hardwood frames; including ironmongery; painting and polishing				
flush; solid core; heavy duty				
single leaf	nr	760.00	to	930.00
double leaf	nr	1075.00	to	1325.00
flush; solid core; heavy duty; plastic laminate faced				
single leaf	nr	900.00	to	1100.00
double leaf	nr	1225.00	to	1500.00
Purpose-made softwood fire doors and hardwood frames; including ironmongery; painting and polishing				
flush; one hour fire-resisting				
single leaf	nr	980.00	to	1175.00
double leaf	nr	1225.00	to	1500.00
flush; one hour fire-resisting; plastic laminated faced				
single leaf	nr	1175.00	to	1450.00
double leaf	nr	1525.00	to	1875.00
purpose-made softwood doors and pressed steel frames;				
flush; half hour fire check; plastic laminate faced	nr	1175.00	to	1425.00

APPROXIMATE ESTIMATING RATES – GENERAL BUILDING WORKS

Item	Unit	Range £		

2.7 INTERNAL WALLS, PARTITIONS AND DOORS – cont

Perimeter treatments

Item	Unit			
Precast concrete lintels; in block walls				
75 mm wide	m	15.50	to	18.90
100 mm wide	m	19.30	to	23.50
Precast concrete lintels; in brick walls				
half brick thick	m	19.30	to	23.50
one brick thick	m	30.00	to	37.00
Purpose-made softwood architraves; painted; including grounds 25 × 50 mm; to both sides of openings				
726 × 2040 mm opening	nr	145.00	to	180.00
826 × 2040 mm opening	nr	100.00	to	120.00

3.1 WALL FINISHES

Item	Unit			
Dry plasterboard lining, taped joints; for direct decoration				
9.5 mm Gyproc Wallboard	m²	10.90	to	13.30
Extra for insulating grade	m²	0.70	to	0.90
Extra for insulating grade; plastic faced	m²	2.30	to	2.80
12.5 mm Gyproc Wallboard (half-hour fire-resisting)	m²	12.50	to	15.30
Extra for insulating grade	m²	0.75	to	0.90
Extra for insulating grade; plastic faced	m²	2.00	to	2.40
two layers of 12.5 mm Gyproc Wallboard (one houre fire-resisting)	m²	21.00	to	25.50
9 mm Supalux (half-hour fire-resisting)	m²	21.00	to	25.50
Dry plasterboard lining, taped joints; for direct decoration; fixed to wall on dabs				
9.5 mm Gyproc Wallboard	m²	12.50	to	15.30
Dry plasterboard lining, taped joints; for direct decoration; including metal tees				
9.5 mm Gyproc Wallboard	m²	24.00	to	29.50
12.5 mm Gyproc Wallboard	m²	25.00	to	31.00
Dry lining/sheet panelling; including battens; plugged to wall				
6.4 mm hardboard	m²	12.20	to	14.90
9.5 mm Gyproc Wallboard	m²	19.20	to	23.50
6 mm birch faced plywood	m²	14.80	to	18.10
6 mm WAM plywood	m²	24.00	to	29.00
15 mm chipboard	m²	175.00	to	210.00
15 mm melamine faced chipboard	m²	31.50	to	38.50
13.2 mm Formica faced chipboard	m²	47.50	to	58.00
In situ wall finishes				
Extra over common brickwork for fair face and pointing both sides	m²	4.30	to	5.30
Comparative finishes				
one mist and two coats emusion	m²	2.70	to	3.30
mutli-coloured gloss paint	m²	5.80	to	7.10
two coats of lightweight plaster	m²	11.20	to	13.70
9.5 mm Gyproc Wallboard and skim coat	m²	14.30	to	17.50
12.5 mm Gyproc Wallboard and skim coat	m²	16.10	to	19.70
two coates of Thistle plaster	m²	15.00	to	18.30
plaster and emulsion	m²	16.00	to	19.60
Extra for gloss paint in lieu of emulsion	m²	2.50	to	3.05
two coat render and emulsion	m²	26.00	to	31.50

APPROXIMATE ESTIMATING RATES – GENERAL BUILDING WORKS

Item	Unit	Range £
Ceramic wall tiles; including backing		
economical	m²	34.00 to 42.00
medium quality	m²	62.00 to 76.00
high quality; to toilet blocks; kitchens and first aid rooms	m²	83.00 to 100.00
Decorations		
Emulsion		
two coats	m²	2.35 to 2.90
one mist and two coats	m²	3.10 to 3.80
Artex plastic compound		
one coat; textured	m²	4.00 to 4.90
Wall paper	m²	6.35 to 7.75
Gloss		
primer and two coats	m²	4.70 to 5.70
primer and three coats	m²	4.00 to 4.90
Comparative steel/metalwork finishes		
primer only	m²	1.20 to 1.45
grit blast and one coat zinc chromate primer	m²	2.15 to 2.65
touch up primer and one coat of two pack epoxy zinc phosphate	m²	2.70 to 3.30
primer gloss three coats	m²	5.95 to 7.25
sprayed mineral fibre; one hour	m²	13.00 to 15.90
sprayed mineral fibre; two hour	m²	14.20 to 17.40
sprayed vermiculite cement; one hour	m²	14.30 to 17.50
sprayed vermiculite cement; two hour	m²	17.10 to 21.00
intumescent coating with decorative top seal; half hour	m²	19.40 to 24.00
intumescent coating with decorative top seal; one hour	m²	32.50 to 39.50
Comparative woodword finishes		
primer only	m²	1.50 to 1.80
two coats gloss; touch up primer	m²	3.00 to 3.70
three coats gloss; touch up primer	m²	4.50 to 5.50
primer and two coat gloss	m²	5.00 to 6.10
primer and three coat gloss	m²	6.00 to 7.35
polyurethene lacquer two coats	m²	6.25 to 7.65
polyurethene lacquer three coats	m²	5.00 to 6.10
flame-retardant paint three coats	m²	7.15 to 8.80
3.2 FLOOR FINISHES		
Chipboard flooring; t & g joints		
18 mm thick	m²	10.10 to 12.40
22 mm thick	m²	12.30 to 15.00
Wrought softwood flooring		
25 mm thick; butt joints; cleaned off and polished	m²	25.00 to 30.00
25 mm thick; t & g joints; cleaned off and polished	m²	28.00 to 34.50
Extra over concrete floor for		
power floating	m²	6.90 to 8.40
power floating; surface hardener	m²	11.00 to 13.50
Latex cement screeds		
3 mm thick; one coat	m²	5.45 to 6.70
5 mm thick; two coat	m²	7.25 to 8.90
Rubber latex non-slip solution and epox sealant	m²	14.50 to 17.70

APPROXIMATE ESTIMATING RATES – GENERAL BUILDING WORKS

Item	Unit	Range £
3.2 FLOOR FINISHES – cont		
Cement and sand (1:3) screeds		
25 mm thick	m²	10.70 to 13.00
50 mm thick	m²	13.20 to 16.20
75 mm thick	m²	18.30 to 22.50
Cement and sand (1:3) paving		
32 mm thick	m²	10.40 to 12.80
32 mm thick; surface hardener	m²	13.80 to 16.80
Screed only (for subsequent finish)	m²	16.80 to 20.50
Screed only (for subsequent finish); allowance for skirtings	m²	19.80 to 24.00
Mastic asphalt paving		
20 mm thick; BS 6925; black	m²	23.00 to 28.00
20 mm thick; BS 6925; red	m²	27.00 to 33.00
Granolithic		
20 mm thick	m²	15.00 to 18.30
25 mm thick	m²	17.80 to 22.00
25 mm thick; including screed	m²	27.00 to 33.00
38 mm thick; including screed	m²	39.00 to 48.00
Synthanite; on and including building paper		
25 mm thick	m²	25.50 to 31.00
50 mm thick	m²	36.00 to 44.00
75 mm thick	m²	4.40 to 5.40
Acrylic polymer floor finish		
10 mm thick	m²	26.00 to 32.00
Epoxy floor finish		
1.5–2 mm thick	m²	28.00 to 34.50
5–6 mm thick	m²	52.00 to 64.00
Polyester resin floor finish		
5–9 mm thick	m²	58.00 to 71.00
Quarry tile flooring		
150 × 150 × 12.5 mm thick; red	m²	30.00 to 36.50
150 × 150 × 12.5 mm thick; brown	m²	38.50 to 47.00
200 × 200 × 19 mm thick; brown	m²	47.00 to 57.00
average tiling	m²	39.00 to 48.00
tiling; including screed	m²	53.00 to 65.00
tiling; including screed and allowance for skirtings	m²	67.00 to 82.00
Glazed ceramic tile flooring		
150 × 150 × 12 mm thick; black	m²	39.50 to 48.50
150 × 150 × 12 mm thick; antislip	m²	44.00 to 54.00
fully vitrified	m²	55.00 to 67.00
fully vitrified; including screed	m²	69.00 to 84.00
fully vitrified; including screed and allowance for skirtings	m²	80.00 to 98.00
Composition block flooring		
174 × 57 mm blocks	m²	68.00 to 84.00
Vinyl tile flooring		
2 mm thick; semi-flexible tiles	m²	11.00 to 13.50
2 mm thick; fully flexible tiles	m²	10.40 to 12.70
2.5 mm thick; semi-flexible tiles	m²	13.20 to 16.20
tiling; including screed	m²	27.00 to 33.50
tiling; including screed and allowance for skirtings	m²	48.50 to 59.00
tiling; antistatic	m²	46.00 to 56.00
tiling; antistatic; including screed	m²	57.00 to 70.00

APPROXIMATE ESTIMATING RATES – GENERAL BUILDING WORKS

Item	Unit	Range £		
Vinyl sheet flooring; heavy duty				
2 mm thick	m²	17.20	to	21.00
2.5 mm thick	m²	19.80	to	24.00
3 mm thick; needle felt backed	m²	13.20	to	16.20
3 mm thick; foam backed	m²	18.50	to	22.50
Sheeting; including screed and allowances for skirtings				
Altro safety flooring	m²	33.50	to	41.00
2 mm thick; Marine T20	m²	26.50	to	32.00
2.5 mm thick; Classic D25	m²	32.00	to	39.00
3.5 mm thick; stonghold	m²	39.50	to	48.50
flooring	m²	33.50	to	41.00
flooring; including screed	m²	47.00	to	57.00
Rubber tile flooring; smooth; ribbed or studded tiles				
2.5 mm thick	m²	32.00	to	39.00
5 mm thick	m²	37.00	to	45.00
5 mm thick; including screed	m²	53.00	to	65.00
Carpet tiles/Underlay				
Underlay	m²	6.60	to	8.10
nylon needlepunch (stick down)	m²	15.90	to	19.40
80% animal hair; 20% wool cord	m²	25.00	to	30.00
100% wool	m²	36.00	to	44.00
80% wool; 20% nylon antistatic	m²	41.50	to	51.00
economical; including screed and allowance for skirtings	m²	38.00	to	46.00
good quality	m²	40.00	to	49.00
good quality; including screed	m²	52.00	to	64.00
good quality; including screed and allowance for skirtings	m²	56.00	to	69.00
Access floors; excluding finish				
600 × 600 mm chipboard panels; faced both sides with galvanized steel sheet; on adjustable steel/aluminium pedestals; cavity height 100–300 mm high				
light grade duty	m²	53.00	to	65.00
medium grade duty	m²	59.00	to	72.00
heavy grade duty	m²	78.00	to	95.00
extra heavy grade duty	m²	83.00	to	100.00
600 × 600 mm chipboard panels; faced both sides with galvanized steel sheet; on adjustable steel/aluminium pedestals; cavity height 300–600 mm high				
medium grade duty	m²	67.00	to	82.00
heavy grade duty	m²	79.00	to	97.00
extra heavy grade duty	m²	83.00	to	100.00
Common floor coverings bonded to access floor panels				
heavy-duty fully flexible vinyl to BS 3261, type A	m²	16.50	to	20.00
anti-static grade sheet PVC to BS 3261	m²	20.00	to	25.00
Comparative skirtings				
25 × 75 mm softwood skirting; painted; including grounds	m	11.40	to	13.90
12.5 × 150 mm Quarry tile skirting; including backing	m	16.50	to	20.00
13 × 75 mm granolithic skirting; including backing	m	23.00	to	28.50
Entrance matting in aluminium-framed				
matwell	m²	430.00	to	520.00

APPROXIMATE ESTIMATING RATES – GENERAL BUILDING WORKS

Item	Unit	Range £		

3.3 CEILING FINISHES

Plastered ceilings
Plaster to soffits
lightweight plaster	m²	12.00	to	14.70
plaster and emulsion	m²	16.80	to	20.50
Extra for gloss paint in lieu of emulsion	m²	2.40	to	2.95

Plasterboard to soffits
9.5 mm Gyproc lath and skim coat	m²	15.50	to	19.00
9.5 mm Gyproc insulating lath and skim coat	m²	16.30	to	19.90
plasterboard, skim and emulsion	m²	18.50	to	22.50
Extra for gloss paint in lieu of emulsion	m²	2.40	to	2.95
plasterboard and Artex	m²	12.30	to	15.10
plasterboard, Artex and emulsion	m²	15.90	to	19.40
plaster and emulsion; including metal lathing	m²	26.50	to	32.00

Other board finishes; with fire-resisting properties; excluding decoration
12.5 Gyproc Fireline; half hour	m²	13.20	to	16.20
6 mm Supalux; half hour	m²	16.30	to	19.90
two layers of 12.5 mm Gyproc Wallboard; half hour	m²	18.10	to	22.00
two layers of 12.5 mm Gyproc Fireline; one hour	m²	21.50	to	26.50
9 mm Supalux; one hour; on fillets	m²	24.00	to	29.50

Specialist plasters; to soffits
sprayed accoustic plaster; self finished	m²	35.50	to	43.00
rendering; Tyrolean finish	m²	36.50	to	45.00

Other ceiling finishes
50 mm wood wool slabs as permanent lining	m²	15.50	to	19.00
12 mm pine tongued and grooved boardings	m²	20.00	to	25.00
16 mm softwood tongued and grooved boardings	m²	24.50	to	30.00

Suspended and intergrated ceilings
Suspended ceilings
economical; exposed grid	m²	27.00	to	33.50
jointless; plasterboard	m²	33.00	to	40.50
semi-concealed grid	m²	36.00	to	43.50
medium quality; Minatone; concealed grid	m²	40.50	to	49.50
high quality; Travertone; concealed grid	m²	45.50	to	56.00
metal linear strip; Dampa/Luxalon	m²	48.50	to	59.00
metal tray	m²	52.00	to	64.00
egg-crate	m²	69.00	to	110.00
open grid; Formalux/Dimension	m²	100.00	to	120.00

Intergrated ceilings
coffered; with steel services	m²	120.00	to	190.00

4.1 FITTINGS AND FURNISHINGS
Comparative wrought softwood shelving
25 × 225 mm; including black japanned brackets	m	13.70	to	16.70
25 mm thick slatted shelving; including bearers	m²	54.00	to	66.00
25 mm thick cross-tongued shelving; including bearers	m²	71.00	to	86.00

APPROXIMATE ESTIMATING RATES – GENERAL BUILDING WORKS

Item	Unit	Range £
Allowances per gross floor area for:		
reception desk, shelves and cupboards for general areas		
economical	m²	5.90 to 9.80
medium quality	m²	11.00 to 13.50
high quality	m²	17.60 to 21.50
Extra for		
high quality finishes to reception areas	m²	8.80 to 10.80
full kitchen equipment	m²	11.90 to 14.60
furniture and fittings to general office areas		
economical	m²	9.90 to 12.10
medium quality	m²	15.00 to 18.30
high quality	m²	25.50 to 31.00
General fittings and equipment		
internal planting	m²	26.50 to 32.00
signs, notice-boards, shelving, fixed seating, curtains and blinds	m²	13.70 to 16.70

5.1 SANITARY INSTALLATIONS

Note: Material prices vary considerably, the following composite rates are based on average prices for mid priced fittings.

Item	Unit	Range £
Lavatory basins; vitreous china; chromium plated taps, waste, chain and plug, cantilever brackets	nr	210.00 to 290.00
Low level WCs; vitreous china pan and cistern; black plastic seat; low pressure ball valve; plastic flush pipe; fixing brackets		
on ground floor	nr	190.00 to 255.00
one of a range; on upper floors	nr	370.00 to 500.00
Bowl type wall urinal; white glazed vitreous china flushing cistern; chromium plated flush pipes and spreaders; fixing brackets		
white	nr	180.00 to 245.00
Shower tray; glazed fireclay; chromium plated waste, chain and plug, riser pipe, rose and mixing valve		
white	nr	510.00 to 690.00
Sink; glazed fireclay; chromium plated waste, chain and plug; fixing brackets		
white	nr	415.00 to 560.00
Sink; stainless steel; chromium plated waste, chain and self coloured plug		
single drainer	nr	225.00 to 305.00
double drainer	nr	260.00 to 350.00
Soil waste stacks; 3.15 m storey height; branch and connection to drain		
110 mm PVC	nr	320.00 to 430.00
Extra for additional floors	nr	165.00 to 225.00
100 mm cast iron; decorated	nr	630.00 to 860.00
Extra for additional floors	nr	325.00 to 440.00
Industrial buildings – Allowance per m² of floor area		
minimum provision	m²	12.50 to 16.90
high provision	m²	17.50 to 23.50
Hot and cold water installations – Allowance per m² of floor area		
complete installations	m²	21.00 to 28.00

APPROXIMATE ESTIMATING RATES – GENERAL BUILDING WORKS

Item	Unit	Range £		
5.6 SPACE HEATING AND AIR CONDITIONING AND 5.7 VENTILATION				
Gas or oil-fired radiator heating				
Gas-fired hot water service and central heating for				
three radiators	nr	2750.00	to	3700.00
Extra for additional radiator	nr	315.00	to	430.00
LPHW radiator system – Allowance per m^2 of floor area	m^2	72.00	to	97.00
Ventilation systems				
Local ventilation to				
WCs	nr	270.00	to	370.00
toilet areas – Allowance per m^2 of floor area	m^2	7.30	to	9.85
Air extract system – Allowance per m^2 of floor area	m^2	45.00	to	61.00
Air supply and extract system – Allowance per m^2 of floor area	m^2	70.00	to	95.00
Heating and ventilation – Allowance per m^2 of floor area	m^2	69.00	to	94.00
Warm air heating and ventilation – Allowance per m^2 of floor area	m^2	140.00	to	185.00
Comfort cooling systems				
Stand-alone air conditioning unit systems				
air supply and extract – Allowance per m^2 of floor area	m^2	260.00	to	350.00
Air conditioning systems				
fully air conditioning with dust and humidity control – Allowance per m^2 of floor area	m^2	230.00	to	310.00
5.8 ELECTRICAL INSTALLATIONS				
Based upon gross internal area serviced				
Mains and sub-mains switchgear and distribution				
mains intake only	m^2	2.60	to	3.55
mains switchgear only	m^2	6.05	to	8.20
Mains and sub-mains distribution				
to floors only	m^2	7.90	to	10.70
to floors; including small power and supplies to equipment	m^2	16.70	to	22.50
to floors; including lighting and power and supplies to equipment	m^2	16.20	to	22.00
Lighting installation				
general lighting; including luminaries	m^2	37.50	to	51.00
emergency lighting	m^2	12.50	to	16.90
standby generators only	m^2	7.10	to	9.60
Lighting and power installations to buildings				
plant area	m^2	62.00	to	85.00
plant are; high provision	m^2	81.00	to	110.00
office area	m^2	110.00	to	150.00
office area; high provision	m^2	150.00	to	205.00
Comparative fittings/rates per each point				
Fittings; excluding lamps or light fittings				
lighting point; PVC cables	nr	57.00	to	77.00
lighting point; PVC cables in screwed conduits	nr	150.00	to	200.00
lighting point; MICC cables	nr	130.00	to	180.00

APPROXIMATE ESTIMATING RATES – GENERAL BUILDING WORKS

Item	Unit	Range £		
Switch socket outlet; PVC cables				
single	nr	52.00	to	71.00
double	nr	74.00	to	100.00
Switch socket outlet; PVC cables in screwed conduit				
single	nr	92.00	to	120.00
double	nr	98.00	to	130.00
Based upon gross internal area serviced				
Fittings; excluding lamps or light fittings				
Switch socket outlet; MICC cables				
single	nr	93.00	to	125.00
double	nr	100.00	to	140.00
Immersion heater point (excluding heater)	nr	90.00	to	120.00
Cooker point; including control unit	nr	155.00	to	210.00
5.10 LIFT AND CONVEYOR INSTALLATIONS				
Goods lifts				
hoist	nr	11000.00	to	33000.00
Electric heavy duty goods lifts				
500 kg 2–3 levels	nr	37500.00	to	51000.00
1000 kg; 2–3 levels	nr	47500.00	to	64000.00
1500 kg; 3 levels	nr	105000.00	to	135000.00
2000 kg; 2 levels	nr	105000.00	to	135000.00
2000 kg; 3 levels	nr	125000.00	to	170000.00
3000 kg; 2 levels	nr	115000.00	to	155000.00
3000 kg; 3 levels	nr	140000.00	to	185000.00
Oil hydraulic heavy duty goods lifts				
500 kg; 3 levels	nr	105000.00	to	140000.00
1000 kg; 3 levels	nr	110000.00	to	150000.00
2000 kg; 3 levels	nr	125000.00	to	170000.00
Dock levellers				
dock levellers	nr	13000.00	to	29000.00
dock levellers and canopy	nr	17000.00	to	35500.00
5.11 FIRE AND LIGHTNING PROTECTION				
Based upon gross internal area serviced				
Fire fighting/protective installations				
smoke detectors, alarms and controls	m^2	5.70	to	8.90
hosereels, dry risers and extinguishers	m^2	8.60	to	13.50
Sprinkler installations				
single level sprinkler systems, alarms and smoke detectors; low hazard	m^2	15.80	to	21.50
single level sprinkler systems, alarms and smoke detectors; ordinary hazard	m^2	23.00	to	28.00
double level sprinkler systems, alarms and smoke detectors; high hazard	m^2	33.50	to	41.00
Lightning protection	m^2	1.00	to	2.95

APPROXIMATE ESTIMATING RATES – GENERAL BUILDING WORKS

Item	Unit	Range £		
5.11 FIRE AND LIGHTNING PROTECTION – cont				
Security/communication installations				
security alarm system	m²	2.50	to	3.00
telephone system	m²	1.65	to	2.50
public address, television ariel and clocks	m²	3.70	to	5.50
closed-circuit television and public address system	m²	5.00	to	6.10
5.14 BUILDER WORK IN CONNECTION WITH SERVICES				
General builders work to				
mains supplies; lighting and power to landlords areas	m²	2.20	to	6.60
central heating and electrical installation	m²	5.40	to	16.20
central heating; electrical and lift installations	m²	7.75	to	18.10
air conditioning	m²	19.60	to	26.50
air conditioning and electrical installation	m²	20.50	to	27.50
air conditioning; electrical and lift installations	m²	22.50	to	30.50
General builders work; including allowance for plant rooms; to				
central heating and electrical installations	m²	31.00	to	38.00
central heating, electrical and lift installations	m²	39.00	to	47.50
air conditioning	m²	53.00	to	65.00
air conditioning and electrical installation	m²	65.00	to	80.00
air conditioning; electrical and lift installations	m²	81.00	to	99.00
8.2 ROADS, PATHS AND SURFACES				
Paved areas				
Precast concrete paving slabs				
50 mm thick	m²	15.40	to	21.00
50 mm thick Texitone slabs	m²	17.20	to	23.00
slabs on subbase; including excavation	m²	28.00	to	38.00
Precast concrete block paviors				
65 mm Keyblock grey paving	m²	21.50	to	29.00
65 mm Mount Sorrel grey paving	m²	21.00	to	29.00
65 mm Intersett paving	m²	23.00	to	31.50
60 mm Pedesta paving	m²	19.80	to	27.00
paviors on subbase; including excavations	m²	33.00	to	45.00
Brick paviors; 229 × 114 × 38 mm paving bricks				
laid flat	m²	36.00	to	48.50
laid to herring-bone pattern	m²	55.00	to	74.00
paviors on subbase; including excavation	m²	64.00	to	87.00
Granite setts				
200 × 100 × 100 mm setts	m²	83.00	to	110.00
setts on subbase; including excavation	m²	100.00	to	135.00
York stone slab paving				
paving on subbase; including excavation	m²	115.00	to	155.00
Cobblestone paving				
50–75 mm	m²	67.00	to	90.00
cobblestones on subbase; including excavation	m²	81.00	to	110.00

APPROXIMATE ESTIMATING RATES – GENERAL BUILDING WORKS

Item	Unit	Range £		
Car parking alternatives				
Surface level parking; including lighting and drainage, forecourts, aprons; not including approach roads				
tarmacadam on subbase	car	2200.00	to	2700.00
concrete interlocking blocks	car	3150.00	to	3850.00
Grasscrete precast concrete units filled with top soil and grass seed	car	1675.00	to	2025.00
Garages etc.				
single car park	nr	940.00	to	1275.00
single; traditional construction; in a block	nr	2800.00	to	3800.00
single; traditional construction; pitched roof	nr	4500.00	to	6100.00
double; traditional construction; pitched roof	nr	8500.00	to	11500.00

8.3 SOFT LANDSCAPING, PLANTING AND IRRIGATION SYSTEMS

Landscaping as a subject matter is sufficiently large to fill a book (see Spons External Works and Landscape Price Book). However there are certain items that arise on a majority of projects and other discrete items for which indicative costs can be produced. The following rates include for normal surface excavation but exclude bulk excavation levelling or earth shifting and land drain provision.

Item	Unit	Range £		
Seeded and planted areas				
Plant supply, planting, maintenance and 12 months guarantee				
seeded areas	m^2	5.20	to	7.10
turfed areas	m^2	6.80	to	9.20
Planted areas (per m^2 of suface area)				
herbaceous plants	m^2	4.60	to	6.20
climbing plants	m^2	7.20	to	9.80
general planting	m^2	13.60	to	18.40
woodland	m^2	36.00	to	48.00
shrubbed planting	m^2	44.00	to	60.00
dense planting	m^2	44.00	to	60.00
shrubbed area including allowances for small trees	m^2	60.00	to	81.00
Trees				
advanced nursery stock trees (12–20 cm girth)	tree	160.00	to	215.00
semi-mature trees; 5–8 m high – coniferous	tree	800.00	to	1075.00
semi-mature trees; 5–8 m high – deciduous	tree	1675.00	to	2250.00

8.4 FENCING, RAILINGS AND WALLS

Item	Unit	Range £		
Chain link fencing; plastic coated				
1.2 m high	m	17.50	to	24.00
1.8 m high	m	24.00	to	33.00
Timber fencing				
1.2 m high chestnut pale fencing	m	19.40	to	26.00
1.8 m high close-boarded fencing	m	52.00	to	70.00
Screen walls; one brick thick; including foundations etc.				
1.8 m high facing brick screen wall	m	260.00	to	350.00
1.8 m high coloured masonry block boundry wall	m	290.00	to	395.00

APPROXIMATE ESTIMATING RATES – GENERAL BUILDING WORKS

Item	Unit	Range £		
8.4 FENCING, RAILINGS AND WALLS – cont				
Guard rails and parking bollards etc.				
open metal post and rail fencing 1 m high	m	140.00	to	190.00
galvanized steel post and rail fencing 2 m high	m	165.00	to	220.00
steel guard rails and vehicle barriers	m	57.00	to	78.00
Parking bollards				
precast concrete	nr	110.00	to	150.00
steel	nr	195.00	to	260.00
cast iron	nr	230.00	to	315.00
Vehicle control barrier				
manual pole	nr	940.00	to	1275.00
automatic barrier equipment	nr	19000.00	to	23000.00
8.5 EXTERNAL FIXTURES				
Street furniture				
Lighting				
reflectorized traffic signs 0.25–0.75 m² area on steel post	nr	120.00	to	200.00
internally illuminated traffic signs (dependent on area)	nr	230.00	to	315.00
externally illuminated traffic signs (up to 4 m²)	nr	550.00	to	1500.00
pedestrian areas and estates roads on 4–6 m columns with up to 70 W lamps	nr	280.00	to	380.00
lighting to pedestrian areas and estates roads on 4–6 m lighting to main roads on 10–12 m columns with 250 W lamps	nr	1075.00	to	1325.00
lighting to main roads on 12–15 m columns with 400 W high pressure sodium lighting	nr	1450.00	to	1750.00
Glavanized steel cycle stand	nr	44.50	to	60.00
Galvanized steel flag staff	nr	1175.00	to	1575.00
Benches – hardwood and precast concrete	nr	230.00	to	310.00
Litter bins				
precast concrete	nr	210.00	to	290.00
hardwood slatted	nr	94.00	to	125.00
cast iron	nr	360.00	to	480.00
large aluminium	nr	630.00	to	850.00
Bus stops	nr	380.00	to	520.00
Bus stops including basic shelter	nr	880.00	to	1175.00
Pillar box on post	nr	320.00	to	430.00
8.6 EXTERNAL DRAINAGE				
Pipework				
100/150 mm clay or PVC-u pipes on granular bed and surround up to 3 m deep	m	42.50	to	58.00
100/150 mm cast iron pipes on concrete beds up to 3 m deep	m	89.00	to	115.00
300 mm clay drain pipe on granular bed and surround up to 3 m deep	m	160.00	to	200.00
450 mm concrete pipe on concrete beds up to 3 m deep	m	130.00	to	160.00
900 mm diameter concrete sewage pipe on granular bed up to 3.5 m deep	m	265.00	to	325.00

APPROXIMATE ESTIMATING RATES – GENERAL BUILDING WORKS

Item	Unit	Range £		
Manholes and gullies				
brick manholes in commons rendered internally clay channel with three branches concrete cover slab cast iron cover and frame 900–1500 mm deep	nr	1350.00	to	1650.00
precast concrete manholes with precast concrete rings up to 1200 mm diameter channel with three branches concrete cover slab and frame 900–1500 mm deep	nr	1475.00	to	1825.00
vitrified clay or precast concrete road gully 450 × 900 mm deep with concrete surround brick seating and cast iron grating	nr	455.00	to	560.00
Land drainage				
Costs include for excavation and backfilling of trenches and laying agricultural clay drain pipes with 75 mm diameter lateral runs average 600 mm deep and 100 mm diameter main runs average 750 mm deep.				
land drainage to parkland with laterals at 30 m centres and main runs at 100 m centres	ha	3350.00	to	3850.00
land drainage to sportsfields with laterals at 10 m centres and main runs at 33 m centres	ha	16000.00	to	18500.00

8.7 EXTERNAL SERVICES

Service runs laid in trenches inlcuding excavation

Water main				
75 mm uPVC main in 225 mm ductile iron pipe as duct	m	85.00	to	115.00
Electric main				
600/1000 volt cables. Two core 25 mm cable including 100 mm clayware duct	m	34.50	to	46.50
Gas main				
150 mm ductile or cast iron gas pipe	m	68.00	to	92.00
Telephone				
British Telecom installation in 100 mm uPVC duct	m	21.00	to	29.00
Residential connection charges				

Costs of services to a site are built up of connection charges and service runs. This can vary significantly depending upon the availability or otherwise of a suitably sized main in the neighbourhood of the site. However typical service charges for an estate of 200 houses might be as follows:

water	unit	600.00	to	1000.00
electric, if all electric supply only	unit	255.00	to	310.00
electric, if gas and electric supplied	unit	470.00	to	580.00
substation if required	nr	14000.00	to	21500.00
gas	unit	540.00	to	660.00
governing station if required	nr	18500.00	to	22500.00
telephone	unit	165.00	to	200.00
sewerage	unit	420.00	to	510.00

ESSENTIAL READING FROM TAYLOR AND FRANCIS

Construction Contracts Questions and Answers

2nd Edition

David Chappell

What they said about the first edition: "A fascinating concept, full of knowledgeable gems put in the most frank of styles... A book to sample when the time is right and to come back to when another time is right, maybe again and again."
– David A Simmonds, Building Engineer magazine

- Is there a difference between inspecting and supervizing?

- What does 'time-barred' mean?

- Is the contractor entitled to take possession of a section of the work even though it is the contractor's fault that possession is not practicable?

Construction law can be a minefield. Professionals need answers which are pithy and straightforward, as well as legally rigorous. The two hundred questions in the book are real questions, picked from the thousands of telephone enquiries David Chappell has received as a Specialist Adviser to the Royal Institute of British Architects. Although the enquiries were originally from architects, the answers to most of them are of interest to project managers, contractors, QSs, employers and others involved in construction.

The material is considerably updated from the first edition – weeded, extended and almost doubled in coverage. The questions range in content from extensions of time, liquidated damages and loss and/or expense to issues of warranties, bonds, novation, practical completion, defects, valuation, certificates and payment, architects' instructions, adjudication and fees. Brief footnotes and a table of cases will be retained for those who may wish to investigate further.

August 2010: 216x138: 344pp
Pb: 978-0-415-56650-6: **£34.99**

To Order: Tel: +44 (0) 1235 400524 **Fax:** +44 (0) 1235 400525
or Post: Taylor and Francis Customer Services,
Bookpoint Ltd, Unit T1, 200 Milton Park, Abingdon, Oxon, OX14 4TA UK
Email: book.orders@tandf.co.uk

For a complete listing of all our titles visit:
www.tandf.co.uk

Taylor & Francis
Taylor & Francis Group

PART 6

Unit Costs – Civil Engineering Works

INTRODUCTORY NOTES

The Unit Costs in this part represent the net cost to the Contractor of executing the work on site; they are not the prices which would be entered in a tender Bill of Quantities.

It must be emphasized that the unit rates are averages calculated on unit outputs for typical site conditions. Costs can vary considerably from contract to contract depending on individual Contractors, site conditions, methods of working and various other factors. Reference should be made to Part 1 for a general discussion on Civil Engineering Estimating.

Guidance prices are included for work normally executed by specialists, with a brief description where necessary of the assumptions upon which the costs have been based. Should the actual circumstances differ, it would be prudent to obtain check prices from the specialists concerned, on the basis of actual/likely quantity of the work, nature of site conditions, geographical location, time constraints, etc.

The method of measurement adopted in this section is the **CESMM3**, *subject to variances where this has been felt to produce more helpful price guidance.*

We have structured this Unit Costs section to cover as many aspects of Civil Engineering works as possible.

The Gang hours column shows the output per measured unit in actual time, not the total labour hours; thus for an item involving a gang of 5 men each for 0.3 hours, the total labour hours would naturally be 1.5, whereas the Gang hours shown would be 0.3.

This section is structured to provide the User with adequate background information on how the rates have been calculated, so as to allow them to be readily adjusted to suit other conditions to the example presented:

- Alternative gang structures as well as the effect of varying bonus levels, travelling costs etc.
- Other types of plant or else different running costs from the medium usage presumed
- Other types of materials or else different discount/waste allowances from the levels presumed

Reference to Part 3 giving basic costs of labour, materials and plant together with Parts 8 and 10 will assist the reader in making adjustments to the unit costs.

ESSENTIAL READING FROM TAYLOR AND FRANCIS

A Guide to Field Instrumentation in Geotechnics

R. Bassett

Geotechnical instrumentation is used for installation, monitoring and assessment on any sizeable project, particularly in urban areas, and is used for recording, controlled remedial work, and safety.

This unique and up-to-date book deals with the conceptual philosophy behind the use of instruments, and then systematically covers their practical use. It is divided into displacement dominated systems and stress recording systems. The limitations are discussed and the theoretical background for data assessment and presentation are covered in some detail, with some relevant background material in theoretical soil mechanics. Relevant advanced electronic techniques such as laser scanning in surveying and fibre-optics are also included and communication and data recovery systems are discussed.

It is written for senior designers, consulting engineers and major contractors who need a major introduction to the general purpose, availability and analysis of field instruments before details of their own project can be progressed, and serves as a text book to any specialist geotechnical MSc or professional seminar course in which instrumentation forms a major part.

November 2011: 232pp
Hb: 978-0-415-67537-6: **£90.00**

To Order: Tel: +44 (0) 1235 400524 **Fax:** +44 (0) 1235 400525
or Post: Taylor and Francis Customer Services,
Bookpoint Ltd, Unit T1, 200 Milton Park, Abingdon, Oxon, OX14 4TA UK
Email: book.orders@tandf.co.uk

For a complete listing of all our titles visit:
www.tandf.co.uk

Taylor & Francis
Taylor & Francis Group

GUIDANCE NOTES

Generally

Adjustments should be made to the rates shown for time, location, local conditions, site constraints and any other factors likely to affect the costs of a specific scheme.

Method of Measurement

Although this part of the book is primarily based on CESMM3, the specific rules have been varied from in cases where it has been felt that an alternative presentation would be of value to the book's main purpose of providing guidance on prices. This is especially so with a number of specialist contractors but also in the cases of work where a more detailed presentation will enable the user to allow for ancillary items.

Materials cost

Materials costs within the rates have been calculated using 'list prices', adjusted to allow for delivery charges (if any) and a 'reasonable' level of discount obtainable by the contractor. This will vary very much depending on the contractor's standing, the potential size of the order and the supplier's eagerness and will vary also between raw traded goods such as timber which will attract a low discount of perhaps 3%, if at all, and manufactured goods where the room for bargaining is much greater and can reach levels of 30% to 40%. High demand for a product at the time of pricing can dramatically reduce the potential discount, as can the world economy in the case of imported goods such as timber and copper. Allowance has also been made for wastage on site (generally 2.5% to 5%), dependent upon the risk of damage, the actual level should take account of the nature of the material and its method of storage and distribution about the site.

Labour cost

The composition of the labour and type of plant is generally stated at the beginning of each section, more detailed information on the calculation of the labour rates is given in Part 4: Resources. In addition is a summary of labour grades and responsibilities extracted from the Working Rule Agreement. Within Parts 6 and 7, each section is prefaced by a detailed build-up of the labour gang assumed for each type of work. This should allow the user to see the cost impact of a different gang as well as different levels of bonus payments, allowances for skills and conditions, travelling allowances etc. The user should be aware that the output constants are based on the gangs shown and would also need to be changed.

Plant cost

A rate build-up of suitable plant is generally stated at the beginning of each section, with more detailed information on alternative machines and their average fuel costs being given in Part 4: Resources. Within Parts 6 and 7, each section is prefaced by a detailed build-up of plant assumed for each type of work. This should allow the user to see the cost impact of using alternative plant as well as different levels of usage. The user should be aware that the output constants are based on the plant shown and would also need to be changed.

Outputs

The user is directed to Part 10: Outputs, which contains a selection of output constants, in particular a chart of haulage times for various capacities of Tippers.

ESSENTIAL READING FROM TAYLOR AND FRANCIS

Failures in Concrete Structures: Case Studies in Reinforced and Prestressed Concrete

R. Whittle

Some lessons are only learned from mistakes. But, it's much cheaper to learn from someone else's mistakes than to have to do so from your own. Drawing on over 50 years of working with concrete structures, Robin Whittle examines the problems which he has seen occur and shows how they could have been avoided.

The first and largest part of the book tells the stories of a number of cases where things have gone wrong with concrete structures. Each case is analysed to identify its cause and how it might have been prevented. It then looks at how failures in structural modelling can lead to big problems if they are not identified before construction is undertaken. Beyond this it examines how contract arrangements can encourage or prevent problems in the designing and building processes. It concludes with an examination of the role research and development in preventing failures.

By identifying the differences between shoddy economisations and genuine efficiency savings, this book offers savings in the short term which won't be at the expense of a structure's long-term performance. Invaluable reading if you're designing or building concrete structures and want to avoid problems which could be expensive or embarrassing further down the line.

July 2012:: 234x156: 154pp
Hb: 978-0-415-56701-5: **£ 60.00**

To Order: Tel: +44 (0) 1235 400524 **Fax:** +44 (0) 1235 400525
or Post: Taylor and Francis Customer Services,
Bookpoint Ltd, Unit T1, 200 Milton Park, Abingdon, Oxon, OX14 4TA UK
Email: book.orders@tandf.co.uk

For a complete listing of all our titles visit:
www.tandf.co.uk

Taylor & Francis
Taylor & Francis Group

CLASS A: GENERAL ITEMS

Item	Gang hours	Labour £	Plant £	Material £	Unit	Total rate £
METHOD RELATED CHARGES						
NOTES Refer to the example calculation of Preliminaries and Oncosts and Profits in Part 2						
CONTRACTUAL REQUIREMENTS						
Performance bond The cost of the bond will relate to the nature and degree of difficulty and risk inherent in the type of works intended, the perceived ability and determination of the Contractor to complete them, his financial status and whether he has a proven track record in this field with the provider. Refer to the discussion of the matter in Part 2, Preliminaries						
Insurance of the Works Refer to the discussion of the matter in Part 2, Preliminaries						
Third party insurance Refer to the discussion of the matter in Part 2, Preliminaries						
SPECIFIED REQUIREMENTS						
General This section entails the listing of services and facilities over and above the 'Permanent Works' which the Contractor would be instructed to provide in the Contract Documents.						
Accommodation for the Engineer's Staff Refer to Resources – Plant section for a list of accommodation types.						
Services for the Engineer's staff Transport vehicles						
4 WD utility short wheelbase	–	–	656.00	–	week	**656.00**
4 WD long wheelbase	–	–	556.00	–	week	**556.00**
(for other vehicles, refer to Resources – Plant) Telephones (allow for connection charges and usage of telephones required for use by the Engineer's Staff)						

CLASS A: GENERAL ITEMS

Item	Gang hours	Labour £	Plant £	Material £	Unit	Total rate £
METHOD RELATED CHARGES – cont						
Equipment for use by the Engineer's staff						
Allow for equipment specifically required; entailing Office Equipment, Laboratory Equipment and Surveying Equipment.						
Attendance upon the Engineer's staff						
Driver	40.00	580.00	–	–	week	**580.00**
Chainmen	40.00	470.00	–	–	week	**470.00**
Laboratory assistants	40.00	380.00	–	–	week	**380.00**
Temporary Works						
Temporary Works relate to other work which the contractor may need to carry out due to his construction method. These are highly specific to the particular project and envisaged construction method. Examples are:						
* access roads and hardstandings/ bases for plant and accommodation as well as to the assembly/ working area generally						
* constructing ramps for access to low excavations						
* steel sheet, diaphragm wall or secant pile cofferdam walling to large excavations subject to strong forces/water penetration						
* bridges						
* temporary support works and decking						
The need for these items should be carefully considered and reference made to the other sections of the book for guidance on what costs should be set against the design assumptions made. Extensive works could well call for the involvement of a contractor's temporary works engineer for realistic advice.						
Accommodation and buildings						
Offices; establishment and removal; Fixed Charge						
80 m^2 mobile unit (10 staff × 8 m^2)	–	–	1750.00	–	sum	**1750.00**
10 m^2 section units; two	–	–	750.00	–	sum	**750.00**
Offices; maintaining; Time Related Charge						
80 m^2 mobile unit (10 staff × 8 m^2)	–	–	450.00	–	week	**450.00**
10 m^2 section units; two	–	–	250.00	–	week	**250.00**
Stores; establishment and removal; Fixed Charge						
22 m^2 section unit	–	–	600.00	–	sum	**600.00**
Stores; maintaining; Time Related Charge						
22 m^2 section unit	–	–	100.00	–	sum	**100.00**
Canteens and messrooms; establishment and removal; Fixed Charge						
70 m^2 mobile unit (70 men)	–	–	1500.00	–	sum	**1500.00**
Canteens and messrooms; maintaining; Time-Related Charge						
70 m^2 mobile unit (70 men)	–	–	425.00	–	week	**425.00**

CLASS A: GENERAL ITEMS

Item	Gang hours	Labour £	Plant £	Material £	Unit	Total rate £
Plant						
General purpose plant not included in Unit Costs:						
Transport						
wheeled tractor	40.00	508.40	1026.00	–	week	1534.40
trailer	–	–	13.86	–	week	13.86
Tractair						
hire charge	–	–	600.00	–	week	600.00
driver (skill rate 4)	40.00	595.20	–	–	week	595.20
fuel and consumables	–	–	588.00	–	week	588.00
Cranes						
10 t crane	40.00	662.00	1704.00	–	week	2366.00
Miscellaneous						
sawbench	–	–	88.00	–	week	88.00
concrete vibrator	–	–	146.00	–	week	146.00
75 mm 750 l/min pump	–	–	148.00	–	week	148.00
compressor	–	–	3151.20	–	week	3151.20
plate compactor; 180 kg	–	–	98.94	–	week	98.94
towed roller; BW6	–	–	1041.60	–	week	1041.60
Excavators etc.						
hydraulic backacter; 14.5 tonne; driver + banksman	80.00	1206.40	1744.40	–	week	2950.80
bulldozer; D6; driver	40.00	724.80	3039.20	–	week	3764.00
loading shovel; CAT 939; driver	40.00	662.00	1543.60	–	week	2205.60
Temporary Works						
Supervision and labour						
Supervision for the duration of construction; Time Related Charge						
Agent	40.00	1360.00	–	–	week	1360.00
Senior Engineer	40.00	1000.00	–	–	week	1000.00
Engineers	40.00	800.00	–	–	week	800.00
General Foreman	40.00	880.00	–	–	week	880.00
Administration for the duration of construction; Time Related Charge						
Office manager/Cost clerk	40.00	500.00	–	–	week	500.00
Timekeeper/Storeman/Checker	40.00	400.00	–	–	week	400.00
Typist/Telephonist	40.00	400.00	–	–	week	400.00
Security guard	40.00	450.00	–	–	week	450.00
Quantity Surveyor	40.00	1120.00	–	–	week	1120.00
Labour teams for the duration of construction; Time Related Charge						
General yard labour (part time); loading and offloading, clearing site rubbish etc.; ganger	40.00	490.00	–	–	week	490.00
General yard labour (part time); loading and offloading, clearing site rubbish etc.; four unskilled operatives	160.00	1360.00	–	–	week	1360.00
Maintenance of contractor's own plant; Fitter	40.00	600.00	–	–	week	600.00
Maintenance of contractor's own plant; Fitter's Mate	40.00	420.00	–	–	week	420.00

CLASS B: GROUND INVESTIGATION

Item	Gang hours	Labour £	Plant £	Material £	Unit	Total rate £
RESOURCES – LABOUR						
Trial hole gang						
1 ganger or chargehand (skill rate 4)		13.61				
1 skilled operative (skill rate 4)		13.61				
2 unskilled operatives (general)		25.42				
1 plant operator (skill rate 3)		16.55				
Total Gang Rate/Hour		£ 69.19				
RESOURCES – PLANT						
Trial holes						
8 tonne wheeled backacter			30.10			
3 tonne dumper			10.50			
3.7 m^3/min compressor, 2 tool			8.58			
two 2.4 m^3/min road breakers			8.75			
extra 50 ft / 15 m hose			0.31			
plate compactors; vibrating compaction; plate 180 kg/600 mm			2.47			
Total Rate/Hour		£	60.72			
TRIAL PITS AND TRENCHES						
The following costs assume the use of mechanical plant and excavating and backfilling on the same day.						
Trial holes measured by number						
Excavating trial hole; plan size 1.0 × 2.0 m; supports, backfilling						
ne 1.0 m deep	0.24	16.61	14.57	–	nr	31.18
1.0–2.0 m deep	0.47	32.52	28.57	–	nr	61.09
over 2.0 m deep	0.53	36.67	32.21	–	nr	68.88
Excavating trial hole in rock or similar; plan size 1.0 × 2.0 m; supports, backfilling						
ne 1.0 m deep	0.28	19.37	17.00	–	nr	36.37
1.0–2.0 m deep	0.51	35.29	31.00	–	nr	66.29
over 2.0 m deep	0.58	40.13	35.22	–	nr	75.35
Trial holes measured by depth						
Excavating trial hole; plan size 1.0 × 2.0 m; supports, backfilling						
ne 1.0 m deep	0.24	16.61	14.57	–	m	31.18
1.0–2.0 m deep	0.53	36.67	32.21	–	m	68.88
2.0–3.0 m deep	0.59	40.82	35.85	–	m	76.67
3.0–5.0 m deep	0.65	44.97	39.50	–	m	84.47
Excavating trial hole in rock or similar; plan size 1.0 × 2.0 m; supports, backfilling						
ne 1.0 m deep	2.95	204.11	179.14	–	m	383.25
1.0–2.0 m deep	3.65	252.54	221.65	–	m	474.19
2.0–3.0 m deep	3.95	273.30	239.86	–	m	513.16
3.0–5.0 m deep	4.50	311.36	273.22	–	m	584.58

Unit Costs – Civil Engineering Works 145

CLASS B: GROUND INVESTIGATION

Item	Gang hours	Labour £	Plant £	Material £	Unit	Total rate £
Sundries in trial holes						
Removal of obstructions from trial holes irrespective of depth	1.00	69.19	60.72	–	hr	129.91
Pumping; maximum depth 4.0 m minimum 750 litres per hour	0.06	1.19	13.62	–	hr	14.81
LIGHT CABLE PERCUSSION BOREHOLES						
The following costs are based on using Specialist Contractors and are for guidance only						
Establishment of standard plant and equipment and removal on completion	–	–	–	–	sum	450.00
Number; 150 mm nominal diameter at base	–	–	–	–	nr	50.00
Depth; 150 mm nominal diameter of base						
in holes of maximum depth not exceeding 5 m	–	–	–	–	m	18.00
in holes of maximum depth 5–10 m	–	–	–	–	m	18.00
in holes of maximum depth 10–20 m	–	–	–	–	m	23.00
in holes of maximum depth 20–30 m	–	–	–	–	m	29.00
Depth backfilled; imported pulverized fuel ash	–	–	–	–	m	5.00
Depth backfilled; imported gravel	–	–	–	–	m	7.00
Depth backfilled; bentonite grout	–	–	–	–	m	10.00
Chiselling to prove rock or to penetrate obstructions	–	–	–	–	h	48.00
Standing time of rig and crew	–	–	–	–	h	48.00
ROTARY DRILLED BOREHOLES						
The following costs are based on using Specialist Contractors and are for guidance only						
Establishment of standard plant and equipment and removal on completion	–	–	–	–	sum	1000.00
Setting up at each borehole position	–	–	–	–	nr	8.00
Depth without core recovery; nominal minimum core diameter 100 mm						
ne 5.0 m deep	–	–	–	–	m	32.00
5–10 m deep	–	–	–	–	m	32.00
10–20 m deep	–	–	–	–	m	33.00
20–30 m deep	–	–	–	–	m	33.00
Depth with core recovery; nominal minimum core diameter 75 mm						
ne 5.0 m deep	–	–	–	–	m	70.00
5–10 m deep	–	–	–	–	m	70.00
10–20 m deep	–	–	–	–	m	78.00
20–30 m deep	–	–	–	–	m	88.00
Depth cased; semi-rigid plastic core barrel liner	–	–	–	–	m	6.00
SAMPLES						
From the surface or from trial pits and trenches						
undisturbed soft material; minimum 200 mm cube	–	–	–	–	nr	16.25
disturbed soft material; minimum 5 kg	–	–	–	–	nr	2.55
rock; minimum 5 kg	–	–	–	–	nr	10.15
groundwater; miniumum 1 l	–	–	–	–	nr	4.50

CLASS B: GROUND INVESTIGATION

Item	Gang hours	Labour £	Plant £	Material £	Unit	Total rate £
SAMPLES – cont						
From boreholes						
open tube; 100 mm diameter × 450 mm long	–	–	–	–	nr	16.00
disturbed; minimum 5 kg	–	–	–	–	nr	2.50
groundwater; minimum 1 l	–	–	–	–	nr	4.00
stationary piston	–	–	–	–	nr	4.00
Swedish foil	–	–	–	–	nr	25.00
Delft	–	–	–	–	nr	25.00
Bishop sand	–	–	–	–	nr	25.00
SITE TESTS AND OBSERVATIONS						
Groundwater level						
Standard penetration						
in light cable percussion boreholes	–	–	–	–	nr	–
Vane in borehole						
Plate bearing						
in pits and trenches; loading table	–	–	–	–	nr	–
in pits and trenches; hydraulic jack and kentledge	–	–	–	–	nr	–
California bearing ratio	–	–	–	–	nr	–
Mackintosh probe						
Hand auger borehole						
mm minimum diameter; 6 m maximum depth					nr	–
INSTRUMENTAL OBSERVATIONS						
General						
Pressure head						
standpipe; 75 mm diameter HDPE pipe	–	–	–	–	m	30.00
piezometer	–	–	–	–	m	42.50
install protective cover	–	–	–	–	nr	82.00
readings	–	–	–	–	nr	20.50
LABORATORY TESTS						
Classification						
moisture content	–	–	–	–	nr	5.00
specific gravity	–	–	–	–	nr	7.00
particle size analysis by sieve	–	–	–	–	nr	28.30
particle size analysis by pipette or hydrometer	–	–	–	–	nr	40.00
Chemical content						
organic matter	–	–	–	–	nr	31.12
sulphate	–	–	–	–	nr	21.22
pH value	–	–	–	–	nr	6.84
contaminants; Comprehensive	–	–	–	–	nr	247.48
contaminants; Abbreviated	–	–	–	–	nr	176.86
contaminants; Mini, Screening	–	–	–	–	nr	141.48
contaminants; nitrogen herbicides	–	–	–	–	nr	127.34
contaminants; organophosphorus pesticides	–	–	–	–	nr	106.12
contaminants; organochlorine pesticides	–	–	–	–	nr	106.12

CLASS B: GROUND INVESTIGATION

Item	Gang hours	Labour £	Plant £	Material £	Unit	Total rate £
Compaction						
standard	–	–	–	–	nr	**127.33**
heavy	–	–	–	–	nr	**127.33**
vibratory	–	–	–	–	nr	**141.48**
Permeability						
falling head	–	–	–	–	nr	**56.59**
Soil strength						
quick undrained triaxial; set of three 38 mm diameter specimens	–	–	–	–	nr	**28.29**
shear box; peak only; size of shearbox 100 × 100 mm	–	–	–	–	nr	**35.38**
California bearing ratio; typical	–	–	–	–	nr	**42.45**
Rock strength						
point load test; minimum 5 kg sample	–	–	–	–	nr	**70.74**
PROFESSIONAL SERVICES						
Technician	–	–	–	–	hr	**35.00**
Technician engineer	–	–	–	–	hr	**47.00**
Engineer or geologist						
graduate	–	–	–	–	hr	**47.00**
chartered	–	–	–	–	hr	**60.00**
principal or consultant	–	–	–	–	hr	**80.50**
Visits to the Site						
technician	–	–	–	–	nr	**35.00**
technician engineer/graduate engineer or geologist	–	–	–	–	nr	**47.00**
chartered engineer	–	–	–	–	nr	**60.00**
principal or consultant	–	–	–	–	nr	**80.50**
Overnight stays in connection with visits to the site						
technician	–	–	–	–	nr	**66.50**
technician engineer/graduate engineer or geologist	–	–	–	–	nr	**66.50**
chartered engineer/senior geologist	–	–	–	–	nr	**86.50**
principal or consultant	–	–	–	–	nr	**86.50**

CLASS C: GEOTECHNICAL AND OTHER SPECIALIST PROCESSES

Item	Gang hours	Labour £	Plant £	Material £	Unit	Total rate £
DRILLING FOR GROUT HOLES						
Notes						
The processes referred to in this section are generally carried out by Specialist Contractors and the costs are therefore an indication of the probable costs based on average site conditions.						
The following unit costs are based on drilling 100 grout holes on a clear site with reasonable access						
Establishment of standard drilling plant and equipment and removal on completion	–	–	–	–	sum	8750.00
Standing time	–	–	–	–	hour	185.00
Drilling through material other than rock or artificial hard material						
vertically downwards						
depth ne 5 m	–	–	–	–	m	17.50
depth 5–10 m	–	–	–	–	m	20.25
depth 10–20 m	–	–	–	–	m	24.00
depth 20–30 m	–	–	–	–	m	29.35
downwards at an angle 0–45° to the vertical						
depth ne 5 m	–	–	–	–	m	17.50
depth 5–10 m	–	–	–	–	m	20.25
depth 10–20 m	–	–	–	–	m	24.00
depth 20–30 m	–	–	–	–	m	29.35
horizontally or downwards at an angle less than 45° to the horizontal						
depth ne 5 m	–	–	–	–	m	17.50
depth 5–10 m	–	–	–	–	m	20.25
depth 10–20 m	–	–	–	–	m	24.00
depth 20–30 m	–	–	–	–	m	29.35
upwards at an angle 0–45° to the horizontal						
depth ne 5 m	–	–	–	–	m	28.50
depth 5–10 m	–	–	–	–	m	32.50
depth 10–20 m	–	–	–	–	m	35.50
depth 20–30 m	–	–	–	–	m	38.50
upwards at an angle less than 45° to the vertical						
depth ne 5 m	–	–	–	–	m	28.50
depth 5–10 m	–	–	–	–	m	32.50
depth 10–20 m	–	–	–	–	m	35.50
depth 20–30 m	–	–	–	–	m	38.50
Drilling through rock or artificial hard material						
Vertically downwards						
depth ne 5 m	–	–	–	–	m	18.75
depth 5–10 m	–	–	–	–	m	23.00
depth 10–20 m	–	–	–	–	m	28.00
depth 20–30 m	–	–	–	–	m	35.00

CLASS C: GEOTECHNICAL AND OTHER SPECIALIST PROCESSES

Item	Gang hours	Labour £	Plant £	Material £	Unit	Total rate £
Downwards at an angle 0–45° to the vertical						
depth ne 5 m	–	–	–	–	m	18.75
depth 5–10 m	–	–	–	–	m	23.00
depth 10–20 m	–	–	–	–	m	28.00
depth 20–30 m	–	–	–	–	m	35.00
Horizontally or downwards at an angle less than 45° to the horizontal						
depth ne 5 m	–	–	–	–	m	18.75
depth 5–10 m	–	–	–	–	m	23.00
depth 10–20 m	–	–	–	–	m	28.00
depth 20–30 m	–	–	–	–	m	35.00
Upwards at an angle 0–45° to the horizontal						
depth ne 5 m	–	–	–	–	m	34.50
depth 5–10 m	–	–	–	–	m	38.80
depth 10–20 m	–	–	–	–	m	44.00
depth 20–30 m	–	–	–	–	m	49.00
Upwards at an angle less than 45° to the horizontal						
depth ne 5 m	–	–	–	–	m	34.50
depth 5–10 m	–	–	–	–	m	38.80
depth 10–20 m	–	–	–	–	m	44.00
depth 20–30 m	–	–	–	–	m	49.00
GROUT HOLES						
The following unit costs are based on drilling 100 grout holes on a clear site with reasonable access						
Grout holes						
number of holes	–	–	–	–	nr	94.50
multiple water pressure tests	–	–	–	–	nr	6.75
GROUT MATERIALS AND INJECTION						
The following unit costs are based on drilling 100 grout holes on a clear site with reasonable access						
Materials						
ordinary Portland cement	–	–	–	–	t	110.00
sulphate-resistant cement	–	–	–	–	t	112.00
cement grout	–	–	–	–	t	137.00
pulverized fuel ash	–	–	–	–	t	16.00
sand	–	–	–	–	t	21.60
pea gravel	–	–	–	–	t	21.60
bentonite (2:1)	–	–	–	–	t	142.00
Injection						
establishment of standard injection plant and removal on completion	–	–	–	–	sum	8750.00
standing time	–	–	–	–	hr	175.00
number of injections	–	–	–	–	nr	57.50
neat cement grout	–	–	–	–	t	100.00
cement/P.F.A. grout	–	–	–	–	t	57.50

CLASS C: GEOTECHNICAL AND OTHER SPECIALIST PROCESSES

Item	Gang hours	Labour £	Plant £	Material £	Unit	Total rate £
DIAPHRAGM WALLS						
Notes						
Diaphragm walls are the construction of vertical walls, cast in place in a trench excavation. They can be formed in reinforced concrete to provide structural elements for temporary or permanent retaining walls. Wall thicknesses of 500 mm to 1.50 m and up to 40 m deep may be constructed. Special equipment such as the Hydrofraise can construct walls up to 100 m deep. Restricted urban sites will significantly increase the costs.						
The following costs are based on constructing a diaphragm wall with an excavated volume of 4000 m³ using standard equipment. Typical progress would be up to 500 m per week						
Establishment of standard plant and equipment including bentonite storage tanks and removal on completion	–	–	–	–	sum	300000.00
Standing time	–	–	–	–	hr	750.00
Excavation, disposal of soil and placing of concrete	–	–	–	–	m³	440.00
Provide and place reinforcement cages	–	–	–	–	t	750.00
Excavate/chisel in hard material/rock	–	–	–	–	hr	975.00
Waterproofed joints	–	–	–	–	m	6.50
Guide walls						
guide walls (twin)	–	–	–	–	m	350.00
GROUND ANCHORAGES						
Notes						
Ground anchorages consist of the installation of a cable or solid bar tendon fixed in the ground by grouting and tensioned to exceed the working load to be carried. Ground anchors may be of a permanent or temporary nature and can be used in conjunction with diaphragm walls or sheet piling to eliminate the use of strutting etc.						
The following costs are based on the installation of 50 nr ground anchors						
Establishment of standard plant and equipment and removal on completion	–	–	–	–	sum	10500.00
Standing time	–	–	–	–	hr	165.00
Ground anchorages; temporary or permanent						
15.0 m maximum depth; in rock, alluvial or clay; 0–50 t load	–	–	–	–	m	37.00
15.0 m maximum depth; in rock or alluvial; 50–90 t load	–	–	–	–	m	52.25
15.0 m maximum depth; in rock only; 90–150 t load	–	–	–	–	m	107.50

CLASS C: GEOTECHNICAL AND OTHER SPECIALIST PROCESSES

Item	Gang hours	Labour £	Plant £	Material £	Unit	Total rate £
Temporary tendons						
in rock, alluvial or clay; 0–50 t load	–	–	–	–	m	26.70
in rock or alluvial; 50–90 t load	–	–	–	–	m	40.00
in rock only; 90–150 t load	–	–	–	–	m	54.25
Permanent tendons						
in rock, alluvial or clay; 0–50 t load	–	–	–	–	m	42.50
in rock or alluvial; 50–90 t load	–	–	–	–	m	55.00
in rock only; 90–150 t load	–	–	–	–	m	70.00

SAND, BAND AND WICK DRAINS

Notes

Vertical drains are a technique by which the rate of consolidation of fine grained soils can be considerably increased by the installation of vertical drainage paths commonly in the form of columns formed by a high-quality plastic material encased in a filter sleeve. Columns of sand are rarely used in this country these days.

Band drains are generally 100 mm wide and 3–5 mm thick. Water is extracted through the drain from the soft soils when the surface is surcharged. The rate of consolidation is dependent on the drain spacing and the height of surcharge.

Drains are usually quickly installed up to depths of 25 m by special lances either pulled or vibrated into the ground. Typical drain spacing would be one per 1–2 m with the rate of installation varying between 1,500 to 6,000 m per day depending on ground conditions and depths.

The following costs are based on the installation of 2,000 nr vertical band drains to a depth of 12 m

Item	Gang hours	Labour £	Plant £	Material £	Unit	Total rate £
Establishment of standard plant and equipment and removal on completion	–	–	–	–	sum	6000.00
Standing time	–	–	–	–	hr	185.00
Set up installation equipment at each drain position	–	–	–	–	nr	1.00
Install drains maximum depth 10–15 m	–	–	–	–	m	0.85
Additional costs in pre-drilling through hard upper strata at each drain position:						
establishment of standard drilling plant and equipment and removal on completion	–	–	–	–	sum	2250.00
set up at each drain position	–	–	–	–	nr	5.15
drilling for vertical band drains up to a maximum depth of 3 m	–	–	–	–	m	11.25

CLASS C: GEOTECHNICAL AND OTHER SPECIALIST PROCESSES

Item	Gang hours	Labour £	Plant £	Material £	Unit	Total rate £
GROUND CONSOLIDATION – VIBRO-REPLACEMENT						
Notes						
Vibro-replacement is a method of considerably increasing the ground bearing pressure and consists of a specifically designed powerful poker vibrator penetrating vertically into the ground hydraulically. Air and water jets may be used to assist penetration. In cohesive soils a hole is formed into which granular backfill is placed and compacted by the poker, forming a dense stone column. In natural sands and gravels the existing loose deposits may be compacted without the addition of extra material other than making up levels after settlement resulting from compaction.						
There are many considerations regarding the soil types to be treated, whether cohesive or non-cohesive, made-up or natural ground, which influence the choice of wet or dry processes, pure densification or stone column techniques with added granular backfill. It is therefore possible only to give indicative costs; a Specialist Contractor should be consulted for more accurate costs for a particular site.						
Testing of conditions after consolidation can be static or dynamic penetration tests, plate bearing tests or zone bearing tests. A frequently adopted specification calls for plate bearing tests at 1 per 1000 stone columns. Allowable bearing pressures of up to 400 kN/m^2 by the installation of stone columns in made or natural ground.						
The following costs are typical rates for this sort of work						
Establishment of standard plant and equipment and removal on completion	–	–	–	–	sum	4000.00
Standing time	–	–	–	–	hr	215.00
Construct stone columns to a depth ne 4 m						
dry formed	–	–	–	–	m	18.50
water jet formed	–	–	–	–	m	27.50
Plate bearing test						
ne 11 t or 2 hour duration	–	–	–	–	nr	450.00
Zone loading test to specification	–	–	–	–	nr	8750.00

CLASS C: GEOTECHNICAL AND OTHER SPECIALIST PROCESSES

Item	Gang hours	Labour £	Plant £	Material £	Unit	Total rate £
GROUND CONSOLIDATION – DYNAMIC COMPACTION						
Notes						
Ground consolidation by dynamic compaction is a technique which involves the dropping of a steel or concrete pounder several times in each location on a grid pattern that covers the whole site. For ground compaction up to 10 m, a 15 t pounder from a free fall of 20 m would be typical. Several passes over the site are normally required to achieve full compaction. The process is recommended for naturally cohesive soils and is usually uneconomic for areas of less than 4,000 m² for sites with granular or mixed granular cohesive soils and 6,000 m for a site with weak cohesive soils. The main considerations to be taken into account when using this method of consolidation are:						
* sufficient area to be viable						
* proximity and condition of adjacent property and services						
* need for blanket layer of granular material for a working surface and as backfill to offset induced settlement						
* water table level						
The final bearing capacity and settlement criteria that can be achieved depend on the nature of the material being compacted. Allowable bearing capacity may be increased by up to twice the pre-treated value for the same settlement. Control testing can be by crater volume measurements, site levelling between passes, penetration tests or plate loading tests.						
The following range of costs are average based on treating an area of about 10,000 m² for a 5–6 m compaction depth. Typical progress would be 1,500–2,000 m² per week						
Establishment of standard plant and equipment and removal on completion	–	–	–	–	sum	21750.00
Ground treatment	–	–	–	–	m²	1.50
Laying free-draining granular blanket layer as both working surface and backfill material (300 mm thickness required of filter material)	–	–	–	–	m²	8.25
Control testing including levelling, piezometers and penetrameter testing	–	–	–	–	m²	3.50
Kentledge load test	–	–	–	–	nr	8000.00

CLASS C: GEOTECHNICAL AND OTHER SPECIALIST PROCESSES

Item	Gang hours	Labour £	Plant £	Material £	Unit	Total rate £
CONSOLIDATION OF ABANDONED MINE WORKINGS						
The following costs are based on using Specialist Contractors						
Transport plant, labour and all equipment to and from site (max. 100 miles)	–	–	–	–	sum	350.00
Drilling bore holes						
move to each separate bore position; erect equipment; dismantle prior to next move	–	–	–	–	nr	36.00
drill 50 mm diameter bore holes	–	–	–	–	m	10.00
drill 100 mm diameter bore holes for pea gravel injection	–	–	–	–	m	17.50
extra for casing, when required	–	–	–	–	m	15.00
standing time for drilling rig and crew	–	–	–	–	hr	95.00
Grouting drilled bore holes						
connecting grout lines	–	–	–	–	m	18.75
injection of grout	–	–	–	–	t	75.00
add for pea gravel injection	–	–	–	–	t	88.15
standing time for grouting rig and crew	–	–	–	–	hr	90.50
Provide materials for grouting						
ordinary Portland cement	–	–	–	–	t	110.00
sulphate-resistant cement	–	–	–	–	t	112.00
pulverized fuel ash (PFA)	–	–	–	–	t	16.00
sand	–	–	–	–	t	21.60
pea gravel	–	–	–	–	t	37.20
bentonite (2:1)	–	–	–	–	t	142.00
cement grout	–	–	–	–	t	137.00
Capping to old shafts or similar; reinforced concrete grade C20P, 20 mm aggregate; thickness						
ne 150 mm	–	–	–	–	m^3	175.00
150–300 mm	–	–	–	–	m^3	170.00
300–500 mm	–	–	–	–	m^3	165.00
over 500 mm	–	–	–	–	m^3	135.00
Mild steel bars BS4449; supplied in bent and cut lengths						
6 mm nominal size	–	–	–	–	t	750.00
8 mm nominal size	–	–	–	–	t	750.00
10 mm nominal size	–	–	–	–	t	750.00
12 mm nominal size	–	–	–	–	t	750.00
16 mm nominal size	–	–	–	–	t	750.00
20 mm nominal size	–	–	–	–	t	750.00
High yield steel bars BS4449 or 4461; supplied in bent and cut lengths						
6 mm nominal size	–	–	–	–	t	900.00
8 mm nominal size	–	–	–	–	t	900.00
10 mm nominal size	–	–	–	–	t	900.00
12 mm nominal size	–	–	–	–	t	900.00
16 mm nominal size	–	–	–	–	t	900.00
20 mm nominal size	–	–	–	–	t	900.00

CLASS D: DEMOLITION AND SITE CLEARANCE

Item	Gang hours	Labour £	Plant £	Material £	Unit	Total rate £
GENERAL CLEARANCE						
The rates for site clearance include for all sundry items, small trees (i.e. under 500 mm diameter), hedges etc., but exclude items that are measured separately; examples of which are given in this section.						
Clear site vegetation						
generally	–	–	–	–	ha	1050.00
wooded areas	–	–	–	–	ha	3150.00
areas below tidal level	–	–	–	–	ha	3850.00
TREES						
The following rates are based on removing a minimum of 100 trees, generally in a group. Cutting down a single tree on a site would be many times these costs						
Remove trees						
girth 500 mm–1 m	–	–	–	–	nr	40.00
girth 1–2 m	–	–	–	–	nr	70.00
girth 2–3 m	–	–	–	–	nr	230.00
girth 3–5 m	–	–	–	–	nr	910.00
girth 7 m	–	–	–	–	nr	1200.00
STUMPS						
Clearance of stumps						
diameter 150–500 mm	–	–	–	–	nr	40.00
diameter 500 mm–1 m	–	–	–	–	nr	80.00
diameter 2 m	–	–	–	–	nr	120.00
Clearance of stumps; backfilling holes with topsoil from site						
diameter 150–500 mm	–	–	–	–	nr	35.00
diameter 500 mm–1 m	–	–	–	–	nr	115.00
diameter 2 m	–	–	–	–	nr	250.00
Clearance of stumps; backfilling holes with imported hardcore						
diameter 150–500 mm	–	–	–	–	nr	55.00
diameter 500 mm–1 m	–	–	–	–	nr	215.00
diameter 2 m	–	–	–	–	nr	700.00

CLASS D: DEMOLITION AND SITE CLEARANCE

Item	Gang hours	Labour £	Plant £	Material £	Unit	Total rate £
BUILDINGS						
The following rates are based on assuming a non urban location where structure does not take up a significant area of the site						
Demolish building to ground level and dispose off site						
brickwork with timber floor and roof	–	–	–	–	m³	6.71
brickwork with concrete floor and roof	–	–	–	–	m³	11.19
masonry with timber floor and roof	–	–	–	–	m³	8.77
reinforced concrete frame with brick infill	–	–	–	–	m³	11.66
steel frame with brick cladding	–	–	–	–	m³	6.35
steel frame with sheet cladding	–	–	–	–	m³	6.04
timber	–	–	–	–	m³	5.43
Demolish buildings with asbestos linings to ground level and dispose off site						
brick with concrete floor and roof	–	–	–	–	m³	27.19
reinforced concrete frame with brick infill	–	–	–	–	m³	28.34
steel frame with brick cladding	–	–	–	–	m³	15.49
steel frame with sheet cladding	–	–	–	–	m³	16.08
OTHER STRUCTURES						
The following rates are based on assuming a non urban location where structure does not take up a significant area of the site						
Demolish walls to ground level and dispose off site						
reinforced concrete wall	–	–	–	–	m³	150.00
brick or masonry wall	–	–	–	–	m³	69.43
brick or masonry retaining wall	–	–	–	–	m³	84.84
PIPELINES						
Removal of redundant services						
electric cable; LV	–	–	–	–	m	2.56
75 mm diameter water main; low pressure	–	–	–	–	m	5.13
150 mm diameter gas main; low pressure	–	–	–	–	m	6.15
earthenware ducts; one way	–	–	–	–	m	2.56
earthenware ducts; two way	–	–	–	–	m	5.13
100 or 150 mm diameter sewer or drain	–	–	–	–	m	11.00
225 mm diameter sewer or drain	–	–	–	–	m	16.00
300 mm diameter sewer or drain	–	–	–	–	m	18.00
450 mm diameter sewer or drain	–	–	–	–	m	22.00
750 mm diameter sewer or drain	–	–	–	–	m	28.00
Extra for breaking up concrete surround	–	–	–	–	m	28.00
Grouting redundant drains or sewers						
100 mm diameter	–	–	–	–	m	6.15
150 mm diameter	–	–	–	–	m	15.00
225 mm diameter	–	–	–	–	m	35.00
manhole chambers	–	–	–	–	m³	850.00

CLASS E: EXCAVATION

Item	Gang hours	Labour £	Plant £	Material £	Unit	Total rate £
NOTES						
Ground conditions						
The following unit costs for 'excavation in material other than topsoil, rock or artificial hard material' are based on excavation in firm sand and gravel soils. For alternative types of soil, multiply the following rates by:						
Scrapers						
Stiff clay × 1.5						
Chalk × 2.5						
Soft rock × 3.5						
Broken rock × 3.7						
Tractor dozers and loaders						
Stiff clay × 2.0						
Chalk × 3.0						
Soft rock × 2.5						
Broken rock × 2.5						
Backacter (minimum bucket size 0.5 m^3)						
Stiff clay × 1.7						
Chalk × 2.0						
Soft rock × 2.0						
Broken rock × 1.7						
Basis of disposal rates						
All pricing and estimating for disposal is based on the volume of solid material excavated and rates for disposal should be adjusted by the following factors for bulkage. Multiply the rates by:						
Sand bulkage × 1.10						
Gravel bulkage × 1.20						
Compacted soil bulkage × 1.30						
Compacted subbase, suitable fill etc. bulkage × 1.30						
Stiff clay bulkage × 1.20						
See also Part 12: Tables and Memoranda						
Basis of rates generally						
To provide an overall cost comparison, rates, prices and outputs have been based on a medium sized Civil Engineering project of £10–12 million, location neither in city centre nor excessively remote, with no abnormal ground conditions that would affect the stated output and consistency of work produced. The rates are optimum rates and assume continuous output with no delays caused by other operations or works.						

CLASS E: EXCAVATION

Item	Gang hours	Labour £	Plant £	Material £	Unit	Total rate £
RESOURCES – LABOUR						
Excavation for cuttings gang						
1 plant operator (skill rate 1) – 33% of time		6.55				
1 plant operator (skill rate 2) – 66% of time		11.96				
1 banksman (skill rate 4)		13.61				
Total Gang Rate/Hour		£ **32.12**				
Excavation for foundations gang						
1 plant operator (skill rate 3) – 33% of time		5.46				
1 plant operator (skill rate 2) – 66% of time		11.96				
1 banksman (skill rate 4)		13.61				
Total Gang Rate/Hour		£ **31.03**				
General excavation gang						
1 plant operator (skill rate 3)		16.55				
1 plant operator (skill rate 3) – 25% of time		4.14				
1 banksman (skill rate 4)		13.61				
Total Gang Rate/Hour		£ **34.30**				
Filling gang						
1 plant operator (skill rate 4)		14.88				
2 unskilled operatives (general)		25.42				
Total Gang Rate/Hour		£ **40.30**				
Treatment of filled surfaces gang						
1 plant operator (skill rate 2)		18.12				
Total Gang Rate/Hour		£ **18.12**				
Geotextiles (light sheets) gang						
1 ganger/chargehand (skill rate 4) – 20% of time		2.91				
2 unskilled operatives (general)		25.42				
Total Gang Rate/Hour		£ **28.33**				
Geotextiles (medium sheets) gang						
1 ganger/chargehand (skill rate 4) – 20% of time		2.91				
3 unskilled operatives (general)		38.13				
Total Gang Rate/Hour		£ **41.04**				
Geotextiles (heavy sheets) gang						
1 ganger/chargehand (skill rate 4) – 20% of time		2.91				
2 unskilled operatives (general)		25.42				
1 plant operator (skill rate 3)		16.55				
Total Gang Rate/Hour		£ **44.88**				
Horticultural works gang						
1 skilled operative (skill rate 4)		13.61				
1 unskilled operative (general)		12.71				
Total Gang Rate/Hour		£ **26.32**				
RESOURCES – PLANT						
Excavation for foundations						
Hydraulic backacter – 21 tonne (33% of time)			18.43			
Hydraulic backacter – 14.5 tonne (33% of time)			14.54			
1000 kg hydraulic breaker (33% of time)			5.81			
Tractor loader – 0.80 m^3 (33% of time)			12.86			
Total Rate/Hour			£ **51.64**			

CLASS E: EXCAVATION

Item	Gang hours	Labour £	Plant £	Material £	Unit	Total rate £
Filling						
1.5 m³ tractor loader			63.41			
6 t vibratory roller			26.04			
Pedestrian Roller, Bomag BW35			6.05			
Total Rate/Hour		£	**95.50**			
Treatment of filled surfaces						
1.5 m³ tractor loader			63.41			
Pedestrian Roller, Bomag BW35			6.05			
Total Rate/Hour		£	**69.46**			
Geotextiles (heavy sheets)						
1.5 m³ tractor loader			63.41			
Total Rate/Hour		£	**63.41**			

EXCAVATION BY DREDGING

Notes

Dredging can be carried out by land-based machines or by floating plant. The cost of the former can be assessed by reference to the excavation costs of the various types of plant given below, suitably adjusted to take account of the type of material to be excavated, depth of water and method of disposal. The cost of the latter is governed by many factors which affect the rates and leads to wide variations. For reliable estimates it is advisable to seek the advice of a Specialist Contractor. The prices included here are for some typical dredging situations and are shown for a cost comparison and EXCLUDE initial mobilization charges which can range widely between £3,000 and £10,000 depending on plant, travelling distance etc. Some clients schedule operations for when the plant is passing so as to avoid the large travelling cost.

Of the factors affecting the cost of floating plant, the matter of working hours is by far one of the most important. The customary practice in the dredging industry is to work 24 hours per day, 7 days per week. Local constraints, particularly noise restrictions, will have a significant impact.

Other major factors affecting the cost of floating plant are:
* type of material to be dredged
* depth of water
* depth of cut
* tidal range
* disposal location
* size and type of plant required
* current location of plant
* method of disposal of dredged material

CLASS E: EXCAVATION

Item	Gang hours	Labour £	Plant £	Material £	Unit	Total rate £
EXCAVATION BY DREDGING – cont						
In tidal locations, creating new channels on approaches to quays and similar locations or within dock systems Backhoe dredger loading material onto two hopper barges with bottom dumping facility maximum water depth 15 m, distance to disposal site less than 20 miles						
approximate daily cost of backhoe dredger and two hopper barges £10,500; average production 100 m³/hr, 1,500 m³/day; locate, load, deposit and relocate	–	–	–	–	m³	**7.50**
For general bed lowering or in maintaining shipping channels in rivers, estuaries or deltas Trailer suction hopper dredger excavating non-cohesive sands, grits or silts; hopper capacity 2000 m³, capable of dredging to depths of 25 m with ability either to dump at disposal site or pump ashore for reclamation. Costs totally dependent on nature of material excavated and method of disposal. Approximate cost per tonne £2.50, which should be adjusted by the relative density of the material for conversion to m³						
locate, load, deposit and relocate	–	–	–	–	tonne	**3.00**
Harbour bed control Maintenance dredging of this nature would in most cases be carried out by trailing suction hopper dredger as detailed above, at similar rates. The majority of the present generation of trailers have the ability to dump at sea or discharge ashore. Floating craft using diesel-driven suction method with a 750 mm diameter flexible pipe for a maximum distance of up to 5000 m from point of suction to point of discharge using a booster (standing alone, the cutter suction craft should be able to pump up to 2000 m). Maximum height of lift 10 m.						
Average pumping capacity of silt/sand type materials containing maximum 30% volume of solids would be about 8,000 m³/day based on 24 hour working, Daily cost (hire basis) in the region of £20,000 including all floating equipment and discharge pipes, maintenance and all labour and plant to service but excluding mobilization/initial set-up and demobilization costs (minimum £10,000).	–	–	–	–	m³	**2.85**

CLASS E: EXCAVATION

Item	Gang hours	Labour £	Plant £	Material £	Unit	Total rate £
For use in lakes, canals, rivers, industrial lagoons and from silted locations in dock systems						
Floating craft using diesel driven suction method with a 200 mm diameter flexible pipe maximum distance from point of suction to point of discharge 1500 m. Maximum dredge depth 5 m.						
Average pumping output 30 m³ per hour and approximate average costs (excluding mobilization and demobilization costs ranging between £3,000 and £6,000 apiece)	–	–	–	–	m³	**5.25**
EXCAVATION FOR CUTTINGS						
The following unit costs are based on backacter and tractor loader machines						
Excavate topsoil						
maximum depth ne 0.25 m	0.03	0.96	1.58	–	m³	**2.54**
Excavate material other than topsoil, rock or artificial hard material						
ne 0.25 m maximum depth	0.03	0.96	1.58	–	m³	**2.54**
0.25–0.5 m maximum depth	0.03	0.96	1.58	–	m³	**2.54**
0.5–1.0 m maximum depth	0.04	1.28	2.04	–	m³	**3.32**
1.0–2.0 m maximum depth	0.06	1.93	3.10	–	m³	**5.03**
2.0–5.0 m maximum depth	0.10	3.21	5.14	–	m³	**8.35**
5.0–10.0 m maximum depth	0.20	6.42	10.36	–	m³	**16.78**
10.0–15.0 m maximum depth	0.29	9.32	14.98	–	m³	**24.30**
The following unit costs are based on backacter machines fitted with hydraulic breakers and tractor loader machines						
Excavate rock (medium hard)						
ne 0.25 m maximum depth	0.31	9.96	16.02	–	m³	**25.98**
0.25–0.5 m maximum depth	0.43	13.81	22.22	–	m³	**36.03**
0.5–1.0 m maximum depth	0.58	18.63	29.94	–	m³	**48.57**
1.0–2.0 m maximum depth	0.80	25.70	41.40	–	m³	**67.10**
Excavate unreinforced concrete exposed at the commencing surface						
ne 0.25 m maximum depth	0.57	18.31	29.48	–	m³	**47.79**
0.25–0.5 m maximum depth	0.60	19.27	31.00	–	m³	**50.27**
0.5–1.0 m maximum depth	0.67	21.52	34.62	–	m³	**56.14**
1.0–2.0 m maximum depth	0.70	22.49	36.14	–	m³	**58.63**
Excavate reinforced concrete exposed at the commencing surface						
ne 0.25 m maximum depth	0.90	28.91	46.50	–	m³	**75.41**
0.25–0.5 m maximum depth	0.90	28.91	46.50	–	m³	**75.41**
0.5–1.0 m maximum depth	0.95	30.52	49.08	–	m³	**79.60**
1.0–2.0 m maximum depth	1.08	34.69	55.80	–	m³	**90.49**

CLASS E: EXCAVATION

Item	Gang hours	Labour £	Plant £	Material £	Unit	Total rate £
EXCAVATION FOR CUTTINGS – cont						
The following unit costs are based on backacter machines fitted with hydraulic breakers and tractor loader machines – cont						
Excavate unreinforced concrete not exposed at the commencing surface						
ne 0.25 m maximum depth	0.65	20.88	33.58	–	m³	**54.46**
0.25–0.5 m maximum depth	0.67	21.52	34.62	–	m³	**56.14**
0.5–1.0 m maximum depth	0.69	22.16	35.69	–	m³	**57.85**
1.0–2.0 m maximum depth	0.72	23.13	37.20	–	m³	**60.33**
EXCAVATION FOR FOUNDATIONS						
The following unit costs are based on the use of backacter machines						
Excavate topsoil						
maximum depth ne 0.25 m	0.03	0.93	1.58	–	m³	**2.51**
Excavate material other than topsoil, rock or artificial hard material						
0.25–0.5 m deep	0.05	1.55	2.62	–	m³	**4.17**
0.5–1.0 m deep	0.06	1.86	3.10	–	m³	**4.96**
1.0–2.0 m deep	0.07	2.17	3.58	–	m³	**5.75**
2.0–5.0 m deep	0.12	3.72	6.20	–	m³	**9.92**
5.0–10.0 m deep	0.23	7.14	11.92	–	m³	**19.06**
The following unit costs are based on backacter machines fitted with hydraulic breakers						
Excavate unreinforced concrete exposed at the commencing surface						
ne 0.25 m maximum depth	0.60	18.62	31.00	–	m³	**49.62**
0.25–0.5 m maximum depth	0.66	20.48	34.10	–	m³	**54.58**
0.5–1.0 m maximum depth	0.72	22.34	37.20	–	m³	**59.54**
1.0–2.0 m maximum depth	0.80	24.82	41.40	–	m³	**66.22**
Excavate reinforced concrete exposed at the commencing surface						
ne 0.25 m maximum depth	0.94	29.17	48.50	–	m³	**77.67**
0.25–0.5 m maximum depth	0.98	30.41	50.70	–	m³	**81.11**
0.5–1.0 m maximum depth	1.00	31.03	51.64	–	m³	**82.67**
1.0–2.0 m maximum depth	1.04	32.27	53.80	–	m³	**86.07**
Excavate tarmacadam exposed at the commencing surface						
ne 0.25 m maximum depth	0.30	9.31	15.50	–	m³	**24.81**
0.25–0.5 m maximum depth	0.34	10.55	17.50	–	m³	**28.05**
0.5–1.0 m maximum depth	0.37	11.48	19.08	–	m³	**30.56**
1.0–2.0 m maximum depth	0.40	12.41	20.60	–	m³	**33.01**
Excavate unreinforced concrete not exposed at the commencing surface						
ne 0.25 m maximum depth	0.70	21.72	36.10	–	m³	**57.82**
0.25–0.5 m maximum depth	0.85	26.38	43.88	–	m³	**70.26**
0.5–1.0 m maximum depth	0.96	29.79	49.60	–	m³	**79.39**
1.0–2.0 m maximum depth	1.02	31.65	52.70	–	m³	**84.35**

CLASS E: EXCAVATION

Item	Gang hours	Labour £	Plant £	Material £	Unit	Total rate £
GENERAL EXCAVATION						
The following unit costs are based on backacter and tractor loader machines						
Excavate topsoil						
maximum depth ne 0.25 m	0.03	1.03	1.58	–	m^3	**2.61**
Excavate material other than topsoil, rock or artificial hard material						
ne 0.25 m maximum depth	0.03	1.03	1.55	–	m^3	**2.58**
0.25–0.5 m maximum depth	0.03	1.03	1.58	–	m^3	**2.61**
0.5–1.0 m maximum depth	0.04	1.37	2.04	–	m^3	**3.41**
1.0–2.0 m maximum depth	0.06	2.06	3.10	–	m^3	**5.16**
2.0–5.0 m maximum depth	0.10	3.43	5.14	–	m^3	**8.57**
5.0–10.0 m maximum depth	0.20	6.86	10.36	–	m^3	**17.22**
10.0–15.0 m maximum depth	0.29	9.95	14.98	–	m^3	**24.93**
The following unit costs are based on backacter machines fitted with hydraulic breakers and tractor loader machines						
Excavate rock (medium hard)						
ne 0.25 m maximum depth	0.31	10.63	16.02	–	m^3	**26.65**
0.25–0.5 m maximum depth	0.43	14.75	22.22	–	m^3	**36.97**
0.5–1.0 m maximum depth	0.58	19.89	29.94	–	m^3	**49.83**
1.0–2.0 m maximum depth	0.80	27.44	41.40	–	m^3	**68.84**
Excavate unreinforced concrete exposed at the commencing surface						
ne 0.25 m maximum depth	0.57	19.55	29.48	–	m^3	**49.03**
0.25–0.5 m maximum depth	0.60	20.58	31.00	–	m^3	**51.58**
0.5–1.0 m maximum depth	0.67	22.98	34.62	–	m^3	**57.60**
1.0–2.0 m maximum depth	0.70	24.01	36.14	–	m^3	**60.15**
Excavate reinforced concrete exposed at the commencing surface						
ne 0.25 m maximum depth	0.90	30.87	46.50	–	m^3	**77.37**
0.25–0.5 m maximum depth	0.90	30.87	46.50	–	m^3	**77.37**
0.5–1.0 m maximum depth	0.95	32.58	49.08	–	m^3	**81.66**
1.0–2.0 m maximum depth	1.08	37.04	55.80	–	m^3	**92.84**
Excavate unreinforced concrete not exposed at the commencing surface						
ne 0.25 m maximum depth	0.65	22.29	33.58	–	m^3	**55.87**
0.25–0.5 m maximum depth	0.67	22.98	34.62	–	m^3	**57.60**
0.5–1.0 m maximum depth	0.69	23.67	35.69	–	m^3	**59.36**
1.0–2.0 m maximum depth	0.72	24.69	37.20	–	m^3	**61.89**

CLASS E: EXCAVATION

Item	Gang hours	Labour £	Plant £	Material £	Unit	Total rate £
EXCAVATION ANCILLARIES						
The following unit costs are for various machines appropriate to the work						
Trimming topsoil; using D4H dozer						
horizontal	0.04	0.60	0.89	–	m²	**1.49**
10–45° to horizontal	0.05	0.75	1.11	–	m²	**1.86**
Trimming material other than topsoil, rock or artificial hard material; using D4H dozer, tractor loader or motor grader average rate						
horizontal	0.04	0.60	1.20	–	m²	**1.80**
10–45° to horizontal	0.04	0.60	1.20	–	m²	**1.80**
45–90° to horizontal	0.06	0.90	1.79	–	m²	**2.69**
Trimming rock; using D6E dozer						
horizontal	0.08	1.21	30.39	–	m²	**31.60**
10–45° to horizontal	0.80	12.06	30.39	–	m²	**42.45**
45–90° to horizontal	1.06	15.98	40.27	–	m²	**56.25**
vertical	1.06	15.98	40.27	–	m²	**56.25**
Preparation of topsoil; using D4H dozer, tractor loader or motor grader average rate						
horizontal	0.06	0.90	1.62	–	m²	**2.52**
10–45° to horizontal	0.06	0.90	1.62	–	m²	**2.52**
Preparation of material other than rock or artificial hard material; using D6E dozer, tractor loader or motor grader average rate						
horizontal	0.06	0.90	2.11	–	m²	**3.01**
10–45° to horizontal	0.06	0.90	2.11	–	m²	**3.01**
45–90° to horizontal	0.06	0.90	2.11	–	m²	**3.01**
Preparation of rock; using D6E dozer						
horizontal	0.80	12.06	30.39	–	m²	**42.45**
10–45° to horizontal	0.60	9.05	22.79	–	m²	**31.84**
45–90° to horizontal	0.60	9.05	22.79	–	m²	**31.84**
The following unit costs for disposal are based on using a 22.5 t ADT for site work and 20 t tipper for off-site work. The distances used in the calculation are quoted to assist estimating, although this goes beyond the specific requirements of CESMM3						
Disposal of excavated topsoil						
storage on site; 100 m maximum distance	0.05	0.91	2.86	–	m³	**3.77**
removal; 5 km distance	0.11	1.99	7.32	–	m³	**9.31**
removal; 15 km distance	0.21	3.81	13.98	–	m³	**17.79**
Disposal of excavated earth other than rock or artificial hard material						
storage on site; 100 m maximum distance; using 22.5 t ADT	0.05	0.91	2.86	–	m³	**3.77**
removal; 5 km distance; using 20 t tipper	0.12	2.17	7.99	–	m³	**10.16**
removal; 15 km distance; using 20 t tipper	0.22	3.99	14.64	–	m³	**18.63**

CLASS E: EXCAVATION

Item	Gang hours	Labour £	Plant £	Material £	Unit	Total rate £
Disposal of excavated rock or artificial hard material						
storage on site; 100 m maximum distance; using						
22.5 t ADT	0.06	1.09	3.43	–	m³	**4.52**
removal; 5 km distance; using 20 t tipper	0.13	2.36	8.65	–	m³	**11.01**
removal; 15 km distance; using 20 t tipper	0.23	4.17	15.31	–	m³	**19.48**
Add to the above rates where tipping charges apply:						
non-hazardous waste	–	–	–	–	m³	**46.50**
hazardous waste	–	–	–	–	m³	**121.50**
special waste	–	–	–	–	m³	**127.50**
contaminated liquid	–	–	–	–	m³	**168.00**
contaminated sludge	–	–	–	–	m³	**210.00**
Add to the above rates where Landfill Tax applies:						
exempted material	–	–	–	–	m³	**–**
inactive or inert material	–	–	–	–	m³	**3.75**
other material	–	–	–	–	m³	**96.00**
The following unit costs for double handling are based on using a 1.5 m³ tractor loader and 22.5 t ADT. A range of distances is listed to assist in estimating, although this goes beyond the specific requirements of CESMM3						
Double handling of excavated topsoil; using 1.5 m³ tractor loader and 22.5 t ADT						
300 m average distance moved	0.08	1.45	4.82	–	m³	**6.27**
600 m average distance moved	0.10	1.81	6.03	–	m³	**7.84**
1000 m average distance moved	0.12	2.17	7.23	–	m³	**9.40**
Double handling of excavated earth other than rock or artificial hard material; using 1.5 m³ tractor loader and 22.5 t ADT						
300 m average distance moved	0.08	1.45	4.82	–	m³	**6.27**
600 m average distance moved	0.10	1.81	6.03	–	m³	**7.84**
1000 m average distance moved	0.12	2.17	7.23	–	m³	**9.40**
Double handling of rock or artificial hard material; using 1.5 m³ tractor loader and 22.5 t ADT						
300 m average distance moved	0.16	2.90	9.64	–	m³	**12.54**
600 m average distance moved	0.18	3.26	10.85	–	m³	**14.11**
1000 m average distance moved	0.20	3.62	12.05	–	m³	**15.67**
The following unit rates for excavation below Final Surface are based on using a 16 t backacter machine						
Excavation of material below the Final Surface and replacement with:						
granular fill	1.20	17.56	28.56	25.29	m³	**71.41**
concrete Grade C7.5P	0.60	8.78	14.28	82.56	m³	**105.62**
Miscellaneous						
Timber supports left in	0.32	6.16	–	8.12	m²	**14.28**
Metal supports left in	0.89	16.74	–	15.00	m²	**31.74**

CLASS E: EXCAVATION

Item	Gang hours	Labour £	Plant £	Material £	Unit	Total rate £
FILLING						
Excavated topsoil; DfT specified type 5 A						
Filling						
to structures	0.07	2.88	6.83	–	m³	9.71
embankments	0.03	1.01	2.39	–	m³	3.40
general	0.02	0.85	2.01	–	m³	2.86
150 mm thick	0.01	0.28	0.67	–	m²	0.95
250 mm thick	0.01	0.48	1.15	–	m²	1.63
400 mm thick	0.02	0.69	1.62	–	m²	2.31
600 mm thick	0.02	0.97	2.29	–	m²	3.26
Imported topsoil DfT specified type 5B;						
Filling						
embankments	0.03	1.01	2.39	22.36	m³	25.76
general	0.02	0.85	2.01	22.36	m³	25.22
150 mm thick	0.01	0.28	0.67	3.35	m²	4.30
250 mm thick	0.01	0.48	1.15	5.59	m²	7.22
400 mm thick	0.02	0.69	1.62	8.94	m²	11.25
600 mm thick	0.02	0.97	2.29	13.41	m²	16.67
Non-selected excavated material other than topsoil or rock						
Filling						
to structures	0.04	1.45	3.44	–	m³	4.89
embankments	0.02	0.60	1.43	–	m³	2.03
general	0.01	0.44	1.05	–	m³	1.49
150 mm thick	–	0.16	0.38	–	m²	0.54
250 mm thick	0.01	0.24	0.57	–	m²	0.81
400 mm thick	0.01	0.36	0.86	–	m²	1.22
600 mm thick	0.01	0.48	1.15	–	m²	1.63
Selected excavated material other than topsoil or rock						
Filling						
to structures	0.04	1.65	3.92	–	m³	5.57
embankments	0.01	0.56	1.34	–	m³	1.90
general	0.01	0.48	1.15	–	m³	1.63
150 mm thick	–	0.16	0.38	–	m²	0.54
250 mm thick	0.01	0.28	0.67	–	m²	0.95
400 mm thick	0.01	0.40	0.95	–	m²	1.35
600 mm thick	0.01	0.52	1.24	–	m²	1.76
Imported natural material other than topsoil or rock; subsoil						
Filling						
to structures	0.04	1.45	3.44	23.25	m³	28.14
embankments	0.02	0.60	1.43	23.25	m³	25.28
general	0.02	0.60	1.43	23.25	m³	25.28
150 mm thick	0.01	0.36	0.86	3.49	m²	4.71
250 mm thick	0.01	0.52	1.24	5.81	m²	7.57
400 mm thick	0.02	0.60	1.43	9.30	m²	11.33
600 mm thick	0.02	0.69	1.62	13.95	m²	16.26

CLASS E: EXCAVATION

Item	Gang hours	Labour £	Plant £	Material £	Unit	Total rate £
Imported natural material other than topsoil or rock; granular graded material						
Filling						
to structures	0.04	1.45	3.44	28.04	m³	**32.93**
embankments	0.02	0.69	1.62	28.04	m³	**30.35**
general	0.02	0.69	1.62	28.04	m³	**30.35**
150 mm thick	0.01	0.48	1.15	4.21	m²	**5.84**
250 mm thick	0.02	0.69	1.62	7.01	m²	**9.32**
400 mm thick	0.02	0.81	1.91	11.22	m²	**13.94**
600 mm thick	0.02	0.89	2.10	16.83	m²	**19.82**
Imported natural material other than topsoil or rock; granular selected material						
Filling						
to structures	0.04	1.45	3.44	24.21	m³	**29.10**
embankments	0.02	0.69	1.62	24.21	m³	**26.52**
general	0.02	0.69	1.62	24.21	m³	**26.52**
150 mm thick	0.01	0.48	1.15	3.63	m²	**5.26**
250 mm thick	0.02	0.69	1.62	6.05	m²	**8.36**
400 mm thick	0.02	0.81	1.91	9.68	m²	**12.40**
600 mm thick	0.02	0.89	2.10	14.53	m²	**17.52**
Excavated rock						
Filling						
to structures	0.04	1.61	3.82	–	m³	**5.43**
embankments	0.05	2.02	4.78	–	m³	**6.80**
general	0.05	2.02	4.78	–	m³	**6.80**
150 mm thick	0.02	0.69	1.62	–	m²	**2.31**
250 mm thick	0.03	1.05	2.48	–	m²	**3.53**
400 mm thick	0.04	1.61	3.82	–	m²	**5.43**
600 mm thick	0.06	2.22	5.25	–	m²	**7.47**
Imported rock						
Filling						
to structures	0.04	1.61	3.82	53.30	m³	**58.73**
embankments	0.02	0.97	2.29	53.30	m³	**56.56**
general	0.02	0.81	1.91	53.30	m³	**56.02**
150 mm thick	0.01	0.48	1.15	7.99	m²	**9.62**
250 mm thick	0.02	0.81	1.91	13.33	m²	**16.05**
400 mm thick	0.03	1.21	2.87	21.32	m²	**25.40**
600 mm thick	0.03	1.21	2.87	31.98	m²	**36.06**

CLASS E: EXCAVATION

Item	Gang hours	Labour £	Plant £	Material £	Unit	Total rate £
FILLING ANCILLARIES						
Trimming of filled surfaces						
Topsoil						
horizontal	0.02	0.38	1.46	–	m²	**1.84**
inclined at an angle of 10–45° to horizontal	0.02	0.38	1.46	–	m²	**1.84**
inclined at an angle of 45–90° to horizontal	0.03	0.49	1.88	–	m²	**2.37**
Material other than topsoil, rock or artificial hard material						
horizontal	0.02	0.38	1.46	–	m²	**1.84**
inclined at an angle of 10–45° to horizontal	0.02	0.38	1.46	–	m²	**1.84**
inclined at an angle of 45–90° to horizontal	0.03	0.49	1.88	–	m²	**2.37**
Rock						
horizontal	0.51	9.24	35.42	–	m²	**44.66**
inclined at an angle of 10–45° to horizontal	0.52	9.42	36.12	–	m²	**45.54**
inclined at an angle of 45–90° to horizontal	0.70	12.68	48.62	–	m²	**61.30**
Preparation of filled surfaces						
Topsoil						
horizontal	0.03	0.54	2.08	–	m²	**2.62**
inclined at an angle of 10–45° to horizontal	0.03	0.54	2.08	–	m²	**2.62**
inclined at an angle of 45–90° to horizontal	0.04	0.67	2.57	–	m²	**3.24**
Material other than topsoil, rock or artificial hard material						
horizontal	0.03	0.54	2.08	–	m²	**2.62**
inclined at an angle of 10–45° to horizontal	0.03	0.54	2.08	–	m²	**2.62**
inclined at an angle of 45–90° to horizontal	0.04	0.67	2.57	–	m²	**3.24**
Rock						
horizontal	0.33	5.98	22.92	–	m²	**28.90**
inclined at an angle of 10–45° to horizontal	0.33	5.98	22.92	–	m²	**28.90**
inclined at an angle of 45–90° to horizontal	0.52	9.42	36.12	–	m²	**45.54**
GEOTEXTILES						
Notes						
The geotextile products mentioned below are not specifically confined to the individual uses stated but are examples of one of many scenarios to which they may be applied. Conversely, the scenarios are not limited to the geotextile used as an example.						
Geotextiles; stabilization applications for reinforcement of granular subbases, capping layers and railway ballast placed over weak and variable soils						
For use over weak soils with moderate traffic intensities e.g. car parks, light access roads; Tensar SS20 Polypropylene Geogrid						
horizontal	0.04	1.19	–	2.07	m²	**3.26**
inclined at an angle of 10 to 45° to the horizontal	0.05	1.50	–	2.07	m²	**3.57**

CLASS E: EXCAVATION

Item	Gang hours	Labour £	Plant £	Material £	Unit	Total rate £
For use over weak soils with high traffic intensities and/or high axle loadings; Tensar SS30 Polypropylene Geogrid						
horizontal	0.05	1.85	–	2.11	m²	**3.96**
inclined at an angle of 10–45° to the horizontal	0.06	2.30	–	2.11	m²	**4.41**
For use over very weak soils e.g. alluvium, marsh or peat or firmer soil subject to exceptionally high axle loadings;Tensar SS40 Polypropylene Geogrid						
horizontal	0.05	2.02	2.85	3.73	m²	**8.60**
inclined at an angle of 10–45° to the horizontal	0.06	2.51	3.55	3.73	m²	**9.79**
For trafficked areas where fill comprises aggregate exceeding 100 mm; Tensar SSLA20 Polypropylene Geogrid						
horizontal	0.04	1.19	–	3.10	m²	**4.29**
inclined at an angle 10–45° to the horizontal	0.05	1.50	–	3.10	m²	**4.60**
Stabilization and separation of granular fill from soft subgrade to prevent intermixing: Terram 1000						
horizontal	0.05	2.09	–	0.55	m²	**2.64**
inclined at an angle of 10–45° to the horizontal	0.06	2.63	–	0.55	m²	**3.18**
Stabilization and separation of granular fill from soft subgrade to prevent intermixing: Terram 2000						
horizontal	0.04	1.89	2.66	1.32	m²	**5.87**
inclined at an angle of 10–45° to the horizontal	0.05	2.38	3.36	1.32	m²	**7.06**
Geotextiles; reinforcement applications for asphalt pavements						
For roads, hardstandings and airfield pavements; Tensar AR-G composite comprising Tensar AR-1 grid bonded to a geotextile, laid within asphalt						
horizontal	0.05	1.85	–	2.79	m²	**4.64**
inclined at an angle of 10–45° to the horizontal	0.06	2.30	–	2.79	m²	**5.09**
Geotextiles; slope reinforcement and embankment support; for use where soils can only withstand limited shear stresses, therefore steep slopes require external support						
Paragrid 30/155; 330 g/m²						
horizontal	0.04	1.19	–	2.24	m²	**3.43**
inclined at an angle of 10–45° to the horizontal	0.05	1.50	–	2.24	m²	**3.74**
Paragrid 100/255; 330 g/m²						
horizontal	0.04	1.19	–	2.92	m²	**4.11**
inclined at an angle of 10–45° to the horizontalhorizontal	0.05	1.50	–	2.92	m²	**4.42**
Paralink 200 s; 1120 g/m²						
horizontal	0.05	2.42	3.42	4.56	m²	**10.40**
inclined at an angle of 10–45° to the horizontal	0.07	3.05	4.31	4.56	m²	**11.92**
Paralink 600 s; 2040 g/m²						
horizontal	0.06	2.83	3.99	10.62	m²	**17.44**
inclined at an angle of 10–45° to the horizontal	0.08	3.55	5.01	10.62	m²	**19.18**
Terram grid 30/30						
horizontal	0.06	2.56	3.61	0.01	m²	**6.18**
inclined at an angle of 10–45° to the horizontal	0.07	3.19	4.50	0.01	m²	**7.70**

CLASS E: EXCAVATION

Item	Gang hours	Labour £	Plant £	Material £	Unit	Total rate £
GEOTEXTILES – cont						
Geotextiles; scour and erosion protection						
For use where erosion protection is required to the surface of a slope once its geotechnical stability has been achieved, and to allow grass establishment; Tensar 'Mat' Polyethelene mesh; fixed with Tensar pegs						
horizontal	0.04	1.68	–	4.30	m²	**5.98**
inclined at an angle of 10–45° to the horizontal	0.05	2.09	–	4.30	m²	**6.39**
For use where hydraulic action exists, such as coastline protection from pressures exerted by waves, currents and tides; Typar SF56						
horizontal	0.05	2.29	3.23	0.49	m²	**6.01**
inclined at an angle of 10–45° to the horizontal	0.06	2.87	4.06	0.49	m²	**7.42**
For protection against puncturing to reservoir liner; Typar SF563						
horizontal	0.05	2.29	3.23	0.49	m²	**6.01**
inclined at an angle of 10–45° to the horizontal	0.06	2.87	4.06	0.49	m²	**7.42**
Geotextiles; temporary parking areas						
For reinforcement of grassed areas subject to wear from excessive pedestrian and light motor vehicle traffic; Netlon CE131 high density polyethelyene geogrid						
horizontal	0.04	1.25	–	3.79	m²	**5.04**
Geotextiles; landscaping applications						
For prevention of weed growth in planted areas by incorporating a geotextile over top soil and below mulch or gravel; Typar SF20						
horizontal	0.08	3.08	–	0.29	m²	**3.37**
inclined at an angle of 10–45° to the horizontal	0.09	3.86	–	0.29	m²	**4.15**
For root growth control, prevention of lateral spread of roots and mixing of road base and humus; Typar SF20						
horizontal	0.08	3.08	–	0.29	m²	**3.37**
inclined at an angle of 10–45° to the horizontal	0.09	3.86	–	0.29	m²	**4.15**
Geotextiles; drainage applications						
For clean installation of pipe support material and to prevent silting of the drainage pipe and minimizing differential settlement; Typar SF10						
horizontal	0.04	1.68	–	0.32	m²	**2.00**
inclined at an angle of 10–45° to the horizontal	0.05	2.09	–	0.32	m²	**2.41**
For wrapping to prevent clogging of drainage pipes surrounded by fine soil; Typar SF10						
sheeting	0.08	3.28	–	0.64	m²	**3.92**

Unit Costs – Civil Engineering Works

CLASS E: EXCAVATION

Item	Gang hours	Labour £	Plant £	Material £	Unit	Total rate £
For wrapping to prevent clogging of drainage pipes surrounded by fine soil; Terram 1000						
sheeting	0.08	3.28	–	0.60	m²	**3.88**
For vertical structure drainage to subsurface walls, roofs and foundations; Filtram 1B1						
sheeting	0.08	3.28	–	3.65	m²	**6.93**
For waterproofing to tunnels, buried structures, etc. where the membrane is buried, forming part of the drainage system; Filtram 1BZ						
sheeting	0.08	3.28	–	6.68	m²	**9.96**
Geotextiles; roofing insulation and protection						
Protection of waterproofing membrane from physical damage and puncturing; Typar SF56						
sheeting	0.05	2.29	3.23	0.49	m²	**6.01**
Internal reinforcement of in situ spread waterproof bitumen emulsion; Typar SF10						
sheeting	0.07	2.71	–	0.32	m²	**3.03**
LANDSCAPING						
Preparatory operations prior to landscaping						
Supply and apply granular cultivation treatments by hand						
35 g/m²	0.50	13.16	–	0.37	100 m²	**13.53**
50 g/m²	0.65	17.11	–	0.53	100 m²	**17.64**
75 g/m²	0.85	22.37	–	0.73	100 m²	**23.10**
100 g/m²	1.00	26.32	–	1.05	100 m²	**27.37**
Supply and apply granular cultivation treatments by machine in suitable economically large areas						
100 g/m²	0.25	3.72	1.57	1.05	100 m²	**6.34**
Supply and incorporate cultivation additives into top 150 mm of topsoil by hand						
1 m³/10 m²	20.00	526.40	–	231.00	100 m²	**757.40**
1 m³/13 m²	20.00	526.40	–	177.64	100 m²	**704.04**
1 m³/20 m²	19.00	500.08	–	115.50	100 m²	**615.58**
1 m³/40 m²	17.00	447.44	–	57.75	100 m²	**505.19**
Supply and incorporate cultivation additives into top 150 mm of topsoil by machine in suitable economically large areas						
1 m³/10 m²	–	–	233.90	231.00	100 m²	**464.90**
1 m³/13 m²	–	–	215.90	177.64	100 m²	**393.54**
1 m³/20 m²	–	–	191.93	115.50	100 m²	**307.43**
1 m³/40 m²	–	–	176.93	57.75	100 m²	**234.68**
Turfing						
Turfing						
horizontal	0.12	3.16	–	6.15	m²	**9.31**
10–45° to horizontal	0.17	4.47	–	6.15	m²	**10.62**
45–90° to horizontal; pegging down	0.19	5.00	–	6.15	m²	**11.15**

CLASS E: EXCAVATION

Item	Gang hours	Labour £	Plant £	Material £	Unit	Total rate £
LANDSCAPING – cont						
Hydraulic mulch grass seeding						
Grass seeding						
horizontal	0.01	0.13	–	1.65	m²	1.78
10–45° to horizontal	0.01	0.18	–	1.65	m²	1.83
45–90° to horizontal	0.01	0.24	–	1.65	m²	1.89
Selected grass seeding						
Grass seeding; sowing at the rate of 0.050 kg/m² in two operations						
horizontal	0.01	0.26	–	0.51	m²	0.77
10–45° to horizontal	0.02	0.39	–	0.51	m²	0.90
45–90° to horizontal	0.02	0.53	–	0.51	m²	1.04
Plants						
Form planting hole in previously cultivated area, supply and plant specified herbaceous plants and backfill with excavated material						
5 plants/m²	0.01	0.26	–	3.72	m²	3.98
10 plants/m²	0.02	0.58	–	13.64	m²	14.22
25 plants/m²	0.05	1.32	–	31.76	m²	33.08
35 plants/m²	0.07	1.84	–	44.47	m²	46.31
50 plants/m²	0.10	2.63	–	63.53	m²	66.16
Supply and fix plant support netting on 50 mm diameter stakes 750 mm long driven into the ground at 1.5 m centres						
1.15 m high green extruded plastic mesh, 125 mm square mesh	0.06	1.58	–	5.00	m²	6.58
Form planting hole in previously cultivated area; supply and plant bulbs and backfill with excavated material						
small	0.01	0.26	–	0.15	each	0.41
medium	0.01	0.26	–	0.24	each	0.50
large	0.01	0.26	–	0.26	each	0.52
Supply and plant bulbs in grassed area using bulb planter and backfill with screened topsoil or peat and cut turf plug						
small	0.01	0.26	–	0.15	each	0.41
medium	0.01	0.26	–	0.24	each	0.50
large	0.01	0.26	–	0.26	each	0.52
Shrubs						
Form planting hole in previously cultivated area, supply and plant specified shrub and backfill with excavated material						
shrub 300 mm high	0.01	0.26	–	6.15	each	6.41
shrub 600 mm high	0.01	0.26	–	6.66	each	6.92
shrub 900 mm high	0.01	0.26	–	7.28	each	7.54
shrub 1 m high and over	0.01	0.26	–	7.69	each	7.95
Supply and fix shrub stake including two ties						
one stake; 1.5 m long, 75 mm diameter	0.12	3.16	–	8.20	each	11.36

CLASS E: EXCAVATION

Item	Gang hours	Labour £	Plant £	Material £	Unit	Total rate £
Trees						
The cost of planting semi-mature trees will depend on the size and species, and on the access to the site for tree handling machines. Prices should be obtained for individual trees and planting.						
Break up subsoil to a depth of 200 mm						
in tree pit	0.05	1.32	–	–	each	**1.32**
Supply and plant tree in prepared pit; backfill with excavated topsoil minimum 600 mm deep						
light standard tree	0.25	6.58	–	52.50	each	**59.08**
standard tree	0.45	11.84	–	65.10	each	**76.94**
selected standard tree	0.75	19.74	–	102.90	each	**122.64**
heavy standard tree	0.85	22.37	–	115.50	each	**137.87**
extra heavy standard tree	1.50	39.48	–	152.25	each	**191.73**
Extra for filling with topsoil from spoil heap ne 100 m distant	0.15	3.95	–	–	m^3	**3.95**
Extra for filling with imported topsoil	0.08	2.11	–	26.25	m^3	**28.36**
Supply tree stake and drive 500 mm into firm ground and trim to approved height, including two tree ties to approved pattern						
one stake; 2.4 m long, 100 mm diameter	0.16	4.21	–	5.78	each	**9.99**
one stake; 3.0 m long, 100 mm diameter	0.20	5.26	–	6.04	each	**11.30**
two stakes; 2.4 m long, 100 mm diameter	0.24	6.32	–	8.40	each	**14.72**
two stakes; 3.0 m long, 100 mm diameter	0.30	7.90	–	8.93	each	**16.83**
Supply and fit tree support comprising three collars and wire guys; including pickets						
galvanized steel 50 × 600 mm	1.50	39.48	–	21.00	each	**60.48**
hardwood 75 × 600 mm	1.50	39.48	–	22.50	each	**61.98**
Supply and fix standard steel tree guard	0.30	7.90	–	21.00	each	**28.90**
Hedges						
Excavate trench by hand for hedge and deposit soil alongside trench						
300 wide × 300 mm deep	0.10	2.63	–	–	m	**2.63**
450 wide × 300 mm deep	0.13	3.42	–	–	m	**3.42**
Excavate trench by machine for hedge and deposit soil alongside trench						
300 wide × 300 mm deep	0.02	0.53	–	–	m	**0.53**
450 wide × 300 mm deep	0.02	0.53	–	–	m	**0.53**
Set out, nick out and excavate trench and break up subsoil to minimum depth of 300 mm						
400 mm minimum deep	0.28	7.34	–	–	m	**7.34**
Supply and plant hedging plants, backfill with excavated topsoil						
single row; plants at 200 mm centres	0.25	6.58	–	15.22	m	**21.80**
single row; plants at 300 mm centres	0.17	4.47	–	10.14	m	**14.61**
single row; plants at 400 mm centres	0.13	3.29	–	7.61	m	**10.90**
single row; plants at 500 mm centres	0.10	2.63	–	6.09	m	**8.72**
single row; plants at 600 mm centres	0.08	2.11	–	5.05	m	**7.16**
double row; plants at 200 mm centres	0.50	13.16	–	30.45	m	**43.61**
double row; plants at 300 mm centres	0.34	8.95	–	20.28	m	**29.23**

CLASS E: EXCAVATION

Item	Gang hours	Labour £	Plant £	Material £	Unit	Total rate £
LANDSCAPING – cont						
Hedges – cont						
Supply and plant hedging plants, backfill with excavated topsoil – cont						
double row; plants at 400 mm centres	0.25	6.58	–	15.22	m	**21.80**
double row; plants at 500 mm centres	0.20	5.26	–	12.18	m	**17.44**
Extra for incorporating manure at the rate of 1 m^3 per 30 m of trench	0.12	3.16	–	0.12	m	**3.28**
COMPARATIVE COSTS – EARTH MOVING						
Notes						
The cost of earth moving and other associated works is dependent on matching the overall quantities and production rate called for by the programme of works with the most appropriate plant and assessing the most suitable version of that plant which will:						
* deal with the site conditions (e.g. type of ground, type of excavation, length of haul, prevailing weather, etc.)						
* comply with the specification requirements (e.g. compaction, separation of materials, surface tolerances, etc.)						
* complete the work economically (e.g. provide surface tolerances which will avoid undue expense of imported materials)						
Labour costs are based on a plant operative skill rate 3 unless otherwise stated.						
Comparative costs of excavation equipment						
The following are comparative costs using various types of excavation equipment and include loading into transport. All costs assume 50 minutes productive work in every hour and adequate disposal transport being available to obviate any delay.						
Dragline (for excavations requiring a long reach and long discharge, mainly in clearing streams and rivers)						
bucket capacity ne 1.15 m^3	0.06	1.09	3.60	–	m^3	**4.69**
bucket capacity ne 2.00 m^3	0.03	0.54	3.12	–	m^3	**3.66**
bucket capacity ne 3.00 m^3	0.02	0.36	2.78	–	m^3	**3.14**
Hydraulic backacter (for all types of excavation and loading, including trenches, breaking hard ground, etc.)						
bucket capacity ne 0.40 m^3	0.05	0.83	1.85	–	m^3	**2.68**
bucket capacity ne 1.00 m^3	0.02	0.36	1.11	–	m^3	**1.47**
bucket capacity ne 1.60 m^3	0.01	0.18	1.09	–	m^3	**1.27**

CLASS E: EXCAVATION

Item	Gang hours	Labour £	Plant £	Material £	Unit	Total rate £
Hydraulic face shovel (predominantly for excavating cuttings and embankments over 2 m high requiring high output)						
bucket capacity ne 1.50 m^3	0.04	0.72	2.96	–	m^3	**3.68**
bucket capacity ne 2.60 m^3	0.02	0.36	2.04	–	m^3	**2.40**
bucket capacity ne 3.40 m^3	0.01	0.18	1.46	–	m^3	**1.64**
Tractor loader (for loading, carrying, placing materials, spreading and levelling, some site clearance operations and reducing levels)						
bucket capacity ne 0.80 m^3	0.02	0.36	0.77	–	m^3	**1.13**
bucket capacity ne 1.50 m^3	0.01	0.18	0.63	–	m^3	**0.81**
bucket capacity ne 3.00 m^3	0.01	0.13	0.88	–	m^3	**1.01**
Multipurpose wheeled loader/backhoe (versatile machine for small to medium excavations, trenches, loading, carrying and back filling)						
bucket capacity ne 0.76 m^3	0.08	1.45	1.83	–	m^3	**3.28**
bucket capacity ne 1.00 m^3	0.06	1.09	1.90	–	m^3	**2.99**
Comparative costs of transportation equipment						
The following are comparative costs for using various types of transportation equipment to transport excavated loose material. The capacity of the transport must be suitable for the output of the loading machine. The cost will vary depending on the number of transport units required to meet the output of the loading unit and the distance to be travelled.						
Loading loose material into transport by wheeled loader						
ne 2.1 m^3 capacity	0.02	0.36	0.99	–	m^3	**1.35**
ne 5.4 m^3 capacity	0.01	0.18	1.50	–	m^3	**1.68**
ne 10.5 m^3 capacity	0.01	0.13	1.29	–	m^3	**1.42**
Transport material within site by dump truck (rear dump)						
ne 24 m^3 heaped capacity, distance travelled ne 0.5 km	0.03	0.54	3.48	–	m^3	**4.02**
Add per 0.5 km additional distance	0.01	0.18	1.16	–	m^3	**1.34**
ne 57 m^3 heaped capacity, distance travelled ne 0.5 km	0.03	0.60	3.26	–	m^3	**3.86**
Add per 0.5 km additional distance	0.03	0.60	1.63	–	m^3	**2.23**
Transport material within site by dump truck (articulated)						
ne 32 t payload, distance travelled ne 0.5 km	0.04	0.79	2.53	–	m^3	**3.32**
Add per 0.5 km additional distance	0.02	0.40	1.27	–	m^3	**1.67**
Transport material within or off site by tipping lorry						
ne 10 t payload, distance travelled ne 1 km	0.05	0.83	1.98	–	m^3	**2.81**
Add per 1 km additional distance	0.03	0.41	0.99	–	m^3	**1.40**
10–15 t payload, distance travelled ne 1 km	0.04	0.66	2.06	–	m^3	**2.72**
Add per 1 km additional distance	0.02	0.33	1.03	–	m^3	**1.36**
15–25 t payload, distance travelled ne 1 km	0.03	0.54	1.84	–	m^3	**2.38**
Add per 1 km additional distance	0.02	0.27	0.92	–	m^3	**1.19**

CLASS E: EXCAVATION

Item	Gang hours	Labour £	Plant £	Material £	Unit	Total rate £
COMPARATIVE COSTS – EARTH MOVING – cont						
Comparative costs of earth moving equipment						
The following are comparative costs using various types of earth moving equipment to include excavation, transport, spreading and levelling. Bulldozer up to 74 kW (CAT D4H sized machine used for site strip, reducing levels or grading and spreading materials over smaller sites)						
average push one way 10 m	0.03	0.50	1.33	–	m^3	1.83
average push one way 30 m	0.08	1.32	3.56	–	m^3	4.88
average push one way 50 m	0.14	2.32	6.23	–	m^3	8.55
average push one way 100 m	0.29	4.80	12.90	–	m^3	17.70
Bulldozer up to 104 kW (CAT D6E sized machine for reducing levels, excavating to greater depths at steeper inclines, grading surfaces, small cut and fill operations; maximum push 100 m)						
average push one way 10 m	0.03	0.54	2.28	–	m^3	2.82
average push one way 30 m	0.07	1.27	5.32	–	m^3	6.59
average push one way 50 m	0.09	1.63	6.84	–	m^3	8.47
average push one way 100 m	0.19	3.44	14.44	–	m^3	17.88
Bull or Angle Dozer up to 212 kW (CAT D8N sized machine for high output, ripping and excavating by reducing levels at steeper inclines or in harder material than with D6E, larger cut and fill operations used in conjunction with towed or S.P. scrapers. Spreading and grading over large areas; maximum push 100 m)						
average push one way 10 m	0.03	0.54	4.22	–	m^3	4.76
average push one way 30 m	0.06	1.09	8.44	–	m^3	9.53
average push one way 50 m	0.07	1.27	9.85	–	m^3	11.12
average push one way 100 m	0.11	1.99	15.47	–	m^3	17.46
Motorized scraper, 15 m^3 capacity (for excavating larger volumes over large haul lengths, excavating to reduce levels and also levelling ground, grading large sites, moving and tipping material including hard material – used in open cast sites)						
average haul one way 500 m	–	0.07	0.54	–	m^3	0.61
average haul one way 1,000 m	0.01	0.11	0.81	–	m^3	0.92
average haul one way 2,000 m	0.01	0.14	1.08	–	m^3	1.22
average haul one way 3,000 m	0.01	0.18	1.35	–	m^3	1.53
Twin engined motorized scraper, 16 m^3 capacity						
average haul one way 500 m	–	0.05	0.53	–	m^3	0.58
average haul one way 1,000 m	–	0.07	0.71	–	m^3	0.78
average haul one way 2,000 m	0.01	0.09	0.88	–	m^3	0.97
average haul one way 3,000 m	0.01	0.13	1.23	–	m^3	1.36
Twin engined motorized scraper, 34 m^3 capacity						
average haul one way 500 m	–	0.04	0.75	–	m^3	0.79
average haul one way 1,000 m	–	0.05	1.12	–	m^3	1.17
average haul one way 2,000 m	–	0.05	1.12	–	m^3	1.17
average haul one way 3,000 m	–	0.07	1.50	–	m^3	1.57

CLASS E: EXCAVATION

Item	Gang hours	Labour £	Plant £	Material £	Unit	Total rate £
Excavation by hand						
Desirable for work around live services or in areas of highly restricted access.						
Excavate and load into skip or dumper bucket						
loose material	2.02	25.67	–	–	m³	**25.67**
compacted soil or clay	3.15	40.04	–	–	m³	**40.04**
mass concrete or sandstone	3.00	38.13	–	–	m³	**38.13**
broken rock	2.95	37.49	–	–	m³	**37.49**
existing subbase or pipe surrounds	3.12	39.66	–	–	m³	**39.66**
Excavate by hand using pneumatic equipment						
Excavate below ground using 1.80 m³/min single tool compressor and pneumatic breaker and load material into skip or dumper bucket						
rock (medium drill)	2.12	26.95	15.39	–	m³	**42.34**
brickwork or mass concrete	2.76	35.08	19.76	–	m³	**54.84**
reinforced concrete	3.87	49.19	26.27	–	m³	**75.46**
asphalt in carriageways	1.15	14.62	11.55	–	m³	**26.17**
Comparative prices for ancillary equipment						
Excavate using 6 t to break out (JCB 3CX and Montalbert 125 breaker)						
medium hard rock	0.43	7.12	16.63	–	m³	**23.75**
brickwork or mass concrete	0.54	8.94	20.25	–	m³	**29.19**
reinforced concrete	0.64	10.59	24.00	–	m³	**34.59**
Load material into skip or dumper						
Load material into skip or dumper bucket using a 11.5 t crawler backacter with 0.80 m³ rock bucket						
medium hard rock	0.07	1.16	2.81	–	m³	**3.97**
brickwork or mass concrete	0.08	1.32	3.20	–	m³	**4.52**
reinforced concrete	0.09	1.49	3.61	–	m³	**5.10**
DRILLING AND BLASTING IN ROCK						
The cost of blasting is controlled by the number of holes and the length of drilling required to achieve the tolerances and degree of shatter required, e.g. line drilling to trenches, depth of drilling to control horizontal overbreak.						
Drilling with rotary percussion drills						
105–110 mm diameter; hard rock	–	–	–	–	m	**14.00**
105–110 mm diameter; sandstone	–	–	–	–	m	**7.05**
125 mm diameter; hard rock	–	–	–	–	m	**11.05**
125 mm diameter; sandstone	–	–	–	–	m	**10.00**
165 mm diameter; hard rock	–	–	–	–	m	**12.65**
165 mm diameter; sandstone	–	–	–	–	m	**12.65**

CLASS E: EXCAVATION

Item	Gang hours	Labour £	Plant £	Material £	Unit	Total rate £
DRILLING AND BLASTING IN ROCK – cont						
Drilling and blasting in open cut for bulk excavation excluding cost of excavatio or trimming						
hard rock	–	–	–	–	m	3.20
sandstone	–	–	–	–	m	4.35
Drilling and blasting for quarry operations with face height exceeding 10 m						
hard rock	–	–	–	–	m	3.05
sandstone	–	–	–	–	m	3.05
Drilling and blasting in trenches excluding cost of excavation or trimming						
trench width ne 1.0 m	–	–	–	–	m	24.01
trench width 1.0–1.5 m	–	–	–	–	m	21.00
trench width over 1.5 m	–	–	–	–	m	17.65
Secondary blasting to boulders						
pop shooting	–	–	–	–	m	7.50
plaster shooting	–	–	–	–	m	4.05
DEWATERING						
The following unit costs are for dewatering pervious ground only and are for sets of equipment comprising:						
1 nr diesel driven pump (WP 150/60 or similar) complete with allowance of £50 for fuel	–	–	–	–	day	92.00
50 m of 150 mm diameter header pipe	–	–	–	–	day	17.50
35 nr of disposable well points	–	–	–	–	buy	280.00
18 m of delivery pipe	–	–	–	–	day	8.00
1 nr diesel driven standby pump	–	–	–	–	day	55.00
1 nr jetting pump with hoses (for installation of wellpoints only)	–	–	–	–	–	92.00
attendant labour and plant (2 hrs per day inclusive of small dumper and bowser)	–	–	–	–	–	27.00
Costs are based on 10 hr shifts with attendant labour and plant (specialist advice)						
Guide price for single set of equipment comprising pump, 150 mm diameter header pipe, 35 nr well points, delivery pipes and attendant labour and plant						
Bring to site equipment and remove upon completion	–	–	–	–	sum	5000.00
Installation costs						
hire of jetting pump with hoses; 1 day	–	–	–	–	sum	92.00
purchase of well points; 35 nr	–	–	–	–	sum	280.00
labour and plant; 10 hours	–	–	–	–	sum	270.00
Operating costs						
hire of pump, header pipe, delivery pipe and standby pump complete with fuel etc. and 2 hours attendant labour and plant	–	–	–	–	day	226.50

CLASS F: IN SITU CONCRETE

Item	Gang hours	Labour £	Plant £	Material £	Unit	Total rate £
NOTES						
The unit rates in this section are based on nett measurements and appropriate adjustments should be made for unmeasured excess (e.g. additional blinding thickness as a result of ground conditions). The unit rates for the provision of concrete are based on ready mixed concrete in full loads delivered to site within 5 miles (8 km) of the concrete mixing plant and include an allowance for wastage prior to placing.						
This section assumes optimum outputs of an efficiently controlled pour with no delays caused by out of sequence working and no abnormal conditions that would affect continuity of work.						
RESOURCES – LABOUR						
Concrete gang						
1 ganger or chargehand (skill rate 4)		14.56				
2 skilled operatives (skill rate 4)		27.22				
4 unskilled operatives (general)		50.84				
1 plant operator (skill rate 3) – 25% of time		4.14				
Total Gang Rate/Hour	£	**96.76**				
RESOURCES – MATERIALS						
The following costs do not reflect in the rates generally shown later and should be considered separately						
Delivery to site for each additional mile from the concrete plant further than 5 miles (8 km)	–	–	–	2.00	m^3	2.00
Mix design, per trial mix	–	–	–	245.00	mix	245.00
Part loads, per m^3 below full load	–	–	–	50.00	m^3	50.00
Waiting time (in excess of 6 mins/m^3 'norm' discharge time)	–	–	–	85.00	hr	85.00
Making and testing concrete cube	0.69	8.74	–	–	nr	8.74
Pumping from ready mix truck to point of placing at the rate of 25 m^3 / hour	0.11	1.33	6.95	–	m^3	8.28
Pumping from ready mix truck to point of placing at the rate of 45 m^3 / hour	0.05	0.64	7.32	–	m^3	7.96
RESOURCES – PLANT						
Concrete						
10 t crane (50% of time)			21.30			
1.00 m^3 concrete skip (50% of time)			1.25			
11.30 m^3/min compressor, 2 tool			34.98			
four 54 mm poker vibrators			5.50			
Total Rate/Hour		£	**63.02**			

CLASS F: IN SITU CONCRETE

Item	Gang hours	Labour £	Plant £	Material £	Unit	Total rate £
PROVISION OF CONCRETE						
Standard mix; cement to BS EN 197–1						
Type ST1	–	–	–	81.71	m³	81.71
Type ST2	–	–	–	82.74	m³	82.74
Type ST3	–	–	–	83.78	m³	83.78
Type ST4	–	–	–	84.81	m³	84.81
Type ST5	–	–	–	85.85	m³	85.85
Standard mix; sulphate-resisting cement to BS 4027						
Type ST1	–	–	–	84.86	m³	84.86
Type ST2	–	–	–	85.89	m³	85.89
Type ST3	–	–	–	86.93	m³	86.93
Type ST4	–	–	–	87.96	m³	87.96
Type ST5	–	–	–	89.00	m³	89.00
Designed mix; cement to BS EN 197–1						
Grade C7.5						
10 mm aggregate	–	–	–	82.95	m³	82.95
20 mm aggregate	–	–	–	81.71	m³	81.71
40 mm aggregate	–	–	–	80.48	m³	80.48
Grade C10						
10 mm aggregate	–	–	–	84.00	m³	84.00
20 mm aggregate	–	–	–	82.74	m³	82.74
40 mm aggregate	–	–	–	81.50	m³	81.50
Grade C15						
10 mm aggregate	–	–	–	85.05	m³	85.05
20 mm aggregate	–	–	–	83.78	m³	83.78
40 mm aggregate	–	–	–	82.52	m³	82.52
Grade C20						
10 mm aggregate	–	–	–	86.10	m³	86.10
20 mm aggregate	–	–	–	84.81	m³	84.81
40 mm aggregate	–	–	–	83.54	m³	83.54
Grade C25						
10 mm aggregate	–	–	–	87.15	m³	87.15
20 mm aggregate	–	–	–	85.85	m³	85.85
40 mm aggregate	–	–	–	84.56	m³	84.56
Grade C30						
10 mm aggregate	–	–	–	90.30	m³	90.30
20 mm aggregate	–	–	–	89.46	m³	89.46
40 mm aggregate	–	–	–	88.12	m³	88.12
Grade C40						
10 mm aggregate	–	–	–	96.60	m³	96.60
20 mm aggregate	–	–	–	95.15	m³	95.15
Grade C50						
10 mm aggregate	–	–	–	100.80	m³	100.80
20 mm aggregate	–	–	–	99.29	m³	99.29
Grade C60						
10 mm aggregate	–	–	–	105.00	m³	105.00
20 mm aggregate	–	–	–	103.42	m³	103.42

CLASS F: IN SITU CONCRETE

Item	Gang hours	Labour £	Plant £	Material £	Unit	Total rate £
Designed mix; sulphate-resisting cement to BS 4027						
Grade C7.5						
10 mm aggregate	–	–	–	86.03	m³	**86.03**
20 mm aggregate	–	–	–	84.79	m³	**84.79**
40 mm aggregate	–	–	–	83.56	m³	**83.56**
Grade C10						
10 mm aggregate	–	–	–	87.13	m³	**87.13**
20 mm aggregate	–	–	–	85.87	m³	**85.87**
40 mm aggregate	–	–	–	84.63	m³	**84.63**
Grade C15						
10 mm aggregate	–	–	–	88.23	m³	**88.23**
20 mm aggregate	–	–	–	86.96	m³	**86.96**
40 mm aggregate	–	–	–	85.70	m³	**85.70**
Grade C20						
10 mm aggregate	–	–	–	89.31	m³	**89.31**
20 mm aggregate	–	–	–	88.02	m³	**88.02**
40 mm aggregate	–	–	–	86.75	m³	**86.75**
Grade C25						
10 mm aggregate	–	–	–	90.47	m³	**90.47**
20 mm aggregate	–	–	–	89.16	m³	**89.16**
40 mm aggregate	–	–	–	87.87	m³	**87.87**
Grade C30						
10 mm aggregate	–	–	–	93.67	m³	**93.67**
20 mm aggregate	–	–	–	92.83	m³	**92.83**
40 mm aggregate	–	–	–	91.48	m³	**91.48**
Prescribed mix; cement to BS EN 197–1						
Grade C7.5						
10 mm aggregate	–	–	–	85.03	m³	**85.03**
20 mm aggregate	–	–	–	83.75	m³	**83.75**
40 mm aggregate	–	–	–	82.49	m³	**82.49**
Grade C10						
10 mm aggregate	–	–	–	86.10	m³	**86.10**
20 mm aggregate	–	–	–	84.81	m³	**84.81**
40 mm aggregate	–	–	–	83.54	m³	**83.54**
Grade C15						
10 mm aggregate	–	–	–	87.18	m³	**87.18**
20 mm aggregate	–	–	–	85.87	m³	**85.87**
40 mm aggregate	–	–	–	84.58	m³	**84.58**
Grade C20						
10 mm aggregate	–	–	–	88.25	m³	**88.25**
20 mm aggregate	–	–	–	86.93	m³	**86.93**
40 mm aggregate	–	–	–	85.63	m³	**85.63**
Grade C25						
10 mm aggregate	–	–	–	89.33	m³	**89.33**
20 mm aggregate	–	–	–	87.99	m³	**87.99**
40 mm aggregate	–	–	–	86.67	m³	**86.67**

CLASS F: IN SITU CONCRETE

Item	Gang hours	Labour £	Plant £	Material £	Unit	Total rate £
PROVISION OF CONCRETE – cont						
Prescribed mix – cont						
Grade C30						
10 mm aggregate	–	–	–	93.09	m^3	93.09
20 mm aggregate	–	–	–	91.70	m^3	91.70
40 mm aggregate	–	–	–	90.32	m^3	90.32
Grade C40						
10 mm aggregate	–	–	–	99.02	m^3	99.02
20 mm aggregate	–	–	–	97.53	m^3	97.53
Grade C50						
10 mm aggregate	–	–	–	103.32	m^3	103.32
20 mm aggregate	–	–	–	101.77	m^3	101.77
Grade C60						
10 mm aggregate	–	–	–	107.63	m^3	107.63
20 mm aggregate	–	–	–	106.01	m^3	106.01
Prescribed mix; sulphate-resisting cement to BS 4027						
Grade C7.5						
10 mm aggregate	–	–	–	88.11	m^3	88.11
20 mm aggregate	–	–	–	86.83	m^3	86.83
40 mm aggregate	–	–	–	85.57	m^3	85.57
Grade C10						
10 mm aggregate	–	–	–	89.23	m^3	89.23
20 mm aggregate	–	–	–	87.94	m^3	87.94
40 mm aggregate	–	–	–	86.66	m^3	86.66
Grade C15						
10 mm aggregate	–	–	–	90.36	m^3	90.36
20 mm aggregate	–	–	–	89.05	m^3	89.05
40 mm aggregate	–	–	–	87.76	m^3	87.76
Grade C20						
10 mm aggregate	–	–	–	91.46	m^3	91.46
20 mm aggregate	–	–	–	90.14	m^3	90.14
40 mm aggregate	–	–	–	88.84	m^3	88.84
Grade C25						
10 mm aggregate	–	–	–	92.65	m^3	92.65
20 mm aggregate	–	–	–	91.31	m^3	91.31
40 mm aggregate	–	–	–	89.98	m^3	89.98
Grade C30						
10 mm aggregate	–	–	–	96.46	m^3	96.46
20 mm aggregate	–	–	–	95.06	m^3	95.06
40 mm aggregate	–	–	–	93.69	m^3	93.69
PLACING OF CONCRETE; MASS						
Blinding; thickness						
ne 150 mm	0.18	17.42	0.68	–	m^3	18.10
150–300 mm	0.16	15.48	1.94	–	m^3	17.42
300–500 mm	0.14	13.55	3.20	–	m^3	16.75
exceeding 500 mm	0.12	11.61	4.46	–	m^3	16.07

CLASS F: IN SITU CONCRETE

Item	Gang hours	Labour £	Plant £	Material £	Unit	Total rate £
Bases, footings, pile caps and ground slabs; thickness						
ne 150 mm	0.20	19.35	12.61	–	m³	**31.96**
150–300 mm	0.17	16.45	10.76	–	m³	**27.21**
300–500 mm	0.15	14.51	9.51	–	m³	**24.02**
exceeding 500 mm	0.14	13.55	8.83	–	m³	**22.38**
ADD to the above for placing against an excavated surface	0.03	2.42	1.55	–	m³	**3.97**
Walls; thickness						
ne 150 mm	0.21	20.32	13.28	–	m³	**33.60**
150–300 mm	0.15	14.51	9.51	–	m³	**24.02**
300–500 mm	0.13	12.58	8.24	–	m³	**20.82**
exceeding 500 mm	0.12	11.61	7.56	–	m³	**19.17**
ADD to the above for placing against an excavated surface	0.03	2.90	1.94	–	m³	**4.84**
Other concrete forms						
plinth 1000 × 1000 × 600 mm	0.33	31.93	20.85	–	m³	**52.78**
plinth 1500 × 1500 × 750 mm	0.25	24.19	15.81	–	m³	**40.00**
plinth 2000 × 2000 × 600 mm	0.20	19.35	12.61	–	m³	**31.96**
surround to precast concrete manhole chambers 200 mm thick	0.29	28.06	18.32	–	m³	**46.38**
PLACING OF CONCRETE; REINFORCED						
Bases, footings, pile caps and ground slabs; thickness						
ne 150 mm	0.21	20.32	13.28	–	m³	**33.60**
150–300 mm	0.18	17.42	11.35	–	m³	**28.77**
300–500 mm	0.16	15.48	10.08	–	m³	**25.56**
exceeding 500 mm	0.15	14.51	9.51	–	m³	**24.02**
Suspended slabs; thickness						
ne 150 mm	0.27	26.12	17.07	–	m³	**43.19**
150–300 mm	0.21	20.32	13.28	–	m³	**33.60**
300–500 mm	0.19	18.38	12.03	–	m³	**30.41**
exceeding 500 mm	0.19	18.38	12.03	–	m³	**30.41**
Walls; thickness						
ne 150 mm	0.29	28.06	18.32	–	m³	**46.38**
150–300 mm	0.22	21.29	13.87	–	m³	**35.16**
300–500 mm	0.20	19.35	12.61	–	m³	**31.96**
exceeding 500 mm	0.20	19.35	12.61	–	m³	**31.96**
Columns and piers; cross-sectional area						
ne 0.03 m²	0.50	48.38	31.52	–	m³	**79.90**
0.03–0.10 m²	0.40	38.70	25.22	–	m³	**63.92**
0.10–0.25 m²	0.35	33.87	22.11	–	m³	**55.98**
0.25–1.00 m²	0.35	33.87	22.11	–	m³	**55.98**
exceeding 1 m²	0.28	27.09	17.65	–	m³	**44.74**
Beams; cross-sectional area						
ne 0.03 m²	0.50	48.38	31.52	–	m³	**79.90**
0.03–0.10 m²	0.40	38.70	25.22	–	m³	**63.92**
0.10–0.25 m²	0.35	33.87	22.11	–	m³	**55.98**
0.25–1.00 m²	0.35	33.87	22.11	–	m³	**55.98**
exceeding 1 m²	0.28	27.09	17.65	–	m³	**44.74**

CLASS F: IN SITU CONCRETE

Item	Gang hours	Labour £	Plant £	Material £	Unit	Total rate £
PLACING OF CONCRETE; REINFORCED – cont						
Casings to metal sections; cross-sectional area						
ne 0.03 m^2	0.47	45.48	29.67	–	m^3	**75.15**
0.03–0.10 m^2	0.47	45.48	29.67	–	m^3	**75.15**
0.10–0.25 m^2	0.40	38.70	25.21	–	m^3	**63.91**
0.25–1.00 m^2	0.40	38.70	25.21	–	m^3	**63.91**
exceeding 1 m^2	0.35	33.87	22.11	–	m^3	**55.98**
PLACING OF CONCRETE; PRESTRESSED						
Suspended slabs; thickness						
ne 150 mm	0.28	27.09	17.65	–	m^3	**44.74**
150–300 mm	0.22	21.29	13.87	–	m^3	**35.16**
300–500 mm	0.20	19.35	12.61	–	m^3	**31.96**
exceeding 500 mm	0.19	18.38	12.03	–	m^3	**30.41**
Beams; cross-sectional area						
ne 0.03 m^2	0.50	48.38	31.52	–	m^3	**79.90**
0.03–0.10 m^2	0.40	38.70	25.22	–	m^3	**63.92**
0.10–0.25 m^2	0.35	33.87	22.11	–	m^3	**55.98**
0.25–1.00 m^2	0.35	33.87	22.11	–	m^3	**55.98**
exceeding 1 m^2	0.28	27.09	17.65	–	m^3	**44.74**

CLASS G: CONCRETE ANCILLARIES

Item	Gang hours	Labour £	Plant £	Material £	Unit	Total rate £
RESOURCES – LABOUR						
Formwork gang – small areas						
1 foreman joiner (craftsman)		22.50				
1 joiner (craftsman)		19.54				
1 unskilled operative (general)		12.71				
1 plant operator (craftsman) – 50% of time		10.55				
Total Gang Rate/Hour		£ 65.30				
Formwork gang – large areas						
1 foreman joiner (craftsman)		22.50				
2 joiners (craftsman)		39.08				
1 unskilled operative (general)		12.71				
1 plant operator (craftsman) – 25% of time		5.28				
Total Gang Rate/Hour		£ 79.56				
Reinforcement gang						
1 foreman steel fixer (craftsman)		22.50				
4 steel fixers (craftsman)		78.16				
1 unskilled operative (general)		12.71				
1 plant operator (craftsman) – 25% of time		5.28				
Total Gang Rate/Hour		£ 118.65				
Reinforcement – on-site bending/baling gang						
1 steel fixer (craftsman)		19.54				
1 unskilled operative (general)		12.71				
1 plant operator (craftsman) – 25% of time		5.28				
Total Gang Rate/Hour		£ 37.52				
Joints gang						
1 ganger/chargehand (skill rate 4)		14.56				
1 skilled operative (skill rate 4)		13.61				
1 unskilled operative (general)		12.71				
Total Gang Rate/Hour		£ 40.88				
Concrete accessories gang						
1 ganger/chargehand (skill rate 4)		14.56				
1 skilled operative (skill rate 4)		13.61				
1 unskilled operative (general)		12.71				
Total Gang Rate/Hour		£ 40.88				
RESOURCES – MATERIALS						
Reinforcement; supply only ex works materials rates						
Grade 500C deformed reinforcing bars						
Reinforcement; high tensile steel bars; straight; T12	–	–	–	650.00	tonne	**650.00**
Bars; stainless steel; straight; lengths; 16 mm nominal size-alloy 1.4301						
Stainless steel bars; straight	–	–	–	3100.00	tonne	**3100.00**
Stainless steel bars; bent	–	–	–	3300.00	tonne	**3300.00**

CLASS G: CONCRETE ANCILLARIES

Item	Gang hours	Labour £	Plant £	Material £	Unit	Total rate £
RESOURCES – MATERIALS – cont						
Reinforcement – cont						
Welded fabric to BS 4483 in sheets 4.80 × 2.40						
A98	–	–	–	1.75	m^2	**1.75**
A142	–	–	–	1.44	m^2	**1.44**
A193	–	–	–	1.93	m^2	**1.93**
A252	–	–	–	2.53	m^2	**2.53**
A393	–	–	–	3.94	m^2	**3.94**
B196	–	–	–	3.68	m^2	**3.68**
B283	–	–	–	2.50	m^2	**2.50**
B385	–	–	–	3.08	m^2	**3.08**
B503	–	–	–	3.91	m^2	**3.91**
B785	–	–	–	5.37	m^2	**5.37**
B1131	–	–	–	7.90	m^2	**7.90**
C283	–	–	–	1.85	m^2	**1.85**
C385	–	–	–	2.40	m^2	**2.40**
C363	–	–	–	3.91	m^2	**3.91**
C503	–	–	–	3.05	m^2	**3.05**
C785	–	–	–	4.50	m^2	**4.50**
D49	–	–	–	0.50	m^2	**0.50**
D98	–	–	–	0.95	m^2	**0.95**
Reinforcement couplers, Lenton type						
Type A; one bar twistable						
12 mm diameter	–	–	–	2.75	nr	**2.75**
16 mm diameter	–	–	–	3.50	nr	**3.50**
20 mm diameter	–	–	–	6.40	nr	**6.40**
25 mm diameter	–	–	–	9.95	nr	**9.95**
32 mm diameter	–	–	–	13.50	nr	**13.50**
40 mm diameter	–	–	–	19.50	nr	**19.50**
Type B; no bar twistable						
12 mm diameter	–	–	–	16.50	nr	**16.50**
16 mm diameter	–	–	–	19.35	nr	**19.35**
20 mm diameter	–	–	–	21.25	nr	**21.25**
25 mm diameter	–	–	–	24.50	nr	**24.50**
32 mm diameter	–	–	–	33.00	nr	**33.00**
40 mm diameter	–	–	–	50.00	nr	**50.00**
extra for threaded bar at factory						
12 mm diameter	–	–	–	1.85	nr	**1.85**
16 mm diameter	–	–	–	2.05	nr	**2.05**
20 mm diameter	–	–	–	2.42	nr	**2.42**
25 mm diameter	–	–	–	2.75	nr	**2.75**
32 mm diameter	–	–	–	3.85	nr	**3.85**
40 mm diameter	–	–	–	4.75	nr	**4.75**

CLASS G: CONCRETE ANCILLARIES

Item	Gang hours	Labour £	Plant £	Material £	Unit	Total rate £
RESOURCES – PLANT						
Formwork – small areas						
20 t crawler crane – 50% of time			25.08			
22" diameter saw bench			2.58			
allowance for small power tools			0.75			
Total Rate/Hour		£	**28.40**			
Formwork – large areas						
20 t crawler crane – 25% of time			12.54			
22" diameter saw bench			2.58			
allowance for small power tools			0.75			
Total Rate/Hour		£	**15.86**			
Reinforcement						
30 t crawler crane – 25% of time			14.11			
bar cropper			3.75			
small power tools			1.50			
support acrows, tirfors, kentledge, etc.			0.70			
Total Rate/Hour		£	**20.06**			
FORMWORK; ROUGH FINISH						
Formwork						
Formwork materials include for shutter, bracing, ties, support, kentledge and all consumables.						
The following unit costs do not include for formwork outside the payline and are based on an optimum of a minimum 8 uses with 10% per use towards the cost of repairs/replacement of components damaged during disassembly						
ADD to formwork material costs generally depending on the number of uses:						
Nr of uses % Addition % Waste						
1 + 90 to 170 + 7						
2 + 50 to 80 + 7						
3 + 15 to 30 + 6						
6 + 5 to 10 + 6						
8 No change + 5						
10 – 5 to 7 + 5						
Plane horizontal, width						
ne 0.1 m	0.10	6.53	2.84	2.19	m	**11.56**
0.1- 0.20 m	0.18	11.75	5.11	2.39	m	**19.25**
0.2–0.40 m	0.40	31.83	6.34	5.90	m^2	**44.07**
0.4–1.22 m	0.40	31.83	6.34	5.90	m^2	**44.07**
exceeding 1.22 m	0.38	30.23	6.03	5.90	m^2	**42.16**
Plane sloping, width						
ne 0.1 m	0.11	7.18	3.13	2.19	m	**12.50**
0.1–0.20 m	0.20	13.06	5.68	2.39	m	**21.13**
0.2–0.40 m	0.43	34.21	6.85	8.48	m^2	**49.54**
0.4–1.22 m	0.43	34.21	6.85	8.48	m^2	**49.54**
exceeding 1.22 m	0.38	30.23	6.03	8.48	m^2	**44.74**

CLASS G: CONCRETE ANCILLARIES

Item	Gang hours	Labour £	Plant £	Material £	Unit	Total rate £
FORMWORK; ROUGH FINISH – cont						
Formwork – cont						
Plane battered, width						
ne 0.1 m	0.12	7.84	3.41	2.83	m	14.08
0.1–0.20 m	0.20	13.06	5.68	3.04	m	21.78
0.2–0.40 m	0.43	34.21	6.95	9.76	m²	50.92
0.4–1.22 m	0.43	34.21	6.95	9.76	m²	50.92
exceeding 1.22 m	0.40	31.83	6.34	9.76	m²	47.93
Plane vertical, width						
ne 0.1 m	0.12	7.84	3.41	2.83	m	14.08
0.1–0.20 m	0.19	12.41	5.40	3.04	m	20.85
0.2–0.40 m	0.70	55.70	11.10	8.48	m²	75.28
0.4–1.22 m	0.51	40.58	7.97	8.48	m²	57.03
exceeding 1.22 m	0.47	37.40	7.58	8.48	m²	53.46
Curved to one radius in one plane, 0.5 m radius, width						
ne 0.1 m	0.19	12.41	5.40	4.12	m	21.93
0.1–0.20 m	0.25	16.32	7.10	4.32	m	27.74
0.2–0.40 m	0.90	71.61	14.28	9.76	m²	95.65
0.4–1.22 m	0.72	57.29	11.42	9.76	m²	78.47
exceeding 1.22 m	0.65	51.72	10.19	9.76	m²	71.67
Curved to one radius in one plane, 2 m radius, width						
ne 0.1 m	0.18	11.75	5.11	4.12	m	20.98
0.1–0.20 m	0.22	14.37	6.25	4.32	m	24.94
0.2–0.40 m	0.84	66.83	13.32	9.76	m²	89.91
0.4–1.22 m	0.66	52.51	10.47	9.76	m²	72.74
exceeding 1.22 m	0.52	41.37	8.25	9.76	m²	59.38
For voids						
small void; depth ne 0.5 m	0.07	4.57	–	3.06	nr	7.63
small void; depth 0.5–1.0 m	0.12	7.84	–	5.88	nr	13.72
small void; depth ne 1.0–2.0 m	0.16	10.45	–	10.86	nr	21.31
large void; depth ne 0.5 m	0.14	9.14	–	9.13	nr	18.27
large void; depth 0.5–1.0 m	0.33	21.55	–	18.83	nr	40.38
large void; depth 1.0–2.0 m	0.58	37.87	–	38.12	nr	75.99
For concrete components of constant cross-section						
beams; 200 × 200 mm	0.48	31.34	13.63	5.11	m	50.08
beams; 500 × 500 mm	0.55	35.91	15.62	8.85	m	60.38
beams; 500 × 800 mm	0.67	43.75	19.03	10.06	m	72.84
columns; 200 × 200 mm	0.55	35.91	15.62	5.51	m	57.04
columns; 300 × 300 mm	0.55	35.91	15.62	7.81	m	59.34
columns; 300 × 500 mm	0.62	40.49	17.61	9.02	m	67.12
to walls; 1.0 m high thickness 250 mm	1.10	87.52	17.45	10.46	m	115.43
to walls; 1.5 m high thickness 300 mm	1.50	119.35	23.79	15.01	m	158.15
box culvert; 2 × 2 m internally and wall thickness 300 mm	4.50	358.04	71.38	86.73	m	516.15
projections (100 mm deep)	0.10	6.53	2.84	2.90	m	12.27
intrusions (100 mm deep)	0.10	6.53	2.84	2.90	m	12.27
Allowance for additional craneage and rub up where required						
ADD to items measures linear	0.04	2.61	1.14	0.03	m	3.78
ADD to items measures m²	0.12	7.84	3.41	0.09	m²	11.34

CLASS G: CONCRETE ANCILLARIES

Item	Gang hours	Labour £	Plant £	Material £	Unit	Total rate £
FORMWORK; FAIR FINISH						
Plane horizontal, width						
ne 0.1 m	0.10	6.53	2.84	3.36	m	**12.73**
0.1–0.20 m	0.18	11.75	5.11	3.80	m	**20.66**
0.2–0.40 m	0.40	31.83	6.34	11.99	m²	**50.16**
0.4–1.22 m	0.40	31.83	6.34	11.99	m²	**50.16**
exceeding 1.22 m	0.38	30.23	6.03	11.99	m²	**48.25**
Plane sloping, width						
ne 0.1 m	0.11	7.18	3.13	3.36	m	**13.67**
0.1–0.20 m	0.20	13.06	3.17	3.80	m	**20.03**
0.2–0.40 m	0.43	34.21	21.13	18.31	m²	**73.65**
0.4–1.22 m	0.43	34.21	21.13	18.31	m²	**73.65**
exceeding 1.22 m	0.38	30.23	6.03	18.31	m²	**54.57**
Plane battered, width						
ne 0.1 m	0.12	7.84	3.41	4.94	m	**16.19**
0.1–0.20 m	0.20	13.06	5.68	5.39	m	**24.13**
0.2–0.40 m	0.43	34.21	6.82	21.48	m²	**62.51**
0.4–1.22 m	0.43	34.21	6.82	21.48	m²	**62.51**
exceeding 1.22 m	0.40	31.83	6.34	21.48	m²	**59.65**
Plane vertical, width						
ne 0.1 m	0.12	7.84	3.41	4.94	m	**16.19**
0.1–0.20 m	0.20	13.06	5.68	5.39	m	**24.13**
0.2–0.40 m	0.72	57.29	11.42	18.31	m²	**87.02**
0.4–1.22 m	0.53	42.17	8.41	18.31	m²	**68.89**
exceeding 1.22 m	0.48	38.19	7.61	18.31	m²	**64.11**
Curved to one radius in one plane, 0.5 m radius, width						
ne 0.1 m	0.20	13.06	5.68	8.10	m	**26.84**
0.1–0.20 m	0.25	16.32	7.10	8.55	m	**31.97**
0.2–0.40 m	0.90	71.61	14.28	21.48	m²	**107.37**
0.4–1.22 m	0.72	57.29	11.42	21.48	m²	**90.19**
exceeding 1.22 m	0.65	51.72	10.31	21.48	m²	**83.51**
Curved to one radius in one plane, 2.0 m radius, width						
ne 0.1 m	0.18	11.75	5.11	8.10	m	**24.96**
0.1–0.20 m	0.22	14.37	6.25	8.55	m	**29.17**
0.2–0.40 m	0.84	66.83	13.32	21.48	m²	**101.63**
0.4–1.22 m	0.66	52.51	10.47	21.48	m²	**84.46**
exceeding 1.22 m	0.52	41.37	8.25	21.48	m²	**71.10**
For voids, using void former						
small void; depth ne 0.5 m	0.07	4.57	–	5.92	nr	**10.49**
small void; depth 0.5–1.0 m	0.12	7.84	–	11.40	nr	**19.24**
small void; depth ne 1.0–2.0 m	0.16	10.45	–	21.73	nr	**32.18**
large void; depth ne 0.5 m	0.14	9.14	–	12.71	nr	**21.85**
large void; depth 0.5–1.0 m	0.33	21.55	–	25.54	nr	**47.09**
large void; depth 1.0–2.0 m	0.58	37.87	–	45.25	nr	**83.12**

CLASS G: CONCRETE ANCILLARIES

Item	Gang hours	Labour £	Plant £	Material £	Unit	Total rate £
FORMWORK; FAIR FINISH – cont						
For concrete components of constant cross-section						
beams; 200 × 200 mm	0.49	32.00	13.92	10.26	m	56.18
beams; 500 × 500 mm	0.57	37.22	16.19	17.31	m	70.72
beams; 500 × 800 mm	0.69	45.06	19.60	21.77	m	86.43
columns; 200 × 200 mm	0.57	37.22	16.19	11.13	m	64.54
columns; 300 × 300 mm	0.57	37.22	16.19	16.02	m	69.43
columns; 300 × 500 mm	0.64	41.79	18.18	19.32	m	79.29
to walls; 1.0 m high thickness 250 mm	1.20	95.48	19.04	22.63	m	137.15
to walls; 1.5 m high thickness 300 mm	1.60	127.30	25.38	33.27	m	185.95
box culvert; 2 × 2 m internally and wall thickness 300 mm	4.60	366.00	72.97	205.30	m	644.27
projections (100 mm deep)	0.10	6.53	2.84	5.39	m	14.76
intrusions (100 mm deep)	0.10	6.53	2.84	5.39	m	14.76
Allowance for additional craneage and rub up where required						
ADD to items measures linear	0.14	9.14	3.98	0.03	m	13.15
ADD to items measures m²	0.12	7.84	3.41	0.12	m²	11.37
FORMWORK; EXTRA SMOOTH FINISH						
Plane horizontal, width						
ne 0.1 m	0.10	6.53	2.84	3.61	m	12.98
0.1–0.20 m	0.18	11.75	5.11	4.26	m	21.12
0.2–0.40 m	0.40	31.83	6.34	14.32	m²	52.49
0.4–1.22 m	0.40	31.83	6.34	14.32	m²	52.49
exceeding 1.22 m	0.38	30.23	6.02	14.32	m²	50.57
Plane sloping, width						
ne 0.1 m	0.11	7.18	3.13	3.61	m	13.92
0.1–0.20 m	0.20	13.06	5.68	4.26	m	23.00
0.2–0.40 m	0.43	34.21	6.85	20.65	m²	61.71
0.4–1.22 m	0.43	34.21	6.85	20.65	m²	61.71
exceeding 1.22 m	0.38	30.23	6.03	20.65	m²	56.91
Plane battered, width						
ne 0.1 m	0.12	7.84	3.41	5.19	m	16.44
0.1–0.20 m	0.20	13.06	5.68	5.84	m	24.58
0.2–0.40 m	0.43	34.21	6.85	23.81	m²	64.87
0.4–1.22 m	0.43	34.21	6.85	23.81	m²	64.87
exceeding 1.22 m	0.40	31.83	6.34	23.81	m²	61.98
Plane vertical, width						
ne 0.1 m	0.12	9.55	3.41	5.19	m	18.15
0.1–0.20 m	0.20	13.06	5.68	5.84	m	24.58
0.2–0.40 m	0.74	58.88	11.74	20.65	m²	91.27
0.4–1.22 m	0.53	42.17	8.44	20.65	m²	71.26
exceeding 1.22 m	0.48	38.19	7.61	20.65	m²	66.45
Curved to one radius in one plane, 0.5 m radius, width						
ne 0.1 m	0.20	13.06	5.68	8.35	m	27.09
0.1–0.20 m	0.25	16.32	7.10	5.84	m	29.26
0.2–0.40 m	0.90	71.61	14.28	23.81	m²	109.70
0.4–1.22 m	0.72	57.29	11.42	23.81	m²	92.52
exceeding 1.22 m	0.65	51.72	10.34	23.81	m²	85.87

CLASS G: CONCRETE ANCILLARIES

Item	Gang hours	Labour £	Plant £	Material £	Unit	Total rate £
Curved to one radius in one plane, 1.0 m radius, width						
ne 0.1 m	0.18	11.75	5.11	8.35	m	25.21
0.1–0.20 m	0.22	14.37	6.25	5.84	m	26.46
0.2–0.40 m	0.84	66.83	23.35	23.81	m²	113.99
0.4–1.22 m	0.66	52.51	10.47	23.81	m²	86.79
exceeding 1.22 m	0.52	41.37	8.25	23.81	m²	73.43
For voids, using void former						
small void; depth ne 0.5 m	0.07	4.57	–	10.26	nr	14.83
small void; depth 0.5–1.0 m	0.12	7.84	–	20.11	nr	27.95
small void; depth ne 1.0–2.0 m	0.16	10.45	–	37.85	nr	48.30
large void; depth ne 0.5 m	0.14	9.14	–	22.57	nr	31.71
large void; depth 0.5–1.0 m	0.33	21.55	–	44.08	nr	65.63
large void; depth 1.0–2.0 m	0.58	37.87	–	76.91	nr	114.78
For concrete components of constant cross-section						
beams; 200 × 200 mm	0.50	32.65	14.20	11.66	m	58.51
beams; 500 × 500 mm	0.57	37.22	9.02	20.81	m	67.05
beams; 500 × 800 mm	0.69	45.06	10.97	25.97	m	82.00
columns; 200 × 200 mm	0.57	37.22	9.02	12.99	m	59.23
columns; 300 × 300 mm	0.57	37.22	9.02	18.81	m	65.05
columns; 300 × 500 mm	0.64	41.79	10.15	23.05	m	74.99
to walls; 1.0 m high thickness 250 mm	1.20	95.48	19.04	27.29	m	141.81
to walls; 1.5 m high thickness 300 mm	1.60	127.30	25.38	40.27	m	192.95
box culvert; 2 × 2 m internally and wall thickness 300 mm	4.60	366.00	72.97	230.06	m	669.03
projections (100 mm deep)	0.10	6.53	2.84	5.84	m	15.21
intrusions (100 mm deep)	0.10	6.53	2.84	5.84	m	15.21
Allowance for additional craneage and rub up where required						
ADD to items measures linear	0.14	9.14	3.98	0.03	m	13.15
ADD to items measures m²	0.12	7.84	3.41	0.15	m²	11.40

REINFORCEMENT
Supply and fit reinforcement materials, include for bars, tying wire, spacers, couplers and steel supports for bottom layer reinforcement (stools, chairs and risers).

Plain round mild steel bars to BS 4449
Bars; supplied in straight lengths

6 mm nominal size	8.00	949.16	160.50	505.00	tonne	1614.66
8 mm nominal size	6.74	799.67	135.22	505.00	tonne	1439.89
10 mm nominal size	6.74	799.67	135.22	505.00	tonne	1439.89
12 mm nominal size	6.74	799.67	135.22	505.00	tonne	1439.89
16 mm nominal size	6.15	729.67	123.39	505.00	tonne	1358.06
20 mm nominal size	4.44	526.78	89.08	505.00	tonne	1120.86
25 mm nominal size	4.44	526.78	89.08	505.00	tonne	1120.86
32 mm nominal size	4.44	526.78	89.08	505.00	tonne	1120.86
40 mm nominal size	4.44	526.78	89.08	505.00	tonne	1120.86

CLASS G: CONCRETE ANCILLARIES

Item	Gang hours	Labour £	Plant £	Material £	Unit	Total rate £
REINFORCEMENT – cont						
Plain round mild steel bars to BS 4449 – cont						
Bars; supplied in bent and cut lengths						
6 mm nominal size	8.00	949.16	160.50	505.00	tonne	1614.66
8 mm nominal size	6.74	799.67	135.22	505.00	tonne	1439.89
10 mm nominal size	6.74	799.67	135.22	505.00	tonne	1439.89
12 mm nominal size	6.74	799.67	135.22	505.00	tonne	1439.89
16 mm nominal size	6.15	729.67	123.39	505.00	tonne	1358.06
20 mm nominal size	4.44	526.78	89.08	505.00	tonne	1120.86
25 mm nominal size	4.44	526.78	89.08	505.00	tonne	1120.86
32 mm nominal size	4.44	526.78	89.08	505.00	tonne	1120.86
40 mm nominal size	4.44	526.78	89.08	505.00	tonne	1120.86
Deformed high yield steel bars to BS 4449						
Bars; supplied in straight lengths						
6 mm nominal size	8.00	949.16	160.50	656.50	tonne	1766.16
8 mm nominal size	6.74	799.67	135.22	656.50	tonne	1591.39
10 mm nominal size	6.74	799.67	135.22	656.50	tonne	1591.39
12 mm nominal size	6.74	799.67	135.22	656.50	tonne	1591.39
16 mm nominal size	6.15	729.67	123.39	656.50	tonne	1509.56
20 mm nominal size	4.44	526.78	89.08	656.50	tonne	1272.36
25 mm nominal size	4.44	526.78	89.08	656.50	tonne	1272.36
32 mm nominal size	4.44	526.78	89.08	656.50	tonne	1272.36
40 mm nominal size	4.44	526.78	89.08	656.50	tonne	1272.36
Bars; supplied in bent and cut lengths						
6 mm nominal size	8.00	949.16	160.50	656.50	tonne	1766.16
8 mm nominal size	6.74	799.67	135.22	656.50	tonne	1591.39
10 mm nominal size	6.74	799.67	135.22	656.50	tonne	1591.39
12 mm nominal size	6.74	799.67	135.22	656.50	tonne	1591.39
16 mm nominal size	6.15	729.67	123.39	656.50	tonne	1509.56
20 mm nominal size	4.44	526.78	89.08	656.50	tonne	1272.36
25 mm nominal size	4.44	526.78	89.08	656.50	tonne	1272.36
32 mm nominal size	4.44	526.78	89.08	656.50	tonne	1272.36
40 mm nominal size	4.44	526.78	89.08	656.50	tonne	1272.36
Stainless steel bars; Alloy 1.4301						
Bars; supplied in straight lengths						
8 mm nominal size	6.74	799.67	135.22	3100.00	tonne	4034.89
10 mm nominal size	6.74	799.67	135.22	3100.00	tonne	4034.89
12 mm nominal size	6.74	799.67	135.22	3100.00	tonne	4034.89
16 mm nominal size	6.15	729.67	123.39	3100.00	tonne	3953.06
20 mm nominal size	4.44	526.78	89.08	3100.00	tonne	3715.86
25 mm nominal size	4.44	526.78	89.08	3100.00	tonne	3715.86
32 mm nominal size	4.44	526.78	89.08	3100.00	tonne	3715.86
Bars; supplied in bent and cut lengths						
8 mm nominal size	6.74	799.67	135.22	3300.00	tonne	4234.89
10 mm nominal size	6.74	799.67	135.22	3300.00	tonne	4234.89
12 mm nominal size	6.74	799.67	135.22	3300.00	tonne	4234.89

CLASS G: CONCRETE ANCILLARIES

Item	Gang hours	Labour £	Plant £	Material £	Unit	Total rate £
16 mm nominal size	6.15	729.67	123.39	3300.00	tonne	**4153.06**
20 mm nominal size	4.44	526.78	89.08	3300.00	tonne	**3915.86**
25 mm nominal size	4.44	526.78	89.08	3300.00	tonne	**3915.86**
32 mm nominal size	4.44	526.78	89.08	3300.00	tonne	**3915.86**
Additional allowances to bar reinforcement						
Add to the above bars						
12–15 m long; mild steel to BS 4449	–	–	–	30.00	tonne	**30.00**
12–15 m long; high yield steel to BS 4449	–	–	–	45.00	tonne	**45.00**
Over 15 m long, per 500 mm increment; mild steel to BS 4449	–	–	–	15.00	tonne	**15.00**
Over 15 m long, per 500 mm increment; high yield steel to BS 4449	–	–	–	50.00	tonne	**50.00**
Add for cutting, bending, tagging and baling reinforcement on site						
6 mm nominal size	4.87	182.75	97.71	2.00	tonne	**282.46**
8 mm nominal size	4.58	171.86	91.89	2.00	tonne	**265.75**
10 mm nominal size	3.42	128.34	68.61	2.00	tonne	**198.95**
12 mm nominal size	2.55	95.69	51.17	2.00	tonne	**148.86**
16 mm nominal size	2.03	76.18	40.73	2.00	tonne	**118.91**
20 mm nominal size	1.68	63.04	33.71	2.00	tonne	**98.75**
25 mm nominal size	1.68	63.04	33.71	2.00	tonne	**98.75**
32 mm nominal size	1.39	52.16	27.89	2.00	tonne	**82.05**
40 mm nominal size	1.39	52.16	27.89	2.00	tonne	**82.05**
Special joints						
Lenton type A couplers; threaded ends on reinforcing bars						
12 mm	0.09	10.68	–	6.56	nr	**17.24**
16 mm	0.09	10.68	–	7.73	nr	**18.41**
20 mm	0.09	10.68	–	11.45	nr	**22.13**
25 mm	0.09	10.68	–	15.75	nr	**26.43**
32 mm	0.09	10.68	–	21.61	nr	**32.29**
40 mm	0.09	10.68	–	29.58	nr	**40.26**
Lenton type B couplers; threaded ends on reinforcing bars						
12 mm	0.09	10.68	–	20.65	nr	**31.33**
16 mm	0.09	10.68	–	23.97	nr	**34.65**
20 mm	0.09	10.68	–	26.67	nr	**37.35**
25 mm	0.09	10.68	–	30.67	nr	**41.35**
32 mm	0.09	10.68	–	41.60	nr	**52.28**
40 mm	0.09	10.68	–	60.84	nr	**71.52**
Steel fabric to BS 4483						
nominal mass 0.77 kg/m^2; ref D49	0.02	2.37	0.40	0.57	m^2	**3.34**
nominal mass 1.54 kg/m^2; ref D98	0.02	2.37	0.40	1.09	m^2	**3.86**
nominal mass 1.54 kg/m^2; ref A98	0.03	3.56	0.61	2.01	m^2	**6.18**
nominal mass 2.22 kg/m^2; ref A142	0.03	3.56	0.61	1.66	m^2	**5.83**
nominal mass 2.61 kg/m^2; ref C283	0.03	3.56	0.61	2.13	m^2	**6.30**
nominal mass 3.02 kg/m^2; ref A193	0.04	4.75	0.80	2.22	m^2	**7.77**
nominal mass 3.05 kg/m^2; ref B196	0.04	4.75	0.80	4.23	m^2	**9.78**

CLASS G: CONCRETE ANCILLARIES

Item	Gang hours	Labour £	Plant £	Material £	Unit	Total rate £
REINFORCEMENT – cont						
Steel fabric to BS 4483 – cont						
nominal mass 3.41 kg/m^2; ref C385	0.04	4.75	0.80	2.76	m^2	**8.31**
nominal mass 3.73 kg/m^2; ref B283	0.04	4.75	0.80	2.88	m^2	**8.43**
nominal mass 3.95 kg/m^2; ref A252	0.04	4.75	0.80	2.91	m^2	**8.46**
nominal mass 4.34 kg/m^2; ref C503	0.05	5.93	1.01	3.51	m^2	**10.45**
nominal mass 4.35 kg/m^2; ref B385	0.05	5.93	1.01	3.54	m^2	**10.48**
nominal mass 5.55 kg/m^2; ref C636	0.05	5.93	1.01	4.50	m^2	**11.44**
nominal mass 5.93 kg/m^2; ref B503	0.05	5.93	1.01	4.50	m^2	**11.44**
nominal mass 6.16 kg/m^2; ref A393	0.07	8.31	1.41	4.53	m^2	**14.25**
nominal mass 6.72 kg/m^2; ref C785	0.07	8.31	1.41	5.17	m^2	**14.89**
nominal mass 8.14 kg/m^2; ref B785	0.08	9.49	1.60	6.18	m^2	**17.27**
nominal mass 10.90 kg/m^2; ref B1131	0.09	10.68	1.81	9.09	m^2	**21.58**
JOINTS						
Open surface plain; average width						
ne 0.5 m; scabbling concrete for subsequent pour	0.04	1.64	1.83	–	m^2	**3.47**
0.5–1 m; scabbling concrete for subsequent pour	0.03	1.23	1.40	–	m^2	**2.63**
Open surface with filler; average width						
ne 0.5 m; 12 mm Flexcell joint filler	0.04	1.64	1.83	3.57	m^2	**7.04**
0.5–1 m; 12 mm Flexcell joint filler	0.04	1.64	1.83	3.57	m^2	**7.04**
ne 0.5 m; 19 mm Flexcell joint filler	0.05	2.04	2.26	6.09	m^2	**10.39**
0.5–1 m; 19 mm Flexcell joint filler	0.05	2.04	2.26	6.09	m^2	**10.39**
Formed surface plain; average width (including formwork)						
ne 0.5 m	0.24	9.81	10.95	7.38	m^2	**28.14**
0.5–1.0 m	0.24	9.81	10.95	7.38	m^2	**28.14**
Formed surface with filler; average width						
ne 0.5 m; 10 mm Flexcell joint filler	0.40	16.35	13.75	10.95	m^2	**41.05**
0.5–1 m; 10 mm Flexcell joint filler	0.41	16.76	18.69	10.95	m^2	**46.40**
ne 0.5 m; 19 mm Flexcell joint filler	0.42	17.17	19.23	13.47	m^2	**49.87**
0.5–1 m; 19 mm Flexcell joint filler	0.43	17.58	19.66	13.47	m^2	**50.71**
ne 0.5 m; 25 mm Flexcell joint filler	0.45	18.40	20.51	15.29	m^2	**54.20**
0.5–1 m; 25 mm Flexcell joint filler	0.47	19.21	21.48	15.29	m^2	**55.98**
PVC						
Plastics or rubber waterstops						
160 mm centre bulb	0.04	1.64	–	3.67	m	**5.31**
junction piece	0.04	1.64	–	7.88	nr	**9.52**
210 mm centre bulb	0.05	2.04	–	5.25	m	**7.29**
junction piece	0.04	1.64	–	8.14	nr	**9.78**
260 mm centre bulb	0.05	2.04	–	6.30	m	**8.34**
junction piece	0.05	2.04	–	8.56	nr	**10.60**
170 mm flat dumbell	0.04	1.64	–	4.72	m	**6.36**
junction piece	0.05	2.04	–	57.75	nr	**59.79**
210 mm flat dumbell	0.04	1.64	–	6.30	m	**7.94**
junction piece	0.07	2.86	–	63.00	nr	**65.86**
250 mm flat dumbell	0.05	2.04	–	7.88	m	**9.92**
junction piece	0.09	3.68	–	120.75	nr	**124.43**

CLASS G: CONCRETE ANCILLARIES

Item	Gang hours	Labour £	Plant £	Material £	Unit	Total rate £
Polysulphide sealant; gun grade						
Sealed rebates or grooves						
10 × 20 mm	0.05	2.04	–	–	m	2.04
20 × 20 mm	0.07	2.86	–	0.01	m	2.87
25 × 20 mm	0.08	3.27	–	0.01	m	3.28
Mild steel						
Dowels, plain or greased						
12 mm diameter × 500 mm long	0.04	1.64	–	0.58	nr	2.22
16 mm diameter × 750 mm long	0.05	1.84	–	1.31	nr	3.15
20 mm diameter × 750 mm long	0.05	1.84	–	2.10	nr	3.94
25 mm diameter × 750 mm long	0.05	1.84	–	2.63	nr	4.47
32 mm diameter × 750 mm long	0.05	1.84	–	3.41	nr	5.25
Dowels, sleeved or capped						
12 mm diameter × 500 mm long, debonding agent for 250 mm and capped with PVC dowel cap	0.05	1.84	–	0.65	nr	2.49
16 mm diameter × 750 mm long, debonding agent for 375 mm and capped with PVC dowel cap	0.05	2.17	–	1.39	nr	3.56
20 mm diameter × 750 mm long, debonding agent for 375 mm and capped with PVC dowel cap	0.05	2.17	–	2.18	nr	4.35
25 mm diameter × 750 mm long, debonding agent for 375 mm and capped with PVC dowel cap	0.06	2.45	–	2.70	nr	5.15
32 mm diameter × 750 mm long, debonding agent for 375 mm and capped with PVC dowel cap	0.07	2.66	–	3.49	nr	6.15
POST-TENSIONED PRESTRESSING						
The design of prestressing is based on standard patented systems, each of which has produced its own method of anchoring, joining and stressing the cables or wires. The companies marketing the systems will either supply all the materials and fittings required together with the sale or hire of suitable jacks and equipment for prestressing and grouting or they will undertake to complete the work on a subcontract basis. The rates given below are therefore indicative only of the probable labour and plant costs and do not include for any permanent materials. The advice of specialist contractors should be sought for more accurate rates based on the design for a particular contract. Pre-tensioned prestressing is normally used only in the manufacture of precast units utilizing special beds set up in the supplier's factory.						
Labour and plant cost in post-tensioning; material cost excluded						
form ducts to profile including supports and fixings; 50 mm internal diameter	0.04	1.41	3.40	–	m	4.81
Extra for grout vents	0.22	7.04	–	–	nr	7.04
form ducts to profile including supports and fixings; 80 mm internal diameter	0.07	2.21	3.99	–	m	6.20

CLASS G: CONCRETE ANCILLARIES

Item	Gang hours	Labour £	Plant £	Material £	Unit	Total rate £
POST-TENSIONED PRESTRESSING – cont						
Extra for grout vents	0.22	7.04	–	–	nr	7.04
form ducts to profile including supports and fixings; 100 mm internal diameter	0.09	3.02	4.20	–	m	7.22
Extra for grout vents	0.22	7.04	–	–	nr	7.04
grout ducts including provision of equipment; 50 mm internal diameter	0.06	2.11	0.59	–	m	2.70
grout ducts including provision of equipment; 80 mm internal diameter	0.06	2.11	0.74	–	m	2.85
grout ducts including provision of equipment; 100 mm internal diameter	0.09	2.86	0.88	–	m	3.74
form tendons including spacers etc. and pull through ducts; 7 Nr strands	0.25	8.13	4.03	–	m	12.16
form tendons including spacers etc. and pull through ducts; 12 Nr strands	0.45	14.63	7.25	–	m	21.88
form tendons including spacers etc. and pull through ducts; 19 Nr strands	0.65	21.13	10.48	–	m	31.61
dead end anchorage; 7 Nr strands	0.75	24.38	12.09	–	nr	36.47
dead end anchorage; 12 Nr strands	0.95	30.88	15.31	–	nr	46.19
dead end anchorage; 19 Nr strands	1.15	37.38	18.54	–	nr	55.92
looped buried dead end anchorage; 7 Nr strands	0.50	16.25	8.06	–	nr	24.31
looped buried dead end anchorage; 12 Nr strands	0.66	21.45	10.64	–	nr	32.09
looped buried dead end anchorage; 19 Nr strands	0.89	28.93	14.35	–	nr	43.28
end anchorage including reinforcement; 7 Nr strands	1.67	54.27	26.92	–	nr	81.19
add to last for anchorage coupling	0.46	14.95	7.42	–	nr	22.37
end anchorage including reinforcement; 12 Nr strands	2.04	66.30	32.88	–	nr	99.18
add to last for anchorage coupling	0.79	25.68	12.73	–	nr	38.41
end anchorage including reinforcement; 19 Nr strands	2.56	83.20	41.27	–	nr	124.47
add to last for anchorage coupling	1.20	39.00	19.34	–	nr	58.34
stress and lock off including multimatic jack; 7 Nr strands	3.21	104.33	51.75	–	nr	156.08
stress and lock off including multimatic jack; 12 Nr strands	4.19	136.18	67.54	–	nr	203.72
stress and lock off including multimatic jack; 19 Nr strands	5.58	181.35	89.95	–	nr	271.30
cut off and seal ends of tendons; 7 Nr strands	0.20	6.50	3.22	–	nr	9.72
cut off and seal ends of tendons; 12 Nr strands	0.35	11.38	5.64	–	nr	17.02
cut off and seal ends of tendons; 19 Nr strands	0.55	17.88	8.87	–	nr	26.75
CONCRETE ACCESSORIES						
Finishing of top surfaces						
wood float; level	0.02	0.82	–	–	m^2	0.82
wood float; falls or cross-falls	0.03	1.23	–	–	m^2	1.23
steel trowel; level	0.03	1.23	–	–	m^2	1.23
wood float; falls or cross-falls	0.03	1.23	–	–	m^2	1.23
steel trowel; falls or cross-falls	0.05	2.04	–	–	m^2	2.04
granolithic finish 20 mm thick laid monolithically	0.07	2.86	0.54	7.88	m^2	11.28

CLASS G: CONCRETE ANCILLARIES

Item	Gang hours	Labour £	Plant £	Material £	Unit	Total rate £
Finishing of formed surfaces						
aggregate exposure using retarder	0.05	2.04	–	1.18	m²	**3.22**
bush hammering; kango hammer	0.28	11.45	3.55	–	m²	**15.00**
rubbing down concrete surfaces after striking shutters	0.02	0.82	–	1.35	m²	**2.17**
Inserts totally within the concrete volume						
HDPE conduit 20 mm diameter	0.10	4.09	–	1.79	m	**5.88**
black enamelled steel conduit 20 mm diameter	0.10	4.09	–	2.56	m	**6.65**
galvanized steel conduit 20 mm diameter	0.10	4.09	–	2.46	m	**6.55**
Unistrut channel type P3270	0.20	8.18	–	8.15	m	**16.33**
Unistrut channel type P3370	0.20	8.18	–	7.00	m	**15.18**
Inserts projecting from surface(s) of the concrete						
expanding bolt; 10 mm diameter × 25 mm deep	0.05	2.04	–	1.67	nr	**3.71**
holding down bolt; 16 mm diameter × 250 mm deep	0.25	10.22	–	2.63	nr	**12.85**
holding down bolt; 16 mm diameter × 350 mm deep	0.25	10.22	–	3.31	nr	**13.53**
holding down bolt; 20 mm diameter × 250 mm deep	0.25	10.22	–	3.67	nr	**13.89**
holding down bolt; 20 mm diameter × 450 mm deep	0.25	10.22	–	3.94	nr	**14.16**
vitrified clay pipe to BS 65; 100 mm diameter × 1000 mm long	0.25	10.22	–	8.60	nr	**18.82**
cast iron pipe to BS 437; 100 mm diameter × 1000 mm long	0.25	10.22	–	23.68	nr	**33.90**
Grouting under plates; cement and sand (1:3)						
area ne 0.1 m²	0.10	4.09	–	0.21	nr	**4.30**
area 0.1–0.5 m²	0.45	18.40	–	0.99	nr	**19.39**
area 0.5–1.0 m²	0.78	31.89	–	1.98	nr	**33.87**
Grouting under plates; non-shrink cementitious grout						
area ne 0.1 m²	0.10	4.09	–	3.43	nr	**7.52**
area 0.1–0.5 m²	0.45	18.40	–	33.00	nr	**51.40**
area 0.5–1.0 m²	0.78	31.89	–	66.00	nr	**97.89**

CLASS H: PRECAST CONCRETE

Item	Gang hours	Labour £	Plant £	Material £	Unit	Total rate £
NOTES						
The cost of precast concrete items is very much dependent on the complexity of the moulds, the number of units to be cast from each mould and the size and weight of the unit to be handled. The unit rates below are for standard precast items that are often to be found on a civil engineering project. It would be misleading to quote for indicative costs for tailor-made precast concrete units and it is advisable to contact Specialist Manufacturers for guide prices.						
BEAMS						
Concrete mix C20						
Beams						
100 × 150 × 1050 mm long	0.19	6.04	–	10.00	nr	**16.04**
225 × 150 × 1200 mm long	0.26	8.38	2.05	19.30	nr	**29.73**
225 × 225 × 1800 mm long	0.33	10.72	4.26	45.00	nr	**59.98**
PRESTRESSED PRE-TENSIONED BEAMS						
Concrete mix C20						
Beams						
100 × 65 × 1050 mm long	0.11	3.58	–	6.75	nr	**10.33**
265 × 65 × 1800 mm long	0.14	4.49	3.95	23.00	nr	**31.44**
Bridge beams						
Inverted T Beams, flange width 495 mm						
section T1; 8 m long, 380 mm deep; mass 1.88 t	–	–	–	–	nr	**850.00**
section T2; 9 m long, 420 mm deep; mass 2.29 t	–	–	–	–	nr	**1050.00**
section T3; 11 m long, 535 mm deep; mass 3.02 t	–	–	–	–	nr	**1250.00**
section T4; 12 m long, 575 mm deep; mass 3.54 t	–	–	–	–	nr	**1300.00**
section T5; 13 m long, 615 mm deep; mass 4.08 t	–	–	–	–	nr	**1440.00**
section T6; 13 m long, 655 mm deep; mass 4.33 t	–	–	–	–	nr	**1440.00**
section T7; 14 m long, 695 mm deep; mass 4.95 t	–	–	–	–	nr	**1550.00**
section T8; 15 m long, 735 mm deep; mass 5.60 t	–	–	–	–	nr	**1750.00**
section T9; 16 m long, 775 mm deep; mass 6.28 t	–	–	–	–	nr	**1800.00**
section T10; 18 m long, 815 mm deep; mass 7.43 t	–	–	–	–	nr	**2100.00**
M beams, flange width 970 mm						
section M2; 17 m long, 720 mm deep; mass 12.95 t	–	–	–	–	nr	**4750.00**
section M3; 18 m long, 800 mm deep; mass 15.11 t	–	–	–	–	nr	**4400.00**
section M6; 22 m long, 1040 mm deep; mass 20.48 t	–	–	–	–	nr	**7150.00**
section M8; 25 m long, 1200 mm deep; mass 23.68 t	–	–	–	–	nr	**9200.00**
U beams, base width 970 mm						
section U3; 16 m long, 900 mm deep; mass 19.24 t	–	–	–	–	nr	**7500.00**
section U5; 20 m long, 1000 mm deep; mass 25.64 t	–	–	–	–	nr	**9500.00**
section U8; 24 m long, 1200 mm deep; mass 34.56 t	–	–	–	–	nr	**10500.00**
section U12; 30 m long, 1600 mm deep; mass 52.74 t	–	–	–	–	nr	**17500.00**

CLASS H: PRECAST CONCRETE

Item	Gang hours	Labour £	Plant £	Material £	Unit	Total rate £
SLABS						
Prestressed precast concrete flooring planks; Bison or similar; cement mortar grout between planks on bearings						
100 mm thick floor						
400 mm wide planks	0.21	11.44	17.03	32.00	m²	**60.47**
1200 mm wide planks	0.12	6.38	9.50	40.15	m²	**56.03**
150 mm thick floor						
400 mm wide planks	0.26	14.30	21.29	34.50	m²	**70.09**
1200 mm wide planks	0.14	7.97	11.87	42.00	m²	**61.84**
COPINGS, SILLS AND WEIR BLOCKS						
Concrete mix C30						
Coping; weathered and throated						
178 × 64 mm	1.00	9.50	5.00	3.25	m	**17.75**
305 × 76 mm	1.00	7.50	3.50	5.75	m	**16.75**

CLASS I: PIPEWORK – PIPES

Item	Gang hours	Labour £	Plant £	Material £	Unit	Total rate £
NOTES						
The rates assume the most efficient items of plant (excavator) and are optimum rates assuming continuous output with no delays caused by other operations or works.						
Ground conditions are assumed to be good easily worked soil with no abnormal conditions that would affect outputs and consistency of work.						
Multiplier Table for labour and plant for various site conditions for working:						
out of sequence × 2.75 (minimum)						
in hard clay × 1.75 to 2.00						
in running sand × 2.75 (minimum)						
in broken rock × 2.75 to 3.50						
below water table × 2.00 (minimum)						
Variance from CESMM3						
Fittings are included with the pipe concerned, for convenience of reference, rather than in Class J.						
RESOURCES – LABOUR						
Drainage/pipework gang (small bore)						
1 ganger/chargehand (skill rate 4) – 50% of time		7.28				
1 skilled operative (skill rate 4)		13.61				
2 unskilled operatives (general)		25.42				
1 plant operator (skill rate 3)		16.55				
1 plant operator (skill rate 3) – 50% of time		8.28				
Total Gang Rate/Hour		£ 71.13				
Drainage/pipework gang (small bore – not in trenches)						
1 ganger/chargehand (skill rate 4) – 50% of time		7.28				
1 skilled operative (skill rate 4)		13.61				
2 unskilled operatives (general)		25.42				
Total Gang Rate/Hour		£ 46.31				
Drainage/pipework gang (large bore)						
Note: relates to pipes exceeding 700 mm diameter						
1 ganger/chargehand (skill rate 4) – 50% of time		7.28				
1 skilled operative (skill rate 4)		13.61				
2 unskilled operatives (general)		25.42				
1 plant operator (skill rate 3)		16.55				
Total Gang Rate/Hour		£ 62.86				
Drainage/pipework gang (large bore – not in trenches)						
Note: relates to pipes exceeding 700 mm diameter						
1 ganger/chargehand (skill rate 4) – 50% of time		7.28				
1 skilled operative (skill rate 4)		13.61				
2 unskilled operatives (general)		25.42				
Total Gang Rate/Hour		£ 46.31				

CLASS I: PIPEWORK – PIPES

Item	Gang hours	Labour £	Plant £	Material £	Unit	Total rate £
RESOURCES – MATERIALS						
Supply only ex works materials rates						
Clayware Pipes and Fittings						
Vitrified clay; socketed pipes to BS EN 295						
Straight pipes						
100 mm diameter	–	–	–	6.64	m	6.64
150 mm diameter	–	–	–	10.10	m	10.10
225 mm diameter	–	–	–	22.67	m	22.67
300 mm diameter	–	–	–	29.00	m	29.00
400 mm diameter	–	–	–	97.64	m	97.64
450 mm diameter	–	–	–	109.25	m	109.25
Vitrified clay; plain ended pipes to BS EN 295						
Straight pipes						
1.6 m lengths; 100 mm diameter	–	–	–	8.39	m	8.39
1.75 m lengths; 150 mm diameter	–	–	–	16.97	m	16.97
1.75 m lengths; 225 mm diameter	–	–	–	54.79	m	54.79
1.75 m lengths; 300 mm diameter	–	–	–	83.88	m	83.88
Couplings						
polypropylene coupling, 100 mm diameter	–	–	–	6.19	nr	6.19
polypropylene coupling, 150 mm diameter	–	–	–	11.25	nr	11.25
polypropylene coupling, 225 mm diameter	–	–	–	21.58	nr	21.58
polypropylene coupling, 300 mm diameter	–	–	–	44.31	nr	44.31
Tapers						
plain end; taper reducer, 225–150 mm	–	–	–	110.85	nr	110.85
plain end; taper reducer, 150–100 mm	–	–	–	49.66	nr	49.66
Field drains to BS 1196, butt joints, nominal bore; Hepworth						
300 mm lengths; 75 mm diameter	–	–	–	2.50	nr	2.50
300 mm lengths; 100 mm diameter	–	–	–	3.50	nr	3.50
300 mm lengths; 150 mm diameter	–	–	–	7.17	nr	7.17
300 mm lengths; 225 mm diameter	–	–	–	18.44	nr	18.44
Perforated vitrified clay sleeved pipe, BS EN295; Hepline						
1.6 m lengths;100 mm diameter	–	–	–	22.50	m	22.50
1.75 m lengths; 225 mm diameter	–	–	–	51.33	m	51.33
1.75 m lengths; 150 mm diameter	–	–	–	44.73	m	44.73
1.75 m lengths; 300 mm diameter	–	–	–	105.32	m	105.32
Concrete pipes						
Concrete pipes, BS5911 Class L						
300 mm diameter	–	–	–	–	m	–
375 mm diameter	–	–	–	21.94	m	21.94
450 mm diameter	–	–	–	25.89	m	25.89
525 mm diameter	–	–	–	33.94	m	33.94
750 mm diameter	–	–	–	86.00	m	86.00
900 mm diameter	–	–	–	115.39	m	115.39
1200 mm diameter	–	–	–	198.00	m	198.00
1500 mm diameter	–	–	–	349.44	m	349.44
1800 mm diameter	–	–	–	486.00	m	486.00
2100 mm diameter	–	–	–	610.00	m	610.00

CLASS I: PIPEWORK – PIPES

Item	Gang hours	Labour £	Plant £	Material £	Unit	Total rate £
RESOURCES – MATERIALS – cont						
Concrete pipes – cont						
Bends						
bend; 300 mm diameter; <45°	–	–	–	177.20	nr	**177.20**
bend; 375 mm diameter; <45°	–	–	–	177.20	nr	**177.20**
bend; 450 mm diameter; <45°	–	–	–	258.90	nr	**258.90**
bend; 525 mm diameter; <45°	–	–	–	339.40	nr	**339.40**
bend; 750 mm diameter; <45°	–	–	–	–	nr	**–**
bend; 900 mm diameter; 45–90°	–	–	–	1153.90	nr	**1153.90**
bend; 1200 mm diameter; <45°	–	–	–	1980.00	nr	**1980.00**
bend; 1500 mm diameter; <45°	–	–	–	3494.00	nr	**3494.00**
bend; 1800 mm diameter; <45°	–	–	–	4862.00	nr	**4862.00**
bend; 2100 mm diameter; <45°	–	–	–	6012.50	nr	**6012.50**
Cast Iron Pipes and Fittings						
Cast iron drain BS EN 877; plain end						
70 mm dia.	–	–	–	19.91	m	**19.91**
100 mm dia.	–	–	–	23.68	m	**23.68**
150 mm dia.	–	–	–	46.92	m	**46.92**
250 mm dia.	–	–	–	109.49	m	**109.49**
Cast iron couplings complete with stainless steel nuts and bolts and synthetic rubber gaskets						
70 mm dia.	–	–	–	8.51	nr	**8.51**
100 mm dia.	–	–	–	11.08	nr	**11.08**
150 mm dia.	–	–	–	22.20	nr	**22.20**
250 mm dia.	–	–	–	71.13	nr	**71.13**
Bends; 45°						
70 mm dia.	–	–	–	15.06	nr	**15.06**
100 mm dia.	–	–	–	17.83	nr	**17.83**
150 mm dia.	–	–	–	32.00	nr	**32.00**
250 mm dia.	–	–	–	18.60	nr	**18.60**
Bends; short radius; 87.5°						
70 mm dia.	–	–	–	15.06	nr	**15.06**
100 mm dia.	–	–	–	17.83	nr	**17.83**
150 mm dia.	–	–	–	32.02	nr	**32.02**
250 mm dia.	–	–	–	186.00	nr	**186.00**
Single junction × 45°						
70 × 70 mm dia.	–	–	–	22.66	nr	**22.66**
100 × 100 mm	–	–	–	28.86	nr	**28.86**
150 × 150 mm	–	–	–	69.08	nr	**69.08**
250 × 250 mm dia.	–	–	–	416.89	nr	**416.89**
Taper reducer pipes						
70 mm dia.	–	–	–	20.58	nr	**20.58**
100 mm dia.	–	–	–	24.20	nr	**24.20**
150 mm dia.	–	–	–	46.46	nr	**46.46**
250 mm dia.	–	–	–	155.84	nr	**155.84**

CLASS I: PIPEWORK – PIPES

Item	Gang hours	Labour £	Plant £	Material £	Unit	Total rate £
RESOURCES – PLANT						
Field drains						
0.4 m³ hydraulic backacter			37.03			
2 t dumper – 30% of time			1.98			
Stihl saw, 12', petrol – 30% of time			0.28			
small pump – 30% of time			1.11			
Total Rate/Hour		£	**40.40**			
Add to the above for trench supports appropriate to trench depth (see below)						
Field drains (not in trenches)						
2 t dumper – 30% of time			1.98			
Stihl saw, 12', petrol – 30% of time			0.28			
Total Rate/Hour		£	**2.26**			
Drains/sewers (small bore)						
1.0 m³ hydraulic backacter			55.30			
2 t dumper – 30% of time			1.98			
2.80 m³/min compressor, 2 tool – 30% of time			2.21			
disc saw – 30% of time			0.30			
extra 50 ft / 15 m hose – 30% of time			0.09			
small pump – 30% of time			1.11			
sundry tools – 30% of time			0.38			
Total Rate/Hour		£	**61.37**			
Add to the above for trench supports appropriate to trench depth (see below)						
Drains/sewers (small bore – not in trenches)						
2 t dumper – 30% of time			1.98			
2.80 m³/min compressor, 2 tool – 30% of time			2.21			
disc saw – 30% of time			0.30			
extra 50 ft / 15 m hose – 30% of time			0.09			
sundry tools – 30% of time			0.38			
Total Rate/Hour		£	**4.96**			
Drains/sewers (large bore)						
1.0 m³ hydraulic backacter			55.30			
20 t crawler crane – 50% of time			25.08			
2 t dumper (30% of time)			1.98			
2.80 m³/min compressor, 2 tool – 30% of time			2.21			
disc saw – 30% of time			0.30			
extra 50 ft / 15 m hose – 30% of time			0.09			
small pump – 30% of time			1.11			
sundry tools – 30% of time			0.38			
Total Rate/Hour		£	**86.44**			
Add to the above for trench supports appropriate to trench depth (see below)						
Drains/sewers (large bore – not in trenches)						
2 t dumper (30% of time)			1.98			
2.80 m³/min compressor, 2 tool – 30% of time			2.21			
disc saw – 30% of time			0.30			
extra 50 ft / 15 m hose – 30% of time			0.09			
sundry tools – 30% of time			0.38			
Total Rate/Hour		£	**4.96**			

CLASS I: PIPEWORK – PIPES

Item	Gang hours	Labour £	Plant £	Material £	Unit	Total rate £
RESOURCES – PLANT – cont						
Trench supports						
In addition to the above, the following allowances for close sheeted trench supports are included in the following unit rates, assuming that the ground conditions warrants it:						
ne 1.50 m deep			3.94		m	
1.50–2.00 m deep			5.49		m	
2.00–2.50 m deep			5.25		m	
2.50–3.00 m deep			6.47		m	
3.00–3.50 m deep			7.86		m	
3.50–4.00 m deep			8.66		m	
4.00–4.50 m deep			9.69		m	
4.50–5.00 m deep			17.15		m	
5.00–5.50 m deep			23.63		m	
CLAY PIPES						
Field drains to BS 1196, butt joints, nominal bore; excavation and supports, backfilling						
75 mm pipes; in trench, depth						
not in trenches	0.06	2.78	0.05	8.54	m	**11.37**
ne 1.50 m deep	0.09	6.40	3.64	8.54	m	**18.58**
1.50–2.00 m deep	0.13	9.25	5.28	8.54	m	**23.07**
2.00–2.50 m deep	0.18	12.80	7.32	8.54	m	**28.66**
2.50–3.00 m deep	0.23	16.36	9.35	8.54	m	**34.25**
100 mm pipes; in trench, depth						
not in trenches	0.06	2.78	0.05	11.96	m	**14.79**
ne 1.50 m deep	0.10	7.11	4.05	11.96	m	**23.12**
1.50–2.00 m deep	0.14	9.96	5.69	11.96	m	**27.61**
2.00–2.50 m deep	0.19	13.52	7.71	11.96	m	**33.19**
2.50–3.00 m deep	0.24	17.07	9.76	11.96	m	**38.79**
150 mm pipes; in trench, depth						
not in trenches	0.06	2.78	0.05	24.49	m	**27.32**
ne 1.50 m deep	0.11	7.82	4.45	24.49	m	**36.76**
1.50–2.00 m deep	0.15	10.67	6.09	24.49	m	**41.25**
2.00–2.50 m deep	0.20	14.23	8.12	24.49	m	**46.84**
2.50–3.00 m deep	0.25	17.78	10.17	24.49	m	**52.44**
225 mm pipes; in trench, depth						
not in trenches	0.06	2.78	0.05	63.00	m	**65.83**
ne 1.50 m deep	0.13	9.25	5.26	63.00	m	**77.51**
1.50–2.00 m deep	0.17	12.09	6.91	63.00	m	**82.00**
2.00–2.50 m deep	0.22	15.65	8.94	63.00	m	**87.59**
2.50–3.00 m deep	0.27	19.21	10.97	63.00	m	**93.18**

CLASS I: PIPEWORK – PIPES

Item	Gang hours	Labour £	Plant £	Material £	Unit	Total rate £
Vitrified clay perforated field drains to BS EN295, sleeved joints; excavation and supports, backfilling						
100 mm pipes; in trench, depth						
not in trenches	0.06	2.78	0.05	23.06	m	**25.89**
ne 1.50 m deep	0.15	10.67	6.06	23.06	m	**39.79**
1.50–2.00 m deep	0.19	13.52	7.71	23.06	m	**44.29**
2.00–2.50 m deep	0.24	17.07	9.74	23.06	m	**49.87**
2.50–3.00 m deep	0.29	20.63	11.79	23.06	m	**55.48**
Extra for bend	0.08	5.69	–	9.81	nr	**15.50**
Extra for single junction	0.09	6.40	–	21.17	nr	**27.57**
150 mm pipes; in trench, depth						
not in trenches	0.06	2.78	0.05	45.85	m	**48.68**
ne 1.50 m deep	0.17	12.09	6.88	45.85	m	**64.82**
1.50–2.00 m deep	0.20	14.23	8.12	45.85	m	**68.20**
2.00–2.50 m deep	0.24	17.07	9.74	45.85	m	**72.66**
2.50–3.00 m deep	0.29	20.63	11.79	45.85	m	**78.27**
Extra for bend	0.13	8.89	–	20.19	nr	**29.08**
Extra for single junction	0.13	9.25	–	29.65	nr	**38.90**
225 mm pipes; in trench, depth						
not in trenches	0.08	3.70	0.05	52.61	m	**56.36**
ne 1.50 m deep	0.18	12.80	7.29	52.61	m	**72.70**
1.50–2.00 m deep	0.22	15.65	8.94	52.61	m	**77.20**
2.00–2.50 m deep	0.26	18.50	10.57	52.61	m	**81.68**
2.50–3.00 m deep	0.30	21.34	12.21	52.61	m	**86.16**
Extra for bend	0.12	8.54	–	102.95	nr	**111.49**
Extra for single junction	0.20	13.87	–	143.73	nr	**157.60**
300 mm pipes; in trench, depth						
not in trenches	0.10	4.63	0.23	107.95	m	**112.81**
ne 1.50 m deep	0.19	13.52	7.68	107.95	m	**129.15**
1.50–2.00 m deep	0.24	17.07	9.75	107.95	m	**134.77**
2.00–2.50 m deep	0.28	19.92	11.37	107.95	m	**139.24**
2.50–3.00 m deep	0.32	22.76	13.01	107.95	m	**143.72**
Extra for bend	0.24	17.07	–	195.53	nr	**212.60**
Extra for single junction	0.23	16.36	–	368.92	nr	**385.28**
Vitrified clay pipes to BS EN295, plain ends with push-fit polypropylene flexible couplings; excavation and supports, backfilling						
100 mm pipes; in trenches, depth						
not in trenches	0.06	2.78	0.31	12.57	m	**15.66**
ne 1.50 m deep	0.15	10.67	9.06	12.57	m	**32.30**
1.50–2.00 m deep	0.19	13.52	11.69	12.57	m	**37.78**
2.00–2.50 m deep	0.24	17.07	14.78	12.57	m	**44.42**
2.50–3.00 m deep	0.29	20.63	17.88	12.57	m	**51.08**
3.00–3.50 m deep	0.36	25.61	22.20	12.57	m	**60.38**
3.50–4.00 m deep	0.44	31.30	27.16	12.57	m	**71.03**
4.00–4.50 m deep	0.55	39.12	33.96	12.57	m	**85.65**
4.50–5.00 m deep	0.70	49.79	43.29	12.57	m	**105.65**
5.00–5.50 m deep	0.90	64.02	55.68	12.57	m	**132.27**
Extra for bend	0.05	3.56	–	26.76	nr	**30.32**
Extra for rest bend	0.06	4.27	–	33.55	nr	**37.82**
Extra for single junction; equal	0.09	6.40	–	44.57	nr	**50.97**
Extra for saddle; oblique	0.23	16.36	–	31.18	nr	**47.54**

CLASS I: PIPEWORK – PIPES

Item	Gang hours	Labour £	Plant £	Material £	Unit	Total rate £
CLAY PIPES – cont						
Vitrified clay pipes to BS EN295, plain ends with push-fit polypropylene flexible couplings – cont						
150 mm pipes; in trenches, depth						
not in trenches	0.06	2.78	0.31	23.98	m	27.07
ne 1.50 m deep	0.17	12.09	10.45	23.98	m	46.52
1.50–2.00 m deep	0.20	14.23	12.32	23.98	m	50.53
2.00–2.50 m deep	0.24	17.07	14.78	23.98	m	55.83
2.50–3.00 m deep	0.29	20.63	17.88	23.98	m	62.49
3.00–3.50 m deep	0.39	27.74	24.04	23.98	m	75.76
3.50–4.00 m deep	0.45	32.01	27.79	23.98	m	83.78
4.00–4.50 m deep	0.58	41.26	35.84	23.98	m	101.08
4.50–5.00 m deep	0.75	53.35	46.36	23.98	m	123.69
5.00–5.50 m deep	0.96	68.29	59.37	23.98	m	151.64
Extra for bend	0.08	5.69	–	54.06	nr	59.75
Extra for rest bend	0.09	6.40	–	54.33	nr	60.73
Extra for single junction; equal	0.11	7.82	–	68.83	nr	76.65
Extra for taper reducer	0.07	4.98	–	69.21	nr	74.19
Extra for saddle; oblique	0.29	20.63	–	48.22	nr	68.85
225 mm pipes; in trenches, depth						
not in trenches	0.08	3.70	0.40	68.12	m	72.22
ne 1.50 m deep	0.18	12.80	11.06	68.12	m	91.98
1.50–2.00 m deep	0.22	15.65	13.56	68.12	m	97.33
2.00–2.50 m deep	0.26	18.50	16.03	68.12	m	102.65
2.50–3.00 m deep	0.30	21.34	18.50	68.12	m	107.96
3.00–3.50 m deep	0.41	29.17	25.29	68.12	m	122.58
3.50–4.00 m deep	0.47	33.43	29.01	68.12	m	130.56
4.00–4.50 m deep	0.65	46.24	40.16	68.12	m	154.52
4.50–5.00 m deep	0.80	56.91	49.45	68.12	m	174.48
5.00–5.50 m deep	1.02	72.56	63.10	68.12	m	203.78
Extra for bend	0.10	7.11	–	93.34	nr	100.45
Extra for rest bend	0.11	7.82	–	178.59	nr	186.41
Extra for single junction; equal	0.16	11.38	–	284.45	nr	295.83
Extra for taper reducer	0.12	8.54	–	148.09	nr	156.63
Extra for saddle; oblique	0.36	25.61	–	158.86	nr	184.47
300 mm pipes; in trenches, depth						
not in trenches	0.10	4.63	0.51	104.14	m	109.28
ne 1.50 m deep	0.19	13.52	11.66	104.14	m	129.32
1.50–2.00 m deep	0.24	17.07	14.78	104.14	m	135.99
2.00–2.50 m deep	0.28	19.92	17.24	104.14	m	141.30
2.50–3.00 m deep	0.32	22.76	19.72	104.14	m	146.62
3.00–3.50 m deep	0.42	29.88	25.92	104.14	m	159.94
3.50–4.00 m deep	0.52	36.99	32.10	104.14	m	173.23
4.00–4.50 m deep	0.70	49.79	43.25	104.14	m	197.18
4.50–5.00 m deep	0.90	64.02	55.65	104.14	m	223.81
5.00–5.50 m deep	1.12	79.67	69.27	104.14	m	253.08
Extra for bend	0.15	10.67	–	142.72	nr	153.39
Extra for rest bend	0.16	11.38	–	445.27	nr	456.65
Extra for single junction; equal	0.19	13.52	–	509.79	nr	523.31
Extra for saddle; oblique	0.44	31.30	–	283.53	nr	314.83

CLASS I: PIPEWORK – PIPES

Item	Gang hours	Labour £	Plant £	Material £	Unit	Total rate £
Vitrified clay pipes to BS EN295, spigot and socket joints with sealing ring; excavation and supports, backfilling						
100 mm pipes; in trenches, depth						
not in trenches	0.06	2.78	0.33	6.80	m	9.91
ne 1.50 m deep	0.15	10.67	9.33	6.80	m	26.80
1.50–2.00 m deep	0.19	13.52	11.69	6.80	m	32.01
2.00–2.50 m deep	0.24	17.07	14.78	6.80	m	38.65
2.50–3.00 m deep	0.29	20.63	17.88	6.80	m	45.31
3.00–3.50 m deep	0.36	25.61	22.20	6.80	m	54.61
3.50–4.00 m deep	0.44	31.30	27.16	6.80	m	65.26
4.00–4.50 m deep	0.55	39.12	33.96	6.80	m	79.88
4.50–5.00 m deep	0.70	49.79	43.29	6.80	m	99.88
5.00–5.50 m deep	0.90	64.02	55.68	6.80	m	126.50
Extra for bend	0.05	3.56	–	17.91	nr	21.47
Extra for rest bend	0.06	4.27	–	27.30	nr	31.57
Extra for single junction; equal	0.09	6.40	–	25.07	nr	31.47
Extra fro saddle; oblique	0.23	16.36	–	24.61	nr	40.97
150 mm pipes; in trenches, depth						
not in trenches	0.06	2.78	0.31	10.35	m	13.44
ne 1.50 m deep	0.17	12.09	10.45	10.35	m	32.89
1.50–2.00 m deep	0.20	14.23	12.32	10.35	m	36.90
2.00–2.50 m deep	0.24	17.07	14.78	10.35	m	42.20
2.50–3.00 m deep	0.29	20.63	17.88	10.35	m	48.86
3.00–3.50 m deep	0.39	27.74	24.04	10.35	m	62.13
3.50–4.00 m deep	0.45	32.01	27.79	10.35	m	70.15
4.00–4.50 m deep	0.58	41.26	35.84	10.35	m	87.45
4.50–5.00 m deep	0.75	53.35	46.36	10.35	m	110.06
5.00–5.50 m deep	0.96	68.29	59.37	10.35	m	138.01
Extra for bend	0.08	5.69	–	23.90	nr	29.59
Extra for rest bend	0.09	6.40	–	30.71	nr	37.11
Extra for single junction; equal	0.11	7.82	–	31.99	nr	39.81
Extra for double junction; equal	0.13	9.25	–	53.12	nr	62.37
Extra for taper reducer	0.07	4.98	–	50.90	nr	55.88
Extra for saddle; oblique	0.29	20.63	–	36.61	nr	57.24
225 mm pipes; in trenches, depth						
not in trenches	0.08	3.70	0.40	23.24	m	27.34
ne 1.50 m deep	0.18	12.80	11.07	23.24	m	47.11
1.50–2.00 m deep	0.22	15.65	13.56	23.24	m	52.45
2.00–2.50 m deep	0.26	18.50	16.03	23.24	m	57.77
2.50–3.00 m deep	0.30	21.34	18.50	23.24	m	63.08
3.00–3.50 m deep	0.41	29.17	25.29	23.24	m	77.70
3.50–4.00 m deep	0.47	33.43	28.98	23.24	m	85.65
4.00–4.50 m deep	0.65	46.24	40.16	23.24	m	109.64
4.50–5.00 m deep	0.80	56.91	49.45	23.24	m	129.60
5.00–5.50 m deep	1.02	72.56	63.10	23.24	m	158.90
Extra for bend	0.10	7.11	–	121.87	nr	128.98
Extra for rest bend	0.11	7.82	–	133.27	nr	141.09
Extra for single junction; equal	0.16	11.38	–	170.12	nr	181.50
Extra for double junction; equal	0.18	12.80	–	162.19	nr	174.99
Extra for taper reducer	0.12	8.54	–	113.22	nr	121.76
Extra for saddle; oblique	0.36	25.61	–	136.20	nr	161.81

CLASS I: PIPEWORK – PIPES

Item	Gang hours	Labour £	Plant £	Material £	Unit	Total rate £
CLAY PIPES – cont						
Vitrified clay pipes to BS EN295, spigot and socket joints with sealing ring – cont						
300 mm pipes; in trenches, depth						
not in trenches	0.10	4.63	0.51	29.73	m	**34.87**
ne 1.50 m deep	0.19	13.52	11.66	29.73	m	**54.91**
1.50–2.00 m deep	0.24	17.07	14.78	29.73	m	**61.58**
2.00–2.50 m deep	0.28	19.92	17.24	29.73	m	**66.89**
2.50–3.00 m deep	0.32	22.76	19.72	29.73	m	**72.21**
3.00–3.50 m deep	0.42	29.88	25.92	29.73	m	**85.53**
3.50–4.00 m deep	0.52	36.99	32.10	29.73	m	**98.82**
4.00–4.50 m deep	0.70	49.79	43.25	29.73	m	**122.77**
4.50–5.00 m deep	0.90	64.02	55.65	29.73	m	**149.40**
5.00–5.50 m deep	1.12	79.67	69.27	29.73	m	**178.67**
Extra for bend	0.15	10.67	–	231.42	nr	**242.09**
Extra for rest bend	0.16	11.38	–	352.22	nr	**363.60**
Extra for single junction; equal	0.19	13.52	–	364.44	nr	**377.96**
Extra for double junction; equal	0.21	14.94	–	372.30	nr	**387.24**
Extra for taper reducer	0.15	10.67	–	223.40	nr	**234.07**
Extra for saddle; oblique	0.44	31.30	–	237.00	nr	**268.30**
400 mm pipes; in trenches, depth						
not in trenches	0.10	4.63	0.51	100.08	m	**105.22**
ne 1.50 m deep	0.23	16.36	14.12	100.08	m	**130.56**
1.50–2.00 m deep	0.29	20.63	17.87	100.08	m	**138.58**
2.00–2.50 m deep	0.32	22.76	19.71	100.08	m	**142.55**
2.50–3.00 m deep	0.38	27.03	23.43	100.08	m	**150.54**
3.00–3.50 m deep	0.46	32.72	28.38	100.08	m	**161.18**
3.50–4.00 m deep	0.58	41.26	35.82	100.08	m	**177.16**
4.00–4.50 m deep	0.75	53.35	46.32	100.08	m	**199.75**
4.50–5.00 m deep	0.95	67.58	58.72	100.08	m	**226.38**
5.00–5.50 m deep	1.20	85.36	74.22	100.08	m	**259.66**
Extra for bend; 90°	0.24	17.07	–	617.58	nr	**634.65**
Extra for bend; 45°	0.24	17.07	–	617.58	nr	**634.65**
Extra for bend; 22.5°	0.24	17.07	–	617.58	nr	**634.65**
450 mm pipes; in trenches, depth						
not in trenches	0.23	10.65	1.14	111.98	m	**123.77**
ne 1.50 m deep	0.23	16.36	14.12	111.98	m	**142.46**
1.50–2.00 m deep	0.30	21.34	18.49	111.98	m	**151.81**
2.00–2.50 m deep	0.32	22.76	19.71	111.98	m	**154.45**
2.50–3.00 m deep	0.38	27.03	23.43	111.98	m	**162.44**
3.00–3.50 m deep	0.47	33.43	28.98	111.98	m	**174.39**
3.50–4.00 m deep	0.60	42.68	37.04	111.98	m	**191.70**
4.00–4.50 m deep	0.77	54.77	47.57	111.98	m	**214.32**
4.50–5.00 m deep	0.97	69.00	59.97	111.98	m	**240.95**
5.00–5.50 m deep	1.20	85.36	74.22	111.98	m	**271.56**
Extra for bend; 90°	0.29	20.63	–	813.25	nr	**833.88**
Extra for bend; 45°	0.29	20.63	–	813.25	nr	**833.88**
Extra for bend; 22.5°	0.29	20.63	–	813.25	nr	**833.88**

CLASS I: PIPEWORK – PIPES

Item	Gang hours	Labour £	Plant £	Material £	Unit	Total rate £
CONCRETE PIPES						
Concrete porous pipes to BS 5911; excavation and supports, backfilling						
150 mm pipes; in trench, depth						
not in trenches	0.06	2.78	0.28	5.85	m	**8.91**
ne 1.50 m deep	0.17	12.09	6.88	5.85	m	**24.82**
1.50–2.00 m deep	0.20	14.23	8.12	5.85	m	**28.20**
2.00–2.50 m deep	0.24	17.07	9.74	5.85	m	**32.66**
2.50–3.00 m deep	0.29	20.63	11.79	5.85	m	**38.27**
3.00–3.50 m deep	0.39	27.74	15.87	5.85	m	**49.46**
3.50–4.00 m deep	0.45	32.01	18.35	5.85	m	**56.21**
4.00–4.50 m deep	0.58	41.26	23.67	5.85	m	**70.78**
4.50–5.00 m deep	0.75	53.35	30.63	5.85	m	**89.83**
5.00–5.50 m deep	0.96	68.29	39.24	5.85	m	**113.38**
225 mm pipes; in trench, depth						
not in trenches	0.08	3.70	0.18	6.30	m	**10.18**
ne 1.50 m deep	0.18	12.80	7.29	6.30	m	**26.39**
1.50–2.00 m deep	0.22	15.65	8.94	6.30	m	**30.89**
2.00–2.50 m deep	0.26	18.50	10.57	6.30	m	**35.37**
2.50–3.00 m deep	0.30	21.34	12.21	6.30	m	**39.85**
3.00–3.50 m deep	0.41	29.17	16.69	6.30	m	**52.16**
3.50–4.00 m deep	0.47	33.43	19.16	6.30	m	**58.89**
4.00–4.50 m deep	0.65	46.24	26.52	6.30	m	**79.06**
4.50–5.00 m deep	0.80	56.91	32.68	6.30	m	**95.89**
5.00–5.50 m deep	1.02	72.56	41.70	6.30	m	**120.56**
300 mm pipes; in trench, depth						
not in trenches	0.10	4.63	0.23	6.94	m	**11.80**
ne 1.50 m deep	0.19	13.52	7.68	6.94	m	**28.14**
1.50–2.00 m deep	0.24	17.07	9.75	6.94	m	**33.76**
2.00–2.50 m deep	0.28	19.92	11.37	6.94	m	**38.23**
2.50–3.00 m deep	0.32	22.76	13.01	6.94	m	**42.71**
3.00–3.50 m deep	0.42	29.88	17.10	6.94	m	**53.92**
3.50–4.00 m deep	0.52	36.99	21.20	6.94	m	**65.13**
4.00–4.50 m deep	0.70	49.79	28.57	6.94	m	**85.30**
4.50–5.00 m deep	0.90	64.02	36.77	6.94	m	**107.73**
5.00–5.50 m deep	1.12	79.67	45.78	6.94	m	**132.39**
Concrete pipes with rebated flexible joints to BS 5911 Class 120; excavation and supports, backfilling						
300 mm pipes; in trenches, depth						
not in trenches	0.12	5.56	0.31	17.72	m	**23.59**
ne 1.50 m deep	0.22	15.65	13.53	17.72	m	**46.90**
1.50–2.00 m deep	0.26	18.50	16.03	17.72	m	**52.25**
2.00–2.50 m deep	0.30	21.34	18.49	17.72	m	**57.55**
2.50–3.00 m deep	0.34	24.19	20.97	17.72	m	**62.88**
3.00–3.50 m deep	0.44	31.30	27.13	17.72	m	**76.15**
3.50–4.00 m deep	0.56	39.84	34.57	17.72	m	**92.13**

CLASS I: PIPEWORK – PIPES

Item	Gang hours	Labour £	Plant £	Material £	Unit	Total rate £
CONCRETE PIPES – cont						
Concrete pipes with rebated flexible joints to BS 5911 Class 120 – cont						
300 mm pipes – cont						
4.00–4.50 m deep	0.73	51.93	45.10	17.72	m	**114.75**
4.50–5.00 m deep	0.92	65.44	56.87	17.72	m	**140.03**
5.00–5.50 m deep	1.19	84.65	73.59	17.72	m	**175.96**
5.50–6.00 m deep	1.42	101.01	87.91	17.72	m	**206.64**
Extra for bend	0.08	5.69	4.52	177.20	nr	**187.41**
375 mm pipes; in trenches, depth						
not in trenches	0.15	6.95	0.74	21.94	m	**29.63**
ne 1.50 m deep	0.24	17.07	14.74	21.94	m	**53.75**
1.50–2.00 m deep	0.29	20.63	17.87	21.94	m	**60.44**
2.00–2.50 m deep	0.33	23.47	20.33	21.94	m	**65.74**
2.50–3.00 m deep	0.38	27.03	23.43	21.94	m	**72.40**
3.00–3.50 m deep	0.46	32.72	28.38	21.94	m	**83.04**
3.50–4.00 m deep	0.58	41.26	35.82	21.94	m	**99.02**
4.00–4.50 m deep	0.75	53.35	46.32	21.94	m	**121.61**
4.50–5.00 m deep	0.95	67.58	58.72	21.94	m	**148.24**
5.00–5.50 m deep	1.20	85.36	74.22	21.94	m	**181.52**
5.50–6.00 m deep	1.45	103.15	89.76	21.94	m	**214.85**
Extra for bend	0.10	7.11	6.31	177.20	nr	**190.62**
450 mm pipes; in trenches, depth						
not in trenches	0.17	7.87	0.85	25.89	m	**34.61**
ne 1.50 m deep	0.25	17.78	15.36	25.89	m	**59.03**
1.50–2.00 m deep	0.31	22.05	19.08	25.89	m	**67.02**
2.00–2.50 m deep	0.35	24.90	21.56	25.89	m	**72.35**
2.50–3.00 m deep	0.40	28.45	24.65	25.89	m	**78.99**
3.00–3.50 m deep	0.48	34.14	29.65	25.89	m	**89.68**
3.50–4.00 m deep	0.63	44.82	38.89	25.89	m	**109.60**
4.00–4.50 m deep	0.80	56.91	49.41	25.89	m	**132.21**
4.50–5.00 m deep	1.00	71.14	61.82	25.89	m	**158.85**
5.00–5.50 m deep	1.25	88.92	77.32	25.89	m	**192.13**
5.50–6.00 m deep	1.51	107.41	93.46	25.89	m	**226.76**
Extra for bend	0.13	9.25	9.04	258.90	nr	**277.19**
525 mm pipes; in trenches, depth						
not in trenches	0.20	9.26	0.99	33.94	m	**44.19**
ne 1.50 m deep	0.27	19.21	16.58	33.94	m	**69.73**
1.50–2.00 m deep	0.33	23.47	20.33	33.94	m	**77.74**
2.00–2.50 m deep	0.37	26.32	22.79	33.94	m	**83.05**
2.50–3.00 m deep	0.42	29.88	25.90	33.94	m	**89.72**
3.00–3.50 m deep	0.49	34.86	30.23	33.94	m	**99.03**
3.50–4.00 m deep	0.65	46.24	40.14	33.94	m	**120.32**
4.00–4.50 m deep	0.83	59.04	51.26	33.94	m	**144.24**
4.50–5.00 m deep	1.03	73.27	63.67	33.94	m	**170.88**
5.00–5.50 m deep	1.28	91.05	79.17	33.94	m	**204.16**
5.50–6.00 m deep	1.55	110.26	95.93	33.94	m	**240.13**
Extra for bend	0.16	11.38	11.74	339.40	nr	**362.52**

CLASS I: PIPEWORK – PIPES

Item	Gang hours	Labour £	Plant £	Material £	Unit	Total rate £
750 mm pipes; in trenches, depth						
not in trenches	0.22	10.19	1.10	86.00	m	**97.29**
ne 1.50 m deep	0.30	18.86	25.96	86.00	m	**130.82**
1.50–2.00 m deep	0.36	22.63	31.20	86.00	m	**139.83**
2.00–2.50 m deep	0.41	25.77	35.54	86.00	m	**147.31**
2.50–3.00 m deep	0.47	29.54	40.74	86.00	m	**156.28**
3.00–3.50 m deep	0.58	36.46	50.33	86.00	m	**172.79**
3.50–4.00 m deep	0.80	50.29	69.45	86.00	m	**205.74**
4.00–4.50 m deep	1.05	66.00	91.19	86.00	m	**243.19**
4.50–5.00 m deep	1.30	81.72	112.98	86.00	m	**280.70**
5.00–5.50 m deep	1.55	97.43	134.72	86.00	m	**318.15**
5.50–6.00 m deep	1.82	114.41	158.31	86.00	m	**358.72**
Extra for bends	0.24	15.09	34.09	860.00	nr	**909.18**
900 mm pipes; in trenches, depth						
not in trenches	0.25	11.58	1.24	115.39	m	**128.21**
ne 1.50 m deep	0.33	20.74	28.55	115.39	m	**164.68**
1.50–2.00 m deep	0.40	25.14	34.66	115.39	m	**175.19**
2.00–2.50 m deep	0.46	28.92	39.88	115.39	m	**184.19**
2.50–3.00 m deep	0.53	33.32	45.96	115.39	m	**194.67**
3.00–3.50 m deep	0.70	44.00	60.74	115.39	m	**220.13**
3.50–4.00 m deep	0.92	57.83	79.87	115.39	m	**253.09**
4.00–4.50 m deep	1.20	75.43	104.21	115.39	m	**295.03**
4.50–5.00 m deep	1.50	94.29	130.36	115.39	m	**340.04**
5.00–5.50 m deep	1.80	113.15	156.46	115.39	m	**385.00**
5.50–6.00 m deep	2.10	132.01	182.66	115.39	m	**430.06**
Extra for bends	0.39	24.52	55.85	1153.90	nr	**1234.27**
1200 mm pipes; in trenches, depth						
not in trenches	0.25	11.58	1.24	198.00	m	**210.82**
ne 1.50 m deep	0.46	28.92	39.80	198.00	m	**266.72**
1.50–2.00 m deep	0.53	33.32	45.93	198.00	m	**277.25**
2.00–2.50 m deep	0.60	37.72	51.99	198.00	m	**287.71**
2.50–3.00 m deep	0.70	44.00	60.71	198.00	m	**302.71**
3.00–3.50 m deep	0.85	53.43	73.74	198.00	m	**325.17**
3.50–4.00 m deep	1.12	70.40	97.23	198.00	m	**365.63**
4.00–4.50 m deep	1.45	91.15	125.93	198.00	m	**415.08**
4.50–5.00 m deep	1.75	110.00	152.06	198.00	m	**460.06**
5.00–5.50 m deep	2.05	128.86	178.20	198.00	m	**505.06**
5.50–6.00 m deep	2.36	148.35	205.26	198.00	m	**551.61**
Extra for bends	0.51	32.06	73.24	1980.00	nr	**2085.30**
1500 mm pipes; in trenches, depth						
not in trenches	0.35	16.21	1.73	349.44	m	**367.38**
ne 1.50 m deep	0.60	37.72	51.90	349.44	m	**439.06**
1.50–2.00 m deep	0.70	44.00	60.67	349.44	m	**454.11**
2.00–2.50 m deep	0.81	50.92	70.20	349.44	m	**470.56**
2.50–3.00 m deep	0.92	57.83	79.77	349.44	m	**487.04**
3.00–3.50 m deep	1.05	66.00	91.09	349.44	m	**506.53**
3.50–4.00 m deep	1.27	79.83	110.25	349.44	m	**539.52**
4.00–4.50 m deep	1.70	106.86	147.65	349.44	m	**603.95**
4.50–5.00 m deep	2.05	128.86	178.14	349.44	m	**656.44**
5.00–5.50 m deep	2.40	150.86	208.62	349.44	m	**708.92**
5.50–6.00 m deep	2.75	172.87	236.77	352.31	m	**761.95**
Extra for bends	0.63	39.60	90.47	3494.00	nr	**3624.07**

CLASS I: PIPEWORK – PIPES

Item	Gang hours	Labour £	Plant £	Material £	Unit	Total rate £
CONCRETE PIPES – cont						
Concrete pipes with rebated flexible joints to BS 5911 Class 120 – cont						
1800 mm pipes; in trenches, depth						
not in trenches	0.40	18.52	1.98	486.00	m	506.50
ne 1.50 m deep	0.77	48.40	66.61	486.00	m	601.01
1.50–2.00 m deep	0.91	57.20	78.85	486.00	m	622.05
2.00–2.50 m deep	1.03	64.75	89.25	486.00	m	640.00
2.50–3.00 m deep	1.12	70.40	97.11	486.00	m	653.51
3.00–3.50 m deep	1.20	75.43	104.09	486.00	m	665.52
3.50–4.00 m deep	1.52	95.55	131.96	486.00	m	713.51
4.00–4.50 m deep	2.00	125.72	173.69	486.00	m	785.41
4.50–5.00 m deep	2.40	150.86	208.54	486.00	m	845.40
5.00–5.50 m deep	2.80	176.01	243.39	486.00	m	905.40
5.50–6.00 m deep	3.15	198.01	273.96	486.00	m	957.97
Extra for bends	0.77	48.40	110.59	4862.00	nr	5020.99
2100 mm pipes; in trenches, depth						
not in trenches	0.45	20.84	2.24	610.00	m	633.08
ne 1.50 m deep	0.98	61.60	84.78	610.00	m	756.38
1.50–2.00 m deep	1.13	71.03	97.93	610.00	m	778.96
2.00–2.50 m deep	1.23	77.32	106.45	610.00	m	793.77
2.50–3.00 m deep	1.30	81.72	112.73	610.00	m	804.45
3.00–3.50 m deep	1.50	94.29	130.13	610.00	m	834.42
3.50–4.00 m deep	1.82	114.41	158.02	610.00	m	882.43
4.00–4.50 m deep	2.35	147.72	204.07	610.00	m	961.79
4.50–5.00 m deep	2.80	176.01	243.30	610.00	m	1029.31
5.00–5.50 m deep	3.20	201.15	278.15	610.00	m	1089.30
5.50–6.00 m deep	3.55	223.15	308.75	610.00	m	1141.90
Extra for bends	0.89	55.95	127.94	6012.50	nr	6196.39
IRON PIPES						
Cast iron pipes to BS 437 plain ended pipe with 'Timesaver' mechanical coupling joints; excavation and supports, backfilling						
75 mm pipes; in trenches, depth						
not in trenches	0.12	5.56	0.59	22.74	m	28.89
ne 1.50 m deep	0.19	13.52	11.47	22.74	m	47.73
1.50–2.00 m deep	0.21	14.94	12.94	22.74	m	50.62
2.00–2.50 m deep	0.28	19.92	17.24	22.74	m	59.90
2.50–3.00 m deep	0.34	24.19	20.97	22.74	m	67.90
3.00–3.50 m deep	0.43	30.59	26.51	22.74	m	79.84
3.50–4.00 m deep	0.55	39.12	33.95	22.74	m	95.81
4.00–4.50 m deep	0.71	50.51	43.85	22.74	m	117.10
4.50–5.00 m deep	0.89	63.31	55.03	22.74	m	141.08
5.00–5.50 m deep	1.15	81.81	71.12	22.74	m	175.67
Extra for bend; 87.5°	0.31	22.05	1.82	32.08	nr	55.95
Extra for bend; 45°	0.31	22.05	1.83	32.08	nr	55.96
Extra for single junction; equal	0.48	34.14	3.62	48.19	nr	85.95
Extra for taper reducer	0.27	19.21	1.83	36.82	nr	57.86

CLASS I: PIPEWORK – PIPES

Item	Gang hours	Labour £	Plant £	Material £	Unit	Total rate £
100 mm pipes; in trenches, depth						
not in trenches	0.13	6.02	0.25	27.38	m	33.65
ne 1.50 m deep	0.21	14.94	12.91	27.38	m	55.23
1.50–2.00 m deep	0.23	16.36	14.16	27.38	m	57.90
2.00–2.50 m deep	0.30	21.34	18.49	27.38	m	67.21
2.50–3.00 m deep	0.37	26.32	22.81	27.38	m	76.51
3.00–3.50 m deep	0.48	34.14	29.60	27.38	m	91.12
3.50–4.00 m deep	0.60	42.68	37.04	27.38	m	107.10
4.00–4.50 m deep	0.75	53.35	46.32	27.38	m	127.05
4.50–5.00 m deep	0.95	67.58	58.72	27.38	m	153.68
5.00–5.50 m deep	1.22	86.78	75.47	27.38	m	189.63
Extra for bend; 87.5°	0.38	27.03	2.76	39.99	nr	69.78
Extra for bend; 45°	0.38	27.03	2.76	39.99	nr	69.78
Extra for bend; long radius	0.38	27.03	2.76	57.16	nr	86.95
Extra for single junction; equal	0.59	41.97	4.54	99.91	nr	146.42
Extra for double junction; equal	0.80	56.91	7.24	87.42	nr	151.57
Extra for taper reducer	0.40	28.45	4.54	43.79	nr	76.78
150 mm pipes; in trenches, depth						
not in trenches	0.14	6.48	0.25	54.32	m	61.05
ne 1.50 m deep	0.24	17.07	14.74	54.32	m	86.13
1.50–2.00 m deep	0.28	19.92	17.24	54.32	m	91.48
2.00–2.50 m deep	0.34	24.19	20.97	54.32	m	99.48
2.50–3.00 m deep	0.40	28.45	24.65	54.32	m	107.42
3.00–3.50 m deep	0.54	38.41	33.32	54.32	m	126.05
3.50–4.00 m deep	0.61	43.39	37.67	54.32	m	135.38
4.00–4.50 m deep	0.79	56.20	48.79	54.32	m	159.31
4.50–5.00 m deep	1.05	74.69	64.92	54.32	m	193.93
5.00–5.50 m deep	1.32	93.90	81.64	54.32	m	229.86
Extra for bend; 87.5°	0.56	39.84	3.61	76.42	nr	119.87
Ectra for bend; 45°	0.56	39.84	3.61	76.40	nr	119.85
Extra for bend; long radius	0.56	39.84	3.61	97.40	nr	140.85
Extra for single junction; equal	0.88	62.60	7.22	135.68	nr	205.50
Extra for taper reducer	0.56	39.84	7.22	117.07	nr	164.13
225 mm pipes; in trenches, depth						
not in trenches	0.15	6.95	0.25	133.19	m	140.39
ne 1.50 m deep	0.25	17.78	15.36	133.19	m	166.33
1.50–2.00 m deep	0.30	21.34	18.49	133.19	m	173.02
2.00–2.50 m deep	0.35	24.90	21.54	133.19	m	179.63
2.50–3.00 m deep	0.41	29.17	25.28	133.19	m	187.64
3.00–3.50 m deep	0.54	38.41	33.32	133.19	m	204.92
3.50–4.00 m deep	0.61	43.39	37.67	133.19	m	214.25
4.00–4.50 m deep	0.77	54.77	47.57	133.19	m	235.53
4.50–5.00 m deep	1.02	72.56	63.07	133.19	m	268.82
5.00–5.50 m deep	1.30	92.48	80.42	133.19	m	306.09
Extra for bend; 87.5°	0.77	54.77	3.61	328.26	nr	386.64
Extra for bend; 45°	0.77	54.77	3.61	160.86	nr	219.24
Extra for single junction; equal	1.21	86.07	8.13	630.28	nr	724.48
Extra for taper reducer	0.77	54.77	8.13	249.17	nr	312.07

CLASS I: PIPEWORK – PIPES

Item	Gang hours	Labour £	Plant £	Material £	Unit	Total rate £
IRON PIPES – cont						
Ductile iron pipes to BS 4772, Tyton joints; excavation and supports, backfilling						
100 mm pipes; in trenches, depth						
not in trenches	0.10	4.63	0.25	71.39	m	**76.27**
ne 1.50 m deep	0.19	13.52	11.07	71.39	m	**95.98**
1.50–2.00 m deep	0.20	14.23	12.32	71.39	m	**97.94**
2.00–2.50 m deep	0.26	18.50	16.03	71.39	m	**105.92**
2.50–3.00 m deep	0.32	22.76	19.72	71.39	m	**113.87**
3.00–3.50 m deep	0.42	29.88	25.92	71.39	m	**127.19**
3.50–4.00 m deep	0.53	37.70	32.73	71.39	m	**141.82**
4.00–4.50 m deep	0.66	46.95	40.78	71.39	m	**159.12**
4.50–5.00 m deep	0.84	59.75	51.93	71.39	m	**183.07**
5.00–5.50 m deep	1.08	76.83	66.80	71.39	m	**215.02**
Extra for bend; 90°	0.38	27.03	17.63	19.72	nr	**64.38**
Extra for single junction; equal	0.60	42.68	5.43	34.87	nr	**82.98**
150 mm pipes; in trenches, depth						
not in trenches	0.11	5.09	0.25	141.13	m	**146.47**
ne 1.50 m deep	0.21	14.94	12.91	141.13	m	**168.98**
1.50–2.00 m deep	0.25	17.78	15.40	141.13	m	**174.31**
2.00–2.50 m deep	0.30	21.34	18.49	141.13	m	**180.96**
2.50–3.00 m deep	0.37	26.32	22.81	141.13	m	**190.26**
3.00–3.50 m deep	0.49	34.86	30.23	141.13	m	**206.22**
3.50–4.00 m deep	0.55	39.12	33.95	141.13	m	**214.20**
4.00–4.50 m deep	0.72	51.22	44.47	141.13	m	**236.82**
4.50–5.00 m deep	0.94	66.87	58.12	141.13	m	**266.12**
5.00–5.50 m deep	1.19	84.65	73.59	141.13	m	**299.37**
Extra for bend; 90°	0.57	40.55	3.61	34.05	nr	**78.21**
Extra for single junction; equal	0.88	62.60	7.22	130.47	nr	**200.29**
250 mm pipes; in trenches, depth						
not in trenches	0.16	7.41	0.25	329.04	m	**336.70**
ne 1.50 m deep	0.24	17.07	14.74	329.04	m	**360.85**
1.50–2.00 m deep	0.30	21.34	18.49	329.04	m	**368.87**
2.00–2.50 m deep	0.35	24.90	21.56	329.04	m	**375.50**
2.50–3.00 m deep	0.42	29.88	25.90	329.04	m	**384.82**
3.00–3.50 m deep	0.56	39.84	34.53	329.04	m	**403.41**
3.50–4.00 m deep	0.64	45.53	39.51	329.04	m	**414.08**
4.00–4.50 m deep	0.81	57.62	50.04	329.04	m	**436.70**
4.50–5.00 m deep	1.02	72.56	63.07	329.04	m	**464.67**
5.00–5.50 m deep	1.27	90.34	78.54	329.04	m	**497.92**
Extra for bend; 90°	0.96	68.29	5.43	188.21	nr	**261.93**
Extra for single junction; equal	1.32	93.90	9.04	247.20	nr	**350.14**
400 mm pipes; in trenches, depth						
not in trenches	0.24	11.11	0.54	779.41	m	**791.06**
ne 1.50 m deep	0.33	23.47	20.28	779.41	m	**823.16**
1.50–2.00 m deep	0.43	30.59	26.47	779.41	m	**836.47**
2.00–2.50 m deep	0.48	34.14	29.56	779.41	m	**843.11**
2.50–3.00 m deep	0.57	40.55	35.14	779.41	m	**855.10**
3.00–3.50 m deep	0.71	50.51	43.78	779.41	m	**873.70**
3.50–4.00 m deep	0.91	64.73	56.17	779.41	m	**900.31**

CLASS I: PIPEWORK – PIPES

Item	Gang hours	Labour £	Plant £	Material £	Unit	Total rate £
4.00–4.50 m deep	1.16	82.52	71.65	779.41	m	933.58
4.50–5.00 m deep	1.47	104.57	90.87	779.41	m	974.85
5.00–5.50 m deep	1.77	125.91	109.48	779.41	m	1014.80
600 mm pipes; in trenches, depth						
not in trenches	0.34	15.75	0.74	1261.49	m	1277.98
ne 1.50 m deep	0.47	33.43	28.86	1261.49	m	1323.78
1.50–2.00 m deep	0.55	39.12	33.86	1261.49	m	1334.47
2.00–2.50 m deep	0.66	46.95	40.66	1261.49	m	1349.10
2.50–3.00 m deep	0.78	55.49	48.09	1261.49	m	1365.07
3.00–3.50 m deep	0.89	63.31	54.89	1261.49	m	1379.69
3.50–4.00 m deep	1.09	77.54	67.30	1261.49	m	1406.33
4.00–4.50 m deep	1.37	97.45	84.63	1261.49	m	1443.57
4.50–5.00 m deep	1.70	120.93	105.11	1261.49	m	1487.53
5.00–5.50 m deep	2.03	144.40	125.54	1261.49	m	1531.43
STEEL PIPES						
Carbon steel pipes to BS EN 10126; welded joints;(for protection and lining refer to manufacturer); excavation and supports, backfilling						
100 mm pipes; in trenches, depth						
not in trenches	0.07	3.24	0.34	25.34	m	28.92
ne 1.50 m deep	0.15	10.67	9.21	25.34	m	45.22
1.50–2.00 m deep	0.17	12.09	10.48	25.34	m	47.91
2.00–2.50 m deep	0.22	15.65	13.56	25.34	m	54.55
2.50–3.00 m deep	0.27	19.21	16.63	25.34	m	61.18
3.00–3.50 m deep	0.35	24.90	21.58	25.34	m	71.82
3.50–4.00 m deep	0.44	31.30	27.16	25.34	m	83.80
4.00–4.50 m deep	0.55	39.12	33.96	25.34	m	98.42
4.50–5.00 m deep	0.70	49.79	60.12	25.34	m	135.25
5.00–5.50 m deep	0.87	61.89	53.80	25.34	m	141.03
Extra for bend; 45°	0.07	4.98	2.70	41.05	nr	48.73
Extra for bend; 90°	0.07	4.98	2.70	40.58	nr	48.26
Extra for single junction; equal	0.11	7.82	4.52	67.52	nr	79.86
150 mm pipes; in trenches, depth						
not in trenches	0.07	3.24	0.34	37.85	m	41.43
ne 1.50 m deep	0.17	12.09	10.45	37.85	m	60.39
1.50–2.00 m deep	0.20	14.23	12.32	37.85	m	64.40
2.00–2.50 m deep	0.24	17.07	14.78	37.85	m	69.70
2.50–3.00 m deep	0.29	20.63	17.88	37.85	m	76.36
3.00–3.50 m deep	0.39	27.74	24.04	37.85	m	89.63
3.50–4.00 m deep	0.44	31.30	27.16	37.85	m	96.31
4.00–4.50 m deep	0.57	40.55	35.22	37.85	m	113.62
4.50–5.00 m deep	0.72	51.22	61.83	37.85	m	150.90
5.00–5.50 m deep	0.90	64.02	55.68	37.85	m	157.55
Extra for bend; 45°	0.09	6.40	3.61	72.82	nr	82.83
Extra for bend; 90°	0.09	6.40	3.61	71.13	nr	81.14
Extra for single junction; equal	0.16	11.38	7.22	116.24	nr	134.84

CLASS I: PIPEWORK – PIPES

Item	Gang hours	Labour £	Plant £	Material £	Unit	Total rate £
STEEL PIPES – cont						
Carbon steel pipes to BS EN 10126 – cont						
200 mm pipes; in trenches, depth						
not in trenches	0.09	4.17	8.88	55.19	m	**68.24**
ne 1.50 m deep	0.18	12.80	11.07	55.19	m	**79.06**
1.50–2.00 m deep	0.21	14.94	12.94	55.19	m	**83.07**
2.00–2.50 m deep	0.25	17.78	15.41	55.19	m	**88.38**
2.50–3.00 m deep	0.30	21.34	18.50	55.19	m	**95.03**
3.00–3.50 m deep	0.40	28.45	24.67	55.19	m	**108.31**
3.50–4.00 m deep	0.46	32.72	28.42	55.19	m	**116.33**
4.00–4.50 m deep	0.59	41.97	36.44	55.19	m	**133.60**
4.50–5.00 m deep	0.74	52.64	45.76	55.19	m	**153.59**
5.00–5.50 m deep	0.92	65.44	56.90	55.19	m	**177.53**
Extra for bend; 45°	0.12	8.54	3.61	103.28	nr	**115.43**
Extra for bend; 90°	0.12	8.54	3.61	111.64	nr	**123.79**
Extra for single junction; equal	0.21	14.94	7.17	177.51	nr	**199.62**
250 mm pipes; in trenches, depth						
not in trenches	0.10	4.63	0.51	78.42	m	**83.56**
ne 1.50 m deep	0.18	12.80	16.26	78.42	m	**107.48**
1.50–2.00 m deep	0.22	15.65	19.29	78.42	m	**113.36**
2.00–2.50 m deep	0.26	18.50	21.66	78.42	m	**118.58**
2.50–3.00 m deep	0.31	22.05	27.15	78.42	m	**127.62**
3.00–3.50 m deep	0.41	29.17	35.93	78.42	m	**143.52**
3.50–4.00 m deep	0.47	33.43	39.94	78.42	m	**151.79**
4.00–4.50 m deep	0.60	42.68	49.42	78.42	m	**170.52**
4.50–5.00 m deep	0.75	53.35	64.39	78.42	m	**196.16**
5.00–5.50 m deep	0.94	66.87	80.73	78.42	m	**226.02**
Extra for bend; 45°	0.13	9.25	3.61	179.47	nr	**192.33**
Extra for bend; 90°	0.13	9.25	3.61	184.00	nr	**196.86**
Extra for single junction; equal	0.23	16.36	7.22	265.88	nr	**289.46**
300 mm pipes; in trenches, depth						
not in trenches	0.11	5.09	0.54	96.69	m	**102.32**
ne 1.50 m deep	0.20	14.23	12.28	96.69	m	**123.20**
1.50–2.00 m deep	0.25	17.78	15.40	96.69	m	**129.87**
2.00–2.50 m deep	0.28	19.92	17.24	96.69	m	**133.85**
2.50–3.00 m deep	0.34	24.19	20.97	96.69	m	**141.85**
3.00–3.50 m deep	0.44	31.30	27.13	96.69	m	**155.12**
3.50–4.00 m deep	0.53	37.70	32.73	96.69	m	**167.12**
4.00–4.50 m deep	0.68	48.37	42.00	96.69	m	**187.06**
4.50–5.00 m deep	0.86	61.18	53.18	96.69	m	**211.05**
5.00–5.50 m deep	1.06	75.40	65.57	96.69	m	**237.66**
Extra for bend; 45°	0.14	9.96	5.43	228.03	nr	**243.42**
Extra for bend; 90°	0.14	9.96	5.43	278.83	nr	**294.22**
Extra for single junction; equal	0.25	17.78	11.74	357.91	nr	**387.43**

CLASS I: PIPEWORK – PIPES

Item	Gang hours	Labour £	Plant £	Material £	Unit	Total rate £
POLYVINYL CHLORIDE PIPES						
Unplasticized PVC perforated pipes; ring seal sockets; excavation and supports, backfilling; 6 m pipe lengths unless stated otherwise						
82 mm pipes; in trench, depth						
not in trenches; 3.00 m pipe lengths	0.06	2.78	0.14	8.50	m	**11.42**
ne 1.50 m deep; 3.00 m pipe lengths	0.10	7.11	4.05	8.50	m	**19.66**
1.50–2.00 m deep	0.13	9.25	5.28	8.50	m	**23.03**
2.00–2.50 m deep	0.16	11.38	6.50	8.50	m	**26.38**
2.50–3.00 m deep	0.19	13.52	7.72	8.50	m	**29.74**
3.00–3.50 m deep	0.22	15.65	8.96	8.50	m	**33.11**
3.50–4.00 m deep	0.25	17.78	10.20	8.50	m	**36.48**
4.00–4.50 m deep	0.28	19.92	11.42	8.50	m	**39.84**
4.50–5.00 m deep	0.32	22.76	13.07	8.50	m	**44.33**
5.00–5.50 m deep	0.35	24.90	14.30	8.50	m	**47.70**
110 mm pipes; in trench, depth						
not in trenches	0.06	2.78	0.07	11.75	m	**14.60**
ne 1.50 m deep	0.10	7.11	4.05	11.75	m	**22.91**
1.50–2.00 m deep	0.13	9.25	5.28	11.75	m	**26.28**
2.00–2.50 m deep	0.16	11.38	6.50	11.75	m	**29.63**
2.50–3.00 m deep	0.19	13.52	7.72	11.75	m	**32.99**
3.00–3.50 m deep	0.22	15.65	8.96	11.75	m	**36.36**
3.50–4.00 m deep	0.25	17.78	10.20	11.75	m	**39.73**
4.00–4.50 m deep	0.28	19.92	11.42	11.75	m	**43.09**
4.50–5.00 m deep	0.32	22.76	13.07	11.75	m	**47.58**
5.00–5.50 m deep	0.36	25.61	14.72	11.75	m	**52.08**
160 mm pipes; in trench, depth						
not in trenches	0.06	2.78	0.14	30.10	m	**33.02**
ne 1.50 m deep	0.11	7.82	4.45	30.10	m	**42.37**
1.50–2.00 m deep	0.14	9.96	5.69	30.10	m	**45.75**
2.00–2.50 m deep	0.18	12.80	7.32	30.10	m	**50.22**
2.50–3.00 m deep	0.22	15.65	8.95	30.10	m	**54.70**
3.00–3.50 m deep	0.23	16.36	9.36	30.10	m	**55.82**
3.50–4.00 m deep	0.26	18.50	10.61	30.10	m	**59.21**
4.00–4.50 m deep	0.30	21.34	12.25	30.10	m	**63.69**
4.50–5.00 m deep	0.35	24.90	14.29	30.10	m	**69.29**
5.00–5.50 m deep	0.40	28.45	16.35	30.10	m	**74.90**
Unplasticized PVC pipes; ring seal sockets; excavation and supports, backfilling; 6 m pipe lengths unless stated otherwise						
82 mm pipes; in trenches, depth						
not in trenches; 3.00 m pipe lengths	0.06	2.78	0.31	10.00	m	**13.09**
ne 1.50 m deep; 3.00 m pipe lengths	0.10	7.11	6.05	10.00	m	**23.16**
1.50–2.00 m deep	0.13	9.25	8.01	10.00	m	**27.26**
2.00–2.50 m deep	0.16	11.38	9.85	10.00	m	**31.23**
2.50–3.00 m deep	0.19	13.52	11.70	10.00	m	**35.22**
3.00–3.50 m deep	0.22	15.65	13.58	10.00	m	**39.23**
3.50–4.00 m deep	0.25	17.78	15.44	10.00	m	**43.22**

CLASS I: PIPEWORK – PIPES

Item	Gang hours	Labour £	Plant £	Material £	Unit	Total rate £
POLYVINYL CHLORIDE PIPES – cont						
Unplasticized PVC pipes – cont						
82 mm pipes – cont						
4.00–4.50 m deep	0.28	19.92	17.30	10.00	m	**47.22**
4.50–5.00 m deep	0.32	22.76	19.78	10.00	m	**52.54**
5.00–5.50 m deep	0.35	24.90	21.64	10.00	m	**56.54**
Extra for bend; short radius (socket/spigot)	0.05	3.56	–	26.93	nr	**30.49**
Extra for branches; equal (socket/spigot)	0.07	4.98	–	34.07	nr	**39.05**
110 mm pipes; in trenches, depth						
not in trenches	0.06	2.78	0.31	13.09	m	**16.18**
ne 1.50 m deep	0.10	7.11	6.16	13.09	m	**26.36**
1.50–2.00 m deep	0.13	9.25	8.01	13.09	m	**30.35**
2.00–2.50 m deep	0.16	11.38	9.85	13.09	m	**34.32**
2.50–3.00 m deep	0.19	13.52	11.70	13.09	m	**38.31**
3.00–3.50 m deep	0.22	15.65	13.58	13.09	m	**42.32**
3.50–4.00 m deep	0.25	17.78	15.44	13.09	m	**46.31**
4.00–4.50 m deep	0.28	19.92	17.30	13.09	m	**50.31**
4.50–5.00 m deep	0.32	22.76	19.78	13.09	m	**55.63**
5.00–5.50 m deep	0.36	25.61	22.27	13.09	m	**60.97**
Extra for bend; short radius (socket/spigot)	0.05	3.56	–	40.96	nr	**44.52**
Extra for bend; adjustable (socket/spigot)	0.05	3.56	–	42.00	nr	**45.56**
Extra for reducer	0.05	3.56	–	15.09	nr	**18.65**
Extra for branches; equal (socket/spigot)	0.07	4.98	–	51.79	nr	**56.77**
160 mm pipes; in trenches, depth						
not in trenches	0.06	2.78	0.31	36.91	m	**40.00**
ne 1.50 m deep	0.11	7.82	6.75	36.91	m	**51.48**
1.50–2.00 m deep	0.14	9.96	8.64	36.91	m	**55.51**
2.00–2.50 m deep	0.18	12.80	11.11	36.91	m	**60.82**
2.50–3.00 m deep	0.22	15.65	13.57	36.91	m	**66.13**
3.00–3.50 m deep	0.23	16.36	14.18	36.91	m	**67.45**
3.50–4.00 m deep	0.26	18.50	16.07	36.91	m	**71.48**
4.00–4.50 m deep	0.30	21.34	18.55	36.91	m	**76.80**
4.50–5.00 m deep	0.35	24.90	21.63	36.91	m	**83.44**
5.00–5.50 m deep	0.40	28.45	24.74	36.91	m	**90.10**
Extra for bend; short radius (socket/spigot)	0.05	3.56	–	158.30	nr	**161.86**
Extra for branches; equal (socket/spigot)	0.07	4.98	–	177.73	nr	**182.71**
225 mm pipes; in trenches, depth						
not in trenches	0.07	3.24	0.34	70.44	m	**74.02**
ne 1.50 m deep	0.12	8.54	7.37	70.44	m	**86.35**
1.50–2.00 m deep	0.15	10.67	9.23	70.44	m	**90.34**
2.00–2.50 m deep	0.20	14.23	12.32	70.44	m	**96.99**
2.50–3.00 m deep	0.23	16.36	14.17	70.44	m	**100.97**
3.00–3.50 m deep	0.24	17.07	14.80	70.44	m	**102.31**
3.50–4.00 m deep	0.27	19.21	16.66	70.44	m	**106.31**
4.00–4.50 m deep	0.32	22.76	19.77	70.44	m	**112.97**
4.50–5.00 m deep	0.36	25.61	22.25	70.44	m	**118.30**
5.00–5.50 m deep	0.45	32.01	27.84	70.44	m	**130.29**
Extra for bend; short radius 45° (double socket)	0.07	4.98	–	318.19	nr	**323.17**
Extra for branches; equal (all socket)	0.09	6.40	–	350.42	nr	**356.82**

CLASS I: PIPEWORK – PIPES

Item	Gang hours	Labour £	Plant £	Material £	Unit	Total rate £
300 mm pipes; in trenches, depth						
not in trenches	0.08	3.70	0.40	109.44	m	**113.54**
ne 1.50 m deep	0.13	9.25	7.99	109.44	m	**126.68**
1.50–2.00 m deep	0.16	11.38	9.85	109.44	m	**130.67**
2.00–2.50 m deep	0.21	14.94	12.94	109.44	m	**137.32**
2.50–3.00 m deep	0.23	16.36	14.17	109.44	m	**139.97**
3.00–3.50 m deep	0.25	17.78	15.42	109.44	m	**142.64**
3.50–4.00 m deep	0.28	19.92	17.29	109.44	m	**146.65**
4.00–4.50 m deep	0.34	24.19	21.02	109.44	m	**154.65**
4.50–5.00 m deep	0.38	27.03	23.51	109.44	m	**159.98**
5.00–5.50 m deep	0.47	33.43	29.06	109.44	m	**171.93**
Extra for bend; short radius 45° (double socket)	0.07	4.98	–	1021.00	nr	**1025.98**
Extra for branches; unequal (all socket)	0.09	6.40	–	610.00	nr	**616.40**
Unplasticized PVC pipes; polypropylene couplings; excavation and supports, backfilling; 6 m pipe lengths unless stated otherwise						
110 mm pipes; in trenches, depth						
not in trenches	0.06	2.78	0.31	8.38	m	**11.47**
ne 1.50 m deep	0.10	7.11	6.16	8.38	m	**21.65**
1.50–2.00 m deep	0.13	9.25	8.01	8.38	m	**25.64**
2.00–2.50 m deep	0.16	11.38	9.85	8.38	m	**29.61**
2.50–3.00 m deep	0.19	13.52	11.70	8.38	m	**33.60**
3.00–3.50 m deep	0.22	15.65	13.58	8.38	m	**37.61**
3.50–4.00 m deep	0.25	17.78	15.44	8.38	m	**41.60**
4.00–4.50 m deep	0.28	19.92	17.30	8.38	m	**45.60**
4.50–5.00 m deep	0.32	22.76	19.78	8.38	m	**50.92**
5.00–5.50 m deep	0.36	25.61	22.27	8.38	m	**56.26**
160 mm pipes; in trenches, depth						
not in trenches	0.06	2.78	0.32	28.17	m	**31.27**
ne 1.50 m deep	0.11	7.82	6.75	28.17	m	**42.74**
1.50–2.00 m deep	0.14	9.96	8.64	28.17	m	**46.77**
2.00–2.50 m deep	0.18	12.80	11.11	28.17	m	**52.08**
2.50–3.00 m deep	0.22	15.65	13.57	28.17	m	**57.39**
3.00–3.50 m deep	0.23	16.36	14.18	28.17	m	**58.71**
3.50–4.00 m deep	0.26	18.50	16.07	28.17	m	**62.74**
4.00–4.50 m deep	0.30	21.34	18.55	28.17	m	**68.06**
4.50–5.00 m deep	0.35	24.90	21.63	28.17	m	**74.70**
5.00–5.50 m deep	0.40	28.45	24.74	28.17	m	**81.36**
225 mm pipes; in trenches, depth						
not in trenches	0.07	3.24	0.34	56.00	m	**59.58**
ne 1.50 m deep	0.12	8.54	7.37	56.00	m	**71.91**
1.50–2.00 m deep	0.15	10.67	9.23	56.00	m	**75.90**
2.00–2.50 m deep	0.20	14.23	12.32	56.00	m	**82.55**
2.50–3.00 m deep	0.23	16.36	14.17	56.00	m	**86.53**
3.00–3.50 m deep	0.24	17.07	14.80	56.00	m	**87.87**
3.50–4.00 m deep	0.27	19.21	16.66	56.00	m	**91.87**
4.00–4.50 m deep	0.32	22.76	19.77	56.00	m	**98.53**
4.50–5.00 m deep	0.36	25.61	22.25	56.00	m	**103.86**
5.00–5.50 m deep	0.45	32.01	27.84	56.00	m	**115.85**

CLASS I: PIPEWORK – PIPES

Item	Gang hours	Labour £	Plant £	Material £	Unit	Total rate £
POLYVINYL CHLORIDE PIPES – cont						
Unplasticized PVC pipes – cont						
300 mm pipes; in trenches, depth						
not in trenches	0.08	3.70	0.40	110.84	m	**114.94**
ne 1.50 m deep	0.13	9.25	7.99	110.84	m	**128.08**
1.50–2.00 m deep	0.16	11.38	9.85	110.84	m	**132.07**
2.00–2.50 m deep	0.21	14.94	12.94	110.84	m	**138.72**
2.50–3.00 m deep	0.23	16.36	14.17	110.84	m	**141.37**
3.00–3.50 m deep	0.25	17.78	15.42	110.84	m	**144.04**
3.50–4.00 m deep	0.28	19.92	17.29	110.84	m	**148.05**
4.00–4.50 m deep	0.34	24.19	21.02	110.84	m	**156.05**
4.50–5.00 m deep	0.38	27.03	23.51	110.84	m	**161.38**
5.00–5.50 m deep	0.47	33.43	29.06	110.84	m	**173.33**
Ultrarib unplasticized PVC pipes; ring seal joints; excavation and supports, backfilling						
150 mm pipes; in trenches, depth						
not in trenches	0.13	6.02	0.65	17.01	m	**23.68**
ne 1.50 m deep	0.16	11.38	9.83	17.01	m	**38.22**
1.50–2.00 m deep	0.19	13.52	11.69	17.01	m	**42.22**
2.00–2.50 m deep	0.22	15.65	13.57	17.01	m	**46.23**
2.50–3.00 m deep	0.24	17.07	14.79	17.01	m	**48.87**
3.00–3.50 m deep	0.24	17.07	14.80	17.01	m	**48.88**
3.50–4.00 m deep	0.27	19.21	16.66	17.01	m	**52.88**
4.00–4.50 m deep	0.32	22.76	19.77	17.01	m	**59.54**
4.50–5.00 m deep	0.36	25.61	22.25	17.01	m	**64.87**
5.00–5.50 m deep	0.45	32.01	27.84	17.01	m	**76.86**
Extra for 45° bends; short radius (socket/spigot)	0.05	3.56	–	37.29	nr	**40.85**
Extra for branches; equal (socket/spigot)	0.09	6.40	–	76.75	nr	**83.15**
225 mm pipes; in trenches, depth						
not in trenches	0.08	3.70	0.40	41.41	m	**45.51**
ne 1.50 m deep	0.13	9.25	7.99	41.41	m	**58.65**
1.50–2.00 m deep	0.16	11.38	9.85	41.41	m	**62.64**
2.00–2.50 m deep	0.21	14.94	12.94	41.41	m	**69.29**
2.50–3.00 m deep	0.23	16.36	14.17	41.41	m	**71.94**
3.00–3.50 m deep	0.25	17.78	15.42	41.41	m	**74.61**
3.50–4.00 m deep	0.28	19.92	17.29	41.41	m	**78.62**
4.00–4.50 m deep	0.34	24.19	21.02	41.41	m	**86.62**
4.50–5.00 m deep	0.38	27.03	23.51	41.41	m	**91.95**
5.00–5.50 m deep	0.47	33.43	29.06	41.41	m	**103.90**
Extra for 45° bends; short radius (socket/spigot)	0.05	3.56	–	156.84	nr	**160.40**
Extra for branches; equal (socket/spigot)	0.09	6.40	–	225.07	nr	**231.47**

CLASS I: PIPEWORK – PIPES

Item	Gang hours	Labour £	Plant £	Material £	Unit	**Total rate £**
300 mm pipes; in trenches, depth						
not in trenches	0.08	3.70	0.40	75.99	m	**80.09**
ne 1.50 m deep	0.13	9.25	7.99	75.99	m	**93.23**
1.50–2.00 m deep	0.16	11.38	9.85	75.99	m	**97.22**
2.00–2.50 m deep	0.21	14.94	12.94	75.99	m	**103.87**
2.50–3.00 m deep	0.23	16.36	14.17	75.99	m	**106.52**
3.00–3.50 m deep	0.25	17.78	15.42	75.99	m	**109.19**
3.50–4.00 m deep	0.28	19.92	17.29	75.99	m	**113.20**
4.00–4.50 m deep	0.34	24.19	21.02	75.99	m	**121.20**
4.50–5.00 m deep	0.38	27.03	23.51	75.99	m	**126.53**
5.00–5.50 m deep	0.47	33.43	29.06	75.99	m	**138.48**
EXtra for 45° bends; short radius (socket/spigot)	0.07	4.98	–	331.85	nr	**336.83**
Extra for branches; equal (socket/spigot)	0.09	6.40	–	483.79	nr	**490.19**

CLASS J: PIPEWORK – FITTINGS AND VALVES

Item	Gang hours	Labour £	Plant £	Material £	Unit	Total rate £
RESOURCES – LABOUR						
Note						
Fittings on pipes shown with the appropriate pipe in Class I						
Fittings and valves gang						
1 ganger/chargehand (skill rate 4) – 50% of time		7.28				
1 skilled operative (skill rate 4)		13.61				
2 unskilled operatives (general)		25.42				
1 plant operator (skill rate 3)		16.55				
Total Gang Rate/Hour	£	**62.86**				
RESOURCES – PLANT						
Fittings and valves						
1.0 m³ hydraulic backacter			55.30			
disc saw – 30% of time			0.30			
2.80 m³/min compressor, 2 tool – 30% of time			2.21			
2 t dumper – 30% of time			1.98			
compressor tools, extra 50 ft / 15 m hose – 30% of time			0.09			
small pump – 30% of time			1.11			
sundry tools – 30% of time			0.38			
Total Rate/Hour		£	**61.37**			
VALVES AND PENSTOCKS						
Valves						
Non-return valves; cast iron; single door; tidal flap						
250 mm	0.40	25.14	12.27	460.00	nr	**497.41**
350 mm	0.40	25.14	12.27	880.00	nr	**917.41**
450 mm	0.70	44.00	12.27	1170.00	nr	**1226.27**
600 mm	0.70	44.00	18.43	1585.00	nr	**1647.43**
800 mm	0.75	47.15	21.47	2200.00	nr	**2268.62**
Penstocks; cast iron; wall mounted; hand operated						
250 mm	1.66	104.35	21.47	405.00	nr	**530.82**
350 mm	2.30	144.58	24.55	570.00	nr	**739.13**
450 mm	2.90	182.29	27.62	680.00	nr	**889.91**
600 × 600 mm	5.40	339.44	49.09	1100.00	nr	**1488.53**
1000 × 1000 mm	9.00	565.74	92.07	1900.00	nr	**2557.81**

CLASS K: PIPEWORK – MANHOLES AND PIPEWORK ACCESSORIES

Item	Gang hours	Labour £	Plant £	Material £	Unit	Total rate £
NOTES						
The rates assume the most efficient items of plant (excavator) and are optimum rates assuming continuous output with no delays caused by other operations or works.						
Ground conditions are assumed to be good easily worked soil with no abnormal conditions that would affect outputs and consistency of work.						
Multiplier Table for labour and plant for various site conditions for working:						
out of sequence × 2.75 (minimum)						
in hard clay × 1.75 to 2.00						
in running sand × 2.75 (minimum)						
in broken rock × 2.75 to 3.50						
below water table × 2.00 (minimum)						
RESOURCES – LABOUR						
Gullies gang						
1 chargehand pipelayer (skill rate 4) – 50% of time		7.28				
1 skilled operative (skill rate 4)		13.61				
2 unskilled operatives (general)		25.42				
1 plant operator (skill rate 3)		16.55				
Total Gang Rate/Hour	£	**62.86**				
French/rubble drains, ditches and trenches gang; ducts and metal culverts gang						
1 chargehand pipelayer (skill rate 4) – 50% of time		7.28				
1 skilled operative (skill rate 4)		13.61				
2 unskilled operatives (general)		25.42				
1 plant operator (skill rate 3)		16.55				
Total Gang Rate/Hour	£	**62.86**				
RESOURCES – MATERIALS						
Supply only ex works materials rates						
Vitrified clay road gullies etc.						
Clay road gulley; 450 mm diameter × 900 mm deep	–	–	–	279.12	nr	**279.12**
C.I. gulley grating and frame; 434 × 434 mm to clay	–	–	–	183.50	nr	**183.50**
Clay yard gulley; 225 mm diameter; including grating	–	–	–	255.36	nr	**255.36**
Clay grease interceptor; 600 × 450 mm	–	–	–	1422.25	nr	**1422.25**
Culvert; Corrugated steel						
1.6 mm thick; 1000 mm diameter	–	–	–	103.30	m	**103.30**
2.0 mm thick; 1600 mm diameter	–	–	–	164.70	m	**164.70**
2.0 mm thick; 2000 mm diameter	–	–	–	355.95	m	**355.95**
2.0 mm thick; 2200 mm diameter	–	–	–	451.85	m	**451.85**

CLASS K: PIPEWORK – MANHOLES AND PIPEWORK ACCESSORIES

Item	Gang hours	Labour £	Plant £	Material £	Unit	Total rate £
RESOURCES – PLANT						
Gullies						
0.4 m³ hydraulic excavator			37.03			
2 t dumper (30% of time)			1.98			
2.80 m³/min compressor, 2 tool (30% of time)			2.21			
compaction plate/roller (30% of time)			0.74			
2.40 m³/min road breaker (30% of time)			1.31			
54 mm poker vibrator (30% of time)			0.41			
extra 15 ft / 50 m hose (30% of time)			0.09			
disc saw (30% of time)			0.28			
small pump (30% of time)			1.11			
Total Rate/Hour		£	45.17			
French/rubble drains, ditches and trenches; ducts and metal culverts						
0.4 m³ hydraulic backacter			37.03			
2 t dumper (30% of time)			1.98			
disc saw (30% of time)			0.28			
compaction plate/roller (30% of time)			1.81			
2.80 m³/min compressor, 2 tool (30% of time)			2.21			
small pump (30% of time)			1.11			
Total Rate/Hour		£	44.42			
MANHOLES						
Brick construction						
Design criteria used in models:						
* class A engineering bricks						
* 215 thick walls generally; 328 thick to chambers exceeding 2.5 m deep						
* 225 mm plain concrete C20/20 base slab						
* 300 mm reinforced concrete C20/20 reducing slab						
* 125 mm reinforced concrete C20/20 top slab						
* maximum height of working chamber 2.0 m above benching						
* 750 × 750 access shaft						
* plain concrete C15/20 benching, 150 mm clay main channel longitudinally and two 100 branch channels						
* step irons at 300 mm centres, doubled if depth to invert exceeds 3000 mm						
* heavy duty manhole cover and frame						
750 × 700 chamber 500 depth to invert						
excavation, support, backfilling and disposal	–	–	–	–	–	28.00
concrete base	–	–	–	–	–	80.00
brickwork chamber	–	–	–	–	–	245.00
concrete cover slab	–	–	–	–	–	85.00
concrete benching, main and branch channels	–	–	–	–	–	20.00
step irons	–	–	–	–	–	15.00
access cover and frame	–	–	–	–	–	350.00
TOTAL	–	–	–	–	£	**823.00**

CLASS K: PIPEWORK – MANHOLES AND PIPEWORK ACCESSORIES

Item	Gang hours	Labour £	Plant £	Material £	Unit	Total rate £
750 × 700 chamber 1000 depth to invert						
excavation, support, backfilling and disposal	–	–	–	–	–	60.00
concrete base	–	–	–	–	–	80.00
brickwork chamber	–	–	–	–	–	495.00
concrete cover slab	–	–	–	–	–	85.00
concrete benching and channels	–	–	–	–	–	20.00
step irons	–	–	–	–	–	25.00
access cover and frame	–	–	–	–	–	350.00
TOTAL	–	–	–	–	£	**1115.00**
750 × 700 chamber 1500 depth to invert						
excavation, support, backfilling and disposal	–	–	–	–	–	80.75
concrete base	–	–	–	–	–	80.00
brickwork chamber	–	–	–	–	–	745.00
concrete cover slab	–	–	–	–	–	85.00
concrete benching and channels	–	–	–	–	–	20.00
step irons	–	–	–	–	–	45.00
access cover and frame	–	–	–	–	–	350.00
TOTAL	–	–	–	–	£	**1405.75**
900 × 700 chamber 500 depth to invert						
excavation, support, backfilling and disposal	–	–	–	–	–	35.00
concrete base	–	–	–	–	–	100.00
brickwork chamber	–	–	–	–	–	265.00
concrete cover slab	–	–	–	–	–	100.00
concrete benching and channels	–	–	–	–	–	30.00
step irons	–	–	–	–	–	15.00
access cover and frame	–	–	–	–	–	350.00
TOTAL	–	–	–	–	£	**895.00**
900 × 700 chamber 1000 depth to invert						
excavation, support, backfilling and disposal	–	–	–	–	–	65.00
concrete base	–	–	–	–	–	100.00
brickwork chamber	–	–	–	–	–	535.00
concrete cover slab	–	–	–	–	–	100.00
concrete benching and channels	–	–	–	–	–	30.00
step irons	–	–	–	–	–	25.00
access cover and frame	–	–	–	–	–	350.00
TOTAL	–	–	–	–	£	**1205.00**
900 × 700 chamber 1500 depth to invert						
excavation, support, backfilling and disposal	–	–	–	–	–	110.25
concrete base	–	–	–	–	–	100.00
brickwork chamber	–	–	–	–	–	800.00
concrete cover slab	–	–	–	–	–	100.00
concrete benching and channels	–	–	–	–	–	30.00
step irons	–	–	–	–	–	45.00
access cover and frame	–	–	–	–	–	350.00
TOTAL	–	–	–	–	£	**1535.25**

CLASS K: PIPEWORK – MANHOLES AND PIPEWORK ACCESSORIES

Item	Gang hours	Labour £	Plant £	Material £	Unit	Total rate £
MANHOLES – cont						
Brick construction – cont						
1050 × 700 chamber 1500 depth to invert						
excavation, support, backfilling and disposal	–	–	–	–	–	141.58
concrete base	–	–	–	–	–	110.00
brickwork chamber	–	–	–	–	–	855.00
concrete cover slab	–	–	–	–	–	115.00
concrete benching and channels	–	–	–	–	–	32.50
step irons	–	–	–	–	–	45.00
access cover and frame	–	–	–	–	–	350.00
TOTAL	–	–	–	–	£	**1649.08**
1050 × 700 chamber 2500 depth to invert						
excavation, support, backfilling and disposal	–	–	–	–	–	255.00
concrete base	–	–	–	–	–	110.00
brickwork chamber	–	–	–	–	–	1425.00
concrete cover slab	–	–	–	–	–	115.00
concrete benching and channels	–	–	–	–	–	32.50
step irons	–	–	–	–	–	82.50
access cover and frame	–	–	–	–	–	350.00
TOTAL	–	–	–	–	£	**2370.00**
1050 × 700 chamber 3500 depth to invert						
excavation, support, backfilling and disposal	–	–	–	–	–	350.00
concrete base	–	–	–	–	–	110.00
brickwork chamber	–	–	–	–	–	1995.00
access shaft	–	–	–	–	–	2040.00
reducing slab	–	–	–	–	–	140.00
concrete cover slab	–	–	–	–	–	80.00
concrete benching and channels	–	–	–	–	–	32.50
step irons	–	–	–	–	–	110.00
access cover and frame	–	–	–	–	–	350.00
TOTAL	–	–	–	–	£	**5207.50**
1350 × 700 chamber 2500 depth to invert						
excavation, support, backfilling and disposal	–	–	–	–	–	308.00
concrete base	–	–	–	–	–	140.00
brickwork chamber	–	–	–	–	–	1600.00
concrete cover slab	–	–	–	–	–	140.00
concrete benching and channels	–	–	–	–	–	40.00
step irons	–	–	–	–	–	82.50
access cover and frame	–	–	–	–	–	350.00
TOTAL	–	–	–	–	£	**2660.50**
1350 × 700 chamber 3500 depth to invert						
excavation, support, backfilling and disposal	–	–	–	–	–	430.00
concrete base	–	–	–	–	–	140.00
brickwork chamber	–	–	–	–	–	2250.00
access shaft	–	–	–	–	–	2040.00
reducing slab	–	–	–	–	–	140.00
concrete cover slab	–	–	–	–	–	80.00
concrete benching and channels	–	–	–	–	–	40.00
step irons	–	–	–	–	–	110.00
access cover and frame	–	–	–	–	–	350.00
TOTAL	–	–	–	–	£	**5580.00**

CLASS K: PIPEWORK – MANHOLES AND PIPEWORK ACCESSORIES

Item	Gang hours	Labour £	Plant £	Material £	Unit	Total rate £
1350 × 700 chamber 4500 depth to invert						
excavation, support, backfilling and disposal	–	–	–	–	–	550.00
concrete base	–	–	–	–	–	140.00
brickwork chamber	–	–	–	–	–	2895.00
access shaft	–	–	–	–	–	2665.00
reducing slab	–	–	–	–	–	140.00
concrete cover slab	–	–	–	–	–	80.00
concrete benching and channels	–	–	–	–	–	40.00
step irons	–	–	–	–	–	145.00
access cover and frame	–	–	–	–	–	350.00
TOTAL	–	–	–	–	£	**7005.00**

Precast concrete construction
Design criteria used in models:
* circular shafts
* 150 mm plain concrete C15/20 surround
* 225 mm plain concrete C20/20 base slab
* precast reducing slab
* precast top slab
* maximum height of working chamber 2.0 m above benching
* 750 mm diameter access shaft
* plain concrete C15/20 benching, 150 mm clay main channel longitudinally and two 100 branch channels
* step irons at 300 mm centres, doubled if depth to invert exceeds 3000 mm
* heavy duty manhole cover and frame
* in manholes over 6 m deep, landings at maximum intervals

Item	Gang hours	Labour £	Plant £	Material £	Unit	Total rate £
675 diameter × 500 depth to invert						
excavation, support, backfilling and disposal	–	–	–	–	–	27.00
concrete base	–	–	–	–	–	50.00
main chamber rings	–	–	–	–	–	20.00
cover slab	–	–	–	–	–	45.00
concrete benching and channels	–	–	–	–	–	12.10
concrete surround	–	–	–	–	–	50.00
step irons	–	–	–	–	–	15.00
access cover and frame	–	–	–	–	–	350.00
TOTAL	–	–	–	–	£	**569.10**
675 diameter × 750 depth to invert						
excavation, support, backfilling and disposal	–	–	–	–	–	35.00
concrete base	–	–	–	–	–	50.00
main chamber rings	–	–	–	–	–	25.00
cover slab	–	–	–	–	–	45.00
concrete benching and channels	–	–	–	–	–	12.10
concrete surround	–	–	–	–	–	70.00
step irons	–	–	–	–	–	–
access cover and frame	–	–	–	–	–	350.00
TOTAL	–	–	–	–	£	**607.10**

CLASS K: PIPEWORK – MANHOLES AND PIPEWORK ACCESSORIES

Item	Gang hours	Labour £	Plant £	Material £	Unit	Total rate £
MANHOLES – cont						
Precast concrete construction – cont						
675 diameter × 1000 depth to invert						
excavation, support, backfilling and disposal	–	–	–	–	–	42.50
concrete base	–	–	–	–	–	50.00
main chamber rings	–	–	–	–	–	32.50
cover slab	–	–	–	–	–	45.00
concrete benching and channels	–	–	–	–	–	12.10
step irons	–	–	–	–	–	–
concrete surround	–	–	–	–	–	100.00
access cover and frame	–	–	–	–	–	350.00
TOTAL	–	–	–	–	£	**657.10**
675 diameter × 1250 depth to invert						
excavation, support, backfilling and disposal	–	–	–	–	–	52.00
concrete base	–	–	–	–	–	50.00
main chamber rings	–	–	–	–	–	37.50
cover slab	–	–	–	–	–	45.00
concrete benching and channels	–	–	–	–	–	12.10
concrete surround	–	–	–	–	–	120.00
step irons	–	–	–	–	–	32.00
access cover and frame	–	–	–	–	–	350.00
TOTAL	–	–	–	–	£	**698.60**
900 diameter × 750 depth to invert						
excavation, support, backfilling and disposal	–	–	–	–	–	115.00
concrete base	–	–	–	–	–	100.00
main chamber rings	–	–	–	–	–	41.30
cover slab	–	–	–	–	–	60.00
concrete benching and channels	–	–	–	–	–	30.00
concrete surround	–	–	–	–	–	100.00
step irons	–	–	–	–	–	–
access cover and frame	–	–	–	–	–	350.00
TOTAL	–	–	–	–	£	**727.80**
900 diameter × 1000 depth to invert						
excavation, support, backfilling and disposal	–	–	–	–	–	150.00
concrete base	–	–	–	–	–	100.00
main chamber rings	–	–	–	–	–	55.07
cover slab	–	–	–	–	–	60.00
concrete benching and channels	–	–	–	–	–	30.00
concrete surround	–	–	–	–	–	130.00
step irons	–	–	–	–	–	–
access cover and frame	–	–	–	–	–	350.00
TOTAL	–	–	–	–	£	**806.57**

CLASS K: PIPEWORK – MANHOLES AND PIPEWORK ACCESSORIES

Item	Gang hours	Labour £	Plant £	Material £	Unit	Total rate £
900 diameter × 1500 depth to invert						
excavation, support, backfilling and disposal	–	–	–	–	–	215.00
concrete base	–	–	–	–	–	100.00
main chamber rings	–	–	–	–	–	65.07
cover slab	–	–	–	–	–	60.00
concrete benching and channels	–	–	–	–	–	30.00
concrete surround	–	–	–	–	–	190.00
step irons	–	–	–	–	–	45.00
access cover and frame	–	–	–	–	–	350.00
TOTAL	–	–	–	–	£	**941.82**
1200 diameter × 1500 depth to invert						
excavation, support, backfilling and disposal	–	–	–	–	–	275.00
concrete base	–	–	–	–	–	170.00
main chamber rings	–	–	–	–	–	135.00
concrete benching and channels	–	–	–	–	–	38.15
cover slab	–	–	–	–	–	78.00
concrete surround	–	–	–	–	–	260.00
step irons	–	–	–	–	–	45.00
access cover and frame	–	–	–	–	–	350.00
TOTAL	–	–	–	–	£	**1351.15**
1200 diameter × 2000 depth to invert						
excavation, support, backfilling and disposal	–	–	–	–	–	380.00
concrete base	–	–	–	–	–	170.00
main chamber rings	–	–	–	–	–	18.00
cover slab	–	–	–	–	–	78.00
concrete benching and channels	–	–	–	–	–	38.15
concrete surround	–	–	–	–	–	340.00
step irons	–	–	–	–	–	60.00
access cover and frame	–	–	–	–	–	350.00
TOTAL	–	–	–	–	£	**1434.15**
1200 diameter × 2500 depth to invert						
excavation, support, backfilling and disposal	–	–	–	–	–	450.00
concrete base	–	–	–	–	–	170.00
main chamber rings	–	–	–	–	–	225.00
cover slab	–	–	–	–	–	78.00
concrete benching and channels	–	–	–	–	–	38.15
concrete surround	–	–	–	–	–	430.00
step irons	–	–	–	–	–	75.00
access cover and frame	–	–	–	–	–	350.00
TOTAL	–	–	–	–	£	**1816.15**
1200 diameter × 3000 depth to invert						
excavation, support, backfilling and disposal	–	–	–	–	–	555.50
concrete base	–	–	–	–	–	170.00
main chamber rings	–	–	–	–	–	270.00
cover slab	–	–	–	–	–	78.00
concrete benching and channels	–	–	–	–	–	38.15
concrete surround	–	–	–	–	–	510.00
step irons	–	–	–	–	–	90.00
access cover and frame	–	–	–	–	–	350.00
TOTAL	–	–	–	–	£	**2061.65**

CLASS K: PIPEWORK – MANHOLES AND PIPEWORK ACCESSORIES

Item	Gang hours	Labour £	Plant £	Material £	Unit	Total rate £
MANHOLES – cont						
Precast concrete construction – cont						
1800 diameter × 1500 depth to invert						
excavation, support, backfilling and disposal	–	–	–	–	–	415.00
concrete base	–	–	–	–	–	380.00
main chamber rings	–	–	–	–	–	405.00
cover slab	–	–	–	–	–	215.00
concrete benching and channels	–	–	–	–	–	85.85
concrete surround	–	–	–	–	–	380.00
step irons	–	–	–	–	–	45.00
access cover and frame	–	–	–	–	–	350.00
TOTAL	–	–	–	–	£	**2275.85**
1800 diameter × 2000 depth to invert						
excavation, support, backfilling and disposal	–	–	–	–	–	550.00
concrete base	–	–	–	–	–	380.00
main chamber rings	–	–	–	–	–	540.00
cover slab	–	–	–	–	–	215.00
concrete benching and channels	–	–	–	–	–	85.85
concrete surround	–	–	–	–	–	510.00
step irons	–	–	–	–	–	60.00
access cover and frame	–	–	–	–	–	350.00
TOTAL	–	–	–	–	£	**2690.85**
1800 diameter × 2500 depth to invert						
excavation, support, backfilling and disposal	–	–	–	–	–	650.00
concrete base	–	–	–	–	–	380.00
main chamber rings	–	–	–	–	–	675.00
cover slab	–	–	–	–	–	215.00
concrete benching and channels	–	–	–	–	–	85.85
concrete surround	–	–	–	–	–	640.00
step irons	–	–	–	–	–	75.00
access cover and frame	–	–	–	–	–	350.00
TOTAL	–	–	–	–	£	**3070.85**
1800 diameter × 3000 depth to invert						
excavation, suport, backfilling and disposal	–	–	–	–	–	830.00
concrete base	–	–	–	–	–	380.00
main chamber rings	–	–	–	–	–	815.00
cover slab	–	–	–	–	–	215.00
concrete surround	–	–	–	–	–	770.00
concrete benching and channels	–	–	–	–	–	85.85
step irons	–	–	–	–	–	90.00
access cover and frame	–	–	–	–	–	350.00
TOTAL	–	–	–	–	£	**3535.85**

CLASS K: PIPEWORK – MANHOLES AND PIPEWORK ACCESSORIES

Item	Gang hours	Labour £	Plant £	Material £	Unit	Total rate £
1800 diameter × 3500 depth to invert						
excavation, support, backfilling and disposal	–	–	–	–	–	950.00
concrete base	–	–	–	–	–	380.00
access shaft	–	–	–	–	–	300.00
main chamber rings	–	–	–	–	–	950.00
reducing slab	–	–	–	–	–	251.00
cover slab	–	–	–	–	–	60.00
concrete benching and channels	–	–	–	–	–	85.85
concrete surround	–	–	–	–	–	900.00
step irons	–	–	–	–	–	115.00
access cover and frame	–	–	–	–	–	350.00
TOTAL	–	–	–	–	£	**4446.85**
1800 diameter × 4000 depth to invert						
excavation, support, backfilling and disposal	–	–	–	–	–	1100.00
concrete base	–	–	–	–	–	380.00
access shaft	–	–	–	–	–	345.00
main chamber rings	–	–	–	–	–	1055.00
reducing slab	–	–	–	–	–	251.00
cover slab	–	–	–	–	–	60.00
concrete benching and channels	–	–	–	–	–	85.85
concrete surround	–	–	–	–	–	1030.00
step irons	–	–	–	–	–	125.00
access cover and frame	–	–	–	–	–	350.00
TOTAL	–	–	–	–	£	**4781.85**
2400 diameter × 1500 depth to invert						
excavation, support, backfilling and disposal	–	–	–	–	–	570.00
concrete base	–	–	–	–	–	680.00
main chamber rings	–	–	–	–	–	1035.00
cover slab	–	–	–	–	–	615.00
concrete benching and channels	–	–	–	–	–	155.00
concrete surround	–	–	–	–	–	510.00
step irons	–	–	–	–	–	45.00
access cover and frame	–	–	–	–	–	350.00
TOTAL	–	–	–	–	£	**3960.00**
2400 diameter × 3000 depth to invert						
excavation, support, backfilling and disposal	–	–	–	–	–	1140.00
concrete base	–	–	–	–	–	680.00
main chamber rings	–	–	–	–	–	2075.00
cover slab	–	–	–	–	–	615.00
concrete benching and channels	–	–	–	–	–	155.00
concrete surround	–	–	–	–	–	1030.00
step irons	–	–	–	–	–	90.00
access cover and frame	–	–	–	–	–	350.00
TOTAL	–	–	–	–	£	**6135.00**

CLASS K: PIPEWORK – MANHOLES AND PIPEWORK ACCESSORIES

Item	Gang hours	Labour £	Plant £	Material £	Unit	Total rate £
MANHOLES – cont						
Precast concrete construction – cont						
2400 diameter × 4500 depth to invert						
excavation, support, backfilling and disposal	–	–	–	–	–	1500.00
concrete base	–	–	–	–	–	680.00
access shaft	–	–	–	–	–	385.00
main chamber rings	–	–	–	–	–	3100.00
reducing slab	–	–	–	–	–	705.00
cover slab	–	–	–	–	–	60.00
concrete benching and channels	–	–	–	–	–	155.00
concrete surround	–	–	–	–	–	1540.00
step irons	–	–	–	–	–	147.00
access cover and frame	–	–	–	–	–	350.00
TOTAL	–	–	–	–	£	**8622.00**
2700 diameter × 1500 depth to invert						
excavation, support, backfilling and disposal	–	–	–	–	–	650.00
concrete base	–	–	–	–	–	870.00
main chamber rings	–	–	–	–	–	1200.00
cover slab	–	–	–	–	–	1100.00
concrete benching and channels	–	–	–	–	–	195.00
concrete surround	–	–	–	–	–	580.00
step irons	–	–	–	–	–	45.00
access cover and frame	–	–	–	–	–	350.00
TOTAL	–	–	–	–	£	**4990.00**
2700 diameter × 3000 depth to invert						
excavation, support, backfilling and disposal	–	–	–	–	–	1310.00
concrete base	–	–	–	–	–	870.00
main chamber rings	–	–	–	–	–	2400.00
cover slab	–	–	–	–	–	1100.00
concrete benching and channels	–	–	–	–	–	195.00
concrete surround	–	–	–	–	–	1150.00
step irons	–	–	–	–	–	90.00
access cover and frame	–	–	–	–	–	350.00
TOTAL	–	–	–	–	£	**7465.00**
2700 diameter × 4500 depth to invert						
excavation, support, backfilling and disposal	–	–	–	–	–	1900.00
concrete base	–	–	–	–	–	870.00
access shaft	–	–	–	–	–	385.00
main chamber rings	–	–	–	–	–	3500.00
reducing slab	–	–	–	–	–	1155.00
cover slab	–	–	–	–	–	60.00
concrete benching and channels	–	–	–	–	–	195.00
concrete surround	–	–	–	–	–	1730.00
step irons	–	–	–	–	–	147.00
access cover and frame	–	–	–	–	–	350.00
TOTAL	–	–	–	–	£	**10292.00**

CLASS K: PIPEWORK – MANHOLES AND PIPEWORK ACCESSORIES

Item	Gang hours	Labour £	Plant £	Material £	Unit	Total rate £
3000 diameter × 3000 depth to invert						
excavation, support, backfilling and disposal	–	–	–	–	–	1500.00
concrete base	–	–	–	–	–	1070.00
main chamber rings	–	–	–	–	–	3375.00
cover slab	–	–	–	–	–	1215.00
concrete benching and channels	–	–	–	–	–	240.00
concrete surround	–	–	–	–	–	1280.00
step irons	–	–	–	–	–	90.00
access cover and frame	–	–	–	–	–	350.00
TOTAL	–	–	–	–	£	**9120.00**
3000 diameter × 4500 depth to invert						
excavation, support, backfilling and disposal	–	–	–	–	–	2250.00
concrete base	–	–	–	–	–	1070.00
access shaft	–	–	–	–	–	385.00
main chamber rings	–	–	–	–	–	4500.00
reducing slab	–	–	–	–	–	1500.00
cover slab	–	–	–	–	–	60.00
concrete benching and channels	–	–	–	–	–	240.00
concrete surround	–	–	–	–	–	1920.00
step irons	–	–	–	–	–	147.00
access cover and frame	–	–	–	–	–	350.00
TOTAL	–	–	–	–	£	**12422.00**
3000 diameter × 6000 depth to invert						
excavation, support, backfilling and disposal	–	–	–	–	–	3000.00
concrete base	–	–	–	–	–	1070.00
access shaft	–	–	–	–	–	515.00
main chamber rings	–	–	–	–	–	6700.00
reducing slab	–	–	–	–	–	1500.00
cover slab	–	–	–	–	–	60.00
concrete benching and channels	–	–	–	–	–	240.00
concrete surround	–	–	–	–	–	2560.00
step irons	–	–	–	–	–	175.00
access cover and frame	–	–	–	–	–	350.00
TOTAL	–	–	–	–	£	**16170.00**
BACKDROPS TO MANHOLES						
Clayware vertical pipe complete with rest bend at base and tumbling bay junction to main drain complete with stopper; concrete grade C20 surround, 150 mm thick; additional excavation and disposal						
100 pipe						
1.15 m to invert	–	–	–	–	nr	**110.00**
2.15 m to invert	–	–	–	–	nr	**140.00**
3.15 m to invert	–	–	–	–	nr	**165.00**
4.15 m to invert	–	–	–	–	nr	**200.50**

CLASS K: PIPEWORK – MANHOLES AND PIPEWORK ACCESSORIES

Item	Gang hours	Labour £	Plant £	Material £	Unit	Total rate £
BACKDROPS TO MANHOLES – cont						
Clayware vertical pipe complete with rest bend at base and tumbling bay junction to main drain complete with stopper – cont						
150 pipe						
1.15 m to invert	–	–	–	–	nr	**165.00**
2.15 m to invert	–	–	–	–	nr	**205.00**
3.15 m to invert	–	–	–	–	nr	**242.00**
4.15 m to invert	–	–	–	–	nr	**285.00**
225 pipe						
1.15 m to invert	–	–	–	–	nr	**335.00**
2.15 m to invert	–	–	–	–	nr	**375.00**
3.15 m to invert	–	–	–	–	nr	**425.00**
4.15 m to invert	–	–	–	–	nr	**495.00**
GULLIES						
Vitrified clay; set in concrete grade C20, 150 mm thick; additional excavation and disposal						
Road gulley						
450 mm diameter × 900 mm deep, 100 mm or 150 mm outlet; cast iron road gulley grating and frame group 4, 434 × 434 mm, on Class B engineering brick seating	0.50	31.43	1.13	525.15	nr	**557.71**
Yard gulley (mud); trapped with rodding eye; galvanized bucket; stopper						
225 mm diameter, 100 mm diameter outlet, cast iron hinged grate and frame	0.30	18.86	0.68	272.47	nr	**292.01**
Grease interceptors; internal access and bucket						
600 × 450 mm, metal tray and lid, square hopper with horizontal inlet	0.35	22.00	0.79	1443.11	nr	**1465.90**
Precast concrete; set in concrete grade C20, 150 mm thick; additional excavation and disposal						
Road gulley; trapped with rodding eye; galvanized bucket; stopper						
450 mm diameter × 750 mm deep; cast iron road gulley grating and frame group 4, 434 × 434 mm, on Class B engineering brick seating	0.50	31.43	1.13	446.56	nr	**479.12**
450 mm diameter × 900 mm deep; cast iron road gulley grating and frame group 4, 434 × 434 mm, on Class B engineering brick seating	0.54	33.94	1.22	461.77	nr	**496.93**
450 mm diameter × 1050 mm deep, cast iron road gulley grating and frame group 4, 434 × 434 mm, on Class B engineering brick seating	0.58	36.46	1.31	471.81	nr	**509.58**

CLASS K: PIPEWORK – MANHOLES AND PIPEWORK ACCESSORIES

Item	Gang hours	Labour £	Plant £	Material £	Unit	Total rate £
FRENCH DRAINS, RUBBLE DRAINS, DITCHES						
The rates assume the most efficient items of plant (excavator) and are optimum rates assuming continuous output with no delays caused by other operations or works.						
Ground conditions are assumed to be good easily worked soil with no abnormal conditions that would affect outputs and consistency of work.						
Multiplier Table for labour and plant for various site conditions for working:						
out of sequence × 2.75 (minimum)						
in hard clay × 1.75 to 2.00						
in running sand × 2.75 (minimum)						
in broken rock × 2.75 to 3.50						
below water table × 2.00 (minimum)						
Excavation of trenches for unpiped rubble drains (excluding trench support); cross-sectional area						
0.25–0.50 m²	0.10	6.29	3.98	–	m	**10.27**
0.50–0.75 m²	0.12	7.54	4.77	–	m	**12.31**
0.75–1.00 m²	0.14	8.80	5.57	–	m	**14.37**
1.00–1.50 m²	0.17	10.69	6.76	–	m	**17.45**
1.50–2.00 m²	0.20	12.57	7.95	–	m	**20.52**
Filling French and rubble drains with graded material						
20 mm stone aggregate; PC £4.45/t	0.30	18.86	11.93	27.26	m³	**58.05**
broken brick/concrete rubble; PC £3.42/t	0.29	18.23	11.53	29.27	m³	**59.03**
Excavation of rectangular section ditches; unlined; cross-sectional area						
0.25–0.50 m²	0.11	6.91	4.29	–	m	**11.20**
0.50–0.75 m²	0.13	8.17	5.07	–	m	**13.24**
0.75–1.00 m²	0.16	10.06	6.24	–	m	**16.30**
1.00–1.50 m²	0.20	12.57	7.80	–	m	**20.37**
1.50–2.00 m²	0.25	15.71	9.75	–	m	**25.46**
Excavation of rectangular ditches; lined with precast concrete slabs; cross-sectional area						
0.25–0.50 m²	0.15	9.43	6.22	14.27	m	**29.92**
0.50–0.75 m²	0.25	15.71	10.38	20.94	m	**47.03**
0.75–1.00 m²	0.36	22.63	14.95	29.70	m	**67.28**
1.00–1.50 m²	0.40	25.14	16.61	41.88	m	**83.63**
1.50–2.00 m²	0.45	28.29	18.69	59.16	m	**106.14**
Excavation of vee section ditches; unlined; cross-sectional area						
0.25–0.50 m²	0.10	6.29	4.46	–	m	**10.75**
0.50–0.75 m²	0.12	7.54	5.33	–	m	**12.87**
0.75–1.00 m²	0.14	8.80	6.24	–	m	**15.04**
1.00–1.50 m²	0.18	11.31	8.01	–	m	**19.32**
1.50–2.00 m²	0.22	13.83	9.79	–	m	**23.62**

CLASS K: PIPEWORK – MANHOLES AND PIPEWORK ACCESSORIES

Item	Gang hours	Labour £	Plant £	Material £	Unit	Total rate £
DUCTS AND METAL CULVERTS						
Galvanized steel culverts; bitumen coated						
Sectional corrugated metal culverts, nominal internal diameter 0.5–1 m; 1000 mm nominal internal diameter, 1.6 mm thick						
not in trenches	0.15	9.43	0.67	110.01	m	**120.11**
in trenches, depth not exceeding 1.5 m	0.31	19.49	13.76	110.01	m	**143.26**
in trenches, depth 1.5–2 m	0.43	27.03	19.10	110.01	m	**156.14**
in trenches, depth 2–2.5 m	0.51	32.06	22.65	110.01	m	**164.72**
in trenches, depth 2.5–3 m	0.60	37.72	26.65	110.01	m	**174.38**
Sectional corrugated metal culverts, nominal internal diameter exceeding 1.5 m; 1600 mm nominal internal diameter, 1.6 mm thick						
not in trenches	0.21	13.20	0.94	176.49	m	**190.63**
in trenches, depth 1.5–2 m	0.44	27.66	19.55	176.49	m	**223.70**
in trenches, depth 2–2.5 m	0.56	35.20	24.88	176.49	m	**236.57**
in trenches, depth 2.5–3 m	0.68	42.74	30.21	176.49	m	**249.44**
in trenches, depth 3–3.5 m	0.82	51.55	36.44	176.49	m	**264.48**
Sectional corrugated metal culverts, nominal internal diameter exceeding 1.5 m; 2000 mm nominal internal diameter, 1.6 mm thick						
not in trenches	0.26	16.34	1.17	373.59	m	**391.10**
in trenches, depth 2–2.5 m	0.46	28.92	20.45	373.59	m	**422.96**
in trenches, depth 2.5–3 m	0.60	37.72	26.65	373.59	m	**437.96**
in trenches, depth 3–3.5 m	0.75	47.15	33.31	373.59	m	**454.05**
in trenches, depth 3.5–4 m	0.93	58.46	41.32	373.59	m	**473.37**
Sectional corrugated metal culverts, nominal internal diameter exceeding 1.5 m; 2200 mm nominal internal diameter, 1.6 mm thick						
not in trenches	0.33	20.74	1.48	472.40	m	**494.62**
in trenches, depth 2.5–3 m	0.64	40.23	28.43	472.40	m	**541.06**
in trenches, depth 3–3.5 m	0.77	48.40	34.21	472.40	m	**555.01**
in trenches, depth 3.5–4 m	1.02	64.12	45.33	472.40	m	**581.85**
in trenches, depth exceeding 4 m	1.32	82.98	58.64	472.40	m	**614.02**
OTHER PIPEWORK ANCILLARIES						
Notes						
Refer to Section G (Concrete and concrete ancillaries) for costs relevant to the construction of Headwall Structure.						
Build Ends in						
Connections to existing manholes and other chambers, pipe bore						
150 mm diameter	0.60	42.68	3.35	8.86	nr	**54.89**
225 mm diameter	0.95	67.58	5.29	16.48	nr	**89.35**
300 mm diameter	1.25	88.92	6.98	25.02	nr	**120.92**
375 mm diameter	1.45	103.15	8.10	32.96	nr	**144.21**
450 mm diameter	1.75	110.00	75.85	41.47	nr	**227.32**

CLASS L: PIPEWORK SUPPORTS AND PIPEWORK ANCILLARIES

Item	Gang hours	Labour £	Plant £	Material £	Unit	Total rate £
NOTES						
The rates assume the most efficient items of plant (excavator) and are optimum rates assuming continuous output with no delays caused by other operations or works.						
Ground conditions are assumed to be good easily worked soil with no abnormal conditions that would affect outputs and consistency of work.						
Multiplier Table for labour and plant for various site conditions for working:						
out of sequence × 2.75 (minimum)						
in hard clay × 1.75 to 2.00						
in running sand × 2.75 (minimum)						
in broken rock × 2.75 to 3.50						
below water table × 2.00 (minimum)						
RESOURCES – LABOUR						
Supports and protection gang						
2 unskilled operatives (general)		25.42				
1 plant operator (skill rate 3)		16.55				
Total Gang Rate/Hour		£ **41.97**				
RESOURCES – PLANT						
Supports and protection						
0.40 m³ hydraulic backacter			37.03			
Bomag BW 65S			7.65			
Total Rate/Hour			£ **44.68**			
EXTRAS TO EXCAVATION AND BACKFILLING						
Drainage sundries						
Extra over any item of drainage for excavation in						
rock	0.65	27.28	34.26	–	m³	**61.54**
mass concrete	0.84	35.25	44.55	–	m³	**79.80**
reinforced concrete	1.18	49.52	62.47	–	m³	**111.99**
Excavation of soft spots, backfilling						
concrete grade C15P	0.30	12.59	15.82	82.91	m³	**111.32**
SPECIAL PIPE LAYING METHODS						
There are many factors, apart from design consideration, which influence the cost of pipe jacking, so that it is only possible to give guide prices for a sample of the work involved. For more reliable estimates it is advisable to seek the advice of a Specialist Contractor.						
The main cost considerations are:						
* the nature of the ground						
* length of drive						
* location						
* presence of water						
* depth below surface						

CLASS L: PIPEWORK SUPPORTS AND PIPEWORK ANCILLARIES

Item	Gang hours	Labour £	Plant £	Material £	Unit	Total rate £
SPECIAL PIPE LAYING METHODS – cont						
Provision of all plant, equipment and labour establishing						
thrust pit; 6 m × 4 m × 8 m deep	–	–	–	–	item	30000.00
reception pit; 4 m × 4 m × 8 m deep	–	–	–	–	item	22000.00
mobilize and set up pipe jacking equipment	–	–	–	–	item	44000.00
Pipe jacking, excluding the cost of non-drainage materials; concrete pipes BS 5911 Part 1 Class 120 with rebated joints, steel reinforcing band; length not exceeding 50 m; in sand and gravel						
900 mm nominal bore	–	–	–	–	m	2070.00
1200 mm nominal bore	–	–	–	–	m	2587.50
1500 mm nominal bore	–	–	–	–	m	1932.50
1800 mm nominal bore	–	–	–	–	m	2185.50
BEDS						
Imported sand						
100 mm deep bed for pipes nominal bore						
100 mm	0.02	0.84	0.74	1.57	m	3.15
150 mm	0.02	0.84	0.74	1.70	m	3.28
225 mm	0.03	1.26	1.11	1.98	m	4.35
300 mm	0.04	1.68	1.48	2.07	m	5.23
150 mm deep bed for pipes nominal bore						
150 mm	0.06	2.52	2.22	2.56	m	7.30
225 mm	0.07	2.94	2.59	2.96	m	8.49
300 mm	0.09	3.78	3.33	3.14	m	10.25
400 mm	0.12	5.04	4.44	3.51	m	12.99
450 mm	0.14	5.88	5.18	4.11	m	15.17
600 mm	0.17	7.13	6.30	5.08	m	18.51
750 mm	0.19	7.97	7.04	5.68	m	20.69
900 mm	0.21	8.81	7.78	6.65	m	23.24
1200 mm	0.25	10.49	9.26	7.82	m	27.57
Imported granular material						
100 mm deep bed for pipes nominal bore						
100 mm	0.02	0.84	0.74	1.62	m	3.20
150 mm	0.03	1.26	1.11	1.75	m	4.12
225 mm	0.04	1.68	1.48	2.01	m	5.17
300 mm	0.05	2.10	1.85	2.15	m	6.10
150 mm deep bed for pipes nominal bore						
150 mm	0.06	2.52	2.22	2.62	m	7.36
225 mm	0.08	3.36	2.96	3.02	m	9.34
300 mm	0.10	4.20	3.70	3.22	m	11.12
400 mm	0.13	5.46	4.81	3.63	m	13.90
450 mm	0.15	6.30	5.55	4.19	m	16.04
600 mm	0.18	7.55	6.67	5.22	m	19.44
750 mm	0.20	8.39	7.41	5.82	m	21.62
900 mm	0.22	9.23	8.15	6.83	m	24.21
1200 mm	0.26	10.91	9.63	8.02	m	28.56

CLASS L: PIPEWORK SUPPORTS AND PIPEWORK ANCILLARIES

Item	Gang hours	Labour £	Plant £	Material £	Unit	Total rate £
Mass concrete						
100 mm deep bed for pipes nominal bore						
100 mm	0.07	2.94	2.59	6.41	m	**11.94**
150 mm	0.08	3.36	2.96	6.95	m	**13.27**
225 mm	0.09	3.78	3.33	8.03	m	**15.14**
300 mm	0.11	4.62	4.07	8.57	m	**17.26**
150 mm deep bed for pipes nominal bore						
100 mm	0.10	4.20	3.70	9.64	m	**17.54**
150 mm	0.12	5.04	4.44	10.43	m	**19.91**
225 mm	0.14	5.88	5.18	12.04	m	**23.10**
300 mm	0.16	6.72	5.92	12.85	m	**25.49**
400 mm	0.19	7.97	7.04	14.42	m	**29.43**
450 mm	0.21	8.81	7.78	16.86	m	**33.45**
600 mm	0.24	10.07	8.89	20.85	m	**39.81**
750 mm	0.26	10.91	9.63	23.25	m	**43.79**
900 mm	0.28	11.75	10.37	27.28	m	**49.40**
1200 mm	0.32	13.43	11.85	32.07	m	**57.35**
HAUNCHES						
The following items allow for dressing the haunching material half-way up the pipe barrel for the full width of the bed and then dressing in triangular fashion to the crown of the pipe. The items exclude the drain bed.						
Mass concrete PC £62.15/m³						
Haunches for pipes nominal bore						
150 mm	0.24	10.07	8.89	5.89	m	**24.85**
225 mm	0.29	12.17	10.74	9.23	m	**32.14**
300 mm	0.36	15.11	13.33	11.64	m	**40.08**
400 mm	0.43	18.05	15.92	15.39	m	**49.36**
450 mm	0.50	20.98	18.52	20.85	m	**60.35**
600 mm	0.56	23.50	20.74	32.28	m	**76.52**
750 mm	0.62	26.02	22.96	39.91	m	**88.89**
900 mm	0.69	28.96	25.55	54.62	m	**109.13**
1200 mm	0.75	31.48	27.77	71.49	m	**130.74**

CLASS L: PIPEWORK SUPPORTS AND PIPEWORK ANCILLARIES

Item	Gang hours	Labour £	Plant £	Material £	Unit	Total rate £
SURROUNDS						
The following items provide for dressing around the pipe above the bed. Sand and granular material is quantified on the basis of the full width of the bed to the stated distance above the crown, concrete as an ellipse from the top corners of the bed to a point at the stated distance above the crown. The items exclude the drain bed.						
Imported sand						
100 mm thick bed for pipes nominal bore						
100 mm	0.04	1.68	1.48	2.85	m	6.01
150 mm	0.05	2.10	1.85	4.09	m	8.04
225 mm	0.06	2.52	2.22	6.55	m	11.29
300 mm	0.08	3.36	2.96	9.13	m	15.45
150 mm thick bed for pipes nominal bore						
100 mm	0.10	4.20	3.70	3.46	m	11.36
150 mm	0.12	5.04	4.44	4.75	m	14.23
225 mm	0.14	5.88	5.18	7.32	m	18.38
300 mm	0.18	7.55	6.67	9.91	m	24.13
400 mm	0.24	10.07	8.89	14.47	m	33.43
450 mm	0.28	11.75	10.37	18.43	m	40.55
600 mm	0.34	14.27	12.59	29.78	m	56.64
750 mm	0.38	15.95	14.07	41.52	m	71.54
900 mm	0.42	17.63	15.55	57.81	m	90.99
1200 mm	0.50	20.98	18.52	92.13	m	131.63
Imported granular material						
100 mm thick bed for pipes nominal bore						
100 mm	0.07	4.98	2.59	2.92	m	10.49
150 mm	0.10	7.11	3.70	4.17	m	14.98
225 mm	0.13	9.25	4.81	6.68	m	20.74
300 mm	0.16	11.38	5.92	9.33	m	26.63
150 mm thick bed for pipes nominal bore						
100 mm	0.10	4.20	3.70	3.53	m	11.43
150 mm	0.12	5.04	4.44	4.85	m	14.33
225 mm	0.14	5.88	5.18	7.47	m	18.53
300 mm	0.18	7.55	6.67	10.16	m	24.38
400 mm	0.24	10.07	8.89	14.82	m	33.78
450 mm	0.28	11.75	10.37	18.91	m	41.03
600 mm	0.34	14.27	12.59	30.40	m	57.26
750 mm	0.38	15.95	14.07	42.52	m	72.54
900 mm	0.42	17.63	15.55	59.18	m	92.36
1200 mm	0.50	20.98	18.52	94.35	m	133.85
Mass concrete						
100 mm thick bed for pipes nominal bore						
100 mm	0.14	5.88	5.18	11.59	m	22.65
150 mm	0.16	6.72	5.92	16.65	m	29.29
225 mm	0.18	7.55	6.67	26.65	m	40.87
300 mm	0.22	9.23	8.15	37.19	m	54.57

CLASS L: PIPEWORK SUPPORTS AND PIPEWORK ANCILLARIES

Item	Gang hours	Labour £	Plant £	Material £	Unit	Total rate £
150 mm thick bed for pipes nominal bore						
150 mm	0.23	9.65	8.52	19.35	m	**37.52**
225 mm	0.26	10.91	9.63	29.79	m	**50.33**
300 mm	0.30	12.59	11.11	40.54	m	**64.24**
400 mm	0.36	15.11	13.33	59.13	m	**87.57**
450 mm	0.40	16.79	14.81	75.36	m	**106.96**
600 mm	0.45	18.89	16.66	121.15	m	**156.70**
750 mm	0.50	20.98	18.52	169.50	m	**209.00**
900 mm	0.55	23.08	20.37	235.77	m	**279.22**
1200 mm	0.61	25.60	22.59	375.77	m	**423.96**
CONCRETE STOOLS AND THRUST BLOCKS						
Mass concrete PC £62.15/m³						
Concrete stools or thrust blocks (nett volume of concrete excluding volume occupied by pipes)						
0.1 m³	0.18	7.55	6.67	9.60	nr	**23.82**
0.1–0.2 m³	0.32	13.43	11.85	19.22	nr	**44.50**
0.2–0.5 m³	0.62	26.02	22.96	48.13	nr	**97.11**
0.5–1.0 m³	0.91	38.19	33.70	96.33	nr	**168.22**
1.0–2.0 m³	1.29	54.14	47.77	192.85	nr	**294.76**
2.0–4.0 m³	3.15	132.21	116.64	385.50	nr	**634.35**

CLASS M: STRUCTURAL METALWORK

Item	Gang hours	Labour £	Plant £	Material £	Unit	Total rate £
NOTES						
The following are guide prices for various structural members commonly found in a Civil Engineering contract. The list is by no means exhaustive and costs are very much dependent on the particular design and will vary greatly according to specific requirements.						
For more detailed prices, reference should be made to Specialist Contractors.						
FABRICATION OF MEMBERS; STEELWORK						
Columns						
universal beams; straight on plan	–	–	–	–	t	1300.00
circular hollow sections; straight on plan	–	–	–	–	t	1600.00
rectangular hollow sections; straight on plan	–	–	–	–	t	1500.00
Beams						
universal beams; straight on plan	–	–	–	–	t	1400.00
universal beams; curved on plan	–	–	–	–	t	2000.00
channels; straight on plan	–	–	–	–	t	1450.00
channels; curved on plan	–	–	–	–	t	2400.00
castellated beams; straight on plan	–	–	–	–	t	1750.00
Portal frames						
straight on plan	–	–	–	–	t	1700.00
Trestles, towers and built-up columns						
straight on plan	–	–	–	–	t	1700.00
Trusses and built-up girders						
straight on plan	–	–	–	–	t	2000.00
curved on plan	–	–	–	–	t	2500.00
Bracings						
angles; straight on plan	–	–	–	–	t	1500.00
circular hollow sections; straight on plan	–	–	–	–	t	2200.00
Purlins and cladding rails						
straight on plan	–	–	–	–	t	1350.00
Cold rolled purlins and rails						
straight on plan	–	–	–	–	t	2000.00
Anchorages and holding down bolt assemblies						
base plate and bolt assemblies complete	–	–	–	–	t	2500.00
ERECTION OF FABRICATED MEMBERS ON SITE						
Trial erection	–	–	–	–	t	550.00
Permanent erection	–	–	–	–	t	550.00
Site bolts						
black	–	–	–	–	t	3200.00
HSFG general grade	–	–	–	–	t	3200.00
HSFG higher	–	–	–	–	t	3600.00
HSFG load indicating or limit types, general grade	–	–	–	–	t	4200.00
HSFG load indicating or limit types, higher grade	–	–	–	–	t	4700.00

CLASS M: STRUCTURAL METALWORK

Item	Gang hours	Labour £	Plant £	Material £	Unit	Total rate £
OFF-SITE SURFACE TREATMENT						
Note: The following preparation and painting systems have been calculated on the basis of 20 m² per tonne.						
Blast cleaning	–	–	–	–	m²	5.00
Galvanizing	–	–	–	–	m²	18.75
Painting						
one coat zinc chromate primer	–	–	–	–	m²	4.85
one coat two pack epoxy zinc phosphate primer (75 microns dry film thickness)	–	–	–	–	m²	7.50
two coats epoxy micaceous iron oxide (100 microns dry film thickness per coat)	–	–	–	–	m²	22.50

CLASS N: MISCELLANEOUS METALWORK

Item	Gang hours	Labour £	Plant £	Material £	Unit	Total rate £
NOTES						
The following are guide prices for various structural members commonly found in a Civil Engineering contract. The list is by no means exhaustive and costs are very much dependent on the particular design and will vary greatly according to specific requirements.						
For more detailed prices, reference should be made to Specialist Contractors.						
Cladding						
CESMM3 N.2.1 requires cladding to be measured in square metres, the item so produced being inclusive of all associated flashings at wall corners and bases, eaves, gables, ridges and around openings. As the relative quantities of these flashings will depend very much on the complexity of the building shape, the guide prices shown below for these items are shown separately to help with the accuracy of the estimate.						
Bridge bearings						
Bridge bearings are manufactured and installed to individual specifications. The following guide prices are for different sizes of simple bridge bearings. If requirements are known, then advice ought to be obtained from specialist manufactureres such as CCL.						
If there is a requirement for testing bridge bearings prior to their being installed then the tests should be enumerated separately.						
Specialist advice should be sought once details are known.						
RESOURCES – LABOUR						
Roofing – cladding gang						
1 ganger/chargehand (skill rate 3) – 50% of time		8.08				
2 skilled operative (skill rate 3)		30.42				
1 unskilled operatives (general) – 50% of time		6.36				
Total Gang Rate/Hour		£ **44.85**				
Bridge bearing gang						
1 skilled operative (skill rate 4)		13.61				
2 unskilled operatives (general)		25.42				
Total Gang Rate/Hour		£ **39.03**				
RESOURCES – PLANT						
Cladding to roofs						
15 m telescopic access platform – 50% of time			11.72			
Total Gang Rate/Hour			£ **11.72**			
Cladding to walls						
15 m telescopic access platform			23.45			
Total Gang Rate/Hour			£ **23.45**			

CLASS N: MISCELLANEOUS METALWORK

Item	Gang hours	Labour £	Plant £	Material £	Unit	Total rate £
MILD STEEL						
Mild steel						
Stairways and landings	–	–	–	–	t	5375.00
Walkways and platforms	–	–	–	–	t	5250.00
Ladders						
cat ladder; 64 × 13 mm bar strings; 19 mm rungs at 250 mm centres; 450 mm wide with safety hoops	–	–	–	–	m	95.00
Miscellaneous framing						
angle section; 200 × 200 × 16 mm (equal)	–	–	–	–	m	265.00
angle section; 150 × 150 × 10 mm (equal)	–	–	–	–	m	122.50
angle section; 100 × 100 × 12 mm (equal)	–	–	–	–	m	220.00
angle section; 200 × 150 × 15 mm (unequal)	–	–	–	–	m	175.00
angle section; 150 × 75 × 10 mm (unequal)	–	–	–	–	m	92.70
universal beams; 914 × 419 mm	–	–	–	–	m	2050.00
universal beams; 533 × 210 mm	–	–	–	–	m	700.00
universal joists; 127 × 76 mm	–	–	–	–	m	75.00
channel section; 381 × 102 mm	–	–	–	–	m	320.00
channel section; 254 × 76 mm	–	–	–	–	m	165.00
channel section; 152 × 76 mm	–	–	–	–	m	105.00
tubular section; 100 × 100 × 10 mm	–	–	–	–	m	150.00
tubular section; 200 × 200 × 15 mm	–	–	–	–	m	500.00
tubular section; 76.1 × 5.0 mm	–	–	–	–	m	52.00
tubular section; 139.7 × 6.3 mm	–	–	–	–	m	145.00
Mild steel; galvanized						
Handrails						
76 mm diameter tubular handrail, 48 mm diameter standards at 750 mm centres, 48 mm diameter middle rail, 1070 mm high overall	–	–	–	–	m	115.00
Plate flooring						
8 mm (on plain) 'Durbar' pattern floor plates, maximum weight each panel 50 kg	–	–	–	–	m^2	300.00
Mild steel; internally and externally acid dipped, rinse and hot dip galvanized, epoxy internal paint						
Uncovered tanks						
1600 litre capacity open top water tank	–	–	–	–	nr	999.38
18180 litre capacity open top water tank	–	–	–	–	nr	9750.00
Covered tanks						
1600 litre capacity open top water tank with loose fitting lid;	–	–	–	–	nr	1450.00
18180 litre capacity open top water tank with loose fitting lid	–	–	–	–	nr	15000.00
Corrugated steel plates to BS 1449 Pt 1, Gr H4, sealed and bolted; BS729 hot dip galvanized, epoxy internal and external paint						
Uncovered tanks						
713 m^3 capacity bolted cylindrical open top tank	–	–	–	–	nr	38500.00

CLASS N: MISCELLANEOUS METALWORK

Item	Gang hours	Labour £	Plant £	Material £	Unit	Total rate £
PROPRIETARY WORK						
Galvanized steel troughed sheeting; 0.70 mm metal thickness, 75 mm deep corrugations; colour coating each side; fixing with plastic capped self-tapping screws to steel purlins or rails						
Cladding						
upper surfaces inclined at an angle ne 30° to the horizontal	0.15	6.73	1.76	43.05	m^2	**51.54**
Extra for:						
galvanized steel inner lining sheet, 0.40 mm thick, Plastisol colour coating	0.06	2.69	1.41	21.00	m^2	**25.10**
galvanized steel inner lining sheet, 0.40 mm thick, Plastisol colour coating; insulation, 80 mm thick	0.08	3.59	0.94	73.50	m^2	**78.03**
surfaces inclined at an angle exceeding 60° to the horizontal	0.16	7.18	3.75	36.75	m^2	**47.68**
Extra for:						
galvanized steel inner lining sheet, 0.40 mm thick, Plastisol colour coating	0.11	4.93	2.58	21.00	m^2	**28.51**
galvanized steel inner lining sheet, 0.40 mm thick, Plastisol colour coating; insulation, 80 mm thick	0.13	5.83	3.05	73.50	m^2	**82.38**
Galvanized steel flashings; 0.90 mm metal thickness; bent to profile; fixing with plastic capped self-tapping screws to steel purlins or rails; mastic sealant						
Flashings to cladding						
250 mm girth	0.12	5.38	2.81	14.44	m	**22.63**
500 mm girth	0.18	8.07	4.22	23.63	m	**35.92**
750 mm girth	0.22	9.87	5.16	32.81	m	**47.84**
Aluminium profiled sheeting; 0.90 mm metal thickness, 75 mm deep corrugations; colour coating each side; fixing with plastic capped self-tapping screws to steel purlins or rails						
Cladding						
upper surfaces inclined at an angle exceeding 60° to the horizontal	0.20	8.97	4.69	42.00	m^2	**55.66**
Extra for:						
aluminium inner lining sheet, 0.70 mm thick, Plastisol colour coating	0.13	5.83	3.13	26.25	m^2	**35.21**
aluminium inner lining sheet, 0.70 mm thick, Plastisol colour coating; insulation, 80 mm thick	0.15	6.73	3.52	78.75	m^2	**89.00**

CLASS N: MISCELLANEOUS METALWORK

Item	Gang hours	Labour £	Plant £	Material £	Unit	Total rate £
Aluminium profiled sheeting; 1.00 mm metal thickness, 75 mm deep corrugations; colour coating each side; fixing with plastic capped self-tapping screws to steel purlins or rails						
Cladding						
upper surfaces inclined at an angle ne 30° to the horizontal	0.17	7.63	1.99	52.50	m²	62.12
Extra for:						
aluminium inner lining sheet, 0.70 mm thick, Plastisol colour coating	0.09	4.04	1.06	26.25	m²	31.35
aluminium inner lining sheet, 0.70 mm thick, Plastisol colour coating; insulation, 80 mm thick	0.11	4.93	1.29	78.75	m²	84.97
Aluminium flashings; 0.90 mm metal thickness; bent to profile; fixing with plastic capped self-tapping screws to steel purlins or rails; mastic sealant						
Flashings to cladding						
250 mm girth	0.12	5.38	2.81	15.75	m	23.94
500 mm girth	0.18	8.07	4.22	26.25	m	38.54
750 mm girth	0.22	9.87	5.16	36.75	m	51.78
Flooring; Eurogrid; galvanized mild steel						
Open grid flooring						
type 41/100; 3 × 25 mm bearer bar; 6 mm diameter transverse bar	–	–	–	–	m²	41.50
type 41/100; 5 × 25 mm bearer bar; 6 mm diameter transverse bar	–	–	–	–	m²	55.30
type 41/100; 3 × 30 mm bearer bar; 6 mm diameter transverse bar	–	–	–	–	m²	47.55
Duct covers; Stelduct; galvanized mild steel						
Duct covers; pedestrian duty						
225 mm clear opening	–	–	–	–	m	90.00
450 mm clear opening	–	–	–	–	m	97.00
750 mm clear opening	–	–	–	–	m	115.00
Duct covers; medium duty						
225 mm clear opening	–	–	–	–	m	365.00
450 mm clear opening	–	–	–	–	m	375.00
750 mm clear opening	–	–	–	–	m	390.00
Duct covers; heavy duty						
225 mm clear opening	–	–	–	–	m	147.00
450 mm clear opening	–	–	–	–	m	220.00
750 mm clear opening	–	–	–	–	m	315.00

CLASS N: MISCELLANEOUS METALWORK

Item	Gang hours	Labour £	Plant £	Material £	Unit	Total rate £
PROPRIETARY WORK – cont						
Bridge bearings						
Supply plain rubber bearings (3 m and 5 m lengths)						
150 × 20 mm	0.35	13.66	–	37.63	m	51.29
150 × 25 mm	0.35	13.66	–	45.15	m	58.81
Supply and place in position laminated elastomeric rubber bearing						
250 × 150 × 19 mm	0.25	9.76	–	27.95	nr	37.71
300 × 200 × 19 mm	0.25	9.76	–	30.64	nr	40.40
300 × 200 × 30 mm	0.27	10.54	–	38.70	nr	49.24
300 × 200 × 41 mm	0.27	10.54	–	58.05	nr	68.59
300 × 250 × 41 mm	0.30	11.71	–	73.10	nr	84.81
300 × 250 × 63 mm	0.30	11.71	–	102.13	nr	113.84
400 × 250 × 19 mm	0.32	12.49	–	53.75	nr	66.24
400 × 250 × 52 mm	0.32	12.49	–	118.25	nr	130.74
400 × 300 × 19 mm	0.32	12.49	–	62.35	nr	74.84
600 × 450 × 24 mm	0.35	13.66	–	129.00	nr	142.66
Adhesive fixings to laminated elastomeric rubber bearings						
2 mm thick epoxy adhesive	1.00	39.03	–	46.98	m^2	86.01
15 mm thick epoxy mortar	1.50	58.55	–	282.13	m^2	340.68
15 mm thick epoxy pourable grout	2.00	78.06	–	281.46	m^2	359.52
Supply and install mechanical guides for laminated elastomeric rubber bearings						
500 kN SLS design load; FP50 fixed pin Type 1	2.00	78.06	–	842.34	nr	920.40
500 kN SLS design load; FP50 fixed pin Type 2	2.00	78.06	–	868.60	nr	946.66
750 kN SLS design load; FP75 fixed pin Type 1	2.10	81.96	–	976.67	nr	1058.63
750 kN SLS design load; FP75 fixed pin Type 2	2.10	81.96	–	1110.00	nr	1191.96
300 kN SLS design load; UG300 Uniguide Type 1	2.00	78.06	–	1010.00	nr	1088.06
300 kN SLS design load; UG300 Uniguide Type 2	2.00	78.06	–	1160.00	nr	1238.06
Supply and install fixed pot bearings						
355 × 355; PF200	2.00	78.06	–	902.94	nr	981.00
425 × 425; PF300	2.10	81.96	–	1076.66	nr	1158.62
Supply and install free sliding pot bearings						
445 × 345; PS200	2.10	81.96	–	1214.02	nr	1295.98
520 × 415; PS300	2.20	85.87	–	1548.33	nr	1634.20
Supply and install guided sliding pot bearings						
455 × 375; PG200	2.20	85.87	–	1612.97	nr	1698.84
545 × 435; PG300	2.30	89.77	–	1920.01	nr	2009.78
Testing; laminated elastomeric bearings						
compression test	–	–	–	56.00	nr	56.00
shear test	–	–	–	75.00	nr	75.00
bond test (Exclusive of cost of bearings as this is a destructive test)	–	–	–	285.00	nr	285.00

Unit Costs – Civil Engineering Works

CLASS O: TIMBER

Item	Gang hours	Labour £	Plant £	Material £	Unit	Total rate £
RESOURCES – LABOUR						
Timber gang						
1 foreman carpenter/joiner (craftsman)		22.50				
1 carpenter/joiner (craftsman)		19.54				
1 unskilled operative (general)		12.71				
1 plant operator (skill rate 3) – 50% of time		8.28				
Total Gang Rate/Hour	£	**63.02**				
Timber fixings gang						
1 foreman carpenter/joiner (craftsman)		22.50				
1 unskilled operative (general)		12.71				
Total Gang Rate/Hour	£	**35.21**				
RESOURCES – MATERIALS						
The timber material prices shown below are averages, actual prices being very much affected by availability of suitably sized forest timbers capable of conversion to the sizes shown. Apart from the practicality of being able to obtain the larger sizes in one timber, normal practice and drive for economy would lead to their being built up using smaller timbers.						
RESOURCES – PLANT						
Timber						
tractor/trailer			32.30			
10 t crawler crane (25% of time)			10.65			
5.6 t rough terrain forklift (25% of time)			8.03			
7.5 kVA diesel generator			8.01			
two K637 rotary hammers			1.20			
two electric screwdrivers			2.85			
Total Gang Rate/Hour		£	**63.03**			
HARDWOOD COMPONENTS						
Greenheart						
100 × 75 mm						
length not exceeding 1.5 m	0.15	9.45	9.50	7.01	m	**25.96**
length 1.5–3 m	0.13	8.19	8.24	7.01	m	**23.44**
length 3–5 m	0.12	7.56	7.57	7.01	m	**22.14**
length 5–8 m	0.11	6.93	6.98	7.37	m	**21.28**
150 × 75 mm						
length not exceeding 1.5 m	0.17	10.71	10.92	8.70	m	**30.33**
length 1.5–3 m	0.15	9.45	9.50	8.50	m	**27.45**
length 3–5 m	0.14	8.82	8.83	8.50	m	**26.15**
length 5–8 m	0.12	7.56	9.81	8.91	m	**26.28**
200 × 100 mm						
length not exceeding 1.5 m	0.24	15.13	15.13	17.83	m	**48.09**
length 1.5–3 m	0.21	13.24	13.44	17.83	m	**44.51**
length 3–5 m	0.20	12.29	12.32	17.83	m	**42.44**
length 5–8 m	0.18	11.34	11.34	17.83	m	**40.51**
length 8–12 m	0.16	10.08	10.08	21.76	m	**41.92**

CLASS O: TIMBER

Item	Gang hours	Labour £	Plant £	Material £	Unit	Total rate £
HARDWOOD COMPONENTS – cont						
Greenheart – cont						
200 × 200 mm						
length not exceeding 1.5 m	0.42	26.47	26.47	30.85	m	83.79
length 1.5–3 m	0.40	25.21	25.21	30.85	m	81.27
length 3–5 m	0.38	23.95	23.95	30.85	m	78.75
length 5–8 m	0.34	21.43	21.43	30.85	m	73.71
length 8–12 m	0.30	18.91	19.28	33.26	m	71.45
225 × 100 mm						
length not exceeding 1.5 m	0.27	17.02	17.22	17.83	m	52.07
length 1.5–3 m	0.24	15.13	15.13	17.83	m	48.09
length 3–5 m	0.22	13.87	14.24	17.83	m	45.94
length 5–8 m	0.20	12.61	16.34	17.83	m	46.78
length 8–12 m	0.18	11.34	11.72	21.76	m	44.82
300 × 100 mm						
length not exceeding 1.5 m	0.36	22.69	22.69	23.32	m	68.70
length 1.5–3 m	0.33	20.80	21.00	23.32	m	65.12
length 3–5 m	0.30	18.91	19.28	23.32	m	61.51
length 5–8 m	0.27	17.02	17.22	23.32	m	57.56
length 8–12 m	0.24	15.13	15.13	31.61	m	61.87
300 × 200 mm						
length not exceeding 1.5 m	0.50	31.51	31.52	46.62	m	109.65
length 1.5–3 m	0.45	28.36	28.56	46.62	m	103.54
length 3–5 m	0.40	25.21	25.21	46.62	m	97.04
length 5–8 m	0.35	22.06	22.26	49.49	m	93.81
length 8–12 m	0.30	18.91	18.91	49.49	m	87.31
300 × 300 mm						
length not exceeding 1.5 m	0.52	32.77	32.78	69.94	m	135.49
length 1.5–3 m	0.48	30.25	30.25	69.94	m	130.44
length 3–5 m	0.44	27.73	27.74	69.94	m	125.41
length 5–8 m	0.40	25.21	25.21	74.37	m	124.79
length 8–12 m	0.36	22.69	22.69	74.37	m	119.75
450 × 450 mm						
length not exceeding 1.5 m	0.98	61.76	62.15	154.05	m	277.96
length 1.5–3 m	0.90	56.72	57.10	154.05	m	267.87
length 3–5 m	0.83	52.31	52.51	154.05	m	258.87
length 5–8 m	0.75	47.27	47.47	157.89	m	252.63
length 8–12 m	0.68	42.86	42.86	165.60	m	251.32
SOFTWOOD COMPONENTS						
Softwood; stress graded SC3/4						
100 × 75 mm						
up to 3.00 m long	0.10	6.30	5.42	2.00	m	13.72
3.00–5.00 m long	0.10	6.30	5.42	3.40	m	15.12
150 × 75 mm						
up to 3.00 m long	0.13	8.19	7.03	4.17	m	19.39
3.00–5.00 m long	0.11	6.93	5.96	6.09	m	18.98

CLASS O: TIMBER

Item	Gang hours	Labour £	Plant £	Material £	Unit	Total rate £
200 × 100 mm						
up to 3.00 m long	0.16	10.08	8.66	7.24	m	**25.98**
3.00–5.00 m long	0.14	8.82	7.59	12.88	m	**29.29**
200 × 200 mm						
up to 3.00 m long	0.25	15.76	13.53	21.45	m	**50.74**
3.00–5.00 m long	0.23	14.50	12.45	31.78	m	**58.73**
5.00–8.00 m long	0.20	12.61	10.83	24.58	m	**48.02**
300 × 200 mm						
up to 3.00 m long	0.27	17.02	14.62	36.03	m	**67.67**
3.00–5.00 m long	0.25	15.76	13.53	52.12	m	**81.41**
5.00–8.00 m long	0.23	14.50	12.45	35.68	m	**62.63**
300 × 300 mm						
up to 3.00 m long	0.30	18.91	16.25	56.59	m	**91.75**
3.00–5.00 m long	0.27	17.02	14.62	81.86	m	**113.50**
5.00–8.00 m long	0.25	15.76	13.53	57.26	m	**86.55**
450 × 450 mm						
up to 3.00 m long	0.35	22.06	18.95	160.52	m	**201.53**
3.00–5.00 m long	0.31	19.54	16.79	220.45	m	**256.78**
5.00–8.00 m long	0.28	17.65	15.16	158.92	m	**191.73**
600 × 600 mm						
3.00–5.00 m long	0.39	24.58	17.26	276.70	m	**318.54**
5.00–8.00 m long	0.37	23.32	20.02	289.84	m	**333.18**
ADD to the above prices for vacuum/pressure impregnating to minimum 5.30 kg/m³ salt retention	–	–	–	35.00	m³	**35.00**
HARDWOOD DECKING						
Greenheart; wrought finish						
Thickness 25–50 mm						
150 × 50 mm	0.58	36.55	31.41	55.16	m²	**123.12**
Thickness 50–75 mm						
200 × 75 mm	0.75	47.27	40.60	70.28	m²	**158.15**
Thickness 75–100 mm						
250 × 100 mm	0.95	59.87	49.91	94.46	m²	**204.24**
SOFTWOOD DECKING						
Douglas Fir						
Thickness 25–50 mm						
150 × 50 mm	0.39	24.58	21.12	28.46	m²	**74.16**
Thickness 50–75 mm						
200 × 75 mm	0.50	31.51	27.08	37.27	m²	**95.86**
Thickness 75–100 mm						
250 × 100 mm	0.65	40.97	46.30	41.14	m²	**128.41**

CLASS O: TIMBER

Item	Gang hours	Labour £	Plant £	Material £	Unit	Total rate £
FITTINGS AND FASTENINGS						
Metalwork						
Spikes; mild steel material rosehead						
14 × 14 × 275 mm long	0.13	4.58	–	1.21	nr	5.79
Metric mild steel bolts, nuts and washers						
M6 × 25 mm long	0.05	1.76	–	0.18	nr	1.94
M6 × 50 mm long	0.05	1.76	–	0.20	nr	1.96
M6 × 75 mm long	0.05	1.76	–	0.23	nr	1.99
M6 × 100 mm long	0.06	2.11	–	0.25	nr	2.36
M6 × 120 mm long	0.06	2.11	–	0.38	nr	2.49
M6 × 150 mm long	0.06	2.11	–	0.44	nr	2.55
M8 × 25 mm long	0.05	1.76	–	0.20	nr	1.96
M8 × 50 mm long	0.05	1.76	–	0.23	nr	1.99
M8 × 75 mm long	0.06	2.11	–	0.28	nr	2.39
M8 × 100 mm long	0.06	2.11	–	0.32	nr	2.43
M8 × 120 mm long	0.07	2.46	–	0.49	nr	2.95
M8 × 150 mm long	0.07	2.46	–	0.58	nr	3.04
M10 × 25 mm long	0.05	1.76	–	0.37	nr	2.13
M10 × 50 mm long	0.06	2.11	–	0.41	nr	2.52
M10 × 75 mm long	0.06	2.11	–	0.46	nr	2.57
M10 × 100 mm long	0.07	2.46	–	0.53	nr	2.99
M10 × 120 mm long	0.07	2.46	–	0.55	nr	3.01
M10 × 150 mm long	0.07	2.46	–	0.90	nr	3.36
M10 × 200 mm long	0.08	2.82	–	1.49	nr	4.31
M12 × 25 mm long	0.06	2.11	–	0.47	nr	2.58
M12 × 50 mm long	0.06	2.11	–	0.51	nr	2.62
M12 × 75 mm long	0.07	2.46	–	0.59	nr	3.05
M12 × 100 mm long	0.07	2.46	–	0.66	nr	3.12
M12 × 120 mm long	0.08	2.82	–	0.72	nr	3.54
M12 × 150 mm long	0.08	2.82	–	0.93	nr	3.75
M12 × 200 mm long	0.08	2.82	–	1.30	nr	4.12
M12 × 240 mm long	0.09	3.17	–	1.95	nr	5.12
M12 × 300 mm long	0.10	3.52	–	2.11	nr	5.63
M16 × 50 mm long	0.07	2.46	–	0.86	nr	3.32
M16 × 75 mm long	0.07	2.46	–	0.95	nr	3.41
M16 × 100 mm long	0.08	2.82	–	1.07	nr	3.89
M16 × 120 mm long	0.08	2.82	–	1.24	nr	4.06
M16 × 150 mm long	0.09	3.17	–	1.38	nr	4.55
M16 × 200 mm long	0.09	3.17	–	1.91	nr	5.08
M16 × 240 mm long	0.10	3.52	–	2.73	nr	6.25
M16 × 300 mm long	0.10	3.52	–	2.96	nr	6.48
M20 × 50 mm long	0.07	2.46	–	0.94	nr	3.40
M20 × 75 mm long	0.08	2.82	–	1.50	nr	4.32
M20 × 100 mm long	0.08	2.82	–	1.69	nr	4.51
M20 × 120 mm long	0.09	3.17	–	2.07	nr	5.24
M20 × 150 mm long	0.09	3.17	–	2.20	nr	5.37
M20 × 200 mm long	0.10	3.52	–	2.87	nr	6.39
M20 × 240 mm long	0.10	3.52	–	3.82	nr	7.34

CLASS O: TIMBER

Item	Gang hours	Labour £	Plant £	Material £	Unit	Total rate £
M20 × 300 mm long	0.11	3.87	–	4.22	nr	8.09
M24 × 50 mm long	0.08	2.82	–	3.00	nr	5.82
M24 × 75 mm long	0.08	2.82	–	3.22	nr	6.04
M24 × 100 mm long	0.09	3.17	–	3.32	nr	6.49
M24 × 120 mm long	0.09	3.17	–	3.56	nr	6.73
M24 × 150 mm long	0.10	3.52	–	3.98	nr	7.50
M24 × 200 mm long	0.11	3.87	–	4.61	nr	8.48
M24 × 240 mm long	0.11	3.87	–	5.87	nr	9.74
M24 × 300 mm long	0.12	4.23	–	6.50	nr	10.73
M30 × 100 mm long	0.09	3.17	–	7.94	nr	11.11
M30 × 120 mm long	0.10	3.52	–	8.25	nr	11.77
M30 × 150 mm long	0.11	3.87	–	8.74	nr	12.61
M30 × 200 mm long	0.11	3.87	–	9.53	nr	13.40
Carriage bolts, nuts and washer						
M6 × 25 mm long	0.05	1.76	–	0.92	nr	2.68
M6 × 50 mm long	0.05	1.76	–	0.50	nr	2.26
M6 × 75 mm long	0.05	1.76	–	0.53	nr	2.29
M6 × 100 mm long	0.06	2.11	–	0.65	nr	2.76
M6 × 150 mm long	0.06	2.11	–	0.82	nr	2.93
M8 × 25 mm long	0.05	1.76	–	0.58	nr	2.34
M8 × 50 mm long	0.05	1.76	–	0.61	nr	2.37
M8 × 75 mm long	0.06	2.11	–	0.63	nr	2.74
M8 × 100 mm long	0.06	2.11	–	0.70	nr	2.81
M8 × 150 mm long	0.07	2.46	–	0.72	nr	3.18
M8 × 200 mm long	0.07	2.46	–	1.14	nr	3.60
M10 × 25 mm long	0.05	1.76	–	0.73	nr	2.49
M10 × 50 mm long	0.06	2.11	–	0.70	nr	2.81
M10 × 75 mm long	0.06	2.11	–	0.72	nr	2.83
M10 × 100 mm long	0.07	2.46	–	0.75	nr	3.21
M10 × 150 mm long	0.07	2.46	–	0.92	nr	3.38
M10 × 200 mm long	0.08	2.82	–	1.32	nr	4.14
M10 × 240 mm long	0.08	2.82	–	2.51	nr	5.33
M10 × 300 mm long	0.09	3.17	–	2.72	nr	5.89
M12 × 25 mm long	0.06	2.11	–	1.08	nr	3.19
M12 × 50 mm long	0.06	2.11	–	1.36	nr	3.47
M12 × 75 mm long	0.07	2.46	–	1.48	nr	3.94
M12 × 100 mm long	0.07	2.46	–	1.54	nr	4.00
M12 × 150 mm long	0.08	2.82	–	1.88	nr	4.70
M12 × 200 mm long	0.08	2.82	–	2.35	nr	5.17
M12 × 240 mm long	0.09	3.17	–	3.44	nr	6.61
M12 × 300 mm long	0.10	3.52	–	3.66	nr	7.18
Galvanized steel						
Straps						
30 × 2.5 × 600 mm girth	0.13	4.58	–	1.94	nr	6.52
30 × 2.5 × 800 mm girth	0.13	4.58	–	2.54	nr	7.12
30 × 2.5 × 1000 mm girth	0.13	4.58	–	3.24	nr	7.82
30 × 2.5 × 1200 mm girth	0.15	5.28	–	3.83	nr	9.11
30 × 2.5 × 1400 mm girth	0.13	4.58	–	4.46	nr	9.04
30 × 2.5 × 1600 mm girth	0.15	5.28	–	5.09	nr	10.37
30 × 2.5 × 1800 mm girth	0.15	5.28	–	5.87	nr	11.15

CLASS O: TIMBER

Item	Gang hours	Labour £	Plant £	Material £	Unit	Total rate £
FITTINGS AND FASTENINGS – cont						
Galvanized steel – cont						
Straps – cont						
30 × 5 × 600 mm girth	0.13	4.58	–	3.78	nr	**8.36**
30 × 5 × 800 mm girth	0.13	4.58	–	5.04	nr	**9.62**
30 × 5 × 1000 mm girth	0.13	4.58	–	6.41	nr	**10.99**
30 × 5 × 1200 mm girth	0.15	5.28	–	7.61	nr	**12.89**
30 × 5 × 1400 mm girth	0.13	4.58	–	8.93	nr	**13.51**
30 × 5 × 1600 mm girth	0.15	5.28	–	10.13	nr	**15.41**
30 × 5 × 1800 mm girth	0.15	5.28	–	11.39	nr	**16.67**
Timber connectors; round toothed plate, single sided for 10 mm or 12 mm bolts						
38 mm diameter	0.01	0.18	–	1.84	nr	**2.02**
50 mm diameter	0.01	0.18	–	1.89	nr	**2.07**
63 mm diameter	0.01	0.28	–	2.05	nr	**2.33**
75 mm diameter	0.01	0.28	–	2.26	nr	**2.54**
Timber connectors; round toothed plate, double sided for 10 mm or 12 mm bolts						
38 mm diameter	0.01	0.18	–	1.86	nr	**2.04**
50 mm diameter	0.01	0.18	–	1.89	nr	**2.07**
63 mm diameter	0.01	0.28	–	2.05	nr	**2.33**
75 mm diameter	0.01	0.28	–	2.26	nr	**2.54**
Split ring connectors						
50 mm diameter	0.06	2.11	–	2.65	nr	**4.76**
63 mm diameter	0.01	0.21	–	2.71	nr	**2.92**
101 mm diameter	0.01	0.21	–	7.09	nr	**7.30**
Shear plate connectors						
67 mm diameter	0.01	0.21	–	4.38	nr	**4.59**
101 mm diameter	0.01	0.21	–	18.06	nr	**18.27**
Flitch plates						
200 × 75 × 10 mm	0.07	2.46	–	6.56	nr	**9.02**
300 × 100 × 10 mm	0.09	3.17	–	12.97	nr	**16.14**
450 × 150 × 12 mm	0.15	5.28	–	35.17	nr	**40.45**
Stainless steel						
Straps						
30 × 2.5 × 600 mm girth	0.13	4.58	–	4.20	nr	**8.78**
30 × 2.5 × 800 mm girth	0.13	4.58	–	5.78	nr	**10.36**
30 × 2.5 × 1000 mm girth	0.13	4.58	–	3.24	nr	**7.82**
30 × 2.5 × 1200 mm girth	0.15	5.28	–	3.83	nr	**9.11**
30 × 2.5 × 1400 mm girth	0.13	4.58	–	4.46	nr	**9.04**
30 × 2.5 × 1600 mm girth	0.15	5.28	–	5.09	nr	**10.37**
30 × 2.5 × 1800 mm girth	0.15	5.28	–	5.87	nr	**11.15**
30 × 5 × 600 mm girth	0.13	4.58	–	6.77	nr	**11.35**
30 × 5 × 800 mm girth	0.13	4.58	–	8.93	nr	**13.51**
30 × 5 × 1000 mm girth	0.13	4.58	–	6.41	nr	**10.99**
30 × 5 × 1200 mm girth	0.15	5.28	–	7.61	nr	**12.89**
30 × 5 × 1400 mm girth	0.13	4.58	–	8.93	nr	**13.51**
30 × 5 × 1600 mm girth	0.15	5.28	–	10.13	nr	**15.41**
30 × 5 × 1800 mm girth	0.15	5.28	–	11.39	nr	**16.67**

CLASS O: TIMBER

Item	Gang hours	Labour £	Plant £	Material £	Unit	Total rate £
Coach screws						
5.0 mm diameter × 75 mm long	0.04	1.41	–	–	nr	**1.41**
7.0 mm diameter × 105 mm long	0.05	1.76	–	–	nr	**1.76**
7.0 mm diameter × 140 mm long	0.06	2.11	–	–	nr	**2.11**
10.0 mm diameter × 95 mm long	0.06	2.11	–	0.01	nr	**2.12**
10.0 mm diameter × 165 mm long	0.07	2.46	–	0.01	nr	**2.47**
Timber connectors; round toothed plate, single sided for 10 mm or 12 mm bolts						
38 mm diameter	0.01	0.18	–	2.77	nr	**2.95**
50 mm diameter	0.01	0.18	–	2.89	nr	**3.07**
63 mm diameter	0.01	0.28	–	3.31	nr	**3.59**
75 mm diameter	0.01	0.28	–	3.67	nr	**3.95**
Timber connectors; round toothed plate, double sided for 10 mm or 12 mm bolts						
38 mm diameter	0.01	0.18	–	2.78	nr	**2.96**
50 mm diameter	0.01	0.18	–	2.89	nr	**3.07**
63 mm diameter	0.01	0.28	–	3.31	nr	**3.59**
75 mm diameter	0.01	0.28	–	3.67	nr	**3.95**
Split ring connectors						
63 mm diameter	0.06	2.11	–	12.07	nr	**14.18**
101 mm diameter	0.06	2.11	–	23.63	nr	**25.74**
Shear plate connectors						
67 mm diameter	0.06	2.11	–	0.47	nr	**2.58**
101 mm diameter	0.06	2.11	–	1.00	nr	**3.11**

CLASS P: PILING

Item	Gang hours	Labour £	Plant £	Material £	Unit	Total rate £
GENERALLY There are a number of different types of piling which are available for use in differing situations. Selection of the most suitable type of piling for a particular site will depend on a number of factors including the physical conditions likely to be encountered during driving, the loads to be carried, the design of superstructure, etc. The most commonly used systems are included in this section. It is essential that a thorough and adequate site investigation is carried out to ascertain details of the ground strata and bearing capacities to enable a proper assessment to be made of the most suitable and economical type of piling to be adopted. There are so many factors, apart from design considerations, which influence the cost of piling that it is not possible to give more than an approximate indication of costs. To obtain reliable costs for a particular contract advice should be sought from a company specializing in the particular type of piling proposed. Some Specialist Contractors will also provide a design service if required. **BORED CAST IN PLACE CONCRETE PILES** **Generally** The items 'number of piles' are calculated based on the following: Allowance for provision of all plant, equipment and labour including transporting to and from site and establishing and dismantling at £6,000 in total. Moving the rig to and setting up at each pile position; preparing to commence driving; £40.00, £60.00 and 75.00 per 300 mm, 450 mm and 600 mm diameter piles using the tripod mounted percussion rig. Standing time is quoted at £100.00 per hour for tripod rig, £195.00 per hour for mobile rig and £120.00 per hour for continuous flight auger. **Disposal of material arising from pile bores** The disposal of excavated material is shown separately, partly as this task is generally carried out by the main contractor rather than the piling specialist, but also to allow for simple adjustment should contaminated ground be envisaged.						

CLASS P: PILING

Item	Gang hours	Labour £	Plant £	Material £	Unit	Total rate £
Disposal of material arising from pile bores; collection from around piling operations						
storage on site; 100 m maximum distance; using 22.5 t ADT	0.05	0.91	2.86	–	m³	3.77
removal; 5 km distance; using 20 t tipper	0.12	2.17	7.99	–	m³	10.16
removal; 15 km distance; using 20 t tipper	0.22	3.99	14.64	–	m³	18.63
Add to the above rates where tipping charges apply (excluding Landfill Tax):						
non-hazardous waste	–	–	–	–	m³	46.50
hazardous waste	–	–	–	–	m³	121.50
special waste	–	–	–	–	m³	127.50
contaminated liquid	–	–	–	–	m³	168.00
contaminated sludge	–	–	–	–	m³	210.00
Add to the above rates where Landfill Tax applies:						
exempted material	–	–	–	–	m³	–
inactive or inert material	–	–	–	–	m³	3.75
other material	–	–	–	–	m³	96.00
Concrete 35 N/mm², 20 mm aggregate; installed by lorry/crawler-mounted rotary rig						
The following unit costs cover the construction of small diameter bored piles using lorry or crawler mounted rotary boring rigs. This type of plant is more mobile and faster in operation than the tripod rigs and is ideal for large contracts in cohesive ground. Construction of piles under bentonite suspension can be carried out to obviate the use of liners. Standard diameters of 450–900 mm diameter can be constructed to depths of 30 m.						
The costs are based on installing 100 piles on a clear site with reasonable access.						
Diameter: 300 mm						
number of piles (see above)	–	–	–	–	nr	156.35
concreted length	–	–	–	–	m	10.50
depth bored to 10 m maximum depth	–	–	–	–	m	32.50
depth bored to 15 m maximum depth	–	–	–	–	m	32.50
depth bored to 20 m maximum depth	–	–	–	–	m	32.50
Diameter: 450 mm						
number of piles (see above)	–	–	–	–	nr	156.35
concreted length	–	–	–	–	m	13.00
depth bored to 10 m maximum depth	–	–	–	–	m	40.65
depth bored to 15 m maximum depth	–	–	–	–	m	40.65
depth bored to 20 m maximum depth	–	–	–	–	m	40.65
Diameter: 600 mm						
number of piles (see above)	–	–	–	–	nr	156.35
concreted length	–	–	–	–	m	33.50
depth bored to 10 m maximum depth	–	–	–	–	m	72.50
depth bored to 15 m maximum depth	–	–	–	–	m	72.50
depth bored to 20 m maximum depth	–	–	–	–	m	72.50

CLASS P: PILING

Item	Gang hours	Labour £	Plant £	Material £	Unit	Total rate £
BORED CAST IN PLACE CONCRETE PILES – cont						
Concrete 35 N/mm², 20 mm aggregate; installed by continuous flight auger						
The following unit costs cover the construction of piles by screwing a continuous flight auger into the ground to a design depth (determined prior to commencement of piling operations and upon which the rates are based and subsequently varied to actual depths). Concrete is then pumped through the hollow stem of the auger to the bottom and the pile formed as the auger is withdrawn. Spoil is removed by the auger as it is withdrawn. This is a fast method of construction without causing disturbance or vibration to adjacent ground. No casing is required even in unsuitable soils. Reinforcement can be placed after grouting is complete.						
The costs are based on installing 100 piles on a clear site with reasonable access.						
Diameter: 300 mm						
number of piles	–	–	–	–	nr	36.50
concreted length	–	–	–	–	m	12.00
depth bored to 10 m maximum depth	–	–	–	–	nr	12.25
depth bored to 15 m maximum depth	–	–	–	–	nr	10.90
depth bored to 20 m maximum depth	–	–	–	–	nr	10.50
Diameter: 450 mm						
number of piles	–	–	–	–	nr	36.50
concreted length	–	–	–	–	m	22.25
depth bored to 10 m maximum depth	–	–	–	–	nr	14.00
depth bored to 15 m maximum depth	–	–	–	–	nr	13.40
depth bored to 20 m maximum depth	–	–	–	–	nr	12.50
Diameter: 600 mm						
number of piles	–	–	–	–	nr	36.50
concreted length	–	–	–	–	m	38.00
depth bored to 10 m maximum depth	–	–	–	–	nr	14.30
depth bored to 15 m maximum depth	–	–	–	–	nr	13.60
depth bored to 20 m maximum depth	–	–	–	–	nr	12.90
DRIVEN CAST IN PLACE CONCRETE PILES						
Generally						
The items 'number of piles' are calculated based on the following:						
Allowance for provision of all plant, equipment and labour including transporting to and from site and establishing and dismantling at £6,250 in total for piles using the Temporary Steel Casing Method and £10,500 in total for piles using the Segmental Casing Method.						

CLASS P: PILING

Item	Gang hours	Labour £	Plant £	Material £	Unit	Total rate £
Moving the rig to and setting up at each pile position; preparing to commence driving; £110.00 per pile. For the Temporary Steel Casing Method, obstructions (where within the capabilities of the normal plant) are quoted at £160.00 per hour and standing time at £152.50 per hour. For the Segmental Steel Casing Method, obstructions (where within the capabilities of the normal plant) are quoted at £275.00 per hour and standing time at £252.00 per hour.						
Temporary steel casing method; concrete 35 N/mm^2; reinforced for 750 kN The following unit costs cover the construction of piles by driving a heavy steel tube into the ground either by using an internal hammer acting on a gravel or concrete plug, as is more usual, or by using an external hammer on a driving helmet at the top of the tube. After driving to the required depth an enlarged base is formed by hammering out sucessive charges of concrete down the tube. The tube is then filled with concrete which is compacted as the tube is vibrated and withdrawn. Piles of 350–500 mm diameter can be constructed with rakes up to 1 in 4 to carry working loads up to 120 t per pile. The costs are based on installing 100 piles on a clear site with reasonable access. Diameter 430 mm; drive shell and form pile						
number of piles	–	–	–	–	nr	205.00
concreted length	–	–	–	–	m	22.50
depth driven; bottom-driven method	–	–	–	–	m	6.50
depth driven; top-driven method	–	–	–	–	m	3.75
Segmental casing method; concrete 35 N/mm^2; nominal reinforcement The following unit costs cover the construction of piles by driving into hard material using a serrated thick walled tube. It is oscillated and pressed into the hard material using a hydraulic attachment to the piling rig. The hard material is broken up using chiselling methods and is then removed by mechanical grab. The costs are based on installing 100 piles on a clear site with reasonable access.						
Diameter 620 mm						
number of piles	–	–	–	–	nr	270.00
concreted length	–	–	–	–	m	130.00
depth bored or driven to 15 m maximum depth	–	–	–	–	m	9.50

CLASS P: PILING

Item	Gang hours	Labour £	Plant £	Material £	Unit	Total rate £
DRIVEN CAST IN PLACE CONCRETE PILES – cont						
Diameter 1180 mm						
number of piles	–	–	–	–	nr	270.00
concreted length	–	–	–	–	m	142.50
depth bored or driven to 15 m maximum depth	–	–	–	–	m	15.00
Diameter 1500 mm						
number of piles	–	–	–	–	nr	270.00
concreted length	–	–	–	–	m	192.00
depth bored or driven to 15 m maximum depth	–	–	–	–	m	23.75

PREFORMED CONCRETE PILES

The following unit costs cover the installation of driven precast concrete piles by using a hammer acting on shoe fitted or cast into the precast concrete pile unit.

Single pile lengths are normally a maximum of 13 m long, at which point, a mechanical interlocking joint is required to extend the pile. These joints are most economically and practically formed at works.

Lengths, sizes of sections, reinforcement details and concrete mixes vary for differing contractors, whose specialist advice should be sought for specific designs.

The following unit costs are based on installing 100 piles on a clear site with reasonable access.

The items 'number of piles' are calculated based on the following:

    Allowance for provision of all plant, equipment and labour including transporting to and from site and establishing and dismantling at £3,150 in total for piles up to 275 × 275 mm and £3,850 in total for piles 350 × 350 mm and over.

    Moving the rig to and setting up at each pile position; preparing to commence driving; piles up to 275 × 275 mm £35.00 each; piles 350 × 350 mm and over, £55.00 each.

    An allowance for the cost of the driving head and shoe; £35.00 for 235 × 235 mm piles, £45.00 for 275 × 275 mm and £55.00 for 350 × 350 mm.

    Cost of providing the pile of the stated length.

Typical allowances for standing time are £138.50 per hour for 235 × 235 mm piles, £157.50 for 275 × 275 mm and £189.50 for 350 × 350 mm.

CLASS P: PILING

Item	Gang hours	Labour £	Plant £	Material £	Unit	Total rate £
Concrete 50 N/mm²; reinforced for 600 kN						
The costs are based on installing 100 piles on a clear site with reasonable access.						
Cross-sectional area: 0.05–0.1 m²; 235 × 235 mm						
number of piles of 10 m length	–	–	–	–	nr	301.90
number of piles of 15 m length	–	–	–	–	nr	401.90
number of piles of 20 m length	–	–	–	–	nr	501.90
number of piles of 25 m length	–	–	–	–	nr	601.90
add for mechanical interlocking joint	–	–	–	–	nr	60.00
depth driven	–	–	–	–	m	2.75
Cross-sectional area: 0.05–0.1 m²; 275 × 275 mm						
number of piles of 10 m length	–	–	–	–	nr	336.90
number of piles of 15 m length	–	–	–	–	nr	454.40
number of piles of 20 m length	–	–	–	–	nr	571.90
number of piles of 25 m length	–	–	–	–	nr	689.40
add for mechanical interlocking joint	–	–	–	–	nr	62.60
depth driven	–	–	–	–	m	2.90
Cross-sectional area: 0.1–0.15 m²; 350 × 350 mm						
number of piles of 10 m length	–	–	–	–	nr	527.60
number of piles of 15 m length	–	–	–	–	nr	712.60
number of piles of 20 m length	–	–	–	–	nr	897.60
number of piles of 25 m length	–	–	–	–	nr	1082.60
number of piles of 30 m length	–	–	–	–	nr	1267.60
add for mechanical interlocking joint	–	–	–	–	nr	120.00
depth driven	–	–	–	–	m	4.10
TIMBER PILES						
The items 'number of piles' are calculated based on the following:						
Allowance for provision of all plant, equipment and labour including transporting to and from site and establishing and dismantling at £6,000 in total. Moving the rig to and setting up at each pile position and preparing to drive at £75.00 per pile. allowance for the cost of the driving head and shoe at £40.00 per 225 × 225 mm pile, £50.00 for 300 × 300 mm, £60.00 for 350 × 350 mm and £70.00 for 450 × 450 mm.						
Cost of providing the pile of the stated length. A typical allowance for standing time is £245.50 per hour.						
Douglas Fir; hewn to mean pile size						
The costs are based on installing 100 piles on a clear site with reasonable access.						
Cross-sectional area: 0.05–0.1 m²; 225 × 225 mm						
number of piles of 10 m length	–	–	–	–	nr	490.00
number of piles of 15 m length	–	–	–	–	nr	640.00
number of piles of 20 m length	–	–	–	–	nr	790.00
depth driven	–	–	–	–	m	3.25

CLASS P: PILING

Item	Gang hours	Labour £	Plant £	Material £	Unit	Total rate £
TIMBER PILES – cont						
Douglas Fir – cont						
Cross-sectional area: 0.05–0.1 m^2; 300 × 300 mm						
number of piles of 10 m length	–	–	–	–	nr	700.00
number of piles of 15 m length	–	–	–	–	nr	950.00
number of piles of 20 m length	–	–	–	–	nr	1200.00
depth driven	–	–	–	–	m	3.75
Cross-sectional area: 0.1–0.15 m^2; 350 × 350 mm						
number of piles of 10 m length	–	–	–	–	nr	920.00
number of piles of 15 m length	–	–	–	–	nr	1270.00
number of piles of 20 m length	–	–	–	–	nr	1620.00
depth driven	–	–	–	–	m	4.50
Cross-sectional area: 0.15–0.25 m^2; 450 × 450 mm						
number of piles of 10 m length	–	–	–	–	nr	1375.00
number of piles of 15 m length	–	–	–	–	nr	1950.00
number of piles of 20 m length	–	–	–	–	nr	2525.00
number of piles of 25 m length	–	–	–	–	nr	3100.00
depth driven	–	–	–	–	m	5.50
Greenheart; hewn to mean pile size						
The costs are based on installing 100 piles on a clear site with reasonable access.						
Cross-sectional area: 0.05–0.1 m^2; 225 × 225 mm						
number of piles of 10 m length	–	–	–	–	nr	540.00
number of piles of 15 m length	–	–	–	–	nr	715.00
number of piles of 20 m length	–	–	–	–	nr	890.00
depth driven	–	–	–	–	m	3.25
Cross-sectional area: 0.05–0.1 m^2; 300 × 300 mm						
number of piles of 10 m length	–	–	–	–	nr	800.00
number of piles of 15 m length	–	–	–	–	nr	1100.00
number of piles of 20 m length	–	–	–	–	nr	1400.00
depth driven	–	–	–	–	m	3.75
Cross-sectional area: 0.1–0.15 m^2; 350 × 350 mm						
number of piles of 10 m length	–	–	–	–	nr	1020.00
number of piles of 15 m length	–	–	–	–	nr	1420.00
number of piles of 20 m length	–	–	–	–	nr	1820.00
depth driven	–	–	–	–	m	4.50
Cross-sectional area: 0.15–0.25 m^2; 450 × 450 mm						
number of piles of 10 m length	–	–	–	–	nr	1575.00
number of piles of 15 m length	–	–	–	–	nr	2250.00
number of piles of 20 m length	–	–	–	–	nr	2925.00
number of piles of 25 m length	–	–	–	–	nr	3600.00
depth driven	–	–	–	–	m	5.50

CLASS P: PILING

Item	Gang hours	Labour £	Plant £	Material £	Unit	Total rate £
ISOLATED STEEL PILES						
Steel bearing piles are commonly carried out by a Specialist Contractor whose advice should be sought to arrive at accurate costing. However the following items can be used to assess a budget cost for such work.						
The following unit costs are based upon driving 100 nr steel bearing piles on a clear site with reasonable access.						
The items 'number of piles' are calculated based on the following:						
Allowance for provision of all plant, equipment and labour including transporting to and from site and establishing and dismantling at £6,600 in total up to a maximum 100 miles radius from base and £16,050 up to a maximum 250 miles radius from base.						
Moving the rig to and setting up at each pile position; preparing to commence driving; £193.15 per pile.						
Cost of providing the pile of the stated length.						
A typical allowance for standing time is £283.90 per hour.						
Steel EN 10025 grade S275; within 100 miles of steel plant						
The costs are based upon installing 100 nr on a clear site with reasonable access.						
Mass 45 kg/m; 203 × 203 mm						
number of piles: length 10 m	–	–	–	5041.44	nr	**5041.44**
number of piles: length 15 m	–	–	–	5316.78	nr	**5316.78**
number of piles: length 20 m	–	–	–	5567.38	nr	**5567.38**
depth driven; vertical	–	–	–	–	m	**3.25**
depth driven; raking	–	–	–	–	m	**5.15**
Mass 54 kg/m; 203 × 203 mm						
number of piles; length 10 m	–	–	–	5136.72	nr	**5136.72**
number of piles; length 15 m	–	–	–	5467.14	nr	**5467.14**
number of piles; length 20 m	–	–	–	5767.85	nr	**5767.85**
depth driven; vertical	–	–	–	–	m	**3.25**
depth driven; raking	–	–	–	–	m	**5.15**
Mass 63 kg/m; 254 × 254 mm						
number of piles; length 10 m	–	–	–	5232.01	nr	**5232.01**
number of piles; length 15 m	–	–	–	5617.49	nr	**5617.49**
number of piles; length 20 m	–	–	–	5968.33	nr	**5968.33**
depth driven; vertical	–	–	–	–	m	**3.50**
depth driven; raking	–	–	–	–	m	**5.35**
Mass 71 kg/m; 254 × 254 mm						
number of piles; length 10 m	–	–	–	5316.71	nr	**5316.71**
number of piles; length 15 m	–	–	–	5751.14	nr	**5751.14**
number of piles; length 20 m	–	–	–	6146.53	nr	**6146.53**
depth driven; vertical	–	–	–	–	m	**3.50**
depth driven; raking	–	–	–	–	m	**5.35**

CLASS P: PILING

Item	Gang hours	Labour £	Plant £	Material £	Unit	Total rate £
ISOLATED STEEL PILES – cont						
Steel EN 10025 grade S275 – cont						
Mass 85 kg/m; 254 × 254 mm						
number of piles; length 10 m	–	–	–	5464.94	nr	5464.94
number of piles; length 15 m	–	–	–	5985.03	nr	5985.03
number of piles; length 20 m	–	–	–	6458.38	nr	6458.38
depth driven; vertical	–	–	–	–	m	3.50
depth driven; raking	–	–	–	–	m	5.35
Mass 79 kg/m; 305 × 305 mm						
number of piles; length 10 m	–	–	–	5401.41	nr	5401.41
number of piles; length 15 m	–	–	–	5884.79	nr	5884.79
number of piles; length 20 m	–	–	–	6324.72	nr	6324.72
depth driven; vertical	–	–	–	–	m	3.50
depth driven; raking	–	–	–	–	m	5.35
Mass 95 kg/m; 305 × 305 mm						
number of piles; length 10 m	–	–	–	5570.81	nr	5570.81
number of piles; length 15 m	–	–	–	6152.09	nr	6152.09
number of piles; length 20 m	–	–	–	6681.13	nr	6681.13
depth driven; vertical	–	–	–	–	m	3.75
depth driven; raking	–	–	–	–	m	5.65
Mass 110 kg/m; 305 × 305 mm						
number of piles; length 10 m	–	–	–	5729.63	nr	5729.63
number of piles; length 15 m	–	–	–	6402.69	nr	6402.69
number of piles; length 20 m	–	–	–	7015.25	nr	7015.25
depth driven; vertical	–	–	–	–	m	3.75
depth driven; raking	–	–	–	–	m	5.65
Mass 109 kg/m; 356 × 368 mm						
number of piles; length 10 m	–	–	–	5958.84	nr	5958.84
number of piles; length 15 m	–	–	–	6745.68	nr	6745.68
number of piles; length 20 m	–	–	–	7472.57	nr	7472.57
depth driven; vertical	–	–	–	–	m	3.75
depth driven; raking	–	–	–	–	m	5.65
Mass 126 kg/m; 305 × 305 mm						
number of piles; length 10 m	–	–	–	5899.03	nr	5899.03
number of piles; length 15 m	–	–	–	6669.99	nr	6669.99
number of piles; length 20 m	–	–	–	7371.65	nr	7371.65
driving piles; vertical	–	–	–	–	m	4.00
driving piles; raking	–	–	–	–	m	6.15
Mass 149 kg/m; 305 × 305 mm						
number of piles; length 10 m	–	–	–	6142.54	nr	6142.54
number of piles; length 15 m	–	–	–	7054.23	nr	7054.23
number of piles; length 20 m	–	–	–	7883.97	nr	7883.97
driving piles; vertical	–	–	–	–	m	4.00
driving piles; raking	–	–	–	–	m	6.15
Mass 186 kg/m; 305 × 305 mm						
number of piles; length 10 m	–	–	–	6534.28	nr	6534.28
number of piles; length 15 m	–	–	–	7672.36	nr	7672.36
number of piles; length 20 m	–	–	–	8708.15	nr	8708.15
driving piles; vertical	–	–	–	–	m	4.67
drive piles; raking	–	–	–	–	m	6.65

CLASS P: PILING

Item	Gang hours	Labour £	Plant £	Material £	Unit	Total rate £
Mass 223 kg/m; 305 × 305 mm						
number of piles; length 10 m	–	–	–	6926.01	nr	6926.01
number of piles; length 15 m	–	–	–	8290.49	nr	8290.49
number of piles; length 20 m	–	–	–	9532.32	nr	9532.32
driving piles; vertical	–	–	–	–	m	4.95
driving piles; raking	–	–	–	–	m	6.85
Mass 133 kg/m; 356 × 368 mm						
number of piles; length 10 m	–	–	–	6265.74	nr	6265.74
number of piles; length 15 m	–	–	–	7225.83	nr	7225.83
number of piles; length 20 m	–	–	–	8112.78	nr	8112.78
driving piles; vertical	–	–	–	–	m	4.00
driving piles; raking	–	–	–	–	m	6.15
Mass 152 kg/m; 356 × 368 mm						
number of piles; length 10 m	–	–	–	6508.70	nr	6508.70
number of piles; length 15 m	–	–	–	7605.95	nr	7605.95
number of piles; length 20 m	–	–	–	8619.60	nr	8619.60
driving piles; vertical	–	–	–	–	m	4.50
driving piles; raking	–	–	–	–	m	6.25
Mass 174 kg/m; 356 × 368 mm						
number of piles; length 10 m	–	–	–	6790.03	nr	6790.03
number of piles; length 15 m	–	–	–	8046.09	nr	8046.09
number of piles; length 20 m	–	–	–	9206.45	nr	9206.45
driving piles; vertical	–	–	–	–	m	4.50
driving piles; raking	–	–	–	–	m	6.25

INTERLOCKING STEEL PILES

Sheet steel piling is commonly carried out by a Specialist Contractor, whose advice should be sought to arrive at accurate costings. However the following items can be used to assess a budget for such work.

Note: area of driven piles will vary from area supplied dependent upon pitch line of piling and provision for such allowance has been made in PC for supply.

The materials cost below includes the manufacturers tarrifs for a 200 mile delivery radius from works, delivery 10–20 t loads and with an allowance of 10% to cover waste/projecting piles etc.

Arcelor Mittal Z section steel piles; EN 10248 grade S270GP steel

The following unit costs are based on driving/extracting 1,500 m² of sheet piling on a clear site with reasonable access.

Provision of all plant, equipment and labour including transport to and from the site and establishing and dismantling for

Item	Gang hours	Labour £	Plant £	Material £	Unit	Total rate £
driving of sheet piling	–	–	–	–	sum	7500.00
extraction of sheet piling	–	–	–	–	sum	500.00
Standing time	–	–	–	–	hr	350.00

CLASS P: PILING

Item	Gang hours	Labour £	Plant £	Material £	Unit	Total rate £
INTERLOCKING STEEL PILES – cont						
Arcelor Mittal Z section steel piles – cont						
Section modulus 800–1200 cm^3/m; section reference AZ 12; mass 98.7 kg/m^2, sectional modulus 1200 cm^3/m; EN 10248 grade S270GP steel						
length of welded corner piles	–	–	–	–	m	68.25
length of welded junction piles	–	–	–	–	m	115.00
driven area	–	–	–	–	m^2	125.00
area of piles of length not exceeding 14 m	–	–	–	87.20	m^2	87.20
area of piles of length 14–24 m	–	–	–	93.47	m^2	93.47
area of piles of length exceeding 24 m	–	–	–	93.47	m^2	93.47
Section modulus 1200–2000 cm^3/m; section reference AZ 17; mass 108.6 kg/m^2; sectional modulus 1665 cm^3/m; EN 10248 grade S270GP steel						
length of welded corner piles	–	–	–	–	m	68.25
length of welded junction piles	–	–	–	–	m	115.00
driven area	–	–	–	–	m^2	145.00
area of piles of length not exceeding 14 m	–	–	–	92.10	m^2	92.10
area of piles of length 14–24 m	–	–	–	93.47	m^2	93.47
area of piles of length exceeding 24 m	–	–	–	93.47	m^2	93.47
Section modulus 2000–3000 cm^3/m; section reference AZ 26; mass 155.2 kg/m^2; sectional modulus 2600 cm^3/m; EN 10248 grade S270GP steel						
driven area	–	–	–	–	m^2	175.00
area of piles of length 6–18 m	–	–	–	104.36	m^2	104.36
area of piles of length 18–24 m	–	–	–	105.90	m^2	105.90
Section modulus 3000–4000 cm^3/m; section reference AZ 36; mass 194.0 kg/m^2; sectional modulus 3600 cm^3/m; EN 10248 grade S270GP steel						
driven area	–	–	–	–	m^2	245.00
area of piles of length 6–18 m	–	–	–	127.43	m^2	127.43
area of piles of length 18–24 m	–	–	–	129.31	m^2	129.31
Straight section modulus ne 500 cm^3/m; section reference AS 500–12 mass 149 kg/m^2; sectional modulus 51 cm^3/m; EN 10248 grade S270GP steel						
driven area	–	–	–	–	m^2	175.00
area of piles of length 6–18 m	–	–	–	158.78	m^2	158.78
area of piles of length 18–24 m	–	–	–	161.12	m^2	161.12
One coat black tar vinyl (PC1) protective treatment applied all surfaces at shop to minimum dry film thickness up to 150 microns to steel piles						
section reference AZ 12; pile area	–	–	–	9.21	m^2	9.21
section reference AZ 17; pile area	–	–	–	9.41	m^2	9.41
section reference AZ 26; pile area	–	–	–	10.29	m^2	10.29
section reference AZ 36; pile area	–	–	–	10.96	m^2	10.96
section reference AS 500–12; pile area	–	–	–	11.45	m^2	11.45

CLASS P: PILING

Item	Gang hours	Labour £	Plant £	Material £	Unit	Total rate £
One coat black high build isocyanate cured epoxy pitch (PC2) protective treatment applied all surfaces at shop to minimum dry film thickness up to 450 microns to steel piles						
section reference AZ 12; pile area	–	–	–	14.46	m^2	**14.46**
section reference AZ 17; pile area	–	–	–	14.79	m^2	**14.79**
section reference AZ 26; pile area	–	–	–	16.17	m^2	**16.17**
section reference AZ 36; pile area	–	–	–	17.21	m^2	**17.21**
section reference AS 500–12; pile area	–	–	–	17.98	m^2	**17.98**
Arcelor Mittal U section steel piles; EN 10248 grade S270GP steel						
The following unit costs are based on driving/extracting 1,500 m^2 of sheet piling on a clear site with reasonable access.						
Provision of plant, equipment and labour including transport to and from the site and establishing and dismantling						
driving of sheet piling	–	–	–	–	sum	**7100.00**
extraction of sheet piling	–	–	–	–	sum	**4750.00**
Standing time	–	–	–	–	hr	**330.00**
Section modulus 500–800 cm^3/m; section reference PU 6; mass 76.0 kg/m^2; sectional modulus 600 cm^3/m						
driven area	–	–	–	–	m^2	**100.00**
area of piles of length 6–18 m	–	–	–	74.88	m^2	**74.88**
area of piles of length 18–24 m	–	–	–	76.05	m^2	**76.05**
Section modulus 800–1200 cm^3/m; section reference PU 8; mass 90.9 kg/m^2; sectional modulus 830 cm^3/m						
driven area	–	–	–	–	m^2	**110.00**
area of piles of length 6–18 m	–	–	–	79.08	m^2	**79.08**
area of piles of length 18–24 m	–	–	–	80.33	m^2	**80.33**
Section modulus 1200–2000 cm^3/m; section reference PU 12; mass 110.1 kg/m^2; sectional modulus 1200 cm^3/m						
driven area	–	–	–	–	m^2	**128.00**
area of piles of length 6–18 m	–	–	–	93.60	m^2	**93.60**
area of piles of length 18–24 m	–	–	–	95.06	m^2	**95.06**
Section modulus 1200–2000 cm^3/m; section reference PU 18; mass 128.2 kg/m^2; sectional modulus 1800 cm^3/m						
driven area	–	–	–	–	m^2	**141.00**
area of piles of length 6–18 m	–	–	–	108.72	m^2	**108.72**
area of piles of length 18–24 m	–	–	–	110.42	m^2	**110.42**
Section modulus 2000–3000 cm^3/m; section reference PU 22; mass 143.6 kg/m^2; sectional modulus 2200 cm^3/m						
driven area	–	–	–	–	m^2	**141.00**
area of piles of length 6–18 m	–	–	–	122.00	m^2	**122.00**
area of piles of length 18–24 m	–	–	–	123.91	m^2	**123.91**

CLASS P: PILING

Item	Gang hours	Labour £	Plant £	Material £	Unit	Total rate £
INTERLOCKING STEEL PILES – cont						
Arcelor Mittal U section steel piles – cont						
Section modulus 3000–4000 cm^3/m; section reference PU 32; mass 190.2 kg/m^2; sectional modulus 3200 cm^3/m						
driven area	–	–	–	–	m^2	176.80
area of piles of length 6–18 m	–	–	–	137.92	m^2	137.92
area of piles of length 18–24 m	–	–	–	140.08	m^2	140.08
One coat black tar vinyl (PC1) protective treatment applied all surfaces at shop to minimum dry film thickness up to 150 microns to steel piles						
section reference PU 6; pile area	–	–	–	11.60	m^2	11.60
section reference PU 8; pile area	–	–	–	11.40	m^2	11.40
section reference PU 12; pile area	–	–	–	12.12	m^2	12.12
section reference PU 18; pile area	–	–	–	13.00	m^2	13.00
section reference PU 22; pile area	–	–	–	13.48	m^2	13.48
section reference PU 32; pile area	–	–	–	13.68	m^2	13.68
One coat black high build isocyanate cured epoxy pitch (PC2) protective treatment applied all surfaces at shop to minimum dry film thickness up to 450 microns to steel piles						
section reference PU 6; pile area	–	–	–	15.95	m^2	15.95
section reference PU 8; pile area	–	–	–	15.68	m^2	15.68
section reference PU 12; pile area	–	–	–	16.66	m^2	16.66
section reference PU 18; pile area	–	–	–	17.88	m^2	17.88
section reference PU 22; pile area	–	–	–	18.54	m^2	18.54
section reference PU 32; pile area	–	–	–	18.81	m^2	18.81
RESOURCE – MATERIALS						
Supply only materials rates						
Interlocking steel sheet piling; all sections hot rolled						
standard lengths to 18 m	–	–	–	775.00	t	775.00
long lengths 18 m to 24 m	–	–	–	787.50	t	787.50
lomg lengths 18 m to 24 m	–	–	–	800.00	t	800.00
One coat tar vinyl (PC1) protection at shop; surface area	–	–	–	4.00	m^2	4.00
One coat epoxy pitch (PC2) protection at shop; surface area	–	–	–	5.50	m^2	5.50
Universal bearing piles						
size 203 × 203 mm to 305 × 305 mm	–	–	–	950.00	t	950.00
size 356 × 368 mm	–	–	–	1150.00	t	1150.00
extra for lengths 15–24 m long (all sizes)	–	–	–	50.00	t	50.00
NB for lengths over 24 m POA						

CLASS Q: PILING ANCILLARIES

Item	Gang hours	Labour £	Plant £	Material £	Unit	Total rate £
CAST IN PLACE CONCRETE PILES						
Bored; lorry/crawler mounted rotary rig						
Backfilling empty bore with selected excavated material						
diameter 500 mm	–	–	–	–	m	3.50
Permanent casings; each length not exceeding 13 m						
diameter 500 mm	–	–	–	–	m	75.00
Permanent casings; each length exceeding 13 m						
diameter 500 mm	–	–	–	–	m	80.00
Enlarged bases						
diameter 1500 mm; to 500 mm diameter pile	–	–	–	–	nr	250.00
Cutting off surplus lengths						
diameter 500 mm	–	–	–	–	m	27.50
Preparing heads						
500 mm diameter	–	–	–	–	nr	40.00
Bored; continuous flight auger						
Backfilling empty bore with selected excavated material						
450 mm diameter piles	–	–	–	–	m	2.75
600 mm diameter piles	–	–	–	–	m	3.75
750 mm diameter piles	–	–	–	–	m	4.10
Permanent casings; each length not exceeding 13 m						
450 mm diameter piles	–	–	–	–	m	66.15
600 mm diameter piles	–	–	–	–	m	90.00
750 mm diameter piles	–	–	–	–	m	112.50
Permanent casings; each length exceeding 13 m						
450 mm diameter piles	–	–	–	–	m	65.00
600 mm diameter piles	–	–	–	–	m	95.50
750 mm diameter piles	–	–	–	–	m	115.00
Enlarged bases						
diameter 1400 mm; to 450 mm diameter piles	–	–	–	–	nr	230.00
diameter 1800 mm; to 600 mm diameter piles	–	–	–	–	nr	280.00
diameter 2100 mm; to 750 mm diameter piles	–	–	–	–	nr	315.00
Cutting off surplus lengths						
450 mm diameter piles	–	–	–	–	m	24.00
600 mm diameter piles	–	–	–	–	m	34.00
750 mm diameter piles	–	–	–	–	m	40.50
Preparing heads						
450 mm diameter piles	–	–	–	–	nr	25.50
600 mm diameter piles	–	–	–	–	nr	45.00
750 mm diameter piles	–	–	–	–	nr	61.00
Collection from around pile heads of spoil accruing from piling operations and depositing in spoil heaps (For final disposal see Class E – Excavation Ancillaries)	–	–	–	–	m^3	3.00

CLASS Q: PILING ANCILLARIES

Item	Gang hours	Labour £	Plant £	Material £	Unit	Total rate £
CAST IN PLACE CONCRETE PILES – cont						
Reinforcement; mild steel						
Straight bars, nominal size						
6 mm	–	–	–	–	t	500.00
8 mm	–	–	–	–	t	500.00
10 mm	–	–	–	–	t	500.00
12 mm	–	–	–	–	t	500.00
16 mm	–	–	–	–	t	500.00
25 mm	–	–	–	–	t	500.00
32 mm	–	–	–	–	t	500.00
50 mm	–	–	–	–	t	500.00
Helical bars, nominal size						
6 mm	–	–	–	–	t	550.00
8 mm	–	–	–	–	t	550.00
10 mm	–	–	–	–	t	550.00
12 mm	–	–	–	–	t	550.00
Reinforcement; high tensile steel						
Straight bars, nominal size						
6 mm	–	–	–	–	t	650.00
8 mm	–	–	–	–	t	650.00
10 mm	–	–	–	–	t	650.00
12 mm	–	–	–	–	t	650.00
16 mm	–	–	–	–	t	650.00
25 mm	–	–	–	–	t	650.00
32 mm	–	–	–	–	t	650.00
50 mm	–	–	–	–	t	650.00
Helical bars, nominal size						
6 mm	–	–	–	–	t	650.00
8 mm	–	–	–	–	t	650.00
10 mm	–	–	–	–	t	650.00
12 mm	–	–	–	–	t	650.00
Couplers; Lenton type A; threaded ends on reinforcing bars						
12 mm	0.09	10.68	–	6.56	nr	17.24
16 mm	0.09	10.68	–	7.73	nr	18.41
20 mm	0.09	10.68	–	11.45	nr	22.13
25 mm	0.09	10.68	–	15.75	nr	26.43
32 mm	0.09	10.68	–	21.61	nr	32.29
40 mm	0.09	10.68	–	29.58	nr	40.26
Couplers; Lenton type B; threaded ends on reinforcing bars						
12 mm	0.09	10.68	–	20.65	nr	31.33
16 mm	0.09	10.68	–	23.97	nr	34.65
20 mm	0.09	10.68	–	26.67	nr	37.35
25 mm	0.09	10.68	–	30.67	nr	41.35
32 mm	0.09	10.68	–	41.60	nr	52.28
40 mm	0.09	10.68	–	60.84	nr	71.52

CLASS Q: PILING ANCILLARIES

Item	Gang hours	Labour £	Plant £	Material £	Unit	Total rate £
PREFORMED CONCRETE PILES						
General						
Preparing heads						
235 × 235 mm piles	–	–	–	–	nr	29.33
275 × 275 mm piles	–	–	–	–	nr	40.05
350 × 350 mm piles	–	–	–	–	nr	61.59
TIMBER PILES						
Douglas Fir						
Cutting off surplus lengths						
cross-sectional area: 0.025–0.05 m^2	–	–	–	–	nr	2.75
cross-sectional area: 0.05–0.1 m^2	–	–	–	–	nr	4.75
cross-sectional area: 0.1–0.15 m^2	–	–	–	–	nr	6.15
cross-sectional area: 0.15–0.25 m^2	–	–	–	–	nr	12.00
Preparing heads						
cross-sectional area: 0.025–0.05 m^2	–	–	–	–	nr	2.75
cross-sectional area: 0.05–0.1 m^2	–	–	–	–	nr	4.75
cross-sectional area: 0.1–0.15 m^2	–	–	–	–	nr	6.15
cross-sectional area: 0.15–0.25 m^2	–	–	–	–	nr	12.00
Greenheart						
Cutting off surplus lengths						
cross-sectional area: 0.025–0.05 m^2	–	–	–	–	nr	5.50
cross-sectional area: 0.05–0.1 m^2	–	–	–	–	nr	10.00
cross-sectional area: 0.1–0.15 m^2	–	–	–	–	nr	12.50
cross-sectional area: 0.15–0.25 m^2	–	–	–	–	nr	22.50
Preparing heads						
cross-sectional area: 0.025–0.05 m^2	–	–	–	–	nr	5.50
cross-sectional area: 0.05–0.1 m^2	–	–	–	–	nr	10.00
cross-sectional area: 0.1–0.15 m^2	–	–	–	–	nr	12.50
cross-sectional area: 0.15–0.25 m^2	–	–	–	–	nr	22.50
ISOLATED STEEL PILES						
Steel bearing piles						
Steel bearing piles are commonly carried out by a Specialist Contractor whose advice should be sought to arrive at accurate costing. However the following items can be used to assess a budget cost for such work.						
The item for number of pile extensions includes for the cost of setting up the rig at the pile position together with welding the extension to the top of the steel bearing pile. The items for length of pile extension cover the material only, the driving cost being included in Class P.						

CLASS Q: PILING ANCILLARIES

Item	Gang hours	Labour £	Plant £	Material £	Unit	Total rate £
ISOLATED STEEL PILES – cont						
Steel bearing piles – cont						
Number of pile extensions						
at each position	–	–	–	–	nr	65.00
Length of pile extensions, each length not exceeding 3 m; steel EN 10025 grade S275						
mass 45 kg/m	–	–	–	50.12	m	50.12
mass 54 kg/m	–	–	–	60.14	m	60.14
mass 63 kg/m	–	–	–	70.17	m	70.17
mass 71 kg/m	–	–	–	79.08	m	79.08
mass 79 kg/m	–	–	–	87.99	m	87.99
mass 85 kg/m	–	–	–	94.67	m	94.67
mass 95 kg/m	–	–	–	105.81	m	105.81
mass 109 kg/m	–	–	–	145.38	m	145.38
mass 110 kg/m	–	–	–	122.51	m	122.51
mass 126 kg/m	–	–	–	140.33	m	140.33
mass 149 kg/m	–	–	–	165.95	m	165.95
mass 133 kg/m	–	–	–	177.39	m	177.39
mass 152 kg/m	–	–	–	202.73	m	202.73
mass 174 kg/m	–	–	–	232.07	m	232.07
mass 186 kg/m	–	–	–	207.16	m	207.16
mass 223 kg/m	–	–	–	248.37	m	248.37
Length of pile extensions, each length exceeding 3 m; steel EN 10025 grade S275						
mass 45 kg/m	–	–	–	50.12	m	50.12
mass 54 kg/m	–	–	–	60.14	m	60.14
mass 63 kg/m	–	–	–	70.17	m	70.17
mass 71 kg/m	–	–	–	79.08	m	79.08
mass 79 kg/m	–	–	–	87.99	m	87.99
mass 85 kg/m	–	–	–	94.67	m	94.67
mass 95 kg/m	–	–	–	105.81	m	105.81
mass 109 kg/m	–	–	–	145.38	m	145.38
mass 110 kg/m	–	–	–	122.51	m	122.51
mass 126 kg/m	–	–	–	140.33	m	140.33
mass 133 kg/m	–	–	–	165.95	m	165.95
mass 149 kg/m	–	–	–	177.39	m	177.39
mass 152 kg/m	–	–	–	202.73	m	202.73
mass 174 kg/m	–	–	–	232.07	m	232.07
mass 186 kg/m	–	–	–	207.16	m	207.16
mass 223 kg/m	–	–	–	248.37	m	248.37
Number of pile extensions						
section size 203 × 203 × any kg/m	–	–	–	–	nr	215.00
section size 254 × 254 × any kg/m	–	–	–	–	nr	255.00
section size 305 × 305 × any kg/m	–	–	–	–	nr	295.00
section size 356 × 368 × any kg/m	–	–	–	–	nr	325.00
Cutting off surplus lengths						
mass 30–60 kg/m	–	–	–	–	nr	75.00
mass 60–120 kg/m	–	–	–	–	nr	105.00
mass 120–250 kg/m	–	–	–	–	nr	145.00

CLASS Q: PILING ANCILLARIES

Item	Gang hours	Labour £	Plant £	Material £	Unit	Total rate £
Burning off tops of piles to level						
mass 30–60 kg/m	–	–	–	–	nr	75.00
mass 60–120 kg/m	–	–	–	–	nr	105.00
mass 120–250 kg/m	–	–	–	–	nr	145.00
INTERLOCKING STEEL PILES						
Arcelor Mittal Z section steel piles; EN 10248 grade S270GP steel						
Steel bearing piles are commonly carried out by a Specialist Contractor whose advice should be sought to arrive at accurate costing. However the following items can be used to assess a budget cost for such work.						
The item for number of pile extensions includes for welding the extension to the top of the steel bearing pile. The items for length of pile extension cover the material only, the driving cost being included in Class P.						
Number of pile extensions						
Cutting off surplus lengths						
section modulus 500–800 cm^3/m	–	–	–	–	m	23.50
section modulus 800–1200 cm^3/m	–	–	–	–	m	23.50
section modulus 1200–2000 cm^3/m	–	–	–	–	m	18.50
section modulus 2000–3000 cm^3/m	–	–	–	–	m	23.50
section modulus 3000–4000 cm^3/m	–	–	–	–	m	18.50
Extract piling and stacking on site						
section modulus 500–800 cm^3/m	–	–	–	–	m^2	37.00
section modulus 800–1200 cm^3/m	–	–	–	–	m^2	18.50
section modulus 1200–2000 cm^3/m	–	–	–	–	m^2	40.75
section modulus 2000–3000 cm^3/m	–	–	–	–	m^2	18.50
section modulus 3000–4000 cm^3/m	–	–	–	–	m^2	18.50
Arcelor Mittal U section steel piles; EN 10248 grade S270GP steel						
Steel bearing piles are commonly carried out by a Specialist Contractor and whose advice should be sought to arrive at accurate costing. However the following items can be used to assess a budget cost for such work.						
The item for number of pile extensions includes for welding the extension to the top of the steel bearing pile. The items for length of pile extension cover the material only, the driving cost being included in Class P.						
Number of pile extensions						
Cutting off surplus lengths						
section modulus 500–800 cm^3/m	–	–	–	–	m	22.10
section modulus 800–1200 cm^3/m	–	–	–	–	m	22.10
section modulus 1200–2000 cm^3/m	–	–	–	–	m	22.10
section modulus 2000–3000 cm^3/m	–	–	–	–	m	22.10
section modulus 3000–4000 cm^3/m	–	–	–	–	m	22.10

CLASS Q: PILING ANCILLARIES

Item	Gang hours	Labour £	Plant £	Material £	Unit	Total rate £
INTERLOCKING STEEL PILES – cont						
Arcelor Mittal U section steel piles – cont						
Extract piling and stack on site						
section modulus 500–800 cm^3/m	–	–	–	–	m^2	18.50
section modulus 800–1200 cm^3/m	–	–	–	–	m^2	18.50
section modulus 800–1200 cm^3/m PU12	–	–	–	–	m^2	18.50
section modulus 800–1200 cm^3/m PU18	–	–	–	–	m^2	18.50
section modulus 2000–3000 cm^3/m	–	–	–	–	m^2	18.50
section modulus 3000–4000 cm^3/m	–	–	–	–	m^2	18.50
OBSTRUCTIONS						
General						
Obstructions	–	–	–	–	hr	95.00
PILE TESTS						
Cast in place						
Pile tests; 500 mm diameter working pile; maximum test load of 600 kN using non-working tension piles as reaction tripod						
first pile	–	–	–	–	nr	4725.00
subsequent pile	–	–	–	–	nr	3675.00
Take and test undisturbed soil samples; tripod	–	–	–	–	nr	165.00
Make, cure and test concrete cubes; tripod	–	–	–	–	nr	12.50
Pile tests; working pile; maximum test load of 1.5 times working load; first pile						
450 mm / 650 kN	–	–	–	–	nr	2500.00
600 mm / 1400 kN	–	–	–	–	nr	5250.00
750 mm / 2200 kN	–	–	–	–	nr	9500.00
Pile tests; working pile; maximum test load of 1.5 times working load; second and subsequent piles						
450 mm / 650 kN	–	–	–	–	nr	2300.00
600 mm / 1400 kN	–	–	–	–	nr	5050.00
750 mm / 2200 kN	–	–	–	–	nr	9100.00
Pile tests; working pile; electronic integrity testing; each pile (minimum 40 piles per visit)	–	–	–	–	nr	11.50
Make, cure and test concrete cubes	–	–	–	–	nr	22.00
Preformed						
Pile tests; working pile; maximum test load of 1.5 times working load	–	–	–	–	nr	4080.00
Pile tests; working pile; dynamic testing with piling hammer	–	–	–	–	nr	270.00

CLASS Q: PILING ANCILLARIES

Item	Gang hours	Labour £	Plant £	Material £	Unit	Total rate £
Steel bearing piles						
Steel bearing piles are commonly carried out by a Specialist Contractor whose advice should be sought to arrive at accurate costing. However the following items can be used to assess a budget cost for such work.						
The following unit costs are based upon driving 100 nr steel bearing piles 15–24 m long on a clear site with reasonable access. Supply is based on delivery 75 miles from works, in loads over 20 t.						
Establishment of pile testing equipment on site preliminary to any piling operation	–	–	–	–	sum	1400.00
Carry out pile test on bearing piles irrespective of section using pile testing equipment on site up to 108 t load	–	–	–	–	nr	4675.00
Driven-temporary casing						
Pile tests; 430 mm diameter working pile; maximum test load of 1125 kN using non-working tension piles as reaction; first piles						
bottom driven	–	–	–	–	nr	3650.00
top driven	–	–	–	–	nr	3000.00
Pile tests; 430 mm diameter working pile; maximum test load of 1125 kN using non-working tension piles as reaction; subsequent piles						
bottom driven	–	–	–	–	nr	1575.00
top driven	–	–	–	–	nr	1575.00
Pile tests; working pile; electronic integrity testing; each pile (minimum 40 piles per visit)	–	–	–	–	nr	15.00
Make cure and test concrete cubes	–	–	–	–	nr	10.50
Driven – segmental casing						
Pile tests; 500 mm diameter working pile; maximum test load of 600 kN using non-working tension piles as reaction						
first pile	–	–	–	–	nr	5500.00
subsequent piles	–	–	–	–	nr	4215.00

CLASS R: ROADS AND PAVINGS

Item	Gang hours	Labour £	Plant £	Material £	Unit	Total rate £
GENERAL						
Notes – Labour and Plant						
All outputs are based on clear runs without undue delay to two pavers with 75% utilization.						
The outputs can be adjusted as follows to take account of space or time influences on the utilization.						
Factors for varying utilization of Labour and Plant:						
1 paver @ 75% utilization = × 2.00	–	–	–	–	–	–
1 paver @ 100% utilization = × 1.50	–	–	–	–	–	–
2 paver @ 100% utilization = × 0.75	–	–	–	–	–	–
RESOURCES – LABOUR						
Subbase laying gang						
1 ganger/chargehand (skill rate 4)		14.56				
1 skilled operative (skill rate 4)		13.61				
2 unskilled operatives (general)		25.42				
1 plant operator (skill rate 2)		18.12				
1 plant operator (skill rate 3)		16.55				
Total Gang Rate/Hour	£	**88.26**				
Flexible paving gang						
1 ganger/chargehand (skill rate 4)		14.56				
2 skilled operatives (skill rate 4)		27.22				
4 unskilled operatives (general)		50.84				
4 plant operators (skill rate 3)		66.20				
Total Gang Rate/Hour	£	**158.82**				
Concrete paving gang						
1 ganger/chargehand (skill rate 4)		14.56				
2 skilled operatives (skill rate 4)		27.22				
4 unskilled operatives (general)		50.84				
1 plant operator (skill rate 2)		18.12				
1 plant operator (skill rate 3)		16.55				
Total Gang Rate/Hour	£	**127.29**				
Road surface spraying gang						
1 plant operator (skill rate 3)		16.55				
Total Gang Rate/Hour	£	**16.55**				
Road chippings gang						
1 ganger/chargehand (skill rate 4) – 50% of time		7.28				
1 skilled operative (skill rate 4)		13.61				
2 unskilled operatives (general)		25.42				
3 plant operators (skill rate 3)		49.65				
Total Gang Rate/Hour	£	**95.96**				

CLASS R: ROADS AND PAVINGS

Item	Gang hours	Labour £	Plant £	Material £	Unit	Total rate £
Cutting slabs gang						
1 unskilled operative (generally)		12.71				
Total Gang Rate/Hour	£	**12.71**				
Concrete filled joints gang						
1 ganger/chargehand (skill rate 4) – 50% of time		7.28				
1 skilled operative (skill rate 4)		13.61				
2 unskilled operatives (general)		25.42				
Total Gang Rate/Hour	£	**46.31**				
Milling gang						
1 ganger/chargehand (skill rate 4)		14.56				
2 skilled operatives (skill rate 4)		27.22				
4 unskilled operative (general)		50.84				
1 plant operators (skill rate 3)		16.55				
1 plant operator (skill rate 2)		18.12				
Total Gang Rate/Hour	£	**127.29**				
Rake and compact planed material gang						
1 ganger/chargehand (skill rate 4)		14.56				
1 skilled operative (skill rate 4)		13.61				
3 unskilled operatives (general)		38.13				
1 plant operator (skill rate 3)		16.55				
1 plant operator (skill rate 4)		14.88				
Total Gang Rate/Hour	£	**97.73**				
Kerb laying gang						
3 skilled operatives (skill rate 4)		40.83				
1 unskilled operative (general)		12.71				
1 plant operator (skill rate 3) – 25% of time		4.14				
Total Gang Rate/Hour	£	**57.68**				
Path subbase, bitmac and gravel laying gang						
1 ganger/chargehand (skill rate 4)		14.56				
2 unskilled operatives (general)		25.42				
1 plant operator (skill rate 3)		16.55				
Total Gang Rate/Hour	£	**56.53**				
Paviors and flagging gang						
1 skilled operative (skill rate 4)		13.61				
1 unskilled operative (general)		12.71				
Total Gang Rate/Hour	£	**26.32**				
Traffic signs gang						
1 ganger/chargehand (skill rate 3)		16.16				
1 skilled operative (skill rate 3)		15.21				
2 unskilled operatives (general)		25.42				
1 plant operator (skill rate 3) – 25% of time		4.14				
Total Gang Rate/Hour	£	**60.93**				

CLASS R: ROADS AND PAVINGS

Item	Gang hours	Labour £	Plant £	Material £	Unit	Total rate £
RESOURCES – PLANT						
Subbase laying						
93 kW motor grader			32.66			
0.80 m³ tractor loader			38.59			
6 t towed roller			26.04			
Total Rate/Hour		£	**97.29**			
Flexible paving						
2 asphalt pavers, 35 kW, 4.0 m			74.54			
2 deadweight rollers, 3 point, 10 t			78.10			
tractor with front bucket and integral 2 tool compressor			29.70			
compressor tools: scabbler			1.50			
tar sprayer, 100 litre			11.30			
self propelled chip spreader			36.30			
channel (heat) iron			2.70			
Total Rate/Hour		£	**234.14**			
Concrete paving						
wheeled loader, 2.60 m²			57.50			
concrete paver, 6.0 m			107.90			
compaction slipform finisher			23.80			
Total Rate/Hour		£	**189.20**			
Road surface spraying						
tar sprayer, 100 litre			11.30			
Total Rate/Hour		£	**11.30**			
Road chippings						
deadweight rollers, 3 point, 10 t			39.05			
tar sprayer, 100 litre			11.30			
self propelled chip spreader			36.30			
channel (heat) iron			2.70			
Total Rate/Hour		£	**88.65**			
Cutting slabs						
compressor, 65 cfm			5.66			
12' disc cutter			1.00			
Total Rate/Hour		£	**6.66**			
Milling						
cold planer, 2.10 m			216.20			
wheeled loader, 2.60 m²			57.50			
Total Rate/ Hour		£	**273.70**			
Heat planing						
heat planer, 4.5 m			84.10			
wheeled loader, 2.60 m²			57.50			
Total Rate/ Hour		£	**141.60**			

CLASS R: ROADS AND PAVINGS

Item	Gang hours	Labour £	Plant £	Material £	Unit	Total rate £
Rake and compact planed material						
deadweight roller, 3 point, 10 t			39.05			
tractor with front bucket and integral 2 tool						
compressor			29.70			
channel (heat) iron			2.00			
Total Rate/Hour		£	**70.75**			
Kerb laying						
backhoe JCB 3CX (25% of time) JCB 3CX (25% of time)			7.50		hr	
12' stihl saw			0.93		hr	
road forms			2.50		hr	
TOTAL		£	**10.85**		hr	
Path subbase, bitmac and gravel laying						
backhoe JCB 3CX			30.00		hr	
2 t dumper			6.60		hr	
pedestrian roller Bomag BW 90S			9.67		hr	
TOTAL		£	**46.27**		hr	
Paviors and flagging						
2 t dumper (33% of time)			2.20			
Total Rate/Hour		£	**2.20**			
Traffic signs						
JCB 3CX backhoe – 50% of time			15.00			
125 cfm compressor – 50% of time			4.29			
compressor tools: hand held hammer drill – 50% of time			0.56			
compressor tools: clay spade – 50% of time			0.89			
compressor tools: extra 15 m hose – 50% of time			0.16			
8 t lorry with hiab lift – 50% of time			15.25			
Total Rate/Hour		£	**36.14**			
SUBBASES, FLEXIBLE ROAD BASES AND SURFACING						
Granular material DfT specified type 1						
Subbase; spread and graded						
75 mm deep	0.04	3.09	3.41	24.35	m^3	**30.85**
100 mm deep	0.04	3.53	3.89	24.35	m^3	**31.77**
150 mm deep	0.05	3.97	4.38	24.35	m^3	**32.70**
200 mm deep	0.05	4.41	4.86	24.35	m^3	**33.62**
Lean concrete DfT specified strength mix C20P/20 mm aggregate						
Subbase; spread and graded						
100 mm deep	0.05	3.97	4.38	61.09	m^3	**69.44**
200 mm deep	0.05	4.41	4.86	61.09	m^3	**70.36**

CLASS R: ROADS AND PAVINGS

Item	Gang hours	Labour £	Plant £	Material £	Unit	Total rate £
SUBBASES, FLEXIBLE ROAD BASES AND SURFACING – cont						
Hardcore						
Subbase; spread and graded						
100 mm deep	0.04	3.53	3.89	19.68	m^3	**27.10**
150 mm deep	0.05	3.97	4.38	19.68	m^3	**28.03**
200 mm deep	0.05	4.41	4.86	19.68	m^3	**28.95**
Geotextiles refer to Class E						
Wet mix macadam; DfT clause 808						
Subbase; spread and graded						
75 mm deep	0.04	3.09	5.16	73.80	m^3	**82.05**
100 mm deep	0.04	3.53	5.35	73.80	m^3	**82.68**
200 mm deep	0.05	4.41	5.64	73.80	m^3	**83.85**
Dense Bitumen Macadam						
Base to DfT clause 903						
100 mm deep	0.02	3.18	4.68	6.06	m^2	**13.92**
150 mm deep	0.03	3.97	5.85	9.09	m^2	**18.91**
200 mm deep	0.03	4.76	7.03	12.12	m^2	**23.91**
Binder Course to DfT clause 906						
50 mm deep	0.02	2.38	3.51	2.52	m^2	**8.41**
100 mm deep	0.02	3.18	4.68	5.05	m^2	**12.91**
Surface Course to DfT clause 912						
30 mm deep	0.01	1.59	2.34	2.12	m^2	**6.05**
50 mm deep	0.02	2.38	3.51	3.54	m^2	**9.43**
Bitumen Macadam						
Binder Course to DfT clause 908						
35 mm deep	0.01	1.59	2.34	2.30	m^2	**6.23**
70 mm deep	0.02	2.38	3.51	4.60	m^2	**10.49**
Dense Tarmacadam						
Base to DfT clause 902						
50 mm deep	0.02	2.38	3.44	3.13	m^2	**8.95**
100 mm deep	0.02	2.38	3.44	6.26	m^2	**12.08**
Binder Course to DfT clause 907						
60 mm deep	0.02	2.38	3.51	4.12	m^2	**10.01**
80 mm deep	0.02	2.38	3.51	5.49	m^2	**11.38**
Dense Tar Surfacing						
Surface Course to DfT clause 913						
30 mm deep	0.01	1.59	2.34	2.36	m^2	**6.29**
50 mm deep	0.02	2.38	3.46	3.94	m^2	**9.78**
Cold Asphalt						
Surface Course to DfT clause 914						
15 mm deep	0.01	1.59	2.34	1.44	m^2	**5.37**
30 mm deep	0.01	1.59	2.34	2.91	m^2	**6.84**

Unit Costs – Civil Engineering Works 281

CLASS R: ROADS AND PAVINGS

Item	Gang hours	Labour £	Plant £	Material £	Unit	Total rate £
Rolled Asphalt						
Binder Course to DfT clause 905						
60 mm deep	0.02	2.38	3.51	5.76	m^2	**11.65**
80 mm deep	0.02	2.38	3.51	7.68	m^2	**13.57**
Surface Course to DfT clause 911						
40 mm deep	0.02	2.38	3.51	3.84	m^2	**9.73**
60 mm deep	0.02	2.38	3.51	5.76	m^2	**11.65**
Slurry sealing; BS 434 class K3						
Sealing to DfT clause 918						
3 mm deep	0.02	0.25	0.17	1.51	m^2	**1.93**
4 mm deep	0.02	0.25	0.17	1.77	m^2	**2.19**
Coated chippings, 9–11 kg/m^2						
Surface dressing to DfT clause 915						
6 mm nominal size	0.01	0.96	0.89	0.98	m^2	**2.83**
8 mm nominal size	0.01	0.96	0.89	1.03	m^2	**2.88**
10 mm nominal size	0.01	0.96	0.89	1.08	m^2	**2.93**
12 mm nominal size	0.01	0.96	0.89	1.13	m^2	**2.98**
Bituminous spray; BS 434 K1–40						
Tack coat to DfT clause 920						
large areas; over 20 m^2	0.02	0.25	0.17	0.27	m^2	**0.69**
small areas; under 20 m^2	0.02	0.25	0.17	0.25	m^2	**0.67**
Removal of flexible surface						
Trimming edges only of existing slabs, floors or similar surfaces (wet or dry); 6 mm cutting width						
50 mm deep	0.02	0.25	0.13	12.50	m	**12.88**
100 mm deep	0.03	0.38	0.21	15.98	m	**16.57**
Cutting existing slabs, floors or similar surfaces (wet or dry); 8 mm cutting width						
50 mm deep	0.03	0.32	0.16	12.50	m	**12.98**
100 mm deep	0.06	0.76	0.40	15.98	m	**17.14**
150 mm deep	0.08	1.02	0.53	18.00	m	**19.55**
Milling pavement (assumes disposal on site or reuse as fill but excludes transport if required)						
75 mm deep	0.03	3.44	7.39	–	m^2	**10.83**
100 mm deep	0.04	4.58	9.85	–	m^2	**14.43**
50 mm deep; scarifying surface	0.02	2.80	6.02	–	m^2	**8.82**
75 mm deep; scarifying surface	0.04	4.71	10.13	–	m^2	**14.84**
25 mm deep; heat planing for reuse	0.03	4.07	4.53	–	m^2	**8.60**
50 mm deep; heat planing for reuse	0.06	7.13	7.93	–	m^2	**15.06**
Raking over scarified or heat planed material; compacting with 10 t roller						
ne 50 mm deep	0.01	0.98	0.71	–	m^2	**1.69**

CLASS R: ROADS AND PAVINGS

Item	Gang hours	Labour £	Plant £	Material £	Unit	Total rate £
CONCRETE PAVEMENTS						
The following unit costs are for jointed reinforced concrete slabs, laid in reasonable areas (over 200 m²) by paver train/slipformer.						
Designed mix; cement to BS EN 197–1; grade C30, 20 mm aggregate						
Carriageway slabs of DfT specified paving quality						
180 mm deep	0.02	1.91	2.84	12.91	m²	17.66
220 mm deep	0.02	2.29	3.41	15.79	m²	21.49
260 mm deep	0.02	2.80	4.16	18.66	m²	25.62
300 mm deep	0.03	3.18	4.73	21.52	m²	29.43
Fabric						
Steel fabric reinforcement to BS 4483						
Ref A142 nominal mass 2.22 kg	0.03	3.82	–	1.66	m²	5.48
Ref A252 nominal mass 3.95 kg	0.04	5.09	–	2.91	m²	8.00
Ref B385 nominal mass 4.53 kg	0.04	5.09	–	3.54	m²	8.63
Ref C636 nominal mass 5.55 kg	0.05	6.36	–	4.50	m²	10.86
Ref B503 nominal mass 5.93 kg	0.05	6.36	–	4.50	m²	10.86
Mild Steel bar reinforcement BS 4449						
Bars; supplied in bent and cut lengths						
6 mm nominal size	8.00	1018.32	–	505.00	t	1523.32
8 mm nominal size	6.74	857.93	–	505.00	t	1362.93
10 mm nominal size	6.74	857.93	–	505.00	t	1362.93
12 mm nominal size	6.74	857.93	–	505.00	t	1362.93
16 mm nominal size	6.15	782.83	–	505.00	t	1287.83
High yield steel bar reinforcement BS 4449 or 4461						
Bars; supplied in bent and cut lengths						
6 mm nominal size	8.00	1018.32	–	656.50	t	1674.82
8 mm nominal size	6.74	857.93	–	656.50	t	1514.43
10 mm nominal size	6.74	857.93	–	656.50	t	1514.43
12 mm nominal size	6.74	857.93	–	656.50	t	1514.43
16 mm nominal size	6.15	782.83	–	656.50	t	1439.33
Sheeting to prevent moisture loss						
Polyethylene sheeting; lapped joints; horizontal below concrete pavements						
250 micron	0.01	1.27	–	0.65	m²	1.92
500 micron	0.01	1.27	–	1.64	m²	2.91
JOINTS IN CONCRETE PAVEMENTS						
Longitudinal joints						
180 mm deep	0.01	1.53	2.27	15.73	m	19.53
220 mm deep	0.01	1.53	2.27	18.50	m	22.30
260 mm deep	0.01	1.53	2.27	22.66	m	26.46
300 mm deep	0.01	1.53	2.27	26.36	m	30.16

CLASS R: ROADS AND PAVINGS

Item	Gang hours	Labour £	Plant £	Material £	Unit	Total rate £
Expansion joints						
180 mm deep	0.01	1.53	2.27	30.97	m	**34.77**
220 mm deep	0.01	1.53	2.27	36.07	m	**39.87**
260 mm deep	0.01	1.53	2.27	41.15	m	**44.95**
300 mm deep	0.01	1.53	2.27	42.17	m	**45.97**
Contraction joints						
180 mm deep	0.01	1.53	2.27	18.07	m	**21.87**
220 mm deep	0.01	1.53	2.27	19.10	m	**22.90**
260 mm deep	0.01	1.53	2.27	20.29	m	**24.09**
300 mm deep	0.01	1.53	2.27	24.19	m	**27.99**
Construction joints						
180 mm deep	0.01	1.53	2.27	11.91	m	**15.71**
220 mm deep	0.01	1.53	2.27	13.00	m	**16.80**
260 mm deep	0.01	1.53	2.27	14.02	m	**17.82**
300 mm deep	0.01	1.53	2.27	15.02	m	**18.82**
Open joints with filler						
ne 0.5 m; 10 mm flexcell joint filler	0.11	5.09	–	3.15	m	**8.24**
0.5–1.00 m; 10 mm flexcell joint filler	0.11	5.09	–	4.45	m	**9.54**
Joint sealants						
10 × 20 mm cold polysulphide sealant	0.14	6.48	–	3.50	m	**9.98**
20 × 20 mm cold polysulphide sealant	0.18	8.34	–	6.00	m	**14.34**
KERBS, CHANNELS AND EDGINGS						
Foundations to kerbs etc.						
Measurement Note: the following are shown separate from their associated kerb etc. to simplify the presentation of cost alternatives.						
Mass concrete						
200 × 100 mm	0.01	0.58	0.10	1.70	m	**2.38**
300 × 150 mm	0.02	0.87	0.15	3.90	m	**4.92**
450 × 150 mm	0.02	1.15	0.22	5.77	m	**7.14**
100 × 100 mm haunching, one side	0.01	0.29	0.06	0.41	m	**0.76**
Precast concrete kerbs; BS 7263: bedded, jointed and pointed in cement mortar						
Kerbs; bullnosed; splayed or half battered; straight or curved over 12 m radius						
125 × 150 mm	0.06	3.46	0.65	9.71	m	**13.82**
125 × 255 mm	0.07	4.04	0.76	10.91	m	**15.71**
150 × 305 mm	0.07	4.04	0.78	16.32	m	**21.14**
Kerbs; bullnosed; splayed or half battered; curved ne 12 m radius						
125 × 150 mm	0.07	3.75	0.71	9.71	m	**14.17**
125 × 255 mm	0.08	4.33	0.81	10.91	m	**16.05**
150 × 305 mm	0.08	4.33	0.84	16.32	m	**21.49**
Quadrants						
305 × 305 × 150 mm	0.08	4.61	0.89	14.56	nr	**20.06**
455 × 455 × 255 mm	0.10	5.77	1.12	21.60	nr	**28.49**

CLASS R: ROADS AND PAVINGS

Item	Gang hours	Labour £	Plant £	Material £	Unit	Total rate £
KERBS, CHANNELS AND EDGINGS – cont						
Precast concrete kerbs – cont						
Drop kerbs						
125 × 255 mm	0.07	4.04	0.78	24.28	m	**29.10**
150 × 305 mm	0.07	4.04	0.78	40.37	m	**45.19**
Channel; straight or curved over 12 m radius						
255 × 125 mm	0.07	4.04	0.78	6.11	m	**10.93**
Channel; curved radius ne 12 m						
255 × 125 mm	0.07	4.04	0.78	6.11	m	**10.93**
Edging; straight or curved over 12 m radius						
50 × 150 mm	0.04	2.31	0.45	2.14	m	**4.90**
Edging; curved ne 12 m radius						
50 × 150 mm	0.05	2.60	0.50	2.14	m	**5.24**
Precast concrete drainage channels; Charcon Safeticurb; channels jointed with plastic rings and bedded; jointed and pointed in cement mortar						
Channel unit; straight; Type DBA/3						
250 × 254 mm; medium duty	0.08	4.33	0.84	48.78	m	**53.95**
305 × 305 mm; heavy duty	0.10	5.48	1.06	102.88	m	**109.42**
Precast concrete Ellis Trief safety kerb; bedded jointed and pointed in cement mortar						
Kerb; straight or curved over 12 m radius						
415 × 380 mm	0.23	12.98	2.42	73.16	m	**88.56**
Kerb; curved ne 12 m radius						
415 × 380 mm	0.25	14.42	2.69	84.50	m	**101.61**
Precast concrete combined kerb and drainage block Beany block system; bedded jointed and pointed in cement mortar						
Kerb; top block, shallow base unit, standard cover plate and frame						
straight or curved over 12 m radius	0.15	8.65	1.62	117.16	m	**127.43**
curved ne 12 m radius	0.20	11.54	2.15	178.90	m	**192.59**
Kerb; top block, standard base unit, standard cover plate and frame						
straight or curved over 12 m radius	0.15	8.65	1.63	117.16	m	**127.44**
curved ne 12 m radius	0.20	11.54	2.15	178.90	m	**192.59**
Kerb; top block, deep base unit, standard cover plate and frame						
straight or curved over 12 m radius	0.15	8.65	1.60	261.53	m	**271.78**
curved ne 12 m radius	0.20	11.54	2.12	334.37	m	**348.03**
base block depth tapers	0.10	5.99	1.09	28.20	m	**35.28**

CLASS R: ROADS AND PAVINGS

Item	Gang hours	Labour £	Plant £	Material £	Unit	Total rate £
Extruded asphalt kerbs to BS 5931; extruded and slip formed						
Kerb; straight or curved over 12 m radius						
75 mm kerb height	–	–	–	3.50	m	**3.50**
100 mm kerb height	–	–	–	7.50	m	**7.50**
125 mm kerb height	–	–	–	10.00	m	**10.00**
Channel; straight or curved over 12 m radius						
300 mm channel width	–	–	–	12.50	m	**12.50**
250 mm channel width	–	–	–	10.00	m	**10.00**
Kerb; curved to radius ne 12 m						
75 mm kerb height	–	–	–	3.33	m	**3.33**
100 mm kerb height	–	–	–	7.13	m	**7.13**
125 mm kerb height	–	–	–	9.50	m	**9.50**
Channel; curved to radius ne 12 m						
300 mm channel width	–	–	–	11.50	m	**11.50**
250 mm channel width	–	–	–	9.50	m	**9.50**
Extruded concrete; slip formed						
Kerb; straight or curved over 12 m radius						
100 mm kerb height	–	–	–	10.00	m	**10.00**
125 mm kerb height	–	–	–	12.50	m	**12.50**
Kerb; curved to radius ne 12 m						
100 mm kerb height	–	–	–	11.00	m	**11.00**
125 mm kerb height	–	–	–	13.00	m	**13.00**
LIGHT DUTY PAVEMENTS						
Subbases						
Measurement Note: the following are shown separate from their associated paving to simplify the presentation of cost alternatives.						
To paved area; sloping not exceeding 10° to the horizontal						
100 mm thick sand	0.01	0.51	0.42	2.67	m^2	**3.60**
150 mm thick sand	0.01	0.68	0.56	4.01	m^2	**5.25**
100 mm thick gravel	0.01	0.51	0.42	2.82	m^2	**3.75**
150 mm thick gravel	0.01	0.68	0.56	4.22	m^2	**5.46**
100 mm thick hardcore	0.01	0.51	0.42	2.42	m^2	**3.35**
150 mm thick hardcore	0.01	0.68	0.56	3.63	m^2	**4.87**
100 mm thick concrete grade 20/20	0.02	1.19	0.97	8.48	m^2	**10.64**
150 mm thick concrete grade 20/20	0.03	1.81	1.48	12.72	m^2	**16.01**

CLASS R: ROADS AND PAVINGS

Item	Gang hours	Labour £	Plant £	Material £	Unit	Total rate £
LIGHT DUTY PAVEMENTS – cont						
Bitumen macadam surfacing; BS 4987; base course of 20 mm open graded aggregate to clause 2.6.1 tables 5–7; wearing course of 6 mm medium graded aggregate to clause 2.7.6 tables 32–33						
Paved area comprising base course 40 mm thick wearing course 20 mm thick						
sloping not exceeding 10° to the horizontal	0.09	4.81	3.93	7.15	m^2	**15.89**
sloping not exceeding 10° to the horizontal; red additives	0.09	4.81	3.93	9.35	m^2	**18.09**
sloping not exceeding 10° to the horizontal; green additives	0.09	4.81	3.93	10.84	m^2	**19.58**
sloping exceeding 10° to the horizontal	0.10	5.38	4.40	7.15	m^2	**16.93**
sloping exceeding 10° to the horizontal; red additives	0.10	5.38	4.40	9.35	m^2	**19.13**
sloping exceeding 10° to the horizontal; green additives	0.10	5.38	4.40	10.84	m^2	**20.62**
Granular base surfacing; Central Reserve Treatments Limestone, graded 10 mm down; laid and compacted						
Paved area 100 mm thick; surface sprayed twice with two coats of cold bituminous emulsion; blinded with 6 mm quartzite fine gravel						
sloping not exceeding 10° to the horizontal	0.02	1.13	0.93	2.75	m^2	**4.81**
Ennstone Johnston Golden gravel; graded 13 mm to fines; rolled wet						
Paved area 50 mm thick; single layer						
sloping not exceeding 10° to the horizontal	0.03	1.70	1.39	8.25	m^2	**11.34**
Precast concrete flags; BS 7263; grey; bedding in cement mortar						
Paved area; sloping not exceeding 10° to the horizontal						
900 × 600 × 63 mm	0.21	5.53	0.46	9.47	m^2	**15.46**
900 × 600 × 50 mm	0.20	5.26	0.44	9.38	m^2	**15.08**
600 × 600 × 63 mm	0.25	6.58	0.55	8.34	m^2	**15.47**
600 × 600 × 50 mm	0.24	6.32	0.53	8.12	m^2	**14.97**
600 × 450 × 50 mm	0.28	7.37	0.62	8.01	m^2	**16.00**
Extra for coloured, 50 mm thick	–	–	–	2.50	m^2	**2.50**
Precast concrete rectangular paving blocks; BS 6717; grey; bedding on 50 mm thick dry sharp sand; filling joints; excluding subbase						
Paved area; sloping not exceeding 10° to the horizontal						
200 × 100 × 80 mm thick	0.30	7.90	0.66	30.26	m^2	**38.82**
200 × 100 × 80 mm thick; coloured blocks	0.30	7.90	0.66	35.51	m^2	**44.07**

CLASS R: ROADS AND PAVINGS

Item	Gang hours	Labour £	Plant £	Material £	Unit	Total rate £
Brick paviors; bedding on 20 mm thick mortar; excluding subbase						
Paved area; sloping not exceeding 10° to the horizontal						
215 × 103 × 65 mm	0.30	7.90	0.66	21.52	m²	**30.08**
Granite setts; bedding on 25 mm cement mortar; excluding subbase						
Paved area; sloping not exceeding 10° to the horizontal						
to random pattern	0.90	23.69	1.98	50.40	m²	**76.07**
to specific pattern	1.20	31.58	2.64	50.40	m²	**84.62**
Cobble paving; 50–75 mm; bedding in 25 mm cement mortar; filling joints; excluding subbase						
Paved area; sloping not exceeding 10° to the horizontal						
50–75 mm diameter cobbles	1.00	26.32	2.18	14.53	m²	**43.03**
ANCILLARIES						
Traffic signs						
In this section prices will vary depending upon the diagram configurations. The following are average costs of signs and bollards. Diagram numbers refer to the Traffic Signs Regulations and General Directions 2002 and the figure numbers refer to the Traffic Sign Manual.						
Examples of Prime Costs for Class 1 (High Intensity) traffic and road signs (ex works).						
600 × 450 mm	–	–	–	32.43	nr	**32.43**
600 mm diameter	–	–	–	126.00	nr	**126.00**
600 mm triangular	–	–	–	105.00	nr	**105.00**
500 × 500 mm	–	–	–	32.38	nr	**32.38**
450 × 450 mm	–	–	–	30.76	nr	**30.76**
450 × 300 mm	–	–	–	29.22	nr	**29.22**
1200 × 400 mm (CHEVRONS)	–	–	–	48.73	nr	**48.73**
Examples of Prime Costs for Class 21 (Engineering Grade) traffic and road signs (ex works).						
600 × 450 mm	–	–	–	34.32	nr	**34.32**
600 mm diameter	–	–	–	34.86	nr	**34.86**
600 mm triangular	–	–	–	29.46	nr	**29.46**
500 × 500 mm	–	–	–	108.76	nr	**108.76**
450 × 450 mm	–	–	–	103.32	nr	**103.32**
450 × 300 mm	–	–	–	98.15	nr	**98.15**
1200 × 400 mm (CHEVRONS)	–	–	–	53.37	nr	**53.37**

CLASS R: ROADS AND PAVINGS

Item	Gang hours	Labour £	Plant £	Material £	Unit	Total rate £
ANCILLARIES – cont						
Standard reflectorized traffic signs						
Note: Unit costs do not include concrete foundations						
Standard one post signs; 600 × 450 mm type						
C1 signs						
fixed back to back to another sign (measured separately) with aluminium clips to existing post (measured separately)	0.04	2.44	1.45	34.05	nr	37.94
Extra for fixing singly with aluminium clips	0.01	0.61	0.28	1.00	nr	1.89
Extra for fixing singly with stainless steel clips	0.01	0.61	0.81	3.50	nr	4.92
fixed back to back to another sign (measured separately) with stainless steel clips to one new 76 mm diameter plastic coated steel post 1.75 m long	0.27	16.45	9.77	84.05	nr	110.27
Extra for fixing singly to one face only	0.01	0.61	0.28	–	nr	0.89
Extra for 76 mm diameter 1.75 m long aluminium post	0.02	1.22	0.59	26.75	nr	28.56
Extra for 76 mm diameter 3.5 m long plastic coated steel post	0.02	1.22	0.59	42.00	nr	43.81
Extra for 76 mm diameter 3.5 m long aluminium post	0.02	1.22	0.59	52.50	nr	54.31
Extra for excavation for post, in hard material	1.10	67.02	31.58	–	nr	98.60
Extra for single external illumination unit with fitted photo cell (excluding trenching and cabling); unit cost per face illuminated	0.33	20.11	9.46	82.50	nr	112.07
Standard two post signs; 1200 × 400 mm;						
fixed back to back to another sign (measured separately) with stainless steel clips to two new 76 mm diameter plastic coated steel posts	0.51	31.07	18.44	152.23	nr	201.74
Extra for fixing singly to one face only	0.02	1.22	0.59	–	nr	1.81
Extra for two 76 mm diameter 1.75 m long aluminium posts	0.04	2.44	1.15	53.50	nr	57.09
Extra for two 76 mm diameter 1.75 m long plastic coated steel posts	0.04	2.44	1.15	84.00	nr	87.59
Extra for two 76 mm diameter 3.5 m long aluminium posts	0.04	2.44	1.15	105.00	nr	108.59
Extra for excavation for post, in hard material	1.10	67.02	31.58	–	nr	98.60
Extra for single external illumination unit with fitted photo cell (including trenching and cabling); unit per face illuminated	0.58	35.34	16.66	127.50	nr	179.50
Standard internally illuminated traffic signs						
Bollard with integral mould-in translucent graphics (excluding trenching and cabling)						
fixing to concrete base	0.48	29.25	17.35	350.00	nr	396.60
Special traffic signs						
Note: Unit costs do not include concrete foundations or trenching and cabling.						

CLASS R: ROADS AND PAVINGS

Item	Gang hours	Labour £	Plant £	Material £	Unit	Total rate £
Externally illuminated relectorized traffic signs manufactured to order						
special signs, surface area 1.50 m² on two 100 mm diameter steel posts	–	–	–	–	nr	**662.50**
special signs, surface area 4.00 m² on three 100 mm diameter steel posts	–	–	–	–	nr	**975.00**
Internally illuminated traffic signs manufactured to order						
special signs, surface area 0.25 m² on one new 76 mm diameter post	–	–	–	–	nr	**165.00**
special signs, surface area 0.75 m² on one new 100 mm diameter steel posts	–	–	–	–	nr	**215.00**
special signs, surface area 4.00 m² on four new 120 mm diameter steel posts	–	–	–	–	nr	**1075.00**
Signs on gantries						
Externally illuminated reflectorized signs						
1.50 m²	1.78	108.76	122.50	950.00	nr	**1181.26**
2.50 m²	2.15	130.99	147.54	1000.00	nr	**1278.53**
3.00 m²	3.07	187.05	210.68	1100.00	nr	**1497.73**
Internally illuminated sign with translucent optical reflective sheeting and remote light source						
0.75 m²	1.56	95.05	107.05	1175.00	nr	**1377.10**
1.00 m²	1.70	103.58	116.66	1560.00	nr	**1780.24**
1.50 m²	2.41	146.84	165.39	2345.00	nr	**2657.23**
Existing signs						
Take from store and re-erect						
3.0 m high road sign	0.28	17.06	19.21	40.00	nr	**76.27**
road sign on two posts	0.50	30.46	34.31	80.00	nr	**144.77**
Surface markings; reflectorized white						
Letters and shapes						
triangles; 1.6 m high	–	–	–	–	nr	**10.00**
triangles; 2.0 m high	–	–	–	–	nr	**12.00**
triangles; 3.75 m high	–	–	–	–	nr	**15.00**
circles with enclosing arrows; 1.6 m diameter	–	–	–	–	nr	**60.00**
arrows; 4.0 m long; straight	–	–	–	–	nr	**25.00**
arrows; 4.0 m long; turning	–	–	–	–	nr	**25.00**
arrows; 6.0 m long; straight	–	–	–	–	nr	**30.00**
arrows; 6.0 m long; turning	–	–	–	–	nr	**30.00**
arrows; 6.0 m long; curved	–	–	–	–	nr	**30.00**
arrows; 6.0 m long; double headed	–	–	–	–	nr	**42.00**
arrows; 8.0 m long; double headed	–	–	–	–	nr	**60.00**
arrows; 16.0 m long; double headed	–	–	–	–	nr	**90.00**
arrows; 32.0 m long; double headed	–	–	–	–	nr	**120.00**
letters or numerals; 1.6 m high	–	–	–	–	nr	**7.75**
letters or numerals; 2.0 m high	–	–	–	–	nr	**11.50**
letters or numerals; 3.75 m high	–	–	–	–	nr	**20.00**

CLASS R: ROADS AND PAVINGS

Item	Gang hours	Labour £	Plant £	Material £	Unit	Total rate £
ANCILLARIES – cont						
Surface markings – cont						
Continuous lines						
150 mm wide	–	–	–	–	m	**1.25**
200 mm wide	–	–	–	–	m	**1.50**
Intermittent lines						
60 mm wide; 0.60 m line and 0.60 m gap	–	–	–	–	m	**0.80**
100 mm wide; 1.0 m line and 5.0 m gap	–	–	–	–	m	**0.82**
100 mm wide; 2.0 m line and 7.0 m gap	–	–	–	–	m	**0.85**
100 mm wide; 4.0 m line and 2.0 m gap	–	–	–	–	m	**0.87**
100 mm wide; 6.0 m line and 3.0 m gap	–	–	–	–	m	**0.89**
150 mm wide; 1.0 m line and 5.0 m gap	–	–	–	–	m	**1.15**
150 mm wide; 6.0 m line and 3.0 m gap	–	–	–	–	m	**1.18**
150 mm wide; 0.60 m line and 0.30 m gap	–	–	–	–	m	**1.20**
200 mm wide; 0.60 m line and 0.30 m gap	–	–	–	–	m	**1.45**
200 mm wide; 1.0 m line and 1.0 m gap	–	–	–	–	m	**1.45**
Surface markings; reflectorized yellow						
Continuous lines						
100 mm wide	–	–	–	–	m	**1.00**
150 mm wide	–	–	–	–	m	**1.25**
Intermittent lines						
kerb marking; 0.25 m long	–	–	–	–	nr	**1.75**
ROAD MARKINGS						
Surface markings; thermoplastic screed or spray						
Note: Unit costs based upon new road with clean surface closed to traffic.						
Continuous line in reflectorized white						
150 mm wide	–	–	–	–	m	**1.25**
200 mm wide	–	–	–	–	m	**1.50**
Continuous line in reflectorized yellow						
100 mm wide	–	–	–	–	m	**1.00**
150 mm wide	–	–	–	–	m	**1.25**
Intermittent line in reflectorized white						
60 mm wide with 0.60 m line and 0.60 m gap	–	–	–	–	m	**0.80**
100 mm wide with 1.0 m line and 5.0 m gap	–	–	–	–	m	**0.82**
100 mm wide with 2.0 m line and 7.0 m gap	–	–	–	–	m	**0.85**
100 mm wide with 4.0 m line and 2.0 m gap	–	–	–	–	m	**0.87**
100 mm wide with 6.0 m line and 3.0 m gap	–	–	–	–	m	**0.89**
150 mm wide with 1.0 m line and 5.0 m gap	–	–	–	–	m	**1.15**
150 mm wide with 6.0 m line and 3.0 m gap	–	–	–	–	m	**1.18**
150 mm wide with 0.6 m line and 0.3 m gap	–	–	–	–	m	**1.20**
200 mm wide with 0.6 m line and 0.3 m gap	–	–	–	–	m	**1.45**
200 mm wide with 1.0 m line and 1.0 m gap	–	–	–	–	m	**1.45**

CLASS R: ROADS AND PAVINGS

Item	Gang hours	Labour £	Plant £	Material £	Unit	Total rate £
Ancillary line in reflectorized white						
150 mm wide in hatched areas	–	–	–	–	m	0.90
200 mm wide in hatched areas	–	–	–	–	m	1.45
Ancillary line in reflectorized yellow						
150 mm wide in hatched areas	–	–	–	–	m	0.95
Triangles in reflectorized white						
1.6 m high	–	–	–	–	nr	10.00
2.0 m high	–	–	–	–	nr	12.00
3.75 m high	–	–	–	–	nr	15.00
Circles with enclosing arrows in reflectorized white						
1.6 m diameter	–	–	–	–	nr	60.00
Arrows in reflectorized white						
4.0 m long straight or turning	–	–	–	–	nr	25.00
6.0 m long straight or turning	–	–	–	–	nr	30.00
6.0 m long curved	–	–	–	–	nr	30.00
6.0 m long double headed	–	–	–	–	nr	42.00
8.0 m long double headed	–	–	–	–	nr	60.00
16.0 m long double headed	–	–	–	–	nr	90.00
32.0 m long double headed	–	–	–	–	nr	120.00
Kerb markings in yellow						
250 mm long	–	–	–	–	nr	1.75
Letters or numerals in reflectorized white						
1.6 m high	–	–	–	–	nr	7.75
2.0 m high	–	–	–	–	nr	11.50
3.75 m high	–	–	–	–	nr	20.00
Surface markings; Verynyl strip markings						
Note: Unit costs based upon new road with clean surface closed to traffic.						
Verynyl strip markings (pedestrian crossings and similar locations)						
200 mm wide line	–	–	–	–	m	7.55
600 × 300 mm single stud tile	–	–	–	–	nr	13.00
Removal of thermoplastic screed or spray markings						
Removal of existing reflectorized thermoplastic markings						
100 mm wide line	–	–	–	–	m	1.25
150 mm wide line	–	–	–	–	m	1.85
200 mm wide line	–	–	–	–	m	2.50
arrow or letter ne 6.0 m long	–	–	–	–	nr	19.15
arrow or letter 6.0–16.00 m long	–	–	–	–	nr	75.00

CLASS R: ROADS AND PAVINGS

Item	Gang hours	Labour £	Plant £	Material £	Unit	Total rate £
REFLECTING ROAD STUDS						
100 × 100 mm square bi-directional reflecting road studs with amber corner cube reflectors	–	–	–	–	nr	**6.50**
140 × 254 mm rectangular one way reflecting road studs with red catseye reflectors	–	–	–	–	nr	**15.20**
140 × 254 mm rectangular one way reflecting road studs with green catseye reflectors	–	–	–	–	nr	**15.25**
140 × 254 mm rectangular bi-directional reflecting road studs with white catseye reflectors	–	–	–	–	nr	**15.25**
140 × 254 mm rectangular bi-directional reflecting road studs with amber catseye reflectors	–	–	–	–	nr	**15.25**
140 × 254 mm rectangular bi-directional reflecting road stud without catseye reflectors	–	–	–	–	nr	**9.50**
REMOVAL OF ROAD STUDS						
100 × 100 mm corner cube type	–	–	–	–	nr	**3.15**
140 × 254 mm cateye type	–	–	–	–	nr	**7.50**
REMOVAL FROM STORE AND REFIX ROAD STUD						
General						
Remove from store and reinstall 100 × 100 mm square bi-directional reflecting road stud with corner cube reflectors.	–	–	–	–	nr	**3.15**
Remove from store and reinstall 140 × 254 mm rectangular one way reflecting road stud with catseye reflectors.	–	–	–	–	nr	**7.50**
TRAFFIC SIGNAL INSTALLATIONS						
Traffic signal installation is carried out exclusively by specialist contractors, although certain items are dealt with by the main contractor or a subcontractor. The following prices are given to assist in the calculation of the total installation cost.						
Installation of signal pedestals, loop detector unit pedestals, controller unit boxes and cable connection pillars						
signal pedestal	–	–	–	–	nr	**45.00**
loop detector unit pedestal	–	–	–	–	nr	**20.00**
controller unit box	–	–	–	–	nr	**45.00**
Excavate trench for traffic signal cable, depth ne 1.50 m; supports, backfilling						
450 mm wide	–	–	–	–	m	**15.00**
Extra for excavating in hard material	–	–	–	–	m^3	**30.00**
Saw cutting grooves in pavement for detector loops and feeder cables; seal with hot bitumen sealant after installation						
25 mm deep	–	–	–	–	m	**22.00**

CLASS S: RAIL TRACK

Item	Gang hours	Labour £	Plant £	Material £	Unit	Total rate £

NOTES
The following unit costs are for guidance only. For more reliable estimates it is advisable to seek the advice of a Specialist Contractor.
These rates are for the supply and laying of track other than in connection with the Permanent Way.
 	
Permanent Way
The following rates would not reflect work carried out on the existing public track infrastructure (Permanent Way), which tends to be more costly not merely due to differences in technology and the level of specification and control standards, but also due to a number of logistical factors, such as:
   * access to the works for personnel, plant and machinery would be via approved access points to the rail followed by travel along the rail to the work area; this calls for the use of additional and expensive transport plant as well as reducing the effective shift time of the works gang
   * effect of track possession periods will dictate when the work can be carried out and could well force night-time or weekend working and perhaps severely reducing the effective shift hours where coupled to long travel to work distances and the need to clear away before the resumption of traffic
   * the labour gang will be composed of more highly paid personnel, reflecting the additional training received; in addition there may well be additional gang members acting as look-outs; this could add 30% to the gang rates shown below
   * plant will tend to cost more, especially if the circumstances of the work call for rail/road plant; this could add 20% to the gang rates shown.
 	
Possession costs
Where the contractor's work is on, over or poses a risk to the safety of the railway, then the contractor normally applies for possession of the track. During the period for which the contractor is given possession, rail traffic stops.
Possessions of the Operational Safety Zone may well be fragmented rather than a single continuous period, dependent upon windows in the pattern of rail traffic.
Costs for working alongside an operational rail system are high, the need for safety demanding a high degree of supervision, look-outs, the induction of labour gangs and may involve temporary works such as safety barriers.						

CLASS S: RAIL TRACK

Item	Gang hours	Labour £	Plant £	Material £	Unit	Total rate £
TRACK FOUNDATIONS						
Imported crushed granite						
Bottom ballast	–	–	–	–	m³	60.00
Top ballast	–	–	–	–	m³	60.00
Imported granular material						
Blankets; 150 mm thick	–	–	–	–	m²	50.00
Imported sand						
Blinding; 100 mm thick	–	–	–	–	m²	30.00
Polythene sheeting						
Waterproof membrane; 1200 gauge	–	–	–	–	m²	3.75
LIFTING, PACKING AND SLEWING						
Maximum distance of slew 300 mm; maximum lift 100 mm; no extra ballast allowed						
Bullhead rail track; fishplated; timber sleepers	–	–	–	–	m	130.00
Bullhead rail track; fishplated; concrete sleepers	–	–	–	–	m	130.00
Bullhead rail track with turnout; timber sleepers	–	–	–	–	nr	130.00
Flat bottom rail track; welded; timber sleepers	–	–	–	–	m	130.00
Flat bottom rail track; welded; concrete sleepers	–	–	–	–	m	130.00
Flat bottom rail track with turnout; concrete sleepers	–	–	–	–	nr	85.00
Buffer stops	–	–	–	–	nr	3750.00
TAKING UP						
Taking up; dismantling into individual components; sorting; storing on site where directed						
Bullhead or flat bottom rails						
plain track; fishplated; timber sleepers	–	–	–	–	m	10.67
plain track; fishplated; concrete sleepers	–	–	–	–	m	10.67
plain track; welded; timber sleepers	–	–	–	–	m	10.67
plain track; welded; concrete sleepers	–	–	–	–	m	10.67
turnouts; fishplated; concrete sleepers	–	–	–	–	nr	1500.00
diamond crossings; fishplated; timber sleepers	–	–	–	–	nr	1500.00
Dock and crane rails						
plain track; welded; base plates	–	–	–	–	m	10.67
turnouts; welded; base plates	–	–	–	–	nr	1500.00
diamonds; welded; base plates	–	–	–	–	nr	1500.00
Check and guard rails						
plain track; fishplated	–	–	–	–	m	10.67
Conductor rails						
plain track; fishplated	–	–	–	–	m	5.60

CLASS S: RAIL TRACK

Item	Gang hours	Labour £	Plant £	Material £	Unit	Total rate £
Sundries						
buffer stops	–	–	–	–	nr	3950.00
retarders	–	–	–	–	nr	750.00
wheel stops	–	–	–	–	nr	750.00
lubricators	–	–	–	–	nr	750.00
switch heaters	–	–	–	–	nr	750.00
switch levers	–	–	–	–	nr	750.00
SUPPLYING (STANDARD GAUGE TRACK)						
Bullhead rails; BS 11; delivered in standard 18.288 m lengths						
BS95R section, 47 kg/m; for jointed track	–	–	–	–	t	655.00
BS95R section, 47 kg/m; for welded track	–	–	–	–	t	655.00
Flat bottom rails; BS 11; delivered in standard 18.288 m lengths						
BS113'A' section, 56 kg/m; for jointed track	–	–	–	–	t	655.00
BS113'A' section, 56 kg/m; for welded track	–	–	–	–	t	655.00
Extra for curved rails to form super elevation; radius over 600 m	–	–	–	–	%	20.00
Check and guard rails; BS 11; delivered in standard 18.288 m lengths; flange planed to allow 50 mm free wheel clearance						
BS113'A' section, 56 kg/m; for bolting	–	–	–	–	t	855.00
Conductor rails; BS 11; delivered in standard 18.288 m lengths						
BS113'A' section, 56 kg/m; for bolting	–	–	–	–	t	855.00
Twist rails; BS 11; delivered in standard 18.288 m lengths						
BS113'A' section, 56 kg/m; for bolting	–	–	–	–	t	855.00
Sleepers; bitumen saturated French Maritime pine						
2600 × 250 × 130 mm	–	–	–	–	nr	26.00
Sleepers; bitumen saturated Douglas fir						
2600 × 250 × 130 mm	–	–	–	–	nr	46.15
Sleepers; prestressed concrete						
2525 × 264 × 204 mm; BR type F27, Pandrol inserts	–	–	–	–	nr	62.00
Fittings						
Cast iron chairs complete with chair screws, plastic ferrules, spring steels and keys; BR type S1	–	–	–	–	nr	52.00
Cast iron chairs complete with chair screws, plastic ferrules, spring steels and keys; BR type CC	–	–	–	–	nr	73.00
Cast iron chairs complete with resilient pad, chair screws, ferrules, rail clips and nylon insulators; BR type PAN 6	–	–	–	–	nr	39.75

CLASS S: RAIL TRACK

Item	Gang hours	Labour £	Plant £	Material £	Unit	Total rate £
SUPPLYING (STANDARD GAUGE TRACK) – cont						
Cast iron chairs complete with resilient pad, chair screws, ferrules, rail clips and nylon insulators; BR type VN	–	–	–	–	nr	45.00
Cast iron chairs complete with resilient pad, chair screws, ferrules, rail clips and nylon insulators; BR type C	–	–	–	–	nr	34.00
Pandrol rail clips and nylon insulator	–	–	–	–	nr	2.81
plain fishplates; for BS 95R section rail, skirted pattern; sets of two; complete with fishbolts, nuts and washers	–	–	–	–	nr	100.00
plain fishplates; for BS 95R section rail, joggled pattern; sets of two; complete with fishbolts, nuts and washers	–	–	–	–	nr	115.00
plain fishplates; for BS 113 'A' section rail, shallow section; sets of two; complete with fishbolts, nuts and washers	–	–	–	–	nr	100.00
insulated fishplates; for BS 95R section rail, steel billet pattern; sets of two; complete with high tensile steel bolts, nuts and washers	–	–	–	–	nr	58.00
insulated fishplates; BS 95R section rail, steel billet pattern; sets of two; complete with high tensile steel bolts, nuts and washers	–	–	–	–	nr	58.00
cast iron spacer block between running and guard rails; for BS 95R section rail; M25 × 220 mm bolt, nut and washers	–	–	–	–	nr	45.00
cast iron spacer block between running and guard rails; for BS 113 'A' section rail; M25 × 220 mm bolt, nut and washers	–	–	–	–	nr	45.00
Turnouts; complete with closures, check rails, fittings, timber sleepers						
Type B8; BS 95R bullhead rail	–	–	–	–	nr	20500.00
Type C10; BS 95R bullhead rail	–	–	–	–	nr	1600.00
Type Bx8; BS 113 'A' section flat bottom rail	–	–	–	–	nr	13500.00
Type Cv9.25; BS 113 'A' section flat bottom rail	–	–	–	–	nr	21600.00
Diamond crossings; complete with closures, check rails, fittings, timber sleepers						
RT standard design, angle 1 in 4; BS 95R bullhead rail	–	–	–	–	nr	90215.33
RT standard design, angle 1 in 4; BS 113 'A' section flat bottom rail	–	–	–	–	nr	97519.36
Sundries						
buffer stops; single raker, steel rail and timber; 2 tonnes approximate weight	–	–	–	–	nr	12500.00
buffer stops; double raker, steel rail and timber; 2.5 tonnes approximate weight	–	–	–	–	nr	14000.00
wheel stops; steel; 100 kg approximate weight	–	–	–	–	nr	600.00
lubricators; single rail	–	–	–	–	nr	1100.00
lubricators; double rail	–	–	–	–	nr	2200.00
switch levers; upright pattern	–	–	–	–	nr	750.00
switch levers; flush type	–	–	–	–	nr	750.00

CLASS S: RAIL TRACK

Item	Gang hours	Labour £	Plant £	Material £	Unit	Total rate £
LAYING (STANDARD GAUGE TRACK)						
Laying						
Bullhead rails; jointed with fishplates; softwood sleepers						
plain track	–	–	–	–	m	67.00
form curve in plain track, radius ne 300 m	–	–	–	–	m	67.00
form curve in plain track, radius over 300 m	–	–	–	–	m	67.00
turnouts; standard, type B8	–	–	–	–	nr	13750.00
turnouts; standard, type C10	–	–	–	–	nr	16750.00
diamond crossings; standard	–	–	–	–	nr	35000.00
welded joints; aluminothermic welding including refractory mould	–	–	–	–	nr	650.00
spot re-sleepering	–	–	–	–	nr	97.00
Bullhead rails; jointed with fishplates; concrete sleepers						
plain track	–	–	–	–	m	67.00
form curve in plain track, radius ne 300 m	–	–	–	–	m	67.00
form curve in plain track, radius over 300 m	–	–	–	–	m	67.00
turnouts; standard, type B8	–	–	–	–	nr	13750.00
turnouts; standard, type C10	–	–	–	–	nr	16750.00
diamond crossings; standard	–	–	–	–	nr	35000.00
welded joints; aluminothermic welding including refractory mould	–	–	–	–	nr	650.00
spot re-sleepering	–	–	–	–	nr	97.00
Bullhead rails; welded joints; softwood sleepers						
plain track	–	–	–	–	m	67.00
form curve in plain track, radius ne 300 m	–	–	–	–	m	67.00
form curve in plain track, radius over 300 m	–	–	–	–	m	67.00
turnouts; standard, type B8	–	–	–	–	nr	13750.00
turnouts; standard, type C10	–	–	–	–	nr	16750.00
diamond crossings; standard	–	–	–	–	nr	35000.00
spot re-sleepering	–	–	–	–	nr	97.00
Bullhead rails; welded joints; concrete sleepers						
plain track	–	–	–	–	m	97.00
form curve in plain track, radius ne 300 m	–	–	–	–	m	97.00
form curve in plain track, radius over 300 m	–	–	–	–	m	97.00
turnouts; standard, type B8	–	–	–	–	nr	13750.00
turnouts; standard, type C10	–	–	–	–	nr	16750.00
diamond crossings; standard	–	–	–	–	nr	35000.00
spot re-sleepering	–	–	–	–	nr	97.00
Flat bottom rails; jointed with fishplates; softwood sleepers						
plain track	–	–	–	–	m	67.00
form curve in plain track, radius not exceeding 300 m	–	–	–	–	m	67.00
form curve in plain track, radius exceeding 300 m	–	–	–	–	m	67.00
turnouts; standard, type Bv8	–	–	–	–	nr	13750.00
turnouts; standard, type Cv9.25	–	–	–	–	nr	16750.00
diamond crossings; standard	–	–	–	–	nr	35000.00
welded joints; aluminothermic welding including refractory mould	–	–	–	–	nr	650.00
spot re-sleepering	–	–	–	–	nr	97.00

CLASS S: RAIL TRACK

Item	Gang hours	Labour £	Plant £	Material £	Unit	Total rate £
LAYING (STANDARD GAUGE TRACK) – cont						
Flat bottom rails; jointed with fishplates; concrete sleepers						
plain track	–	–	–	–	m	97.00
form curve in plain track, radius ne 300 m	–	–	–	–	m	97.00
form curve in plain track, radius over 300 m	–	–	–	–	m	97.00
turnouts; standard, type Bv8	–	–	–	–	nr	13750.00
turnouts; standard, type Cv9.25	–	–	–	–	nr	16750.00
diamond crossings; standard	–	–	–	–	nr	35000.00
welded joints; aluminothermic welding including refractory mould	–	–	–	–	nr	650.00
spot re-sleepering	–	–	–	–	nr	97.00
Flat bottom rails; welded joints; softwood sleepers						
plain track	–	–	–	–	m	97.00
form curve in plain track, radius ne 300 m	–	–	–	–	m	97.00
form curve in plain track, radius over 300 m	–	–	–	–	m	97.00
turnouts; standard, type Bv8	–	–	–	–	nr	13750.00
turnouts; standard, type Cv9.25	–	–	–	–	nr	16750.00
diamond crossings; standard	–	–	–	–	nr	35000.00
spot re-sleepering	–	–	–	–	nr	97.00
Flat bottom rails; welded joints; concrete sleepers						
plain track	–	–	–	–	m	650.00
form curve in plain track, radius ne 300 m	–	–	–	–	m	97.00
form curve in plain track, radius over 300 m	–	–	–	–	m	97.00
turnouts; standard, type Bv8	–	–	–	–	nr	13750.00
turnouts; standard, type Cv9.25	–	–	–	–	nr	16750.00
diamond crossings; standard	–	–	–	–	nr	35000.00
spot re-sleepering	–	–	–	–	nr	97.00
Check rails, flat bottom; jointed with fishplates						
rail	–	–	–	–	m	97.00
welded joints; aluminothermic welding including refractory mould	–	–	–	–	nr	97.00
Guard rails, bullhead; jointed with fishplates						
rail	–	–	–	–	m	97.00
welded joints; aluminothermic welding including refractory mould	–	–	–	–	nr	650.00
Guard rails, flat bottom; jointed with fishplates						
rail	–	–	–	–	m	97.00
welded joints; alumino-thermic welding including refractory mould	–	–	–	–	nr	650.00
Conductor rails, bullhead; jointed with fishplates						
rail	–	–	–	–	m	97.00
welded joints; aluminothermic welding including refractory mould	–	–	–	–	nr	97.00
Sundries						
buffer stops; single raker, steel rail and timber; 2 tonnes approximate weight	–	–	–	–	nr	3750.00
buffer stops; double raker, steel rail and timber; 2.5 tonnes approximate weight	–	–	–	–	nr	3750.00
wheel stops; steel; 100 kg approximate weight	–	–	–	–	nr	3750.00

CLASS S: RAIL TRACK

Item	Gang hours	Labour £	Plant £	Material £	Unit	Total rate £
lubricators; single rail	–	–	–	–	nr	960.00
lubricators; double rail	–	–	–	–	nr	960.00
switch levers; upright pattern	–	–	–	–	nr	1250.00
switch levers; flush type	–	–	–	–	nr	1250.00
conductor rail guard boards	–	–	–	–	m	2.50
DECAUVILLE TRACK						
Supplying						
Dock and crane rails; for welded track						
section 56 crane rail; 12.2 m lengths	–	–	–	–	nr	145.00
section 101 crane rail; 9.144 m lengths	–	–	–	–	nr	1280.00
Fittings						
20 mm mild steel sole plate 400 mm wide; drilled with two bolt holes at 1200 mm centres	–	–	–	–	m	65.00
M20 × 250 mm holding down bolt, nut and washers	–	–	–	–	nr	2.75
rail clips, spring type, adjustable; complete with M20 × 60 mm stud welded to sole plate	–	–	–	–	nr	16.00
Sundries						
wheel stops; 200 kg each	–	–	–	–	nr	194.00
Laying						
Crane rails, section 56; continuous sole plate; welded						
plain track	–	–	–	–	m	55.00
form curve in plain track, radius ne 300 m	–	–	–	–	m	10.20
Crane rails, section 101; continuous sole plate; welded						
plain track	–	–	–	–	m	69.50
form curve in plain track, radius ne 300 m	–	–	–	–	m	10.15
Sundries						
wheel stops	–	–	–	–	nr	105.00

CLASS T: TUNNELS

Item	Gang hours	Labour £	Plant £	Material £	Unit	Total rate £
NOTES						
There are so many factors, apart from design considerations, which influence the cost of tunnelling that it is only possible to give guide prices for a sample of the work involved. For more reliable estimates it is advisable to seek the advice of a Specialist Contractor. The main cost considerations are:						
* the nature of the ground						
* size of tunnel						
* length of drive						
* depth below surface						
* anticipated overbreak						
* support of face and roof of tunnel (rock bolting etc.)						
* necessity for pre-grouting						
* ventilation						
* presence of water						
* use of compressed air working						
The following rates for mass concrete work cast in primary and secondary linings to tunnels and access shafts are based on a 5.0 m depth of shaft and 15.0 m head of tunnel.						
Apply the following factors for differing depths and lengths:						
HEAD LENGTH: 15 m 30 m 60 m 90 m						
Shaft depth 5 m +0% +10% +20% +32½%						
Shaft depth 10 m +5% +12½% +27½% +35%						
Shaft depth 15 m +10% +17½% +32½% +40%						
Shaft depth 20 m +15% +20% +37½% +42½%						
EXCAVATION						
NB: Rates +25% or −10% Dependent on Ground Conditions						
Excavating tunnels in rock						
1.5 m diameter	–	–	–	–	m^3	**528.00**
3.0 m diameter	–	–	–	–	m^3	**330.00**
Excavating tunnels in soft material						
1.5 m diameter	–	–	–	–	m^3	**250.25**
3.0 m diameter	–	–	–	–	m^3	**137.50**
Excavating shafts in rock						
3.0 m diameter	–	–	–	–	m^3	**203.50**
4.5 m diameter	–	–	–	–	m^3	**170.50**
Excavating shafts in soft material						
3.0 m diameter	–	–	–	–	m^3	**118.25**
4.5 m diameter	–	–	–	–	m^3	**101.75**
Excavating other cavities in rock						
1.5 m diameter	–	–	–	–	m^3	**528.00**
3.0 m diameter	–	–	–	–	m^3	**330.00**
Excavating other cavities in soft material						
1.5 m diameter	–	–	–	–	m^3	**250.25**
3.0 m diameter	–	–	–	–	m^3	**137.50**
Excavated surfaces in rock	–	–	–	–	m^2	**19.25**
Excavated surfaces in soft material	–	–	–	–	m^2	**19.25**

Unit Costs – Civil Engineering Works

CLASS T: TUNNELS

Item	Gang hours	Labour £	Plant £	Material £	Unit	Total rate £
IN SITU LINING TO TUNNELS						
Notes						
The following rates for mass concrete work cast in primary and secondary linings to tunnels and access shafts are based on a 5.0 m depth of shaft and 15.0 m head of tunnel.						
See above for additions for differing shaft depths and tunnel lengths.						
NB: Rates +10% or −10% Dependent on Ground Conditions						
Mass concrete; grade C30, 20 mm aggregate						
Cast primary lining to tunnels						
1.5 m diameter	–	–	–	–	m³	**333.50**
3.0 m diameter	–	–	–	–	m³	**287.50**
Secondary lining to tunnels						
1.5 m diameter	–	–	–	–	m³	**388.13**
3.0 m diameter	–	–	–	–	m³	**307.63**
Formwork; rough finish						
Tunnel lining						
1.5 m diameter	–	–	–	–	m²	**54.05**
3.0 m diameter	–	–	–	–	m²	**54.05**
IN SITU LINING TO ACCESS SHAFTS						
Notes						
The following rates for mass concrete work cast in primary and secondary linings to tunnels and access shafts are based on a 5.0 m depth of shaft and 15.0 m head of tunnel.						
See above for additions for differing shaft depths and tunnel lengths.						
NB: Rates +10% or −10% Dependent on Ground Conditions						
Mass concrete; grade C30, 20 mm aggregate						
Secondary linings to shafts						
3.0 m int diameter	–	–	–	–	m³	**307.63**
4.5 m int diameter	–	–	–	–	m³	**299.00**
Cast primary lining to shafts						
3.0 m int diameter	–	–	–	–	m³	**327.75**
4.5 m int diameter	–	–	–	–	m³	**313.38**
Formwork; rough finish						
Shaft lining						
3.0 m int diameter	–	–	–	–	m²	**81.65**
4.5 m int diameter	–	–	–	–	m²	**55.20**

CLASS T: TUNNELS

Item	Gang hours	Labour £	Plant £	Material £	Unit	Total rate £
IN SITU LINING TO OTHER CAVITIES						
Notes						
The following rates for mass concrete work cast in primary and secondary linings to tunnels and access shafts are based on a 5.0 m depth of shaft and 15.0 m head of tunnel.						
See above for additions for differing shaft depths and tunnel lengths.						
NB: Rates +10% or −10% Dependent on Ground Conditions						
Mass concrete; grade C30, 20 mm aggregate						
Cast primary lining to other cavities						
1.5 m int diameter	–	–	–	–	m^3	336.38
3.0 m int diameter	–	–	–	–	m^3	287.50
Secondary linings to other cavities						
1.5 m int diameter	–	–	–	–	m^3	388.13
3.0 m int diameter	–	–	–	–	m^3	307.63
Formwork; rough finish						
Other cavities lining						
1.5 m int diameter	–	–	–	–	m^2	54.05
3.0 m int diameter	–	–	–	–	m^2	54.05
PREFORMED SEGMENTAL LININGS TO TUNNELS						
Precast concrete bolted rings; flanged; including packing; guide price/ring based upon standard bolted concrete segmental rings; ring width 610 mm						
Under current market conditions the reader is currently advised to contact specialist contractors directly for costs associated with these works.						
PREFORMED SEGMENTAL LININGS TO SHAFTS						
Precast concrete bolted rings; flanged; including packing; guide price/ring based upon standard bolted concrete segmental rings; ring width 610 mm						
Under current market conditions the reader is currently advised to contact specialist contractors directly for costs associated with these works.						

CLASS T: TUNNELS

Item	Gang hours	Labour £	Plant £	Material £	Unit	Total rate £
PREFORMED SEGMENTAL LININGS TO OTHER						
Precast concrete bolted rings; flanged; including packing; guide price/ring based upon standard bolted concrete segmental rings; ring width 610 mm						
NB: Rates +10% or −10% Dependent on Ground Conditions						
Linings to tunnels						
1.5 m int diameter; Ring 1.52 m ID × 1.77 m OD; 6 segments, maximum piece weight 139 kg	–	–	–	–	nr	589.39
3.0 m int diameter; Ring 3.05 m ID × 3.35 m OD; 7 segments, maximum piece weight 247 kg	–	–	–	–	nr	1055.37
Linings to shafts						
3.0 m int diameter; Ring 3.05 m ID × 3.35 m OD; 7 segments, maximum piece weight 247 kg	–	–	–	–	nr	835.18
Lining ancillaries; bitumen impregnated fibreboard						
Parallel circumferential packing						
1.5 m int diameter	–	–	–	–	nr	5.60
3.0 m int diameter	–	–	–	–	nr	11.19
Lining ancillaries; PC4 AF caulking compound						
Caulking						
1.5 m int diameter	–	–	–	–	m	10.11
3.0 m int diameter	–	–	–	–	m	10.11
SUPPORT AND STABILIZATION						
NB: Rates +15% or −5% Dependent on Ground Conditions						
Rock bolts						
mechanical	–	–	–	–	m	28.00
mechanical grouted	–	–	–	–	m	43.73
pre-grouted impacted	–	–	–	–	m	42.07
chemical end anchor	–	–	–	–	m	42.07
chemical grouted	–	–	–	–	m	29.92
chemically filled	–	–	–	–	m	47.07
Internal support						
steel arches; supply	–	–	–	–	t	1284.26
steel arches; erection	–	–	–	–	t	485.59
timber supports; supply	–	–	–	–	m^3	358.41
timber supports; erection	–	–	–	–	m^3	309.03
lagging	–	–	–	–	m^2	26.50
sprayed concrete	–	–	–	–	m^2	30.31
mesh or link	–	–	–	–	m^2	9.79
Pressure grouting						
sets of drilling and grouting plant	–	–	–	–	nr	1328.56
face packers	–	–	–	–	nr	58.85
deep packers	–	–	–	–	nr	100.90
drilling and flushing to 40 mm diameter	–	–	–	–	m	22.29
re-drilling and flushing	–	–	–	–	m	17.68
injection of grout materials; chemical grout	–	–	–	–	t	706.33
Forward probing	–	–	–	–	m	22.97

CLASS U: BRICKWORK, BLOCKWORK AND MASONRY

Item	Gang hours	Labour £	Plant £	Material £	Unit	Total rate £
NOTES						
Apply the following multipliers to both labour and plant for rubble walls:						
height 2 to 5 m × 1.21						
height 5 to 10 m × 1.37						
wall to radius small × 1.75						
wall to radius large × 1.50						
wall to rake or batter × 1.15						
wall in piers or stanchion × 1.50						
wall in butresses × 1.15						
RESOURCES – LABOUR						
Brickwork, blockwork and masonry gang						
1 foreman bricklayer (craftsman)		22.50				
4 bricklayers (craftsman)		78.16				
1 unskilled operative (general)		12.71				
Total Gang Rate/Hour		£ 113.37				
RESOURCES – MATERIAL						
Supply only materials rates						
Clay bricks						
Common bricks; PC £ per thousand	–	–	–	440.00	Nr	440.00
Facing bricks; PC £ per thousand	–	–	–	630.00	Nr	630.00
Class A engineering bricks; PC £ per thousand	–	–	–	940.00	Nr	940.00
Class B engineering bricks; PC £ per thousand	–	–	–	450.00	Nr	450.00
Concrete blocks to BS EN 771–3						
Lightweight						
solid; 3.5 N/mm^2; 100 mm thick	–	–	–	7.50	m^2	7.50
solid; 3.5 N/mm^2; 140 mm thick	–	–	–	10.25	m^2	10.25
solid 3.5 N/mm^2; 215 mm thick	–	–	–	15.00	m^2	15.00
Dense						
solid 7 N/mm^2; 100 mm thick	–	–	–	5.00	m^2	5.00
solid 7 N/mm^2; 140 mm thick	–	–	–	7.50	m^2	7.50
solid 7 N/mm^2; 215 mm thick	–	–	–	10.00	m^2	10.00
RESOURCES – PLANT						
Brickwork, blockwork and masonry						
2 t dumper (50% of time)			3.30			
mixer (50% of time)			1.46			
small power tools			0.75			
scaffold, etc.			1.15			
Total Rate/Hour		£	6.66			

CLASS U: BRICKWORK, BLOCKWORK AND MASONRY

Item	Gang hours	Labour £	Plant £	Material £	Unit	Total rate £
COMMON BRICkWORK						
Common bricks in cement mortar designation (ii)						
Thickness 103 mm						
vertical straight walls	0.23	25.85	1.52	29.56	m²	**56.93**
vertical curved walls	0.30	33.56	1.95	29.56	m²	**65.07**
battered straight walls	0.33	37.30	2.19	29.56	m²	**69.05**
battered curved walls	0.37	41.95	2.46	29.56	m²	**73.97**
vertical facing to concrete	0.25	28.00	1.65	35.19	m²	**64.84**
battered facing to concrete	0.37	41.95	2.46	35.21	m²	**79.62**
casing to metal sections	0.32	35.71	2.10	35.21	m²	**73.02**
Thickness 215 mm						
vertical straight walls	0.44	50.11	2.94	59.60	m²	**112.65**
vertical curved walls	0.57	64.62	3.80	59.60	m²	**128.02**
battered straight walls	0.63	71.42	4.20	59.60	m²	**135.22**
battered curved walls	0.71	80.49	4.70	59.60	m²	**144.79**
vertical facing to concrete	0.48	54.42	3.19	65.22	m²	**122.83**
battered facing to concrete	0.71	80.49	4.70	65.22	m²	**150.41**
casing to metal sections	0.61	69.16	4.03	65.22	m²	**138.41**
Thickness 328 mm						
vertical straight walls	0.64	72.56	4.29	89.99	m²	**166.84**
vertical curved walls	0.82	92.96	5.48	89.99	m²	**188.43**
battered straight walls	0.91	103.17	6.04	89.99	m²	**199.20**
battered curved walls	1.01	114.50	6.73	89.99	m²	**211.22**
vertical facing to concrete	0.69	78.23	4.62	95.61	m²	**178.46**
battered facing to concrete	1.01	114.50	6.73	95.61	m²	**216.84**
casing to metal sections	0.87	98.63	5.80	95.61	m²	**200.04**
Thickness 440 mm						
vertical straight walls	0.84	95.23	5.56	120.23	m²	**221.02**
vertical curved walls	1.06	120.17	7.05	120.23	m²	**247.45**
battered straight walls	1.16	131.51	7.74	120.23	m²	**259.48**
battered curved walls	1.29	146.25	8.58	120.23	m²	**275.06**
vertical facing to concrete	0.90	102.03	5.98	125.79	m²	**233.80**
battered facing to concrete	1.29	146.25	8.58	125.86	m²	**280.69**
casing to metal sections	1.12	126.97	7.45	125.86	m²	**260.28**
Thickness 890 mm						
vertical straight walls	1.50	170.06	10.00	241.40	m²	**421.46**
vertical curved walls	1.85	209.73	12.34	241.40	m²	**463.47**
battered straight walls	2.00	226.74	13.38	241.40	m²	**481.52**
battered curved walls	2.20	249.41	14.62	241.40	m²	**505.43**
vertical facing to concrete	1.60	181.39	10.67	247.03	m²	**439.09**
battered facing to concrete	2.20	249.41	14.62	247.03	m²	**511.06**
casing to metal sections	1.94	219.94	12.95	247.03	m²	**479.92**
Thickness exceeding 1 m						
vertical straight walls	1.64	185.93	10.90	270.16	m²	**466.99**
vertical curved walls	2.00	226.74	13.38	270.16	m²	**510.28**
battered straight walls	2.17	246.01	14.47	270.16	m²	**530.64**
battered curved walls	2.37	268.69	15.76	270.16	m²	**554.61**
vertical facing to concrete	1.74	197.26	11.61	275.79	m²	**484.66**
battered facing to concrete	2.37	268.69	15.76	275.79	m²	**560.24**
casing to metal sections	2.10	238.08	14.01	275.79	m²	**527.88**

CLASS U: BRICKWORK, BLOCKWORK AND MASONRY

Item	Gang hours	Labour £	Plant £	Material £	Unit	Total rate £
COMMON BRICKWORK – cont						
Columns and piers						
215 × 215 mm	0.13	14.74	0.87	13.21	m	**28.82**
440 × 215 mm	0.24	27.21	1.60	26.62	m	**55.43**
665 × 328 mm	0.44	49.88	2.93	60.15	m	**112.96**
890 × 890 mm	1.10	124.71	7.33	215.07	m	**347.11**
Surface features						
copings; standard header-on-edge; 215 mm wide × 103 mm high	0.10	11.34	0.70	6.64	m	**18.68**
sills; standard header-on-edge; 215 mm wide × 103 mm high	0.13	14.74	0.83	6.50	m	**22.07**
rebates	0.30	34.01	2.00	–	m	**36.01**
chases	0.35	39.68	2.33	–	m	**42.01**
band courses; flush; 215 mm wide	0.05	5.67	0.30	–	m	**5.97**
band courses; projection 103 mm; 215 mm wide	0.05	5.67	0.33	–	m	**6.00**
corbels; maximum projection 103 mm; 215 mm wide	0.15	17.01	1.00	6.43	m	**24.44**
pilasters; 328 mm wide × 103 mm projection	–	–	0.47	13.21	m	**13.68**
pilasters; 440 mm wide × 215 mm projection	0.12	13.60	0.80	29.77	m	**44.17**
plinths; projection 103 mm × 900 mm wide	0.19	21.54	1.27	26.96	m	**49.77**
fair facing	0.06	6.80	0.38	–	m²	**7.18**
Ancillaries						
bonds to existing work; to brickwork	1.50	170.06	10.00	30.38	m²	**210.44**
built-in pipes and ducts, cross-sectional area not exceeding 0.05 m²; excluding supply; brickwork 103 mm thick	0.06	6.80	0.40	0.97	nr	**8.17**
built-in pipes and ducts, cross-sectional area not exceeding 0.05 m²; excluding supply; brickwork 215 mm thick	0.12	13.60	0.77	1.41	nr	**15.78**
built-in pipes and ducts, cross-sectional area 0.05–0.25 m²; excluding supply; brickwork 103 mm thick	0.15	17.01	1.00	1.38	nr	**19.39**
built-in pipes and ducts, cross-sectional area 0.05–0.25 m²; excluding supply; brickwork 215 mm thick	0.29	32.88	1.93	2.02	nr	**36.83**
built-in pipes and ducts, cross-sectional area 0.25–0.5 m²; excluding supply; brickwork 103 mm thick	0.18	20.41	1.17	1.58	nr	**23.16**
built-in pipes and ducts, cross-sectional area 0.25–0.5 m²; excluding supply; brickwork 215 mm thick	0.34	38.55	2.27	2.29	nr	**43.11**
FACING BRICKWORK						
Facing bricks; in plasticized cement mortar designation (iii)						
Thickness 103 mm						
vertical straight walls	0.34	38.55	2.27	45.50	m²	**86.32**
vertical curved walls	0.45	51.02	3.00	45.50	m²	**99.52**
battered straight walls	0.45	51.02	3.00	45.50	m²	**99.52**

Unit Costs – Civil Engineering Works

CLASS U: BRICKWORK, BLOCKWORK AND MASONRY

Item	Gang hours	Labour £	Plant £	Material £	Unit	Total rate £
battered curved walls	0.56	63.49	3.75	45.50	m²	**112.74**
vertical facing to concrete	0.37	41.95	2.30	51.13	m²	**95.38**
battered facing to concrete	0.56	63.49	3.75	51.13	m²	**118.37**
casing to metal sections	0.49	55.55	3.26	51.13	m²	**109.94**
Thickness 215 mm						
vertical straight walls	0.57	64.62	5.17	91.48	m²	**161.27**
vertical curved walls	0.84	95.23	5.60	91.48	m²	**192.31**
battered straight walls	0.84	95.23	5.60	91.48	m²	**192.31**
battered curved walls	1.02	115.64	6.82	91.48	m²	**213.94**
vertical facing to concrete	0.66	74.82	4.40	97.11	m²	**176.33**
battered facing to concrete	1.02	115.64	6.82	97.11	m²	**219.57**
casing to metal sections	0.82	92.96	5.46	97.11	m²	**195.53**
Thickness 328 mm						
vertical straight walls	0.83	94.10	5.51	137.81	m²	**237.42**
vertical curved walls	1.12	126.97	7.43	137.81	m²	**272.21**
battered straight walls	1.12	126.97	7.43	137.81	m²	**272.21**
battered curved walls	1.36	154.18	9.06	137.81	m²	**301.05**
vertical facing to concrete	0.88	99.77	5.84	143.43	m²	**249.04**
battered facing to concrete	1.36	154.18	9.06	143.43	m²	**306.67**
casing to metal sections	1.18	133.78	7.86	143.43	m²	**285.07**
Thickness 440 mm						
vertical straight walls	1.08	122.44	7.20	183.99	m²	**313.63**
vertical curved walls	1.39	157.58	9.26	183.99	m²	**350.83**
battered straight walls	1.39	157.58	9.26	183.99	m²	**350.83**
vertical facing to concrete	1.09	123.57	7.28	189.62	m²	**320.47**
battered facing to concrete	1.39	157.58	9.26	189.62	m²	**356.46**
casing to metal sections	1.39	157.58	9.26	189.62	m²	**356.46**
Columns and piers						
215 × 215 mm	0.17	19.27	1.11	20.40	m	**40.78**
440 × 215 mm	0.29	32.88	1.93	40.97	m	**75.78**
665 × 328 mm	0.58	65.75	3.84	92.43	m	**162.02**
Surface features						
copings; standard header-on-edge; standard bricks; 215 mm wide × 103 mm high	0.13	14.74	0.89	9.55	m	**25.18**
flat arches; standard stretcher-on-end; 215 mm wide × 103 mm high	0.21	23.81	1.40	9.34	m	**34.55**
flat arches; standard stretcher-on-end; bullnosed special bricks; 103 mm × 215 mm high	0.22	24.94	1.47	55.37	m	**81.78**
segmental arches; single ring; standard bricks; 103 mm wide × 215 mm high	0.37	41.95	2.46	9.34	m	**53.75**
segmental arches; two ring; standard bricks; 103 mm wide × 440 mm high	0.49	55.55	3.26	18.82	m	**77.63**
segmental arches; cut voussoirs; 103 mm wide × 215 mm high	0.39	44.21	2.60	22.96	m	**69.77**
rebates	0.33	37.41	2.20	–	m	**39.61**
chases	0.37	41.95	2.46	–	m	**44.41**
cornices; maximum projection 103 mm; 215 mm wide	0.37	41.95	2.46	9.30	m	**53.71**
band courses; projection 113 mm; 215 mm wide	0.06	6.80	0.37	–	m	**7.17**
corbels; maximum projection 113 mm; 215 mm wide	0.37	41.95	2.46	9.30	m	**53.71**

CLASS U: BRICKWORK, BLOCKWORK AND MASONRY

Item	Gang hours	Labour £	Plant £	Material £	Unit	Total rate £
FACING BRICkWORK – cont						
Facing bricks – cont						
Surface features – cont						
pilasters; 328 mm wide × 113 mm projection	0.05	5.67	0.33	20.39	m	**26.39**
pilasters; 440 mm wide × 215 mm projection	0.06	6.80	0.37	45.91	m	**53.08**
plinths; projection 113 mm × 900 mm wide	0.24	27.21	1.60	41.30	m	**70.11**
fair facing; pointing as work proceeds	0.06	6.80	0.40	–	m²	**7.20**
Ancillaries						
bonds to existing work; to brickwork	1.50	170.06	10.00	10.96	m	**191.02**
built-in pipes and ducts, cross-sectional area not exceeding 0.05 m²; excluding supply; brickwork half brick thick	0.10	11.34	0.63	0.95	nr	**12.92**
built-in pipes and ducts, cross-sectional area not exceeding 0.05 m²; excluding supply; brickwork one brick thick	0.15	17.01	0.68	1.70	nr	**19.39**
built-in pipes and ducts, cross-sectional area 0.05–0.25 m²; excluding supply; brickwork half brick thick	0.19	21.54	1.27	1.50	nr	**24.31**
built-in pipes and ducts, cross-sectional area 0.05–0.25 m²; excluding supply; brickwork one brick thick	0.33	37.41	2.20	2.38	nr	**41.99**
built-in pipes and ducts, cross-sectional area 0.25–0.5 m²; excluding supply; brickwork half brick thick	0.23	26.08	1.53	1.70	nr	**29.31**
built-in pipes and ducts, cross-sectional area 0.25–0.5 m²; excluding supply; brickwork one brick thick	0.40	45.35	2.63	2.72	nr	**50.70**
ENGINEERING BRICkWORK						
Class A engineering bricks, solid; in cement mortar designation (ii)						
Thickness 103 mm						
vertical straight walls	0.27	30.61	1.79	60.67	m²	**93.07**
vertical curved walls	0.37	41.95	2.46	60.67	m²	**105.08**
battered straight walls	0.37	41.95	2.46	60.67	m²	**105.08**
battered curved walls	0.41	46.48	2.75	60.67	m²	**109.90**
vertical facing to concrete	0.32	36.28	2.10	66.30	m²	**104.68**
battered facing to concrete	0.46	52.15	3.09	66.30	m²	**121.54**
casing to metal sections	0.41	46.48	2.75	66.30	m²	**115.53**
Thickness 215 mm						
vertical straight walls	0.52	58.95	3.46	121.82	m²	**184.23**
vertical curved walls	0.71	80.49	4.70	121.82	m²	**207.01**
battered straight walls	0.71	80.49	4.70	121.82	m²	**207.01**
battered curved walls	0.78	88.43	5.20	121.82	m²	**215.45**
vertical facing to concrete	0.61	69.16	4.03	127.45	m²	**200.64**
battered facing to concrete	0.87	98.63	5.80	127.45	m²	**231.88**
casing to metal sections	0.78	88.43	5.20	127.45	m²	**221.08**

CLASS U: BRICKWORK, BLOCKWORK AND MASONRY

Item	Gang hours	Labour £	Plant £	Material £	Unit	Total rate £
Thickness 328 mm						
vertical straight walls	0.75	85.03	5.02	183.32	m²	**273.37**
vertical curved walls	1.01	114.50	6.73	183.32	m²	**304.55**
battered straight walls	1.01	114.50	6.73	183.32	m²	**304.55**
battered curved walls	1.11	125.84	7.40	183.32	m²	**316.56**
vertical facing to concrete	0.87	98.63	5.80	188.95	m²	**293.38**
battered facing to concrete	1.24	140.58	8.23	188.95	m²	**337.76**
casing to metal sections	1.11	125.84	7.40	188.95	m²	**322.19**
Thickness 440 mm						
vertical straight walls	0.97	109.97	6.47	244.68	m²	**361.12**
vertical curved walls	1.29	146.25	8.58	244.68	m²	**399.51**
battered straight walls	1.29	146.25	8.58	244.68	m²	**399.51**
battered curved walls	1.41	159.85	9.40	244.68	m²	**413.93**
vertical facing to concrete	1.12	126.97	7.45	250.30	m²	**384.72**
battered facing to concrete	1.56	176.86	10.39	250.30	m²	**437.55**
casing to metal sections	1.41	159.85	9.40	250.30	m²	**419.55**
Thickness 890 mm						
vertical straight walls	1.72	195.00	11.45	490.09	m²	**696.54**
vertical curved walls	2.20	249.41	14.62	490.09	m²	**754.12**
battered straight walls	2.20	249.41	14.62	490.09	m²	**754.12**
battered curved walls	2.37	268.69	15.79	490.09	m²	**774.57**
vertical facing to concrete	1.94	219.94	12.95	495.72	m²	**728.61**
battered facing to concrete	2.58	292.49	17.17	495.72	m²	**805.38**
casing to metal sections	2.37	268.69	15.79	495.72	m²	**780.20**
Thickness exceeding 1 m						
vertical straight walls	1.87	212.00	12.43	548.50	m³	**772.93**
vertical curved walls	2.37	268.69	15.76	548.50	m³	**832.95**
battered straight walls	2.37	268.69	15.76	548.50	m³	**832.95**
battered curved walls	2.55	289.09	16.98	548.50	m³	**854.57**
vertical facing to concrete	2.10	238.08	14.01	554.06	m³	**806.15**
battered facing to concrete	2.76	312.90	18.39	554.06	m³	**885.35**
casing to metal sections	2.55	289.09	16.98	554.06	m³	**860.13**
Columns and piers						
215 × 215 mm	0.16	18.14	1.06	27.21	m	**46.41**
440 × 215 mm	0.28	31.74	1.85	54.62	m	**88.21**
665 × 328 mm	0.56	63.49	3.71	123.15	m	**190.35**
890 × 890 mm	1.78	201.80	11.85	439.08	m	**652.73**
Surface features						
copings; standard header-on-edge; 215 mm wide × 103 mm high	0.11	12.47	0.70	13.64	m	**26.81**
sills; standard header-on-edge; 215 mm wide × 103 mm high	0.13	14.74	0.83	13.50	m	**29.07**
rebates	0.33	37.41	2.20	–	m	**39.61**
chases	0.37	41.95	2.46	–	m	**44.41**
band courses; flush; 215 mm wide	0.05	5.67	0.30	–	m	**5.97**
band courses; projection 103 mm; 215 mm wide	0.05	5.67	0.33	–	m	**6.00**
corbels; maximum projection 103 mm; 215 mm wide	0.15	17.01	1.00	13.43	m	**31.44**
pilasters; 328 mm wide × 103 mm projection	0.07	7.94	0.47	27.21	m	**35.62**
pilasters; 440 mm wide × 215 mm projection	0.12	13.60	0.80	61.27	m	**75.67**
plinths; projection 103 mm × 900 mm wide	0.19	21.54	1.27	54.96	m	**77.77**
fair facing	0.06	6.80	0.40	–	m²	**7.20**

CLASS U: BRICKWORK, BLOCKWORK AND MASONRY

Item	Gang hours	Labour £	Plant £	Material £	Unit	Total rate £
ENGINEERING BRICkWORK – cont						
Class A engineering bricks, solid – cont						
Ancillaries						
bonds to existing brickwork; to brickwork	0.29	32.88	1.93	15.41	m²	**50.22**
built-in pipes and ducts, cross-sectional area not exceeding 0.05 m²; excluding supply; brickwork half brick thick	0.10	11.34	0.63	1.65	nr	**13.62**
built-in pipes and ducts, cross-sectional area not exceeding 0.05 m²; excluding supply; brickwork one brick thick	0.15	17.01	1.00	2.25	nr	**20.26**
built-in pipes and ducts, cross-sectional area 0.05–0.25 m²; excluding supply; brickwork half brick thick	0.19	21.54	1.27	2.11	nr	**24.92**
built-in pipes and ducts, cross-sectional area 0.05–0.25 m²; excluding supply; brickwork one brick thick	0.33	37.41	2.20	3.02	nr	**42.63**
built-in pipes and ducts, cross-sectional area 0.025–0.5 m²; excluding supply; brickwork half brick thick	0.23	26.08	1.53	2.33	nr	**29.94**
built-in pipes and ducts, cross-sectional area 0.025–0.5 m²; excluding supply; brickwork one brick thick	0.40	45.35	2.63	3.39	nr	**51.37**
Class B engineering bricks, perforated; in cement mortar designation (ii)						
Thickness 103 mm						
vertical straight walls	0.27	30.61	1.79	30.18	m²	**62.58**
vertical curved walls	0.37	41.95	2.46	30.18	m²	**74.59**
battered straight walls	0.37	41.95	2.46	30.18	m²	**74.59**
vertical facing to concrete	0.32	36.28	2.10	35.81	m²	**74.19**
casings to metal sections	0.41	46.48	2.75	35.81	m²	**85.04**
Thickness 215 mm						
vertical straight walls	0.52	58.95	3.46	60.84	m²	**123.25**
vertical curved walls	0.71	80.49	4.70	60.84	m²	**146.03**
battered straight walls	0.71	80.49	4.70	60.84	m²	**146.03**
vertical facing to concrete	0.61	69.16	4.03	66.47	m²	**139.66**
casings to metal sections	0.78	88.43	5.20	66.47	m²	**160.10**
Thickness 328 mm						
vertical straight walls	0.76	86.16	5.04	91.85	m²	**183.05**
vertical curved walls	1.01	114.50	6.73	91.85	m²	**213.08**
battered straight walls	1.01	114.50	6.73	91.85	m²	**213.08**
vertical facing to concrete	0.87	98.63	5.80	97.48	m²	**201.91**
casings to metal sections	1.11	125.84	7.40	97.48	m²	**230.72**
Thickness 440 mm						
vertical straight walls	0.97	109.97	6.47	122.72	m²	**239.16**
vertical curved walls	1.29	146.25	8.58	122.72	m²	**277.55**
battered straight walls	1.29	146.25	8.58	122.72	m²	**277.55**
vertical facing to concrete	1.12	126.97	7.45	128.34	m²	**262.76**
casings to metal sections	1.41	159.85	9.40	128.34	m²	**297.59**

CLASS U: BRICKWORK, BLOCKWORK AND MASONRY

Item	Gang hours	Labour £	Plant £	Material £	Unit	Total rate £
Thickness 890 mm						
vertical straight walls	1.72	195.00	11.45	246.18	m²	**452.63**
vertical curved walls	2.20	249.41	14.62	246.18	m²	**510.21**
battered straight walls	2.20	249.41	14.62	246.18	m²	**510.21**
battered curved walls	2.37	268.69	15.79	246.18	m²	**530.66**
vertical facing to concrete	1.94	219.94	12.95	251.81	m²	**484.70**
battered facing to concrete	2.58	292.49	17.17	251.81	m²	**561.47**
casing to metal sections	2.37	268.69	15.79	251.81	m²	**536.29**
Thickness exceeding 1 m						
vertical straight walls	1.87	212.00	12.43	275.81	m³	**500.24**
vertical curved walls	2.37	268.69	15.76	275.81	m³	**560.26**
battered straight walls	2.37	268.69	15.76	275.81	m³	**560.26**
battered curved walls	2.55	289.09	16.98	275.81	m³	**581.88**
vertical facing to concrete	2.10	238.08	14.01	281.37	m³	**533.46**
battered facing to concrete	2.76	312.90	18.39	281.37	m³	**612.66**
casing to metal sections	2.55	289.09	16.98	281.37	m³	**587.44**
Columns and piers						
215 × 215 mm	0.16	18.14	1.06	13.49	m	**32.69**
440 × 215 mm	0.28	31.74	1.85	27.18	m	**60.77**
665 × 328 mm	0.56	63.49	3.71	61.41	m	**128.61**
890 × 890 mm	1.78	201.80	11.85	219.55	m	**433.20**
Surface features						
copings; standard header-on-edge; 215 mm wide × 103 mm high	0.11	12.47	0.70	6.78	m	**19.95**
sills; standard header-on-edge; 215 mm wide × 103 mm high	0.13	14.74	0.83	6.64	m	**22.21**
rebates	0.33	37.41	2.20	–	m	**39.61**
chases	0.37	41.95	2.46	–	m	**44.41**
band courses; flush; 215 mm wide	0.05	5.67	0.30	–	m	**5.97**
band courses; projection 103 mm; 215 mm wide	0.05	5.67	0.33	–	m	**6.00**
corbels; maximum projection 103 mm; 215 mm wide	0.15	17.01	1.00	6.57	m	**24.58**
pilasters; 328 mm wide × 103 mm projection	0.07	7.94	0.47	13.49	m	**21.90**
pilasters; 440 mm wide × 215 mm projection	0.12	13.60	0.80	30.40	m	**44.80**
plinths; projection 103 mm × 900 mm wide	0.19	21.54	1.27	27.52	m	**50.33**
fair facing	0.06	6.80	0.40	–	m²	**7.20**
Ancillaries						
bonds to existing brickwork; to brickwork	0.29	32.88	1.93	14.89	m²	**49.70**
built-in pipes and ducts, cross-sectional area not exceeding 0.05 m²; excluding supply; brickwork half brick thick	0.10	11.34	0.63	1.65	nr	**13.62**
built-in pipes and ducts, cross-sectional area not exceeding 0.05 m²; excluding supply; brickwork one brick thick	0.15	17.01	1.00	2.25	nr	**20.26**

CLASS U: BRICKWORK, BLOCKWORK AND MASONRY

Item	Gang hours	Labour £	Plant £	Material £	Unit	Total rate £
ENGINEERING BRICkWORK – cont						
Class B engineering bricks, perforated – cont						
Ancillaries – cont						
built-in pipes and ducts, cross-sectional area 0.05–0.25 m^2; excluding supply; brickwork half brick thick	0.19	21.54	1.27	2.11	nr	24.92
built-in pipes and ducts, cross-sectional area 0.05–0.25 m^2; excluding supply; brickwork one brick thick	0.33	37.41	2.20	3.02	nr	42.63
built-in pipes and ducts, cross-sectional area 0.025–0.5 m^2; excluding supply; brickwork half brick thick	0.23	26.08	1.53	2.33	nr	29.94
built-in pipes and ducts, cross-sectional area 0.025–0.5 m^2; excluding supply; brickwork one brick thick	0.40	45.35	2.63	3.34	nr	51.32
LIGHTWEIGHT BLOCkWORK						
Lightweight concrete blocks; 3.5 N/mm^2; in cement-lime mortar designation (iii)						
Thickness 100 mm;						
vertical straight walls	0.17	19.27	1.16	8.12	m^2	28.55
vertical curved walls	0.23	26.08	1.54	8.12	m^2	35.74
vertical facework to concrete	0.18	20.41	1.19	10.07	m^2	31.67
casing to metal sections	0.21	23.81	1.40	10.07	m^2	35.28
Thickness 140 mm;						
vertical straight walls	0.23	26.08	1.50	10.80	m^2	38.38
vertical curved walls	0.30	34.01	1.99	10.80	m^2	46.80
vertical facework to concrete	0.23	26.08	1.55	12.67	m^2	40.30
casing to metal sections	0.27	30.61	1.80	12.67	m^2	45.08
Thickness 215 mm;						
vertical straight walls	0.28	31.74	1.83	16.27	m^2	49.84
vertical curved walls	0.37	41.95	2.44	16.27	m^2	60.66
vertical facework to concrete	0.28	31.74	1.88	18.09	m^2	51.71
casing to metal sections	0.31	35.14	2.07	18.09	m^2	55.30
Columns and piers						
440 × 100 mm	0.08	9.07	0.53	3.59	m	13.19
890 × 140 mm	0.22	24.94	1.47	9.69	m	36.10
Surface features						
fair facing	0.06	6.80	0.40	–	m^2	7.20
Ancillaries						
built-in pipes and ducts, cross-sectional area not exceeding 0.05 m^2; excluding supply; blockwork 100 mm thick	0.04	4.53	0.27	0.27	nr	5.07
built-in pipes and ducts, cross-sectional area not exceeding 0.05 m^2; excluding supply; blockwork 140 mm thick	0.09	10.20	0.60	0.76	nr	11.56
built-in pipes and ducts, cross-sectional area not exceeding 0.05 m^2; excluding supply; blockwork 215 mm thick	0.13	14.74	0.84	0.97	nr	16.55

CLASS U: BRICKWORK, BLOCKWORK AND MASONRY

Item	Gang hours	Labour £	Plant £	Material £	Unit	Total rate £
built-in pipes and ducts, cross-sectional area 0.05–0.25 m^2; excluding supply; blockwork 100 mm thick	0.13	14.74	0.84	1.03	nr	**16.61**
built-in pipes and ducts, cross-sectional area 0.05–0.25 m^2; excluding supply; blockwork 140 mm thick	0.18	20.41	1.17	1.57	nr	**23.15**
built-in pipes and ducts, cross-sectional area 0.05–0.25 m^2; excluding supply; blockwork 215 mm thick	0.25	28.34	1.66	2.02	nr	**32.02**
built-in pipes and ducts, cross-sectional area 0.25–0.5 m^2; excluding supply; blockwork 100 mm thick	0.15	17.01	1.00	1.31	nr	**19.32**
built-in pipes and ducts, cross-sectional area 0.25–0.5 m^2; excluding supply; blockwork 140 mm thick	0.21	23.81	1.40	1.57	nr	**26.78**
built-in pipes and ducts, cross-sectional area 0.25–0.5 m^2; excluding supply; blockwork 215 mm thick	0.30	34.01	1.97	2.41	nr	**38.39**
DENSE CONCRETE BLOCkWORK						
Dense concrete blocks; 7 N/mm^2; in cement mortar designation (iii)						
Walls, built vertical and straight						
100 mm thick	0.19	21.54	1.27	5.53	m^2	**28.34**
140 mm thick	0.25	28.34	1.65	8.21	m^2	**38.20**
215 mm thick	0.34	38.55	2.27	11.15	m^2	**51.97**
Walls, built vertical and curved						
100 mm thick	0.25	28.34	1.69	5.56	m^2	**35.59**
140 mm thick	0.33	37.41	2.20	8.21	m^2	**47.82**
215 mm thick	0.45	51.02	3.01	11.17	m^2	**65.20**
Walls, built vertical in facework to concrete						
100 mm thick	0.20	22.67	1.31	6.87	m^2	**30.85**
140 mm thick	0.26	29.48	1.70	9.65	m^2	**40.83**
215 mm thick	0.35	39.68	2.33	12.35	m^2	**54.36**
Walls, as casings to metal sections, built vertical and straight						
100 mm thick	0.23	26.08	1.53	6.87	m^2	**34.48**
140 mm thick	0.30	34.01	1.97	9.65	m^2	**45.63**
Columns and piers						
440 × 100 mm	0.08	9.07	0.53	2.37	m	**11.97**
890 × 140 mm thick	0.22	24.94	1.47	7.38	m	**33.79**
Ancillaries						
built-in pipes and ducts, cross-sectional area not exceeding 0.05 m^2; excluding supply; blockwork 100 mm thick	0.05	5.67	0.30	0.26	nr	**6.23**
built-in pipes and ducts, cross-sectional area not exceeding 0.05 m^2; excluding supply; blockwork 140 mm thick	0.10	11.34	0.67	0.62	nr	**12.63**
built-in pipes and ducts, cross-sectional area not exceeding 0.05 m^2; excluding supply; blockwork 215 mm thick	0.14	15.87	0.90	0.74	nr	**17.51**

CLASS U: BRICKWORK, BLOCKWORK AND MASONRY

Item	Gang hours	Labour £	Plant £	Material £	Unit	Total rate £
DENSE CONCRETE BLOCkWORK – cont						
Dense concrete blocks – cont						
Ancillaries – cont						
built-in pipes and ducts, cross-sectional area 0.05–0.25 m^2; excluding supply; blockwork 100 mm thick	0.14	15.87	0.90	0.80	nr	**17.57**
built-in pipes and ducts, cross-sectional area 0.05–0.25 m^2; excluding supply; blockwork 140 mm thick	0.19	21.54	1.27	1.25	nr	**24.06**
built-in pipes and ducts, cross-sectional area 0.05–0.25 m^2; excluding supply; blockwork 215 mm thick	0.27	30.61	1.80	1.47	nr	**33.88**
built-in pipes and ducts, cross-sectional area 0.25–0.5 m^2; excluding supply; blockwork 100 mm thick	0.16	18.14	1.08	0.99	nr	**20.21**
built-in pipes and ducts, cross-sectional area 0.25–0.5 m^2; excluding supply; blockwork 140 mm thick	0.23	26.08	1.53	1.28	nr	**28.89**
built-in pipes and ducts, cross-sectional area 0.25–0.5 m^2; excluding supply; blockwork 215 mm thick	0.32	36.28	2.13	1.77	nr	**40.18**
ARTIFICIAL STONE BLOCkWORK						
Reconstituted stone masonry blocks; Bradstone 100 bed weathered Cotswold or North Cerney rough hewn rockfaced blocks; in coloured cement-lime mortar designation (iii)						
Thickness 100 mm vertical facing						
vertical straight walls	0.30	34.01	2.00	51.36	m^2	**87.37**
vertical curved walls	0.39	44.21	2.60	51.36	m^2	**98.17**
vertical facework to concrete	0.31	35.14	2.08	51.39	m^2	**88.61**
casing to metal sections	0.40	45.35	2.67	51.39	m^2	**99.41**
Ancillaries						
built-in pipes and ducts, cross-sectional area not exceeding 0.05 m^2; excluding supply	0.07	7.94	0.47	0.51	nr	**8.92**
built-in pipes and ducts, cross-sectional area 0.05–0.25 m^2; excluding supply	0.22	24.94	1.47	5.35	nr	**31.76**
built-in pipes and ducts, cross-sectional area 0.25–0.5 m^2; excluding supply	0.26	29.48	1.73	6.71	nr	**37.92**
Reconstituted stone masonry blocks; Bradstone Architectural dressing in weathered Cotswold or North Cerney shades; in coloured cement-lime mortar designation (iii)						
Surface features; Pier Caps						
305 × 305 mm, weathered and throated	0.09	10.20	0.60	19.12	nr	**29.92**
381 × 381 mm, weathered and throated	0.11	12.47	0.73	27.10	nr	**40.30**
457 × 457 mm, weathered and throated	0.13	14.74	0.87	37.12	nr	**52.73**
533 × 533 mm, weathered and throated	0.15	17.01	1.00	51.67	nr	**69.68**

CLASS U: BRICKWORK, BLOCKWORK AND MASONRY

Item	Gang hours	Labour £	Plant £	Material £	Unit	Total rate £
Surface features; Copings						
152 × 76 mm, twice weathered and throated	0.08	9.07	0.53	8.33	m	**17.93**
152 × 76 mm, curved on plan, twice weathered and throated	0.11	12.47	0.71	28.49	m	**41.67**
305 × 76 mm, twice weathered and throated	0.10	11.34	2.67	14.01	m	**28.02**
305 × 76 mm, curved on plan, twice weathered and throated	0.13	14.74	0.89	76.20	m	**91.83**
Surface features; Pilasters						
440 × 100 mm	0.14	15.87	0.93	22.61	m	**39.41**
Surface features; Corbels						
479 × 100 × 215 mm, splayed	0.49	55.55	3.26	128.69	nr	**187.50**
665 × 100 × 215 mm, splayed	0.55	62.35	3.67	92.81	nr	**158.83**
Surface features; Lintels						
100 × 140 mm	0.11	12.47	0.73	39.87	m	**53.07**
100 × 215 mm	0.16	18.14	1.07	51.95	m	**71.16**
ASHLAR MASONRY						
Portland Whitbed limestone; in cement-lime mortar designation (iv); pointed one side cement-lime mortar designation (iii) incorporating stone dust						
Thickness 50 mm						
vertical facing to concrete	0.85	96.36	5.66	97.95	m²	**199.97**
Thickness 75 mm						
vertical straight walls	0.95	107.70	6.33	173.50	m²	**287.53**
vertical curved walls	1.45	164.39	9.66	307.88	m²	**481.93**
Surface features						
copings; weathered and twice throated; 250 × 150 mm	0.45	51.02	3.00	129.66	m	**183.68**
copings; weathered and twice throated; 250 × 150 mm; curved on plan	0.45	51.02	3.00	155.57	m	**209.59**
copings; weathered and twice throated; 400 × 150 mm	0.49	55.55	3.26	190.60	m	**249.41**
copings; weathered and twice throated; 400 × 150 mm; curved on plan	0.49	55.55	3.26	228.70	m	**287.51**
string courses; shaped and dressed; 75 mm projection × 150 mm high	0.45	51.02	3.00	103.63	m	**157.65**
corbel; shaped and dressed; 500 × 450 × 300 mm	0.55	62.35	3.67	171.62	nr	**237.64**
keystone; shaped and dressed; 750 × 900 × 300 mm (extreme)	1.30	147.38	8.66	543.44	nr	**699.48**
RUBBLE MASONRY						
Rubble masonry; random stones; in cement-lime mortar designation (iii)						
Walls, built vertical and straight; not exceeding 2 m high						
300 mm thick	1.25	141.71	8.33	259.01	m²	**409.05**
450 mm thick	1.80	204.07	11.99	447.38	m²	**663.44**
600 mm thick	2.40	272.09	15.99	517.99	m²	**806.07**

CLASS U: BRICKWORK, BLOCKWORK AND MASONRY

Item	Gang hours	Labour £	Plant £	Material £	Unit	Total rate £
RUBBLE MASONRY – cont						
Rubble masonry; random stones – cont						
Walls, built vertical, curved on plan; not exceeding 2 m high						
300 mm thick	1.40	158.72	3.33	265.06	m^2	**427.11**
450 mm thick	2.00	226.74	13.32	388.54	m^2	**628.60**
600 mm thick	2.65	300.43	17.65	517.99	m^2	**836.07**
Walls, built with battered face; not exceeding 2 m high						
300 mm thick	1.40	158.72	9.33	259.01	m^2	**427.06**
450 mm thick	2.00	226.74	13.32	388.54	m^2	**628.60**
600 mm thick	2.65	300.43	17.65	517.99	m^2	**836.07**
Rubble masonry; squared stones; in cement-lime mortar designation (iii)						
Walls, built vertical and straight; not exceeding 2 m high						
300 mm thick	1.25	141.71	8.33	467.90	m^2	**617.94**
450 mm thick	1.80	204.07	11.99	701.84	m^2	**917.90**
600 mm thick	2.40	272.09	15.99	935.80	m^2	**1223.88**
Walls, built vertical, curved on plan; not exceeding 2 m high						
300 mm thick	1.40	158.72	9.33	467.90	m^2	**635.95**
450 mm thick	2.00	226.74	13.32	701.84	m^2	**941.90**
600 mm thick	2.40	272.09	15.99	985.28	m^2	**1273.36**
Walls, built with battered face; not exceeding 2 m high						
300 mm thick	1.40	158.72	9.33	467.90	m^2	**635.95**
450 mm thick	2.00	226.74	13.32	701.84	m^2	**941.90**
600 mm thick	2.40	272.09	15.99	985.28	m^2	**1273.36**
Dry stone walling; random stones						
Average thickness 300 mm						
battered straight walls	1.15	130.38	3.79	239.20	m^2	**373.37**
Average thickness 450 mm						
battered straight walls	1.65	187.06	5.45	358.76	m^2	**551.27**
Average thickness 600 mm						
battered straight walls	2.15	243.75	7.10	478.37	m^2	**729.22**
Surface features						
copings; formed of rough stones 275 × 200 mm (average) high	0.45	51.02	3.00	61.69	m	**115.71**
copings; formed of rough stones 500 × 200 mm (average) high	0.55	62.35	3.67	112.16	m	**178.18**

CLASS U: BRICKWORK, BLOCKWORK AND MASONRY

Item	Gang hours	Labour £	Plant £	Material £	Unit	Total rate £
ANCILLARIES COMMON TO ALL DIVISIONS						
Expamet joint reinforcement						
Ancillaries						
joint reinforcement; 65 mm wide	0.01	1.13	0.04	1.55	m	**2.72**
joint reinforcement; 115 mm wide	0.01	1.13	0.05	2.66	m	**3.84**
joint reinforcement; 175 mm wide	0.01	1.13	0.07	4.18	m	**5.38**
joint reinforcement; 225 mm wide	0.01	1.13	0.09	5.37	m	**6.59**
Hyload pitch polymer damp-proof course; lapped joints; in cement mortar						
Ancillaries						
103 mm wide; horizontal	0.01	1.13	0.07	6.00	m	**7.20**
103 mm wide; vertical	0.02	2.27	0.11	6.00	m	**8.38**
215 mm wide; horizontal	0.03	3.40	0.17	13.47	m	**17.04**
215 mm wide; vertical	0.04	4.53	0.27	13.47	m	**18.27**
328 mm wide; horizontal	0.04	4.53	0.25	19.42	m	**24.20**
328 mm wide; vertical	0.06	6.80	0.38	19.42	m	**26.60**
Preformed closed cell joint filler; pointing with polysulphide sealant						
Ancillaries						
movement joints; 12 mm filler 90 mm wide; 12 × 12 sealant one side	0.06	6.80	0.37	0.71	m	**7.88**
movement joints; 12 mm filler 200 mm wide; 12 × 12 sealant one side	0.07	7.94	0.43	1.07	m	**9.44**
Dritherm cavity insulation						
Infills						
50 mm thick	0.04	4.53	0.27	39.56	m^2	**44.36**
75 mm thick	0.05	5.67	0.30	41.82	m^2	**47.79**
Concrete						
Infills						
50 mm thick	0.06	6.80	0.38	4.54	m^2	**11.72**
Galvanized steel wall ties						
Fixings and ties						
vertical twist strip type; 900 mm horizontal and 450 mm vertical staggered spacings	0.02	2.27	0.10	1.38	m^2	**3.75**
Stainless steel wall ties						
Fixings and ties						
vertical twist strip type; 900 mm horizontal and 450 mm vertical staggered spacings	0.02	2.27	0.10	3.04	m^2	**5.41**

CLASS V: PAINTING

Item	Gang hours	Labour £	Plant £	Material £	Unit	Total rate £
RESOURCES – LABOUR						
Painting gang						
1 ganger (skill rate 3)		16.16				
3 painters (skill rate 3)		45.63				
1 unskilled operative (general)		12.71				
Total Gang Rate/Hour		£ 74.50				
RESOURCES – PLANT						
Painting						
1.5 kVA diesel generator			4.43			
transformers/cables; junction box			0.64			
4.5' electric grinder			0.40			
transit van (50% of time)			6.20			
ladders			1.15			
Total Rate/Hour			£ 12.82			
LEAD BASED PRIMER PAINT						
One coat calcium plumbate primer						
Metal						
upper surfaces inclined at an angle ne 30° to the horizontal	0.03	1.86	0.32	0.66	m²	2.84
upper surfaces inclined at an angle 30–60° to the horizontal	0.03	1.86	0.32	0.66	m²	2.84
surfaces inclined at an angle exceeding 60° to the horizontal	0.03	1.86	0.32	0.66	m²	2.84
soffit surfaces and lower surfaces inclined at an angle ne 60° to the horizontal	0.03	2.23	0.39	0.66	m²	3.28
surfaces of width ne 300 mm	0.01	0.75	0.14	0.20	m	1.09
surfaces of width 300 mm–1 m	0.03	1.86	0.32	0.43	m	2.61
Metal sections	0.03	2.23	0.38	0.66	m²	3.27
Pipework	0.03	2.23	0.38	0.66	m²	3.27
IRON BASED PRIMER PAINT						
One coat iron oxide primer						
Metal						
upper surfaces inclined at an angle ne 30° to the horizontal	0.03	1.86	0.32	0.55	m²	2.73
upper surfaces inclined at an angle 30–60° to the horizontal	0.03	1.86	0.32	0.55	m²	2.73
surfaces inclined at an angle exceeding 60° to the horizontal	0.03	1.86	0.32	0.55	m²	2.73
soffit surfaces and lower surfaces inclined at an angle ne 60° to the horizontal	0.03	2.23	0.39	0.55	m²	3.17
surfaces of width ne 300 mm	0.01	0.75	0.14	0.17	m	1.06
surfaces of width 300 mm–1 m	0.03	1.86	0.32	0.37	m	2.55
Metal sections	0.03	2.23	0.38	0.55	m²	3.16
Pipework	0.03	2.23	0.38	0.55	m²	3.16

CLASS V: PAINTING

Item	Gang hours	Labour £	Plant £	Material £	Unit	Total rate £
ACRYLIC PRIMER PAINT						
One coat acrylic wood primer						
Timber						
upper surfaces inclined at an angle ne 30° to the horizontal	0.02	1.49	0.26	0.44	m^2	**2.19**
upper surfaces inclined at an angle 30–60° to the horizontal	0.02	1.49	0.26	0.44	m^2	**2.19**
surfaces inclined at an angle exceeding 60° to the horizontal	0.02	1.49	0.24	0.44	m^2	**2.17**
soffit surfaces and lower surfaces inclined at an angle ne 60° to the horizontal	0.03	1.86	0.32	0.44	m^2	**2.62**
surfaces of width ne 300 mm	0.01	0.75	0.13	0.13	m	**1.01**
surfaces of width 300 mm–1 m	0.03	1.86	0.32	0.29	m	**2.47**
GLOSS PAINT						
One coat calcium plumbate primer; one undercoat and one finishing coat of gloss paint						
Metal						
upper surfaces inclined at an angle ne 30° to the horizontal	0.07	5.21	0.91	1.63	m^2	**7.75**
upper surfaces inclined at an angle 30–60° to the horizontal	0.08	5.59	0.97	1.63	m^2	**8.19**
surfaces inclined at an angle exceeding 60° to the horizontal	0.08	5.96	1.03	1.63	m^2	**8.62**
soffit surfaces and lower surfaces inclined at an angle ne 60° to the horizontal	0.09	6.33	1.08	1.63	m^2	**9.04**
surfaces of width ne 300 mm	0.03	1.86	0.32	0.49	m	**2.67**
surfaces of width 300 mm–1 m	0.05	3.73	0.65	1.06	m	**5.44**
Metal sections	0.09	6.71	1.13	1.63	m^2	**9.47**
Pipework	0.09	6.71	1.13	1.63	m^2	**9.47**
One coat acrylic wood primer; one undercoat and one finishing coat of gloss paint						
Timber						
upper surfaces inclined at an angle ne 30° to the horizontal	0.07	4.84	0.78	1.41	m^2	**7.03**
upper surfaces inclined at an angle 30–60° to the horizontal	0.07	5.21	0.91	1.41	m^2	**7.53**
surfaces inclined at an angle exceeding 60° to the horizontal	0.08	5.59	0.97	1.42	m^2	**7.98**
soffit surfaces and lower surfaces inclined at an angle ne 60° to the horizontal	0.08	5.96	1.03	1.41	m^2	**8.40**
surfaces of width ne 300 mm	0.03	1.86	0.18	0.42	m	**2.46**
surfaces of width 300 mm–1 m	0.05	3.73	0.65	0.92	m	**5.30**

CLASS V: PAINTING

Item	Gang hours	Labour £	Plant £	Material £	Unit	Total rate £
GLOSS PAINT – cont						
One coat calcium plumbate primer; two undercoats and one finishing coat of gloss paint						
Metal						
upper surfaces inclined at an angle ne 30° to the horizontal	0.06	4.47	0.77	1.94	m^2	**7.18**
upper surfaces inclined at an angle 30–60° to the horizontal	0.07	4.84	0.83	1.94	m^2	**7.61**
surfaces inclined at an angle exceeding 60° to the horizontal	0.07	5.21	0.91	1.94	m^2	**8.06**
soffit surfaces and lower surfaces inclined at an angle ne 60° to the horizontal	0.08	5.59	0.97	1.89	m^2	**8.45**
surfaces of width ne 300 mm	0.03	2.23	0.39	0.58	m	**3.20**
surfaces of width 300 mm–1 m	0.06	4.47	0.77	1.26	m	**6.50**
One coat acrylic wood primer; two undercoats and one finishing coat of gloss paint						
Timber						
upper surfaces inclined at an angle ne 30° to the horizontal	0.07	4.84	0.83	1.72	m^2	**7.39**
upper surfaces inclined at an angle 30–60° to the horizontal	0.07	5.21	0.91	1.72	m^2	**7.84**
surfaces inclined at an angle exceeding 60° to the horizontal	0.08	5.59	0.97	1.72	m^2	**8.28**
soffit surfaces and lower surfaces inclined at an angle ne 60° to the horizontal	0.08	5.96	0.94	1.72	m^2	**8.62**
surfaces of width ne 300 mm	0.03	2.23	0.91	0.52	m	**3.66**
surfaces of width 300 mm–1 m	0.06	4.47	0.77	1.12	m	**6.36**
One coat alkali resisting primer; two undercoats and one finishing coat of gloss paint						
Smooth concrete						
upper surfaces inclined at an angle ne 30° to the horizontal	0.04	2.61	0.45	1.71	m^2	**4.77**
upper surfaces inclined at an angle 30–60° to the horizontal	0.04	2.98	0.51	1.71	m^2	**5.20**
surfaces inclined at an angle exceeding 60° to the horizontal	0.05	3.35	0.57	1.72	m^2	**5.64**
soffit surfaces and lower surfaces inclined at an angle ne 60° to the horizontal	0.05	3.73	0.65	1.71	m^2	**6.09**
surfaces of width ne 300 mm	0.02	1.12	0.20	0.51	m	**1.83**
surfaces of width 300 mm–1 m	0.03	2.23	0.39	1.11	m	**3.73**
Brickwork and rough concrete						
upper surfaces inclined at an angle ne 30° to the horizontal	0.04	2.98	0.51	1.73	m^2	**5.22**
upper surfaces inclined at an angle 30–60° to the horizontal	0.05	3.35	0.57	1.74	m^2	**5.66**
surfaces inclined at an angle exceeding 60° to the horizontal	0.05	3.73	0.65	1.78	m^2	**6.16**

Unit Costs – Civil Engineering Works

CLASS V: PAINTING

Item	Gang hours	Labour £	Plant £	Material £	Unit	Total rate £
soffit surfaces and lower surfaces inclined at an angle ne 60° to the horizontal	0.06	4.10	0.71	1.78	m²	6.59
surfaces of width ne 300 mm	0.02	1.49	0.26	0.53	m	2.28
surfaces of width 300 mm–1 m	0.04	2.61	0.45	1.15	m	4.21
Blockwork						
upper surfaces inclined at an angle ne 30° to the horizontal	0.06	4.10	0.71	1.78	m²	6.59
soffit surfaces and lower surfaces inclined at an angle ne 60° to the horizontal	0.06	4.47	0.77	1.78	m²	7.02
surfaces of width ne 300 mm	0.03	1.86	0.32	0.53	m	2.71
surfaces of width 300 mm–1 m	0.04	2.98	0.51	1.16	m	4.65
Two coats anti-condensation paint						
Metal						
upper surfaces inclined at an angle ne 30° to the horizontal	0.08	5.59	0.97	2.05	m²	8.61
upper surfaces inclined at an angle 30–60° to the horizontal	0.08	5.59	0.97	2.05	m²	8.61
surfaces inclined at an angle exceeding 60° to the horizontal	0.08	5.59	0.97	2.05	m²	8.61
soffit surfaces and lower surfaces inclined at an angle ne 60° to the horizontal	0.08	5.96	1.03	2.05	m²	9.04
surfaces of width ne 300 mm	0.03	2.23	0.39	0.61	m	3.23
surfaces of width 300 mm–1 m	0.05	3.73	0.65	1.33	m	5.71
Metal sections	0.09	6.71	1.16	2.05	m²	9.92

EMULSION PAINT

One thinned coat, two coats vinyl emulsion paint
Smooth concrete

Item	Gang hours	Labour £	Plant £	Material £	Unit	Total rate £
upper surfaces inclined at an angle ne 30° to the horizontal	0.07	4.84	0.83	1.15	m²	6.82
upper surfaces inclined at an angle 30–60° to the horizontal	0.07	5.21	0.91	1.15	m²	7.27
surfaces inclined at an angle exceeding 60° to the horizontal	0.08	5.59	0.97	1.15	m²	7.71
soffit surfaces and lower surfaces inclined at an angle ne 60° to the horizontal	0.08	5.96	1.03	1.15	m²	8.14
surfaces of width ne 300 mm	0.03	2.23	0.39	0.34	m	2.96
surfaces of width 300 mm–1 m	0.07	5.21	0.91	0.75	m	6.87
Brickwork and rough concrete						
upper surfaces inclined at an angle ne 30° to the horizontal	0.08	5.59	0.97	1.18	m²	7.74
upper surfaces inclined at an angle 30–60° to the horizontal	0.08	5.96	1.03	1.18	m²	8.17
surfaces inclined at an angle exceeding 60° to the horizontal	0.09	6.33	1.08	1.18	m²	8.59
soffit surfaces and lower surfaces inclined at an angle ne 60° to the horizontal	0.09	6.71	1.16	1.18	m²	9.05
surfaces of width ne 300 mm	0.03	2.23	0.39	0.35	m	2.97
surfaces of width 300 mm–1 m	0.08	5.96	1.03	0.81	m	7.80

CLASS V: PAINTING

Item	Gang hours	Labour £	Plant £	Material £	Unit	Total rate £
EMULSION PAINT – cont						
One thinned coat, two coats vinyl emulsion paint – cont						
Blockwork						
surfaces inclined at an angle exceeding 60° to the horizontal	0.09	6.71	1.16	1.21	m²	9.08
soffit surfaces and lower surfaces inclined at an angle ne 60° to the horizontal	0.10	7.45	1.28	1.21	m²	9.94
surfaces of width ne 300 mm	0.04	2.98	0.51	0.36	m	3.85
surfaces of width 300 mm–1 m	0.08	5.96	1.03	0.79	m	7.78
CEMENT PAINT						
Two coats masonry paint						
Smooth concrete						
upper surfaces inclined at an angle ne 30° to the horizontal	0.07	4.84	0.83	0.93	m²	6.60
upper surfaces inclined at an angle 30–60° to the horizontal	0.07	5.21	0.91	0.93	m²	7.05
surfaces inclined at an angle exceeding 60° to the horizontal	0.08	5.59	0.97	0.93	m²	7.49
soffit surfaces and lower surfaces inclined at an angle ne 60° to the horizontal	0.08	5.96	1.03	0.93	m²	7.92
surfaces of width ne 300 mm	0.03	2.23	0.39	0.28	m	2.90
surfaces of width 300 mm–1 m	0.07	5.21	0.91	0.61	m	6.73
Brickwork and rough concrete						
upper surfaces inclined at an angle ne 30° to the horizontal	0.08	5.59	0.97	1.08	m²	7.64
upper surfaces inclined at an angle 30–60° to the horizontal	0.08	5.96	1.03	1.08	m²	8.07
surfaces inclined at an angle exceeding 60° to the horizontal	0.09	6.33	1.08	1.08	m²	8.49
soffit surfaces and lower surfaces inclined at an angle ne 60° to the horizontal	0.09	6.71	1.16	1.08	m²	8.95
surfaces of width ne 300 mm	0.03	2.23	0.39	0.32	m	2.94
surfaces of width 300 mm–1 m	0.08	5.96	1.03	0.70	m	7.69
Blockwork						
surfaces inclined at an angle exceeding 60° to the horizontal	0.09	6.71	1.16	1.40	m²	9.27
soffit surfaces and lower surfaces inclined at an angle ne 60° to the horizontal	0.10	7.45	1.28	1.40	m²	10.13
surfaces of width ne 300 mm	0.04	2.98	0.51	0.42	m	3.91
surfaces of width 300 mm–1 m	0.08	5.96	1.03	0.91	m	7.90

CLASS V: PAINTING

Item	Gang hours	Labour £	Plant £	Material £	Unit	Total rate £
One thinned coat, two coats concrete floor paint						
Smooth concrete						
upper surfaces inclined at an angle ne 30° to the horizontal	0.07	4.84	0.83	3.29	m²	**8.96**
upper surfaces inclined at an angle 30–60° to the horizontal	0.07	5.21	0.91	3.29	m²	**9.41**
surfaces inclined at an angle exceeding 60° to the horizontal	0.08	5.59	0.97	3.29	m²	**9.85**
soffit surfaces and lower surfaces inclined at an angle ne 60° to the horizontal	0.08	5.96	1.03	2.87	m²	**9.86**
surfaces of width ne 300 mm	0.03	2.23	0.39	1.02	m	**3.64**
surfaces of width 300 mm–1 m	0.07	5.21	0.91	2.14	m	**8.26**
Additional coats						
width exceeding 1000 mm	0.03	2.23	0.39	1.32	m²	**3.94**
surfaces of width ne 300 mm	0.01	0.75	0.14	0.40	m	**1.29**
surfaces of width 300 mm–1 m	0.02	1.12	0.32	0.86	m	**2.30**
EPOXY OR POLYURETHANE PAINT						
Blast clean to BS 7079; one coat zinc chromate etc.h primer, two coats zinc phosphate CR/alkyd undercoat off site; one coat MIO CR undercoat and one coat CR finish on site						
Metal sections	–	–	–	–	m²	**21.68**
Blast clean to BS 7079; one coat zinc rich 2 pack primer, one coat MIO high build epoxy 2 pack paint off site; one coat polyurethane 2 pack undercoat and one coat polyurethane 2 pack finish on site						
Metal sections	–	–	–	–	m²	**14.20**
Two coats clear polyurethane varnish						
Timber						
upper surfaces inclined at an angle ne 30° to the horizontal	0.04	2.98	0.51	0.70	m²	**4.19**
upper surfaces inclined at an angle 30–60° to the horizontal	0.05	3.35	0.57	0.70	m²	**4.62**
surfaces inclined at an angle exceeding 60° to the horizontal	0.05	3.73	0.65	0.70	m²	**5.08**
soffit surfaces and lower surfaces inclined at an angle ne 60° to the horizontal	0.06	4.47	0.77	0.70	m²	**5.94**
surfaces of width ne 300 mm	0.03	1.86	0.32	0.21	m	**2.39**
surfaces of width 300 mm–1 m	0.05	3.73	0.65	0.45	m	**4.83**

CLASS V: PAINTING

Item	Gang hours	Labour £	Plant £	Material £	Unit	Total rate £
EPOXY OR POLYURETHANE PAINT – cont						
Two coats colour stained polyurethane varnish						
Timber						
upper surfaces inclined at an angle ne 30° to the horizontal	0.04	2.98	0.51	1.10	m²	**4.59**
upper surfaces inclined at an angle 30–60° to the horizontal	0.05	3.35	0.57	1.10	m²	**5.02**
surfaces inclined at an angle exceeding 60° to the horizontal	0.05	3.73	0.65	1.10	m²	**5.48**
soffit surfaces and lower surfaces inclined at an angle ne 60° to the horizontal	0.06	4.47	0.77	1.10	m²	**6.34**
surfaces of width ne 300 mm	0.03	1.86	0.32	0.33	m	**2.51**
surfaces of width 300 mm–1 m	0.05	3.73	0.65	0.71	m	**5.09**
Three coats colour stained polyurethane varnish						
Timber						
upper surfaces inclined at an angle ne 30° to the horizontal	0.06	4.10	0.71	1.64	m²	**6.45**
upper surfaces inclined at an angle 30–60° to the horizontal	0.07	4.84	0.83	1.64	m²	**7.31**
surfaces inclined at an angle exceeding 60° to the horizontal	0.07	5.21	0.91	1.64	m²	**7.76**
soffit surfaces and lower surfaces inclined at an angle ne 60° to the horizontal	0.09	6.33	1.08	1.64	m²	**9.05**
surfaces of width ne 300 mm	0.04	2.61	0.45	0.49	m	**3.55**
surfaces of width 300 mm–1 m	0.07	5.21	0.91	1.07	m	**7.19**
One coat hardwood stain basecoat and two coats hardwood woodstain						
Timber						
upper surfaces inclined at an angle ne 30° to the horizontal	0.06	4.47	0.77	1.47	m²	**6.71**
upper surfaces inclined at an angle 30–60° to the horizontal	0.07	4.84	0.83	1.47	m²	**7.14**
surfaces inclined at an angle exceeding 60° to the horizontal	0.07	5.21	0.91	1.47	m²	**7.59**
soffit surfaces and lower surfaces inclined at an angle ne 60° to the horizontal	0.08	5.96	1.03	1.58	m²	**8.57**
surfaces of width ne 300 mm	0.04	2.61	0.45	0.44	m	**3.50**
surfaces of width 300 mm–1 m	0.07	5.21	0.91	0.95	m	**7.07**

CLASS V: PAINTING

Item	Gang hours	Labour £	Plant £	Material £	Unit	**Total rate £**
BITUMINOUS OR PRESERVATIVE PAINT						
Two coats golden brown wood preservative						
Timber						
upper surfaces inclined at an angle ne 30° to the horizontal	0.05	3.35	0.57	1.39	m^2	**5.31**
upper surfaces inclined at an angle 30–60° to the horizontal	0.05	3.73	0.65	1.39	m^2	**5.77**
surfaces inclined at an angle exceeding 60° to the horizontal	0.06	4.10	0.71	1.39	m^2	**6.20**
soffit surfaces and lower surfaces inclined at an angle ne 60° to the horizontal	0.07	4.84	0.83	1.39	m^2	**7.06**
surfaces of width ne 300 mm	0.03	2.23	0.39	0.42	m	**3.04**
surfaces of width 300 mm–1 m	0.06	4.10	0.71	0.90	m	**5.71**
Two coats bituminous paint						
Metal sections	0.08	5.96	1.03	0.83	m^2	**7.82**
Pipework	0.09	6.71	1.22	0.83	m^2	**8.76**

CLASS W: WATERPROOFING

Item	Gang hours	Labour £	Plant £	Material £	Unit	Total rate £
RESOURCES – LABOUR						
Roofing – cladding gang						
1 ganger/chargehand (skill rate 3) – 50% of time		8.08				
2 skilled operatives (skill rate 3)		30.42				
1 unskilled operative (general) – 50% of time		6.36				
Total Gang Rate/Hour		£ 44.85				
Damp-proofing gang						
1 ganger (skill rate 4)		14.56				
1 skilled operative (skill rate 4)		13.61				
1 unskilled labour (general)		12.71				
Total Gang Rate/Hour		£ 40.88				
Roofing – asphalt gang						
1 ganger/chargehand (skill rate 4) – 50% of time		7.28				
1 skilled operative (skill rate 4)		13.61				
1 unskilled operative (general)		12.71				
Total Gang Rate/Hour		£ 33.60				
Tanking – asphalt gang						
1 ganger/chargehand (skill rate 4) – 50% of time		7.28				
1 skilled operative (skill rate 4)		13.61				
1 unskilled operative (general)		12.71				
Total Gang Rate/Hour		£ 33.60				
Tanking – waterproof sheeting gang						
1 skilled operative (skill rate 4)		13.61				
Total Gang Rate/Hour		£ 13.61				
Tanking – rendering gang						
1 ganger/chargehand (skill rate 4)		14.56				
1 skilled operative (skill rate 4)		13.61				
1 unskilled operative (generally)		12.71				
Total Gang Rate/Hour		£ 40.88				
Potective layers – flexible sheeting, sand and pea gravel coverings gang						
1 unskilled operative (generally)		12.71				
Total Gang Rate/Hour		£ 12.71				
Protective layers – screed gang						
1 ganger/chargehand (skill rate 4)		14.56				
1 skilled operative (skill rate 4)		13.61				
1 unskilled operative (general)		12.71				
Total Gang Rate/Hour		£ 40.88				
Sprayed or brushed waterproofing gang						
1 ganger/chargehand (skill rate 4) – 30% of time		4.37				
1 skilled operative (skill rate 4)		13.61				
Total Gang Rate/Hour		£ 17.98				
RESOURCES – PLANT						
Damp-proofing						
2 t dumper (50% of time)			3.30			
Total Rate/Hour		£	3.30			

Unit Costs – Civil Engineering Works

CLASS W: WATERPROOFING

Item	Gang hours	Labour £	Plant £	Material £	Unit	Total rate £
Tanking – asphalt						
tar boiler (50% of time)			18.35			
2 t dumper (50% of time)			3.30			
Total Rate/Hour		£	**21.65**			
Tanking – waterproof sheeting						
2 t dumper (50% of time)			3.30			
Total Rate/Hour		£	**3.30**			
Tanking – rendering						
mixer			2.08			
2 t dumper (50% time)			3.30			
Total Rate/Hour		£	**5.38**			
Protective layers – flexible sheeting, sand and pea gravel coverings						
2 t dumper (50% of time)			3.30			
Total Rate/Hour		£	**3.30**			
Protective layers – screed						
mixer			2.08			
2 t dumper (50% time)			3.30			
Total Rate/Hour		£	**5.38**			
Sprayed or brushed waterproofing						
2 t dumper (50% of time)			3.30			
Total Rate/Hour		£	**3.30**			
DAMP-PROOFING						
Waterproof sheeting						
0.3 mm polythene sheet						
ne 300 mm wide	0.01	0.12	0.03	0.39	m	**0.54**
300 mm–1 m wide	0.01	0.14	0.03	0.77	m	**0.94**
on horizontal or included surfaces	0.01	0.16	0.04	1.18	m^2	**1.38**
on vertical surfaces	0.01	0.14	0.03	1.18	m^2	**1.35**

NOTES

Asphalt roofing
Work has been presented in more detail than CESMM3 to allow the user greater freedom to access an appropriate rate to suit the complexity of their work.

CLASS W: WATERPROOFING

Item	Gang hours	Labour £	Plant £	Material £	Unit	Total rate £
TANKING						
Asphalt						
13 mm Mastic asphalt to BS 6925, Type T 1097; two coats; on concrete surface						
upper surfaces inclined at an angle ne 30° to the horizontal	0.45	15.12	10.62	10.17	m²	**35.91**
upper surfaces inclined at an angle 30–60° to the horizontal	0.60	20.16	14.16	10.17	m²	**44.49**
upper surfaces inclined at an angle exceeding 60° to the horizontal	1.00	33.60	23.60	10.17	m²	**67.37**
curved surfaces	1.20	40.32	28.32	10.17	m²	**78.81**
domed surfaces	1.50	50.40	35.40	10.17	m²	**95.97**
ne 300 mm wide	0.20	6.72	4.72	3.07	m	**14.51**
300 mm–1 m wide	0.45	15.12	10.62	10.17	m	**35.91**
20 mm Mastic asphalt to BS 6925, Type T 1097; two coats; on concrete surface						
upper surfaces inclined at an angle ne 30° to the horizontal	0.50	16.80	11.80	15.62	m²	**44.22**
upper surfaces inclined at an angle 30–60° to the horizontal	0.70	23.52	16.52	15.62	m²	**55.66**
upper surfaces inclined at an angle exceeding 60° to the horizontal	1.20	40.32	28.32	15.62	m²	**84.26**
curved surfaces	1.40	47.04	33.04	15.62	m²	**95.70**
domed surfaces	1.75	58.80	41.30	15.62	m²	**115.72**
ne 300 mm wide	0.23	7.73	5.43	4.68	m	**17.84**
300 mm–1 m wide	0.50	16.80	11.80	15.62	m	**44.22**
13 mm Mastic asphalt to BS 6925, Type T 1097; two coats; on brickwork surface; raking joints to form key						
upper surfaces inclined at an angle 30–60° to the horizontal	0.90	30.24	14.23	15.69	m²	**60.16**
upper surfaces inclined at an angle exceeding 60° to the horizontal	1.30	43.68	30.68	15.69	m²	**90.05**
curved surfaces	1.50	50.40	35.40	15.69	m²	**101.49**
domed surfaces	1.80	60.48	42.48	15.69	m²	**118.65**
ne 300 mm wide	0.30	10.08	4.72	4.71	m	**19.51**
300 mm–1 m wide	0.75	25.20	17.79	15.69	m	**58.68**
Waterproof sheeting						
Bituthene 3000; lapped joints						
upper surfaces inclined at an angle ne 30° to the horizontal	0.05	2.04	0.17	9.37	m²	**11.58**
upper surfaces inclined at an angle 30–60° to the horizontal	0.05	2.04	0.17	9.37	m²	**11.58**
ne 300 mm wide	0.03	1.02	0.08	2.47	m	**3.57**
300 mm–1 m wide	0.04	1.64	0.13	5.26	m	**7.03**
Bituthene 3000; lapped joints; primer coat						
upper surfaces inclined at an angle exceeding 60° to the horizontal	0.06	2.45	0.20	10.21	m²	**12.86**

CLASS W: WATERPROOFING

Item	Gang hours	Labour £	Plant £	Material £	Unit	Total rate £
Rendering in waterproof cement mortar						
19 mm render in waterproof cement mortar (1:3); two coat work						
upper surfaces inclined at an angle exceeding 60° to the horizontal	0.11	4.50	0.59	7.82	m²	**12.91**
ne 300 mm wide	0.07	2.86	0.38	2.90	m	**6.14**
300 mm–1 m wide	0.11	4.50	0.59	5.27	m	**10.36**
32 mm render in waterproof cement mortar (1:3); one coat work						
upper surfaces inclined at an angle ne 30° to the horizontal	0.11	4.50	0.59	11.66	m²	**16.75**
ne 300 mm wide	0.07	2.86	0.38	3.39	m	**6.63**
300 mm–1 m wide	0.11	4.50	0.59	5.91	m	**11.00**
ROOFING						
Asphalt						
13 mm Mastic asphalt to BS 6925 Type R 988; two coats; on concrete surface						
ne 300 mm wide	0.20	6.72	4.72	2.83	m	**14.27**
20 mm Mastic asphalt to BS 6925 Type R 988; two coats; on concrete surface						
upper surfaces inclined at an angle ne 30° to the horizontal	0.50	16.80	11.80	15.12	m²	**43.72**
upper surfaces inclined at an angle 30–60° to the horizontal	0.70	23.52	16.52	15.12	m²	**55.16**
surfaces inclined at an angle exceeding 60° to the horizontal	1.20	40.32	28.32	15.12	m²	**83.76**
curved surfaces	1.50	50.40	35.40	15.12	m²	**100.92**
domed surfaces	1.75	58.80	41.30	15.12	m²	**115.22**
ne 300 mm wide	0.23	7.73	5.43	4.54	m	**17.70**
300 mm–1 m wide	0.50	16.80	11.80	15.12	m	**43.72**
Extra for:						
10 mm thick limestone chippings bedded in hot bitumen	0.05	1.68	0.47	2.11	m²	**4.26**
dressing with solar reflective paint	0.05	1.68	0.47	5.87	m²	**8.02**
300 × 300 × 8 mm GRP tiles bedded in hot bitumen	0.30	10.08	2.82	120.45	m²	**133.35**
PROTECTIVE LAYERS						
Flexible sheeting						
3 mm Servi-pak protection board to Bituthene						
upper surfaces inclined at an angle ne 30° to the horizontal	0.20	2.54	0.66	18.56	m²	**21.76**
ne 300 mm wide	0.10	1.27	0.33	7.92	m	**9.52**
300 mm–1 m wide	0.20	2.54	0.66	18.56	m	**21.76**

CLASS W: WATERPROOFING

Item	Gang hours	Labour £	Plant £	Material £	Unit	Total rate £
PROTECTIVE LAYERS – cont						
Flexible sheeting – cont						
3 mm Servi-pak protection board to Bituthene; fixing with adhesive dabs						
upper surfaces inclined at an angle exceeding 60° to the horizontal	0.25	3.18	0.82	19.65	m²	**23.65**
ne 300 mm wide	0.12	1.53	0.40	8.34	m	**10.27**
300 mm–1 m wide	0.25	3.18	0.82	19.65	m	**23.65**
6 mm Servi-pak protection board to Bituthene						
upper surfaces inclined at an angle ne 30° to the horizontal	0.20	2.54	0.66	34.44	m²	**37.64**
ne 300 mm wide	0.10	1.27	0.33	14.18	m	**15.78**
300 mm–1 m wide	0.20	2.54	0.66	34.44	m	**37.64**
6 mm Servi-pak protection board to Bituthene; fixing with adhesive dabs						
upper surfaces inclined at an angle exceeding 60° to the horizontal	0.25	3.18	0.82	36.12	m²	**40.12**
ne 300 mm wide	0.12	1.53	0.40	15.21	m	**17.14**
300 mm–1 m wide	0.25	3.18	0.82	36.12	m	**40.12**
Sand covering						
25 mm thick						
upper surfaces inclined at an angle ne 30° to the horizontal	0.02	0.25	0.03	3.55	m²	**3.83**
Pea gravel covering						
50 mm thick						
upper surfaces inclined at an angle ne 30° to the horizontal	0.02	0.25	0.03	11.56	m²	**11.84**
Sand and cement screed						
50 mm screed in cement mortar (1:4); one coat work						
upper surfaces inclined at an angle ne 30° to the horizontal	0.13	5.31	0.62	17.70	m²	**23.63**
SPRAYED OR BRUSHED WATERPROOFING						
Two coats RIW liquid asphaltic composition						
on horizontal or vertical surfaces	0.06	1.08	0.20	70.35	m²	**71.63**
Two coats Aquaseal						
on horizontal or vertical surfaces	0.06	1.08	0.20	1.03	m²	**2.31**
One coat Ventrot primer; one coat Ventrot hot applied damp-proof membrane						
on horizontal or vertical surfaces	0.05	0.90	0.17	5.47	m²	**6.54**

Unit Costs – Civil Engineering Works

CLASS X: MISCELLANEOUS WORK

Item	Gang hours	Labour £	Plant £	Material £	Unit	Total rate £
RESOURCES – LABOUR						
Fencing/barrier gang						
1 ganger/chargehand (skill rate 4) – 50% of time		7.28				
1 skilled operative (skill rate 4) – 50% of time		6.80				
1 unskilled operative (general)		12.71				
1 plant operator (skill rate 4)		14.88				
Total Gang Rate/Hour	£	**41.68**				
Safety fencing gang						
1 ganger/chargehand (skill rate 4)		14.56				
1 skilled operative (skill rate 4)		13.61				
2 unskilled operatives (general)		25.42				
1 plant operator (skill rate 4)		14.88				
Total Gang Rate/Hour	£	**68.47**				
Guttering gang						
2 skilled operatives (skill rate 4)		27.22				
Total Gang Rate/Hour	£	**27.22**				
Rock filled gabions gang						
1 ganger/chargehand (skill rate 4)		14.56				
4 unskilled operatives (general)		50.84				
1 plant operator (skill rate 3) – 50% of time		8.28				
Total Gang Rate/Hour	£	**73.68**				
RESOURCES – PLANT						
Fencing/barriers						
agricultural type tractor; fencing auger			21.80			
gas oil for ditto			6.30			
drop sided trailer; two axles			0.35			
power tools (fencing)			1.15			
Total Rate/Hour		£	**29.60**			
Guttering						
ladders			1.15			
Total Rate/Hour		£	**1.15**			
Rock filled gabions						
16 t crawler backacter (50% of time)			23.80			
Total Rate/Hour		£	**23.80**			
FENCES						
Timber fencing						
Timber post and wire						
1.20 m high; DfT type 3; timber posts, driven; cleft chestnut paling	0.07	2.92	2.07	9.22	m	**14.21**
0.90 m high; DfT type 4; galvanized rectangular wire mesh	0.13	5.42	3.85	9.22	m	**18.49**
1.275 m high; DfT type 1; galvanized wire, 2 barbed, 4 plain	0.06	2.50	1.78	5.13	m	**9.41**
1.275 m high; DfT type 2; galvanized wire, 2 barbed, 4 plain	0.06	2.50	1.78	7.69	m	**11.97**

CLASS X: MISCELLANEOUS WORK

Item	Gang hours	Labour £	Plant £	Material £	Unit	Total rate £
FENCES – cont						
Timber fencing – cont						
Concrete post and wire						
1.20 m high; DfT type 3; timber posts, driven; cleft chestnut paling	0.07	2.92	2.07	9.22	m	**14.21**
0.90 m high; DfT type 4; galvanized rectangular wire mesh	0.13	5.42	3.85	9.22	m	**18.49**
1.275 m high; DfT type 1; galvanized wire, 2 barbed, 4 plain	0.06	2.50	1.78	5.13	m	**9.41**
1.275 m high; DfT type 2; galvanized wire, 2 barbed, 4 plain	0.06	2.50	1.78	7.69	m	**11.97**
Metal post and wire						
1.20 m high; DfT type 3; timber posts, driven; cleft chestnut paling	0.07	2.92	2.07	9.22	m	**14.21**
0.90 m high; DfT type 4; galvanized rectangular wire mesh	0.13	5.42	3.85	9.22	m	**18.49**
1.275 m high; DfT type 1; galvanized wire, 2 barbed, 4 plain	0.06	2.50	1.78	5.13	m	**9.41**
1.275 m high; DfT type 2; galvanized wire, 2 barbed, 4 plain	0.06	2.50	1.78	7.69	m	**11.97**
Timber post and wire						
1.20 m high; DfT type 3; timber posts, driven; cleft chestnut paling	0.07	2.92	2.07	9.22	m	**14.21**
0.90 m high; DfT type 4; galvanized rectangular wire mesh	0.13	5.42	3.85	9.22	m	**18.49**
1.275 m high; DfT type 1; galvanized wire, 2 barbed, 4 plain	0.06	2.50	1.78	5.13	m	**9.41**
1.275 m high; DfT type 2; galvanized wire, 2 barbed, 4 plain	0.06	2.50	1.78	7.69	m	**11.97**
Timber closeboarded; concrete posts						
Timber closeboarded						
1.80 m high; 125 × 125 mm posts	0.30	12.50	8.88	41.00	m	**62.38**
Wire rope safety fencing to BS 5750; based on 600 m lengths						
Metal crash barriers						
600 mm high; 4 wire ropes; long line posts at 2.40 m general spacings, driven	0.16	10.96	4.74	57.98	m	**73.68**
600 mm high; 4 wire ropes; short line posts at 2.40 m general spacings, 400 × 400 × 600 mm concrete footing	0.27	18.49	7.99	60.08	m	**86.56**
600 mm high; 4 wire ropes; short line posts at 2.40 m general spacings, 400 × 400 × 600 mm concrete footing, socketed	0.32	21.91	9.47	60.27	m	**91.65**
600 mm high; 4 wire ropes; short line posts at 2.40 m general spacings, bolted to structure	0.20	13.69	5.92	57.29	m	**76.90**

Unit Costs – Civil Engineering Works

CLASS X: MISCELLANEOUS WORK

Item	Gang hours	Labour £	Plant £	Material £	Unit	Total rate £
Pedestrian guard rails						
Metal guard rails						
1000 mm high; tubular galvanized mild steel to BS 3049, mesh infill (105 swg, 50 × 50 mm mesh; steel posts with concrete footing	0.80	10.80	4.74	71.75	m	**87.29**
Beam safety fencing; based on 600 m lengths						
Metal crash barriers						
600 mm high; untensioned corrugated beam, single sided; long posts at 3.20 m general spacings, driven	0.10	6.85	2.96	69.60	m	**79.41**
600 mm high; untensioned corrugated beam, double sided; long posts at 3.20 m general spacings, driven	0.26	17.80	7.69	128.12	m	**153.61**
600 mm high; untensioned open box beam, single sided; long posts at 3.20 m general spacings, driven	0.10	6.85	2.96	65.06	m	**74.87**
600 mm high; untensioned open box beam, double sided; long posts at 3.20 m general spacings, driven	0.26	17.80	7.69	119.34	m	**144.83**
600 mm high; untensioned open box beam, double height; long posts at 3.20 m general spacings, driven	0.30	20.54	8.88	139.58	m	**169.00**
600 mm high; tensioned corrugated beam, single sided; long posts at 3.20 m general spacings, driven	0.13	8.90	3.85	61.24	m	**73.99**
600 mm high; tensioned corrugated beam, double sided; long posts at 3.20 m general spacings, driven	0.37	25.33	10.95	111.40	m	**147.68**
GATES AND STILES						
Notes						
Refer to Part 7 Series 300.						
Timber field gates						
single; 3.00 m wide × 1.27 m high	1.84	76.68	53.82	251.13	nr	**381.63**
single; 3.60 m wide × 1.27 m high	1.90	79.18	55.58	266.50	nr	**401.26**
single; 4.10 m wide × 1.27 m high	2.00	83.35	58.50	281.88	nr	**423.73**
single; 4.71 m wide × 1.27 m high	2.00	83.35	58.50	307.50	nr	**449.35**
Timber wicket gates						
single; 1.20 m wide × 1.20 m high; DfT Type 1	1.20	50.01	35.10	128.13	nr	**213.24**
single; 1.20 m wide × 1.02 m high; DfT Type 2	1.20	50.01	35.10	179.38	nr	**264.49**
Metal field gates						
single; steel tubular; 3.60 m wide × 1.175 m high	0.30	12.50	8.78	143.50	nr	**164.78**
single; steel tubular; 4.50 m wide × 1.175 m high	0.30	12.50	8.78	164.00	nr	**185.28**
single; steel tubular; half mesh; 3.60 m wide × 1.175 m high	0.30	12.50	8.78	169.13	nr	**190.41**
single; steel tubular; half mesh; 4.50 m wide × 1.175 m high	0.30	12.50	8.78	189.63	nr	**210.91**
single; steel tubular; extra wide; 4.88 m wide × 1.175 m high	0.30	12.50	8.78	194.75	nr	**216.03**
double; steel tubular; 5.02 m wide × 1.175 m high	0.60	25.00	17.55	210.13	nr	**252.68**
Stiles						
1.00 m wide × 1.45 m high; DfT Type 1	1.50	62.51	43.88	107.63	nr	**214.02**
1.00 m wide × 1.45 m high; DfT Type 2	1.40	58.34	40.95	76.88	nr	**176.17**

CLASS X: MISCELLANEOUS WORK

Item	Gang hours	Labour £	Plant £	Material £	Unit	Total rate £
DRAINAGE TO STRUCTURES ABOVE GROUND						
Note						
Outputs are based on heights up to 3 m above ground and exclude time spent on erecting access equipment, but include marking, cutting, drilling to wood, brick or concrete and all fixings. Testing of finished work is not included.						
Output multipliers for labour and plant for heights over 3 m:						
3–6 m × 1.25						
6–9 m × 1.50						
9–12 m × 1.75						
12–15 m × 2.00						
Cast iron gutters and fittings; BS 460 2002						
100 × 75 mm gutters; support brackets	0.25	6.80	0.29	55.20	m	62.29
stop end	0.03	0.73	0.03	43.26	nr	44.02
running outlet	0.04	0.95	0.04	44.28	nr	45.27
angle	0.04	0.95	0.04	69.10	nr	70.09
125 × 75 mm gutters; support brackets	0.30	8.17	0.34	68.39	m	76.90
stop end	0.03	0.87	0.04	50.57	nr	51.48
running outlet	0.04	1.09	0.05	51.99	nr	53.13
angle	0.04	1.09	0.05	83.57	nr	84.71
Cast iron rainwater pipes and fittings; BS 460 2002						
65 mm diameter; support brackets	0.26	7.08	0.30	38.31	m	45.69
bend	0.28	7.62	0.32	20.36	nr	28.30
offset, 75 mm projection	0.30	8.17	0.34	30.58	nr	39.09
offset, 225 mm projection	0.30	8.17	0.34	35.59	nr	44.10
offset, 455 mm projection	0.30	8.17	0.34	97.40	nr	105.91
shoe	0.18	4.90	0.21	32.64	nr	37.75
75 mm diameter; support brackets	0.28	7.62	0.32	51.03	m	58.97
bend	0.28	7.62	0.32	34.24	nr	42.18
offset, 75 mm projection	0.30	8.17	0.34	30.58	nr	39.09
offset, 225 mm projection	0.30	8.17	0.34	69.86	nr	78.37
offset, 455 mm projection	0.30	8.17	0.34	140.69	nr	149.20
shoe	0.18	4.90	0.21	43.33	nr	48.44
PVC-u gutters and fittings; Marley						
116 × 75 mm gutters; support brackets	0.18	4.90	0.21	12.97	m	18.08
stop end	0.05	1.36	0.06	4.25	nr	5.67
running outlet	0.05	1.36	0.06	15.20	nr	16.62
angle	0.14	3.81	0.16	18.55	nr	22.52
150 mm half round gutters; support brackets	0.18	4.90	0.21	15.30	m	20.41
stop end	0.05	1.36	0.06	6.75	nr	8.17
running outlet	0.05	1.36	0.06	18.72	nr	20.14
angle	0.14	3.81	0.16	17.09	nr	21.06

CLASS X: MISCELLANEOUS WORK

Item	Gang hours	Labour £	Plant £	Material £	Unit	Total rate £
PVC-u external rainwater pipes and fittings; Marley						
68 mm diameter; support brackets	0.15	4.08	0.17	7.81	m	**12.06**
bend	0.16	4.36	0.18	8.17	nr	**12.71**
offset bend	0.18	4.90	0.21	4.76	nr	**9.87**
shoe	0.10	2.72	0.12	7.43	nr	**10.27**
110 mm diameter; support brackets	0.15	4.08	0.17	17.80	m	**22.05**
bend	0.16	4.36	0.18	21.09	nr	**25.63**
offset bend	0.18	4.90	0.21	20.40	nr	**25.51**
shoe	0.10	2.72	0.12	17.05	nr	**19.89**
ROCK FILLED GABIONS						
Gabions						
PVC coated galvanized wire mesh box gabions, wire laced; graded broken stone filling						
1.0 × 1.0 m module sizes	0.65	47.89	15.47	61.39	m^3	**124.75**
1.0 × 0.5 m module sizes	0.80	58.94	19.04	74.20	m^3	**152.18**
Heavily galvanized woven wire mesh box gabions, wire laced; graded broken stone filling						
1.0 × 1.0 m module sizes	0.65	47.89	15.47	56.16	m^3	**119.52**
1.0 × 0.5 m module sizes	0.80	58.94	19.04	67.03	m^3	**145.01**
Reno mattresses						
PVC coated woven wire mesh mattresses, wire tied; graded broken stone filling						
230 mm deep	0.15	11.05	3.57	20.62	m^2	**35.24**
Heavily galvanized woven wire mesh, wire tied; graded broken stone filling						
300 mm deep	0.15	11.05	3.57	23.95	m^2	**38.57**

CLASS Y: SEWER AND WATER MAIN RENOVATION AND ANCILARY WORKS

Item	Gang hours	Labour £	Plant £	Material £	Unit	Total rate £
RESOURCES – LABOUR						
Drain repair gang						
1 chargehand pipelayer (skill rate 4) – 50% of time		7.28				
1 skilled operative (skill rate 4)		13.61				
2 unskilled operatives (generally)		25.42				
1 plant operator (skill rate 3)		16.55				
Total Gang Rate/Hour		£ 62.86				
RESOURCES – PLANT						
Drainage						
0.40 m³ hydraulic excavator			37.03			
trench sheets, shores, props etc.			28.88			
2 t dumper (30% of time)			1.98			
compaction/plate roller (30% of time)			0.57			
7.30 m³/min compressor			6.40			
small pump			1.11			
Total Rate/Hour			£ 75.96			
PREPARATION OF EXISTING SEWERS						
Cleaning						
eggshape sewer 1300 mm high	–	–	–	–	m	17.00
Removing intrusions						
brickwork	–	–	–	–	m³	84.80
concrete	–	–	–	–	m³	146.25
reinforced concrete	–	–	–	–	m³	186.75
Plugging laterals with concrete plug						
bore not exceeding 300 mm	–	–	–	–	nr	82.05
bore 450 mm	–	–	–	–	nr	126.40
Plugging laterals with brickwork plug						
bore 750 mm	–	–	–	–	nr	380.20
Local internal repairs to brickwork						
area not exceeding 0.1 m²	–	–	–	–	nr	16.93
area 0.1–0.25 m²	–	–	–	–	nr	44.05
area 1 m²	–	–	–	–	nr	107.05
area 10 m²	–	–	–	–	nr	493.17
Grouting ends of redundant drains and sewers						
100 mm diameter	0.03	1.89	2.26	5.03	nr	9.18
300 mm diameter	0.13	7.86	9.47	20.11	nr	37.44
450 mm diameter	0.26	16.34	19.79	43.88	nr	80.01
600 mm diameter	0.50	31.43	38.03	88.81	nr	158.27
1200 mm diameter	1.70	106.86	129.17	251.34	nr	487.37
STABILIZATION OF EXISTING SEWERS						
Pointing, cement mortar (1:3)						
faces of brickwork	–	–	–	–	m²	36.00

CLASS Y: SEWER AND WATER MAIN RENOVATION AND ANCILARY WORKS

Item	Gang hours	Labour £	Plant £	Material £	Unit	Total rate £
RENOVATION OF EXISTING SEWERS						
Sliplining						
GRP one piece unit; eggshape sewer 1300 mm high	–	–	–	–	m	382.15
Renovation of existing sewers						
sliplining	–	–	–	–	hr	92.50
LATERALS TO RENOVATED SEWERS						
Jointing						
bore not exceeding 150 mm	–	–	–	–	nr	59.25
bore 150–300 mm	–	–	–	–	nr	105.15
bore 450 mm	–	–	–	–	nr	147.15
EXISTING MANHOLES						
Abandonment						
Sealing redundant road gullies with grade C15 concrete	0.02	1.45	1.77	15.92	nr	19.14
Sealing redundant chambers with grade C15 concrete						
ne 1.0 m deep to invert	0.09	5.66	6.86	64.10	nr	76.62
1.0–2.0 m deep to invert	0.21	13.20	15.97	103.47	nr	132.64
2.0–3.0 m deep to invert	0.55	34.57	41.76	172.17	nr	248.50
Alteration						
100 × 100 mm water stop tap boxes on 100 × 100 mm brick chambers						
raising the level by 150 mm or less	0.06	3.77	4.60	20.54	nr	28.91
lowering the level by 150 mm or less	0.04	2.51	3.04	12.16	nr	17.71
420 × 420 mm cover and frame on 420 × 420 mm in situ concrete chamber						
raising the level by 150 mm or less	0.10	6.29	7.64	37.10	nr	51.03
lowering the level by 150 mm or less	0.06	3.77	4.60	25.30	nr	33.67
Raising the level of 700 × 700 mm cover and frame on 700 × 500 mm in situ concrete chamber						
by 150 mm or less	0.17	10.69	12.93	53.23	nr	76.85
600 × 600 mm grade 'A' heavy duty manhole cover and frame on 600 × 600 mm brick chamber						
raising the level by 150 mm or less	0.17	10.69	12.93	53.23	nr	76.85
raising the level by 150–300 mm	0.21	13.20	15.97	65.80	nr	94.97
lowering the level by 150 mm or less	0.10	6.29	7.64	34.38	nr	48.31
INTERRUPTIONS						
Preparation of existing sewers						
cleaning	–	–	–	–	hr	393.50
Stabilization of existing sewers						
pointing	–	–	–	–	hr	79.50

CLASS Z: SIMPLE BUILDING WORKS INCIDENTAL TO CIVIL ENGINEERING WORKS

Item	Gang hours	Labour £	Plant £	Material £	Unit	Total rate £
NOTES						
The user should refer to Unit Costs – General Civil Engineering Works for cost guidance on the matters listed in this section.						
RESOURCES – LABOUR						
Carpentry and joinery gang						
1 foreman carpenter (craftsman)		22.50				
5 carpenters (craftsmen)		97.70				
1 unskilled operative (general)		12.71				
Total Gang Rate/Hour	£	**132.91**				
Ironmongery gang						
1 carpenter (craftsman)		19.54				
Total Gang Rate/Hour	£	**19.54**				
Glazing gang						
1 glazier (craftsman)		19.54				
Total Gang Rate/Hour	£	**19.54**				
Finishings gang						
2 plasterers/ceramic tilers (craftsmen)		39.08				
1 unskilled operative (general)		12.71				
Total Gang Rate/Hour	£	**51.79**				
Vinyl tiling gang						
1 tiler (craftsman)		19.54				
1 unskilled operative (general)		12.71				
Total Gang Rate/Hour	£	**32.25**				
Plumbing gang						
1 plumber (craftsman)		19.54				
1 unskilled operative (general)		12.71				
Total Gang Rate/Hour	£	**32.25**				
CARPENTRY AND JOINERY						
Softwood; structural grade SC3; sawn; tanalized						
Structural and carcassing timber; floors						
38 × 100	0.02	2.92	–	1.56	m	**4.48**
38 × 125	0.02	3.06	–	1.73	m	**4.79**
38 × 150	0.03	3.32	–	2.36	m	**5.68**
50 × 100	0.03	3.32	–	2.07	m	**5.39**
50 × 125	0.03	3.46	–	2.60	m	**6.06**
50 × 150	0.03	3.59	–	3.10	m	**6.69**
50 × 175	0.03	3.59	–	3.63	m	**7.22**
50 × 200	0.03	3.85	–	4.14	m	**7.99**
50 × 225	0.03	3.85	–	4.67	m	**8.52**
75 × 200	0.03	4.12	–	6.21	m	**10.33**
75 × 225	0.03	4.12	–	6.98	m	**11.10**
100 × 200	0.04	5.58	–	7.76	m	**13.34**
100 × 225	0.05	5.98	–	8.63	m	**14.61**

CLASS Z: SIMPLE BUILDING WORKS INCIDENTAL TO CIVIL ENGINEERING WORKS

Item	Gang hours	Labour £	Plant £	Material £	Unit	Total rate £
Structural and carcassing timber; walls and partitions						
38 × 100	0.03	3.59	–	1.56	m	**5.15**
38 × 125	0.03	3.99	–	1.73	m	**5.72**
38 × 150	0.03	4.25	–	2.36	m	**6.61**
50 × 100	0.03	4.39	–	2.07	m	**6.46**
50 × 125	0.04	4.65	–	2.60	m	**7.25**
Structural and carcassing timber; flat roofs						
38 × 100	0.03	3.32	–	1.56	m	**4.88**
38 × 125	0.02	3.06	–	1.73	m	**4.79**
38 × 150	0.03	3.32	–	2.36	m	**5.68**
50 × 100	0.03	3.32	–	2.07	m	**5.39**
50 × 125	0.03	3.46	–	2.60	m	**6.06**
50 × 150	0.03	3.59	–	3.10	m	**6.69**
50 × 175	0.03	3.59	–	3.63	m	**7.22**
50 × 200	0.03	3.85	–	4.14	m	**7.99**
50 × 225	0.03	3.85	–	4.67	m	**8.52**
75 × 200	0.03	4.12	–	6.21	m	**10.33**
75 × 225	0.03	4.12	–	6.98	m	**11.10**
100 × 200	0.04	5.58	–	7.76	m	**13.34**
100 × 225	0.05	5.98	–	8.63	m	**14.61**
Structural and carcassing timber; pitched roofs						
38 × 100	0.03	3.59	–	1.56	m	**5.15**
38 × 125	0.02	3.06	–	1.73	m	**4.79**
38 × 150	0.03	3.59	–	2.36	m	**5.95**
50 × 100	0.03	4.39	–	2.07	m	**6.46**
50 × 125	0.03	3.46	–	2.60	m	**6.06**
50 × 150	0.04	5.05	–	3.10	m	**8.15**
50 × 175	0.04	5.05	–	3.63	m	**8.68**
50 × 200	0.04	5.05	–	4.14	m	**9.19**
50 × 225	0.04	5.05	–	4.67	m	**9.72**
75 × 125	0.05	5.98	–	3.88	m	**9.86**
75 × 150	0.05	5.98	–	4.68	m	**10.66**
Structural and carcassing timber; plates and bearers						
38 × 100	0.01	1.46	–	1.56	m	**3.02**
50 × 75	0.01	1.46	–	1.56	m	**3.02**
50 × 100	0.01	1.73	–	2.07	m	**3.80**
75 × 100	0.01	1.73	–	3.10	m	**4.83**
75 × 125	0.02	1.99	–	3.88	m	**5.87**
75 × 150	0.02	1.99	–	4.68	m	**6.67**
Structural and carcassing timber; struts						
38 × 100	0.06	7.31	–	1.56	m	**8.87**
50 × 75	0.06	7.31	–	1.56	m	**8.87**
50 × 100	0.06	7.31	–	2.07	m	**9.38**
Structural and carcassing timber; cleats						
225 mm × 100 mm × 75 mm	0.04	4.78	–	0.71	nr	**5.49**
Structural and carcassing timber; trussed rafters and roof trusses						

CLASS Z: SIMPLE BUILDING WORKS INCIDENTAL TO CIVIL ENGINEERING WORKS

Item	Gang hours	Labour £	Plant £	Material £	Unit	Total rate £
CARPENTRY AND JOINERY – cont						
Softwood; joinery quality; wrought; tanalized						
Strip boarding; walls						
18 mm nominal thick	0.14	18.61	–	18.66	m²	37.27
18 mm nominal thick; not exceeding 100 mm wide	0.05	5.98	–	1.87	m²	7.85
18 mm nominal thick; 100–200 mm wide	0.05	6.65	–	3.73	m²	10.38
18 mm nominal thick; 200–300 mm wide	0.06	7.97	–	5.60	m²	13.57
Softwood; joinery quality; wrought						
Strip boarding; walls						
18 mm nominal thick	0.14	18.61	–	18.66	m²	37.27
18 mm nominal thick; not exceeding 100 mm wide	0.05	5.98	–	1.87	m²	7.85
18 mm nominal thick; 100–200 mm wide	0.05	6.65	–	3.73	m²	10.38
18 mm nominal thick; 200–300 mm wide	0.06	7.97	–	5.60	m²	13.57
Miscellaneous joinery; skirtings						
19 × 100	0.02	2.66	–	2.56	m	5.22
25 × 150	0.03	3.32	–	3.94	m	7.26
Miscellaneous joinery; architraves						
12 × 50	0.02	3.19	–	1.84	m	5.03
19 × 63	0.02	3.19	–	2.10	m	5.29
Miscellaneous joinery; trims						
12 × 25	0.02	3.19	–	1.44	m	4.63
12 × 50	0.02	3.19	–	1.84	m	5.03
16 × 38	0.02	3.19	–	1.42	m	4.61
19 × 19	0.02	3.19	–	1.35	m	4.54
Plywood; marine quality						
Sheet boarding; walls						
18 mm nominal tihck	0.14	18.61	–	21.81	m²	40.42
Plywood; external quality						
Sheet boarding; floors						
18 mm nominal thick	0.11	14.62	–	18.66	m²	33.28
18 mm nominal thick; not exceeding 100 mm wide	0.04	4.65	–	1.87	m²	6.52
18 mm nominal thick; 100–200 mm wide	0.04	5.05	–	3.73	m²	8.78
18 mm nominal thick; 200–300 mm wide	0.05	5.98	–	5.60	m²	11.58
Sheet boarding; walls						
18 mm nominal thick	0.14	18.61	–	18.66	m²	37.27
18 mm nominal thick; not exceeding 100 mm wide	0.05	5.98	–	1.87	m²	7.85
18 mm nominal thick; 100–200 mm wide	0.05	6.65	–	3.73	m²	10.38
18 mm nominal thick; 200–300 mm wide	0.06	7.97	–	5.60	m²	13.57
Sheet boarding; soffits						
18 mm nominal thick	0.16	20.60	–	18.66	m²	39.26
18 mm nominal thick; not exceeding 100 mm wide	0.05	6.65	–	1.87	m²	8.52
18 mm nominal thick; 100–200 mm wide	0.06	7.31	–	3.73	m²	11.04
18 mm nominal thick; 200–300 mm wide	0.07	8.77	–	5.60	m²	14.37

CLASS Z: SIMPLE BUILDING WORKS INCIDENTAL TO CIVIL ENGINEERING WORKS

Item	Gang hours	Labour £	Plant £	Material £	Unit	Total rate £
INSULATION						
Vapour barrier; Sisalkraft 728 building paper (Class A1F); 150 mm laps; fixed to softwood						
Sheets						
floors	0.01	1.73	–	2.12	m²	**3.85**
sloping upper surfaces	0.01	1.86	–	2.12	m²	**3.98**
walls	0.02	2.26	–	2.12	m²	**4.38**
soffits	0.02	2.66	–	2.12	m²	**4.78**
Insulation quilt; Isover glass fibre; laid loose between members at 600 mm centres						
Quilts; floors						
60 mm thick	0.03	3.72	–	1.44	m²	**5.16**
80 mm thick	0.03	3.72	–	1.89	m²	**5.61**
100 mm thick	0.03	3.72	–	2.25	m²	**5.97**
150 mm thick	0.03	3.72	–	3.44	m²	**7.16**
Quilts; sloping upper surfaces						
60 mm thick	0.03	4.12	–	1.44	m²	**5.56**
80 mm thick	0.03	4.52	–	1.89	m²	**6.41**
100 mm thick	0.04	4.78	–	2.25	m²	**7.03**
150 mm thick	0.04	5.32	–	3.44	m²	**8.76**
Insulation quilt; Isover glass fibre; laid between members at 600 mm centres; fixing with staples						
Quilts; walls						
60 mm thick	0.04	4.65	–	1.62	m²	**6.27**
80 mm thick	0.04	5.05	–	2.08	m²	**7.13**
100 mm thick	0.04	5.58	–	2.43	m²	**8.01**
150 mm thick	0.05	6.65	–	3.63	m²	**10.28**
Quilts; soffits						
60 mm thick	0.04	5.58	–	1.62	m²	**7.20**
80 mm thick	0.05	6.25	–	2.08	m²	**8.33**
100 mm thick	0.05	6.65	–	2.43	m²	**9.08**
150 mm thick	0.05	7.18	–	3.63	m²	**10.81**
Insulation board; Jablite or similar expanded polystyrene standard grade; fixing with adhesive						
Boards; floors						
25 mm thick	0.06	7.44	–	1.68	m²	**9.12**
40 mm thick	0.06	7.44	–	2.02	m²	**9.46**
50 mm thick	0.06	7.44	–	2.35	m²	**9.79**
Boards; sloping upper surfaces						
25 mm thick	0.07	8.64	–	1.68	m²	**10.32**
40 mm thick	0.07	8.64	–	2.02	m²	**10.66**
50 mm thick	0.07	8.64	–	2.35	m²	**10.99**
Boards; walls						
25 mm thick	0.08	9.97	–	1.68	m²	**11.65**
40 mm thick	0.08	9.97	–	2.02	m²	**11.99**
50 mm thick	0.08	9.97	–	2.35	m²	**12.32**
Boards; soffits						
25 mm thick	0.08	11.03	–	1.68	m²	**12.71**
40 mm thick	0.08	11.03	–	2.02	m²	**13.05**
50 mm thick	0.08	11.03	–	2.35	m²	**13.38**

CLASS Z: SIMPLE BUILDING WORKS INCIDENTAL TO CIVIL ENGINEERING WORKS

Item	Gang hours	Labour £	Plant £	Material £	Unit	Total rate £
WINDOWS, DOORS AND GLAZING						
Timber windows; treated planed softwood; Jeld-Wen; plugged and screwed to masonry						
High performance top hung reversible windows; weather stripping; opening panes hung on rustproof hinges with aluminized lacquered espagnolette bolts; low E 24 mm double glazing						
630 mm × 900 mm; ref LEC109 AR	0.16	21.80	–	330.21	nr	**352.01**
630 mm × 1200 mm; ref LEC112 AR	0.21	27.78	–	368.14	nr	**395.92**
1200 mm × 1050 mm; ref LEC210 AR	0.25	32.56	–	461.42	nr	**493.98**
1770 mm × 1050 mm; ref LEC310 AER	0.31	41.07	–	787.04	nr	**828.11**
Metal windows; steel fixed light; factory finished polyester powder coating; Crittal Homelight range; fixing lugs plugged and screwed to masonry						
One piece composites; glazing 4 mm OQ glass; easy-glaze beads and weather stripping						
628 mm × 923 mm; ref ZNC5	0.22	28.97	–	202.44	nr	**231.41**
1237 mm × 923 mm; ref ZNC13	0.32	42.27	–	390.79	nr	**433.06**
1237 mm × 1218 mm; ref ZND13	0.32	42.27	–	498.30	nr	**540.57**
1846 mm × 1513 mm; ref ZNDV14	0.32	42.27	–	558.30	nr	**600.57**
Plastics windows; PVC-u, reinforced where appropriate with aluminium alloy; standard ironmongery; fixing lugs plugged and screwed to masonry						
Casement/fixed light; glazing 4 mm OQ glass; e.p.d.m. glazing gaskets and weather seals						
600 mm × 1200 mm; single glazed	0.32	42.27	–	–	nr	**42.27**
600 mm × 1200 mm; double glazed	0.32	42.27	–	79.47	nr	**121.74**
1200 mm × 1200 mm; single glazed	0.36	48.38	–	–	nr	**48.38**
1200 mm × 1200 mm; double glazed	0.36	48.38	–	122.86	nr	**171.24**
1800 mm × 1200 mm; single glazed	0.41	54.76	–	–	nr	**54.76**
1800 mm × 1200 mm; double glazed	0.41	54.76	–	226.36	nr	**281.12**
Timber doors; treated planed softwood						
Matchboarded, ledged and braced doors; 25 mm thick ledges and braces; 19 mm thick tongued, grooved and v-jointed one side vertical boarding						
762 mm × 1981 mm	0.27	36.28	–	62.33	nr	**98.61**
838 mm × 1981 mm	0.27	36.28	–	62.33	nr	**98.61**
Panelled doors; one open panel for glass; including beads						
762 × 1981 × 44 mm	0.32	42.53	–	105.75	nr	**148.28**
Panelled doors; two open panels for glass; including beads						
762 × 1981 × 44 mm	0.32	42.53	–	137.41	nr	**179.94**
838 × 1981 × 44 mm	0.32	42.53	–	137.41	nr	**179.94**

CLASS Z: SIMPLE BUILDING WORKS INCIDENTAL TO CIVIL ENGINEERING WORKS

Item	Gang hours	Labour £	Plant £	Material £	Unit	Total rate £
Timber doors; standard flush pattern						
Flush door; internal quality; skeleton or cellular core; hardboard faced both sides; lipped on two long edges; primed						
626 × 2040 × 40 mm	0.22	29.64	–	22.72	nr	**52.36**
726 × 2040 × 40 mm	0.22	29.64	–	22.72	nr	**52.36**
826 × 2040 × 40 mm	0.22	29.64	–	22.29	nr	**51.93**
Flush door; internal quality; skeleton or cellular core; chipboard faced both sides; lipped all edges; primed						
626 × 2040 × 40 mm	0.22	29.64	–	29.45	nr	**59.09**
726 × 2040 × 40 mm	0.22	29.64	–	29.45	nr	**59.09**
826 × 2040 × 40 mm	0.22	29.64	–	29.45	nr	**59.09**
Flush door; internal quality; skeleton or cellular core; Sapele veneered both sides; lipped all edges; primed						
626 × 2040 × 40 mm	0.34	45.19	–	45.23	nr	**90.42**
726 × 2040 × 40 mm	0.34	45.19	–	45.23	nr	**90.42**
826 × 2040 × 40 mm	0.34	45.19	–	45.23	nr	**90.42**
Flush door; half-hour fire check (30/20); solid core; chipboard faced both sides; lipped all edges; primed						
626 × 2040 × 44 mm	0.32	42.53	–	54.45	nr	**96.98**
726 × 2040 × 44 mm	0.32	42.53	–	54.45	nr	**96.98**
826 × 2040 × 44 mm	0.32	42.53	–	54.45	nr	**96.98**
Timber frames or lining sets; treated planed softwood						
Internal door frame or lining; 30 × 107 mm lining with 12 × 38 mm door stop						
for 726 × 2040 mm door	0.14	19.27	–	39.54	nr	**58.81**
for 826 × 2040 mm door	0.14	19.27	–	39.54	nr	**58.81**
Internal door frame or lining; 30 × 133 mm lining with 12 × 38 mm door stop						
for 726 × 2040 mm door	0.14	19.27	–	44.26	nr	**63.53**
for 826 × 2040 mm door	0.14	19.27	–	44.26	nr	**63.53**
Ironmongery						
Hinges						
100 mm; light steel	0.17	3.42	–	0.88	nr	**4.30**
S/D rising butts; R/L hand; 102 × 67 mm; BRS	0.17	3.42	–	24.63	nr	**28.05**
Door closers						
light duty surface fixed door closer; L/R hand; SIL	1.30	25.40	–	101.15	nr	**126.55**
door selector; face fixing; SAA	0.80	15.63	–	86.57	nr	**102.20**
floor spring; single and double action; ZP	3.33	65.07	–	212.57	nr	**277.64**
Locks						
mortice dead lock; 63 × 108 mm; SSS	1.00	19.54	–	14.01	nr	**33.55**
rim lock; 140 × 73 mm; GYE	0.56	10.94	–	10.65	nr	**21.59**
upright mortice lock; 103 × 82 mm; 3 lever	1.11	21.69	–	14.44	nr	**36.13**
Bolts						
flush; 152 × 25 mm; SCP	0.80	15.63	–	18.99	nr	**34.62**
indicating; 76 × 41 mm; SAA	0.89	17.39	–	13.89	nr	**31.28**
panic; single; SVE	3.33	65.07	–	87.62	nr	**152.69**
panic; double; SVE	4.67	91.25	–	119.81	nr	**211.06**
necked tower; 203 mm; BJ	0.40	7.82	–	5.19	nr	**13.01**

CLASS Z: SIMPLE BUILDING WORKS INCIDENTAL TO CIVIL ENGINEERING WORKS

Item	Gang hours	Labour £	Plant £	Material £	Unit	Total rate £
WINDOWS, DOORS AND GLAZING – cont						
Ironmongery – cont						
Handles						
pull; 225 mm; back fixing; PAA	0.23	4.49	–	8.41	nr	**12.90**
pull; 225 mm; face fixing with cover rose; PAA	0.44	8.60	–	45.18	nr	**53.78**
lever; PAA	0.44	8.60	–	31.48	nr	**40.08**
Plates						
finger plate; 300 × 75 × 3 mm; SAA	0.23	4.49	–	5.50	nr	**9.99**
kicking plate; 1000 × 150 × 3 mm; PAA	0.44	8.60	–	15.99	nr	**24.59**
letter plate; 330 × 76 mm; aluminium finish	1.77	34.59	–	11.45	nr	**46.04**
Brackets						
head bracket; open; side fixing; bolting to masonry	0.50	9.77	–	9.47	nr	**19.24**
head bracket; open; soffit fixing; bolting to masonry	0.50	9.77	–	4.66	nr	**14.43**
Sundries						
rubber door stop; SAA	0.11	2.15	–	1.60	nr	**3.75**
Glazing; standard plain glass to BS 952, clear float; glazing with putty or bradded beads						
Glass						
3 mm thick	0.95	18.56	–	25.56	m^2	**44.12**
4 mm thick	0.95	18.56	–	27.03	m^2	**45.59**
5 mm thick	0.95	18.56	–	32.99	m^2	**51.55**
6 mm thick	0.95	18.56	–	36.21	m^2	**54.77**
Hermetically sealed units, factory-made						
two panes 4 mm thick, 6 mm air space	0.95	18.56	–	105.00	m^2	**123.56**
Glazing; standard plain glass to BS 952, rough cast; glazing with putty or bradded beads						
Glass						
6 mm thick	0.95	18.56	–	35.95	m^2	**54.51**
Glazing; standard plain glass to BS 952, Georgian wired cast; glazing with putty or bradded beads						
Glass						
7 mm thick	0.95	18.56	–	37.74	m^2	**56.30**
Glazing; standard plain glass to BS 952, Georgian wired polished; glazing with putty or bradded beads						
Glass						
6 mm thick	0.95	18.56	–	58.03	m^2	**76.59**

CLASS Z: SIMPLE BUILDING WORKS INCIDENTAL TO CIVIL ENGINEERING WORKS

Item	Gang hours	Labour £	Plant £	Material £	Unit	Total rate £
Glazing; special glass to BS 952, toughened clear float; glazing with putty or bradded beads						
Glass						
4 mm thick	0.95	18.56	–	30.92	m^2	**49.48**
5 mm thick	0.95	18.56	–	40.97	m^2	**59.53**
6 mm thick	0.95	18.56	–	45.07	m^2	**63.63**
10 mm thick	1.05	20.52	–	74.06	m^2	**94.58**
Glazing; special glass to BS 952, clear laminated safety; glazing with putty or bradded beads						
Glass						
4.4 mm thick	0.95	18.56	–	37.59	m^2	**56.15**
5.4 mm thick	0.95	18.56	–	40.79	m^2	**59.35**
6.4 mm thick	0.95	18.56	–	43.98	m^2	**62.54**
Glazing; standard plain glass to BS 952, clear float; glazing with putty or bradded beads						
Glass						
4.4 mm thick	0.95	18.56	–	37.59	m^2	**56.15**
5.4 mm thick	0.95	18.56	–	40.79	m^2	**59.35**
6.4 mm thick	0.95	18.56	–	43.98	m^2	**62.54**
Patent glazing; aluminium alloy bars 2.55 m long at 622 mm centres; fixed to supports						
Patent glazing						
roofs	–	–	–	395.00	m^2	**395.00**
SURFACE FINISHES, LININGS AND PARTITIONS						
In situ finishes; cement and sand (1:3); steel trowelled						
Floors						
30 mm thick	0.12	6.21	–	2.69	m^2	**8.90**
40 mm thick	0.12	6.21	–	3.58	m^2	**9.79**
50 mm thick	0.14	7.25	–	4.47	m^2	**11.72**
60 mm thick	0.15	7.77	–	5.36	m^2	**13.13**
70 mm thick	0.16	8.29	–	6.26	m^2	**14.55**
Sloping upper surfaces						
30 mm thick	0.16	8.29	–	2.69	m^2	**10.98**
40 mm thick	0.16	8.29	–	3.58	m^2	**11.87**
50 mm thick	0.18	9.32	–	4.47	m^2	**13.79**
60 mm thick	0.19	9.84	–	5.36	m^2	**15.20**
70 mm thick	0.20	10.36	–	6.26	m^2	**16.62**
Walls						
12 mm thick	0.31	16.05	–	1.07	m^2	**17.12**
15 mm thick	0.34	17.61	–	1.31	m^2	**18.92**
20 mm thick	0.38	19.68	–	1.79	m^2	**21.47**

CLASS Z: SIMPLE BUILDING WORKS INCIDENTAL TO CIVIL ENGINEERING WORKS

Item	Gang hours	Labour £	Plant £	Material £	Unit	Total rate £
SURFACE FINISHES, LININGS AND PARTITIONS – cont						
In situ finishes – cont						
Surfaces of width not exceeding 300 mm						
12 mm thick	0.15	7.77	–	1.07	m	**8.84**
15 mm thick	0.13	6.73	–	1.34	m	**8.07**
20 mm thick	0.12	6.21	–	1.79	m	**8.00**
30 mm thick	0.08	4.14	–	2.69	m	**6.83**
40 mm thick	0.08	4.14	–	3.58	m	**7.72**
50 mm thick	0.09	4.66	–	4.47	m	**9.13**
60 mm thick	0.09	4.66	–	5.36	m	**10.02**
70 mm thick	0.10	5.18	–	6.26	m	**11.44**
Surfaces of width 300 mm–1 m						
12 mm thick	0.23	11.91	–	1.07	m	**12.98**
15 mm thick	0.26	13.47	–	1.34	m	**14.81**
20 mm thick	0.26	13.47	–	1.79	m	**15.26**
30 mm thick	0.12	6.21	–	2.69	m	**8.90**
40 mm thick	0.12	6.21	–	3.58	m	**9.79**
50 mm thick	0.13	6.73	–	4.47	m	**11.20**
60 mm thick	0.14	7.25	–	5.36	m	**12.61**
70 mm thick	0.15	7.77	–	6.26	m	**14.03**
In situ finishes; lightweight plaster; Thistle; steel trowelled						
Walls						
12 mm thick	0.22	11.39	–	1.09	m^2	**12.48**
Surfaces of width not exceeding 300 mm						
12 mm thick	0.22	11.39	–	0.33	m	**11.72**
Surfaces of width 300 mm–1 m						
12 mm thick	0.16	8.29	–	1.09	m	**9.38**
Beds and backings; cement and sand (1:3)						
Floors						
50 mm thick	0.14	7.25	–	4.47	m^2	**11.72**
70 mm thick	0.16	8.29	–	6.26	m^2	**14.55**
Sloping upper surfaces						
30 mm thick	0.16	8.29	–	2.69	m^2	**10.98**
40 mm thick	0.16	8.29	–	3.58	m^2	**11.87**
Walls						
12 mm thick	0.31	16.05	–	1.07	m^2	**17.12**
20 mm thick	0.38	19.68	–	1.79	m^2	**21.47**
Surfaces of width not exceeding 300 mm						
12 mm thick	0.15	7.77	–	1.07	m	**8.84**
20 mm thick	0.12	6.21	–	1.79	m	**8.00**
30 mm thick	0.08	4.14	–	2.69	m	**6.83**
50 mm thick	0.09	4.66	–	4.47	m	**9.13**
70 mm thick	0.10	5.18	–	6.26	m	**11.44**

CLASS Z: SIMPLE BUILDING WORKS INCIDENTAL TO CIVIL ENGINEERING WORKS

Item	Gang hours	Labour £	Plant £	Material £	Unit	Total rate £
Surfaces of width 300 mm–1 m						
12 mm thick	0.23	11.91	–	1.07	m	12.98
20 mm thick	0.26	13.47	–	1.79	m	15.26
30 mm thick	0.12	6.21	–	2.69	m	8.90
50 mm thick	0.13	6.73	–	4.47	m	11.20
70 mm thick	0.15	7.77	–	6.26	m	14.03
Tiles; red clay; bedding 10 mm thick and jointing in cement mortar (1:3); grouting with cement mortar (1:1)						
Floors						
150 × 150 × 12.5 mm	0.32	16.57	–	23.99	m²	40.56
200 × 200 × 19 mm	0.32	16.57	–	33.32	m²	49.89
Surfaces of width not exceeding 300 mm						
150 × 150 × 12.5 mm	0.16	8.29	–	7.20	m	15.49
200 × 200 × 19 mm	0.13	6.73	–	10.00	m	16.73
Surfaces of width 300 mm–1 m						
150 × 150 × 12.5 mm	0.24	12.43	–	23.99	m	36.42
200 × 200 × 19 mm	0.24	12.43	–	33.34	m	45.77
Tiles; brown clay; bedding 10 mm thick and jointing in cement mortar (1:3); grouting with cement mortar (1:1)						
Floors						
150 × 150 × 12.5 mm	0.32	16.57	–	28.00	m²	44.57
200 × 200 × 19 mm	0.32	16.57	–	39.03	m²	55.60
Surfaces of width not exceeding 300 mm						
150 × 150 × 12.5 mm	0.16	8.29	–	8.40	m	16.69
200 × 200 × 19 mm	0.13	6.73	–	11.71	m	18.44
Surfaces of width 300 mm–1 m						
150 × 150 × 12.5 mm	0.24	12.43	–	28.00	m	40.43
200 × 200 × 19 mm	0.24	12.43	–	39.05	m	51.48
Tiles; glazed ceramic wall tiles; BS 6431; white; fixing with adhesive; pointing joints with grout						
Sloping upper surfaces						
152 × 152 × 5.5 mm thick	0.28	14.50	–	21.72	m²	36.22
200 × 100 × 6.5 mm thick	0.28	14.50	–	21.72	m²	36.22
Walls						
152 × 152 × 5.5 mm thick	0.24	12.43	–	21.72	m²	34.15
200 × 100 × 6.5 mm thick	0.24	12.43	–	21.72	m²	34.15
Soffit						
152 × 152 × 5.5 mm thick	0.28	14.50	–	21.72	m²	36.22
200 × 100 × 6.5 mm thick	0.28	14.50	–	21.72	m²	36.22
Surfaces of width not exceeding 300 mm						
152 × 152 × 5.5 mm thick	0.12	6.21	–	6.52	m	12.73
200 × 100 × 6.5 mm thick	0.12	6.21	–	6.52	m	12.73
Surfaces of width 300 mm–1 m						
152 × 152 × 5.5 mm thick	0.18	9.32	–	21.72	m	31.04
200 × 100 × 6.5 mm thick	0.18	9.32	–	21.72	m	31.04

CLASS Z: SIMPLE BUILDING WORKS INCIDENTAL TO CIVIL ENGINEERING WORKS

Item	Gang hours	Labour £	Plant £	Material £	Unit	Total rate £
SURFACE FINISHES, LININGS AND PARTITIONS – cont						
Tiles; slate; Riven Welsh; bedding 10 mm thick and jointing in cement mortar (1:3); grouting with cement mortar (1:1)						
Floors						
250 × 250 × 12–15 mm	0.24	12.43	–	67.32	m^2	**79.75**
Surfaces of width not exceeding 300 mm						
250 × 250 × 12–15 mm	0.12	6.21	–	20.20	m	**26.41**
Surfaces of width 300 mm–1 m						
250 × 250 × 12–15 mm	0.18	9.32	–	67.32	m	**76.64**
Tiles; vinyl; Accoflex; fixing with adhesive						
Floors						
300 × 300 × 2 mm	0.17	5.39	–	7.68	m^2	**13.07**
300 × 300 × 2.5 mm	0.17	5.39	–	8.74	m^2	**14.13**
Surfaces of width not exceeding 300 mm						
300 × 300 × 2 mm	0.08	2.68	–	2.31	m	**4.99**
300 × 300 × 2.5 mm	0.08	2.68	–	2.63	m	**5.31**
Surfaces of width 300 mm–1 m						
300 × 300 × 2 mm	0.13	4.03	–	7.68	m	**11.71**
300 × 300 × 2.5 mm	0.13	4.03	–	8.74	m	**12.77**
Tiles; vinyl; Marley HD; fixing with adhesive						
Floors						
300 × 300 × 2 mm	0.23	7.51	–	10.71	m^2	**18.22**
Surfaces of width not exceeding 300 mm						
300 × 300 × 2 mm	0.12	3.77	–	3.21	m	**6.98**
Surfaces of width 300 mm–1 m						
300 × 300 × 2 mm	0.17	5.64	–	10.71	m	**16.35**
Tiles; rubber studded; Altro Mondopave; type MRB; black; fixing with adhesive						
Floors						
500 × 500 × 2.5 mm	0.40	12.90	–	29.72	m^2	**42.62**
500 × 500 × 4.0 mm	0.40	12.90	–	28.46	m^2	**41.36**
Surfaces of width not exceeding 300 mm						
500 × 500 × 2.5 mm	0.20	6.45	–	8.90	m	**15.35**
500 × 500 × 4.0 mm	0.20	6.45	–	8.54	m	**14.99**
Surfaces of width 300 mm–1 m						
500 × 500 × 2.5 mm	0.30	9.68	–	29.72	m	**39.40**
500 × 500 × 4.0 mm	0.30	9.68	–	28.46	m	**38.14**

CLASS Z: SIMPLE BUILDING WORKS INCIDENTAL TO CIVIL ENGINEERING WORKS

Item	Gang hours	Labour £	Plant £	Material £	Unit	Total rate £
Tiles; rubber studded; Altro Mondopave; type MRB; colour; fixing with adhesive						
Floors						
500 × 500 × 2.5 mm	0.40	12.90	–	32.99	m^2	**45.89**
500 × 500 × 4.0 mm	0.40	12.90	–	32.70	m^2	**45.60**
Surfaces of width not exceeding 300 mm						
500 × 500 × 2.5 mm	0.20	6.45	–	9.88	m	**16.33**
500 × 500 × 4.0 mm	0.20	6.45	–	9.81	m	**16.26**
Surfaces of width 300 mm–1 m						
500 × 500 × 2.5 mm	0.30	9.68	–	32.99	m	**42.67**
500 × 500 × 4.0 mm	0.30	9.68	–	32.70	m	**42.38**
Tiles; linoleum; Forbo Nairn; Marmoleum Dual; level; fixing with adhesive						
Floors						
2.50 mm thick; marbled	0.27	8.61	–	15.59	m^2	**24.20**
Surfaces of width not exceeding 300 mm						
2.50 mm thick; marbled	0.13	4.29	–	4.68	m	**8.97**
Surfaces of width 300 mm–1 m						
2.50 mm thick; marbled	0.20	6.45	–	1.80	m	**8.25**
Flexible sheer; linoleum; Forbo Nairn; Marmoleum Real; level; fixing with adhesive						
Floors						
2.00 mm thick; marbled	0.33	10.75	–	11.66	m^2	**22.41**
2.50 mm thick; marbled	0.33	10.75	–	12.47	m^2	**23.22**
3.20 mm thick; marbled	0.33	10.75	–	14.97	m^2	**25.72**
4.00 mm thick; marbled	0.33	10.75	–	18.62	m^2	**29.37**
Surfaces of width not exceeding 300 mm						
2.00 mm thick; marbled	0.17	5.39	–	3.50	m	**8.89**
2.50 mm thick; marbled	0.17	5.39	–	3.74	m	**9.13**
3.20 mm thick; marbled	0.17	5.39	–	4.49	m	**9.88**
4.00 mm thick; marbled	0.17	5.39	–	5.45	m	**10.84**
Surfaces of width 300 mm – 1 m						
2.00 mm thick; marbled	0.25	8.06	–	11.66	m	**19.72**
2.50 mm thick; marbled	0.25	8.06	–	12.47	m	**20.53**
3.20 mm thick; marbled	0.25	8.06	–	14.97	m	**23.03**
4.00 mm thick; marbled	0.25	8.06	–	18.62	m	**26.68**
Flexible sheet; vinyl; slip-resistant; Forbo Nairn; Surestep PUR; fixing with adhesive						
Floors						
2.00 mm thick	0.33	10.74	–	15.55	m^2	**26.29**
Surfaces of width not exceeding 300 mm						
2.00 mm thick	0.17	5.39	–	4.67	m	**10.06**
Surfaces of width 300 mm–1 m						
2.00 mm thick	0.25	8.06	–	15.55	m	**23.61**

CLASS Z: SIMPLE BUILDING WORKS INCIDENTAL TO CIVIL ENGINEERING WORKS

Item	Gang hours	Labour £	Plant £	Material £	Unit	Total rate £
SURFACE FINISHES, LININGS AND PARTITIONS – cont						
Flexible sheet; vinyl; Marley HD; fixing with adhesive						
Floors						
2.00 mm thick	0.30	9.68	–	10.71	m^2	**20.39**
2.50 mm thick	0.33	10.75	–	11.97	m^2	**22.72**
Surfaces of width not exceeding 300 mm						
2.00 mm thick	0.15	4.84	–	3.21	m	**8.05**
2.50 mm thick	0.17	5.39	–	3.59	m	**8.98**
Surfaces of width 300 mm–1 m						
2.00 mm thick	0.23	7.26	–	10.71	m	**17.97**
2.50 mm thick	0.25	8.06	–	11.97	m	**20.03**
Flexible sheet; vinyl; Armstrong Contract Interior; fixing with adhesive						
Floors						
2.00 mm thick	0.33	10.75	–	13.59	m^2	**24.34**
Surfaces of width not exceeding 300 mm						
2.00 mm thick	0.17	5.39	–	4.08	m	**9.47**
Surfaces of width 300 mm–1 m						
2.00 mm thick	0.25	8.06	–	13.59	m	**21.65**
Flexible sheet; carpet; Armstrong Strong; fixing with adhesive						
Floors						
5.00 mm thick	0.17	5.39	–	44.62	m^2	**50.01**
Surfaces of width not exceeding 300 mm						
5.00 mm thick	0.08	2.68	–	13.39	m	**16.07**
Surfaces of width 300 mm–1 m						
5.00 mm thick	0.13	4.03	–	44.62	m	**48.65**
Suspended ceilings; mineral fibre tiles in exposed grid; suspension system and wire hangers to structural soffit						
Suspended ceiling						
150–500 mm depth of suspension	–	–	–	–	m^2	**22.19**
Suspended ceilings; mineral fibre tiles in concealed grid; suspension system and wire hangers to structural soffit						
Suspended ceiling						
150–500 mm depth of suspension	–	–	–	–	m^2	**36.44**
Bulkheads						
250 mm girth	–	–	–	–	m	**13.16**
500 mm girth	–	–	–	–	m	**26.32**

CLASS Z: SIMPLE BUILDING WORKS INCIDENTAL TO CIVIL ENGINEERING WORKS

Item	Gang hours	Labour £	Plant £	Material £	Unit	Total rate £
PIPED BUILDING SERVICES						
Pipework; copper pipes EN 1057; capillary fittings; pipe clips screwed to background						
Pipes						
15 mm	0.13	4.03	–	3.50	m	**7.53**
22 mm	0.13	4.19	–	5.41	m	**9.60**
28 mm	0.14	4.68	–	9.27	m	**13.95**
35 mm	0.17	5.32	–	28.51	m	**33.83**
42 mm	0.19	6.13	–	34.49	m	**40.62**
Fittings						
15 mm made bend	0.08	2.42	–	9.80	nr	**12.22**
15 mm elbow	0.08	2.42	–	1.41	nr	**3.83**
15 mm equal tee	0.13	4.03	–	1.59	nr	**5.62**
15 mm straight coupling	0.09	2.74	–	0.57	nr	**3.31**
22 mm made bend	0.10	3.23	–	19.26	nr	**22.49**
22 mm elbow	0.11	3.55	–	4.05	nr	**7.60**
22 mm equal tee	0.17	5.32	–	4.44	nr	**9.76**
22 mm straight coupling	0.11	3.55	–	1.33	nr	**4.88**
28 mm made bend	0.13	4.03	–	24.60	nr	**28.63**
28 mm elbow	0.14	4.51	–	11.09	nr	**15.60**
28 mm equal tee	0.20	6.61	–	10.15	nr	**16.76**
28 mm straight coupling	0.14	4.51	–	2.64	nr	**7.15**
35 mm made bend	0.15	4.84	–	41.01	nr	**45.85**
35 mm elbow	0.17	5.32	–	24.91	nr	**30.23**
35 mm equal tee	0.23	7.42	–	25.60	nr	**33.02**
35 mm straight coupling	0.17	5.32	–	7.98	nr	**13.30**
42 mm made bend	0.20	6.45	–	44.31	nr	**50.76**
42 mm elbow	0.20	6.45	–	46.16	nr	**52.61**
42 mm equal tee	0.26	8.38	–	39.99	nr	**48.37**
42 mm straight coupling	0.20	6.45	–	12.68	nr	**19.13**
Pipework; 19 mm thick rigid mineral glass fibre sectional pipe lagging; plain finish; fixing with aluminium bands						
Insulation						
around 15 mm pipes	0.04	1.13	–	2.98	m	**4.11**
around 22 mm pipes	0.05	1.61	–	3.17	m	**4.78**
around 28 mm pipes	0.06	1.77	–	3.41	m	**5.18**
around 35 mm pipes	0.06	1.94	–	3.81	m	**5.75**
around 42 mm pipes	0.07	2.10	–	4.07	m	**6.17**
Equipment; polyethylene cold water feed and expansion cistern to BS 4213, with cover; placing in position						
Cisterns and tanks						
68 litres; ref SC15	0.63	20.16	–	30.39	nr	**50.55**
114 litres; ref SC25	0.72	23.38	–	39.33	nr	**62.71**

CLASS Z: SIMPLE BUILDING WORKS INCIDENTAL TO CIVIL ENGINEERING WORKS

Item	Gang hours	Labour £	Plant £	Material £	Unit	Total rate £
PIPED BUILDING SERVICES – cont						
Equipment; grp cold water storage cistern, with cover; placing in position						
Cisterns and tanks						
68 litres	0.63	20.16	–	88.98	nr	**109.14**
114 litres	0.72	23.38	–	93.05	nr	**116.43**
Equipment; copper single feed coil indirect cylinder to BS 1566 Part 2 grade 3; placing in position						
Cisterns and tanks						
96 litres; ref 2	1.00	32.25	–	162.80	nr	**195.05**
114 litres; ref 3	1.13	36.28	–	183.80	nr	**220.08**
Equipment; combination copper coil direct hot water storage units to BS 3198; placing in position						
Cisterns and tanks						
400 × 900 mm; 65/20 litres	1.40	45.15	–	385.17	nr	**430.32**
450 × 1075 mm; 115/25 litres	2.45	79.01	–	442.72	nr	**521.73**
Sanitary appliances and fittings						
Sink; white glazed fireclay to BS 1206 with pair of cast iron cantilever brackets						
610 × 455 × 205 mm	1.50	48.38	–	127.21	nr	**175.59**
610 × 455 × 205 mm	1.50	48.38	–	206.07	nr	**254.45**
Sink; stainless steel combined bowl and drainer; pair 19 mm chromium plated high neck pillar taps; chain and self colour plug to BS 3380; setting on base unit						
1050 × 500, single drainer, single bowl						
420 × 350 × 175 mm	0.88	28.22	–	134.25	nr	**162.47**
1550 × 500, double drainer, single bowl						
420 × 350 × 200 mm	0.88	28.22	–	222.20	nr	**250.42**
Lavatory basin; white vitreous china to BS 1188; pair 12 mm chromium plated pillar taps; chain and self colour plug to BS 3380; trap; painted cast iron brackets plugged and screwed to masonry						
560 × 405 mm	1.15	37.09	–	155.77	nr	**192.86**
635 × 455 mm	1.15	37.09	–	163.21	nr	**200.30**
Add for coloured	–	–	–	–	nr	**–**
Add for pedestal in lieu of brackets	0.10	3.23	–	64.31	nr	**67.54**
WC suite; low level; white glazed vitreous china pan; black plastic seat; 9 litre white glazed vitreous china cistern and brackets; low pressure ball valve; plastic flush pipe; building bracket into masonry; plugging and screwing pipe brackets and pan; bedding pan in mastic						
P trap outlet	1.50	48.38	–	266.90	nr	**315.28**
Add for coloured	–	–	–	–	nr	**–**

CLASS Z: SIMPLE BUILDING WORKS INCIDENTAL TO CIVIL ENGINEERING WORKS

Item	Gang hours	Labour £	Plant £	Material £	Unit	Total rate £
Bowl type wall urinal; white glazed vitreous china; white glazed vitreous china automatic flushing cistern and brackets; chromium plated flush pipe and spreaders; building brackets into masonry; plugging and screwing pipe brackets						
single; 455 × 380 × 330 mm	2.00	64.50	–	296.73	nr	**361.23**
range of two; 455 × 380 × 330 mm	5.50	177.38	–	457.45	nr	**634.83**
Add for each additional urinal	1.60	51.60	–	176.86	nr	**228.46**
Add for division between urinals	0.38	12.09	–	84.46	nr	**96.55**

ESSENTIAL READING FROM TAYLOR AND FRANCIS

Performance Based Fire Engineering of Structures

Yong Wang et al

This authoritative book summarizes the latest knowledge on performance-based approaches to the fire safety engineering of structures. Going beyond codes of practice, it addresses issues that have become more important in recent years, including the integrity of fire-resistant compartmentation, the progressive collapse of structures in fire and provisions for connection robustness, and accurate information on material properties at high temperatures. Written by core members of the COST C26 working group on fire resistance, it explains how to use advanced fire engineering design to improve structural safety and reduce construction costs.

June 2012: 234x156: 394pp
Hb: 978-0-415-55733-7: **£80.00**

To Order: Tel: +44 (0) 1235 400524 **Fax:** +44 (0) 1235 400525
or Post: Taylor and Francis Customer Services,
Bookpoint Ltd, Unit T1, 200 Milton Park, Abingdon, Oxon, OX14 4TA UK
Email: book.orders@tandf.co.uk

**For a complete listing of all our titles visit:
www.tandf.co.uk**

Taylor & Francis
Taylor & Francis Group

PART 7

Unit Costs – Highway Works

INTRODUCTORY NOTES

The Unit Costs in this part represent the net cost to the Contractor of executing the work on site; they are not the prices which would be entered in a tender Bill of Quantities.

It must be emphasized that the unit rates are averages calculated on unit outputs for typical site conditions. Costs can vary considerably from contract to contract depending on individual Contractors, site conditions, working methods and other factors. Reference should be made to Part 1 for a general discussion on Civil Engineering Estimating.

Guidance prices are included for work normally executed by specialists, with a brief description where necessary of the assumptions upon which the costs have been based. Should the actual circumstances differ, it would be prudent to obtain check prices from the specialists concerned, on the basis of actual/likely quantity of the work, nature of site conditions, geographical location, time constraints, etc.

The method of measurement adopted in this section is the **Method of Measurement for Highway Works**, subject to variances where this has been felt to be of advantage to produce more helpful price guidance.

We have structured this Unit Cost section to cover as many aspects of Civil and Highway works as possible.
The Gang hours column shows the output per measured unit in actual time, not the total labour hours; thus for an item involving a gang of 5 men each for 0.3 hours, the total labour hours would naturally be 1.5, whereas the Gang hours shown would be 0.3.

This section is structured in such a manner as to provide the User with adequate background information on how the rates have been calculated, so as to allow them to be readily adjusted to suit other conditions to the example presented:

- Alternative gang structures as well as the effect of varying bonus levels, travelling costs etc.
- Other types of plant or else different running costs from the medium usage presumed
- Other types of materials or else different discount/waste allowances from the levels presumed

ESSENTIAL READING FROM TAYLOR AND FRANCIS

Damage to Concrete Structures

De Schutter

Unlike the more specialist books which deal with diagnosis techniques or non-destructive testing, with repair and strengthening of concrete structures or with specific degradation mechanisms such as sulfate attack or alkali-silica reaction, this book describes and explains the different types of damage to concrete structures comprehensively. It is written as a textbook for undergraduate and masters students, and is also helpful to practitioners such as design engineers, architects or consultants.

November 2012: 234x156: 204pp
Pb: 978-0-415-60388-1: **£35.00**

To Order: Tel: +44 (0) 1235 400524 **Fax:** +44 (0) 1235 400525
or Post: Taylor and Francis Customer Services,
Bookpoint Ltd, Unit T1, 200 Milton Park, Abingdon, Oxon, OX14 4TA UK
Email: book.orders@tandf.co.uk

For a complete listing of all our titles visit:
www.tandf.co.uk

Taylor & Francis
Taylor & Francis Group

GUIDANCE NOTES

Generally

Adjustments should be made to the rates shown for time, location, local conditions, site constraints and any other factors likely to affect the costs of a specific scheme.

Materials cost

Materials costs within the rates have been calculated using the 'list prices' contained in Part 4: Resources, adjusted to allow for delivery charges (if any) and a 'reasonable' level of discount obtainable by the contractor, this will vary very much depending on the contractor's standing, the potential size of the order and the supplier's eagerness and will vary also between raw traded goods such as timber which will attract a low discount of perhaps 3%, if at all, and manufactured goods where the room for bargaining is much greater and can reach levels of 30% to 40%. High demand for a product at the time of pricing can dramatically reduce the potential discount, as can the world economy in the case of imported goods such as timber and copper. Allowance has also been made for wastage on site (generally 2.5% to 5%) dependent upon the risk of damage, the actual level should take account of the nature of the material and its method of storage and distribution about the site.

Labour cost

The composition of the labour and type of plant is generally stated at the beginning of each section, more detailed information on the calculation of the labour rates is given in Part 4: Resources. In addition is a summary of labour grades and responsibilities extracted from the Working Rule Agreement. Within Parts 6 and 7, each section is prefaced by a detailed build-up of the labour gang assumed for each type of work. This should allow the user to see the cost impact of a different gang as well as different levels of bonus payments, allowances for skills and conditions, travelling allowances etc. The user should be aware that the output constants are based on the gangs shown and would also need to be changed.

Plant cost

A rate build-up of suitable plant is generally stated at the beginning of each section, with more detailed information on alternative machines and their average fuel costs being given in Part 4: Resources. Within Parts 6 and 7, each section is prefaced by a detailed build-up of plant assumed for each type of work This should allow the user to see the cost impact of using alternative plant as well as different levels of usage. The user should be aware that the output constants are based on the plant shown and would also need to be changed.

Outputs

The user is directed to Part 10: Outputs, which contains a selection of output constants and in particular a chart of haulage times for various capacities of Tippers.

Method of Measurement

A keynote to bills of quantities for highway works is the brevity of descriptions due to a strong emphasis being placed on the estimator pricing the work described in the Specification and shown on the Drawings.

Although this part of the book is primarily based on MMHW, the specific rules have been varied from in cases where it has been felt that an alternative presentation would be of value to the book's main purpose of providing guidance on prices. This is especially so with a number of specialist contractors but also in the cases of work where a more detailed presentation will enable the user to allow for ancillary items.

LEVEL 1 DIVISION		LEVEL 2 CONSTRUCTION HEADING	LEVEL 3 MMHW SERIES HEADINGS
(i) Preliminaries		Preliminaries	Series 100
(ii) Roadworks		Roadworks General	Series 200 Series 300 Series 400 Series 600
		Main Carriageway	Series 500 Series 700 Series 1100
		Interchanges	Series 500 Series 700 Series 1100
		Side Roads	Series 500 Series 700 Series 1100
		Signs, Motorway Communications and Lighting	Series 1200 Series 1300 Series 1400 Series 1500
		Landscape and Ecology	Series 3000
(iii) Structures	Structure in form of Bridge or Viaduct; Name or Reference	Special Preliminaries Piling Substructure – End Supports	Series 2700 Series 1600 Series 500 Series 600 Series 1100 Series 1700 Series 1800 Series 1900 Series 2300 Series 2400
		Substructure – Intermediate Supports Substructure – Main Span Substructure – Approach Spans	As for End Supports
		Superstructure – Main Span Superstructure – Approach Spans Superstructure – Arch Ribs	Series 500 Series 1700 Series 1800 Series 1900 Series 2100 Series 2300 Series 2400
		Finishings	Series 400 Series 600 Series 700 Series 1100 Series 2000 Series 2400 Series 5000

LEVEL 1 DIVISION		LEVEL 2 CONSTRUCTION HEADING	LEVEL 3 MMHW SERIES HEADINGS
	Retaining Wall, Culvert, Subway, Gantry, Large Headwall, Gabion Wall, Diaphragm Wall, Pocket Type Reinforced Brickwork, Retaining Wall and the like; Name or Reference	Special Preliminaries Main Construction	Series 500 Series 600 Series 1100 Series 1600 Series 1700 Series 1800 Series 1900 Series 2300 Series 2400
		Finishings	Series 400 Series 600 Series 700 Series 1100 Series 2000 Series 2400 Series 5000
(iv) Structures where a choice of designs is offered	Structure Designed by the Overseeing Organization; Name or Reference Structure Designed by the Contractor; Name or Reference	To comply with the principles set down above for Structures	
(v) Structures Designed by the Contractor	Structure; Name or Reference		
(vi) Service Areas		Roadworks Structures	To comply with the principles set down above for Roadworks and Structures
(vii) Maintenance Compounds		Roadworks Structures	
(viii) Accommodation Works		Interest; Name or Reference	
(ix) Works for Statutory or Other Bodies		Body; Name or Reference	To comply with the principles set down above for Roadworks and Structures
(x) Daywork		Daywork	
(xi) PC & Provisional Sum		PC & Provisional Sum	

ESSENTIAL READING FROM TAYLOR AND FRANCIS

A Guide to Field Instrumentation in Geotechnics

R. Bassett

Geotechnical instrumentation is used for installation, monitoring and assessment on any sizeable project, particularly in urban areas, and is used for recording, controlled remedial work, and safety.

This unique and up-to-date book deals with the conceptual philosophy behind the use of instruments, and then systematically covers their practical use. It is divided into displacement dominated systems and stress recording systems. The limitations are discussed and the theoretical background for data assessment and presentation are covered in some detail, with some relevant background material in theoretical soil mechanics. Relevant advanced electronic techniques such as laser scanning in surveying and fibre-optics are also included and communication and data recovery systems are discussed.

It is written for senior designers, consulting engineers and major contractors who need a major introduction to the general purpose, availability and analysis of field instruments before details of their own project can be progressed, and serves as a text book to any specialist geotechnical MSc or professional seminar course in which instrumentation forms a major part.

November 2011: 232pp
Hb: 978-0-415-67537-6: **£90.00**

To Order: Tel: +44 (0) 1235 400524 **Fax:** +44 (0) 1235 400525
or Post: Taylor and Francis Customer Services,
Bookpoint Ltd, Unit T1, 200 Milton Park, Abingdon, Oxon, OX14 4TA UK
Email: book.orders@tandf.co.uk

For a complete listing of all our titles visit:
www.tandf.co.uk

Taylor & Francis
Taylor & Francis Group

SERIES 100: PRELIMINARIES

Item	Gang hours	Labour £	Plant £	Material £	Unit	Total rate £
NOTES						
General						
Refer to the example calculation of Preliminaries and Oncosts and Profits in Part 2.						
TEMPORARY ACCOMMODATION						
Erection of principal offices for the Overseeing Organization						
prefabricated unit; connect to services	–	–	500.00	–	nr	500.00
Erection of offices and messes for the Contractor						
prefabricated unit; connect to services	–	–	500.00	–	nr	500.00
Erection of stores and workshops for the Contractor						
prefabricated unit; connect to services	–	–	500.00	–	nr	500.00
Servicing of principal offices for the Overseeing Organization						
jack leg hutment; 3.7 m × 2.6 m	–	–	55.00	–	week	55.00
jack leg hutment; 7.35 m × 3.1 m	–	–	120.00	–	week	120.00
jack leg hutment; 14.7 m × 3.7 m	–	–	275.00	–	week	275.00
jack leg toilet unit; 4.9 m × 2.6 m	–	–	35.00	–	week	35.00
Servicing of portable offices for the Overseeing Organization						
wheeled cabin; 3.7 m × 2.3 m	–	–	92.50	–	week	92.50
wheeled cabin; 6.7 m × 2.3 m	–	–	120.00	–	week	120.00
Servicing of offices and messes for the Overseeing Organization						
jack leg hutment; 3.7 m × 2.6 m	–	–	55.00	–	week	55.00
jack leg hutment; 7.35 m × 3.1 m	–	–	120.00	–	week	120.00
jack leg hutment; 14.7 m × 3.7 m	–	–	275.00	–	week	275.00
wheeled cabin; 3.7 m × 2.3 m	–	–	92.50	–	week	92.50
wheeled cabin; 6.7 m × 2.3 m	–	–	120.00	–	week	120.00
jack leg toilet unit; 4.9 m × 2.6 m	–	–	35.00	–	week	35.00
canteen unit; 9.8 m × 2.6 m	–	–	115.00	–	week	115.00
Servicing of stores and workshops for the Contractor						
jack leg hutment; 3.7 m × 2.6 m	–	–	55.00	–	week	55.00
jack leg hutment; 7.35 m × 3.1 m	–	–	120.00	–	week	120.00
wheeled cabin; 3.7 m × 2.3 m	–	–	92.50	–	week	92.50
wheeled cabin; 6.7 m × 2.3 m	–	–	120.00	–	week	120.00
pollution decontamination unit; 6.7 m × 2.3 m	–	–	175.00	–	week	175.00
Dismantling of principal offices for the Overseeing Organization						
prefabricated unit; disconnect from services; removing	–	–	500.00	–	nr	500.00
Dismantling of offices and messes for the Contractor						
prefabricated unit; disconnect from services; removing	–	–	500.00	–	nr	500.00

SERIES 100: PRELIMINARIES

Item	Gang hours	Labour £	Plant £	Material £	Unit	Total rate £
TEMPORARY ACCOMMODATION – cont						
Dismantling of stores and workshops for the Contractor						
prefabricated unit; disconnect from services; removing	–	–	500.00	–	nr	500.00
VEHICLES FOR THE OVERSEEING ORGANIZATION						
Land Rover or similar, short wheelbase	–	–	556.00	–	week	556.00
Land Rover or similar, long wheelbase	–	–	656.00	–	week	656.00
For other types of transport vehicles refer to Resources – Plant	–	–	–	–	–	–
OPERATIVES FOR THE ENGINEER						
OPERATIVES FOR THE OVERSEEING ORGANIZATION						
Chainman for the Overseeing Organization	40.00	470.00	–	–	week	470.00
Driver for the Overseeing Organization	40.00	580.00	–	–	week	580.00
Laboratory assistant for the Overseeing Organization	40.00	380.00	–	–	week	380.00
INFORMATION BOARD						
Roadworks sign ref. 7002	–	–	–	–	nr	78.55
Roadworks sign ref. 7003	–	–	–	–	nr	98.75
Roadworks sign fref. 7004	–	–	–	–	nr	86.65

SERIES 200: SITE CLEARANCE

Item	Gang hours	Labour £	Plant £	Material £	Unit	Total rate £
NOTES						
Notes						
The prices in this section are to include for the removal of superficial obstructions down to existing ground level.						
Demolition of individual or groups of buildings or structures						
MMHW states that individual structures should be itemized. The following rates are given as £ per m^3 to simplify the pricing of different sized structures. (Refer also to Part 6 of this book Class D)						
RESOURCES – LABOUR						
Clearance gang						
1 ganger/chargehand (skill rate 4)		14.56				
1 skilled operative (skill rate 4)		13.61				
2 unskilled operatives (general)		25.42				
1 plant operator (skill rate 3)		16.55				
Total Gang Rate/Hour		£ **70.14**				
RESOURCES – PLANT						
Clearance						
0.8 m^3 tractor loader – 50% of time			14.83			
8 t lorry with hiab – 25% of time			6.31			
4 t dumper – 50% of time			5.95			
20 t mobile crane – 25% of time			26.69			
compressor, 11.3 m^3/min (450 cfm), 4 tool			6.75			
compressor tools: two brick hammers/picks – 50% of time			1.50			
compressor tools: chipping hammer			0.66			
compressor tools: medium rock drill 30			0.80			
compressor tools: road breaker			2.08			
compressor tools: two 15 m lengths hose			0.31			
Total Gang Rate/Hour		£	**65.88**			
SITE CLEARANCE						
General site clearance						
open field site	9.91	695.09	647.74	–	ha	**1342.83**
medium density wooded	20.61	1445.59	1346.97	–	ha	**2792.56**
heavy density wooded	32.09	2250.79	2097.51	–	ha	**4348.30**
urban areas (town centre)	30.60	2146.28	2000.07	–	ha	**4146.35**
live dual carriageway	30.60	2146.28	2000.07	–	ha	**4146.35**

SERIES 200: SITE CLEARANCE

Item	Gang hours	Labour £	Plant £	Material £	Unit	Total rate £
SITE CLEARANCE – cont						
Demolition of building or structure						
building; brick construction with timber floor and roof	–	–	–	–	m^3	7.42
building; brick construction with concrete floor and roof	–	–	–	–	m^3	13.09
building; masonry construction with timber floor and roof	–	–	–	–	m^3	9.65
building; reinforced concrete frame construction with brick infill	–	–	–	–	m^3	12.83
building; steel frame construction with brick infill	–	–	–	–	m^3	6.99
building; steel frame construction with cladding	–	–	–	–	m^3	6.66
building; timber	–	–	–	–	m^3	6.00
reinforced concrete bridge deck or superstructure	–	–	–	–	m^3	60.50
reinforced concrete bridge abutment or bank seat	–	–	–	–	m^3	55.00
reinforced concrete retaining wall	–	–	–	–	m^3	165.00
brick or masonry retaining wall	–	–	–	–	m^3	92.95
brick or masonry boundary wall	–	–	–	–	m^3	77.00
dry stone boundary wall	–	–	–	–	m^3	82.50
TAKE UP OR TAKE DOWN AND SET ASIDE FOR REUSE OR REMOVE TO STORE OR TIP OFF SITE						
Take up or down and set aside for reuse						
precast concrete kerbs and channels	0.02	1.19	2.54	–	m	3.73
precast concrete edgings	0.01	0.90	1.95	–	m	2.85
precast concrete drainage and kerb blocks	0.02	1.49	3.14	–	m	4.63
precast concrete drainage channel systems	0.02	1.49	3.14	–	m	4.63
tensioned single sided corrugated beam safety fence	0.14	9.82	9.24	–	m	19.06
timber post and 4 rail fence	0.08	5.61	5.27	–	m	10.88
bench seat	0.13	9.12	7.59	–	nr	16.71
cattle trough	0.16	10.94	9.10	–	nr	20.04
permanent bollard	0.13	9.12	7.59	–	nr	16.71
pedestrian crossing lights; pair	0.26	18.24	17.15	–	nr	35.39
lighting column including bracket arm and lantern; 5 m high	0.58	40.68	38.23	–	nr	78.91
lighting column including bracket arm and lantern; 10 m high	0.61	42.79	40.18	–	nr	82.97
traffic sign	0.26	18.24	17.15	–	nr	35.39
timber gate	0.13	9.12	7.59	–	nr	16.71
stile	0.13	9.12	7.59	–	nr	16.71
road stud	0.03	1.82	1.50	–	nr	3.32
chamber cover and frame	0.03	1.82	1.50	–	nr	3.32
gully grating and frame	0.03	1.82	1.50	–	nr	3.32
feeder pillars	0.03	1.82	1.50	–	nr	3.32
Take up or down and remove to store off site						
precast concrete kerbs and channels	0.02	1.19	5.35	–	m	6.54
precast concrete edgings	0.01	0.90	4.06	–	m	4.96
precast concrete drainage and kerb blocks	0.02	1.49	6.66	–	m	8.15

SERIES 200: SITE CLEARANCE

Item	Gang hours	Labour £	Plant £	Material £	Unit	Total rate £
precast concrete drainage channel systems	0.02	1.49	6.66	–	m	8.15
tensioned single sided corrugated beam safety fence	0.14	9.82	11.56	–	m	21.38
timber post and 4 rail fence	0.08	5.61	6.60	–	m	12.21
bench seat	0.13	9.12	16.20	–	nr	25.32
cattle trough	0.16	10.94	19.43	–	nr	30.37
permanent bollard	0.13	9.12	16.20	–	nr	25.32
pedestrian crossing lights; pair	0.26	18.24	21.45	–	nr	39.69
lighting column including bracket arm and lantern; 5 m high	0.58	40.68	47.83	–	nr	88.51
lighting column including bracket arm and lantern; 10 m high	0.61	42.79	50.28	–	nr	93.07
traffic sign	0.26	18.24	21.45	–	nr	39.69
timber gate	0.13	9.12	16.20	–	nr	25.32
stile	0.13	9.12	16.20	–	nr	25.32
road stud	0.03	1.82	3.23	–	nr	5.05
chamber cover and frame	0.03	1.82	3.23	–	nr	5.05
gully grating and frame	0.03	1.82	3.23	–	nr	5.05
feeder pillars	0.03	1.82	3.23	–	nr	5.05
Take up or down and remove to tip off site						
tensioned single sided corrugated beam safety fence	0.17	11.92	11.42	–	m	23.34
timber post and 4 rail fence	0.09	6.31	6.04	–	m	12.35
low pressure gas mains up to 150 mm diameter	0.04	2.81	2.69	–	m	5.50
low pressure water mains up to 75 mm diameter	0.03	2.10	2.03	–	m	4.13
power cable laid singly	0.03	2.10	2.03	–	m	4.13
lighting column including bracket arm and lantern; 5 m high	0.82	57.51	55.12	–	nr	112.63
lighting column including bracket arm and lantern; 10 m high	0.85	59.62	57.10	–	nr	116.72
traffic sign including posts	0.38	26.65	25.03	–	nr	51.68
Removal of existing reflectorized thermoplastic road markings						
100 mm wide line	–	–	–	–	m	1.25
150 mm wide line	–	–	–	–	m	1.85
200 mm wide line	–	–	–	–	m	2.50
arrow or letter ne 6.0 m long	–	–	–	–	nr	18.75
arrow or letter 6.0–16.0 m long	–	–	–	–	nr	78.50

SERIES 300: FENCING

Item	Gang hours	Labour £	Plant £	Material £	Unit	Total rate £
NOTES						
This section is restricted to those fences and barriers which are most commonly found on Highway Works. Hedges have been included, despite not being specifically catered for in the MMHW.						
The re-erection cost for fencing taken from store assumes that major components are in good condition; the prices below allow a sum of 20% of the value of new materials to cover minor repairs, new fixings and touching up any coatings.						
RESOURCES – LABOUR						
Fencing/barrier gang						
1 ganger/chargehand (skill rate 4) – 50% of time		7.28				
1 skilled operative (skill rate 4) – 50% of time		6.80				
1 unskilled operative (general)		12.71				
1 plant operator (skill rate 4)		14.88				
Total Gang Rate/Hour	£	**41.68**				
Horticultural works gang						
1 skilled operative (skill rate 4)		13.61				
1 unskilled operative (general)		12.71				
Total Gang Rate/Hour	£	**26.32**				
RESOURCES – PLANT						
Fencing/Barriers						
agricultural type tractor; fencing auger			21.80			
gas oil for ditto			6.30			
drop sided trailer, two axles			0.35			
power tools etc. (fencing)			1.15			
Total Gang Rate/Hour		£	**29.60**			
RESOURCES – MATERIALS						
All rates for materials are based on the most economically available materials with a minimum waste allowance of 2.5% and supplier's discount of 15%.						
ENVIRONMENTAL BARRIERS						
Hedges						
Set out, nick out and excavate trench minimum 400 mm deep and break up subsoil to minimum depth 300 mm	0.15	3.95	–	–	m	**3.95**
Supply and plant hedging plants ; backfilling with excavated topsoil						
single row plants at 200 mm centres	0.25	6.58	–	4.08	m	**10.66**
single row plants at 300 mm centres	0.17	4.47	–	2.71	m	**7.18**
single row plants at 400 mm centres	0.13	3.29	–	2.04	m	**5.33**

SERIES 300: FENCING

Item	Gang hours	Labour £	Plant £	Material £	Unit	Total rate £
single row plants at 500 mm centres	0.10	2.63	–	1.63	m	4.26
single row plants at 600 mm centres	0.08	2.11	–	1.35	m	3.46
double row plants at 200 mm centres	0.50	13.16	–	8.15	m	21.31
double row plants at 300 mm centres	0.34	8.95	–	5.43	m	14.38
double row plants at 400 mm centres	0.25	6.58	–	4.08	m	10.66
double row plants at 500 mm centres	0.20	5.26	–	3.26	m	8.52
double row plants at 600 mm centres	0.16	4.21	–	2.71	m	6.92
Extra for incorporating manure at $1\,m^3/30\,m^3$	0.60	7.63	–	0.24	m^3	7.87

Noise barriers
Noise barriers consist of the erection of reflective or absorptive acoustical screening to reduce nuisance from noise. Due to the divergence in performance requirements and specification for various locations it is not practical to state all inclusive unit costs. Therefore advice should be sought from Specialist Contractors in order to obtain accurate costings. However listed below are examples of sample specifications together with approximate costings in order to obtain budget prices.
NB: The following unit costs are based upon a 2.0 m high barrier

Noise reflective barriers

Item	Gang hours	Labour £	Plant £	Material £	Unit	Total rate £
Barrier with architectural precast concrete panels and integral posts	–	–	–	–	m^2	167.00
Barrier with acoustical timber planks post support system	–	–	–	–	m^2	149.75

Sound absorptive barriers

Item	Gang hours	Labour £	Plant £	Material £	Unit	Total rate £
Barrier with architectural precast wood fibre concrete panels and integral posts	–	–	–	–	m^2	198.50
Barrier with perforated steel and mineral wool blankets in self-supporting system	–	–	–	–	m^2	220.00

FENCING, GATES AND STILES
Temporary fencing

Item	Gang hours	Labour £	Plant £	Material £	Unit	Total rate £
Type 1; 1.275 m high, timber posts and two strands of galvanized barbed wire and four strands of galvanized plain wire	0.06	2.50	1.78	5.13	m	9.41
Type 2; 1.275 m high, timber posts and two strands of galvanized barbed wire and four strands of galvanized plain wire	0.06	2.50	1.78	7.69	m	11.97
Type 3; 1.2 m high, timber posts and cleft chestnut paling	0.07	2.92	2.07	9.22	m	14.21
Type 4; 0.9 m high, timber posts and galvanized rectangular wire mesh	0.13	5.42	3.85	9.22	m	18.49

SERIES 300: FENCING

Item	Gang hours	Labour £	Plant £	Material £	Unit	Total rate £
FENCING, GATES AND STILES – cont						
Timber rail fencing						
1.4 m high, timber posts and four rails	0.13	5.42	3.85	12.30	m	**21.57**
Plastic coated heavy pattern chain link fencing						
1.40 m high with 125 × 125 mm concrete posts	0.05	2.08	1.48	7.94	m	**11.50**
1.80 m high with 125 × 125 mm concrete posts	0.06	2.50	1.78	15.12	m	**19.40**
Plastic coated strained wire fencing						
1.35 m high, nine strand with 40 × 40 × 3 mm plastic coated RHS posts	0.16	6.67	4.74	14.86	m	**26.27**
1.80 m high, eleven strand with 50 × 50 × 3 mm plastic coated RHS posts	0.20	8.34	5.92	19.99	m	**34.25**
2.10 m high, fifteen strand with 50 × 50 × 3 mm plastic coated RHS posts	0.22	9.17	6.51	23.06	m	**38.74**
Woven wire fencing						
1.23 m high, galvanized wire with 75 × 150 mm timber posts	0.06	2.50	1.78	6.66	m	**10.94**
Closeboarded fencing						
1.80 m high with 125 × 125 mm concrete posts	0.30	12.50	8.88	41.00	m	**62.38**
Concrete foundation						
to main posts	0.09	3.75	0.14	4.88	nr	**8.77**
to straining posts	0.09	3.75	0.14	4.88	nr	**8.77**
to struts	0.09	3.75	0.14	4.88	nr	**8.77**
to intermediate posts	0.09	3.75	0.14	4.88	nr	**8.77**
Steel tubular frame single field gates						
1.175 m high 3.60 m wide	0.30	12.50	8.88	143.50	nr	**164.88**
1.175 m high 4.50 m wide	0.30	12.50	8.88	164.00	nr	**185.38**
Steel tubular frame half mesh single field gates						
1.175 m high 3.60 m wide	0.30	12.50	8.88	169.13	nr	**190.51**
1.175 m high 4.50 m wide	0.30	12.50	8.88	189.63	nr	**211.01**
Steel tubular frame extra wide single field gates						
1.175 m high 4.88 m wide	0.30	12.50	8.88	194.75	nr	**216.13**
Steel tubular frame double field gates						
1.175 m high 5.02 m wide	0.60	25.00	17.76	210.13	nr	**252.89**
Timber single field gates						
1.27 m high 3.00 m wide	1.84	76.68	54.46	251.13	nr	**382.27**
1.27 m high 3.60 m wide	1.90	79.18	56.23	266.50	nr	**401.91**
1.27 m high 4.10 m wide	2.00	83.35	59.19	281.88	nr	**424.42**
1.27 m high 4.71 m wide	2.00	83.35	59.19	307.50	nr	**450.04**
Timber Type 1 wicket gates						
1.27 m high 1.20 m wide	1.20	50.01	35.52	128.13	nr	**213.66**
Timber Type 2 wicket gates						
1.27 m high 1.02 m wide	1.20	50.01	35.52	179.38	nr	**264.91**
Timber kissing gates						
1.27 m high 1.77 m wide	2.00	83.35	59.19	230.63	nr	**373.17**
Timber stiles Type 1						
1.45 m high 1.00 m wide	1.50	62.51	44.40	107.63	nr	**214.54**
Timber stiles Type 2						
1.45 m high 1.00 m wide	1.40	58.34	41.43	76.88	nr	**176.65**
Extra for						
sheep netting on post and wire	0.04	1.46	0.05	0.77	m	**2.28**
pig netting on post and wire	0.04	1.46	0.05	0.77	m	**2.28**

SERIES 300: FENCING

Item	Gang hours	Labour £	Plant £	Material £	Unit	Total rate £
REMOVE FROM STORE AND RE-ERECT FENCING, GATES AND STILES						
Timber rail fencing						
1.4 m high, timber posts and four rails	0.13	5.42	3.85	2.82	m	**12.09**
Plastic coated heavy pattern chain link fencing						
1.40 m high with 125 × 125 mm concrete posts	0.05	2.08	1.48	1.79	m	**5.35**
1.80 m high with 125 × 125 mm concrete posts	0.06	2.50	1.78	2.31	m	**6.59**
Plastic coated strained wire fencing						
1.35 m high, nine strand with 40 × 40 × 3 mm plastic coated RHS posts	0.16	6.67	4.73	3.23	m	**14.63**
1.80 m high, eleven strand with 50 × 50 × 3 mm plastic coated RHS posts	0.20	8.34	5.92	4.25	m	**18.51**
2.10 m high, fifteen strand with 50 × 50 × 3 mm plastic coated RHS posts	0.22	9.17	6.51	4.56	m	**20.24**
Woven wire fencing						
1.23 m high, galvanized wire with 75 × 150 mm timber posts	0.06	2.50	1.78	1.40	m	**5.68**
Closeboarded fencing						
1.80 m high with 125 × 125 mm concrete posts	0.30	12.50	8.88	7.69	m	**29.07**
Steel tubular frame single field gates						
1.175 m high 3.60 m wide	0.30	12.50	8.88	12.81	nr	**34.19**
1.175 m high 4.50 m wide	0.30	12.50	8.88	16.45	nr	**37.83**
Steel tubular frame half mesh single field gates						
1.175 m high 3.60 m wide	0.30	12.50	8.88	16.45	nr	**37.83**
1.175 m high 4.50 m wide	0.30	12.50	8.88	19.99	nr	**41.37**
Steel tubular frame extra wide single field gates						
1.175 m high 4.88 m wide	0.30	12.50	8.88	18.30	nr	**39.68**
Steel tubular frame double field gates						
1.175 m high 5.02 m wide	0.60	25.00	17.76	27.42	nr	**70.18**
Timber single field gates						
1.27 m high 3.00 m wide	1.84	76.68	54.46	32.29	nr	**163.43**
1.27 m high 3.60 m wide	1.90	79.18	56.23	37.41	nr	**172.82**
1.27 m high 4.10 m wide	2.00	83.35	59.19	45.10	nr	**187.64**
1.27 m high 4.71 m wide	2.00	83.35	59.19	46.48	nr	**189.02**
Timber Type 1 wicket gates						
1.27 m high 1.20 m wide	1.20	50.01	35.52	10.25	nr	**95.78**
Timber Type 2 wicket gates						
1.27 m high 1.02 m wide	1.20	50.01	35.52	11.79	nr	**97.32**
Timber kissing gates						
1.27 m high 1.77 m wide	2.00	83.35	59.19	32.29	nr	**174.83**
Timber stiles Type 1						
1.45 m high 1.00 m wide	1.50	62.51	44.40	26.65	nr	**133.56**
Timber stiles Type 2						
1.45 m high 1.00 m wide	1.40	58.34	41.43	19.99	nr	**119.76**
Excavation in hard material						
Extra over excavation for excavation in Hard Material in fencing works	0.50	20.84	–	–	m^3	**20.84**

SERIES 400: ROAD RESTRAINT SYSTEMS (VEHICLE AND PEDESTRIAN)

Item	Gang hours	Labour £	Plant £	Material £	Unit	Total rate £
NOTES						
The re-erection cost for safety fencing taken from store assumes that major components are in good condition; the prices below allow a sum of 20% of the value of new materials to cover minor repairs, new fixings and touching up any coatings.						
The heights of the following parapets are in accordance with the Standard Designs and DfT requirements. The rates include for all anchorages and fixings and in the case of steel, galvanizing at works and painting four coat paint system on site together with etc.hing the galvanized surface, as necessary.						
RESOURCES – LABOUR						
Safety barrier gang						
1 ganger/chargehand (skill rate 4)		14.56				
1 skilled operative (skill rate 4)		13.61				
2 unskilled operatives (general)		25.42				
1 plant operator (skill rate 4)		14.88				
Total Gang Rate/Hour		£ **68.47**				
Parapet gang						
1 ganger/chargehand (skill rate 4)		14.56				
1 skilled operative (skill rate 4)		13.61				
2 unskilled operatives (general)		25.42				
1 plant operator (skill rate 4)		14.88				
Total Gang Rate/Hour		£ **68.47**				
RESOURCES – PLANT						
Safety barriers						
agricultural type tractor; fencing auger			21.80			
gas oil for ditto			6.30			
drop sided trailer, two axles			0.35			
small tools (part time)			1.15			
Total Gang Rate/Hour			£ **29.60**			
Parapets						
agricultural type tractor; front bucket – 50% of time			6.00			
gas oil for ditto			3.15			
2.80 m³/min (100 cfm) compressor; two tool			3.50			
gas oil for ditto			4.20			
compressor tools: heavy rock drill 33, 84 cfm			1.88			
compressor tools: rotary drill, 10 cfm			1.75			
8 t lorry with 1 t hiab – 50% of time			10.00			
gas oil for ditto			2.63			
Total Gang Rate/Hour			£ **33.10**			

SERIES 400: ROAD RESTRAINT SYSTEMS (VEHICLE AND PEDESTRIAN)

Item	Gang hours	Labour £	Plant £	Material £	Unit	Total rate £
BEAM SAFETY BARRIERS						
Prices generally are for beams 'straight or curved exceeding 120 m radius', for work to a tighter radius						
50–120 m radius – Add 15%						
not exceeding 50 m radius – Add 40%						
Untensioned beams						
single sided corrugated beam	0.07	4.79	2.07	19.15	m	26.01
double sided corrugated beam	0.22	15.06	6.51	38.35	m	59.92
single sided open box beam	0.07	4.79	2.07	38.95	m	45.81
single sided double rail open box beam	0.22	15.06	6.51	77.50	m	99.07
double height open box beam	0.24	16.43	7.10	80.15	m	103.68
Tensioned beams						
single sided corrugated beam	0.10	6.85	2.96	20.25	m	30.06
double sided corrugated beam	0.12	8.22	3.55	41.75	m	53.52
Long driven post						
for single sided tensioned corrugated beam	0.06	3.77	1.63	45.00	nr	50.40
for double sided tensioned corrugated beam	0.06	3.77	1.63	45.00	nr	50.40
for single sided open box beam	0.06	3.77	1.63	45.00	nr	50.40
for double sided open box beam	0.06	3.77	1.63	45.00	nr	50.40
for double height open box beam	0.06	3.77	1.63	45.00	nr	50.40
Short post for setting in concrete or socket						
for single sided tensioned corrugated beam	0.06	4.11	1.78	17.50	nr	23.39
for double sided tensioned corrugated beam	0.06	4.11	1.78	18.00	nr	23.89
for single sided open box beam	0.06	4.11	1.78	21.25	nr	27.14
for double sided open box beam	0.06	4.11	1.78	25.00	nr	30.89
Mounting bracket fixed to structure						
for single sided open box beam	0.16	10.96	4.74	70.00	nr	85.70
Terminal section						
for untensioned single sided corrugated beam	0.71	48.61	21.02	490.00	nr	559.63
for untensioned double sided corrugated beam	1.25	85.59	37.00	520.00	nr	642.59
for untensioned single sided open box beam	1.01	69.15	29.89	525.00	nr	624.04
for untensioned double sided open box beam	1.78	121.88	52.68	875.00	nr	1049.56
for tensioned single sided corrugated beam	0.96	65.73	28.41	350.00	nr	444.14
for tensioned double sided corrugated beam	1.70	116.40	50.31	550.00	nr	716.71
Full height anchorage						
for single sided tensioned corrugated beam	3.95	270.46	116.91	750.00	nr	1137.37
for double sided tensioned corrugated beam	4.35	297.84	128.74	880.00	nr	1306.58
for single sided open box beam	3.80	260.19	112.46	655.00	nr	1027.65
for double sided open box beam	4.20	287.57	124.31	775.00	nr	1186.88
Expansion joint anchorage						
for single sided open box beam	4.52	309.48	133.78	1450.00	nr	1893.26
for double sided open box beam	5.15	352.62	152.42	1600.00	nr	2105.04
Type 048 connection to bridge parapet						
for single sided open box beam	0.70	47.93	20.72	172.50	nr	241.15
Connection piece for single sided open box beam to single sided corrugated beam	0.78	53.41	23.08	175.00	nr	251.49
Standard concrete foundation						
for post for corrugated beam	0.23	15.75	6.69	7.15	nr	29.59
for post for open box beam	0.23	15.75	6.80	8.85	nr	31.40

SERIES 400: ROAD RESTRAINT SYSTEMS (VEHICLE AND PEDESTRIAN)

Item	Gang hours	Labour £	Plant £	Material £	Unit	Total rate £
BEAM SAFETY BARRIERS – cont						
Concrete foundation Type 1 spanning filter drain						
for post for corrugated beam	0.25	17.12	7.40	12.30	nr	**36.82**
for post for open box beam	0.25	17.12	7.40	10.25	nr	**34.77**
Standard socketed foundation for post for open box beam	0.30	20.54	8.88	11.15	nr	**40.57**
CONCRETE SAFETY BARRIERS						
Permanent vertical concrete safety barrier; TRL Design – DfT Approved						
Intermediate units Type V01 & V02; 3 m long						
straight or curved exceeding 50 m radius	0.16	10.96	4.74	427.50	nr	**443.20**
curved not exceeding 50 m radius	0.20	13.69	5.92	427.50	nr	**447.11**
Make up units Type V05 & V06; 1 m long						
straight or curved exceeding 50 m radius	0.30	20.54	8.88	547.50	nr	**576.92**
Termination units Type V03 & V04; 3 m long	0.50	34.23	14.80	537.50	nr	**586.53**
Transition to single sided open box beam unit Type V08 & V09; 1.5 m long	0.37	25.33	10.95	639.50	nr	**675.78**
Transition to rectangular hollow section beam unit Type V10, V11 & V12; 1.5 m long	0.37	25.33	10.95	547.50	nr	**583.78**
Transition to double sided open box beam; 1.5 m long unit Type V07	0.37	25.33	10.95	617.50	nr	**653.78**
Anchor plate sets (normally two plates per first and last three units in any run)	0.15	10.27	9.25	42.50	nr	**62.02**
Temporary concrete safety barrier; TRL Design – DfT Approved						
Intermediate units Type V28; 3 m long						
straight or curved exceeding 50 m radius	0.16	10.96	4.74	385.88	nr	**401.58**
curved not exceeding 50 m radius	0.20	13.69	5.92	385.88	nr	**405.49**
Termination units Type V29; 3 m long	0.50	34.23	14.80	435.88	nr	**484.91**
WIRE ROPE SAFETY FENCES						
Brifen wire rope safety fencing DfT approved; based on 600 m lengths; 4 rope system; posts at 2.40 m general spacings						
Wire rope	0.03	2.05	0.89	12.15	m	**15.09**
Long driven line posts	0.05	3.42	1.48	22.50	nr	**27.40**
Long driven deflection posts	0.05	3.42	1.48	25.75	nr	**30.65**
Long driven height restraining posts	0.05	3.42	1.48	27.00	nr	**31.90**
Short line post for setting in concrete or socket	0.06	4.11	1.78	20.50	nr	**26.39**
Short deflection post for setting in concrete or socket	0.06	4.11	1.78	20.85	nr	**26.74**
Short height restraining post for setting in concrete or socket	0.06	4.11	1.78	22.50	nr	**28.39**
Fixed height surface mounted post fixed to structure or foundation	0.09	6.16	2.67	28.50	nr	**37.33**
Standard intermediate anchorage	2.00	136.94	59.19	325.00	nr	**521.13**
Standard end anchorage	2.00	136.94	59.19	118.50	nr	**314.63**
In situ standard concrete foundation for post	0.23	15.75	6.80	7.05	nr	**29.60**
In situ standard socketed foundation for post	0.33	22.60	9.77	7.15	nr	**39.52**
Concrete foundation Type 1 spanning filter drain for post	0.37	25.33	10.95	10.15	nr	**46.43**

SERIES 400: ROAD RESTRAINT SYSTEMS (VEHICLE AND PEDESTRIAN)

Item	Gang hours	Labour £	Plant £	Material £	Unit	Total rate £
VEHICLE PARAPETS The heights of the following parapets are in accordance with the Standard Designs and DfT requirements. The rates include for all anchorages and fixings and in the case of steel, galvanizing at works and painting four coat paint system on site, together with etc.hing the galvanized surface as necessary						
Steel Parapets						
Metal parapet Group P1; 1.0 m high; comprising steel yielding posts and steel horizontal rails						
straight or curved exceeding 50 m radius	–	–	–	164.00	m	**164.00**
curved not exceeding 50 m radius	–	–	–	169.13	m	**169.13**
Metal parapet Group P2 (48 Kph); 1.0 m high; comprising steel yielding posts and steel horizontal rails with vertical infill bars						
straight or curved exceeding 50 m radius	–	–	–	215.25	m	**215.25**
curved not exceeding 50 m radius	–	–	–	230.63	m	**230.63**
Metal parapet Group P2 (80 Kph); 1.0 m high; comprising steel yielding posts and steel horizontal rails						
straight or curved exceeding 50 m radius	–	–	–	235.75	m	**235.75**
curved not exceeding 50 m radius	–	–	–	240.88	m	**240.88**
Metal parapet Group P2 (113 Kph); 1.0 m high; comprising steel yielding posts and steel horizontal rails						
straight or curved exceeding 50 m radius	–	–	–	256.25	m	**256.25**
curved not exceeding 50 m radius	–	–	–	240.88	m	**240.88**
Metal parapet Group P4; 1.15 m high; comprising steel yielding posts and steel horizontal rails						
straight or curved exceeding 50 m radius	–	–	–	333.13	m	**333.13**
curved not exceeding 50 m radius	–	–	–	358.75	m	**358.75**
Metal parapet Group P5; 1.25 m high; comprising steel yielding posts and steel horizontal rails						
straight or curved exceeding 50 m radius	–	–	–	390.24	m	**390.24**
curved not exceeding 50 m radius	–	–	–	461.25	m	**461.25**
Metal parapet Group P5; 1.50 m high; comprising steel yielding posts and steel horizontal rails						
straight or curved exceeding 50 m radius	–	–	–	666.25	m	**666.25**
curved not exceeding 50 m radius	–	–	–	717.50	m	**717.50**
Metal parapet Group P6; 1.50 m high; comprising steel yielding posts and steel horizontal rails						
straight or curved exceeding 50 m radius	–	–	–	768.75	m	**768.75**
curved not exceeding 50 m radius	–	–	–	820.00	m	**820.00**

SERIES 400: ROAD RESTRAINT SYSTEMS (VEHICLE AND PEDESTRIAN)

Item	Gang hours	Labour £	Plant £	Material £	Unit	Total rate £
VEHICLE PARAPETS – cont						
Aluminium Parapets						
Metal parapet Group P1; 1.0 m high; comprising aluminium yielding/frangible posts and aluminium horizontal rails						
straight or curved exceeding 50 m radius	–	–	–	166.05	m	**166.05**
curved not exceeding 50 m radius	–	–	–	174.25	m	**174.25**
Metal parapet Group P2 (80 Kph); 1.0 m high; comprising aluminium yielding/frangible posts and aluminium horizontal rails with mesh infill						
straight or curved exceeding 50 m radius	–	–	–	179.38	m	**179.38**
curved not exceeding 50 m radius	–	–	–	181.43	m	**181.43**
Metal parapet Group P2 (113 Kph); 1.0 m high; comprising aluminium yielding/frangible posts and aluminium horizontal rails with mesh infill						
straight or curved exceeding 50 m radius	–	–	–	177.39	m	**177.39**
curved not exceeding 50 m radius	–	–	–	190.00	m	**190.00**
Metal parapet Group P4; 1.15 m high; comprising aluminium yielding/frangible posts and aluminium horizontal rails						
straight or curved exceeding 50 m radius	–	–	–	174.25	m	**174.25**
curved not exceeding 50 m radius	–	–	–	181.94	m	**181.94**
Metal parapet Group P5; 1.25 m high; comprising aluminium yielding/frangible posts and aluminium horizontal rails with solid sheet infill, anti-access panels						
straight or curved exceeding 50 m radius	–	–	–	205.00	m	**205.00**
curved not exceeding 50 m radius	–	–	–	220.38	m	**220.38**
Metal parapet Group P5; 1.50 m high; comprising aluminium yielding/frangible posts and aluminium horizontal rails with solid sheet infill, anti-access panels						
straight or curved exceeding 50 m radius	–	–	–	235.75	m	**235.75**
curved not exceeding 50 m radius	–	–	–	240.88	m	**240.88**
Crash cushions						
Static crash cushion system to BS EN1317-3, Class 110 km/hr	–	–	–	–	nr	**38000.00**
REMOVE FROM STORE AND RE-ERECT BEAM SAFETY BARRIERS						
Prices generally are for beams straight or curved exceeding 120 m radius, for work to a tighter radius						
50–120 m radius – Add 15%						
not exceeding 50 m radius – Add 25%						
Untensioned beams						
single sided corrugated beam	0.07	4.79	2.07	3.85	m	**10.71**
double sided corrugated beam	0.22	15.06	6.51	7.65	m	**29.22**
single sided open box beam	0.07	4.79	2.07	7.75	m	**14.61**
single sided double rail open box beam	0.22	15.06	6.51	15.85	m	**37.42**
double height open box beam	0.24	16.43	7.10	16.05	m	**39.58**

SERIES 400: ROAD RESTRAINT SYSTEMS (VEHICLE AND PEDESTRIAN)

Item	Gang hours	Labour £	Plant £	Material £	Unit	Total rate £
Tensioned beams						
single sided corrugated beam	0.10	6.85	2.96	4.05	m	**13.86**
double sided corrugated beam	0.12	8.22	3.55	8.35	m	**20.12**
Long driven post						
for single sided tensioned corrugated beam	0.06	3.77	1.63	9.50	nr	**14.90**
for double sided tensioned corrugated beam	0.06	3.77	1.63	10.00	nr	**15.40**
for single sided open box beam	0.06	3.77	1.63	9.15	nr	**14.55**
for double sided open box beam	0.06	3.77	1.63	10.75	nr	**16.15**
for double height open box beam	0.06	3.77	1.63	27.00	nr	**32.40**
Short post for setting in concrete or socket						
for single sided tensioned corrugated beam	0.06	4.11	1.78	17.50	nr	**23.39**
for double sided tensioned corrugated beam	0.06	4.11	1.78	18.00	nr	**23.89**
for single sided open box beam	0.06	4.11	1.78	21.50	nr	**27.39**
for double sided open box beam	0.06	4.11	1.78	26.00	nr	**31.89**
Mounting bracket fixed to structure						
for single sided open box beam	0.16	10.96	4.73	6.75	nr	**22.44**
Terminal section						
for untensioned single sided corrugated beam	0.71	48.61	21.02	100.00	nr	**169.63**
for untensioned double sided corrugated beam	1.25	85.59	37.00	105.00	nr	**227.59**
for untensioned single sided open box beam	1.01	69.15	29.89	107.50	nr	**206.54**
for untensioned double sided open box beam	1.78	121.88	52.68	175.00	nr	**349.56**
for tensioned single sided corrugated beam	0.96	65.73	28.41	65.00	nr	**159.14**
for tensioned double sided corrugated beam	1.70	116.40	50.31	105.00	nr	**271.71**
Full height anchorage						
for single sided tensioned corrugated beam	3.95	270.46	116.91	130.00	nr	**517.37**
for double sided tensioned corrugated beam	4.35	297.84	128.74	170.00	nr	**596.58**
for single sided open box beam	3.80	260.19	112.46	130.00	nr	**502.65**
for double sided open box beam	4.20	287.57	124.31	155.00	nr	**566.88**
Expansion joint anchorage						
for single sided open box beam	4.52	309.48	133.78	270.00	nr	**713.26**
for double sided open box beam	5.15	352.62	152.42	350.00	nr	**855.04**
Type 048 connection to bridge parapet						
for single sided open box beam	0.70	47.93	20.72	7.57	nr	**76.22**
Connection piece for single sided open box beam to single sided corrugated beam	0.78	53.41	23.08	175.00	nr	**251.49**
Standard concrete foundation						
for post for corrugated beam	0.23	15.75	6.69	7.15	nr	**29.59**
for post for open box beam	0.23	15.75	6.80	8.85	nr	**31.40**
Concrete foundation Type 1 spanning filter drain						
for post for corrugated beam	0.25	17.12	7.40	12.30	nr	**36.82**
for post for open box beam	0.25	17.12	7.40	10.25	nr	**34.77**
Standard socketed foundation for post for open box beam	0.30	20.54	8.88	11.15	nr	**40.57**

SERIES 400: ROAD RESTRAINT SYSTEMS (VEHICLE AND PEDESTRIAN)

Item	Gang hours	Labour £	Plant £	Material £	Unit	Total rate £
PEDESTRIAN GUARD RAILS AND HANDRAILS						
New work						
Tubular galvanized mild steel pedestrian guard rails to BS 7818 with mesh infill (105 swg, 50 × 50 mm mesh); 1.0 m high						
mounted on posts with concrete footing	0.16	10.96	4.74	201.00	m	**216.70**
mounted on posts bolted to structure or ground beam	0.14	9.59	4.14	257.00	m	**270.73**
Solid section galvanized steel pedestrian guard rails with vertical rails (group P4 parapet); 1.0 m high						
mounted on posts with concrete footing	0.19	13.01	5.62	33.50	m	**52.13**
mounted on posts bolted to structure or ground beam	0.17	11.64	5.03	42.83	m	**59.50**
Tubular double ball galvanized steel handrail						
50 mm diameter; 1.20 m high posts	0.15	10.27	4.44	148.00	m	**162.71**
63 mm diameter; 1.20 m high posts	0.15	10.27	4.44	155.00	m	**169.71**
Extra for concrete footings for handrail support posts	0.05	3.42	–	4.50	m	**7.92**
Existing guard rails						
Take from store and re-erect						
pedestrian guard railing, 3.0 m long × 1.0 m high panels	0.15	10.27	4.44	5.15	nr	**19.86**

SERIES 500: DRAINAGE AND SERVICE DUCTS

Item	Gang hours	Labour £	Plant £	Material £	Unit	Total rate £
RESOURCES – LABOUR						
Note						
The re-erection cost for covers and grating complete with frames taken from store assumes that major components are in good condition; the prices below allow a sum of 10% of the value of new materials to cover minor repairs, new fixings and touching up any coatings.						
Drains/sewers/culverts gang (small bore)						
1 ganger/chargehand (skill rate 4) – 50% of time		7.28				
1 skilled operative (skill rate 4)		13.61				
2 unskilled operatives (general)		25.42				
1 plant operator (skill rate 3)		16.55				
Total Gang Rate/Hour		£ 62.86				
Drains/sewers/culverts gang (large bore)						
1 ganger/chargehand (skill rate 4) – 50% of time		7.28				
1 skilled operative (skill rate 4)		13.61				
2 unskilled operatives (general)		25.42				
1 plant operator (skill rate 3)		16.55				
1 plant operator (skill rate 3) – 30% of time		4.96				
Total Gang Rate/Hour		£ 67.83				
Gullies gang						
1 ganger/chargehand (skill rate 4) – 50% of time		7.28				
1 skilled operative (skill rate 4)		13.61				
2 unskilled operatives (general)		25.42				
Total Gang Rate/Hour		£ 46.31				
RESOURCES – PLANT						
Drains/sewers/culverts gang (small bore)						
0.4 m^3 hydraulic excavator			37.65			
2 t dumper – 30% of time			1.68			
360 mm compaction plate – 30% of time			1.19			
2.80 m^3/min compressor, 2 tool – 30% of time			2.31			
disc saw – 30% of time			0.30			
extra 15 ft/50 m hose – 30% of time			0.09			
small pump – 30% of time			0.97			
Total Gang Rate/Hour		£	44.20			
Note: in addition to the above are the following allowances for trench struts/props/sheeting, assuming the need for closeboarded earth support:						
average 1.00 m deep			4.33			
average 1.50 m deep			4.83			
average 2.00 m deep			5.09			
average 2.50 m deep			5.20			
average 3.00 m deep			7.47			
average 3.50 m deep			9.23			

SERIES 500: DRAINAGE AND SERVICE DUCTS

Item	Gang hours	Labour £	Plant £	Material £	Unit	Total rate £
RESOURCES – PLANT – cont						
Drains/sewers/culverts gang (small bore) – cont						
Note: excavation in hard materials as above but with the addition of breaker attachments to the excavator as follows:						
generally: BRH91 (141 kg)			3.75			
reinforced concrete; BRH125 (310 kg)			7.50			
Drains/sewers/culverts gang (large bore i.e. greater than 700 mm bore)						
0.4 m³ hydraulic excavator			37.65			
2 t dumper 30% of time			1.68			
2.80 m³/min compressor, 2 tool 30% of time			2.31			
compaction plate/roller 30% of time			1.19			
disc saw 30% of time			0.28			
small pump 30% of time			0.97			
10 t crawler crane 30% of time			12.78			
Total Gang Rate/Hour		£	56.86			
Gullies						
2 t dumper 30% of time			1.68			
Stihl disc saw 30% of time			0.28			
Total Gang Rate/Hour		£	1.96			1.96
RESOURCES – MATERIALS						
For the purposes of bedding widths for pipe bedding materials, trenches have been taken as exceeding 1.50 m in depth; trenches to lesser depths are generally 150 mm narrower than those given here so that the rates need to be reduced proportionately.						
DRAINS AND SERVICE DUCTS (EXCLUDING FILTER, NARROW FILTER DRAINS AND FIN DRAINS)						
Vitrified clay pipes to BS 65, plain ends with push-fit polypropylene flexible couplings						
150 mm diameter drain or sewer in trench, depth to invert						
average 1.00 m deep	0.17	10.69	14.93	23.50	m	49.12
average 1.50 m deep	0.19	11.94	16.52	23.50	m	51.96
average 2.00 m deep	0.22	13.83	18.91	23.50	m	56.24
average 2.50 m deep	0.26	16.34	22.14	23.50	m	61.98
average 3.00 m deep	0.34	21.37	28.83	23.50	m	73.70
average 3.50 m deep	0.42	26.40	35.60	23.50	m	85.50

SERIES 500: DRAINAGE AND SERVICE DUCTS

Item	Gang hours	Labour £	Plant £	Material £	Unit	Total rate £
Extra for						
Type N sand bed 650 × 100 mm	0.04	2.51	3.53	1.71	m	7.75
Type T sand surround 650 wide × 100 mm	0.08	5.03	7.05	4.10	m	16.18
Type F granular bed 650 × 100 mm	0.05	3.14	4.42	1.82	m	9.38
Type S granular surround 650 wide × 100 mm	0.16	10.06	14.11	4.36	m	28.53
Type A concrete bed 650 × 100 mm	0.11	6.91	9.27	6.95	m	23.13
Type B 100 mm concrete bed and haunch	0.24	15.09	20.23	5.94	m	41.26
Type Z concrete surround 650 wide × 100 mm	0.22	13.83	18.55	16.66	m	49.04
225 mm diameter drain or sewer in trench, depth to invert						
average 1.50 m deep	0.20	12.57	12.02	66.76	m	91.35
average 2.00 m deep	0.24	15.09	20.61	66.76	m	102.46
average 2.50 m deep	0.28	17.60	22.97	66.76	m	107.33
average 3.00 m deep	0.35	22.00	29.64	66.76	m	118.40
average 3.50 m deep	0.44	27.66	37.28	66.76	m	131.70
average 4.00 m deep	0.56	35.20	47.44	66.76	m	149.40
Extra for						
Type N sand bed 750 × 150 mm	0.12	7.54	10.12	2.96	m	20.62
Type T sand surround 750 wide × 150 mm	0.24	15.09	20.23	7.31	m	42.63
Type F granular bed 750 × 150 mm	0.13	8.17	10.96	3.14	m	22.27
Type S granular surround 750 wide × 150 mm	0.24	15.09	20.23	7.78	m	43.10
Type A concrete bed 750 × 150 mm	0.19	11.94	16.01	12.04	m	39.99
Type B 150 mm concrete bed and haunch	0.29	18.23	24.45	9.27	m	51.95
Type Z concrete surround 750 wide × 150 mm	0.36	22.63	30.35	29.85	m	82.83
300 mm diameter drain or sewer in trench, depth to invert						
average 1.50 m deep	0.21	13.20	18.53	102.08	m	133.81
average 2.00 m deep	0.26	16.34	22.35	102.08	m	140.77
average 2.50 m deep	0.30	18.86	24.63	102.08	m	145.57
average 3.00 m deep	0.37	23.26	31.39	102.08	m	156.73
average 4.00 m deep	0.61	38.34	50.66	102.08	m	191.08
average 5.00 m deep	1.01	63.49	84.20	102.08	m	249.77
Extra for						
Type N sand bed 800 × 150 mm	0.13	8.17	10.96	3.15	m	22.28
Type T sand surround 800 × 150 mm	0.26	16.34	19.25	10.16	m	45.75
Type F granular bed 800 × 150 mm	0.14	8.80	11.81	3.33	m	23.94
Type S granular surround 800 wide × 150 mm	0.26	16.34	21.92	10.59	m	48.85
Type A concrete bed 800 × 150 mm	0.20	12.57	16.86	12.85	m	42.28
Type B 150 mm concrete bed and haunch	0.36	22.63	30.35	11.70	m	64.68
Type Z concrete surround 800 wide × 150 mm	0.22	13.83	46.73	38.33	m	98.89

SERIES 500: DRAINAGE AND SERVICE DUCTS

Item	Gang hours	Labour £	Plant £	Material £	Unit	Total rate £
DRAINS AND SERVICE DUCTS (EXCLUDING FILTER, NARROW FILTER DRAINS AND FIN DRAINS) – cont						
Vitrified clay pipes to BS 65, spigot and socket joints with sealing ring						
400 mm diameter drain or sewer in trench, depth to invert						
average 2.00 m deep	0.31	19.49	26.61	87.49	m	**133.59**
average 2.50 m deep	0.35	22.00	28.70	87.49	m	**138.19**
average 3.00 m deep	0.42	26.40	35.60	87.49	m	**149.49**
average 4.00 m deep	0.66	41.49	54.80	87.49	m	**183.78**
average 5.00 m deep	1.07	67.26	89.38	87.49	m	**244.13**
average 6.00 m deep	1.85	116.29	154.57	87.49	m	**358.35**
Extra for						
Type N sand bed 900 × 150 mm	0.14	8.80	11.81	3.55	m	**24.16**
Type T sand surround 900 wide × 150 mm	0.28	17.60	23.60	14.53	m	**55.73**
Type F granular bed 900 × 150 mm	0.15	9.43	12.64	3.79	m	**25.86**
Type S granular surround 900 × 150 mm	0.28	17.60	23.60	15.46	m	**56.66**
Type A concrete bed 900 × 150 mm	0.21	13.20	17.70	14.43	m	**45.33**
Type B, 150 mm concrete bed and haunch	0.43	27.03	36.24	15.42	m	**78.69**
Type Z concrete surround 900 wide × 150 mm	0.40	25.14	33.72	59.24	m	**118.10**
Concrete pipes with rebated flexible joints to BS 5911-1						
450 mm diameter piped culvert in trench, depth to invert						
average 2.00 m deep	0.33	20.74	28.35	26.54	m	**75.63**
average 3.00 m deep	0.44	27.66	37.14	26.54	m	**91.34**
average 4.00 m deep	0.72	45.26	59.53	26.54	m	**131.33**
average 6.00 m deep	1.65	103.72	132.80	26.54	m	**263.06**
Extra for						
Type A concrete bed 1050 × 150 mm	0.24	15.09	20.23	16.84	m	**52.16**
Type Z concrete surround 1050 wide × 150 mm	0.45	28.29	37.93	75.48	m	**141.70**
750 mm diameter piped culvert in trench, depth to invert						
average 2.00 m deep	0.38	25.77	42.18	88.15	m	**156.10**
average 3.00 m deep	0.52	35.27	57.12	88.15	m	**180.54**
average 4.00 m deep	0.92	62.40	99.53	88.15	m	**250.08**
average 6.00 m deep	2.05	139.04	216.55	88.15	m	**443.74**
Extra for						
Type A concrete bed 1250 × 150 mm	0.26	17.63	21.92	23.33	m	**62.88**
Type Z concrete surround 1250 wide × 150 mm	0.50	33.91	42.15	169.62	m	**245.68**
900 mm diameter piped culvert in trench, depth to invert						
average 2.00 m deep	0.43	29.16	47.70	118.27	m	**195.13**
average 3.00 m deep	0.62	42.05	68.35	118.27	m	**228.67**
average 4.00 m deep	1.06	71.89	114.62	118.27	m	**304.78**
average 6.00 m deep	2.25	152.61	237.61	118.27	m	**508.49**
Extra for						
Type A concrete bed 1500 × 150 mm	0.28	18.99	23.60	27.31	m	**69.90**
Type Z concrete surround 1500 wide × 150 mm	0.55	37.30	46.36	235.86	m	**319.52**

SERIES 500: DRAINAGE AND SERVICE DUCTS

Item	Gang hours	Labour £	Plant £	Material £	Unit	Total rate £
Corrugated steel pipes galvanized, hot dip bitumen coated (Armco type)						
1000 mm diameter piped culvert in trench, Type S granular surround, depth to invert						
average 2.00 m deep	0.99	67.15	109.60	120.43	m	**297.18**
1600 mm diameter piped culvert in trench, Type S granular surround, depth to invert						
average 2.00 m deep	2.13	144.47	235.87	196.71	m	**577.05**
2000 mm diameter piped culvert in trench, Type S granular surround, depth to invert						
average 3.00 m deep	2.97	201.44	328.83	394.03	m	**924.30**
2200 mm diameter piped culvert in trench, Type S granular surround, depth to invert						
average 3.00 m deep	3.29	223.14	364.11	445.22	m	**1032.47**
Clay cable ducts; Hepduct						
100 mm diameter service duct in trench, Type S granular surround, depth to invert						
average 1.00 m deep	0.14	8.80	11.70	16.93	m	**37.43**
Two 100 mm diameter service ducts in trench, Type S granular surround, depth to invert						
average 1.00 m deep	0.24	15.09	20.28	33.63	m	**69.00**
Three 100 mm diameter service ducts in trench, Type S granular surround, depth to invert						
average 1.00 m deep	0.32	20.12	26.88	49.04	m	**96.04**
Four 100 mm diameter service ducts in trench, Type S granular surround, depth to invert						
average 1.00 m deep	0.40	25.14	33.72	66.94	m	**125.80**
Six 100 mm diameter service ducts in trench, Type S granular surround, depth to invert						
average 1.00 m deep	0.60	37.72	50.34	94.61	m	**182.67**
Extra for						
Type Z concrete surround on single duct	0.08	5.03	6.74	9.41	m	**21.18**
Type Z concrete surround on additional ways	0.08	5.03	6.74	9.41	m	**21.18**
150 mm diameter conduit, per way	0.01	0.63	0.85	18.39	m	**19.87**
225 mm diameter conduit, per way	0.01	0.63	0.85	51.18	m	**52.66**
FILTER DRAINS						
Vitrified clay perforated pipes to BS 65, sleeved joints						
150 mm diameter filter drain in trench with Type A bed and Type A fill filter material						
average 1.00 m deep	0.26	16.34	21.33	37.09	m	**74.76**
average 2.00 m deep	0.30	18.86	24.44	45.13	m	**88.43**
average 3.00 m deep	0.35	22.00	30.09	53.18	m	**105.27**
150 mm pipes with Type A bed and Type B fill, depth						
average 1.00 m deep	0.26	16.34	21.33	45.46	m	**83.13**
average 2.00 m deep	0.30	18.86	24.44	61.31	m	**104.61**
average 3.00 m deep	0.35	22.00	30.09	77.07	m	**129.16**

SERIES 500: DRAINAGE AND SERVICE DUCTS

Item	Gang hours	Labour £	Plant £	Material £	Unit	Total rate £
FILTER DRAINS – cont						
Vitrified clay perforated pipes to BS 65, sleeved joints – cont						
225 mm pipes with Type A bed and Type A fill, depth						
average 1.00 m deep	0.27	16.97	21.40	63.93	m	102.30
average 2.00 m deep	0.31	19.49	25.04	73.75	m	118.28
average 3.00 m deep	0.36	22.63	29.69	82.88	m	135.20
225 mm pipes with Type A bed and Type B fill, depth						
average 1.00 m deep	0.27	16.97	21.40	72.63	m	111.00
average 2.00 m deep	0.31	19.49	25.04	91.57	m	136.10
average 3.00 m deep	0.36	22.63	29.69	109.67	m	161.99
300 mm pipes with Type A bed and Type A fill, depth						
average 1.00 m deep	0.28	17.60	22.14	117.65	m	157.39
average 2.00 m deep	0.32	20.12	26.11	128.93	m	175.16
average 3.00 m deep	0.37	23.26	30.30	139.71	m	193.27
300 mm pipes with Type A bed and Type B fill, depth						
average 1.00 m deep	0.28	17.60	22.14	125.16	m	164.90
average 2.00 m deep	0.32	20.12	26.11	146.42	m	192.65
average 3.00 m deep	0.37	23.26	30.30	167.46	m	221.02
Concrete porous pipe BS 5911-114, sleeved joints						
150 mm pipes with Type A bed and Type A fill, depth						
average 1.00 m deep	0.26	16.34	21.33	17.05	m	54.72
average 2.00 m deep	0.30	18.86	24.44	25.09	m	68.39
average 3.00 m deep	0.35	22.00	30.09	33.14	m	85.23
150 mm pipes with Type A bed and Type B fill, depth						
average 1.00 m deep	0.26	16.34	21.33	25.42	m	63.09
average 2.00 m deep	0.30	18.86	24.44	41.27	m	84.57
average 3.00 m deep	0.35	22.00	30.09	57.03	m	109.12
225 mm pipes with Type A bed and Type A fill, depth						
average 1.00 m deep	0.27	16.97	21.40	19.70	m	58.07
average 2.00 m deep	0.31	19.49	25.04	29.52	m	74.05
average 3.00 m deep	0.36	22.63	29.69	38.65	m	90.97
225 mm pipes with Type A bed and Type B fill, depth						
average 1.00 m deep	0.27	16.97	21.40	28.40	m	66.77
average 2.00 m deep	0.31	19.49	25.04	47.34	m	91.87
average 3.00 m deep	0.36	22.63	29.69	65.44	m	117.76
300 mm pipes with Type A bed and Type A fill, depth						
average 1.00 m deep	0.28	17.60	22.14	22.98	m	62.72
average 2.00 m deep	0.32	20.12	26.11	34.26	m	80.49
average 3.00 m deep	0.37	23.26	30.30	45.04	m	98.60
300 mm pipes with Type A bed and Type B fill, depth						
average 1.00 m deep	0.28	17.60	22.14	30.49	m	70.23
average 2.00 m deep	0.32	20.12	26.11	51.75	m	97.98
average 3.00 m deep	0.37	23.26	30.30	72.79	m	126.35

SERIES 500: DRAINAGE AND SERVICE DUCTS

Item	Gang hours	Labour £	Plant £	Material £	Unit	Total rate £
Filter material contiguous with filter drains, subbase material and lightweight aggregate infill						
Type A	0.07	4.40	4.70	18.16	m^3	**27.26**
Type B	0.07	4.40	5.12	36.06	m^3	**45.58**
Excavate and replace filter material contiguous with filter drain						
Type A	0.47	14.58	22.71	18.16	m^3	**55.45**
Type B	0.47	14.58	22.71	36.06	m^3	**73.35**
FIN DRAINS AND NARROW FILTER DRAINS						
Fin Drain DfT type 6 using Trammel drainage fabrics and perforated clay drain; surrounding pipe with sand and granular fill						
100 mm clay perforated pipes, depth						
average 1.00 m deep	0.17	10.69	14.73	21.24	m	**46.66**
average 2.00 m deep	0.23	14.46	19.84	27.44	m	**61.74**
Fin Drain DfT type 7 using Trammel drainage fabrics and slotted uPVC drain; surrounding pipe with sand and granular fill, backfilling with selected suitable material						
100 mm uPVC slotted pipes, depth						
average 1.00 m deep	0.17	10.69	14.73	24.03	m	**49.45**
average 2.00 m deep	0.23	14.46	19.84	30.23	m	**64.53**
Narrow Filter Drain DfT type 8 using Trammel drainage fabrics and perforated uPVC drain; surrounding pipe with sand and granular fill, backfilling with granular material						
110 mm uPVC perforated pipes, depth						
average 1.00 m deep	0.17	10.69	14.73	15.90	m	**41.32**
average 2.00 m deep	0.22	13.83	19.14	19.71	m	**52.68**
Narrow Filter Drain DfT type 9 using Trammel drainage fabrics and perforated uPVC drain; surrounding pipe with sand and granular fill, backfilling with granular material						
110 mm uPVC perforated pipes, depth						
average 1.00 m deep	0.17	10.69	14.73	18.09	m	**43.51**
average 2.00 m deep	0.23	14.46	19.84	23.79	m	**58.09**

SERIES 500: DRAINAGE AND SERVICE DUCTS

Item	Gang hours	Labour £	Plant £	Material £	Unit	Total rate £
CONNECTIONS						
Note: excavation presumed covered by new trench						
Connection of pipe to existing drain, sewer or piped culvert						
150 mm	0.42	26.40	24.84	34.81	nr	**86.05**
225 mm	0.60	37.72	42.85	130.22	nr	**210.79**
300 mm	1.15	72.29	85.91	226.60	nr	**384.80**
Connection of pipes to existing chambers						
150 mm to one brick	1.20	75.43	103.69	4.86	nr	**183.98**
150 mm to precast	0.60	37.72	51.82	4.86	nr	**94.40**
300 mm to one and a half brick	2.40	150.86	207.15	7.92	nr	**365.93**
CHAMBERS AND GULLIES						
Notes						
The rates assume the most efficient items of plant (excavator) and are optimum rates, assuming continuous output with no delays caused by other operations or works. Ground conditions are assumed to be good soil with no abnormal conditions that would affect outputs and consistency of work. Multiplier Table for Labour and Plant for various site conditions and for working:						
out of sequence with other works × 2.75 minimum						
in hard clay × 1.75–2.00						
in running sand × 2.75 minimum						
in broken rock × 2.75–3.50						
below water table × 2.00 minimum						
Brick construction						
Design criteria used in models:						
class A engineering bricks						
215 thick walls generally; 328 thick to chambers exceeding 2.5 m deep						
225 plain concrete C20/20 base slab						
300 reinforced concrete C20/20 reducing slab						
125 reinforced concrete C20/20 top slab						
maximum height of working chamber 2.0 m above benching						
750 × 750 access shaft						
plain concrete C15/20 benching, 150 clay main channel longitudinally and two 100 branch channels						
step irons at 300 mm centres, doubled if depth to invert exceeds 3000 mm						
heavy duty manhole cover and frame						
750 × 700 chamber 500 depth to invert						
excavation, support, backfilling and disposal						28.00
concrete base						80.00
brickwork chamber						**245.00**

SERIES 500: DRAINAGE AND SERVICE DUCTS

Item	Gang hours	Labour £	Plant £	Material £	Unit	Total rate £
concrete cover slab						80.00
concrete benching, main and branch channels						20.00
step irons						15.00
access cover and frame						350.00
TOTAL					£	**818.00**
750 × 700 chamber 1000 depth to invert						
excavation, support, backfilling and disposal						60.00
concrete base						80.00
brickwork chamber						495.00
concrete cover slab						80.00
concrete benching; main and branch channels						20.00
step irons						25.00
access cover and frame						350.00
TOTAL					£	**1110.00**
750 × 700 chamber 1500 depth to invert						
excavation, support, backfilling and disposal						80.75
concrete base						80.00
brickwork chamber						745.00
concrete cover slab						80.00
concrete benching; main and branch channels						20.00
step irons						45.00
access cover and frame						350.00
TOTAL					£	**1400.75**
900 × 700 chamber 500 depth to invert						
excavation, support, backfilling and disposal						35.00
concrete base						100.00
brickwork chamber						265.00
concrete cover slab						100.00
concrete benching; main and branch channels						30.00
step irons						15.00
access cover and frame						350.00
TOTAL					£	**895.00**
900 × 700 chamber 1000 depth to invert						
excavation, support, backfilling and disposal						65.00
concrete base						100.00
brickwork chamber						535.00
concrete cover slab						100.00
concrete benching; main and branch channels						30.00
step irons						25.00
access cover and frame						350.00
TOTAL					£	**1205.00**

SERIES 500: DRAINAGE AND SERVICE DUCTS

Item	Gang hours	Labour £	Plant £	Material £	Unit	Total rate £
CHAMBERS AND GULLIES – cont						
Brick construction – cont						
900 × 700 chamber 1500 depth to invert						
excavation, support, backfilling and disposal						110.25
concrete base						100.00
brickwork chamber						800.00
concrete cover slab						100.00
concrete benching; main and branch channels						30.00
step irons						45.00
access cover and frame						350.00
TOTAL					£	**1535.25**
1050 × 700 chamber 1500 depth to invert						
excavation, support, backfilling and disposal						141.50
concrete base						110.00
brickwork chamber						855.00
concrete cover slab						115.00
concrete benching; main and branch channels						32.50
step irons						45.00
access cover and frame						350.00
TOTAL					£	**1649.00**
1050 × 700 chamber 2500 depth to invert						
excavation, support, backfilling and disposal						255.00
concrete base						110.00
brickwork chamber						1425.00
concrete cover slab						115.00
concrete benching; main and branch channels						32.50
step irons						82.50
access cover and frame						350.00
TOTAL					£	**2370.00**
1050 × 700 chamber 3500 depth to invert						
excavation, support, backfilling and disposal						350.00
concrete base						110.00
brickwork chamber						1995.00
concrete reducing slab						140.00
brickwork access shaft						2040.00
concrete cover slab						80.00
concrete benching; main and branch channels						32.50
step irons						110.00
access cover and frame						350.00
TOTAL					£	**5207.50**
1350 × 700 chamber 2500 depth to invert						
excavation, support, backfilling and disposal						308.00
concrete base						140.00
brickwork chamber						1600.00
concrete cover slab						140.00
concrete benching; main and branch channels						40.00
step irons						82.50
access cover and frame						350.00
TOTAL					£	**2660.50**

SERIES 500: DRAINAGE AND SERVICE DUCTS

Item	Gang hours	Labour £	Plant £	Material £	Unit	Total rate £
1350 × 700 chamber 3500 depth to invert						
excavation, support, backfilling and disposal						430.00
concrete base						140.00
brickwork chamber						2250.00
concrete reducing slab						150.00
brickwork access shaft						2040.00
concrete cover slab						80.00
concrete benching; main and branch channels						40.00
step irons						110.00
access cover and frame						350.00
TOTAL					£	**5590.00**
1350 × 700 chamber 4500 depth to invert						
excavation, support, backfilling and disposal						550.00
concrete base						140.00
brickwork chamber						2895.00
concrete reducing slab						150.00
brickwork access shaft						2665.00
concrete cover slab						80.00
concrete benching; main and branch channels						40.00
step irons						145.00
access cover and frame						350.00
TOTAL					£	**7015.00**

Precast concrete construction
Design criteria used in models:
* circular shafts
* 150 plain concrete surround
* 225 plain concrete C20/20 base slab
* precast reducing slab
* precast top slab
* maximum height of working chamber 2.0 m above benching
* 750 diameter access shaft
* plain concrete C15/20 benching, 150 clay main channel longitudinally and two 100 branch channels
* step irons at 300 mm centres, doubled if depth to invert exceeds 3000 mm
* heavy duty manhole cover and frame
* in manholes over 6 m deep, landings at maximum intervals

Item	Gang hours	Labour £	Plant £	Material £	Unit	Total rate £
675 diameter × 500 depth to invert						
excavation, support, backfilling and disposal						27.00
concrete base						50.00
main chamber rings						20.00
cover slab						45.00
concrete surround						50.00
concrete benching, main and branch channels						12.10
step irons						15.00
access cover and frame						350.00
TOTAL					£	**569.10**

SERIES 500: DRAINAGE AND SERVICE DUCTS

Item	Gang hours	Labour £	Plant £	Material £	Unit	Total rate £
CHAMBERS AND GULLIES – cont						
Precast concrete construction – cont						
675 diameter × 750 depth to invert						
excavation, support, backfilling and disposal						35.00
concrete base						50.00
main chamber rings						25.00
cover slab						45.00
concrete surround						70.00
concrete benching, main and branch channels						12.10
step irons						20.00
access cover and frame						350.00
TOTAL					£	**607.10**
675 diameter × 1000 depth to invert						
excavation, support, backfilling and disposal						42.50
concrete base						50.00
main chamber rings						32.50
cover slab						45.00
concrete surround						100.00
concrete benching, main and branch channels						12.10
step irons						25.00
access cover and frame						350.00
TOTAL					£	**657.10**
675 diameter × 1250 depth to invert						
excavation, support, backfilling and disposal						52.00
concrete base						50.00
main chamber rings						37.50
cover slab						45.00
concrete surround						120.00
concrete benching, main and branch channels						12.10
step irons						32.00
access cover and frame						350.00
TOTAL					£	**698.60**
900 diameter × 750 depth to invert						
excavation, support, backfilling and disposal						115.00
concrete base						100.00
main chamber rings						41.30
cover slab						60.00
concrete surround						100.00
concrete benching, main and branch channels						21.50
step irons						20.00
access cover and frame						350.00
TOTAL					£	**807.80**
900 diameter × 1000 depth to invert						
excavation, support, backfilling and disposal						150.00
concrete base						100.00
main chamber rings						55.07
cover slab						60.00
concrete surround						130.00
concrete benching, main and branch channels						21.50
step irons						25.00
access cover and frame						350.00
TOTAL					£	**891.57**

SERIES 500: DRAINAGE AND SERVICE DUCTS

Item	Gang hours	Labour £	Plant £	Material £	Unit	Total rate £
900 diameter × 1500 depth to invert						
excavation, support, backfilling and disposal						215.00
concrete base						100.00
main chamber rings						65.07
cover slab						60.00
concrete surround						190.00
concrete benching, main and branch channels						21.50
step irons						45.00
access cover and frame						350.00
TOTAL					£	1046.57
1200 diameter × 1500 depth to invert						
excavation, support, backfilling and disposal						275.00
concrete base						170.00
main chamber rings						135.00
cover slab						78.00
concrete surround						260.00
concrete benching, main and branch channels						38.15
step irons						45.00
access cover and frame						350.00
TOTAL					£	1351.15
1200 diameter × 2000 depth to invert						
excavation, support, backfilling and disposal						380.00
concrete base						170.00
main chamber rings						180.00
cover slab						78.00
concrete surround						340.00
concrete benching, main and branch channels						38.15
step irons						60.00
access cover and frame						350.00
TOTAL					£	1596.15
1200 diameter × 2500 depth to invert						
excavation, support, backfilling and disposal						450.00
concrete base						170.00
main chamber rings						225.00
cover slab						78.00
concrete surround						430.00
concrete benching, main and branch channels						38.15
step irons						75.00
access cover and frame						350.00
TOTAL					£	1816.15
1200 diameter × 3000 depth to invert						
excavation, support, backfilling and disposal						555.00
concrete base						170.00
main chamber rings						270.00
cover slab						78.00
concrete surround						510.00
concrete benching, main and branch channels						38.15
step irons						90.00
access cover and frame						350.00
TOTAL					£	2061.15

SERIES 500: DRAINAGE AND SERVICE DUCTS

Item	Gang hours	Labour £	Plant £	Material £	Unit	Total rate £
CHAMBERS AND GULLIES – cont						
Precast concrete construction – cont						
1800 diameter × 1500 depth to invert						
excavation, support, backfilling and disposal						415.00
concrete base						380.00
main chamber rings						405.00
cover slab						215.00
concrete surround						380.00
concrete benching, main and branch channels						85.85
step irons						45.00
access cover and frame						350.00
TOTAL					£	**2275.85**
1800 diameter × 2000 depth to invert						
excavation, support, backfilling and disposal						555.00
concrete base						380.00
main chamber rings						540.00
cover slab						215.00
concrete surround						510.00
concrete benching, main and branch channels						85.85
step irons						60.00
access cover and frame						350.00
TOTAL					£	**2695.85**
1800 diameter × 2500 depth to invert						
excavation, support, backfilling and disposal						650.00
concrete base						380.00
main chamber rings						675.00
cover slab						215.00
concrete surround						640.00
concrete benching, main and branch channels						85.85
step irons						75.00
access cover and frame						350.00
TOTAL					£	**3070.85**
1800 diameter × 3000 depth to invert						
excavation, support, backfilling and disposal						830.00
concrete base						380.00
main chamber rings						815.00
cover slab						215.00
concrete surround						770.00
concrete benching, main and branch channels						85.85
step irons						90.00
access cover and frame						350.00
TOTAL					£	**3535.85**

SERIES 500: DRAINAGE AND SERVICE DUCTS

Item	Gang hours	Labour £	Plant £	Material £	Unit	Total rate £
1800 diameter × 3500 depth to invert						
excavation, support, backfilling and disposal						950.00
concrete base						380.00
main chamber rings						950.00
reducing slab						251.00
access shaft						300.00
cover slab						60.00
concrete surround						900.00
concrete benching, main and branch channels						85.85
step irons						115.00
access cover and frame						350.00
TOTAL					£	**4341.85**
1800 diameter × 4000 depth to invert						
excavation, support, backfilling and disposal						1100.00
concrete base						380.00
main chamber rings						1055.00
reducing slab						251.00
access shaft						345.00
cover slab						60.00
concrete surround						1030.00
concrete benching, main and branch channels						85.85
step irons						125.00
access cover and frame						350.00
TOTAL					£	**4781.85**
2400 diameter × 1500 depth to invert						
excavation, support, backfilling and disposal						570.00
concrete base						680.00
main chamber rings						1035.00
cover slab						615.00
concrete surround						510.00
concrete benching, main and branch channels						155.00
step irons						45.00
access cover and frame						350.00
TOTAL					£	**3960.00**
2400 diameter × 3000 depth to invert						
excavation, support, backfilling and disposal						1140.00
concrete base						680.00
main chamber rings						2075.00
cover slab						615.00
concrete surround						1030.00
concrete benching, main and branch channels						155.00
step irons						90.00
access cover and frame						350.00
TOTAL					£	**6135.00**

SERIES 500: DRAINAGE AND SERVICE DUCTS

Item	Gang hours	Labour £	Plant £	Material £	Unit	Total rate £
CHAMBERS AND GULLIES – cont						
Precast concrete construction – cont						
2400 diameter × 4500 depth to invert						
excavation, support, backfilling and disposal						1500.00
concrete base						680.00
main chamber rings						3100.00
reducing slab						705.00
access shaft						385.00
cover slab						60.00
concrete surround						1540.00
concrete benching, main and branch channels						155.00
step irons						147.00
access cover and frame						350.00
TOTAL					£	**8622.00**
2700 diameter × 1500 depth to invert						
excavation, support, backfilling and disposal						650.00
concrete base						870.00
main chamber rings						1200.00
cover slab						1100.00
concrete surround						580.00
concrete benching, main and branch channels						195.00
step irons						45.00
access cover and frame						350.00
TOTAL					£	**4990.00**
2700 diameter × 3000 depth to invert						
excavation, support, backfilling and disposal						1310.00
concrete base						870.00
main chamber rings						2400.00
cover slab						1100.00
concrete surround						1150.00
concrete benching, main and branch channels						195.00
step irons						90.00
access cover and frame						350.00
TOTAL					£	**7465.00**
2700 diameter × 4500 depth to invert						
excavation, support, backfilling and disposal						1900.00
concrete base						870.00
main chamber rings						3500.00
reducing slab						1155.00
access shaft						385.00
cover slab						60.00
concrete surround						1730.00
concrete benching, main and branch channels						195.00
step irons						147.00
access cover and frame						350.00
TOTAL					£	**10292.00**

SERIES 500: DRAINAGE AND SERVICE DUCTS

Item	Gang hours	Labour £	Plant £	Material £	Unit	Total rate £
3000 diameter × 3000 depth to invert						
excavation, support, backfilling and disposal						1500.00
concrete base						1070.00
main chamber rings						3375.00
cover slab						1215.00
concrete surround						1280.00
concrete benching, main and branch channels						240.00
step irons						90.00
access cover and frame						350.00
TOTAL					£	9120.00
3000 diameter × 4500 depth to invert						
excavation, support, backfilling and disposal						2250.00
concrete base						1070.00
main chamber rings						4500.00
reducing slab						1500.00
access shaft						385.00
cover slab						60.00
concrete surround						1920.00
concrete benching, main and branch channels						240.00
step irons						147.00
access cover and frame						350.00
TOTAL					£	12422.00
3000 diameter × 6000 depth to invert						
excavation, support, backfilling and disposal						3000.00
concrete base						1070.00
main chamber rings						6700.00
reducing slab						1500.00
access shaft						515.00
cover slab						60.00
concrete surround						2560.00
concrete benching, main and branch channels						240.00
step irons						175.00
access cover and frame						350.00
TOTAL					£	16170.00
Clayware vertical pipe complete with rest bend at base and tumbling bay junction to main drain complete with stopper; concrete grade C20 surround, 150 mm thick; additional excavation and disposal						
100 pipe						
1.15 m to invert	–	–	–	–	nr	110.00
2.15 m to invert	–	–	–	–	nr	140.00
3.15 m to invert	–	–	–	–	nr	165.00
4.15 m to invert	–	–	–	–	nr	200.00
150 pipe						
1.15 m to invert	–	–	–	–	nr	165.00
2.15 m to invert	–	–	–	–	nr	205.00
3.15 m to invert	–	–	–	–	nr	242.00
4.15 m to invert	–	–	–	–	nr	285.00

SERIES 500: DRAINAGE AND SERVICE DUCTS

Item	Gang hours	Labour £	Plant £	Material £	Unit	Total rate £
CHAMBERS AND GULLIES – cont						
Clayware vertical pipe complete with rest bend at base and tumbling bay junction to main drain complete with stopper – cont						
225 pipe						
1.15 m to invert	–	–	–	–	nr	335.00
2.15 m to invert	–	–	–	–	nr	375.00
3.15 m to invert	–	–	–	–	nr	425.00
4.15 m to invert	–	–	–	–	nr	495.00
Vitrified clay; set in concrete grade C20, 150 mm thick; additional excavation and disposal						
Road gulley						
450 mm diameter × 900 mm deep, 100 mm or 150 mm outlet; cast iron road gulley grating and frame group 4, 434 × 434 mm, on Class B engineering brick seating	0.50	23.16	0.98	522.41	nr	546.55
Yard gulley (mud); trapped with rodding eye; galvanized bucket; stopper						
225 mm diameter, 100 mm diameter outlet, cast iron hinged grate and frame	0.30	13.89	0.62	456.58	nr	471.09
Grease interceptors; internal access and bucket						
600 × 450 mm, metal tray and lid, square hopper with horizontal inlet	0.35	16.21	0.69	1441.80	nr	1458.70
Precast concrete; set in concrete grade C20P, 150 mm thick; additional excavation and disposal						
Road gulley; trapped with rodding eye; galvanized bucket; stopper						
450 mm diameter × 900 mm deep, cast iron road gulley grating and frame group 4, 434 × 434 mm, on Class B engineering brick seating	0.54	25.01	1.06	458.90	nr	484.97
SOFT SPOTS AND OTHER VOIDS						
Excavation of soft spots and other voids in bottom of trenches, chambers and gullies	0.07	4.40	5.84	–	m^3	10.24
Filling of soft spots and other voids in bottom of trenches, chambers and gullies with imported selected sand	0.09	5.66	7.17	25.63	m^3	38.46
Filling of soft spots and other voids in bottom of trenches, chambers and gullies with imported natural gravel	0.09	5.66	7.17	30.44	m^3	43.27
Filling of soft spots and other voids in bottom of trenches, chambers and gullies with concrete Grade C15, 40 mm aggregate	0.09	5.66	7.10	82.52	m^3	95.28
Filling of soft spots and other voids in bottom of trenches, chambers and gullies with concrete Grade C20, 20 mm aggregate	0.09	5.66	7.10	84.81	m^3	97.57

SERIES 500: DRAINAGE AND SERVICE DUCTS

Item	Gang hours	Labour £	Plant £	Material £	Unit	Total rate £
SUPPORTS LEFT IN EXCAVATION						
Timber closeboarded supports left in						
trench	–	–	7.50	–	m²	7.50
pits	–	–	7.50	–	m²	7.50
Steel trench sheeting supports left in						
trench	–	–	35.00	–	m²	35.00
pits	–	–	35.00	–	m²	35.00
RENEWAL, RAISING OR LOWERING OF COVERS AND GRATINGS ON EXISTING CHAMBERS						
Raising the level of 100 × 100 mm water stop tap boxes on 100 × 100 mm brick chambers						
by 150 mm or less	0.06	2.78	0.43	25.00	nr	28.21
Lowering the level of 100 × 100 mm water stop tap boxes on 100 × 100 mm brick chambers						
by 150 mm or less	0.04	1.85	0.27	12.50	nr	14.62
Raising the level of 420 × 420 mm cover and frame on 420 × 420 mm in situ concrete chamber						
by 150 mm or less	0.10	4.63	0.70	27.50	nr	32.83
Lowering the level of 420 × 420 mm British Telecom cover and frame on 420 × 420 mm in situ concrete chamber						
by 150 mm or less	0.08	3.47	2.37	16.50	nr	22.34
Raising the level of 700 × 500 mm cover and frame on 700 × 500 mm in situ concrete chamber						
by 150 mm or less	0.17	7.87	1.16	39.38	nr	48.41
Raising the level of 600 × 600 mm grade A heavy duty manhole cover and frame on 600 × 600 mm brick chamber						
by 150 mm or less	0.17	7.87	1.16	30.00	nr	39.03
by 150–300 mm	0.21	9.73	1.44	34.50	nr	45.67
Lowering the level of 600 × 600 mm grade A heavy duty manhole cover and frame on 600 × 600 mm brick chamber						
by 150 mm or less	0.10	4.63	0.70	17.15	nr	22.48
REMOVE FROM STORE AND REINSTALL COVERS						
600 × 600 mm; Group 5; super heavy duty E600 cast iron	0.25	11.58	0.56	201.90	nr	214.04
600 × 600 mm; Group 4; heavy duty triangular D400 cast iron	0.25	11.58	0.56	204.90	nr	217.04
600 × 600 × 75 mm; Group 2; light duty single seal B125 cast iron	0.25	11.58	0.56	213.15	nr	225.29
600 × 600 × 100 mm; Group 2; medium duty single seal B125 cast iron	0.25	11.58	0.56	240.90	nr	253.04

SERIES 500: DRAINAGE AND SERVICE DUCTS

Item	Gang hours	Labour £	Plant £	Material £	Unit	Total rate £
GROUTING UP OF EXISTING DRAINS AND SERVICE DUCTS						
Concrete Grade C15						
Sealing redundant road gullies	0.02	0.93	0.11	12.99	m³	**14.03**
Filling redundant chambers with						
ne 1.0 m deep to invert	0.09	5.66	5.52	52.11	nr	**63.29**
1.0–2.0 m deep to invert	0.21	13.20	12.87	84.22	nr	**110.29**
2.0–3.0 m deep to invert	0.55	34.57	33.70	140.33	nr	**208.60**
Grouting up of existing drains and service ducts						
100 mm diameter	0.03	1.89	1.86	3.33	m	**7.08**
300 mm diameter	0.13	8.17	8.10	14.35	m	**30.62**
450 mm diameter	0.26	16.34	16.20	29.73	m	**62.27**
600 mm diameter	0.50	31.43	31.14	60.32	m	**122.89**
1200 mm diameter	1.70	106.86	105.84	168.61	m	**381.31**
EXCAVATION IN HARD MATERIAL						
Extra over excavation for excavation in Hard Material in drainage						
existing pavement, brickwork, concrete, masonry and the like	0.15	9.43	12.79	–	m³	**22.22**
rock	0.35	22.00	29.84	–	m³	**51.84**
reinforced concrete	0.60	37.72	53.40	–	m³	**91.12**
Reinstatement of pavement construction; extra over excavation for breaking up and subsequently reinstating 150 mm flexible surfacing and 280 mm subbase						
100 mm diameter sewer, drain or service duct	0.09	5.66	7.97	14.08	m	**27.71**
150 mm diameter sewer, drain or service duct	0.10	6.29	8.86	15.01	m	**30.16**
225 mm diameter sewer, drain or service duct	0.10	6.29	8.86	16.42	m	**31.57**
300 mm diameter sewer, drain or service duct	0.10	6.29	8.86	16.89	m	**32.04**
375 mm diameter sewer, drain or service duct	0.10	6.29	8.86	17.36	m	**32.51**
450 mm diameter sewer, drain or service duct	0.12	7.54	9.72	18.30	m	**35.56**
two way 100 mm diameter service ducts	0.10	6.29	8.86	14.54	m	**29.69**
three way 100 mm diameter service ducts	0.10	6.29	8.86	15.01	m	**30.16**
four way 100 mm diameter service ducts	0.10	6.29	8.86	16.42	m	**31.57**
MANHOLES						
Materials						
In order to arrive at an all in rate for enumerated manholes the various constituent elements have been broken down and unit rates given against each item.						
Earthworks						
Excavation by machine						
ne 2.0 m deep	0.22	13.83	9.13	–	m³	**22.96**
2.0–3.0 m deep	0.24	15.09	9.96	–	m³	**25.05**
3.0–4.0 m deep	0.26	16.34	10.80	–	m³	**27.14**

SERIES 500: DRAINAGE AND SERVICE DUCTS

Item	Gang hours	Labour £	Plant £	Material £	Unit	Total rate £
Backfiling						
ne 2.0 m deep	0.11	6.91	4.56	–	m³	**11.47**
2.0–3.0 m deep	0.11	6.91	4.56	–	m³	**11.47**
3.0–4.0 m deep	0.11	6.91	4.56	–	m³	**11.47**
Earthwork support						
ne 2.0 m deep	0.01	0.44	3.17	–	m³	**3.61**
2.0–3.0 m deep	0.01	0.69	4.82	–	m³	**5.51**
3.0–4.0 m deep	0.01	0.88	8.15	–	m³	**9.03**
Disposal of excavated material						
off site	0.55	22.00	6.88	–	m³	**28.88**
to spoil heap average 50 m distant	0.19	2.83	1.06	–	m³	**3.89**
Mass concrete grade C15						
Blinding; thickness						
ne 150 mm deep	1.05	13.72	4.66	83.78	m³	**102.16**
Surrounds to manholes; thickness						
ne 150 mm deep	2.30	30.06	10.16	83.78	m³	**124.00**
150–500 mm deep	1.70	22.22	7.51	83.78	m³	**113.51**
Benching to bottom of manhole						
650 mm diameter	1.00	13.07	3.58	6.96	nr	**23.61**
900 mm diameter	2.00	26.14	7.38	13.33	nr	**46.85**
1050 mm diameter	3.60	47.05	12.91	18.14	nr	**78.10**
1200 mm diameter	4.75	62.08	17.04	23.69	nr	**102.81**
1500 mm diameter	5.75	75.15	20.63	37.03	nr	**132.81**
rectangular	3.50	45.74	12.55	20.94	m²	**79.23**
Reinforced concrete grade C20						
Bases; thickness						
150–300 mm deep	2.30	30.06	8.25	84.81	m³	**123.12**
300–500 mm deep	2.50	32.67	8.96	84.81	m³	**126.44**
Suspended slabs; thickness						
ne 150 mm deep	3.45	45.09	12.38	84.81	m³	**142.28**
150–300 mm deep	2.70	35.29	9.68	84.81	m³	**129.78**
Formwork; GRP forms						
Plain curved, vertical	–	–	–	–	m²	–
to base and benching	0.60	7.84	2.15	9.38	m²	**19.37**
to surrounds	0.40	5.23	1.43	6.92	hr	**13.58**
Bar reinforcement						
See Series 1700						
Precast concrete circular manhole rings; BS 5911 Part 1						
Shaft rings						
675 mm; plain/reinforced	3.20	45.18	34.65	59.10	m	**138.93**
900 mm; plain/reinforced	3.80	53.66	41.13	78.68	m	**173.47**
1050 mm; plain/reinforced	4.20	59.30	55.43	86.23	m	**200.96**
1200 mm; plain/reinforced	4.68	66.08	50.68	103.87	m	**220.63**
1500 mm; reinforced	5.28	74.55	57.14	159.51	m	**291.20**
1800 mm; reinforced	6.00	84.72	64.98	229.73	m	**379.43**
2100 mm; reinforced	6.60	100.70	180.42	435.43	m	**716.55**

SERIES 500: DRAINAGE AND SERVICE DUCTS

Item	Gang hours	Labour £	Plant £	Material £	Unit	Total rate £
MANHOLES – cont						
Precast concrete circular manhole rings – cont						
Shaft rings – cont						
2400 mm; reinforced	7.20	109.85	202.33	552.53	m	**864.71**
2700 mm; reinforced	8.00	122.06	224.81	638.64	m	**985.51**
3000 mm; reinforced	8.60	131.21	241.67	874.05	m	**1246.93**
Reducing slabs + landing slabs; 900 mm diameter access						
1050 mm; plain/reinforced	2.52	35.58	27.28	103.51	nr	**166.37**
1200 mm; plain/reinforced	2.80	39.54	30.33	124.91	nr	**194.78**
1500 mm; reinforced	3.16	44.62	34.20	205.64	nr	**284.46**
1800 mm; reinforced	3.60	50.83	38.99	286.32	nr	**376.14**
2100 mm; reinforced	4.00	61.03	112.41	605.03	nr	**778.47**
2400 mm; reinforced	4.40	67.13	123.66	789.91	nr	**980.70**
2700 mm; reinforced	4.80	73.24	134.89	1256.67	nr	**1464.80**
3000 mm; reinforced	5.20	79.34	146.14	1595.30	nr	**1820.78**
Cover slabs						
675 mm plain/reinforced	1.98	27.76	19.94	54.93	nr	**102.63**
900 mm; plain/reinforced	2.16	30.50	23.40	73.82	nr	**127.72**
1050 mm; plain/reinforced	2.52	35.58	27.27	80.71	nr	**143.56**
1200 mm; plain/reinforced	2.80	39.54	30.33	99.28	nr	**169.15**
1500 mm; reinforced	3.16	44.62	34.20	163.10	nr	**241.92**
1800 mm; reinforced	3.60	50.83	38.99	245.32	nr	**335.14**
2100 mm; reinforced	4.00	61.03	112.41	512.78	nr	**686.22**
2400 mm; reinforced	4.40	67.13	123.66	687.41	nr	**878.20**
2700 mm; reinforced	4.80	73.24	134.89	1154.17	nr	**1362.30**
3000 mm; reinforced	5.20	79.34	146.14	1333.92	nr	**1559.40**
Engineering bricks, Class B; PC £330.00/1000; in cement mortar (1:3)						
Walls, built vertical and straight						
one brick thick; depth ne 1.0 m	2.63	47.28	3.97	57.51	m²	**108.76**
one brick thick; depth 1.0–2.0 m	2.80	50.43	4.23	57.51	m²	**112.17**
one brick thick; depth 2.0–3.0 m	3.04	54.84	4.60	57.51	m²	**116.95**
one brick thick; depth 3.0–4.0 m	3.29	59.26	4.97	57.51	m²	**121.74**
one brick thick; depth 4.0–5.0 m	3.68	66.19	5.55	57.51	m²	**129.25**
one and a half bricks thick; depth ne 1.0 m	3.60	64.93	5.45	85.39	m²	**155.77**
one and a half bricks thick; depth 1.0–2.0 m	4.02	72.50	6.09	85.39	m²	**163.98**
one and a half bricks thick; depth 2.0–3.0 m	4.41	79.43	6.67	85.39	m²	**171.49**
one and a half bricks thick; depth 3.0–4.0 m	4.79	86.36	7.25	85.39	m²	**179.00**
one and a half bricks thick; depth 4.0–5.0 m	5.18	93.30	7.83	85.39	m²	**186.52**
Step irons; BS 1247; malleable; galvanized						
Built into joints; at depths						
ne 1.0 m	0.11	1.89	0.16	13.13	nr	**15.18**
1.0–2.0 m	0.11	1.89	0.16	13.13	nr	**15.18**
2.0–3.0 m	0.14	2.52	0.21	13.13	nr	**15.86**
3.0–4.0 m	0.18	3.15	0.26	13.13	nr	**16.54**
4.0–5.0 m	0.21	3.78	0.51	13.13	nr	**17.42**

SERIES 500: DRAINAGE AND SERVICE DUCTS

Item	Gang hours	Labour £	Plant £	Material £	Unit	Total rate £
Cast into concrete; at depths						
ne 1.0 m	0.09	1.20	0.16	13.13	nr	14.49
1.0–2.0 m	0.09	1.20	0.16	13.13	nr	14.49
2.0–3.0 m	0.12	1.60	0.21	13.13	nr	14.94
3.0–4.0 m	0.15	2.00	0.26	13.13	nr	15.39
4.0–5.0 m	0.18	2.40	0.51	13.13	nr	16.04
Vitrified clayware channels; bedding in cement mortar in bottom of manhole						
Half section straight						
100 mm diameter	0.54	7.19	0.81	27.89	nr	35.89
150 mm diameter	0.69	9.18	1.04	45.67	nr	55.89
225 mm diameter	0.96	12.78	1.44	151.54	nr	165.76
300 mm diameter	1.29	17.17	1.94	231.27	nr	250.38
Channel bend						
100 mm diameter	0.42	5.59	0.63	12.35	nr	18.57
150 mm diameter	0.57	7.59	0.86	20.31	nr	28.76
225 mm diameter	0.84	11.18	1.26	67.47	nr	79.91
300 mm diameter	1.20	15.97	1.81	137.49	nr	155.27
Channel branch						
100 mm diameter	0.48	6.43	0.72	23.26	nr	30.41
150 mm diameter	0.78	10.38	1.18	38.43	nr	49.99
225 mm diameter	2.01	26.75	3.03	110.61	nr	140.39
300 mm diameter	3.21	42.73	4.83	270.13	nr	317.69
Half section straight taper						
150 mm diameter	0.66	8.78	0.99	29.71	nr	39.48
225 mm diameter	0.99	13.18	1.49	92.47	nr	107.14
300 mm diameter	1.44	19.17	2.17	254.99	nr	276.33
Half section branch channel bend						
100 mm diameter	0.42	5.59	0.63	25.11	nr	31.33
150 mm diameter	0.57	7.59	0.86	41.11	nr	49.56
225 mm diameter	0.84	11.18	1.26	136.45	nr	148.89
300 mm diameter	1.20	15.97	1.81	208.24	nr	226.02
Three quarter section branch channel bend						
100 mm diameter	0.42	5.59	0.63	27.70	nr	33.92
150 mm diameter	0.57	7.59	0.86	46.42	nr	54.87
225 mm diameter	0.84	11.18	1.26	169.15	nr	181.59
300 mm diameter	1.20	15.97	1.81	240.95	nr	258.73
Spun concrete channels; bedding in cement mortar in bottom of manhole						
Half section straight						
300 mm diameter	1.14	15.17	2.01	9.88	nr	27.06
450 mm diameter	2.25	29.95	3.97	16.80	nr	50.72
600 mm diameter	3.15	41.93	4.85	30.41	nr	77.19
900 mm diameter	5.91	78.66	10.42	58.50	nr	147.58
1200 mm diameter	9.42	125.38	16.61	102.34	nr	244.33

SERIES 500: DRAINAGE AND SERVICE DUCTS

Item	Gang hours	Labour £	Plant £	Material £	Unit	Total rate £
MANHOLES – cont						
Labours						
Build in pipes; to brickwork						
300 mm diameter	0.28	5.04	0.41	1.14	m²	**6.59**
450 mm diameter	0.39	6.93	0.56	1.60	m²	**9.09**
600 mm diameter	0.52	9.46	0.77	2.15	m²	**12.38**
900 mm diameter	0.70	12.61	1.02	3.80	m²	**17.43**
1200 mm diameter	0.77	13.87	1.12	7.65	m²	**22.64**
Build in pipes; to precast concrete						
300 mm diameter	1.35	17.97	2.29	1.60	nr	**21.86**
450 mm diameter	2.01	26.75	3.41	2.29	nr	**32.45**
600 mm diameter	2.61	34.74	4.43	3.21	nr	**42.38**
900 mm diameter	3.84	51.11	6.52	6.92	nr	**64.55**
1200 mm diameter	4.83	64.29	8.20	12.13	nr	**84.62**
Access covers and frames; coated; BS EN 124; bed frame in cement mortar; set cover in grease and sand						
Group 5; super heavy duty; solid top						
600 × 600 mm; E600	1.80	23.96	2.36	249.20	nr	**275.52**
Group 4; heavy duty; solid top						
600 × 600 mm; double trianglar, hinged; D400	1.44	19.17	1.89	253.10	nr	**274.16**
Group 2; medium duty						
600 × 600 mm; single seal; B125	1.17	15.57	1.54	301.10	nr	**318.21**
Group 2; light duty						
600 × 600 mm; single seal; B125	0.87	11.58	1.14	264.10	nr	**276.82**
Manhole ancillaries						
Manhole interceptors						
100 mm diameter	0.54	7.19	0.92	139.51	nr	**147.62**
150 mm diameter	0.60	7.99	1.02	201.16	nr	**210.17**
225 mm diameter	0.90	11.98	1.53	627.70	nr	**641.21**
300 mm diameter	1.20	15.97	2.04	1004.50	nr	**1022.51**
Rodding eye points (excluding pipework)						
100 mm diameter	1.20	15.97	2.04	174.49	nr	**192.50**
150 mm diameter	1.35	17.97	2.03	178.69	nr	**198.69**

SERIES 600: EARTHWORKS

Item	Gang hours	Labour £	Plant £	Material £	Unit	Total rate £
NOTES The cost of earth moving and other associated works is dependent on matching the overall quantities and production rates called for by the programme of works with the most appropriate plant and assessing the most suitable version of that plant that will: deal with the site conditions (e.g. type of ground, type of excavation, length of haul, prevailing weather, etc.); comply with the specification requirements (e.g. compaction, separation of materials, surface tolerances, etc.); complete the work economically (e.g. provide surface tolerances which will avoid undue excessive thickness of expensive imported materials). **Excavation rates** Unless stated otherwise the units for excavation in material other than topsoil, rock or artificial hard materials are based on excavation in firm gravel soils. Factors for alternative types of soil multiply the rates by: Scrapers Stiff clay × 1.5 Chalk × 2.5 Soft rock × 3.5 Broken rock × 3.7 Tractor dozers and loaders Stiff clay × 2.0 Chalk × 3.0 Soft rock × 2.5 Broken rock × 2.5 Backacter (minimum bucket size 0.5 m^3) Stiff clay × 1.7 Chalk × 2.0 Soft rock × 2.0 Broken rock × 1.7						

SERIES 600: EARTHWORKS

Item	Gang hours	Labour £	Plant £	Material £	Unit	Total rate £

NOTES – cont
 Disposal rates
 The other important consideration in excavation of material is bulkage of material after it is dug and loaded onto transport.
 All pricing and estimating for disposal is based on the volume of solid material excavated and rates for disposal should be adjusted by the following factors for bulkage:
  Sand bulkage × 1.10
  Gravel bulkage × 1.20
  Compacted soil bulkage × 1.30
  Compacted subbase, acceptable fill etc. bulkage × 1.30
  Stiff clay bulkage × 1.20						

Fill rates The price for filling presume the material being supplied in loose state.						
RESOURCES – LABOUR						
Scraper gang 1 plant operator (skill rate 2) Total Gang Rate/Hour	 £	 18.12 **18.12**				
Scraper and ripper bulldozer gang (hard material) 2 plant operator (skill rate 2) Total Gang Rate/Hour	 £	 36.24 **36.24**				
General excavation gang 1 plant operator (skill rate 3) 1 plant operator (skill rate 3) – 25% of time 1 banksman (skill rate 4) Total Gang Rate/Hour	 £	 16.55 4.14 13.61 **34.30**				
Shore defences (armour stones) gang 1 plant operator (skill rate 2) 1 banksman (skill rate 4) Total Gang Rate/Hour	 £	 18.12 13.61 **31.73**				
Filling gang 1 plant operator (skill rate 4) 2 unskilled operatives (general) Total Gang Rate/Hour	 £	 14.88 25.42 **40.30**				
Treatment of filled surfaces gang 1 plant operator (skill rate 2) Total Gang Rate/Hour	 £	 18.12 **18.12**				

SERIES 600: EARTHWORKS

Item	Gang hours	Labour £	Plant £	Material £	Unit	Total rate £
Geotextiles (light sheets) gang						
1 ganger/chargehand (skill rate 4) – 50% of time		7.28				
2 unskilled operatives (general)		25.42				
Total Gang Rate/Hour		£ **32.70**				
Geotextiles (medium sheets) gang						
1 ganger/chargehand (skill rate 4) – 20% of time		2.91				
3 unskilled operatives (general)		38.13				
Total Gang Rate/Hour		£ **41.04**				
Geotextiles (heavy sheets) gang						
1 ganger/chargehand (skill rate 4) – 50% of time		7.28				
2 unskilled operatives (general)		25.42				
1 plant operator (skill rate 4)		14.88				
Total Gang Rate/Hour		£ **47.58**				
Horticultural works gang						
1 skilled operative (skill rate 4)		13.61				
1 unskilled operative (general)		12.71				
Total Gang Rate/Hour		£ **26.32**				
RESOURCES – PLANT						
Scraper excavation						
motor scraper, 16.80 m^3, elevating			147.20			
Total Gang Rate/Hour			£ **147.20**			
Scraper and ripper bulldozer excavation						
motor scraper, 16.80 m^3, elevating			147.20			
D8 tractor dozer			131.60			
dozer attachment: triple shank ripper			7.50			
Total Gang Rate/Hour			£ **286.30**			
General excavation						
hydraulic crawler backacter, 0.40 m^3			37.65			
backacter attachments: percussion breaker (25% of time)			1.89			
tractor loader, 1.50 m^3 (25% of time)			11.01			
loader attachments: ripper (25% of time)			1.88			
Total Gang Rate/Hour			£ **52.43**			
Shore defences (armour stones)						
hydraulic crawler backacter, 1.20 m^3			69.80			
backacter attachments: rock bucket			3.00			
backacter attachments: clamshell grab			4.00			
Total Gang Rate/Hour			£ **76.80**			
Geotextiles (heavy sheets)						
tractor loader, 1.5 m^3			44.05			
Total Gang Rate/Hour			£ **44.05**			

SERIES 600: EARTHWORKS

Item	Gang hours	Labour £	Plant £	Material £	Unit	Total rate £
RESOURCES – PLANT – cont						
Filling						
tractor loader, 1.5 m^3			44.05			
Total Gang Rate/Hour		£	44.05			
Treatment of filled surfaces						
tractor loader, 1.50 m^3			44.05			
3 wheel deadweight roller, 10 tonne			36.20			
Total Gang Rate/Hour		£	80.25			
EXCAVATION						
Typical motorway cutting generally using motorized scrapers and/or dozers on an average haul of 2000 m (one way)						
Excavation of acceptable material Class 5A	0.01	0.14	1.18	–	m^3	1.32
Excavation of acceptable material excluding Class 5A in cutting and other excavation	0.01	0.23	2.79	–	m^3	3.02
General excavation using backacters						
Excavation of acceptable material excluding Class 5A in new watercourses	0.06	2.06	3.49	–	m^3	5.55
Excavation of unacceptable material Class U1 / U2 in new watercourses	0.07	2.23	3.65	–	m^3	5.88
General excavation using backacters						
Excavation of unacceptable material Class U1/U2 in clearing abandoned watercourses	0.06	1.89	3.10	–	m^3	4.99
General excavation using backacters and tractor loaders						
Excavation of acceptable material Class 5A	0.04	1.37	2.18	–	m^3	3.55
Excavation of acceptable material excluding Class 5A in structural foundations						
ne 3.0 m deep	0.08	2.57	4.20	–	m^3	6.77
ne 6.0 m deep	0.20	6.86	10.98	–	m^3	17.84
Excavation of acceptable material excluding Class 5A in foundations for corrugated steel buried structures and the like						
ne 3.0 m deep	0.08	2.57	4.20	–	m^3	6.77
ne 6.0 m deep	0.20	6.86	10.98	–	m^3	17.84
Excavation of unacceptable material Class U1 / U2 in structural foundations						
ne 3.0 m deep	0.09	2.92	4.78	–	m^3	7.70
ne 6.0 m deep	0.21	7.20	11.60	–	m^3	18.80

SERIES 600: EARTHWORKS

Item	Gang hours	Labour £	Plant £	Material £	Unit	Total rate £
Excavation of unacceptable material Class U1 / U2 in foundations for corrugated steel buried structures and the like						
ne 3.0 m deep	0.09	2.92	4.78	–	m³	**7.70**
ne 6.0 m deep	0.21	7.20	11.60	–	m³	**18.80**
General excavation using backacters and tractor loader with ripper						
Excavation of acceptable material Class 5 A	0.03	1.03	1.63	–	m³	**2.66**
Excavation of acceptable material excluding Class 5 A in cutting and other excavation	0.04	1.37	2.18	–	m³	**3.55**
Excavate unacceptable material Class U1 / U2 in cutting and other excavation	0.04	1.37	2.18	–	m³	**3.55**
EXCAVATION IN HARD MATERIAL						
Typical motorway cutting generally using motorized scrapers and/or dozers on an average haul of 2000 m (one way)						
Excavate in hard material; using scraper and ripper bulldozer						
mass concrete/medium hard rock	0.09	3.26	25.77	–	m³	**29.03**
reinforced concrete/hard rock	0.15	5.44	42.95	–	m³	**48.39**
tarmacadam	0.16	5.80	4.58	–	m³	**10.38**
General excavation using backacters						
Extra over excavation in new watercourses for excavation in hard material						
rock	0.74	25.38	39.96	–	m³	**65.34**
pavements, brickwork, concrete and masonry	0.69	23.67	37.23	–	m³	**60.90**
reinforced concrete	1.12	38.41	60.41	–	m³	**98.82**
General excavation using backacters and tractor loaders						
Extra over excavation in structural foundations for excavation in hard material						
rock	0.74	25.38	39.96	–	m³	**65.34**
pavements, brickwork, concrete and masonry	0.69	23.67	37.23	–	m³	**60.90**
reinforced concrete	1.12	38.41	60.41	–	m³	**98.82**
Extra over excavation for excavation in foundations for corrugated steel buried structures and the like for excavation in hard material						
rock	0.74	25.38	39.96	–	m³	**65.34**
pavements, brickwork, concrete and masonry	0.69	23.67	37.23	–	m³	**60.90**
reinforced concrete	1.12	38.41	60.41	–	m³	**98.82**

SERIES 600: EARTHWORKS

Item	Gang hours	Labour £	Plant £	Material £	Unit	Total rate £
EXCAVATION IN HARD MATERIAL – cont						
General excavation using backacters and tractor loader with ripper						
Extra over excavation for excavation in cutting and other excavation for excavation in hard material						
rock	0.71	24.35	38.30	–	m^3	**62.65**
pavements, brickwork, concrete and masonry	0.66	22.64	35.62	–	m^3	**58.26**
reinforced concrete	1.04	35.67	56.11	–	m^3	**91.78**
DEPOSITION OF FILL						
Deposition of acceptable material Class 1C/6B in						
embankments and other areas of fill	0.01	0.44	0.48	–	m^3	**0.92**
strengthened embankments	0.01	0.44	0.48	–	m^3	**0.92**
reinforced earth structures	0.01	0.44	0.48	–	m^3	**0.92**
anchored earth structures	0.01	0.44	0.48	–	m^3	**0.92**
landscaped areas	0.01	0.44	0.48	–	m^3	**0.92**
environmental bunds	0.01	0.44	0.48	–	m^3	**0.92**
fill to structures	0.01	0.44	0.48	–	m^3	**0.92**
fill above structural concrete foundations	0.01	0.44	0.48	–	m^3	**0.92**
DISPOSAL OF MATERIAL						
Disposal of acceptable material excluding Class 5A						
using 10 tonnes capacity tipping lorry for on-site or off-site use; haul distance to tip not exceeding						
1 km	0.06	1.05	2.38	–	m^3	**3.43**
ADD per further km haul	0.03	0.53	1.19	–	m^3	**1.72**
using 10–15 tonnes capacity tipping lorry for on-site or off-site use; haul distance to tip not exceeding 1 km	0.05	0.88	1.98	–	m^3	**2.86**
ADD per further km haul	0.03	0.44	0.99	–	m^3	**1.43**
using 15–25 tonnes capacity tipping lorry for on-site or off-site use; haul distance to tip not exceeding 1 km	0.05	0.79	1.78	–	m^3	**2.57**
ADD per further km haul	0.02	0.39	0.87	–	m^3	**1.26**
Disposal of acceptable material Class 5A (excluding resale value of soil)						
using 10 tonnes capacity tipping lorry for on-site or off-site use; haul distance to tip not exceeding						
1 km	0.06	1.05	2.38	–	m^3	**3.43**
ADD per further km haul	0.03	0.53	1.19	–	m^3	**1.72**
using 10–15 tonnes capacity tipping lorry for on-site or off-site use; haul distance to tip not exceeding 1 km	0.05	0.88	1.98	–	m^3	**2.86**
ADD per further km haul	0.03	0.44	0.99	–	m^3	**1.43**
using 15–25 tonnes capacity tipping lorry for on-site or off-site use; haul distance to tip not exceeding 1 km	0.05	0.79	1.78	–	m^3	**2.57**
ADD per further km haul	0.02	0.39	0.87	–	m^3	**1.26**

SERIES 600: EARTHWORKS

Item	Gang hours	Labour £	Plant £	Material £	Unit	Total rate £
Disposal of unacceptable material Class U1						
using 10 tonnes capacity tipping lorry for on-site or off-site use; haul distance to tip not exceeding 1 km	0.06	1.05	2.38	–	m^3	3.43
ADD per further km haul	0.03	0.53	1.19	–	m^3	1.72
using 10–15 tonnes capacity tipping lorry for on-site or off-site use; haul distance to tip not exceeding 1 km	0.05	0.88	1.98	–	m^3	2.86
ADD per further km haul	0.03	0.44	0.99	–	m^3	1.43
using 15–25 tonnes capacity tipping lorry for on-site or off-site use; haul distance to tip not exceeding 1 km	0.05	0.79	1.78	–	m^3	2.57
ADD per further km haul	0.02	0.39	0.87	–	m^3	1.26
Disposal of unacceptable material Class U2						
using 10 tonnes capacity tipping lorry for on-site or off-site use; haul distance to tip not exceeding 1 km	0.06	1.05	2.38	–	m^3	3.43
ADD per further km haul	0.03	0.53	1.19	–	m^3	1.72
using 10–15 tonnes capacity tipping lorry for on-site or off-site use; haul distance to tip not exceeding 1 km	0.05	0.88	1.98	–	m^3	2.86
ADD per further km haul	0.03	0.44	0.99	–	m^3	1.43
using 15–25 tonnes capacity tipping lorry for on-site or off-site use; haul distance to tip not exceeding 1 km	0.05	0.79	1.78	–	m^3	2.57
ADD per further km haul	0.02	0.39	0.87	–	m^3	1.26
Add to the above rates where tipping charges apply:						
non-hazardous waste	–	–	–	–	m^3	31.00
hazardous waste	–	–	–	–	m^3	81.00
special waste	–	–	–	–	m^3	85.00
contaminated liquid	–	–	–	–	m^3	112.50
contaminated sludge	–	–	–	–	m^3	147.00
Add to the above rates where Landfill Tax applies:						
exempted material	–	–	–	–	m^3	–
inactive or inert material	–	–	–	–	m^3	2.50
other material	–	–	–	–	m^3	64.00

SERIES 600: EARTHWORKS

Item	Gang hours	Labour £	Plant £	Material £	Unit	Total rate £
IMPORTED FILL						
Imported graded granular fill, natural gravels DfT Class 1 A/B/C (1.9 t/m^3)						
Imported acceptable material in						
embankments and other areas of fill	0.02	0.73	0.79	19.96	m^3	**21.48**
extra for Aggregate Tax	–	–	–	4.75	m^3	**4.75**
Imported graded granular fill, crushed gravels or rock DfT Class 1 A/B/C (1.9 t/m^3); using tractor loader						
Imported acceptable material in						
embankments and other areas of fill	0.02	0.73	0.79	22.40	m^3	**23.92**
extra for Aggregate Tax	–	–	–	4.75	m^3	**4.75**
Cohesive material DfT Class 2 A/B/C/D (1.8 t/m^3); using tractor loader						
Imported acceptable material in						
embankments and other areas of fill	0.02	0.73	0.79	20.76	m^3	**22.28**
landscaped areas	0.02	0.73	0.79	20.76	m^3	**22.28**
environmental bunds	0.02	0.85	0.93	20.76	m^3	**22.54**
fill to structures	0.02	0.97	1.06	20.76	m^3	**22.79**
fill above structural concrete foundations	0.02	0.77	0.84	20.76	m^3	**22.37**
Reclaimed pulverized fuel ash DfT Class 2E [2.8 t/m^3]; using tractor loader						
Imported acceptable material in						
embankments and other areas of fill	0.02	0.73	0.79	22.40	m^3	**23.92**
Reclaimed quarry waste DfT Class 2E (1.8 t/m^3); using tractor loader						
Imported acceptable material in						
embankments and other areas of fill	0.02	0.73	0.79	14.49	m^3	**16.01**
extra for Aggregate Tax (fill other than from acceptable process)	–	–	–	4.50	m^3	**4.50**
Imported topsoil DfT Class 5B (1.44 t/m^3); using tractor loader						
Imported topsoil Class 5B	0.02	0.73	0.79	19.44	m^3	**20.96**
Imported selected well graded granular fill DfT Class 6 A (1.90 t/m^3); using tractor loader						
Imported acceptable material						
embankments and other areas of fill	–	–	0.93	25.80	m^3	**26.73**
landscape areas	0.02	0.85	0.93	25.80	m^3	**27.58**
environmental bunds	0.02	0.73	0.79	25.80	m^3	**27.32**
fill to structures	0.02	0.97	1.06	25.80	m^3	**27.83**
fill above structural concrete foundations	0.02	0.77	0.84	25.80	m^3	**27.41**
extra for Aggregate Tax	–	–	–	4.75	m^3	**4.75**

SERIES 600: EARTHWORKS

Item	Gang hours	Labour £	Plant £	Material £	Unit	Total rate £
Imported selected granular fill, DfT Class 6F (1.9 t/m^3); using tractor loader						
Imported acceptable material in						
embankments and other areas of fill	0.10	3.83	4.18	25.65	m^3	**33.66**
extra for Aggregate Tax	–	–	–	4.75	m^3	**4.75**
Imported selected well graded fill DfT Class 6I (1.90 t/m^3); using tractor loader						
Imported acceptable material						
reinforced earth structures	0.02	0.89	0.97	28.24	m^3	**30.10**
extra for Aggregate Tax	–	–	–	4.75	m^3	**4.75**
Imported rock fill (1.90 t/m^3); using tractor loader						
Imported acceptable material						
in embankments and other areas of fill	0.02	0.81	0.88	34.08	m^3	**35.77**
extra for Aggregate Tax	–	–	–	4.75	m^3	**4.75**
Imported well graded granular material (1.90 t/m^3); (bedding/free draining materials under shore protection) using tractor loader						
Imported acceptable material						
in embankments and other areas of fill	0.02	0.81	0.88	26.29	m^3	**27.98**
Imported rock fill (as a core embankment) (1.90 t/m^3); using tractor loader						
Imported acceptable material						
in embankments and other areas of fill	0.03	1.33	1.45	34.08	m^3	**36.86**
extra for Aggregate Tax	–	–	–	4.75	m^3	**4.75**
Imported Armour Stones (1.90 t/m^3) (shore protection of individual rocks up to 0.5 t each); using backacter						
Imported acceptable material						
in embankments and other areas of fill	0.06	1.78	4.30	38.95	m^3	**45.03**
extra for Aggregate Tax	–	–	–	4.75	m^3	**4.75**
Imported Armour Stones (1.90 t/m^3) (shore protection of individual rocks up to 1.0 t each); using backacter						
Imported acceptable material						
in embankments and other areas of fill	0.03	1.02	2.46	43.82	m^3	**47.30**
extra for Aggregate Tax	–	–	–	4.75	m^3	**4.75**
Imported Armour Stones (1.90 t/m^3) (shore protection of individual rocks up to 3.0 t each); using backacter						
Imported acceptable material						
in embankments and other areas of fill	0.03	0.82	2.01	48.69	m^3	**51.52**
extra for Aggregate Tax	–	–	–	4.75	m^3	**4.75**

SERIES 600: EARTHWORKS

Item	Gang hours	Labour £	Plant £	Material £	Unit	Total rate £
COMPACTION OF FILL						
Compaction of granular fill material						
in embankments and other areas of fill	0.01	0.09	0.40	–	m³	**0.49**
adjacent to structures	0.01	0.14	0.64	–	m³	**0.78**
above structural concrete foundations	0.01	0.22	0.96	–	m³	**1.18**
Compaction of fill material						
in subbase or capping layers under verges, central reserves or side slopes	0.02	0.36	1.60	–	m³	**1.96**
adjacent to structures	0.04	0.63	2.81	–	m³	**3.44**
above structural concrete foundations	0.04	0.72	3.21	–	m³	**3.93**
Compaction of graded fill material						
in embankments and other areas of fill	0.01	0.11	0.48	–	m³	**0.59**
adjacent to structures	0.01	0.18	0.80	–	m³	**0.98**
above structural concrete foundations	0.02	0.27	1.20	–	m³	**1.47**
Compaction of rock fill materials						
in embankments and other areas of fill	0.01	0.16	0.72	–	m³	**0.88**
adjacent to structures	0.01	0.25	1.12	–	m³	**1.37**
above structural concrete foundations	0.02	0.31	1.36	–	m³	**1.67**
Compaction of clay fill material						
in embankments and other areas of fill	0.03	0.54	2.41	–	m³	**2.95**
adjacent to structures	0.05	0.87	3.85	–	m³	**4.72**
above structural concrete foundations	0.05	0.87	3.85	–	m³	**4.72**
GEOTEXTILES						
Notes						
The geotextile products mentioned below are not specifically confined to the individual uses stated but are examples of one of many scenarios to which they may be applied. Conversely, the scenarios are not limited to the geotextile used as an example. The heavier grades of sheeting will need to be manipulated into place by machine and cutting would be by hacksaw rather than knife. Care should be taken in assessing the wastage of the more expensive sheeting. The prices include for preparing surfaces, overlaps and turnups, jointing and sealing, fixing material in place if required and reasonable waste between 5% and 10%.						
Stabilization applications for reinforcement of granular subbases and capping layers placed over weak and variable soils.						
For use in weak soils with moderate traffic intensities e.g. light access roads: Tensar SS20 Polypropylene Geogrid						
horizontal	0.04	1.37	–	2.18	m²	**3.55**
inclined at an angle 10–45° to the horizontal	0.05	1.73	–	2.18	m²	**3.91**

SERIES 600: EARTHWORKS

Item	Gang hours	Labour £	Plant £	Material £	Unit	Total rate £
For use in weak soils with high traffic intensities and/or high axle loadings: Tensar SS30 Polypropylene Geogrid						
horizontal	0.05	1.85	–	2.61	m²	4.46
inclined at an angle 10–45° to the horizontal	0.06	2.30	–	2.61	m²	4.91
For construction over very weak soils e.g. alluvium, marsh or peat, or firmer soil subject to exceptionally high axle loadings: Tensar SS40 Polypropylene Geogrid						
horizontal	0.05	2.14	1.98	4.04	m²	8.16
inclined at an angle 10–45° to the horizontal	0.06	2.66	2.47	4.04	m²	9.17
For trafficked areas where fill comprises of aggregate exceeding 100 mm: Tensar SSLA20 Polypropylene Geogrid						
horizontal	0.04	1.37	–	3.03	m²	4.40
inclined at an angle 10–45° to the horizontal	0.05	1.73	–	3.03	m²	4.76
For stabilization and separation of granular fill from soft subgrade to prevent intermixing: Terram 1000						
horizontal	0.05	2.09	–	0.55	m²	2.64
inclined at an angle 10–45° to the horizontal	0.06	2.63	–	0.55	m²	3.18
For stabilization and separation of granular fill from soft subgrade to prevent intermixing: Terram 2000						
horizontal	0.04	2.00	1.85	1.32	m²	5.17
inclined at an angle 10–45° to the horizontal	0.05	2.52	2.33	1.32	m²	6.17
Reinforcement applications for asphalt pavements						
For roads, hardstandings and airfield pavements: Tensar AR-G grid bonded to a geotextile						
horizontal	0.05	1.85	–	2.72	m²	4.57
inclined at an angle 10–45° to the horizontal	0.06	2.30	–	2.72	m²	5.02
Slope Reinforcement and Embankment Support; For use where soils can only withstand limited shear stresses, therefore steep slopes require external support						
Paragrid 30/155; 330 g/m²						
horizontal	0.04	1.37	–	1.93	m²	3.30
inclined at an angle 10–45° to the horizontal	0.05	1.73	–	1.93	m²	3.66
Paragrid 100/255; 330 g/m²						
horizontal	0.04	1.37	–	2.83	m²	4.20
inclined at an angle 10–45° to the horizontal	0.05	1.73	–	2.83	m²	4.56
Paralink 200; 1120 g/m²						
horizontal	0.05	2.57	2.38	5.70	m²	10.65
inclined at an angle 10–45° to the horizontal	0.07	3.24	3.00	5.70	m²	11.94
Paralink 600; 2040 g/m²						
horizontal	0.06	3.00	2.78	12.04	m²	17.82
inclined at an angle 10–45° to the horizontal	0.08	3.76	3.48	12.04	m²	19.28

SERIES 600: EARTHWORKS

Item	Gang hours	Labour £	Plant £	Material £	Unit	Total rate £
GEOTEXTILES – cont						
Slope Reinforcement and Embankment Support – cont						
TerramGrid 3/3 W						
horizontal	0.06	2.71	2.51	2.62	m²	**7.84**
inclined at an angle 10–45° to the horizontal	0.07	3.38	3.13	2.62	m²	**9.13**
Scour and Erosion Protection						
For use where erosion protection is required to the surface of a slope once its geotechnical stability has been achieved, and to allow grass establishment: Tensar 'Mat' Polyethylene mesh, fixed with Tensar pegs						
horizontal	0.04	1.68	–	4.17	m²	**5.85**
inclined at an angle 10–45° from the horizontal	0.05	2.09	–	4.49	m²	**6.58**
For use where hydraulic action exists, such as coastline protection from pressures exerted by waves, currents and tides: Typar SF56						
horizontal	0.05	2.43	2.25	0.49	m²	**5.17**
inclined at an angle 10–45° from the horizontal	0.06	3.05	2.82	0.49	m²	**6.36**
For protection against puncturing to reservoir liner: Typar SF56						
horizontal	0.05	2.43	2.25	0.49	m²	**5.17**
inclined at an angle 10–45° from the horizontal	0.06	3.05	2.82	0.49	m²	**6.36**
Temporary parking areas						
For reinforcement of grassed areas subject to wear from excessive pedestrian and light motor vehicle traffic: Netlon CE131 high density polyethylene geogrid, including fixing pegs						
sheeting	0.04	1.44	–	4.65	m²	**6.09**
Landscaping applications						
For prevention of weed growth in planted areas by incorporating a geotextile over topsoil: Typar SF20, including pegs						
horizontal	0.08	3.08	–	0.61	m²	**3.69**
inclined at an angle 10–45° to the horizontal	0.09	3.86	–	0.61	m²	**4.47**
For root growth control, prevention of lateral spread of roots and mixing of road base and humus: Typar SF20						
horizontal	0.08	3.08	–	0.29	m²	**3.37**
inclined at an angle 10–45° to the horizontal	0.09	3.86	–	0.29	m²	**4.15**

SERIES 600: EARTHWORKS

Item	Gang hours	Labour £	Plant £	Material £	Unit	Total rate £
SOFT SPOTS AND OTHER VOIDS						
Excavation of soft spots and other voids using motorized scrapers and/or dozers						
Excavate below cuttings or under embankments	0.05	0.91	7.36	–	m^3	**8.27**
Excavate in side slopes	0.06	1.09	8.83	–	m^3	**9.92**
Excavation of soft spots and other voids using backacters and tractor loader						
Excavate below structural foundations and foundations for corrugated steel buried structures	0.09	1.47	2.18	–	m^3	**3.65**
Imported graded granular fill; deposition using tractor loader and towed roller						
Filling of soft spots and other voids below cuttings or under embankments	0.04	0.57	1.44	20.17	m^3	**22.18**
Filling of soft spots and other voids in side slopes	0.04	0.70	1.77	20.17	m^3	**22.64**
Filling of soft spots and other voids below structural foundations and foundations for corrugated steel buried structures	0.04	0.57	1.44	20.17	m^3	**22.18**
Imported rock fill; 1.9 t/m^3; using tractor loader						
Deposition into soft areas (rock punching)	0.03	1.01	1.10	34.08	m^3	**36.19**
DISUSED SEWERS, DRAINS, CABLES, DUCTS, PIPELINES AND THE LIKE						
Removal of disused sewer or drain						
100 mm internal diameter; with less than 1 m of cover to formation level	0.16	2.41	4.82	–	m	**7.23**
150 mm internal diameter; with 1 to 2 m of cover to formation level	0.20	2.96	5.91	–	m	**8.87**
Backfilling with acceptable material of disused sewer or drain						
100 mm internal diameter; with less than 1 m of cover to formation level	0.23	3.47	6.93	–	m^3	**10.40**
150 mm internal diameter; with 1 to 2 m of cover to formation level	0.23	3.47	6.93	–	m^3	**10.40**
Backfilling of disused basements, cellars and the like with acceptable material	0.26	3.92	7.84	–	m^3	**11.76**
Backfilling of disused gullies with concrete Grade C15	0.10	1.51	3.02	10.05	nr	**14.58**

SERIES 600: EARTHWORKS

Item	Gang hours	Labour £	Plant £	Material £	Unit	Total rate £
SUPPORTS LEFT IN EXCAVATION						
Timber closeboarded supports left in excavation	–	–	7.50	–	m²	7.50
Steel trench sheeting supports left in excavation	–	–	35.00	–	m²	35.00
TOPSOILING AND STORAGE OF TOPSOIL						
Topsoiling 150 mm thick to surfaces						
at 10° or less to horizontal	0.04	0.63	5.15	–	m²	5.78
more than 10° to horizontal	0.04	0.72	5.89	–	m²	6.61
Topsoiling 350 mm thick to surfaces						
at 10° or less to horizontal	0.05	0.82	6.62	–	m²	7.44
more than 10° to horizontal	0.05	0.91	7.36	–	m²	8.27
Topsoiling 450 mm thick to surfaces						
at 10° or less to horizontal	0.05	0.91	7.36	–	m²	8.27
more than 10° to horizontal	0.06	1.00	8.10	–	m²	9.10
Topsoiling 600 mm thick to surfaces						
at 10° or less to horizontal	0.05	0.91	7.36	–	m²	8.27
more than 10° to horizontal	0.06	1.00	8.10	–	m²	9.10
Permanent storage of topsoil	0.06	1.09	8.83	–	m³	9.92
COMPLETION OF FORMATION AND SUBFORMATION						
Completion of subformation on material other than Class 1C, 6B or rock in cuttings	0.01	0.20	0.96	–	m²	1.16
Completion of formation on material other than Class 1C, 6B or rock in cuttings	0.01	0.21	0.64	–	m²	0.85
LINING OF WATERCOURSES						
Lining new watercourse invert with precast concrete units 63 mm thick	0.21	5.53	0.39	20.59	m²	26.51
Lining new watercourse side slopes with precast concrete units 63 mm thick	0.25	6.58	0.44	19.34	m²	26.36
Lining enlarged watercourse invert with precast concrete units 63 mm thick	0.24	6.32	0.45	20.59	m²	27.36
Lining enlarged watercourse side slopes with precast concrete units 63 mm thick	0.29	7.63	0.54	19.34	m²	27.51

SERIES 600: EARTHWORKS

Item	Gang hours	Labour £	Plant £	Material £	Unit	Total rate £
GROUND IMPROVEMENT – DYNAMIC COMPACTION						
Ground consolidation by dynamic compaction is a technique which involves the dropping of a steel or concrete pounder several times in each location in a grid pattern that covers the whole site. For a ground compaction of up to 10 m, a 15 t pounder with a free fall of 20 m would be typical. Several passes over the site are normally required to achieve full compaction. The process is recommended for naturally cohesive soils and is usually uneconomic for areas of less than 4000 m^2, for sites with granular or mixed granular cohesive soils and 6000 m^2 for a site with weak cohesive soils. The main considerations to be taken into account when using this method of consolidation are:						
* sufficient area to be viable						
* proximity and condition of adjacent property and services						
* need for blanket layer of granular material for a working surface and as backfill to offset induced settlement						
* water table level						
The granular blanket layer performs the dual functions of working surface and backfill material. Generally 300 mm thickness is required.						
The final bearing capacity and settlement criteria that can be achieved depend on the nature of the material being compacted. Allowable bearing capacity may be increased by up to twice the pretreated value for the same settlement. Control testing can be by crater volume measurements, site levelling between passes, penetration tests or plate loading tests.						
DYNAMIC COMPACTION						
Dynamic compaction in main compaction with a 15 t pounder	–	–	–	–	m^2	5.00
Dynamic compaction plant standing time	–	–	–	–	hr	15.00
Free-draining granular material in granular blanket	–	–	–	–	t	15.50
Control testing including levelling, piezometers and penetrameter testing	–	–	–	–	m^2	5.00
Kentledge load test	–	–	–	–	nr	10000.00
ESTABLISHMENT OF PLANT						
Establishment of dynamic compaction plant	–	–	–	–	sum	21750.00

SERIES 600: EARTHWORKS

Item	Gang hours	Labour £	Plant £	Material £	Unit	Total rate £
GROUND IMPROVEMENT – DYNAMIC COMPACTION – cont						
VIBRATED STONE COLUMNS						
Refer to Civil Engineering Works, Class C: Geotechnical and Other Specialist Services, for an explanation and items for vibrated stone columns.						
GABION WALLING AND MATTRESSES						
Gabion walling with plastic coated galvanized wire mesh, wire laced; filled with 50 mm Class 6G material						
2.0 × 1.0 × 1.0 m module sizes	3.98	46.75	15.73	48.20	m^3	**110.68**
2.0 × 1.0 × 0.5 m module sizes	5.20	58.94	38.72	80.70	m^3	**178.36**
Gabion walling with heavily galvanized woven wire mesh, wire laced; filled with 50 mm Class 6G material						
2.0 × 1.0 × 1.0 m module sizes	4.22	47.89	31.46	39.94	m^3	**119.29**
2.0 × 1.0 × 0.5 m module sizes	5.20	58.94	38.72	67.67	m^3	**165.33**
Mattress with plastic coated galvanized wire mesh; filled with 50 mm Class 6G material installed at 10° or less to the horizontal						
6.0 × 2.0 × 0.23 m module sizes	4.22	47.89	31.46	24.50	m^3	**103.85**
CRIB WALLING						
Notes						
There are a number of products specially designated for large scale earth control. Crib walling consists of a rigid unit built of rectangular interlocking timber or precast concrete members forming a skeleton of cells laid on top of each other and filled with earth or rock. Prices for these items depend on quantity, difficulty of access to the site and availability of suitable filling material; estimates should be obtained from the manufacturer when the site conditions have been determined.						
Crib walling of timber components laid with a battering face; stone infill						
ne 1.5 m high	5.28	63.13	17.74	182.14	m^2	**263.01**
ne 3.7 m high	7.11	85.01	20.49	187.14	m^2	**292.64**
ne 4.2 m high	8.13	97.21	22.02	192.14	m^2	**311.37**
ne 5.9 m high	10.71	128.06	25.89	202.14	m^2	**356.09**
ne 7.4 m high	17.16	205.18	35.56	212.14	m^2	**452.88**
Crib walling of precast concrete crib units, laid dry jointed with a battered face; stone infill						
1.0 m high; no dowels	6.75	80.71	45.39	192.14	m	**318.24**
1.5 m high; no dowels	9.36	111.91	62.95	288.21	m	**463.07**
2.5 m high; no dowels	22.11	264.36	153.85	580.35	m	**998.56**
4.0 m high; no dowels	34.11	407.84	237.35	928.56	m	**1573.75**

SERIES 600: EARTHWORKS

Item	Gang hours	Labour £	Plant £	Material £	Unit	Total rate £
GROUND ANCHORAGES Ground anchorages consist of the installation of a cable or solid bar tendon fixed in the ground by grouting and tensioned to exceed the working load to be carried. Ground anchors may be of a permanent or temporary nature and can be used in conjunction with diaphragm walling or sheet piling to eliminate the use of strutting etc. The following costs are based on the installation of 50 nr ground anchors.						
GROUND ANCHORAGE PLANT						
Establishment of ground anchorage plant	–	–	–	–	sum	10500.00
Ground anchorage plant standing time	–	–	–	–	hr	175.00
GROUND ANCHORAGES Ground anchorages; temporary or permanent						
15.0 m maximum depth; in rock, alluvial or clay; 0–50 t load	–	–	–	–	nr	90.00
15.0 m maximum depth; in rock or alluvial; 50–90 t load	–	–	–	–	nr	108.00
15.0 m maximum depth; in rock only; 90–150 t load	–	–	–	–	nr	145.00
Temporary tendons						
in rock, alluvial or clay; 0–50 t load	–	–	–	–	nr	67.50
in rock or alluvial; 50–90 t load	–	–	–	–	nr	105.00
in rock only; 90–150 t load	–	–	–	–	nr	142.00
Permanent tendons						
in rock, alluvial or clay; 0–50 t load	–	–	–	–	nr	105.00
in rock or alluvial; 50–90 t load	–	–	–	–	nr	145.00
in rock only; 90–150 t load	–	–	–	–	nr	182.00
GROUND WATER LOWERING The following unit costs are for dewatering pervious ground only and are for sets of equipment comprising:						
* hire of 1 nr diesel driven pump (WP 150/60 or similar) complete with allowance of £50 for fuel	–	–	–	–	day	110.00
* hire of 50 m of 150 mm diameter header pipe	–	–	–	–	day	20.00
* purchase of 35 nr of disposable well points	–	–	–	–	sum	9625.00
* hire of 18 m of delivery pipe	–	–	–	–	day	7.75
* hire of 1 nr diesel driven standby pump	–	–	–	–	day	62.00
* hire of 1 nr jetting pump with hoses (for installation of wellpoints only)	–	–	–	–	day	65.00
* cost of attendant labour and plant (2 hrs per day) inclusive of small dumper and bowser	–	–	–	–	day	105.00
Costs are based on 24 hr operation in 12 hr shifts with attendant operators (specialist advice)						

SERIES 600: EARTHWORKS

Item	Gang hours	Labour £	Plant £	Material £	Unit	Total rate £
GROUND WATER LOWERING – cont						
Guide price for single set of equipment comprising pump, 150 mm diameter header pipe, 35 nr well points, delivery pipe and attendant labour and plant						
Bring to site equipment and remove upon completion	–	–	–	–	sum	2500.00
Installation: 3 day hire of jetting pump	–	–	–	–	sum	195.00
Operating costs						
purchase of well points	–	–	–	–	nr	275.00
hire of pump, header pipe, delivery pipe and standby pump complete with fuel etc. and attendant labour and plant	–	–	–	–	day	304.75
TRIAL PITS						
The following costs assume the use of mechanical plant and excavating and backfilling on the same day						
Trial pit						
ne 1.0 m deep	1.44	17.22	14.22	–	nr	31.44
1.0–2.0 m deep	2.62	31.18	27.85	–	nr	59.03
over 2.0 m deep	3.18	38.02	31.41	–	nr	69.43
BREAKING UP AND PERFORATION OF REDUNDANT PAVEMENTS						
Using scraper and ripper bulldozer						
Breaking up of redundant concrete slab						
ne 100 mm deep	0.05	0.91	14.32	–	m^2	15.23
100 to 200 mm deep	0.10	1.81	28.63	–	m^2	30.44
Breaking up of redundant flexible pavement						
ne 100 mm deep	0.02	0.36	5.73	–	m^2	6.09
100 to 200 mm deep	0.03	0.54	8.59	–	m^2	9.13
Using backacters and tractor loader with ripper						
Breaking up of redundant reinforced concrete pavement						
ne 100 mm deep	0.28	4.28	6.30	–	m^2	10.58
100 to 200 mm deep	0.49	7.49	11.03	–	m^2	18.52
200 to 300 mm deep	0.70	10.70	15.78	–	m^2	26.48
Using scraper and ripper bulldozer						
Breaking up of redundant flexible pavement using scraper and ripper bulldozer						
ne 100 mm deep	0.02	0.36	5.73	–	m^2	6.09
100 to 200 mm deep	0.03	0.54	8.59	–	m^2	9.13

SERIES 600: EARTHWORKS

Item	Gang hours	Labour £	Plant £	Material £	Unit	Total rate £
Using backacters and breakers						
Perforation of redundant reinforced concrete pavement						
ne 100 mm deep	0.02	0.33	0.78	–	m²	1.11
100 to 200 mm deep	0.03	0.50	1.19	–	m²	1.69
200 to 300 mm deep	0.05	0.83	1.97	–	m²	2.80
Perforation of redundant flexible pavement						
ne 100 mm deep	0.01	0.17	0.41	–	m²	0.58
100 to 200 mm deep	0.02	0.25	0.59	–	m²	0.84
PERFORATION OF REDUNDANT SLABS, BASEMENTS AND THE LIKE						
Using backacters and breakers						
Perforation of redundant reinforced concrete slab						
ne 100 mm deep	0.02	0.33	0.78	–	m²	1.11
100 to 200 mm deep	0.03	0.50	1.19	–	m²	1.69
200 to 300 mm deep	0.05	0.83	1.97	–	m²	2.80
Perforation of redundant reinforced concrete basement						
ne 100 mm deep	0.02	0.35	0.82	–	m²	1.17
100 to 200 mm deep	0.03	0.52	1.25	–	m²	1.77
200 to 300 mm deep	0.05	0.87	2.07	–	m²	2.94
REINFORCED AND ANCHORED EARTH STRUCTURES						
Specialist advice						
As each structure is different, it is virtually impossible to give accurate unit cost prices, as they will vary with the following parameters:						
* Type of structure						
* Where located (in water, dry condition)						
* Where geographically in the country						
* Type of fill						
* Duration of structure						
* Size of structure, etc.						
To arrive at the unit costs below assumptions have been made for a structure with the following characteristics:						
* Structure – retaining wall 6 m high × 150 m in length						
* Construction – as DfT Specification BE 3/78						
* Site conditions – good foundations						
* Fill – 5 m³ per m² of wall face						
* Fill costs – DfT Specification average £9.50 / tonne						
Therefore specialist advice should be sought in order to give accurate budget costings for individual projects.						

SERIES 600: EARTHWORKS

Item	Gang hours	Labour £	Plant £	Material £	Unit	Total rate £
REINFORCED AND ANCHORED EARTH STRUCTURES – cont						
Specialist advice – cont						
Retaining wall (per m² of face)						
concrete faced using ribbed strip	1.00	12.00	4.45	170.00	m²	**186.45**
concrete faced using flat strip	1.00	12.00	4.45	190.00	m²	**206.45**
concrete faced using polyester strip	1.00	30.00	9.25	185.00	m²	**224.25**
concrete faced using geogrid reinforcement	1.00	25.00	10.15	190.00	m²	**225.15**
preformed mesh using ribbed strip	1.00	15.00	7.00	170.00	m²	**192.00**

SERIES 700: PAVEMENTS

Item	Gang hours	Labour £	Plant £	Material £	Unit	Total rate £
RESOURCES – LABOUR						
Subbase laying gang						
1 ganger/chargehand (skill rate 4)		14.56				
1 skilled operative (skill rate 4)		13.61				
2 unskilled operatives (general)		25.42				
1 plant operator (skill rate 2)		18.12				
1 plant operator (skill rate 3)		16.55				
Total Gang Rate/Hour	£	**88.26**				
Flexible paving gang						
1 ganger/chargehand (skill rate 4)		14.56				
2 skilled operatives (skill rate 4)		27.22				
4 unskilled operatives (general)		50.84				
4 plant operators (skill rate 3)		66.20				
Total Gang Rate/Hour	£	**158.82**				
Concrete paving gang						
1 ganger/chargehand (skill rate 4)		14.56				
2 skilled operatives (skill rate 4)		27.22				
4 unskilled operatives (general)		50.84				
1 plant operator (skill rate 2)		18.12				
1 plant operator (skill rate 3)		16.55				
Total Rate/Hour	£	**127.29**				
Road surface spraying gang						
1 plant operator (skill rate 3)		16.55				
Total Gang Cost/Hour	£	**16.55**				
Road chippings gang						
1 ganger/chargehand (skill rate 4) – 50% of time		7.28				
1 skilled operative (skill rate 4)		13.61				
2 unskilled operatives (general)		25.42				
3 plant operators (skill rate 3)		49.65				
Total Gang Rate/Hour	£	**95.96**				
Cutting slabs gang						
1 unskilled operative (general)		12.71				
Total Gang Rate/Hour	£	**12.71**				
Concrete filled joints gang						
1 ganger/chargehand – 50% of time		7.28				
1 skilled operative (skill rate 4)		13.61				
2 unskilled operatives (general)		25.42				
Total Gang Rate/Hour	£	**46.31**				
Milling gang						
1 ganger/chargehand (skill rate 4)		14.56				
2 skilled operatives (skill rate 4)		27.22				
4 unskilled operatives (general)		50.84				
1 plant operator (skill rate 2)		18.12				
1 plant operator (skill rate 3)		16.55				
Total Gang Rate/Hour	£	**127.29**				

SERIES 700: PAVEMENTS

Item	Gang hours	Labour £	Plant £	Material £	Unit	Total rate £
RESOURCES – LABOUR – cont						
Rake and compact planed material gang						
1 ganger/chargehand (skill rate 4)		14.56				
1 skilled operative (skill rate 4)		13.61				
3 unskilled operatives (general)		38.13				
1 plant operator (skill rate 4)		14.88				
1 plant operator (skill rate 3)		16.55				
Total Gang Rate/Hour		£ 97.73				
RESOURCES – PLANT						
Subbase laying						
93 kW motor grader			50.15			
0.8 m³ tractor loader			29.66			
6 t towed vibratory roller			17.75			
Total Gang Rate/Hour			£ 97.56			
Flexible paving						
2 asphalt pavers, 35 kW, 4.0 m			74.83			
2 deadweight rollers, 3 point, 10 t			72.40			
tractor with front bucket and integral 2 tool compressor			29.15			
compressor tools: rammer			0.50			
compressor tools: poker vibrator			1.88			
compressor tools: extra 15 m hose			0.31			
tar sprayer, 100 litre			11.30			
self propelled chip spreader			9.15			
channel (heat) iron			2.00			
Total Gang Rate/Hour			£ 201.51			
Concrete paving						
wheeled loader, 2.60 m³			75.10			
concrete paver, 6.0 m			107.90			
concrete slipform finisher			23.80			
Total Gang Rate/Hour			£ 206.80			
Road surface spraying						
tar sprayer; 100 litre			11.30			
Total Gang Rate/Hour			£ 11.30			
Road chippings						
deadweight roller, 3 point, 10 t			36.20			
tar sprayer, 100 litre			11.30			
self propelled chip spreader			9.15			
channel (heat) iron			2.00			
Total Gang Rate/Hour			£ 58.65			
Cutting slabs						
compressor, 65 cfm			5.80			
12' disc cutter			1.00			
Total Gang Rate/Hour			£ 6.80			

SERIES 700: PAVEMENTS

Item	Gang hours	Labour £	Plant £	Material £	Unit	Total rate £
Cold milling						
cold planer, 2.10 m			216.21			
wheeled loader, 2.60 m^3			75.10			
Total Gang Rate/Hour		£	291.31			
Heat planing						
heat planer, 4.5 m			84.10			
wheeled loader, 2.60 m^3			75.10			
Total Gang Rate/Hour		£	159.20			
Rake and compact planed material						
deadweight roller, 3 point, 10 t			36.20			
tractor with front bucket and integral 2 tool						
compressor			29.15			
channel (heat) iron			2.00			
Total Gang Rate/Hour		£	67.35			
SUBBASE						
The following unit costs are generally based on the Highways Agency Specification for Highway Works and reference is made throughout this Section to clauses within that specification.						
Granular material DfT Type 1						
Subbase in carriageway, hardshoulder and hardstrip						
75 mm deep	0.04	3.09	3.41	24.35	m^3	30.85
100 mm deep	0.04	3.53	3.90	24.35	m^3	31.78
150 mm deep	0.05	3.97	4.39	24.35	m^3	32.71
200 mm deep	0.05	4.41	4.88	24.35	m^3	33.64
Granular material DfT Type 2						
Subbase; spread and graded						
75 mm deep	0.04	3.09	3.41	26.16	m^3	32.66
100 mm deep	0.04	3.53	3.90	26.16	m^3	33.59
150 mm deep	0.05	3.97	4.39	26.16	m^3	34.52
200 mm deep	0.05	4.41	4.88	26.16	m^3	35.45
Wet lean concrete DfT specified strength mix C20, 20 mm aggregate						
Subbase; spread and graded						
100 mm deep	0.05	3.97	4.39	90.69	m^3	99.05
200 mm deep	0.05	4.41	4.88	90.69	m^3	99.98
Hardcore						
Subbase; spread and graded						
100 mm deep	0.04	3.53	3.90	36.08	m^3	43.51
150 mm deep	0.05	3.97	4.39	36.08	m^3	44.44
200 mm deep	0.05	4.41	4.88	36.08	m^3	45.37
Wet mix macadam; DfT Series 900						
Subbase; spread and graded						
75 mm deep	0.04	3.09	5.17	25.42	m^3	33.68
100 mm deep	0.04	3.53	5.37	25.42	m^3	34.32
200 mm deep	0.05	4.41	5.66	25.42	m^3	35.49

SERIES 700: PAVEMENTS

Item	Gang hours	Labour £	Plant £	Material £	Unit	Total rate £
PAVEMENT (FLEXIBLE)						
Notes – Labour and Plant						
All outputs are based on clear runs without undue delay to two pavers with a 75% utilization. The outputs can be adjusted as follows to take account of space or time influences on the utilization. Factors for varying utilization of Labour and Plant:						
1 paver @ 75% utilization = × 2.00						
1 paver @ 100% utilization = × 1.50						
2 pavers @ 100% utilization = × 0.75						
Dense Bitumen Macadam						
Base to DfT Clause 903						
100 mm deep	0.02	3.18	4.03	5.91	m^2	13.12
150 mm deep	0.03	3.97	5.04	8.86	m^2	17.87
200 mm deep	0.03	4.76	6.05	11.82	m^2	22.63
Binder Course to DfT Clause 904						
50 mm deep	0.02	2.38	3.03	2.58	m^2	7.99
100 mm deep	0.02	3.18	4.03	5.15	m^2	12.36
Surface Course to DfT Clause 909						
30 mm deep	0.01	1.59	2.02	2.20	m^2	5.81
50 mm deep	0.02	2.38	3.03	3.66	m^2	9.07
Bitumen Macadam						
Binder Course to DfT Clause 901						
35 mm deep	0.01	1.59	2.02	2.34	m^2	5.95
70 mm deep	0.02	2.38	3.03	4.67	m^2	10.08
Dense Tarmacadam						
Base to DfT Clause 903						
50 mm deep	0.02	2.38	3.03	3.21	m^2	8.62
100 mm deep	0.02	2.38	3.03	6.41	m^2	11.82
Binder Course to DfT Clause 907						
60 mm deep	0.02	2.38	3.03	4.18	m^2	9.59
80 mm deep	0.02	2.38	3.03	5.58	m^2	10.99
Dense Tar Surfacing						
Surface Course to DfT Series 900						
30 mm deep	0.01	1.59	2.02	2.24	m^2	5.85
50 mm deep	0.02	2.38	3.03	3.74	m^2	9.15
Cold Asphalt						
Surface Course to DfT Series 900						
15 mm deep	0.01	1.59	2.02	1.14	m^2	4.75
30 mm deep	0.01	1.59	2.02	2.27	m^2	5.88
Rolled Asphalt						
Base to DfT Clause 904						
60 mm deep	0.02	2.38	3.03	3.94	m^2	9.35
80 mm deep	0.02	2.38	3.03	5.25	m^2	10.66
Surface Course to DfT Clause 905						
40 mm deep	0.02	2.38	3.03	3.56	m^2	8.97
60 mm deep	0.02	2.38	3.03	5.33	m^2	10.74

SERIES 700: PAVEMENTS

Item	Gang hours	Labour £	Plant £	Material £	Unit	Total rate £
PAVEMENT (CONCRETE)						
The following unit costs are for jointed reinforced concrete slabs, laid in reasonable areas (over 200 m^2) by paver train/slipformer.						
Designed mix; cement to BS EN 197–1; grade C30, 20 mm aggregate						
Slab, runway, access roads or similar						
180 mm deep	0.02	1.91	3.10	12.91	m^2	17.92
220 mm deep	0.02	2.29	3.72	15.79	m^2	21.80
260 mm deep	0.02	2.80	4.55	18.66	m^2	26.01
300 mm deep	0.03	3.18	5.17	21.52	m^2	29.87
Fabric reinforcement						
Steel fabric reinforcement to BS4483						
Ref A142 nominal mass 2.22 kg	0.03	3.82	–	1.66	m^2	5.48
Ref A252 nominal mass 3.95 kg	0.04	5.09	–	2.91	m^2	8.00
Ref B385 nominal mass 4.53 kg	0.04	5.09	–	3.54	m^2	8.63
Ref C636 nominal mass 5.55 kg	0.05	6.36	–	4.50	m^2	10.86
Ref B503 nominal mass 5.93 kg	0.05	6.36	–	4.50	m^2	10.86
Mild Steel bar reinforcement BS 4449						
Bars; supplied in bent and cut lengths						
6 mm nominal size	8.00	1018.32	–	505.00	t	1523.32
8 mm nominal size	6.74	857.93	–	505.00	t	1362.93
10 mm nominal size	6.74	857.93	–	505.00	t	1362.93
12 mm nominal size	6.74	857.93	–	505.00	t	1362.93
16 mm nominal size	6.15	782.83	–	504.50	t	1287.33
High yield steel bar reinforcement BS 4449 or 4461						
Bars; supplied in bent and cut lengths						
6 mm nominal size	8.00	1018.32	–	656.50	t	1674.82
8 mm nominal size	6.74	857.93	–	656.50	t	1514.43
10 mm nominal size	6.74	857.93	–	656.50	t	1514.43
12 mm nominal size	6.74	857.93	–	656.50	t	1514.43
16 mm nominal size	6.15	782.83	–	656.50	t	1439.33
Sheeting to prevent moisture loss						
Polyethelene sheeting; lapped joints; horizontal below concrete pavements						
1000 gauge	0.01	1.27	–	0.89	m^2	2.16
2000 gauge	0.01	1.27	–	1.63	m^2	2.90
Joints in concrete slabs						
Longitudinal joints						
180 mm deep concrete	0.01	1.53	2.48	15.73	m	19.74
220 mm deep concrete	0.01	1.53	2.48	18.50	m	22.51
260 mm deep concrete	0.01	1.53	2.48	22.60	m	26.61
300 mm deep concrete	0.01	1.53	2.48	26.36	m	30.37

SERIES 700: PAVEMENTS

Item	Gang hours	Labour £	Plant £	Material £	Unit	Total rate £
PAVEMENT (CONCRETE) – cont						
Joints in concrete slabs – cont						
Expansion joints						
180 mm deep concrete	0.01	1.53	2.48	30.97	m	**34.98**
220 mm deep concrete	0.01	1.53	2.48	36.07	m	**40.08**
260 mm deep concrete	0.01	1.53	2.48	41.15	m	**45.16**
300 mm deep concrete	0.01	1.53	2.48	42.17	m	**46.18**
Contraction joints						
180 mm deep concrete	0.01	1.53	2.48	18.07	m	**22.08**
220 mm deep concrete	0.01	1.53	2.48	19.10	m	**23.11**
260 mm deep concrete	0.01	1.53	2.48	20.29	m	**24.30**
300 mm deep concrete	0.01	1.53	2.48	24.19	m	**28.20**
Construction joints						
180 mm deep concrete	0.01	1.53	2.48	11.91	m	**15.92**
220 mm deep concrete	0.01	1.53	2.48	13.00	m	**17.01**
260 mm deep concrete	0.01	1.53	2.48	14.02	m	**18.03**
300 mm deep concrete	0.01	1.53	2.48	15.02	m	**19.03**
Open joints with filler						
ne 0.5 m; 10 mm flexcell joint filler	0.11	5.09	–	3.15	m	**8.24**
0.5–1 m; 10 mm flexcell joint filler	0.11	5.09	–	4.45	m	**9.54**
Joint sealants						
10 × 20 mm hot bitumen sealant	0.14	6.48	–	3.50	m	**9.98**
20 × 20 mm cold polysulphide sealant	0.18	8.34	–	6.00	m	**14.34**
Trimming edges only of existing slabs, floors or similar surfaces (wet or dry); 6 mm cutting width						
50 mm deep	0.02	0.25	0.14	12.50	m	**12.89**
100 mm deep	0.03	0.38	0.21	17.00	m	**17.59**
Cutting existing slabs, floors or similar surfaces (wet or dry); 8 mm cutting width						
50 mm deep	0.03	0.32	0.17	12.50	m	**12.99**
100 mm deep	0.06	0.76	0.41	15.98	m	**17.15**
150 mm deep	0.08	1.02	0.54	18.00	m	**19.56**
SURFACE TREATMENT						
Slurry sealing; BS 434 class K3						
Slurry sealing to DfT Clause 918						
3 mm deep	0.02	0.25	0.17	1.26	m²	**1.68**
4 mm deep	0.02	0.25	0.17	1.77	m²	**2.19**
Coated chippings, 9–11 kg/m²						
Surface dressing to DfT Clause 915						
6 mm nominal size	0.01	0.96	0.59	0.76	m²	**2.31**
8 mm nominal size	0.01	0.96	0.59	0.78	m²	**2.33**
10 mm nominal size	0.01	0.96	0.59	0.81	m²	**2.36**
12 mm nominal size	0.01	0.96	0.59	0.91	m²	**2.46**
Anti Skid Surfacing System						
High friction surfacing to DfT Clause 924						
Proprietary resin bonded surfacing system, colours (Buff, Grey, Red, Green)	–	–	–	–	m²	**19.25**

SERIES 700: PAVEMENTS

Item	Gang hours	Labour £	Plant £	Material £	Unit	Total rate £
TACK COAT						
Bituminous spray; BS 434 K1–40						
Tack coat to DfT Clause 920						
large areas; over 20 m^2	0.02	0.25	0.17	0.25	m^2	**0.67**
COLD MILLING (PLANING)						
Milling pavement (assumes disposal on site or reuse as fill but excludes transport if required)						
75 mm deep	0.03	3.44	7.87	–	m^2	**11.31**
100 mm deep	0.04	4.58	10.49	–	m^2	**15.07**
50 mm deep; scarifying surface	0.02	2.80	6.41	–	m^2	**9.21**
75 mm deep; scarifying surface	0.04	4.71	10.78	–	m^2	**15.49**
25 mm deep; heat planing for reuse	0.03	4.07	5.09	–	m^2	**9.16**
50 mm deep; heat planing for reuse	0.06	7.13	8.92	–	m^2	**16.05**
IN SITU RECYCLING						
Raking over scarified or heat planed material; compacting with 10 t roller						
50 mm deep	0.01	0.98	0.67	–	m^2	**1.65**

SERIES 1100: KERBS, FOOTWAYS AND PAVED AREAS

Item	Gang hours	Labour £	Plant £	Material £	Unit	Total rate £
NOTES						
Measurement Note: subbases are shown separate from their associated paving to simplify the presentation of cost alternatives.						
Measurement Note: bases are shown separate from their associated kerb etc. to simplify the presentation of cost alternatives.						
Kerb quadrants, droppers are shown separately						
The re-erection cost for kerbs, channels and edgings etc. taken from store assumes that major components are in good condition; the prices below allow a sum of 20% of the value of new materials to cover minor repairs together with an allowance for replacing a proportion of units.						
RESOURCES – LABOUR						
Kerb laying gang						
3 skilled operatives (skill rate 4)		40.83				
1 unskilled operative (general)		12.71				
1 plant operator (skill rate 3) – 25% of time		4.14				
Total Gang Rate/Hour		£ 57.68				
Path subbase, bitmac and gravel laying gang						
1 skilled operative (skill rate 4)		13.61				
2 unskilled operatives (general)		25.42				
1 plant operator (skill rate 3)		16.55				
Total Gang Rate/Hour		£ 55.58				
Paviors and flagging gang						
1 skilled operative (skill rate 4)		13.61				
1 unskilled operative (general)		12.71				
Total Gang Rate/Hour		£ 26.32				
RESOURCES – PLANT						
Kerb laying						
backhoe JCB 3CX (25% of time)			8.19			
12' Stihl saw			0.93			
road forms			2.50			
Total Gang Rate/Hour			£ 11.62			
Path subbase, bitmac and gravel laying						
backhoe JCB3CX			32.75			
2 t dumper			5.60			
pedestrian roller `Bomag BW90S			9.68			
Total Gang Rate/Hour			£ 48.03			
Paviors and flagging						
2 t dumper (33% of time)			1.87			
Total Gang Rate/Hour			£ 1.87			

SERIES 1100: KERBS, FOOTWAYS AND PAVED AREAS

Item	Gang hours	Labour £	Plant £	Material £	Unit	Total rate £
KERBS, CHANNELS, EDGINGS, COMBINED DRAINAGE AND KERB BLOCKS AND LINEAR DRAINAGE						
Foundations to kerbs etc.						
Mass concrete						
200 × 100 mm	0.01	0.58	0.11	1.70	m	**2.39**
300 × 150 mm	0.02	0.87	0.17	3.90	m	**4.94**
450 × 150 mm	0.02	1.15	0.23	5.77	m	**7.15**
100 × 100 mm haunching, per side	0.01	0.29	0.06	0.41	m	**0.76**
Precast concrete units; BS 7263; bedded jointed and pointed in cement mortar						
Kerbs; bullnosed, splayed or half battered; laid straight or curved exceeding 12 m radius						
125 × 150 mm	0.06	3.46	0.70	10.12	m	**14.28**
125 × 255 mm	0.07	4.04	0.81	11.32	m	**16.17**
150 × 305 mm	0.07	4.04	0.81	16.73	m	**21.58**
Kerbs; bullnosed, splayed or half battered; laid to curves not exceeding 12 m radius						
125 × 150 mm	0.07	3.75	0.76	9.22	m	**13.73**
125 × 255 mm	0.08	4.33	0.87	10.30	m	**15.50**
150 × 305 mm	0.08	4.33	0.87	15.17	m	**20.37**
Quadrants (normally included in general rate for kerbs; shown separately for estimating purposes)						
305 × 305 × 150 mm	0.08	4.61	0.93	17.05	nr	**22.59**
455 × 455 × 255 mm	0.10	5.77	1.16	22.00	nr	**28.93**
Drop kerbs (normally included in general rate for kerbs; shown separately for estimating purposes)						
125 × 255 mm	0.07	4.04	0.81	22.68	nr	**27.53**
150 × 305 mm	0.07	4.04	0.81	37.40	nr	**42.25**
Channels; laid straight or curved exceeding 12 m radius						
125 × 255 mm	0.07	4.04	0.81	46.18	m	**51.03**
Channels; laid to curves not exceeding 12 m radius						
255 × 125 mm	0.07	4.04	0.81	47.39	m	**52.24**
Edgings; laid straight or curved exceeding 12 m radius						
150 × 50 mm	0.04	2.31	0.46	2.85	m	**5.62**
Edgings; laid to curves not exceeding 12 m radius						
150 × 50 mm	0.05	2.60	0.52	3.70	m	**6.82**
Precast concrete drainage channels; Charcon Safeticurb; channels jointed with plastic rings and bedded, jointed and pointed in cement mortar						
Channel unit; Type DBA/3; laid straight or curved exceeding 12 m radius						
250 × 250 mm; medium duty	0.08	4.33	0.87	47.10	m	**52.30**
305 × 305 mm; heavy duty	0.10	5.48	1.11	93.89	m	**100.48**

SERIES 1100: KERBS, FOOTWAYS AND PAVED AREAS

Item	Gang hours	Labour £	Plant £	Material £	Unit	Total rate £
KERBS, CHANNELS, EDGINGS, COMBINED DRAINAGE AND KERB BLOCKS AND LINEAR DRAINAGE – cont						
Precast concrete Ellis Trief safety kerb; bedded jointed and pointed in cement mortar						
Kerbs; laid straight or curved exceeding 12 m radius						
415 × 380 mm	0.23	12.98	2.62	80.78	m	**96.38**
Kerbs; laid to curves not exceeding 12 m radius						
415 × 380 mm	0.25	14.42	2.91	97.48	m	**114.81**
Precast concrete combined kerb and drainage block Beany Block System; bedded jointed and pointed in cement mortar						
Kerb; top block, shallow base unit, standard cover plate and frame						
laid straight or curved exceeding 12 m radius	0.15	8.65	1.74	117.57	m	**127.96**
laid to curves not exceeding 12 m radius	0.20	11.54	2.32	179.30	m	**193.16**
Kerb; top block, standard base unit, standard cover plate and frame						
laid straight or curved exceeding 12 m radius	0.15	8.65	1.74	117.57	m	**127.96**
laid to curves not exceeding 12 m radius	0.20	11.54	2.32	179.30	m	**193.16**
Kerb; top block, deep base unit, standard cover plate and frame						
Straight or curved over 12 m radius	0.15	8.65	1.74	261.93	m	**272.32**
laid to curves not exceeding 12 m radius	0.20	11.54	2.32	334.78	m	**348.64**
Base block depth tapers	0.10	5.77	1.16	28.60	m	**35.53**
Extruded asphalt edgings to pavings; slip formed BS 5931						
Kerb; laid straight or curved exceeding 12 m radius						
75 mm kerb height	–	–	–	7.50	m	**7.50**
100 mm kerb height	–	–	–	10.00	m	**10.00**
125 mm kerb height	–	–	–	12.50	m	**12.50**
Channel; laid straight or curved exceeding 12 m radius						
300 mm channel width	–	–	–	13.75	m	**13.75**
250 mm channel width	–	–	–	13.00	m	**13.00**
Kerb; laid to curves not exceeding 12 m radius						
75 mm kerb height	–	–	–	8.50	m	**8.50**
100 mm kerb height	–	–	–	11.00	m	**11.00**
125 mm kerb height	–	–	–	13.00	m	**13.00**
Channel; laid to curves not exceeding 12 m radius						
300 mm channel width	–	–	–	14.75	m	**14.75**
250 mm channel width	–	–	–	14.00	m	**14.00**

SERIES 1100: KERBS, FOOTWAYS AND PAVED AREAS

Item	Gang hours	Labour £	Plant £	Material £	Unit	Total rate £
Extruded concrete; slip formed						
Kerb; laid straight or curved exceeding 12 m radius						
100 mm kerb height	–	–	–	8.50	m	**8.50**
125 mm kerb height	–	–	–	13.00	m	**13.00**
Kerb; laid to curves not exceeding 12 m radius						
100 mm kerb height	–	–	–	10.75	m	**10.75**
125 mm kerb height	–	–	–	13.85	m	**13.85**
Additional concrete for kerbs, channels, edgings, combined drainage and kerb blocks and linear drainage channel systems						
Additional in situ concrete						
kerbs	0.50	28.84	5.25	84.81	m³	**118.90**
channels	0.50	28.84	5.25	84.81	m³	**118.90**
edgings	0.50	28.84	5.25	84.81	m³	**118.90**
Remove from store and relay precast concrete units; bedded jointed and pointed in cement mortar						
Kerbs; laid straight or curved exceeding 12 m radius						
125 × 150 mm	0.06	3.46	0.70	2.91	m	**7.07**
125 × 255 mm	0.07	4.04	0.81	3.15	m	**8.00**
150 × 305 mm	0.07	4.04	0.81	4.23	m	**9.08**
Kerbs; laid to curves not exceeding 12 m radius						
125 × 150 mm	0.07	3.75	0.76	2.73	m	**7.24**
125 × 255 mm	0.08	4.33	0.87	2.94	m	**8.14**
150 × 305 mm	0.08	4.33	0.87	3.92	m	**9.12**
Quadrants (normally included in general rate for kerbs; shown separately for estimating purposes)						
305 × 305 × 150 mm	0.08	4.61	0.93	4.29	nr	**9.83**
455 × 455 × 255 mm	0.10	5.77	1.16	5.28	nr	**12.21**
Drop kerbs (normally included in general rate for kerbs; shown separately for estimating purposes)						
125 × 255 mm	0.07	4.04	0.81	5.42	nr	**10.27**
150 × 305 mm	0.07	4.04	0.81	8.36	nr	**13.21**
Channels; laid straight or curved exceeding 12 m radius						
125 × 255 mm	0.07	4.04	0.81	10.12	m	**14.97**
Channels; laid to curves not exceeding 12 m radius						
255 × 125 mm	0.07	4.04	0.81	10.36	m	**15.21**
Edgings; laid straight or curved exceeding 12 m radius						
150 × 50 mm	0.04	2.31	0.46	1.45	m	**4.22**
Edgings; laid to curves not exceeding 12 m radius						
150 × 50 mm	0.05	2.60	0.52	1.45	m	**4.57**

SERIES 1100: KERBS, FOOTWAYS AND PAVED AREAS

Item	Gang hours	Labour £	Plant £	Material £	Unit	Total rate £
KERBS, CHANNELS, EDGINGS, COMBINED DRAINAGE AND KERB BLOCKS AND LINEAR DRAINAGE – cont						
Remove from store and relay precast concrete drainage channels; Charcon Safeticurb; channels jointed with plastic rings and bedded, jointed and pointed in cement mortar						
Channel unit; Type DBA/3; laid straight or curved exceeding 12 m radius						
250 × 254 mm; medium duty	0.08	4.33	0.87	10.30	m	**15.50**
305 × 305 mm; heavy duty	0.10	5.48	1.11	19.66	m	**26.25**
Remove from store and relay precast concrete Ellis Trief safety kerb; bedded jointed and pointed in cement mortar						
Kerbs; laid straight or curved exceeding 12 m radius						
415 × 380 mm	0.23	12.98	2.62	17.04	m	**32.64**
Kerbs; laid to curves not exceeding 12 m radius						
415 × 380 mm	0.25	14.42	2.91	20.38	m	**37.71**
Remove from store and relay precast concrete combined kerb and drainage block Beany Block System; bedded jointed and pointed in cement mortar						
Kerb; top block, shallow base unit, standard cover plate and frame						
laid straight or curved exceeding 12 m radius	0.15	8.65	1.74	24.40	m	**34.79**
laid to curves not exceeding 12 m radius	0.20	11.54	2.32	36.74	m	**50.60**
Kerb; top block, standard base unit, standard cover plate and frame						
laid straight or curved exceeding 12 m radius	0.15	8.65	1.74	24.40	m	**34.79**
laid to curves not exceeding 12 m radius	0.20	11.54	2.32	36.74	m	**50.60**
Kerb; top block, deep base unit, standard cover plate and frame						
Straight or curved over 12 m radius	0.15	8.65	1.74	53.27	m	**63.66**
laid to curves not exceeding 12 m radius	0.20	11.54	2.32	67.84	m	**81.70**
Base block depth tapers	0.10	5.77	1.16	6.60	m	**13.53**
FOOTWAYS AND PAVED AREAS						
Subbases						
To paved area; sloping not exceeding 10° to the horizontal						
100 mm thick sand	0.01	0.50	0.43	2.67	m²	**3.60**
150 mm thick sand	0.01	0.67	0.58	4.01	m²	**5.26**
100 mm thick gravel	0.01	0.50	0.43	2.82	m²	**3.75**
150 mm thick gravel	0.01	0.67	0.58	4.22	m²	**5.47**
100 mm thick hardcore	0.01	0.50	0.43	1.32	m²	**2.25**
150 mm thick hardcore	0.01	0.67	0.58	1.98	m²	**3.23**
100 mm thick concrete grade 20/20	0.02	1.17	1.01	9.16	m²	**11.34**
150 mm thick concrete grade 20/20	0.03	1.78	1.54	13.74	m²	**17.06**

SERIES 1100: KERBS, FOOTWAYS AND PAVED AREAS

Item	Gang hours	Labour £	Plant £	Material £	Unit	Total rate £
Bitumen macadam surfacing; BS 4987; binder course of 20 mm open graded aggregate to clause 2.6.1 tables 5–7; surface course of 6 mm medium graded aggregate to clause 2.7.6 tables 32–33; excluding subbase						
Paved area 60 mm thick; comprising binder course 40 mm thick surface course 20 mm thick						
sloping at 10° or less to the horizontal	0.09	4.72	4.08	10.72	m²	**19.52**
sloping at more than 10° to the horizontal	0.10	5.28	4.56	10.72	m²	**20.56**
Bitumen macadam surfacing; red additives; BS 4987; binder course of 20 mm open graded aggregate to clause 2.6.1 tables 5–7; surface course of 6 mm medium graded aggregate to clause 2.7.6 tables 32–33; excluding subbase						
Paved area 60 mm thick;comprising binder course 40 mm thick surface course 20 mm thick						
sloping at 10° or less to the horizontal	0.09	4.72	4.08	12.98	m²	**21.78**
sloping at more than 10° to the horizontal	0.10	5.28	4.56	12.98	m²	**22.82**
Bitumen macadam surfacing; green additives; BS 4987; binder course of 20 mm open graded aggregate to clause 2.6.1 tables 5–7; surface course of 6 mm medium graded aggregate to clause 2.7.6 tables 32–33; excluding subbase						
Paved area 60 mm thick; comprising binder course 40 mm thick surface course 20 mm thick						
sloping at 10° or less to the horizontal	0.09	4.72	4.08	14.57	m²	**23.37**
sloping at more than 10° to the horizontal	0.10	5.28	4.56	14.57	m²	**24.41**
Granular base surfacing; Central Reserve Treatments Limestone, graded 10 mm down laid and compacted; excluding subbase						
Paved area 100 mm thick; surface sprayed with two coats of cold bituminous emulsion; blinded with 6 mm quarzite fine gravel						
sloping not exceeding 10° to the horizontal	0.02	1.11	0.96	6.19	m²	**8.26**
Ennstone Johnston Golden gravel; graded 13 mm to fines; rolled wet						
Paved area 50 mm thick; single layer						
sloping not exceeding 10° to the horizontal	0.03	1.67	1.44	8.25	m²	**11.36**
Precast concrete slabs; BS 7263; grey; 5 point bedding and pointing joints in cement mortar; excluding subbase						
Paved area 50 mm thick; comprising 600 × 450 × 50 mm units						
sloping at 10° or less to the horizontal	0.28	7.37	0.52	8.01	m²	**15.90**

SERIES 1100: KERBS, FOOTWAYS AND PAVED AREAS

Item	Gang hours	Labour £	Plant £	Material £	Unit	Total rate £
FOOTWAYS AND PAVED AREAS – cont						
Precast concrete slabs – cont						
Paved area 50 mm thick; comprising						
600 × 600 × 50 mm units						
sloping at 10° or less to the horizontal	0.24	6.32	0.45	8.12	m²	14.89
Paved area 50 mm thick; comprising						
900 × 600 × 50 mm units						
sloping at 10° or less to the horizontal	0.20	5.26	0.37	9.38	m²	15.01
Extra for coloured, 50 mm thick	–	–	–	2.69	m²	2.69
Paved area 63 mm thick; comprising						
600 × 600 × 63 mm units						
sloping at 10° or less to the horizontal	0.25	6.58	0.46	8.34	m²	15.38
Paved area 63 mm thick; comprising						
900 × 600 × 63 mm units						
sloping at 10° or less to the horizontal	0.21	5.53	0.39	9.47	m²	15.39
Precast concrete rectangular paving blocks; BS 6717; grey; bedding on 50 mm thick dry sharp sand; filling joints; excluding subbase						
Paved area 80 mm thick; comprising						
200 × 100 × 80 mm units						
sloping at 10° or less to the horizontal	0.30	7.90	0.56	14.32	m²	22.78
Precast concrete rectangular paving blocks; BS 6717; coloured; bedding on 50 mm thick dry sharp sand; filling joints; excluding subbase						
Paved area 80 mm thick; comprising						
200 × 100 × 80 mm units						
sloping at 10° or less to the horizontal	0.30	7.90	0.56	15.21	m²	23.67
Brick paviors delivered to site; bedding on 20 mm thick mortar; excluding subbase						
Paved area 85 mm thick; comprising						
215 × 103 × 65 mm units						
sloping at 10° or less to the horizontal	0.30	7.90	0.56	25.64	m²	34.10
Granite setts (2.88 kg/mm thickness/m²); bedding on 25 mm cement mortar; excluding subbase						
Paved area 100 mm thick; comprising						
100 × 100 × 100 mm units; laid to random pattern						
sloping at 10° or less to the horizontal	0.90	23.69	1.68	60.27	m²	85.64
Paved area 100 mm thick; comprising						
100 × 100 × 100 mm units; laid to specific pattern						
sloping at 10° or less to the horizontal	1.20	31.58	2.24	60.27	m²	94.09

SERIES 1100: KERBS, FOOTWAYS AND PAVED AREAS

Item	Gang hours	Labour £	Plant £	Material £	Unit	Total rate £
Cobble paving; 50–75 mm stones; bedding on 25 mm cement mortar; filling joints; excluding subbase Paved area; comprising 50–75 mm stones; laid to random pattern sloping at 10° or less to the horizontal	1.00	26.32	1.85	18.01	m^2	**46.18**
REMOVE FROM STORE AND RELAY PAVING FLAGS, SLABS AND BLOCKS						
Note An allowance of 20% of the cost of providing new pavings has been included in the rates below to allow for units which have to be replaced through unacceptable damage.						
Remove from store and relay precast concrete slabs; 5 point bedding and pointing joints in cement mortar; excluding subbase Paved area 50 mm thick; comprising 600 × 450 × 50 mm units sloping at 10° or less to the horizontal	0.28	7.37	0.52	2.11	m^2	**10.00**
Paved area 50 mm thick; comprising 600 × 600 × 50 mm units sloping at 10° or less to the horizontal	0.24	6.32	0.45	2.13	m^2	**8.90**
Paved area 50 mm thick; comprising 900 × 600 × 50 mm units sloping at 10° or less to the horizontal	0.20	5.26	0.37	2.38	m^2	**8.01**
Paved area 63 mm thick; comprising 600 × 600 × 63 mm units sloping at 10° or less to the horizontal	0.25	6.58	0.46	2.17	m^2	**9.21**
Paved area 63 mm thick; comprising 900 × 600 × 63 mm units sloping at 10° or less to the horizontal	0.21	5.53	0.39	2.40	m^2	**8.32**
Remove from store and relay precast concrete rectangular paving blocks; bedding on 50 mm thick dry sharp sand; filling joints; excluding subbase Paved area 80 mm thick; comprising 200 × 100 × 80 mm units sloping at 10° or less to the horizontal	0.30	7.90	0.56	3.96	m^2	**12.42**
Remove from store and relay brick paviors; bedding on 20 mm thick mortar; excluding subbase Paved area 85 mm thick; comprising 215 × 103 × 65 mm units sloping at 10° or less to the horizontal	0.30	7.90	0.56	6.58	m^2	**15.04**

SERIES 1200: TRAFFIC SIGNS AND ROAD MARKINGS

Item	Gang hours	Labour £	Plant £	Material £	Unit	Total rate £
NOTES						
The re-erection cost for traffic signs taken from store assumes that major components are in good condition; the prices below allow a sum of 20% of the value of new materials to cover minor repairs, new fixings and touching up any coatings.						
RESOURCES – LABOUR						
Traffic signs gang						
1 ganger/chargehand (skill rate 3)		16.16				
1 skilled operative (skill rate 3)		15.21				
2 unskilled operatives (general)		25.42				
1 plant operator (skill rate 3) – 25% of time		4.14				
Total Gang Rate/Hour		£ 60.93				
Bollards, furniture gang						
1 ganger/chargehand (skill rate 4)		14.56				
1 skilled operative (skill rate 4)		13.61				
2 unskilled operatives (general)		25.42				
Total Gang Rate/Hour		£ 53.59				
RESOURCES – PLANT						
Traffic signs						
JCB 3CX backhoe – 50% of time			16.38			
125 cfm compressor – 50% of time			4.41			
compressor tools: hand held hammer drill – 50% of time			0.56			
compressor tools: clay spade – 50% of time			0.76			
compressor tools: extra 15 m hose – 50% of time			0.16			
8 t lorry with hiab lift – 50% of time			12.63			
Total Rate/Hour		£	34.89			
Bollards, furniture						
125 cfm compressor – 50% of time			4.41			
compressor tools: hand held hammer drill – 50% of time			0.56			
compressor tools: clay spade – 50% of time			0.76			
compressor tools: extra 15 m hose – 50% of time			0.16			
8 t lorry with hiab lift – 25% of time			6.31			
Total Rate/Hour		£	12.20			
TRAFFIC SIGNS						
In this section prices will vary depending upon the diagram configurations. The following are average costs of signs and bollards. Diagram numbers refer to the Traffic Signs Regulations and General Directions 2002 and the figure numbers refer to the Traffic Signs Manual. Examples of Prime Costs for Class 1 (High Intensity) traffic and road signs (ex works) for orders exceeding £1,000.						

SERIES 1200: TRAFFIC SIGNS AND ROAD MARKINGS

Item	Gang hours	Labour £	Plant £	Material £	Unit	Total rate £
Permanent traffic sign as non-lit unit on						
600 × 450 mm	–	–	–	32.43	nr	**32.43**
600 mm diameter	–	–	–	126.00	nr	**126.00**
600 mm triangular	–	–	–	105.00	nr	**105.00**
500 × 500 mm	–	–	–	32.38	nr	**32.38**
450 × 450 mm	–	–	–	30.76	nr	**30.76**
450 × 300 mm	–	–	–	29.22	nr	**29.22**
1200 × 400 mm (CHEVRONS)	–	–	–	48.73	nr	**48.73**
Examples of Prime Costs for Class 2 (Engineering Grade) traffic and road signs (ex works) for orders exceeding £1,000.						
600 × 450 mm	–	–	–	30.75	nr	**30.75**
600 mm diameter	–	–	–	31.25	nr	**31.25**
600 mm triangular	–	–	–	26.29	nr	**26.29**
500 × 500 mm	–	–	–	97.43	nr	**97.43**
450 × 450 mm	–	–	–	92.56	nr	**92.56**
450 × 300 mm	–	–	–	87.93	nr	**87.93**
1200 × 400 mm (CHEVRONS)	–	–	–	47.81	nr	**47.81**
Standard reflectorized traffic signs						
Note: Unit costs do not include concrete foundations (see Series 1700).						
Standard one post signs; 600 × 450 mm type						
C1 signs						
fixed back to back to another sign (measured separately) with aluminium clips to existing post (measured separately)	0.04	2.44	1.40	34.05	nr	**37.89**
Extra for fixing singly with aluminium clips	0.01	0.61	0.29	1.00	nr	**1.90**
Extra for fixing singly with stainless steel clips	0.01	0.61	0.83	1.00	nr	**2.44**
fixed back to back to another sign (measured separately) with stainless steel clips to one new 76 mm diameter plastic coated steel posts 1.75 m long	0.27	16.45	9.43	84.05	nr	**109.93**
Extra for fixing singly to one face only	0.01	0.61	0.29	–	nr	**0.90**
Extra for 76 mm diameter 1.75 m long aluminium post	0.02	1.22	0.62	26.00	nr	**27.84**
Extra for 76 mm diameter 3.5 m long plastic coated steel post	0.02	1.22	0.62	42.00	nr	**43.84**
Extra for 76 mm diameter 3.5 m long aluminium post	0.02	1.22	0.62	52.00	nr	**53.84**
Extra for excavation for post, in hard material	1.10	67.02	33.08	–	nr	**100.10**
Extra for single external illumination unit with fitted photo cell (excluding trenching and cabling – see Series 1400); unit cost per face illuminated	0.33	20.11	9.91	82.50	nr	**112.52**

SERIES 1200: TRAFFIC SIGNS AND ROAD MARKINGS

Item	Gang hours	Labour £	Plant £	Material £	Unit	Total rate £
TRAFFIC SIGNS – cont						
Standard reflectorized traffic signs – cont						
Standard two post signs; 1200 × 400 mm, signs fixed back to back to another sign (measured separately) with stainless steel clips to two new 76 mm diameter plastic coated steel posts 1.75 m long	0.51	31.07	17.80	153.73	nr	202.60
Extra for fixing singly to one face only	0.02	1.22	0.62	–	nr	1.84
Extra for two 76 mm diameter 1.75 m long aluminium posts	0.04	2.44	1.20	52.00	nr	55.64
Extra for two 76 mm diameter 3.5 m long plastic coated steel posts	0.04	2.44	1.20	84.00	nr	87.64
Extra for two 76 mm diameter 3.5 m long aluminium post	0.04	2.44	1.20	104.00	nr	107.64
Extra for excavation for post, in hard material	1.10	67.02	33.08	–	nr	100.10
Extra for single external illumination unit with fitted photo cell (excluding trenching and cabling – see Series 1400); unit cost per face illuminated	0.58	35.34	17.45	127.50	nr	180.29
Standard internally illuminated traffic signs						
Bollard with integral mould-in translucent graphics (excluding trenching and cabling)						
fixing to concrete base	0.48	29.25	16.75	370.00	nr	416.00
Special traffic signs						
Note: Unit costs do not include concrete foundations (see Series 1700) or trenching and cabling (see Series 1400).						
Externally illuminated reflectorized traffic signs manufactured to order						
special signs, surface area 1.50 m² on two 100 mm diameter steel posts	–	–	–	–	nr	550.00
special signs, surface area 4.00 m² on three 100 mm diameter steel posts	–	–	–	–	nr	785.00
Internally illuminated traffic signs manufactured to order						
special signs, surface area 0.25 m² on one new 76 mm diameter steel post	–	–	–	–	nr	165.00
special signs, surface area 0.75 m² on one new 100 mm diameter steel post	–	–	–	–	nr	215.00
special signs, surface area 4.00 m² on four new 120 mm diameter steel posts	–	–	–	–	nr	1050.00
Signs on gantries						
Externally illuminated reflectorized signs						
1.50 m²	1.78	108.76	117.81	950.00	nr	1176.57
2.50 m²	2.15	130.99	141.90	1000.00	nr	1272.89
3.00 m²	3.07	187.05	202.62	1100.00	nr	1489.67

SERIES 1200: TRAFFIC SIGNS AND ROAD MARKINGS

Item	Gang hours	Labour £	Plant £	Material £	Unit	Total rate £
Internally illuminated sign with translucent optical reflective sheeting and remote light source						
0.75 m²	1.56	95.05	102.96	1500.00	nr	1698.01
1.00 m²	1.70	103.58	112.20	1850.00	nr	2065.78
1.50 m²	2.41	146.84	159.06	2500.00	nr	2805.90
Remove from Store and Re-erect Traffic Signs						
Take from store and re-erect						
3.0 m high road sign	0.28	17.06	18.48	80.00	nr	115.54
road sign on two posts	0.50	30.46	33.00	160.00	nr	223.46
ROAD MARKINGS						
Thermoplastic screed or spray						
Note: Unit costs based upon new road with clean surface closed to traffic).						
Continuous line in reflectorized white						
150 mm wide	–	–	–	–	m	1.25
200 mm wide	–	–	–	–	m	1.50
Continuous line in reflectorized yellow						
100 mm wide	–	–	–	–	m	1.00
150 mm wide	–	–	–	–	m	1.25
Intermittent line in reflectorized white						
60 mm wide with 0.60 m line and 0.60 m gap	–	–	–	–	m	0.80
100 mm wide with 1.0 m line and 5.0 m gap	–	–	–	–	m	0.81
100 mm wide with 2.0 m line and 7.0 m gap	–	–	–	–	m	0.82
100 mm wide with 4.0 m line and 2.0 m gap	–	–	–	–	m	0.83
100 mm wide with 6.0 m line and 3.0 m gap	–	–	–	–	m	0.84
150 mm wide with 1.0 m line and 5.0 m gap	–	–	–	–	m	1.05
150 mm wide with 6.0 m line and 3.0 m gap	–	–	–	–	m	1.07
150 mm wide with 0.60 m line and 0.30 m gap	–	–	–	–	m	1.07
200 mm wide with 0.60 m line and 0.30 m gap	–	–	–	–	m	1.45
200 mm wide with 1.0 m line and 1.0 m gap	–	–	–	–	m	1.45
Ancillary line in reflectorized white						
150 mm wide in hatched areas	–	–	–	–	m	0.95
200 mm wide in hatched areas	–	–	–	–	m	1.45
Ancillary line in yellow						
150 mm wide in hatched areas	–	–	–	–	m	0.95
Triangles in reflectorized white						
1.6 m high					nr	10.00
2.0 m high					nr	12.00
3.75 m high					nr	15.50
Circles with enclosing arrows in reflectorized white						
1.6 m diameter	–	–	–	–	nr	62.15
Arrows in reflectorized white						
4.0 m long straight or turning	–	–	–	–	nr	24.50
6.0 m long straight or turning	–	–	–	–	nr	30.00
6.0 m long curved	–	–	–	–	nr	30.00
6.0 m long double headed	–	–	–	–	nr	45.00
8.0 m long double headed	–	–	–	–	nr	62.15
16.0 m long double headed	–	–	–	–	nr	90.00
32.0 m long double headed	–	–	–	–	nr	131.25

SERIES 1200: TRAFFIC SIGNS AND ROAD MARKINGS

Item	Gang hours	Labour £	Plant £	Material £	Unit	Total rate £
ROAD MARKINGS – cont						
Thermoplastic screed or spray – cont						
Kerb markings in yellow						
250 mm long	–	–	–	–	nr	1.75
Letters or numerals in reflectorized white						
1.6 m high	–	–	–	–	nr	8.15
2.0 m high	–	–	–	–	nr	12.00
3.75 m high	–	–	–	–	nr	20.55
Verynyl strip markings						
Note: Unit costs based upon new road with clean surface closed to traffic.						
Verynyl strip markings (pedestrian crossings and similar locations)						
200 mm wide line	–	–	–	–	m	8.15
600 × 300 mm single stud tile	–	–	–	–	nr	13.50
REFLECTING ROAD STUDS						
100 × 100 mm square bi-directional reflecting road stud with amber corner cube reflectors	–	–	–	–	nr	7.50
140 × 254 mm rectangular one way reflecting road stud with red catseye reflectors	–	–	–	–	nr	15.00
140 × 254 mm rectangular one way reflecting road stud with green catseye reflectors	–	–	–	–	nr	15.00
140 × 254 mm rectangular bi-directional reflecting road stud with white catseye reflectors	–	–	–	–	nr	15.00
140 × 254 mm rectangular bi-directional reflecting road stud with amber catseye reflectors	–	–	–	–	nr	15.00
140 × 254 mm rectangular bi-directional reflecting road stud without catseye reflectors	–	–	–	–	nr	9.00
Remove from store and reinstall road studs						
100 × 100 mm square bi-directional reflecting road stud with corner cube reflectors	–	–	–	–	nr	3.00
140 × 254 mm rectangular one way reflecting road stud with catseye reflectors	–	–	–	–	nr	7.00
TRAFFIC SIGNAL INSTALLATIONS						
Traffic signal installation is carried out exclusively by specialist contractors, although certain items are dealt with by the main contractor or a subcontractor. The following detailed prices are given to assist in the calculation of the total installation cost.						
Installation of signal pedestals, loop detector unit pedestals, controller unit boxes and cable connection pillars						
signal pedestal	–	–	–	–	nr	45.00
loop detector unit pedestal	–	–	–	–	nr	20.00
controller unit box	–	–	–	–	nr	45.00

SERIES 1200: TRAFFIC SIGNS AND ROAD MARKINGS

Item	Gang hours	Labour £	Plant £	Material £	Unit	Total rate £
Excavate trench for traffic signal cable, depth ne 1.50 m; supports, backfilling						
450 mm wide	–	–	–	–	m	7.50
Extra for excavating in hard material	–	–	–	–	m^3	30.00
Saw cutting grooves in pavement for detector loops and feeder cables; seal with hot bitumen sealant after installation						
25 mm deep	–	–	–	–	m	4.50
MARKER POSTS						
Glass reinforced plastic marker posts						
types 1, 2, 3 or 4	–	–	–	–	nr	15.00
types 5, 6, 7 or 8	–	–	–	–	nr	12.50
Line posts for emergency crossing	–	–	–	–	nr	8.00
Standard reflectorized traffic cylinder 1000 mm high 125 mm diameter; mounted in cats eye base (deliniator)	–	–	–	–	nr	25.00
PERMANENT BOLLARDS						
Permanent bollard; non-illuminated; precast concrete						
150 mm minimum diameter 750 mm high	0.80	42.87	9.54	132.44	nr	184.85
300 mm minimum diameter 750 mm high	0.80	42.87	9.54	155.44	nr	207.85
Extra for exposed aggregate finish	–	–	–	11.50	nr	11.50
Permanent bollard; non-illuminated; galvanized steel removable and lockable pattern	0.80	42.87	16.09	116.94	nr	175.90
MISCELLANEOUS FURNITURE						
Galvanized steel lifting traffic barrier						
4.0 m wide	2.40	128.62	28.61	834.81	nr	992.04
Precast concrete seats						
bench seat 2.0 m long	0.75	40.19	8.95	869.81	nr	918.95
bench seat with concrete ends and timber slats 2.0 m long	0.75	40.19	8.95	664.81	nr	713.95
Timber seat fixed to concrete base						
bench seat 2.0 m long	0.45	24.12	4.45	669.81	nr	698.38
Metal seat						
bench seat 2.0 m long	0.75	40.19	8.95	642.40	nr	691.54

SERIES 1300: ROAD LIGHTING COLUMNS AND BRACKETS, CCTV MASTS AND CANTILEVER MASTS

Item	Gang hours	Labour £	Plant £	Material £	Unit	Total rate £
NOTES						
For convenience in pricing this section departs from Series 1300 requirements and shows lighting column costs broken down into main components of columns, brackets, and lamps and cabling.						
The outputs assume operations are continuous and are based on at least 10 complete units and do not include any allowance for on-site remedial works after erection of the columns.						
Painting and protection of the columns apart from galvanizing is not included in the following prices or outputs.						
The re-erection cost for lighting columns taken from store assumes that major components are in good condition; the prices below allow a sum of 20% of the value of new materials to cover minor repairs, new fixings and touching up any coatings.						
RESOURCES – LABOUR						
Column erection gang						
1 ganger/chargehand (skill rate 4)		14.56				
1 skilled operative (skill rate 4)		13.61				
1 unskilled operative (general)		12.71				
1 plant operator (craftsman) – 50% of time		10.55				
Total Gang Rate/Hour	£	**51.43**				
Bracket erection gang						
1 ganger/chargehand (skill rate 4)		14.56				
1 skilled operative (skill rate 4)		13.61				
1 plant operator (skill rate 4)		14.88				
1 plant operator (craftsman)		21.10				
Total Gang Rate/Hour	£	**64.15**				
Lanterns gang						
1 skilled operative (skill rate 3)		15.21				
1 skilled operative (skill rate 4)		13.61				
1 plant operator (skill rate 4)		14.88				
1 plant operator (craftsman)		21.10				
Total Gang Rate/Hour	£	**64.80**				
RESOURCES – PLANT						
Columns and bracket arms						
15 t mobile crane – 50% of time			40.93			
125 cfm compressor – 50% of time			4.41			
compressor tools: 2 single head scabbler – 50% of time			1.55			
2 t dumper – 50% of time			2.80			
Total Rate/Hour		£	**49.69**			

SERIES 1300: ROAD LIGHTING COLUMNS AND BRACKETS, CCTV MASTS AND CANTILEVER MASTS

Item	Gang hours	Labour £	Plant £	Material £	Unit	Total rate £
Bracket arms						
15 t mobile crane			81.87			
access platform, Simon hoist (50 ft)			23.45			
Total rate/Hour		£	105.32			
Lanterns						
15 t mobile crane			81.87			
access platform, Simon hoist (50 ft)			23.45			
Total Rate/Hour		£	105.32			
ROAD LIGHTING COLUMNS, BRACKETS, WALL MOUNTINGS, CCTV MASTS AND CANTILEVER MASTS						
Galvanized steel road lighting columns to BS EN 40 with flange plate base (including all control gear, switching, fuses and internal wiring)						
4.0 m nominal height	0.75	38.57	37.28	117.88	nr	**193.73**
6.0 m nominal height	0.80	41.14	52.85	322.88	nr	**416.87**
8.0 m nominal height	0.96	49.37	63.42	410.00	nr	**522.79**
10.0 m nominal height	1.28	65.83	84.57	512.50	nr	**662.90**
12.0 m nominal height	1.44	74.06	95.14	789.25	nr	**958.45**
15.0 m nominal height	1.76	90.52	116.28	1230.00	nr	**1436.80**
3.0 m cast iron column (pedestrian/landscape area)	0.75	38.57	49.56	1435.00	nr	**1523.13**
Precast concrete lighting columns to BS EN 40 with flange plate base (including all control gear, switching, fuses and internal wiring)						
5.0 m nominal height	0.75	38.57	49.56	235.75	nr	**323.88**
10.0 m nominal height	1.28	65.83	84.60	599.63	nr	**750.06**
Galvanized steel bracket arm to BS EN 40; with 5° uplift						
0.5 m projection, single arm	0.16	10.26	40.34	256.25	nr	**306.85**
1.0 m projection, single arm	0.19	12.19	20.01	358.75	nr	**390.95**
1.5 m projection, single arm	0.21	10.80	21.82	410.00	nr	**442.62**
2.0 m projection, single arm	0.27	17.32	28.44	461.25	nr	**507.01**
1.0 m projection, double arm	0.29	18.60	30.54	461.25	nr	**510.39**
2.0 m projection, double arm	0.32	20.53	33.70	512.50	nr	**566.73**
Precast concrete bracket arm to BS EN 40; with 5° uplift						
1.0 m projection, single arm	0.24	15.40	25.28	143.50	nr	**184.18**
2.0 m projection, single arm	0.32	20.53	33.70	169.13	nr	**223.36**
1.0 m projection, double arm	0.35	22.45	36.86	205.00	nr	**264.31**
2.0 m projection, doube arm	0.37	23.74	38.97	256.25	nr	**318.96**
Lantern unit with photoelectric control set to switch on at 100 lux; lamps						
55 W SON (P226); to suit 4 m and 5 m columns	0.40	25.92	25.75	256.25	nr	**307.92**
70 W SON (P236); to suit 5 m and 6 m columns	0.40	25.92	25.75	271.63	nr	**323.30**
250 W SON (P426); to suit 8 m, 10 m and 12 m columns	0.50	32.40	32.19	399.75	nr	**464.34**
Sphere 70 W SON; to suit 3 m columns (P456)	0.50	32.40	32.19	476.63	nr	**541.22**
400 W SON High pressure sodium; to suit 12 m and 15 m columns	0.50	32.40	32.19	476.63	nr	**541.22**

SERIES 1300: ROAD LIGHTING COLUMNS AND BRACKETS, CCTV MASTS AND CANTILEVER MASTS

Item	Gang hours	Labour £	Plant £	Material £	Unit	Total rate £
REMOVE FROM STORE AND RE-ERECT ROAD LIGHTING COLUMNS AND BRACKETS AND MASTS						
Re-erection of galvanized steel road lighting columns with flange plate base; including all control gear, switching, fuses and internal wiring						
4.0 m nominal height	0.75	38.57	37.29	19.22	nr	**95.08**
6.0 m nominal height	0.80	41.14	39.78	25.11	nr	**106.03**
8.0 m nominal height	0.96	49.37	47.70	31.01	nr	**128.08**
10.0 m nominal height	1.28	65.83	63.64	36.90	nr	**166.37**
12.0 m nominal height	1.44	74.06	71.59	42.18	nr	**187.83**
15.0 m nominal height	1.76	90.52	87.49	47.41	nr	**225.42**
3.0 m cast iron column (pedestrian/landscape area)	0.75	38.57	37.31	16.14	nr	**92.02**
Re-erection of precast concrete lighting columns with flange plate base; including all control gear, switching, fuses and internal wiring						
5.0 m nominal height	0.75	38.57	37.29	22.29	nr	**98.15**
10.0 m nominal height	1.28	65.83	63.64	36.90	nr	**166.37**
Re-erection of galvanized steel bracket arms						
0.5 m projection, single arm	0.16	10.26	16.85	30.75	nr	**57.86**
1.0 m projection, single arm	0.19	12.19	20.01	30.75	nr	**62.95**
1.5 m projection, single arm	0.21	13.47	22.12	30.75	nr	**66.34**
2.0 m projection, single arm	0.27	17.32	28.44	30.75	nr	**76.51**
1.0 m projection, double arm	0.29	18.60	30.54	30.75	nr	**79.89**
2.0 m projection, double arm	0.32	20.53	33.70	30.75	nr	**84.98**
Re-erection of precast concrete bracket arms						
1.0 m projection, single arm	0.24	15.40	25.28	30.75	nr	**71.43**
2.0 m projection, single arm	0.32	20.53	33.70	30.75	nr	**84.98**
1.0 m projection, double arm	0.35	22.45	37.84	30.75	nr	**91.04**
2.0 m projection, double arm	0.37	23.74	36.86	30.75	nr	**91.35**
Reinstalling lantern unit with photoelectric control set to switch on at 100 lux; lamps						
55 W SON (P226); to suit 4 m and 5 m columns	0.40	25.92	25.75	–	nr	**51.67**
70 W SON (P236); to suit 5 m and 6 m columns	0.22	14.26	25.75	–	nr	**40.01**
250 W SON (P426); to suit 8 m, 10 m and 12 m columns	0.50	32.40	32.19	–	nr	**64.59**
Sphere 70 W SON; to suit 3 m (P456)	0.50	32.40	32.19	–	nr	**64.59**
400 W SON High pressure sodium; to suit 12 m and 15 m columns	0.50	32.40	32.19	–	nr	**64.59**

SERIES 1400: ELECTRICAL WORK FOR ROAD LIGHTING AND TRAFFIC SIGNS

Item	Gang hours	Labour £	Plant £	Material £	Unit	Total rate £
RESOURCES – LABOUR						
Trenching gang						
1 ganger/chargehand (skill rate 4)		14.56				
1 skilled operative (skill rate 4)		13.61				
2 unskilled operatives (general)		25.42				
1 plant operator (skill rate 3) – 75% of time		12.41				
Total Gang Rate/Hour	£	**66.00**				
Cable laying gang						
1 ganger/chargehand (skill rate 4)		14.56				
1 skilled operative (skill rate 4)		13.61				
2 skilled operatives (skill rate 3)		30.42				
Total Gang Rate/Hour	£	**58.59**				
RESOURCES – PLANT						
Service trenching						
JCB 3CX backhoe – 50% of time			16.38			
125 cfm compressor – 50% of time			4.41			
compressor tools: 2 single head scabbler – 50% of time			1.55			
2 t dumper – 50% of time			2.80			
trench excavator – 25% of time			14.56			
Total Rate/Hour		£	**39.70**			
Cable laying						
8 t IVECO chassis or similar – 50% of time			7.80			
Total Rate/Hour		£	**7.80**			
LOCATING BURIED ROAD LIGHTING AND TRAFFIC SIGNS CABLES						
Locating buried road lighting and traffic signs cable						
in carriageways, footways, bridge decks and paved areas	0.25	16.50	9.92	–	m	**26.42**
in verges and central reserves	0.20	13.20	7.94	–	m	**21.14**
in side slopes of cuttings or side slopes of embankments	0.15	9.90	5.96	–	m	**15.86**
TRENCH FOR CABLE OR DUCT						
300 to 450 mm wide; depth not exceeding 1.5 m	0.15	9.90	5.96	–	m	**15.86**
450 to 600 mm wide; depth not exceeding 1.5 m	0.20	13.20	7.94	–	m	**21.14**
Extra for excavating rock or reinforced concrete in trench	0.50	33.00	19.86	–	m³	**52.86**
Extra for excavating brickwork or mass concrete in trench	0.40	26.40	15.88	–	m³	**42.28**
Extra for backfilling with pea gravel	0.02	1.32	0.80	22.00	m³	**24.12**
Extra for 450 × 100 mm sand cable bedding and covering	0.02	1.32	0.80	1.46	m	**3.58**
Extra for PVC marker tape	0.01	0.33	0.21	2.05	m	**2.59**
Extra for 150 × 300 clay cable tiles	0.05	3.30	1.98	4.61	m	**9.89**
Extra for 150 × 900 concrete cable tiles	0.03	1.98	1.20	7.69	m	**10.87**

SERIES 1400: ELECTRICAL WORK FOR ROAD LIGHTING AND TRAFFIC SIGNS

Item	Gang hours	Labour £	Plant £	Material £	Unit	Total rate £
CABLE AND DUCT						
600/1000 V 2 core, PVC/SWA/PVC cable with copper conductors						
Cable; in trench not exceeding 1.5 m deep						
2.5 mm^2	0.01	0.59	0.08	2.41	m	3.08
4 mm^2	0.02	1.17	0.16	3.23	m	4.56
6 mm^2	0.02	1.17	0.16	3.84	m	5.17
10 mm^2	0.02	1.17	0.16	5.28	m	6.61
16 mm^2	0.02	1.17	0.16	6.66	m	7.99
25 mm^2	0.03	1.76	0.23	8.46	m	10.45
600/1000 V 4 core, PVC/SWA/PVC cable with copper conductors						
Cable; in trench not exceeding 1.5 m deep						
16 mm^2	0.03	1.76	0.23	9.74	m	11.73
35 mm^2	0.13	7.62	1.01	11.79	m	20.42
70 mm^2	0.15	8.79	1.17	15.89	m	25.85
600/1000 V 2 core, PVC/SWA/PVC cable drawn into ducts, pipe bays or troughs						
Cable; in trench not exceeding 1.5 m deep						
2.5 mm^2	0.01	0.59	0.08	2.41	m	3.08
4 mm^2	0.02	1.17	0.16	3.23	m	4.56
6 mm^2	0.02	1.17	0.16	3.84	m	5.17
10 mm^2	0.02	1.17	0.16	5.28	m	6.61
16 mm^2	0.03	1.76	0.23	6.66	m	8.65
35 mm^2	0.12	7.03	0.94	8.46	m	16.43
70 mm^2	0.14	8.20	1.09	12.81	m	22.10
CABLE JOINTS AND CABLE TERMINATIONS						
Straight joint in 2 core PVC/SWA/PVC cable						
2.5 mm^2	0.30	17.58	2.34	48.48	nr	68.40
4 mm^2	0.30	17.58	2.34	48.48	nr	68.40
6 mm^2	0.32	18.75	2.50	48.48	nr	69.73
10 mm^2	0.42	24.61	3.28	49.49	nr	77.38
16 mm^2	0.50	29.30	3.90	51.91	nr	85.11
35 mm^2	0.80	46.87	6.24	66.02	nr	119.13
70 mm^2	1.15	67.38	8.97	73.05	nr	149.40
Tee joint in 2 core PVC/SWA/PVC cable						
2.5 mm^2	0.46	26.95	3.59	68.91	nr	99.45
4 mm^2	0.46	26.95	3.59	68.91	nr	99.45
6 mm^2	0.48	28.12	3.74	68.91	nr	100.77
10 mm^2	0.62	36.33	4.84	70.68	nr	111.85
16 mm^2	0.74	43.36	5.77	73.79	nr	122.92
25 mm^2	0.91	53.32	7.10	79.04	nr	139.46
Tee joint in 4 core PVC/SWA/PVC cable						
16 mm^2	1.10	64.45	8.58	81.16	nr	154.19
35 mm^2	1.30	76.17	10.14	90.88	nr	177.19
70 mm^2	1.60	93.74	12.48	130.50	nr	236.72

SERIES 1400: ELECTRICAL WORK FOR ROAD LIGHTING AND TRAFFIC SIGNS

Item	Gang hours	Labour £	Plant £	Material £	Unit	Total rate £
Looped terminations of 2 core PVC/SWA/PVC cable in lit sign units, traffic signals installation control unit, pedestrian crossing control unit, road lighting column, wall mounting, subway distribution box, gantry distribution box or feeder pillar						
2.5 mm^2	0.15	8.79	1.17	6.76	nr	**16.72**
6 mm^2	0.15	8.79	1.17	9.15	nr	**19.11**
10 mm^2	0.16	9.37	1.25	11.90	nr	**22.52**
16 mm^2	0.25	14.65	1.95	12.15	nr	**28.75**
25 mm^2	0.30	17.58	2.34	16.94	nr	**36.86**
Terminations of 4 core PVC/SWA/PVC cable in lit sign units, traffic signals installation control unit, pedestrian crossing control unit, road lighting column, wall mounting, subway distribution box, gantry distribution box or feeder pillar						
16 mm^2	0.35	20.51	2.73	13.13	nr	**36.37**
35 mm^2	0.55	32.22	4.29	23.94	nr	**60.45**
70 mm^2	0.68	39.84	5.30	45.43	nr	**90.57**
FEEDER PILLARS						
Galvanized steel feeder pillars						
411 × 610 mm	4.64	271.86	36.19	410.00	nr	**718.05**
611 × 810 mm	4.24	248.42	33.07	461.25	nr	**742.74**
811 × 1110 mm	4.96	290.61	38.69	922.50	nr	**1251.80**
1111 × 1203 mm	4.19	245.49	32.76	1230.00	nr	**1508.25**
EARTH ELECTRODES						
Earth electrodes providing minimal protection using earth rods, plates or stops and protective tape and joint						
to suit columns ne 12.0 m	0.40	23.44	–	261.38	nr	**284.82**
to suit columns ne 15.0 m	0.40	23.44	–	343.38	nr	**366.82**
to suit Superstructure or Buildings using copper lead conductor (per 23 m height)	1.00	58.59	–	579.13	nr	**637.72**
CHAMBERS						
Brick chamber with galvanized steel cover and frame; depth to uppermost surface of base slab						
ne 1.0 m deep	–	–	–	–	nr	**755.00**

SERIES 1500: MOTORWAY COMMUNICATIONS

Item	Gang hours	Labour £	Plant £	Material £	Unit	Total rate £
RESOURCES – LABOUR						
Service trenching gang						
1 ganger/chargehand (skill rate 4)		14.56				
1 skilled operative (skill rate 4)		13.61				
2 unskilled operatives (general)		25.42				
1 plant operator (skill rate 3) – 75% of time		12.41				
Total Gang Rate/Hour	£	**66.00**				
RESOURCES – PLANT						
Trenching						
JCB 3CX backhoe (50% of time)			16.38			
125 cfm compressor (50% of time)			4.41			
compressor tools: 2 single head scabbler (50% of time)			1.55			
2 t dumper (50% of time)			2.80			
trench excavator (25% of time)			14.56			
Total Gang Rate/Hour		£	**39.70**			
LOCATING BURIED COMMUNICATIONS CABLING						
Locating buried road lighting and traffic signs cable						
in carriageways, footways, bridge decks and paved areas	0.25	16.50	9.92	–	m	**26.42**
in verges and central reserves	0.20	13.20	7.94	–	m	**21.14**
in side slopes of cuttings or side slopes of embankments	0.15	9.90	5.96	–	m	**15.86**
TRENCH FOR COMMUNICATIONS CABLE OR DUCT						
Trench for cable						
300 to 450 mm wide; depth not exceeding 1.5 m	0.15	9.90	5.96	–	m	**15.86**
450 to 600 mm wide; depth not exceeding 1.5 m	0.20	13.20	7.94	–	m	**21.14**
Extra for excavating rock or reinforced concrete in trench	0.50	33.00	19.86	–	m^3	**52.86**
Extra for excavating brickwork or mass concrete in trench	0.40	26.40	15.88	–	m^3	**42.28**
Extra for backfilling with pea gravel	0.02	1.32	0.80	22.00	m^3	**24.12**
Extra for 450 × 100 mm sand cable bedding and covering	0.02	1.32	0.80	1.46	m	**3.58**
Extra for PVC marker tape	0.01	0.33	0.21	2.05	m	**2.59**
Extra for 150 × 300 clay cable tiles	0.05	3.30	1.98	4.61	m	**9.89**
Extra for 150 × 900 concrete cable tiles	0.03	1.98	1.20	7.69	m	**10.87**
COMMUNICATIONS CABLING AND DUCT						
Communication cables laid in trench						
type A1 2 pair 0.9 mm² armoured multi-pair	–	–	–	–	m	**4.62**
type A2 20 pair 0.9 mm² armoured multi-pair	–	–	–	–	m	**8.29**
type A3 30 pair 0.9 mm² armoured multi-pair	–	–	–	–	m	**10.19**

Unit Costs – Highway Works 449

SERIES 1500: MOTORWAY COMMUNICATIONS

Item	Gang hours	Labour £	Plant £	Material £	Unit	Total rate £
Power cables laid in trench						
type A4 10.0 mm² armoured split concentric	–	–	–	–	m	7.07
Detector feeder cables laid in trench						
type A6 50/0.25 m² single core detector feeder cable	–	–	–	–	m	8.93
type A7 50/0.25 m² single core detector feeder cable	–	–	–	–	m	8.93
COMMUNICATIONS CABLE JOINTS AND TERMINATIONS						
Cable terminations						
of type 1 cable	–	–	–	–	nr	65.94
of type 2 cable	–	–	–	–	nr	294.63
of type 3 cable	–	–	–	–	nr	422.63
of type 4 cable	–	–	–	–	nr	70.67
of type 6 cable	–	–	–	–	nr	42.10
of type 7 cable	–	–	–	–	nr	42.10
COMMUNICATIONS EQUIPMENT						
Cabinet bases						
foundation plinth	–	–	–	–	nr	133.46
Matrix signal post bases						
foundation plinth	–	–	–	–	nr	138.18
CCTV camera bases						
foundation plinth	–	–	–	–	nr	133.46
Wall mounted brackets						
at maximum 15.0 m height	–	–	–	–	nr	44.52
Fix only the following equipment						
communication equipment cabinet, 600 type series	–	–	–	–	nr	140.32
terminator type II	–	–	–	–	nr	148.73
emergency telephone post	–	–	–	–	nr	74.52
telephone housing	–	–	–	–	nr	23.27
matrix signal post	–	–	–	–	nr	74.52
Motorwarn/fogwarn	–	–	–	–	nr	148.93
distributor on gantry	–	–	–	–	nr	189.63
isolator switch for gantry	–	–	–	–	nr	412.97
heater unit mounted on gantry; Henleys' 65 W type 22501	–	–	–	–	nr	28.29
Terminal blocks						
Klippon BK6	–	–	–	–	nr	7.67
Klippon BK12	–	–	–	–	nr	8.61
Work to pavement for loop detection circuits						
cut or form grooves in pavement for detector loops and feeders	–	–	–	–	m	5.00
additional cost for sealing with hot bitumen sealant	–	–	–	–	m	0.51
CHAMBERS						
Brick chamber with galvanized steel cover and frame; depth to uppermost surface of base slab						
ne 1.0 m deep	–	–	–	–	nr	755.00

SERIES 1600: PILING AND EMBEDDED RETAINING WALLS

Item	Gang hours	Labour £	Plant £	Material £	Unit	Total rate £
NOTES						
There are a number of different types of piling which are available for use in differing situations. Selection of the most suitable type of piling for a particular site will depend on a number of factors including the physical conditions likely to be encountered during driving, the loads to be carried, the design of superstructure, etc.						
The most commonly used systems are included in this section.						
It is essential that a thorough and adequate site investigation is carried out to ascertain details of the ground strata and bearing capacities to enable a proper assessment to be made of the most suitable and economical type of piling to be adopted.						
There are so many factors, apart from design considerations, which influence the cost of piling that it is not possible to give more than an approximate indication of costs. To obtain reliable costs for a particular contract advice should be sought from a company specializing in the particular type of piling proposed. Some Specialist Contractors will also provide a design service if required.						
PILING PLANT						
Driven precast concrete reinforced piles						
Establishment of piling plant for						
235 × 235 mm precast reinforced and prestressed concrete piles in main piling	–	–	–	–	item	**4500.00**
275 × 275 mm precast reinforced and prestressed concrete piles in main piling	–	–	–	–	item	**4500.00**
350 × 350 mm precast reinforced and prestressed concrete piles in main piling	–	–	–	–	item	**4800.00**
Moving piling plant for						
235 × 235 mm precast reinforced and prestressed concrete piles in main piling	–	–	–	–	nr	**50.00**
275 × 275 mm precast reinforced and prestressed concrete piles in main piling	–	–	–	–	nr	**50.00**
350 × 350 mm precast reinforced and prestressed concrete piles in main piling	–	–	–	–	nr	**55.00**
Bored in situ reinforced concrete piling (tripod rig)						
Establishment of piling plant for 500 mm diameter cast-in-place concrete piles (tripod rig) in main piling	–	–	–	–	item	**6825.00**
Moving piling plant for 500 mm diameter cast-in-place concrete piles (tripod rig) in main piling	–	–	–	–	nr	**175.00**

SERIES 1600: PILING AND EMBEDDED RETAINING WALLS

Item	Gang hours	Labour £	Plant £	Material £	Unit	Total rate £
Bored in situ reinforced concrete piling (mobile rig)						
Establishment of piling plant for 500 mm diameter cast-in-place concrete piles (mobile rig) in main piling	–	–	–	–	item	15000.00
Moving piling plant for 500 mm diameter cast-in-place concrete piles in (mobile rig) main piling	–	–	–	–	nr	141.50
Concrete injected piles (continuous flight augered)						
Establishment of piling plant for cast-in-place concrete piles (CFA) in main piling						
450 mm diameter; 650 kN	–	–	–	–	item	8000.00
600 mm diameter; 1400 kN	–	–	–	–	item	8000.00
750 mm diameter; 2200 kN	–	–	–	–	item	8000.00
Moving piling plant for cast-in-place concrete piles (CFA) in main piling						
450 mm diameter; 650 kN	–	–	–	–	nr	40.00
600 mm diameter 1400 kN	–	–	–	–	nr	40.00
750 mm diameter 2200 kN	–	–	–	–	nr	40.00
Driven cast in place piles; segmental casing method						
Establishment of piling plant for cast-in-place concrete piles in main piling	–	–	–	–	item	11500.00
Moving piling plant for cast-in-place concrete piles in main piling	–	–	–	–	nr	25.00
Establishment of piling plant for cast-in-place concrete piles in main piling						
bottom driven	–	–	–	–	item	6000.00
top driven	–	–	–	–	item	6000.00
Moving piling plant for 430 mm diameter cast-in-place concrete piles in main piling						
bottom driven	–	–	–	–	nr	90.76
top driven	–	–	–	–	nr	62.80
Steel bearing piles						
Establishment of piling plant for steel bearing piles in main piling						
maximum 100 miles radius from base	–	–	–	–	item	4500.00
maximum 250 miles radius from base	–	–	–	–	item	10000.00
Moving piling plant for steel bearing piles in main piling	–	–	–	–	nr	65.00
Z section sheet steel piles						
Provision of all plant, equipment and labour including transport to and from the site and establishing and dismantling for						
driving of sheet piling	–	–	–	–	item	7100.00
extraction of sheet piling	–	–	–	–	item	4750.00

SERIES 1600: PILING AND EMBEDDED RETAINING WALLS

Item	Gang hours	Labour £	Plant £	Material £	Unit	Total rate £
PILING PLANT – cont						
U section sheet steel piles						
Provision of plant, equipment and labour including transport to and from the site and establishing and dismantling for						
driving of sheet piling	–	–	–	–	item	6500.00
extraction of sheet piling	–	–	–	–	item	5500.00
Establishment of piling plant for steel tubular piles in main piling						
maximum 100 miles radius from base	–	–	–	–	item	6750.00
maximum 250 miles radius from base	–	–	–	–	item	16750.00
Moving piling plant for steel tubular piles in main piling	–	–	–	–	nr	248.06
PRECAST CONCRETE PILES						
Driven precast reinforced concrete piles						
The following unit costs cover the installation of driven precast concrete piles by using a hammer acting on a shoe fitted onto or cast into the pile unit. The costs are based installing 100 piles of nominal sizes stated, and a concrete strength of 50 N/mm² suitably reinforced for a working load not exceeding 600 kN, with piles average 15 m long, on a clear site with reasonable access.						
Single pile lengths are normally a maximum of 13 m long, at which point, a mechanical interlocking joint is required to extend the pile. These joints are most economically and practically formed at works.						
Lengths, sizes of sections, reinforcement details and concrete mixes vary for differing contractors, whose specialist advice should be sought for specific designs.						
Precast concrete piles; concrete 50 N/mm²						
235 × 235 mm; 5–10 m in length; main piling	–	–	–	–	m	25.00
275 × 275 mm; 5–10 m in length; main piling	–	–	–	–	m	25.00
350 × 350 mm; 5–10 m in length; main piling	–	–	–	–	m	45.00
Mechanical Interlocking joint						
235 × 235 mm	–	–	–	–	nr	60.00
275 × 275 mm	–	–	–	–	nr	62.50
350 × 350 mm	–	–	–	–	nr	120.00
Driving vertical precast piles						
235 × 235 mm; 5–10 m in length; in main piling	–	–	–	–	m	3.25
275 × 275 mm; 5–10 m in length; in main piling	–	–	–	–	m	3.75
350 × 350 mm; 5–10 m in length; in main piling	–	–	–	–	m	4.50

SERIES 1600: PILING AND EMBEDDED RETAINING WALLS

Item	Gang hours	Labour £	Plant £	Material £	Unit	Total rate £
Stripping vertical precast concrete pile heads						
235 × 235 mm piles in main piling	–	–	–	–	nr	34.34
275 × 275 mm piles in main piling	–	–	–	–	nr	46.13
350 × 350 mm piles in main piling	–	–	–	–	nr	67.14
Standing time						
275 × 275 mm	–	–	–	–	hr	175.00
350 × 350 mm	–	–	–	–	hr	175.00

CAST-IN-PLACE PILES

Bored in situ reinforced concrete piling (tripod rig)
The following unit costs cover the construction of small diameter bored piling using light and compact tripod rigs requiring no expensive site levelling or access ways. Piling can be constructed in very restricted headroom or on confined and difficult sites. Standard diameters are between 400 and 600 mm with a normal maximum depth of 30 m.
The costs are based on installing 100 piles of 500 mm nominal diameter, a concrete strength of 20 N/mm² with nominal reinforcement, on a clear site with reasonable access. Disposal of excavated material is included separately.

Item	Gang hours	Labour £	Plant £	Material £	Unit	Total rate £
Vertical 500 mm diameter cast-in-place piles; 20 N/mm² concrete; nominal reinforcement; in main piling	–	–	–	–	m	190.00
Vertical 500 mm diameter empty bores in main piling	–	–	–	–	m	40.00
Add for boring through obstructions	–	–	–	–	hr	95.00
Standing time	–	–	–	–	hr	115.00

Bored in situ reinforced concrete piling (mobile rig)
The following unit costs cover the construction of small diameter bored piles using lorry or crawler mounted rotary boring rigs. This type of plant is more mobile and faster in operation than the tripod rigs and is ideal for large contracts in cohesive ground. Construction of piles under bentonite suspension can be carried out to obviate the use of liners. Standard diameters of 450 to 900 mm diameter can be constructed to depths of 30 m.
The costs are based on installing 100 piles of 500 mm nominal diameter, a concrete strength of 20 N/mm² with nominal reinforcement, on a clear site with reasonable access. Disposal of excavated material is included separately.

Item	Gang hours	Labour £	Plant £	Material £	Unit	Total rate £
Vertical 500 mm diameter cast-in-place piles; 20 N/mm² concrete; nominal reinforcement	–	–	–	–	m	125.00
Vertical 500 mm diameter empty bores	–	–	–	–	m	55.00
Add for boring through obstructions	–	–	–	–	hr	450.00
Standing time	–	–	–	–	hr	350.00

SERIES 1600: PILING AND EMBEDDED RETAINING WALLS

Item	Gang hours	Labour £	Plant £	Material £	Unit	Total rate £
CAST-IN-PLACE PILES – cont						
Concrete injected piles (continuous flight augered)						
The following unit costs cover the construction of piles by screwing a continuous flight auger into the ground to a design depth (determined prior to commencement of piling operations and upon which the rates are based and subsequently varied to actual depths). Concrete is then pumped through the hollow stem of the auger to the bottom and the pile formed as the auger is withdrawn. Spoil is removed by the auger as it is withdrawn. This is a fast method of construction without causing disturbance or vibration to adjacent ground. No casing is required even in unsuitable soils. Reinforcement can be placed after grouting is complete.						
The costs are based on installing 100 piles on a clear site with reasonable access. Disposal of excavated material is included separately.						
Vertical cast-in-place piles; 20 N/mm² concrete						
450 mm diameter; 650 kN; 10–15 m in length; main piling	–	–	–	–	m	21.35
600 mm diameter; 1400 kN; 10–15 m in length; main piling	–	–	–	–	m	50.00
750 mm diameter; 2200 kN; 10–15 m in length; main piling	–	–	–	–	m	54.50
Vertical empty bores						
450 mm diameter	–	–	–	–	m	25.85
600 mm diameter	–	–	–	–	m	26.15
750 mm diameter	–	–	–	–	m	30.50
Standing time/Boring through obstructions time						
450 mm diameter	–	–	–	–	hr	350.00
600 mm diameter	–	–	–	–	hr	350.00
750 mm diameter	–	–	–	–	hr	350.00
DRIVEN CAST-IN-PLACE PILES						
Driven cast-in-place piles; segmental casing method						
The following unit costs cover the construction of piles by driving into hard material using a serrated thick wall tube. It is oscillated and pressed into the hard material using a hydraulic attachment to the piling rig. The hard material is broken up using chiselling methods and is then removed by mechanical grab.						

SERIES 1600: PILING AND EMBEDDED RETAINING WALLS

Item	Gang hours	Labour £	Plant £	Material £	Unit	Total rate £
Vertical cast-in-place piles; 20 N/mm² concrete						
620 mm diameter; 10–15 m in length; main piling	–	–	–	–	m	170.00
1180 mm diameter; 10–15 m in length; main piling	–	–	–	–	m	185.00
1500 mm diameter; 10–15 m in length; main piling	–	–	–	–	m	255.00
Standing time	–	–	–	–	hr	275.00
Add for driving through obstructions	–	–	–	–	hr	295.00
Driven in situ reinforced concrete piling						
The following unit costs cover the construction of piles by driving a tube into the ground either by using an internal hammer acting on a gravel or concrete plug or, as is more usual, by using an external hammer on a driving helmet at the top of the tube. After driving to the required depth an enlarged base is formed by hammering out sucessive charges of concrete down the tube. The tube is then filled with concrete which is compacted as the tube is vibrated and withdrawn. Piles of 350 to 500 mm diameter can be constructed with rakes up to 1 in 4 to carry working loads up to 120 t per pile.						
The costs are based on installing 100 piles of 430 mm nominal diameter, a concrete strength of 20 N/mm² suitably reinforced for a working load not exceeding 750 kN, on a clear site with reasonable access.						
SEE SECTION PILING PLANT Establishment of piling plant for cast-in-place concrete piles in main piling						
bottom driven	–	–	–	–	item	8000.00
top driven	–	–	–	–	item	8000.00
SEE SECTION PILING PLANT Moving piling plant for 430 mm diameter cast-in-place concrete piles in main piling						
bottom driven	–	–	–	–	nr	40.00
top driven	–	–	–	–	nr	40.00
Vertical 430 mm diameter cast-in-place piles 20 N/mm² concrete; reinforcement for 750 kN maximum load						
bottom driven	–	–	–	–	m	20.36
top driven	–	–	–	–	m	20.36
Standing time	–	–	–	–	hr	175.00
Add for driving through obstructions where within the capabilities of the normal plant	–	–	–	–	hr	185.00
Stripping vertical concrete pile heads						
430 mm diameter heads	–	–	–	–	nr	75.00

SERIES 1600: PILING AND EMBEDDED RETAINING WALLS

Item	Gang hours	Labour £	Plant £	Material £	Unit	Total rate £
REINFORCEMENT FOR CAST-IN-PLACE PILES						
Mild steel						
Steel bar reinforcement						
6 mm nominal size; not exceeding 12 in length	–	–	–	–	t	500.00
8 mm nominal size; not exceeding 12 in length	–	–	–	–	t	500.00
10 mm nominal size; not exceeding 12 in length	–	–	–	–	t	500.00
12 mm nominal size; not exceeding 12 in length	–	–	–	–	t	500.00
16 mm nominal size; not exceeding 12 in length	–	–	–	–	t	500.00
25 mm nominal size; not exceeding 12 in length	–	–	–	–	t	500.00
32 mm nominal size; not exceeding 12 in length	–	–	–	–	t	500.00
40 mm nominal size; not exceeding 12 in length	–	–	–	–	t	500.00
Steel helical reinforcement						
6 mm nominal size; not exceeding 12 in length	–	–	–	–	t	550.00
8 mm nominal size; not exceeding 12 in length	–	–	–	–	t	550.00
10 mm nominal size; not exceeding 12 in length	–	–	–	–	t	500.00
12 mm nominal size; not exceeding 12 in length	–	–	–	–	t	550.00
High tensile steel						
Steel bar reinforcement						
6 mm nominal size; not exceeding 12 in length	–	–	–	–	t	650.00
8 mm nominal size; not exceeding 12 in length	–	–	–	–	t	650.00
10 mm nominal size; not exceeding 12 in length	–	–	–	–	t	650.00
12 mm nominal size; not exceeding 12 in length	–	–	–	–	t	650.00
16 mm nominal size; not exceeding 12 in length	–	–	–	–	t	500.00
25 mm nominal size; not exceeding 12 in length	–	–	–	–	t	500.00
32 mm nominal size; not exceeding 12 in length	–	–	–	–	t	650.00
40 mm nominal size; not exceeding 12 in length	–	–	–	–	t	650.00
Steel helical reinforcement						
6 mm nominal size; not exceeding 12 in length	–	–	–	–	t	650.00
8 mm nominal size; not exceeding 12 in length	–	–	–	–	t	650.00
10 mm nominal size; not exceeding 12 in length	–	–	–	–	t	650.00
12 mm nominal size; not exceeding 12 in length	–	–	–	–	t	650.00
STEEL BEARING PILES						
Steel bearing piles are commonly carried out by a Specialist Contractor whose advice should be sought to arrive at accurate costing. However the following items can be used to assess a budget cost for such work.						
The following unit costs are based upon driving 100 nr steel bearing piles on a clear site with reasonable access. Supply is based on delivery 75 miles from works, in loads over 20 t.						
Steel bearing piles						
Standing time	–	–	–	–	hr	350.00

Unit Costs – Highway Works

SERIES 1600: PILING AND EMBEDDED RETAINING WALLS

Item	Gang hours	Labour £	Plant £	Material £	Unit	Total rate £
203 × 203 × 45 kg/m steel bearing piles; Grade S275						
not exceeding 5 m in length in main piling	–	–	–	48.51	m	**48.51**
5–10 m in length in main piling	–	–	–	47.77	m	**47.77**
10–15 m in length in main piling	–	–	–	47.77	m	**47.77**
15–20 m in length in main piling	–	–	–	50.24	m	**50.24**
203 × 203 × 54 kg/m steel bearing piles; Grade S275						
not exceeding 5 m in length in main piling	–	–	–	58.21	m	**58.21**
5–10 m in length in main piling	–	–	–	57.32	m	**57.32**
10–15 m in length in main piling	–	–	–	57.32	m	**57.32**
15–20 m in length in main piling	–	–	–	60.29	m	**60.29**
254 × 254 × 63 kg/m steel bearing piles; Grade S275						
ne 5 m in length	–	–	–	67.91	m	**67.91**
5–10 m in length	–	–	–	66.87	m	**66.87**
10–15 m in length	–	–	–	66.87	m	**66.87**
15–20 m in length	–	–	–	70.34	m	**70.34**
254 × 254 × 71 kg/m steel bearing piles; Grade S275						
ne 5 m in length	–	–	–	76.54	m	**76.54**
5–10 m in length	–	–	–	75.37	m	**75.37**
10–15 m in length	–	–	–	75.37	m	**75.37**
15–20 m in length	–	–	–	79.27	m	**79.27**
254 × 254 × 85 kg/m steel bearing piles; Grade S275						
ne 5 m in length	–	–	–	91.63	m	**91.63**
5–10 m in length	–	–	–	90.23	m	**90.23**
10–15 m in length	–	–	–	90.23	m	**90.23**
15–20 m in length	–	–	–	94.90	m	**94.90**
305 × 305 × 79 kg/m steel bearing piles; Grade S275						
ne 5 m in length	–	–	–	85.16	m	**85.16**
5–10 m in length	–	–	–	83.86	m	**83.86**
10–15 m in length	–	–	–	83.86	m	**83.86**
15–20 m in length	–	–	–	88.20	m	**88.20**
305 × 305 × 95 kg/m steel bearing piles; Grade S275						
ne 5 m in length	–	–	–	102.41	m	**102.41**
5–10 m in length	–	–	–	100.84	m	**100.84**
10–15 m in length	–	–	–	100.84	m	**100.84**
15–20 m in length	–	–	–	106.07	m	**106.07**
305 × 305 × 110 kg/m steel bearing piles; Grade S275						
ne 5 m in length	–	–	–	118.58	m	**118.58**
5–10 m in length	–	–	–	116.77	m	**116.77**
10–15 m in length	–	–	–	116.77	m	**116.77**
15–20 m in length	–	–	–	122.81	m	**122.81**
305 × 305 × 126 kg/m steel bearing piles; Grade S275						
ne 5 m in length	–	–	–	135.83	m	**135.83**
5–10 m in length	–	–	–	133.75	m	**133.75**
10–15 m in length	–	–	–	133.75	m	**133.75**
15–20 m in length	–	–	–	140.68	m	**140.68**
305 × 305 × 149 kg/m steel bearing piles; Grade S275						
ne 5 m in length	–	–	–	160.62	m	**160.62**
5–10 m in length	–	–	–	158.16	m	**158.16**
10–15 m in length	–	–	–	158.16	m	**158.16**
15–20 m in length	–	–	–	166.36	m	**166.36**

SERIES 1600: PILING AND EMBEDDED RETAINING WALLS

Item	Gang hours	Labour £	Plant £	Material £	Unit	Total rate £
STEEL BEARING PILES – cont						
Steel bearing piles – cont						
305 × 305 × 186 kg/m steel bearing piles; Grade S275						
ne 5 m in length	–	–	–	200.51	m	**200.51**
5–10 m in length	–	–	–	197.44	m	**197.44**
10–15 m in length	–	–	–	197.44	m	**197.44**
15–20 m in length	–	–	–	207.67	m	**207.67**
305 × 305 × 233 kg/m steel bearing piles; Grade S275						
ne 5 m in length	–	–	–	240.39	m	**240.39**
5–10 m in length	–	–	–	236.71	m	**236.71**
10–15 m in length	–	–	–	236.71	m	**236.71**
15–20 m in length	–	–	–	248.98	m	**248.98**
356 × 368 × 109 kg/m steel bearing piles; Grade S275						
ne 5 m in length	–	–	–	141.48	m	**141.48**
5–10 m in length	–	–	–	139.68	m	**139.68**
10–15 m in length	–	–	–	139.68	m	**139.68**
15–20 m in length	–	–	–	145.68	m	**145.68**
356 × 368 × 133 kg/m steel bearing piles; Grade S275						
ne 5 m in length	–	–	–	172.63	m	**172.63**
5–10 m in length	–	–	–	170.44	m	**170.44**
10–15 m in length	–	–	–	170.44	m	**170.44**
15–20 m in length	–	–	–	177.75	m	**177.75**
356 × 368 × 152 kg/m steel bearing piles; Grade S275						
ne 5 m in length	–	–	–	197.30	m	**197.30**
5–10 m in length	–	–	–	194.79	m	**194.79**
10–15 m in length	–	–	–	194.79	m	**194.79**
15–20 m in length	–	–	–	203.15	m	**203.15**
356 × 368 × 174 kg/m steel bearing piles; Grade S275						
ne 5 m in length	–	–	–	225.85	m	**225.85**
5–10 m in length	–	–	–	222.98	m	**222.98**
10–15 m in length	–	–	–	222.98	m	**222.98**
15–20 m in length	–	–	–	232.55	m	**232.55**
Driving vertical steel bearing piles						
section weight not exceeding 70 kg/m	–	–	–	–	m	**3.00**
section weight 70–90 kg/m	–	–	–	–	m	**3.25**
section weight 90–110 kg/m	–	–	–	–	m	**3.50**
section weight 90–110 kg/m	–	–	–	–	m	**3.80**
section weight 110–130 kg/m	–	–	–	–	m	**3.80**
section weight 150–170 kg/m	–	–	–	–	m	**4.10**
Driving raking steel bearing piles						
section weight not exceeding 70 kg/m	–	–	–	–	m	**4.75**
section weight 70–90 kg/m	–	–	–	–	m	**5.00**
section weight 90–110 kg/m	–	–	–	–	m	**5.30**
section weight 110–130 kg/m	–	–	–	–	m	**5.30**
section weight 130–150 kg/m	–	–	–	–	m	**3.80**
section weight 150–170 kg/m	–	–	–	–	m	**4.10**
section weight 170–190 kg/m	–	–	–	–	m	**4.35**
section weight 190–210 kg/m	–	–	–	–	m	**4.60**

SERIES 1600: PILING AND EMBEDDED RETAINING WALLS

Item	Gang hours	Labour £	Plant £	Material £	Unit	Total rate £
Allow 30% of the respective item above for the lengthened section only.						
Welding on lengthening pieces to vertical steel bearing piles						
203 × 203 × any kg/m	–	–	–	–	nr	200.00
254 × 254 × any kg/m	–	–	–	–	nr	240.00
305 × 305 × any kg/m	–	–	–	–	nr	275.00
356 × 368 × any kg/m	–	–	–	–	nr	305.00
Cutting or burning off surplus length of vertical steel bearing piles						
203 × 203 × any kg/m	–	–	–	–	nr	70.00
254 × 254 × any kg/m	–	–	–	–	nr	100.00
305 × 305 × any kg/m	–	–	–	–	nr	120.00
356 × 368 × any kg/m	–	–	–	–	nr	135.00
STEEL TUBULAR PILES						
Steel tubular piles are commonly carried out by a Specialist Contractor and whose advice should be sought to arrive at accurate costings. However the following items can be used to assess a budget cost for each work.						
The following unit costs are based upon driving 100 nr steel tubular piles on a clear site with reasonable access.						
Standing time	–	–	–	–	hr	309.19
Steel Grade 275; delivered in 10–20 t loads; mass 60–120 kg/m						
section 508 mm × 8 mm × 98.6 kg/m	–	–	–	128.18	m	128.18
section 559 mm × 8 mm × 109 kg/m	–	–	–	141.70	m	141.70
Steel Grade 275; delivered in 10–20 t loads; mass 120–250 kg/m						
section 508 mm × 10 mm × 123 kg/m	–	–	–	159.90	m	159.90
section 508 mm × 12.5 mm × 153 kg/m	–	–	–	198.90	m	198.90
section 508 mm × 16 mm × 194 kg/m	–	–	–	252.20	m	252.20
section 508 mm × 20 mm × 241 kg/m	–	–	–	313.30	m	313.30
section 559 mm × 10 mm × 135 kg/m	–	–	–	175.50	m	175.50
section 559 mm × 12.5 mm × 168 kg/m	–	–	–	218.40	m	218.40
section 559 mm × 16 mm × 214 kg/m	–	–	–	278.20	m	278.20
section 610 mm × 8 mm × 119 kg/m	–	–	–	154.70	m	154.70
section 610 mm × 10 mm × 148 kg/m	–	–	–	192.40	m	192.40
section 610 mm × 12.5 mm × 184 kg/m	–	–	–	239.20	m	239.20
section 610 mm × 16 mm × 234 kg/m	–	–	–	304.20	m	304.20
section 660 mm × 8 mm × 129 kg/m	–	–	–	167.70	m	167.70
section 660 mm × 10 mm × 160 kg/m	–	–	–	208.00	m	208.00
section 660 mm × 12.5 mm × 200 kg/m	–	–	–	260.00	m	260.00
section 711 mm × 8 mm × 134 kg/m	–	–	–	174.20	m	174.20
section 711 mm × 10 mm × 173 kg/m	–	–	–	224.90	m	224.90
section 711 mm × 12 mm × 215 kg/m	–	–	–	279.50	m	279.50
section 762 mm × 8 mm × 149 kg/m	–	–	–	193.70	m	193.70
section 762 mm × 10 mm × 185 kg/m	–	–	–	240.50	m	240.50
section 762 mm × 12.5 mm × 231 kg/m	–	–	–	300.30	m	300.30

SERIES 1600: PILING AND EMBEDDED RETAINING WALLS

Item	Gang hours	Labour £	Plant £	Material £	Unit	Total rate £
STEEL TUBULAR PILES – cont						
Steel Grade 275; delivered in 10–20 t loads; mass 250–500 kg/m						
section 559 mm × 20 mm × 266 kg/m	–	–	–	345.80	m	**345.80**
section 610 mm × 20 mm × 291 kg/m	–	–	–	378.30	m	**378.30**
section 660 mm × 16 mm × 254 kg/m	–	–	–	330.20	m	**330.20**
section 660 mm × 20 mm × 316 kg/m	–	–	–	410.80	m	**410.80**
section 660 mm × 25 mm × 392 kg/m	–	–	–	509.60	m	**509.60**
section 711 mm × 16 mm × 274 kg/m	–	–	–	356.20	m	**356.20**
section 711 mm × 20 mm × 341 kg/m	–	–	–	443.30	m	**443.30**
section 711 mm × 25 mm × 423 kg/m	–	–	–	549.90	m	**549.90**
section 762 mm × 16 mm × 294 kg/m	–	–	–	382.20	m	**382.20**
section 762 mm × 20 mm × 366 kg/m	–	–	–	475.80	m	**475.80**
section 762 mm × 25 mm × 454 kg/m	–	–	–	590.20	m	**590.20**
Driving vertical steel tubular piles						
mass 60–120 kg/m	–	–	–	–	m	**7.25**
mass 120–150 kg/m	–	–	–	–	m	**7.40**
mass 150–160 kg/m	–	–	–	–	m	**7.45**
mass 160–190 kg/m	–	–	–	–	m	**7.60**
mass 190–220 kg/m	–	–	–	–	m	**7.75**
mass 220–250 kg/m	–	–	–	–	m	**7.85**
mass 250–280 kg/m	–	–	–	–	m	**7.95**
mass 280–310 kg/m	–	–	–	–	m	**8.07**
mass 310–340 kg/m	–	–	–	–	m	**8.15**
mass 340–370 kg/m	–	–	–	–	m	**8.30**
mass 370–400 kg/m	–	–	–	–	m	**8.55**
mass 400–430 kg/m	–	–	–	–	m	**8.65**
mass 430–460 kg/m	–	–	–	–	m	**8.85**
Driving raking steel tubular piles						
mass 60–120 kg/m	–	–	–	–	m	**7.65**
mass 120–150 kg/m	–	–	–	–	m	**7.75**
mass 150–160 kg/m	–	–	–	–	m	**7.85**
mass 160–190 kg/m	–	–	–	–	m	**7.95**
mass 190–220 kg/m	–	–	–	–	m	**8.15**
mass 220–250 kg/m	–	–	–	–	m	**8.25**
mass 250–280 kg/m	–	–	–	–	m	**8.35**
mass 280–310 kg/m	–	–	–	–	m	**8.45**
mass 310–340 kg/m	–	–	–	–	m	**8.60**
mass 340–370 kg/m	–	–	–	–	m	**8.75**
mass 370–400 kg/m	–	–	–	–	m	**8.95**
mass 400–430 kg/m	–	–	–	–	m	**9.10**
mass 430–460 kg/m	–	–	–	–	m	**9.35**
Driving lengthened vertical steel tubular piles						
mass 60–120 kg/m	–	–	–	–	m	**9.40**
mass 120–150 kg/m	–	–	–	–	m	**9.70**
mass 150–160 kg/m	–	–	–	–	m	**9.75**
mass 160–190 kg/m	–	–	–	–	m	**9.85**
mass 190–220 kg/m	–	–	–	–	m	**10.00**
mass 220–250 kg/m	–	–	–	–	m	**10.15**
mass 250–280 kg/m	–	–	–	–	m	**10.30**

SERIES 1600: PILING AND EMBEDDED RETAINING WALLS

Item	Gang hours	Labour £	Plant £	Material £	Unit	Total rate £
mass 280–310 kg/m	–	–	–	–	m	10.45
mass 310–340 kg/m	–	–	–	–	m	10.58
mass 340–370 kg/m	–	–	–	–	m	10.75
mass 370–400 kg/m	–	–	–	–	m	11.00
mass 400–430 kg/m	–	–	–	–	m	11.15
mass 430–460 kg/m	–	–	–	–	m	11.45
Driving lengthened raking steel tubular piles						
mass 60–120 kg/m	–	–	–	–	m	9.85
mass 120–150 kg/m	–	–	–	–	m	10.00
mass 150–160 kg/m	–	–	–	–	m	10.20
mass 160–190 kg/m	–	–	–	–	m	10.35
mass 190–220 kg/m	–	–	–	–	m	10.50
mass 220–250 kg/m	–	–	–	–	m	10.65
mass 250–280 kg/m	–	–	–	–	m	10.85
mass 280–310 kg/m	–	–	–	–	m	10.95
mass 310–340 kg/m	–	–	–	–	m	11.15
mass 340–370 kg/m	–	–	–	–	m	11.30
mass 370–400 kg/m	–	–	–	–	m	11.50
mass 400–430 kg/m	–	–	–	–	m	11.75
mass 430–460 kg/m	–	–	–	–	m	12.00
Welding on lengthening piece to steel tubular piles						
section diameter 508 × any thickness	–	–	–	–	nr	160.00
section diameter 559 × any thickness	–	–	–	–	nr	170.00
section diameter 610 × any thickness	–	–	–	–	nr	180.00
section diameter 660 × any thickness	–	–	–	–	nr	188.00
section diameter 711 × any thickness	–	–	–	–	nr	198.00
section diameter 762 × any thickness	–	–	–	–	nr	207.50
Cutting or burning off surplus length of steel tubular piles						
section diameter 508 × any thickness	–	–	–	–	nr	10.50
section diameter 559 × any thickness	–	–	–	–	nr	10.60
section diameter 610 × any thickness	–	–	–	–	nr	10.70
section diameter 660 × any thickness	–	–	–	–	nr	10.75
section diameter 711 × any thickness	–	–	–	–	nr	10.80
section diameter 762 × any thickness	–	–	–	–	nr	10.85

SERIES 1600: PILING AND EMBEDDED RETAINING WALLS

Item	Gang hours	Labour £	Plant £	Material £	Unit	Total rate £
PROOF LOADING OF PILES						
Driven precast concrete piles						
Establishment of proof loading equipment; proof loading of vertical precast concrete piles with maintained load to 900 kN	–	–	–	–	item	3000.00
Establishment of proof loading equipment; proof loading of vertical precast concrete piles by dynamic testing with piling hammer	–	–	–	–	nr	700.00
Bored in situ reinforced concrete piling (tripod rig)						
Establishment of proof loading equipment for bored cast-in-place piles	–	–	–	–	item	1000.00
Proof loading of vertical cast-in-place piles with maximum test load of 600 kN on a working pile 500 mm diameter using tension piles as reaction	–	–	–	–	nr	3500.00
Bored in situ reinforced concrete piling (mobile rig)						
Establishment of proof loading equipment for bored cast-in-place piles	–	–	–	–	item	1400.00
Proof loading of vertical cast-in-place piles with maximum test load of 600 kN on a working pile 500 mm diameter using tension piles as reaction	–	–	–	–	nr	3500.00
Concrete injected piles (continuous flight augered)						
Establishment of proof loading equipment for bored cast-in-place piles						
450 mm diameter; 650 kN	–	–	–	–	item	1750.00
600 mm diameter; 1400 kN	–	–	–	–	item	1750.00
750 mm diameter; 2200 kN	–	–	–	–	item	1750.00
Proof loading of vertical cast-in-place piles to 1.5 times working load						
450 mm diameter; 650 kN	–	–	–	–	nr	2700.00
600 mm diameter; 1400 kN	–	–	–	–	nr	5500.00
750 mm diameter; 2200 kN	–	–	–	–	nr	10000.00
Electronic integrity testing						
cost per pile (minimum 40 piles per visit)	–	–	–	–	nr	12.00
Segmental casing method piles						
Establishment of proof loading equipment for driven cast-in-place piles	–	–	–	–	item	1400.00
Proof loading of vertical cast-in-place piles with maximum test load of 600 kN on a working pile 500 mm diameter using non-working tension piles as reaction	–	–	–	–	nr	4000.00

SERIES 1600: PILING AND EMBEDDED RETAINING WALLS

Item	Gang hours	Labour £	Plant £	Material £	Unit	Total rate £
Driven in situ reinforced concrete piling						
Establishment of proof loading equipment for driven cast-in-place piles						
bottom driven	–	–	–	–	item	1000.00
top driven	–	–	–	–	item	1000.00
Proof loading of vertical cast-in-place piles with maximum test load of 1125 kN on a working pile 430 mm diameter using non-working tension piles as reaction						
bottom driven	–	–	–	–	nr	3500.00
top driven	–	–	–	–	nr	3500.00
Electronic integrity testing						
Cost per pile (minimum 40 piles per visit)	–	–	–	–	nr	25.00
Steel bearing piles						
Establishment of proof loading equipment for steel bearing piles in main piling	–	–	–	–	item	1360.00
Proof loading of vertical steel bearing piles with maximum test load of 108 t load on a working pile using non-working tension piles as reaction	–	–	–	–	nr	4350.00
Steel tubular piles						
Establishment of proof loading equipment for steel tubular piles	–	–	–	–	item	7000.00
Proof loading of steel tubular piles with maximum test load of 108 t load on a working pile using non-working tension piles as reaction	–	–	–	–	nr	6500.00
STEEL SHEET PILES						
Sheet steel piling is commonly carried out by a Specialist Contractor, whose advice should be sought to arrive at accurate costings. However, the following items can be used to assess a budget for such work.						
The following unit costs are based on driving/extracting 1,500 m² of sheet piling on a clear site with reasonable access.						
Note: area of driven piles will vary from area supplied dependent upon pitch line of piling and provision for such allowance has been made in PC for supply.						
The materials cost below includes the manufacturers tariffs for a 200 mile delivery radius from works, delivery in 5–10 t loads and with an allowance of 10% to cover waste/projecting piles etc.						

SERIES 1600: PILING AND EMBEDDED RETAINING WALLS

Item	Gang hours	Labour £	Plant £	Material £	Unit	Total rate £
STEEL SHEET PILES – cont						
Arcelor Mittal Z section steel piles; EN 10248 grade S270GP steel						
The following unit costs are based on driving/extracting 1,500 m² of sheet piling on a clear site with reasonable access.						
Provision of all plant, equipment and labour including transport to and from the site and establishing and dismantling for						
driving of sheet piling	–	–	–	–	sum	**7100.00**
extraction of sheet piling	–	–	–	–	sum	**4750.00**
Standing time	–	–	–	–	hr	**350.00**
Section modulus 800–1200 cm³/m; section reference AZ 12; mass 98.7 kg/m², sectional modulus 1200 cm³/m; EN 10248 grade S270GP steel						
length of welded corner piles	–	–	–	–	m	**94.57**
length of welded junction piles	–	–	–	–	m	**132.39**
driven area	–	–	–	–	m²	**35.00**
area of piles of length not exceeding 14 m	–	–	–	–	m²	**78.49**
length 14– 24 m	–	–	–	–	m²	**84.36**
area of piles of length exceeding 24 m	–	–	–	–	m²	**87.18**
Section modulus 1200–2000 cm³/m; section reference AZ 17; mass 108.6 kg/m²; sectional modulus 1665 cm³/m; EN 10248 grade S270GP steel						
length of welded corner piles	–	–	–	–	m	**94.57**
length of welded junction piles	–	–	–	–	m	**132.39**
driven area	–	–	–	–	m²	**37.50**
area of piles of length not exceeding 14 m	–	–	–	–	m²	**82.96**
length 14– 24 m	–	–	–	–	m²	**84.36**
area of piles of length exceeding 24 m	–	–	–	–	m²	**87.18**
Section modulus 2000–3000 cm³/m; section reference AZ 26; mass 155.2 kg/m²; sectional modulus 2600 cm³/m; EN 10248 grade S270GP steel						
driven area	–	–	–	–	m²	**39.50**
area of piles of length 6–18 m	–	–	–	–	m²	**94.00**
area of piles of length 18–24 m	–	–	–	–	m²	**95.59**
Section modulus 3000–4000 cm³/m; section reference AZ 36; mass 194.0 kg/m²; sectional modulus 3600 cm³/m; EN 10248 grade S270GP steel						
driven area	–	–	–	–	m²	**40.15**
area of piles of length 6–18 m	–	–	–	–	m²	**114.77**
area of piles of length 18–24 m	–	–	–	–	m²	**116.72**
Straight section modulus ne 500 cm³/m; section reference AS 500–12 mass 149 kg/m²; sectional modulus 51 cm³/m; EN 10248 grade S270GP steel						
driven area	–	–	–	–	m²	**30.60**
area of piles of length 6–18 m	–	–	–	–	m²	**155.49**
area of piles of length 18–24 m	–	–	–	–	m²	**159.12**

SERIES 1600: PILING AND EMBEDDED RETAINING WALLS

Item	Gang hours	Labour £	Plant £	Material £	Unit	Total rate £
One coat black tar vinyl (PC1) protective treatment applied all surfaces at shop to minimum dry film thickness up to 150 microns to steel piles						
section reference AZ 12; pile area	–	–	–	–	m^2	11.05
section reference AZ 17; pile area	–	–	–	–	m^2	11.35
section reference AZ 26; pile area	–	–	–	–	m^2	12.35
section reference AZ 36; pile area	–	–	–	–	m^2	13.15
section reference AS 500–12; pile area	–	–	–	–	m^2	14.00
One coat black high build isocyanate cured epoxy pitch (PC2) protective treatment applied all surfaces at shop to minimum dry film thickness up to 450 microns to steel piles						
section reference AZ 12; pile area	–	–	–	–	m^2	17.35
section reference AZ 17; pile area	–	–	–	–	m^2	17.75
section reference AZ 26; pile area	–	–	–	–	m^2	19.50
section reference AZ 36; pile area	–	–	–	–	m^2	20.50
section reference AS 500–12; pile area	–	–	–	–	m^2	21.50
Arcelor Mittal U section steel piles; EN 10248 grade S270GP steel						
The following unit costs are based on driving/extracting 1,500 m^2 of sheet piling on a clear site with reasonable access.						
Provision of plant, equipment and labour including transport to and from the site and establishing and dismantling						
driving of sheet piling	–	–	–	–	sum	6500.00
extraction of sheet piling	–	–	–	–	sum	5500.00
Standing time	–	–	–	–	hr	350.00
Section modulus 500–800 cm^3/m; section reference PU 6; mass 76.0 kg/m^2; sectional modulus 600 cm^3/m						
driven area	–	–	–	–	m^2	50.00
area of piles of length 6–18 m	–	–	–	–	m^2	65.00
area of piles of length 18–24 m	–	–	–	–	m^2	70.00
Section modulus 800–1200 cm^3/m; section reference PU 8; mass 90.9 kg/m^2; sectional modulus 830 cm^3/m						
driven area	–	–	–	–	m^2	45.00
area of piles of length 6–18 m	–	–	–	–	m^2	69.00
area of piles of length 18–24 m	–	–	–	–	m^2	74.00
Section modulus 1200–2000 cm^3/m; section reference PU 12; mass 110.1 kg/m^2; sectional modulus 1200 cm^3/m						
driven area	–	–	–	–	m^2	38.50
area of piles of length 6–18 m	–	–	–	–	m^2	73.00
area of piles of length 18–24 m	–	–	–	–	m^2	87.50
Section modulus 1200–2000 cm^3/m; section reference PU 18; mass 128.2 kg/m^2; sectional modulus 1800 cm^3/m						
driven area	–	–	–	–	m^2	35.00
area of piles of length 6–18 m	–	–	–	–	m^2	86.00
area of piles of length 18–24 m	–	–	–	–	m^2	92.40

SERIES 1600: PILING AND EMBEDDED RETAINING WALLS

Item	Gang hours	Labour £	Plant £	Material £	Unit	Total rate £
STEEL SHEET PILES – cont						
Arcelor Mittal U section steel piles – cont						
Section modulus 2000–3000 cm^3/m; section reference PU 22; mass 143.6 kg/m^2; sectional modulus 2200 cm^3/m						
driven area	–	–	–	–	m^2	**31.50**
area of piles of length 6–18 m	–	–	–	–	m^2	**112.50**
area of piles of length 18–24 m	–	–	–	–	m^2	**114.05**
Section modulus 3000–4000 cm^3/m; section reference PU 32; mass 190.2 kg/m^2; sectional modulus 3200 cm^3/m						
driven area	–	–	–	–	m^2	**27.50**
area of piles of length 6–18 m	–	–	–	–	m^2	**127.00**
area of piles of length 18–24 m	–	–	–	–	m^2	**128.50**
One coat black tar vinyl (PC1) protective treatment applied all surfaces at shop to minimum dry film thickness up to 150 microns to steel piles						
section reference PU 6; pile area	–	–	–	–	m^2	**10.15**
section reference PU 8; pile area	–	–	–	–	m^2	**10.57**
section reference PU 12; pile area	–	–	–	–	m^2	**10.61**
section reference PU 18; pile area	–	–	–	–	m^2	**11.38**
section reference PU 22; pile area	–	–	–	–	m^2	**11.79**
section reference PU 32; pile area	–	–	–	–	m^2	**11.97**
One coat black high build isocyanate cured epoxy pitch (PC2) protective treatment applied all surfaces at shop to minimum dry film thickness up to 450 microns to steel piles						
section reference PU 6; pile area	–	–	–	–	m^2	**15.65**
section reference PU 8; pile area	–	–	–	–	m^2	**15.95**
section reference PU 12; pile area	–	–	–	–	m^2	**16.66**
section reference PU 18; pile area	–	–	–	–	m^2	**17.88**
section reference PU 22; pile area	–	–	–	–	m^2	**18.54**
section reference PU 32; pile area	–	–	–	–	m^2	**18.81**

SERIES 1600: PILING AND EMBEDDED RETAINING WALLS

Item	Gang hours	Labour £	Plant £	Material £	Unit	Total rate £
EMBEDDED RETAINING WALL PLANT Diaphgram walls are the construction of vertical walls, cast in place in a trench excavation. They can be formed in reinforced concrete to provide structural elements for temporary or permanent retaining walls. Wall thicknesses of 500 to 1,500 mm up to 40 m deep may be constructed. Special equipment such as the Hydrofraise can construct walls up to 100 m deep. Restricted urban sites will significantly increase the costs. The following costs are based on constructing a diaphgram wall with an excavated volume of 4000 m^3 using a grab. Typical progress would be up to 500 m per week.						
Establishment of standard diaphragm walling plant, including bentonite storage tanks	–	–	–	–	item	300000.00
standing time	–	–	–	–	hr	750.00
guide walls (twin)	–	–	–	–	m	350.00
waterproofed joints	–	–	–	–	m	6.50
DIAPHRAGM WALLS						
Excavation for walls 1000 mm thick, disposal of soil and placing of concrete	–	–	–	–	m^3	440.00
Provide and place reinforcement cages	–	–	–	–	t	750.00
Excavate/chisel in hard materials/rock	–	–	–	–	hr	975.00

SERIES 1700: STRUCTURAL CONCRETE

Item	Gang hours	Labour £	Plant £	Material £	Unit	Total rate £
NOTES						
Refer also to Civil Engineering – Concrete, Formwork, Reinforcement and Precast Concrete, although this section is fundamentally different in that the provision of concrete of different classes and its placement is combined in the unit costs.						
RESOURCES – LABOUR						
Concreting gang						
1 ganger/chargehand (skill rate 4)		14.56				
2 skilled operatives (skill rate 4)		27.22				
4 unskilled operatives (general)		50.84				
1 plant operator (skill rate 3) – 25% of time		4.14				
Total Gang Rate/Hour	£	**96.76**				
Formwork gang						
1 foreman (craftsman)		22.50				
2 joiners (craftsman)		39.08				
1 unskilled operative (general)		12.71				
1 plant operator (craftsman) – 25% of time		5.28				
Total Gang Rate/Hour	£	**79.56**				
Reinforcement gang						
1 foreman (craftsman)		22.50				
4 steel fixers (craftsman)		78.16				
1 unskilled operative (general)		12.71				
1 plant operator (craftsman) – 25% of time		5.28				
Total Gang Rate/Hour	£	**118.65**				
RESOURCES – PLANT						
Concreting						
10 t crane- 25% of time			10.13			
gas oil for ditto			0.53			
0.76 m^3 concrete skip – 25% of time			0.63			
11.3 m^3/min compressor, 4 tool			3.00			
gas oil for ditto			10.50			
4 poker vibrators P5475 mm or less in thickness			5.50			
Total Gang Rate/Hour		£	**30.28**			
Formwork						
20 t crawler crane – 25% of time			11.93			
gas oil for ditto			0.61			
22' saw bench			1.88			
gas oil for ditto			0.70			
small power tools (formwork)			0.75			
Total Gang Rate/Hour		£	**15.86**			

SERIES 1700: STRUCTURAL CONCRETE

Item	Gang hours	Labour £	Plant £	Material £	Unit	Total rate £
Reinforcement						
30 t crawler crane (25% of time)			13.50			
gas oil for ditto			0.61			
bar cropper			4.13			
small power tools (reinforcement)			1.50			
tirfors, kentledge etc.			0.75			
Total Gang Rate/Hour		£	20.49			
RESOURCES – MATERIALS						
Wastage allowances 5% onto delivered ready mixed concrete.						
IN SITU CONCRETE						
In situ concrete Grade C10						
Blinding						
75 mm or less in thickness	0.18	17.42	5.45	86.10	m³	**108.97**
Blinding; in narrow widths up to 1.0 m wide or in bottoms of trenches up to 2.5 m wide; excluding formwork						
75 mm or less in thickness	0.20	19.35	6.06	86.10	m³	**111.51**
In situ concrete Grade C15						
Blinding; excluding formwork						
75 mm or less in thickness	0.16	15.48	4.84	87.15	m³	**107.47**
Blinding; in narrow widths up to 1.0 m wide or in bottoms of trenches up to 2.5 m wide; excluding formwork						
75 mm or less in thickness	0.18	17.42	5.45	87.15	m³	**110.02**
In situ concrete Grade C20/20						
Bases, footings, pile caps and ground beams; thickness						
ne 150 mm	0.20	19.35	6.06	86.91	m³	**112.32**
150–300 mm	0.17	16.45	5.16	86.91	m³	**108.52**
300–500 mm	0.15	14.51	4.54	86.91	m³	**105.96**
exceeding 500 mm	0.14	13.55	4.24	86.91	m³	**104.70**
Walls; thickness						
ne 150 mm	0.21	20.32	6.36	86.91	m³	**113.59**
150–300 mm	0.15	14.51	4.54	86.91	m³	**105.96**
300–500 mm	0.13	12.58	3.94	86.91	m³	**103.43**
exceeding 500 mm	0.12	11.61	3.64	86.91	m³	**102.16**
Suspended slabs; thickness						
ne 150 mm	0.27	26.12	8.18	86.91	m³	**121.21**
150–300 mm	0.21	20.32	6.36	86.91	m³	**113.59**
300–500 mm	0.19	18.38	5.76	86.91	m³	**111.05**
exceeding 500 mm	0.19	18.38	5.76	86.91	m³	**111.05**

SERIES 1700: STRUCTURAL CONCRETE

Item	Gang hours	Labour £	Plant £	Material £	Unit	Total rate £
IN SITU CONCRETE – cont						
In situ concrete Grade C20/20 – cont						
Columns, piers and beams; cross-sectional area						
ne 0.03 m²	0.50	48.38	15.13	86.91	m³	**150.42**
0.03–0.10 m²	0.40	38.70	12.11	86.91	m³	**137.72**
0.10–0.25 m²	0.35	33.87	10.60	86.91	m³	**131.38**
0.25–1.00 m²	0.35	33.87	10.60	86.91	m³	**131.38**
exceeding 1 m²	0.28	27.09	8.48	86.91	m³	**122.48**
ADD to the above prices for						
sulphate-resisting cement	–	–	–	10.50	m³	**10.50**
air entrained concrete	–	–	–	5.78	m³	**5.78**
water repellant concrete	–	–	–	5.51	m³	**5.51**
In situ concrete Grade C30/20						
Bases, footings, pile caps and ground beams; thickness						
ne 150 mm	0.21	20.32	6.36	91.56	m³	**118.24**
150–300 mm	0.18	17.42	5.45	91.56	m³	**114.43**
300–500 mm	0.15	14.51	4.54	91.56	m³	**110.61**
exceeding 500 mm	0.14	13.55	4.24	91.56	m³	**109.35**
Walls; thickness						
ne 150 mm	0.22	21.29	6.66	91.56	m³	**119.51**
150–300 mm	0.16	15.48	6.65	91.56	m³	**113.69**
300–500 mm	0.13	12.58	5.41	91.56	m³	**109.55**
exceeding 500 mm	0.12	11.61	4.99	91.56	m³	**108.16**
Suspended slabs; thickness						
ne 150 mm	0.28	27.09	11.63	91.56	m³	**130.28**
150–300 mm	0.22	21.29	9.15	91.56	m³	**122.00**
300–500 mm	0.19	18.38	7.90	91.56	m³	**117.84**
exceeding 500 mm	0.18	17.42	7.48	91.56	m³	**116.46**
Columns, piers and beams; cross-sectional area						
ne 0.03 m²	0.53	51.28	22.03	91.56	m³	**164.87**
0.03–0.10 m²	0.42	40.64	17.46	91.56	m³	**149.66**
0.10–0.25 m²	0.36	34.83	14.96	91.56	m³	**141.35**
0.25–1.00 m²	0.35	33.87	14.55	91.56	m³	**139.98**
exceeding 1 m²	0.28	27.09	11.63	91.56	m³	**130.28**
ADD to the above prices for						
sulphate-resisting cement	–	–	–	10.50	m³	**10.50**
air entrained concrete	–	–	–	5.78	m³	**5.78**
water-repellant concrete	–	–	–	5.51	m³	**5.51**

SERIES 1700: STRUCTURAL CONCRETE

Item	Gang hours	Labour £	Plant £	Material £	Unit	Total rate £
PRECAST CONCRETE The cost of precast concrete item is very much dependent on the complexity of the moulds, the number of units to be cast from each mould and the size and the weight of the unit to be handled. The unit rates below are for standard precast items that are often to be found on a Civil Engineering project. It would be misleading to quote for indicative costs for tailor-made precast concrete units and it is advisable to contact specialist maunfacturers for guide prices.						
Pre-tensioned prestressed beams; concrete Grade C20						
Beams						
100 × 65 × 1050 mm long	1.00	3.75	1.60	10.00	nr	15.35
265 × 65 × 1800 mm long	1.00	4.35	3.35	25.00	nr	32.70
Inverted T beams, flange width 495 mm						
section T1; 8 m long, 380 mm deep; mass 1.88 t	–	–	–	–	nr	750.00
section T2; 9 m long, 420 mm deep; mass 2.29 t	–	–	–	–	nr	900.00
section T3; 11 m long, 535 mm deep; mass 3.02 t	–	–	–	–	nr	1050.00
section T4; 12 m long, 575 mm deep; mass 3.54 t	–	–	–	–	nr	1150.00
section T5; 13 m long, 615 mm deep; mass 4.08 t	–	–	–	–	nr	1250.00
section T6; 13 m long, 655 mm deep; mass 4.33 t	–	–	–	–	nr	1250.00
section T7; 12 m long, 695 mm deep; mass 4.95 t	–	–	–	–	nr	1500.00
section T8; 15 m long, 735 mm deep; mass 5.60 t	–	–	–	–	nr	1700.00
section T9; 16 m long, 775 mm deep; mass 6.28 t	–	–	–	–	nr	1800.00
section T10; 18 m long, 815 mm deep; mass 7.43 t	–	–	–	–	nr	2000.00
M beams, flange width 970 mm						
section M2; 17 m long, 720 mm deep; mass 12.95 t	–	–	–	–	nr	4000.00
section M3; 18 m long, 800 mm deep; mass 15.11 t	–	–	–	–	nr	4250.00
section M6; 22 m long, 1040 mm deep; mass 20.48 t	–	–	–	–	nr	7200.00
section M8; 25 m long, 1200 mm deep; mass 23.68 t	–	–	–	–	nr	8200.00
U beams, base width 970 mm						
section U3; 16 m long, 900 mm deep; mass 19.24 t	–	–	–	–	nr	6765.00
section U5; 20 m long, 1000 mm deep; mass 25.64 t	–	–	–	–	nr	8700.00
section U8; 24 m long, 1200 mm deep; mass 34.56 t	–	–	–	–	nr	12000.00
section U12; 30 m long, 1600 mm deep; mass 52.74 t	–	–	–	–	nr	16450.00

SERIES 1700: STRUCTURAL CONCRETE

Item	Gang hours	Labour £	Plant £	Material £	Unit	Total rate £
PRECAST CONCRETE – cont						
Precast concrete culverts, cattle creeps and subway units; rebated joints						
Rectangular cross section						
500 mm high × 1000 mm wide	1.00	4.55	4.55	290.00	m	**299.10**
1000 mm high × 1500 mm wide	1.00	10.50	10.14	475.00	m	**495.64**
1500 mm high × 1500 mm wide	1.00	18.50	18.75	550.00	m	**587.25**
2000 mm high × 2750 mm wide	1.00	35.00	29.83	1600.00	m	**1664.83**
2750 mm high × 3000 mm wide	1.00	45.25	59.88	1785.00	m	**1890.13**
Extra for units curved on plan to less than 20 m radius	–	–	–	295.00	m	**295.00**

SERIES 1700: STRUCTURAL CONCRETE

Item	Gang hours	Labour £	Plant £	Material £	Unit	Total rate £
SURFACE FINISH OF CONCRETE – FORMWORK						
Materials						
Formwork materials include for shutter, bracing, ties, support, kentledge and all consumables.						
These unit costs are based upon those outputs and prices detailed in Civil Engineering – Concrete Formwork but are referenced to The Specification for Highway Works, clause 1708.						
The following unit rates do not include for formwork outside the payline and are based on an optimum of a minimum 8 uses with 10% per use towards the cost of repairs/replacement of components damaged during disassembly.						
ADJUST formwork material costs generally depending on the number of uses:						

Nr of Uses	% Adjustment	Inclusion for waste
1	Add 90–170%	7%
2	Add 50–180%	7%
3	Add 15–30%	6%
6	Add 5–10%	5%
8	No change	5%
10	Deduct 5–7%	5%

Definitions
Class F1 formwork is rough finish
Class F2 formwork is fair finish
Class F3 formwork is extra smooth finish

Item	Gang hours	Labour £	Plant £	Material £	Unit	Total rate £
Formwork Class F1						
Horizontal more than 300 mm wide	0.52	41.37	8.25	5.90	m^2	55.52
Inclined more than 300 mm wide	0.55	43.76	8.73	8.50	m^2	60.99
Vertical more than 300 mm wide	0.61	48.53	9.68	8.48	m^2	66.69
300 mm wide or less at any inclination	0.72	57.29	11.42	8.48	m^2	77.19
Curved of both girth and width more than 300 mm at any inclination	0.95	75.59	15.07	9.76	m^2	100.42
Curved of girth or width of 300 mm or less at any inclination	0.72	57.29	11.42	9.76	m^2	78.47
Domed	1.20	95.48	19.04	12.13	m^2	126.65
Void former cross-section 100 × 100 mm	0.07	5.57	0.24	3.01	m	8.82
Void former cross-section 250 × 250 mm	0.12	9.55	0.40	6.46	m	16.41
Void former cross-section 500 × 500 mm	0.30	23.87	1.00	12.29	m	37.16
Formwork Class F2						
Horizontal more than 300 mm wide	0.54	42.97	8.57	12.00	m^2	63.54
Inclined more than 300 mm wide	0.57	45.35	9.05	18.33	m^2	72.73
Vertical more than 300 mm wide	0.63	50.13	10.00	18.33	m^2	78.46

SERIES 1700: STRUCTURAL CONCRETE

Item	Gang hours	Labour £	Plant £	Material £	Unit	Total rate £
SURFACE FINISH OF CONCRETE – FORMWORK – cont						
Formwork Class F2 – cont						
300 mm wide or less at any inclination	0.74	58.88	11.74	18.33	m²	**88.95**
Curved of both girth and width more than 300 mm at any inclination	0.98	77.97	15.55	21.49	m²	**115.01**
Curved of girth or width of 300 mm or less at any inclination	0.75	59.67	11.90	21.49	m²	**93.06**
Domed	1.40	111.39	22.21	27.30	m²	**160.90**
Formwork Class F3						
Horizontal more than 300 mm wide	0.56	44.56	8.88	14.34	m²	**67.78**
Inclined more than 300 mm wide	0.59	46.94	9.36	20.67	m²	**76.97**
Vertical more than 300 mm wide	0.65	51.72	10.31	20.67	m²	**82.70**
300 mm wide or less at any inclination	0.76	60.47	12.06	20.67	m²	**93.20**
Curved of both girth and width more than 300 mm at any inclination	0.99	78.77	15.71	23.83	m²	**118.31**
Curved of girth or width of 300 mm or less at any inclination	0.77	61.27	12.22	23.83	m²	**97.32**
Domed	1.45	115.37	23.00	29.64	m²	**168.01**
Formwork ancillaries						
Allowance for additional craneage and rub up where required	0.13	10.34	2.07	0.25	m²	**12.66**
PATTERNED PROFILE FORMWORK						
Extra over formwork for patterned profile formliners, INSITEX or similar	–	–	–	–	m²	**34.85**
STEEL REINFORCEMENT FOR STRUCTURES						
Stainless steel bars						
Bar reinforcement nominal size 16 mm and under not exceeding 12 m in length	6.74	799.67	138.09	3100.00	t	**4037.76**
Bar reinforcement nominal size 20 mm and over not exceeding 12 m in length						
20 mm nominal size	4.44	526.78	90.96	3100.00	t	**3717.74**
25 mm nominal size	4.44	526.78	90.96	3100.00	t	**3717.74**
32 mm nominal size	4.44	526.78	90.96	3100.00	t	**3717.74**
ADD to the above for bars						
12–13.5 m long	–	–	–	42.50	t	**42.50**
13.5–15 m long	–	–	–	42.50	t	**42.50**
over 15 m long; per 500 mm increment	–	–	–	5.00	t	**5.00**

SERIES 1700: STRUCTURAL CONCRETE

Item	Gang hours	Labour £	Plant £	Material £	Unit	Total rate £
High yield steel bars BS 4449; deformed, Grade 500C						
Bar reinforcement nominal size 16 mm and under not exceeding 12 m in length	6.74	799.67	138.09	650.00	t	1587.76
Bar reinforcement nominal size 20 mm and over not exceeding 12 m in length						
20 mm nominal size	4.44	526.78	90.96	650.00	t	1267.74
25 mm nominal size	4.44	526.78	90.96	650.00	t	1267.74
32 mm nominal size	4.44	526.78	90.96	650.00	t	1267.74
40 mm nominal size	4.44	526.78	90.96	650.00	t	1267.74
ADD to the above for bars						
12–13.5 m long	–	–	–	45.00	t	45.00
13.5–15 m long	–	–	–	45.00	t	45.00
over 15 m long; per 500 mm increment	–	–	–	5.00	t	5.00
Helical reinforcement nominal size 16 mm and under ne 12 m in length	6.74	799.67	138.09	676.91	t	1614.67
Helical reinforcement nominal size 20 mm and over						
20 mm nominal size	4.44	526.78	90.96	656.50	t	1274.24
25 mm nominal size	4.44	526.78	90.96	656.50	t	1274.24
32 mm nominal size	4.44	526.78	90.96	656.50	t	1274.24
40 mm nominal size	4.44	526.78	90.96	656.50	t	1274.24
Dowels						
16 mm diameter × 600 mm long	0.10	11.86	2.05	1.71	nr	15.62
20 mm diameter × 600 mm long	0.10	11.86	2.05	3.10	nr	17.01
25 mm diameter × 600 mm long	0.10	11.86	2.05	3.55	nr	17.46
32 mm diameter × 600 mm long	0.10	11.86	2.05	4.34	nr	18.25
Mild steel bars BS 4449; Grade 250						
Bar reinforcement nominal size 16 mm and under ne 12 m in length	6.74	799.67	138.09	570.05	t	1507.81
Bar reinforcement nominal size 20 mm and over; not exceeding 12 m in length						
20 mm nominal size	4.44	526.78	90.96	504.50	t	1122.24
25 mm nominal size	4.44	526.78	90.96	504.50	t	1122.24
32 mm nominal size	4.44	526.78	90.96	505.00	t	1122.74
40 mm nominal size	4.44	526.78	90.96	508.40	t	1126.14
ADD to the above for bars						
12–13.5 m long	–	–	–	45.00	t	45.00
13.5–15 m long	–	–	–	45.00	t	45.00
over 15 m long, per 500 mm increment	–	–	–	5.00	t	5.00
ADD for cutting, bending, tagging and baling reinforcement on site						
6 mm nominal size	4.87	157.06	99.78	2.00	t	258.84
8 mm nominal size	4.58	147.71	93.83	2.00	t	243.54
10 mm nominal size	3.42	110.30	70.07	2.00	t	182.37
12 mm nominal size	2.55	82.24	52.25	2.00	t	136.49
16 mm nominal size	2.03	65.47	41.60	2.00	t	109.07
20 mm nominal size	1.68	54.18	34.42	2.00	t	90.60
25 mm nominal size	1.68	54.18	34.42	2.00	t	90.60
32 mm nominal size	1.39	44.83	28.49	2.00	t	75.32
40 mm nominal size	1.39	44.83	28.49	2.00	t	75.32

SERIES 1700: STRUCTURAL CONCRETE

Item	Gang hours	Labour £	Plant £	Material £	Unit	Total rate £
STEEL REINFORCEMENT FOR STRUCTURES – cont						
Fabric reinforcement; high yield steel BS 4483						
Fabric reinforcement						
BS ref A98; nominal mass 1.54 kg/m^2	0.03	3.56	0.62	2.01	m^2	**6.19**
BS ref A142; nominal mass 2.22 kg/m^2	0.03	3.56	0.62	1.66	m^2	**5.84**
BS ref A193; nominal mass 3.02 kg/m^2	0.04	4.75	0.82	2.22	m^2	**7.79**
BS ref A252; nominal mass 3.95 kg/m^2	0.04	4.75	0.82	2.91	m^2	**8.48**
BS ref A393; nominal mass 6.16 kg/m^2	0.07	8.31	1.44	4.53	m^2	**14.28**
BS ref B196; nominal mass 3.05 kg/m^2	0.04	4.75	0.82	4.23	m^2	**9.80**
BS ref B283; nominal mass 3.73 kg/m^2	0.04	4.75	0.82	2.88	m^2	**8.45**
BS ref B385; nominal mass 4.53 kg/m^2	0.05	5.93	1.03	3.54	m^2	**10.50**
BS ref B503; nominal mass 5.93 kg/m^2	0.05	5.93	1.03	4.50	m^2	**11.46**
BS ref B785; nominal mass 8.14 kg/m^2	0.08	9.49	1.64	6.18	m^2	**17.31**
BS ref B1131; nominal mass 10.90 kg/m^2	0.09	10.68	1.85	8.27	m^2	**20.80**
BS ref C282; nominal mass 2.61 kg/m^2	0.03	3.56	0.62	2.13	m^2	**6.31**
BS ref C385; nominal mass 3.41 kg/m^2	0.04	4.75	0.82	2.76	m^2	**8.33**
BS ref C503; nominal mass 4.34 kg/m^2	0.05	5.93	1.03	3.51	m^2	**10.47**
BS ref C636; nominal mass 5.55 kg/m^2	0.05	5.93	1.03	4.50	m^2	**11.46**
BS ref C785; nominal mass 6.72 kg/m^2	0.07	8.31	1.44	5.17	m^2	**14.92**
BS ref D49; nominal mass 0.77 kg/m^2	0.02	2.37	0.41	0.57	m^2	**3.35**
BS ref D98; nominal mass 1.54 kg/m^2	0.02	2.37	0.41	1.09	m^2	**3.87**

SERIES 1800: STEELWORK FOR STRUCTURES

Item	Gang hours	Labour £	Plant £	Material £	Unit	Total rate £
FABRICATION OF STEELWORK						
Steelwork to BS EN10025; Grade S275						
Fabrication of main members						
rolled sections	–	–	–	–	t	1500.00
plated rolled sections	–	–	–	–	t	4000.00
plated girders	–	–	–	–	t	3500.00
box girders	–	–	–	–	t	1500.00
Fabrication of deck panels						
rolled sections	–	–	–	–	t	2600.00
plated rolled sections	–	–	–	–	t	2600.00
plated girders	–	–	–	–	t	2600.00
Fabrication of subsidiary steelwork						
rolled sections	–	–	–	–	t	2750.00
plated rolled sections	–	–	–	–	t	2750.00
plated girders	–	–	–	–	t	2750.00
ERECTION OF STEELWORK						
Trial erection at the place of fabrication	–	–	–	–	t	320.00
Permanent erection of steelwork; substructure	–	–	–	–	t	250.00
Permanent erection of steelwork; superstructure	–	–	–	–	t	250.00
MISCELLANEOUS METALWORK						
Mild steel						
Ladders						
Cat ladder; 64 × 13 mm bar strings; 19 mm rungs at 250 mm centres; 450 mm wide with safety hoops	–	–	–	–	m	95.00
Handrails						
Galvanized tubular metal; 76 mm diameter handrail, 48 mm diameter standards at 750 mm centres, 48 mm diameter rail; 1070 mm high overall	–	–	–	–	m	115.00
Metal access cover and frame						
Group 4, ductile iron, single seal 610 × 610 × 100 mm depth; D400	–	–	–	–	nr	190.00
Group 2, ductile iron, double seal single piece cover 600 × 450 mm; B125	–	–	–	–	nr	155.00

SERIES 1900: PROTECTION OF STEELWORK AGAINST CORROSION

Item	Gang hours	Labour £	Plant £	Material £	Unit	Total rate £
RESOURCES – LABOUR						
Protective painting gang						
1 ganger/chargehand (skill rate 4)		14.56				
2 skilled operatives (skill rate 4)		27.22				
2 unskilled operatives (general)		25.42				
Total Gang rate/Hour		£ 67.20				
RESOURCES – PLANT						
Protective painting						
power tools (protection of steelwork)			2.85			
access scaffolding, trestles and ladders			6.58			
5 t transit van (50% of time)			3.75			
gas oil for ditto			2.45			
Total Gang Rate/Hour			£ 15.63			
RESOURCES – MATERIALS						
All coats applied off site except as noted. The external environment has been taken as Inland 'B' Exposed.						
PROTECTIVE SYSTEM						
Galvanizing to BS EN ISO 1461; apply protective coatings comprising: 1st coat: Mordant T wash; 2nd coat: Zinc rich epoxy primer; 3rd coat: Zinc phosphate, CR/Alkyd Undercoat; 4th coat; MIO CR Undercoat-on site externally; 5th coat: CR coloured finish-on site externally						
To metal parapets and fencing, lighting columns, brackets						
by brush or airless spray to dry film thickness 200 microns	0.20	13.44	3.13	19.91	m^2	**36.48**
Blast clean to BS 7079 (surface preparation); apply protective coatings comprising: 1st coat: Zinc Chromate, Red Oxide Blast Primer; 2nd coat: Zinc Phosphate, Epoxy Ester Undercoat; 3rd coat: MIO Undercoat; 4th coat: MIO coloured finish-on site externally						
To subsiduary steelwork, interior finishes						
By brush or airless spray to dry film thickness 175 microns	0.15	10.08	2.36	15.00	m^2	**27.44**

SERIES 1900: PROTECTION OF STEELWORK AGAINST CORROSION

Item	Gang hours	Labour £	Plant £	Material £	Unit	Total rate £
Blast clean to BS 7079 (surface preparation); apply metal coating of aluminium spray at works; apply protective coatings comprising: 1st coat: Zinc Chromate Etch Primer (2 pack); 2nd coat: Zinc Phosphate, CR/Aalkyd Undercoat; 3rd coat: Zinc Phosphate, CR/alkyd Undercoat; 4th coat: MIO CR Undercoat; 5th coat: CR finish-on site externally						
To main steel members						
By brush or airless spray to dry film thickness 250 microns	0.18	12.10	2.81	16.62	m^2	31.53
Blast clean to BS 7079 (second quality surface preparation); remove all surface defects to BS EN10025; apply protective coatings comprising: 1st coat: Zinc rich primer (2 pack); 2nd coat: Epoxy High Build M10 (2 pack); 3rd coat: Polyurethane Undercoat (2 pack)-on site internally; 4th coat: Finish coat polyurethane (2 pack)-on site externally						
To internal steel members						
by brush or airless spray	0.15	10.08	2.36	22.86	m^2	35.30
ALTERNATIVE SURFACE TREATMENTS						
Galvanizing (Hot dip) to BS EN ISO 1461, assuming average depth 20 m^2 per tonne of steel	–	–	–	–	m^2	15.00
Shot blasting (at works)	–	–	–	–	m^2	3.00
Grit blasting (at works)	–	–	–	–	m^2	4.35
Sand blasting (at works)	–	–	–	–	m^2	6.50
Shot blasting (on site)	–	–	–	–	m^2	4.25
Grit blasting (on site)	–	–	–	–	m^2	6.15
Sand blasting (on site)	–	–	–	–	m^2	9.15

SERIES 2000: WATERPROOFING FOR STRUCTURES

Item	Gang hours	Labour £	Plant £	Material £	Unit	Total rate £
NOTES						
This section is based around the installation of proprietary systems to new/recently completed works as part of major scheme, for minor works/repairs outputs will be many times more. Outputs are also based on use of skilled labour, therefore effieciency is high and wastage low (5–7% only allowed) excepting laps where required.						
RESOURCES – LABOUR						
Asphalting gang						
1 ganger/chargehand (skill rate 4)		14.56				
2 unskilled operative (general)		25.42				
Total Gang rate/Hour	£	**39.98**				
Damp-proofing gang						
1 ganger/chargehand (skill rate 4)		14.56				
1 skilled operative (skill rate 4)		13.61				
1 unskilled operative (general)		12.71				
Total Gang rate/Hour	£	**40.88**				
Sprayed/brushed waterproofing gang						
1 ganger/chargehand (skill rate 4) – 30% of time		4.37				
1 skilled operative (skill rate 4)		13.61				
Total Gang rate/Hour	£	**17.98**				
Protective layers – screed gang						
1 ganger/chargehand (skill rate 4)		14.56				
1 skilled operative (skill rate 4)		13.61				
1 unskilled operative (general)		12.71				
Total Gang rate/Hour	£	**40.88**				
RESOURCES – PLANT						
Asphalting						
45 litre portable tar boiler including sprayer (50% of time)			0.50			
2 t dumper (50% of time)			1.75			
gas oil for ditto			1.05			
Total Gang Rate/Hour		£	**3.65**			
Damp-proofing						
2 t dumper (50% of time)			1.75			
gas oil for ditto			1.05			
Total Gang Rate/Hour		£	**2.80**			

SERIES 2000: WATERPROOFING FOR STRUCTURES

Item	Gang hours	Labour £	Plant £	Material £	Unit	Total rate £
WATERPROOFING						
Mastic asphalt; BS 6925 Type T 1097; 20 mm thick; two coats						
over 300 mm wide; ne 30° to horizontal	0.33	13.19	1.83	14.97	m²	**29.99**
over 300 mm wide; 30–90° to horizontal	0.50	19.99	1.20	14.97	m²	**36.16**
ne 300 mm wide; at any inclination	0.60	23.99	1.20	14.97	m²	**40.16**
to domed surfaces	0.75	29.98	1.20	14.97	m²	**46.15**
Bituthene 1000; lapped joints						
over 300 mm wide; ne 30° to horizontal	0.05	2.04	0.14	8.26	m²	**10.44**
over 300 mm wide; 30–90° to horizontal	0.06	2.45	0.17	10.54	m²	**13.16**
ne 300 mm wide; at any inclination	0.08	3.27	0.22	10.54	m²	**14.03**
Extra; one coat primer on vertical surfaces	0.03	1.23	0.08	0.03	m²	**1.34**
Bituthene 4000; lapped joints						
over 300 mm wide; ne 30° to horizontal	0.05	2.04	0.14	8.49	m²	**10.67**
over 300 mm wide; 30–90° to horizontal	0.06	2.45	0.92	11.32	m²	**14.69**
ne 300 mm wide; at any inclination	0.08	3.27	1.06	11.32	m²	**15.65**
Famguard (hot applied) with Fam-primer						
over 300 mm wide; ne 30° to horizontal	0.32	13.08	0.90	18.87	m²	**32.85**
over 300 mm wide; 30–90° to horizontal	0.35	14.31	0.98	19.23	m²	**34.52**
ne 300 mm wide; at any inclination	0.40	16.35	1.12	22.07	m²	**39.54**
Famflex (hot applied) with Fam-primer						
over 300 mm wide; ne 30° to horizontal	0.32	13.08	0.90	14.57	m²	**28.55**
over 300 mm wide; 30–90° to horizontal	0.34	13.90	0.95	14.85	m²	**29.70**
ne 300 mm wide; at any inclination	0.38	15.53	1.06	16.98	m²	**33.57**
Two coats of RIW liquid asphaltic composition sprayed or brushed on						
over 300 mm wide; ne 30° to horizontal	0.03	0.54	0.08	52.18	m²	**52.80**
over 300 mm wide; 30–90° to horizontal	0.03	0.54	0.08	52.18	m²	**52.80**
ne 300 mm wide; at any inclination	0.04	0.72	0.11	52.18	m²	**53.01**
Two coats of Mulseal sprayed or brushed on						
any inclination	0.07	1.26	0.20	22.26	m²	**23.72**
20 mm thick red tinted sand asphalt layer						
onto bridge deck	0.02	0.82	0.34	10.78	m²	**11.94**
SURFACE IMPREGNATION OF CONCRETE						
Silane waterproofing						
Surface impregnation to plain surfaces	–	–	–	–	m²	**3.75**
REMOVAL OF EXISTING ASPHALT WATERPROOFING						
over 300 mm wide; ne 30° to horizontal	0.13	5.20	–	–	m²	**5.20**
over 300 mm wide; 30–90° to horizontal	0.18	7.20	–	–	m²	**7.20**
over 300 mm wide; at any inclination	0.20	8.00	–	–	m²	**8.00**
to domed surfaces	0.25	9.99	–	–	m²	**9.99**

SERIES 2100: BRIDGE BEARINGS

Item	Gang hours	Labour £	Plant £	Material £	Unit	Total rate £
NOTES						
Notes						
Bridge bearings are manufactured and installed to individual specifications. The following guide prices are for different sizes of simple bridge bearings. If requirements are known, then advice ought to be obtained from specialist maunfacturers such as CCL.						
RESOURCES – LABOUR						
Bridge bearing gang						
1 ganger/chargehand (skill rate 4)		14.56				
2 unskilled operatives (general)		25.42				
Total Gang Rate/Hour	£	**39.98**				
BEARINGS						
Supply plain rubber bearings (3 m and 5 m lengths)						
150 × 20 mm	0.35	13.99	–	37.63	m	**51.62**
150 × 25 mm	0.35	13.99	–	45.15	m	**59.14**
Supply and place in position laminated elastomeric rubber bearing						
250 × 150 × 19 mm	0.25	9.99	–	27.95	nr	**37.94**
300 × 200 × 19 mm	0.25	9.99	–	30.64	nr	**40.63**
300 × 200 × 30 mm	0.27	10.79	–	38.70	nr	**49.49**
300 × 200 × 41 mm	0.27	10.79	–	58.18	nr	**68.97**
300 × 250 × 41 mm	0.30	11.99	–	73.10	nr	**85.09**
300 × 250 × 63 mm	0.30	11.99	–	102.13	nr	**114.12**
400 × 250 × 19 mm	0.32	12.79	–	53.75	nr	**66.54**
400 × 250 × 52 mm	0.32	12.79	–	118.25	nr	**131.04**
400 × 300 × 19 mm	0.32	12.79	–	62.35	nr	**75.14**
600 × 450 × 24 mm	0.35	13.99	–	129.00	nr	**142.99**
Adhesive fixings to laminated elastomeric rubber bearings						
2 mm thick epoxy adhesive	1.00	39.98	–	46.98	m²	**86.96**
15 mm thick epoxy mortar	1.50	59.97	–	262.45	m²	**322.42**
15 mm thick epoxy pourable grout	2.00	79.96	–	281.46	m²	**361.42**
Supply and install mechanical guides for laminated elastomeric rubber bearings						
500 kN SLS design load; FP50 fixed pin Type 1	2.00	79.96	–	842.35	nr	**922.31**
500 kN SLS design load; FP50 fixed pin Type 2	2.00	79.96	–	868.60	nr	**948.56**
750 kN SLS design load; FP75 fixed pin Type 1	2.10	83.96	–	976.67	nr	**1060.63**
750 kN SLS design load; FP75 fixed pin Type 2	2.10	83.96	–	1110.00	nr	**1193.96**
300 kN SLS design load; UG300 Uniguide Type 1	2.00	79.96	–	1010.00	nr	**1089.96**
300 kN SLS design load; UG300 Uniguide Type 2	2.00	79.96	–	1160.00	nr	**1239.96**
Supply and install fixed pot bearings						
355 × 355; PF200	2.00	79.96	–	902.94	nr	**982.90**
425 × 425; PF300	2.10	83.96	–	1076.66	nr	**1160.62**

SERIES 2100: BRIDGE BEARINGS

Item	Gang hours	Labour £	Plant £	Material £	Unit	Total rate £
Supply and install free sliding pot bearings						
445 × 345; PS200	2.10	83.96	–	1215.00	nr	**1298.96**
520 × 415; PS300	2.20	87.96	–	1548.00	nr	**1635.96**
Supply and install guided sliding pot bearings						
455 × 375; PG200	2.20	87.96	–	1612.97	nr	**1700.93**
545 × 435; PG300	2.30	91.95	–	1920.01	nr	**2011.96**
TESTING BEARINGS						
If there is a requirement for testing bridge bearings prior to their being installed then the tests should be enumerated separately. Specialist advice should be sought once details are known.						
Compression test for laminated elastomeric bearings generally	–	–	–	–	nr	**60.00**
Shear test for laminated elastomeric bearings generally	–	–	–	–	nr	**75.00**

SERIES 2300: BRIDGE EXPANSION JOINTS AND SELAING OF GAPS

Item	Gang hours	Labour £	Plant £	Material £	Unit	Total rate £
NOTES						
Notes						
Major movement joints to bridge and viaduct decks are manufactured and installed to individual specifications determined by the type of structure location in the deck, amount of movement to be expected, and many other variables. The following unit rates are for other types of movement joints found in structures.						
RESOURCES – LABOUR						
Bridge jointing gang						
1 ganger/chargehand (skill rate 4)		14.56				
1 skilled operative (skill rate 4)		13.61				
1 unskilled operative		12.71				
TOTAL	£	40.88				
Jointing gang						
1 ganger/chargehand (skill rate 3)		16.16				
1 skilled operative (skill rate 3)		15.21				
1 unskilled operative		12.71				
TOTAL	£	44.08				
SEALING OF GAPS						
Flexcell joint filler board						
10 mm thick	0.10	4.09	–	3.48	m²	**7.57**
19 mm thick	0.16	6.54	–	5.97	m²	**12.51**
25 mm thick	0.16	6.54	–	7.45	m²	**13.99**
Building paper slip joint to abutment toe	0.01	0.45	–	18.25	m²	**18.70**
Bond breaking agent	0.03	1.02	–	3.50	m²	**4.52**
Hot poured rubber bitumen joint sealant						
10 × 20 mm	0.03	1.35	–	8.46	m	**9.81**
20 × 20 mm	0.04	1.64	–	3.31	m	**4.95**
25 × 15 mm	0.07	2.66	–	3.16	m	**5.82**
25 × 25 mm	0.07	2.98	–	5.17	m	**8.15**
Cold applied polysulphide joint sealant						
20 × 20 mm	0.07	2.66	–	5.57	m	**8.23**
Gun grade cold applied elastomeric joint sealant						
25 × 25 mm on 3 mm foam strip	0.07	2.66	–	8.50	m	**11.16**
50 × 25 mm on 3 mm foam strip	0.09	3.68	–	16.50	m	**20.18**
PVC centre bulb waterstop						
160 mm wide	0.08	3.27	–	8.45	m	**11.72**
210 mm wide	0.09	3.68	–	8.67	m	**12.35**
260 mm wide	0.11	4.50	–	13.46	m	**17.96**
PVC flat dumbell waterstop						
170 mm wide	0.08	3.27	–	39.05	m	**42.32**
210 mm wide	0.10	4.09	–	46.16	m	**50.25**
250 mm wide	0.12	4.91	–	93.51	m	**98.42**

SERIES 2300: BRIDGE EXPANSION JOINTS AND SELAING OF GAPS

Item	Gang hours	Labour £	Plant £	Material £	Unit	Total rate £
Dowels, plain or greased						
12 mm mild steel 450 mm long	0.04	1.64	–	1.75	nr	**3.39**
16 mm mild steel 750 mm long	0.05	1.84	–	2.76	nr	**4.60**
16 mm mild steel 750 mm long with debonding agent for 375 mm	0.05	2.17	–	3.36	nr	**5.53**
Dowels, sleeved or capped						
12 mm mild steel 450 mm long with debonding agent for 225 mm and PVC dowel cap	0.05	1.84	–	1.68	nr	**3.52**

SERIES 2400: BRICKWORK, BLOCKWORK AND STONEWORK

Item	Gang hours	Labour £	Plant £	Material £	Unit	Total rate £
RESOURCES – LABOUR						
Masonry gang						
1 foreman bricklayer (craftsman)		22.50				
4 bricklayers (craftsman)		78.16				
1 unskilled operative (general)		12.71				
Total Gang Rate/Hour		£ 113.37				
RESOURCES – PLANT						
Masonry						
dumper 2 t (50% of time)			1.75			
gas oil for ditto			1.05			
cement mixer 4/3 (50% of time)			0.41			
petrol for ditto			1.05			
small power tools (masonry)			0.75			
minor scaffolding (masonry)			1.05			
Total Rate/Hour		£	6.06			

RESOURCES – MATERIALS

Half brick thick walls are in stretcher bond, thicker than this in English bond (3 stretchers: 1 header) unless otherwise stated.

DfT Table 24/1: Mortar Proportions by Volume:

Mortar Type	Cement: Lime: sand	Masonry Cement: sand	Cement: Sand
(i)	1.0 to 1/4.3	–	–
(ii)	1: ½:4 to 4.5	1:2.5 to 3.5	1:3 to 4
(iii)	1:1:5 to 6	1:4.5	1:5 to 6

BRICKWORK

Item	Gang hours	Labour £	Plant £	Material £	Unit	Total rate £
Common bricks; in stretcher bond; in cement mortar designation (ii)						
Walls						
half brick thick	0.23	25.85	1.38	36.53	m²	63.76
one brick thick	0.44	50.11	2.68	73.70	m²	126.49
one and a half bricks thick	0.64	73.01	3.91	111.34	m²	188.26
two bricks thick	0.83	94.49	5.06	148.78	m²	248.33
Walls, curved on plan						
half brick thick	0.30	33.56	1.77	36.53	m²	71.86
one brick thick	0.57	64.62	3.45	73.70	m²	141.77
one and a half bricks thick	0.82	92.96	4.99	111.34	m²	209.29
two bricks thick	1.06	119.95	6.41	148.78	m²	275.14
Walls, with a battered face						
half brick thick	0.33	37.30	1.99	36.53	m²	75.82
one brick thick	0.63	71.42	3.82	73.70	m²	148.94
one and a half bricks thick	0.91	102.83	5.50	111.34	m²	219.67
two bricks thick	1.16	131.74	7.04	148.78	m²	287.56

SERIES 2400: BRICKWORK, BLOCKWORK AND STONEWORK

Item	Gang hours	Labour £	Plant £	Material £	Unit	Total rate £
Facework to concrete						
half brick thick	0.25	28.00	1.50	42.16	m^2	71.66
one brick thick	0.48	54.19	2.90	79.32	m^2	136.41
one and a half bricks thick	0.69	78.68	4.21	116.96	m^2	199.85
two bricks thick	0.90	101.81	5.44	154.34	m^2	261.59
In alteration work						
half brick thick	0.30	33.56	1.79	42.16	m^2	77.51
one brick thick	0.57	64.62	3.45	79.32	m^2	147.39
one and a half bricks thick	0.82	93.30	4.99	116.96	m^2	215.25
two bricks thick	1.06	119.95	6.41	154.34	m^2	280.70
ADD or DEDUCT to materials costs for variation of £10.00/1000 in PC of common bricks						
half brick thick	–	–	–	0.63	m^2	0.63
one brick thick	–	–	–	1.26	m^2	1.26
one and a half bricks thick	–	–	–	1.89	m^2	1.89
two bricks thick	–	–	–	2.52	m^2	2.52
Copings; standard header-on-edge; 215 mm wide × 103 mm high	0.11	11.90	0.64	8.20	m	20.74
ADD or DEDUCT to copings for variation of £1.00/100 in PC of common bricks	–	–	–	0.13	m	0.13
Class A engineering bricks, perforated; in cement mortar designation (ii)						
Walls						
half brick thick	0.27	30.50	1.63	55.20	m^2	87.33
one brick thick	0.52	58.95	3.15	111.03	m^2	173.13
one and a half bricks thick	0.75	85.37	4.56	167.34	m^2	257.27
two bricks thick	0.97	110.08	5.89	223.45	m^2	339.42
Walls, curved on plan						
half brick thick	0.37	41.95	2.24	55.20	m^2	99.39
one brick thick	0.70	79.93	4.27	111.03	m^2	195.23
one and a half bricks thick	1.01	114.50	6.12	167.34	m^2	287.96
two bricks thick	1.29	146.02	7.81	223.45	m^2	377.28
Walls, with a battered face						
half brick thick	0.37	41.95	2.24	55.20	m^2	99.39
one brick thick	0.70	79.93	4.27	111.03	m^2	195.23
one and a half bricks thick	1.01	114.50	6.12	167.34	m^2	287.96
two bricks thick	1.29	146.02	7.81	223.45	m^2	377.28
Facework to concrete						
half brick thick	0.32	35.71	1.91	60.82	m^2	98.44
one brick thick	0.60	68.59	3.67	116.66	m^2	188.92
ADD or DEDUCT to materials costs for variation of £10.00/1000 in PC of engineering bricks						
half brick thick	–	–	–	0.63	m^2	0.63
one brick thick	–	–	–	1.26	m^2	1.26
one and a half bricks thick	–	–	–	1.89	m^2	1.89
two bricks thick	–	–	–	2.52	m^2	2.52
Brick coping in standard bricks in headers on edge; 215 mm wide × 103 mm high	0.29	32.88	1.76	14.40	m	49.04
ADD or DEDUCT to copings for variation of £1.00/100 in PC of Class A engineering bricks	–	–	–	0.13	m	0.13

SERIES 2400: BRICKWORK, BLOCKWORK AND STONEWORK

Item	Gang hours	Labour £	Plant £	Material £	Unit	Total rate £
BRICKWORK – cont						
Class B engineering bricks, perforated; in cement mortar designation (ii)						
Walls						
half brick thick	0.29	32.88	1.76	22.36	m²	**57.00**
one brick thick	0.48	54.98	2.94	46.11	m²	**104.03**
one and a half bricks thick	0.69	78.79	4.21	67.26	m²	**150.26**
two bricks thick	0.91	102.60	5.49	86.85	m²	**194.94**
Walls, curved on plan						
half brick thick	0.42	47.16	2.52	22.36	m²	**72.04**
one brick thick	0.78	88.43	4.73	46.11	m²	**139.27**
one and a half bricks thick	1.04	117.90	6.30	67.26	m²	**191.46**
two bricks thick	1.30	147.38	7.88	86.85	m²	**242.11**
Walls, with battered face						
half brick thick	0.42	47.16	2.52	22.36	m²	**72.04**
one brick thick	0.78	88.43	4.73	46.11	m²	**139.27**
one and a half bricks thick	1.04	117.90	6.30	67.26	m²	**191.46**
two bricks thick	1.30	147.38	7.88	86.85	m²	**242.11**
Facework to concrete						
half brick thick	0.32	36.05	1.93	27.93	m²	**65.91**
one brick thick	0.56	63.49	3.40	51.69	m²	**118.58**
ADD or DEDUCT to materials costs for variation of £10.00/1000 in PC of engineering bricks						
half brick thick	–	–	–	0.63	m²	**0.63**
one brick thick	–	–	–	1.26	m²	**1.26**
one and a half bricks thick	–	–	–	1.89	m²	**1.89**
two bricks thick	–	–	–	2.52	m²	**2.52**
Brick coping in standard bricks in headers on edge;						
215 mm wide × 103 mm high	0.13	15.19	0.81	8.61	m	**24.61**
ADD or DEDUCT to copings for variation of £1.00/100 in PC of Class B engineering bricks	–	–	–	0.13	m	**0.13**
Facing bricks; in lime mortar designation (ii)						
Walls						
half brick thick	0.34	38.55	2.06	38.60	m²	**79.21**
one brick thick	0.57	65.07	4.57	77.85	m²	**147.49**
one and a half bricks thick	0.83	93.76	5.01	117.56	m²	**216.33**
two bricks thick	1.08	122.44	6.55	157.08	m²	**286.07**
Walls, curved on plan						
half brick thick	0.45	51.02	2.73	38.60	m²	**92.35**
one brick thick	0.84	95.23	5.09	77.85	m²	**178.17**
one and a half bricks thick	1.11	126.41	6.76	117.56	m²	**250.73**
two bricks thick	1.39	157.58	8.43	157.08	m²	**323.09**
Walls, with a battered face						
half brick thick	0.45	51.02	2.73	38.60	m²	**92.35**
one brick thick	0.84	95.23	5.09	77.85	m²	**178.17**
one and a half bricks thick	1.11	126.41	6.76	117.56	m²	**250.73**
two bricks thick	1.39	157.58	8.43	157.08	m²	**323.09**

SERIES 2400: BRICKWORK, BLOCKWORK AND STONEWORK

Item	Gang hours	Labour £	Plant £	Material £	Unit	Total rate £
Facework to concrete						
half brick thick	0.37	41.72	2.08	44.23	m²	**88.03**
one brick thick	0.66	74.82	4.00	83.47	m²	**162.29**
Extra over common brickwork in English bond for facing with facing bricks in lime mortar designation (ii)	0.11	12.70	0.73	25.40	m²	**38.83**
ADD or DEDUCT to materials costs for variation of £10.00/1000 in PC of facing bricks						
half brick thick	–	–	–	0.63	m²	**0.63**
one brick thick	–	–	–	1.26	m²	**1.26**
one and a half bricks thick	–	–	–	1.89	m²	**1.89**
two bricks thick	–	–	–	2.52	m²	**2.52**
Brick coping in standard bricks in headers on edge						
215 mm wide × 103 mm high	0.13	15.19	0.81	8.11	m	**24.11**
Flat arches in standard stretchers on end						
103 mm wide × 215 mm high	0.21	23.81	1.27	7.83	m	**32.91**
Flat arches in bullnose stretchers on end						
103 mm × 215 mm high	0.22	24.94	1.34	23.33	m	**49.61**
Segmental arches in single ring stretchers on end						
103 mm wide × 215 mm high	0.37	41.95	2.24	7.83	m	**52.02**
Segmental arches in double ring stretchers on end						
103 mm wide × 440 mm high	0.49	55.55	2.97	31.91	m	**90.43**
Segmental arches; cut voussoirs						
103 mm wide × 215 mm high	0.39	44.21	2.37	49.79	m	**96.37**
ADD or DEDUCT to copings and arches for variation of £1.00/100 in PC of facing bricks						
header-on-edge	–	–	–	0.13	m	**0.13**
stretcher-on-end	–	–	–	0.13	m	**0.13**
stretcher-on-end bullnose specials	–	–	–	0.13	m	**0.13**
single ring	–	–	–	0.13	m	**0.13**
two ring	–	–	–	0.27	m	**0.27**
BLOCKWORK AND STONEWORK						
Lightweight concrete blocks; solid; 3.5 N/mm²; in cement-lime mortar						
Walls						
100 mm thick	0.17	19.73	1.06	10.61	m²	**31.40**
140 mm thick	0.23	25.51	1.36	13.64	m²	**40.51**
215 mm thick	0.28	31.18	1.67	17.79	m²	**50.64**
Walls, curved on plan						
100 mm thick	0.23	26.19	1.40	10.61	m²	**38.20**
140 mm thick	0.30	33.90	1.81	13.64	m²	**49.35**
215 mm thick	0.37	41.49	2.22	17.79	m²	**61.50**
Facework to concrete						
100 mm thick	0.18	20.29	1.08	16.24	m²	**37.61**
140 mm thick	0.23	26.30	1.41	19.26	m²	**46.97**
215 mm thick	0.28	32.08	1.71	23.41	m²	**57.20**

SERIES 2400: BRICKWORK, BLOCKWORK AND STONEWORK

Item	Gang hours	Labour £	Plant £	Material £	Unit	Total rate £
BLOCKWORK AND STONEWORK – cont						
Lightweight concrete blocks – cont						
In alteration work						
100 mm thick	0.17	19.73	1.06	10.61	m^2	**31.40**
140 mm thick	0.23	25.51	1.36	13.64	m^2	**40.51**
215 mm thick	0.28	31.18	1.67	17.79	m^2	**50.64**
Dense concrete blocks; solid; 3.5 or 7 N/mm²; in cement-lime mortar						
Walls						
100 mm thick	0.17	19.27	1.03	11.10	m^2	**31.40**
140 mm thick	0.20	22.67	1.21	14.37	m^2	**38.25**
215 mm thick	0.24	27.21	1.45	19.84	m^2	**48.50**
Walls, curved on plan						
100 mm thick	0.23	25.62	1.37	11.10	m^2	**38.09**
140 mm thick	0.23	25.62	1.61	14.37	m^2	**41.60**
215 mm thick	0.32	36.16	1.93	19.24	m^2	**57.33**
Facework to concrete						
100 mm thick	0.17	19.84	1.06	16.73	m^2	**37.63**
140 mm thick	0.21	23.35	1.25	19.99	m^2	**44.59**
215 mm thick	0.35	39.68	1.68	24.87	m^2	**66.23**
In alteration work						
100 mm thick	0.27	30.16	1.37	11.10	m^2	**42.63**
140 mm thick	0.26	29.48	1.61	14.37	m^2	**45.46**
215 mm thick	0.32	36.16	1.93	19.24	m^2	**57.33**
Reconstituted stone; Bradstone 100 bed weathered Cotswold or North Cerney masonary blocks; rough hewn rockfaced blocks; in coloured cement-lime mortar designation (1:2:9) (iii)						
Walls, thickness 100 mm						
vertical and straight	0.30	34.01	1.82	53.49	m^2	**89.32**
curved on plan	0.39	44.21	2.37	53.49	m^2	**100.07**
with a battered face	0.34	39.11	2.09	53.49	m^2	**94.69**
in arches	0.57	65.19	3.49	53.49	m^2	**122.17**
Facing to concrete; wall ties						
vertical and straight	0.24	26.87	1.44	81.92	m^2	**110.23**
curved on plan	0.32	35.71	1.91	81.92	m^2	**119.54**
with a battered face	0.36	40.36	2.16	81.92	m^2	**124.44**
Reconstituted stone; Bradstone Architectural dressings in weathered Cotswold or North Cerney shades; in coloured cement-lime mortar designation (1:2:9) (iii)						
Copings; twice weathered and throated						
152 × 76 mm	0.08	9.07	0.49	15.78	m	**25.34**
152 × 76 mm; curved on plan	0.11	12.02	0.64	56.23	m	**68.89**
305 × 76 mm	0.10	11.34	0.61	36.04	m	**47.99**
305 × 76 mm; curved on plan	0.13	15.08	0.81	72.89	m	**88.78**

SERIES 2400: BRICKWORK, BLOCKWORK AND STONEWORK

Item	Gang hours	Labour £	Plant £	Material £	Unit	Total rate £
Corbels						
479 × 100 × 215 mm, splayed	0.49	55.55	2.97	273.22	nr	331.74
665 × 100 × 215 mm, splayed	0.55	62.35	3.34	272.52	nr	338.21
Pier caps						
305 × 305 mm	0.09	10.20	0.54	19.86	nr	30.60
381 × 381 mm	0.11	12.47	0.67	28.92	nr	42.06
457 × 457 mm	0.13	14.74	0.79	38.70	nr	54.23
533 × 533 mm	0.15	17.01	0.91	53.76	nr	71.68
Lintels						
100 × 140 mm	0.11	12.47	0.67	41.16	m	54.30
100 × 215 mm	0.16	18.14	0.97	55.33	m	74.44
Natural stone ashlar; Portland Whitbed limestone; in cement-lime mortar designation (iii)						
Walls						
vertical and straight	12.70	1439.80	72.23	2393.72	m³	3905.75
curved on plan	19.30	2188.04	117.01	3572.47	m³	5877.52
with a battered face	19.30	2188.04	117.01	3572.47	m³	5877.52
Facing to concrete; wall ties						
vertical and straight	17.00	1927.29	103.06	2423.72	m³	4454.07
curved on plan	25.80	2924.95	156.41	3602.47	m³	6683.83
with a battered face	25.80	2924.95	156.41	3602.47	m³	6683.83
Copings; twice weathered and throated						
250 × 150 mm	0.45	51.02	2.73	129.80	m	183.55
250 × 150 mm; curved on plan	0.45	51.02	2.73	155.70	m	209.45
400 × 150 mm	0.49	55.55	2.97	190.74	m	249.26
400 × 150 mm; curved on plan	0.49	55.55	2.97	228.84	m	287.36
Shaped and dressed string courses						
75 mm projection × 150 mm high	0.45	51.02	2.73	116.51	m	170.26
Corbel						
500 × 450 × 300 mm	0.55	62.35	3.34	173.53	nr	239.22
Keystone						
750 × 900 × 300 mm (extreme)	1.30	147.38	7.88	545.23	nr	700.49
Random rubble uncoursed, weighing 2.0 t/m³ of wall; in cement-lime mortar designation (iii)						
Walls						
vertical and straight	4.17	472.75	25.28	265.80	m³	763.83
curved on plan	4.67	529.44	28.31	332.22	m³	889.97
with a battered face	4.67	529.44	28.31	332.22	m³	889.97
in arches	8.53	967.05	51.71	332.22	m³	1350.98
Facework to concrete						
vertical and straight	4.17	472.75	25.28	337.84	m³	835.87
curved on plan	4.67	529.44	28.31	337.84	m³	895.59
with a battered face	4.67	529.44	28.31	337.84	m³	895.59
in arches	8.53	967.05	51.71	337.84	m³	1356.60
Copings						
500 × 125 mm	0.49	55.55	2.97	220.51	m	279.03

SERIES 2400: BRICKWORK, BLOCKWORK AND STONEWORK

Item	Gang hours	Labour £	Plant £	Material £	Unit	Total rate £
BLOCKWORK AND STONEWORK – cont						
Squared random rubble uncoursed, weighing 2.0 t/m^3 of wall; in cement-lime mortar designation (iii)						
Walls						
vertical and straight	4.17	472.75	25.28	453.06	m^3	**951.09**
curved on plan	4.67	529.44	28.31	453.06	m^3	**1010.81**
with a battered face	4.67	529.44	28.31	453.06	m^3	**1010.81**
in arches	8.53	967.05	51.71	453.06	m^3	**1471.82**
Facework to concrete						
vertical and straight	4.17	472.75	25.28	458.69	m^3	**956.72**
curved on plan	4.67	529.44	28.31	458.69	m^3	**1016.44**
with a battered face	4.67	529.44	28.31	458.69	m^3	**1016.44**
in arches	8.53	967.05	51.71	458.69	m^3	**1477.45**
Copings						
500 × 125 mm	0.49	55.55	2.97	220.51	m	**279.03**
Dry rubble, weighing 2.0 t/m^3 of wall						
Walls						
vertical and straight	3.83	434.21	23.22	344.00	m^3	**801.43**
curved on plan	4.33	490.89	26.25	344.00	m^3	**861.14**
with a battered face	4.33	490.89	26.25	344.00	m^3	**861.14**
Copings formed of rough stones						
275 × 200 mm (average) high	0.45	51.02	2.73	29.15	m	**82.90**
500 × 200 mm	0.55	62.35	3.34	50.14	m	**115.83**

Unit Costs – Highway Works 493

SERIES 2500: SPECIAL STRUCTURES

Item	Gang hours	Labour £	Plant £	Material £	Unit	Total rate £

SPECIAL STRUCTURES DESIGNED BY CONTRACTOR
 	
 Notes
 This section envisages the following types of structure which may be required to be designed by the Contractor based on stipulated performance criteria:
   * Buried structures
   * Earth retaining structures
   * Environmental barriers
   * Underbridges up to 8 m span
   * Footbridges
   * Piped culverts
   * Box culverts
   * Drainage exceeding 900 mm diameter
   * Other structures
 Naturally, this work cannot be catered for directly in this section and will require the preparation of a sketch solution and approximate quantities to allow pricing using the various other Unit Costs sections as well as the Approximate Estimates section.
 An allowance must be added to such an estimate to cover the Contractor's design fee(s) and expenses.						

SERIES 2700: ACCOMMODATION WORKS, WORKS FOR STATUTORY UNDERTAKERS

Item	Gang hours	Labour £	Plant £	Material £	Unit	Total rate £
ACCOMMODATION WORKS, WORKS FOR STATUTORY UNDERTAKERS **Note** Cost items in this series will be specific to individual contract agreements and it is felt that inclusion of prices in this publication would not provide useful guidance.						

SERIES 3000: LANDSCAPING AND ECOLOGY

Item	Gang hours	Labour £	Plant £	Material £	Unit	Total rate £
GROUND PREPARATION AND CULTIVATION						
Supply and apply granular cultivation treatments by hand						
35 g/m^2	0.50	13.16	–	12.25	100 m^2	**25.41**
50 g/m^2	0.65	17.11	–	17.50	100 m^2	**34.61**
75 g/m^2	0.85	22.37	–	26.25	100 m^2	**48.62**
100 g/m^2	1.00	26.32	–	35.00	100 m^2	**61.32**
150 g/m^2	1.20	31.58	–	52.50	100 m^2	**84.08**
Supply and apply granular cultivation treatments by machine in suitable economically large areas						
100 g/m^2	–	–	6.75	35.00	100 m^2	**41.75**
ADD to above for:						
granular treatments per £0.10/kg PC variation +10%						
selective weedkiller +171%						
herbicide +567%						
fertilizer +100%						
Supply and incorporate cultivation additives into top 150 mm topsoil by hand						
1 m^3 / 10 m^2	20.00	526.40	–	12.50	100 m^2	**538.90**
1 m^3 / 13 m^2	20.00	526.40	–	9.61	100 m^2	**536.01**
1 m^3 / 20 m^2	19.00	500.08	–	6.25	100 m^2	**506.33**
1 m^3 / 40 m^2	17.00	447.44	–	3.13	100 m^2	**450.57**
Supply and incorporate cultivation additives into top 150 mm topsoil by machine in suitable economically large areas						
1 m^3 / 10 m^2	–	–	247.17	12.50	100 m^2	**259.67**
1 m^3 / 13 m^2	–	–	228.16	9.61	100 m^2	**237.77**
1 m^3 / 20 m^2	–	–	202.80	6.25	100 m^2	**209.05**
1 m^3 / 40 m^2	–	–	186.96	3.13	100 m^2	**190.09**
ADD to above for:						
cultivation additives per £0.10/m^2 PC variation + 10%						
compost + 300%						
manure +166%						
peat +1800%						
SEEDING AND TURFING						
Selected grass seed; at the rate of 0.050 kg/m^2 in two operations						
Grass seeding; by conventional sowing						
at 10° or less to the horizontal	0.01	0.26	–	0.49	m^2	**0.75**
more than 10° to horizontal	0.02	0.39	–	0.49	m^2	**0.88**
Wild Flora mixture BSH ref WF1 combined with low maintenance conservation grass BSH ref A4 (20%:80%); at the rate of 80 g/m^2						
Wildflower seeding; by conventional sowing						
at 10° or less to the horizontal	0.50	13.16	–	1376.82	100 m^2	**1389.98**
more than 10° to horizontal	0.02	0.39	–	0.49	m^2	**0.88**

SERIES 3000: LANDSCAPING AND ECOLOGY

Item	Gang hours	Labour £	Plant £	Material £	Unit	Total rate £
SEEDING AND TURFING – cont						
Hydraulic mulch grass seed						
Grass seeding by hydraulic seeding						
at 10° or less to the horizontal	0.01	0.13	0.17	2.39	m²	**2.69**
at more than 10° to the horizontal	0.01	0.18	0.17	2.39	m²	**2.74**
Imported turf						
Turfing to surfaces						
at 10° or less to the horizontal	0.12	3.16	–	8.57	m²	**11.73**
more than 10° to the horizontal; pegging down	0.17	4.47	–	8.57	m²	**13.04**
PLANTING						
Trees						
The cost of planting semi-mature trees will depend on the size and species, and on the access to the site for tree handling machines. Prices should be obtained for individual trees and planting.						
Break up subsoil to a depth of 200 mm in treepit	0.05	1.32	–	–	nr	**1.32**
Supply and plant tree in prepared pit; backfill with excavated topsoil minimum 600 mm deep						
light standard; in pits	0.25	6.58	–	52.50	nr	**59.08**
standard tree	0.45	11.84	–	65.10	nr	**76.94**
selected standard tree	0.75	19.74	–	102.90	nr	**122.64**
heavy standard tree	0.85	22.37	–	115.50	nr	**137.87**
extra heavy standard tree	1.50	39.48	–	152.25	nr	**191.73**
extra for filling with topsoil from spoil heap ne 100 m distant	0.15	3.95	0.84	–	m³	**4.79**
extra for filling with imported topsoil	0.08	2.11	0.45	14.18	m³	**16.74**
Extra for incorporating manure or compost into top soil at the rate of 1 m³ per 5 m³ +60%						
Supply tree stake and drive 500 mm into firm ground and trim to approved height, including two tree ties to approved pattern						
one stake; 2.4 m long, 100 mm diameter	0.16	4.21	–	5.78	nr	**9.99**
one stake; 3.0 m long, 100 mm diameter	0.20	5.26	–	6.04	nr	**11.30**
two stakes; 2.4 m long, 100 mm diameter	0.24	6.32	–	8.40	nr	**14.72**
two stakes; 3.0 m long, 100 mm diameter	0.30	7.90	–	8.93	nr	**16.83**
Supply and fit tree support comprising three collars and wire guys; including pickets						
galvanized steel 50 × 600 mm	1.50	39.48	–	21.00	nr	**60.48**
hardwood 75 × 600 mm	1.50	39.48	–	22.50	nr	**61.98**
Supply and fix standard steel tree guard	0.30	7.90	–	21.00	nr	**28.90**
Hedge plants						
Excavate trench by hand for hedge and deposit soil alongside trench						
300 wide × 300 mm deep	0.10	2.63	–	–	m	**2.63**
450 wide × 300 mm deep	0.13	3.42	–	–	m	**3.42**

SERIES 3000: LANDSCAPING AND ECOLOGY

Item	Gang hours	Labour £	Plant £	Material £	Unit	Total rate £
Excavate trench by machine for hedge and deposit soil alongside trench						
300 wide × 300 mm deep	0.02	0.30	0.66	–	m	0.96
450 wide × 300 mm deep	0.02	0.30	0.66	–	m	0.96
Set out, nick out and excavate trench and break up subsoil to minimum depth of 300 mm						
400 mm minimum deep	0.15	3.95	–	–	m	3.95
Supply and plant hedging plants; backfill with excavated topsoil						
single row plants at 200 mm centres	0.25	6.58	–	4.08	m	10.66
single row plants at 300 mm centres	0.17	4.47	–	2.71	m	7.18
single row plants at 400 mm centres	0.13	3.29	–	2.04	m	5.33
single row plants at 500 mm centres	0.10	2.63	–	1.63	m	4.26
single row plants at 600 mm centres	0.08	2.11	–	1.35	m	3.46
double row plants at 200 mm centres	0.50	13.16	–	8.15	m	21.31
double row plants at 300 mm centres	0.34	8.95	–	5.43	m	14.38
double row plants at 400 mm centres	0.25	6.58	–	4.08	m	10.66
double row plants at 500 mm centres	0.20	5.26	–	3.26	m	8.52
double row plants at 600 mm centres	0.16	4.21	–	2.71	m	6.92
Extra for incorporating manure at 1 m^3 / 30 m^3	0.60	7.63	–	0.24	–	7.87
Shrubs						
Form planting hole in previously cultivated area, supply and plant specified shrub and backfill with excavated material						
shrub 300 mm high	0.10	2.63	–	6.15	each	8.78
shrub 600 mm high	0.10	2.63	–	6.66	each	9.29
shrub 900 mm high	0.10	2.63	–	7.28	each	9.91
shrub 1.0 m high and over	0.10	2.63	–	7.69	each	10.32
Supply and fix shrub stake including two ties						
one stake; 1.5 m long, 75 mm diameter	0.12	3.16	–	8.20	each	11.36
Extra for the above items for planting in prefabricated or in situ planters	–	–	–	–	%	20.00
Herbaceous plants						
Form planting hole in previously cultivated area, supply and plant specified herbaceous plants and backfill with excavated material						
5 plants/m^2	0.05	1.32	–	6.35	m^2	7.67
10 plants/m^2	0.16	4.21	–	12.71	m^2	16.92
25 plants/m^2	0.42	11.05	–	31.76	m^2	42.81
35 plants/m^2	0.58	15.27	–	44.47	m^2	59.74
50 plants/m^2	0.83	21.85	–	63.52	m^2	85.37
Supply and fix plant support netting on 50 mm diameter stakes 750 mm long driven into the ground at 1.5 m centres						
1.15 m high green extruded plastic mesh, 125 mm square mesh	0.06	1.58	–	37.63	m^2	39.21
Extra to the above items for planting in prefabricated or in situ planters	–	–	–	–	%	20.00

SERIES 3000: LANDSCAPING AND ECOLOGY

Item	Gang hours	Labour £	Plant £	Material £	Unit	Total rate £
PLANTING – cont						
Herbaceous plants – cont						
Form planting hole in previously cultivated area; supply and plant bulbs and backfill with excavated material						
small	0.01	0.26	–	0.15	each	**0.41**
medium	0.01	0.26	–	0.24	each	**0.50**
large	0.01	0.26	–	0.26	each	**0.52**
Supply and plant bulb in grassed area using bulb planter and backfill with screened topsoil or peat and cut turf plug						
small	0.01	0.26	–	0.15	each	**0.41**
medium	0.01	0.26	–	0.24	each	**0.50**
large	0.01	0.26	–	0.26	each	**0.52**
Extra to the above items for planting in prefabricated or in situ planters	–	–	–	–	%	**15.00**
MULCHING						
Organic mulching of medium bark mulch to a depth of 50 mm in planting areas to surfaces						
at 10° or less to the horizontal	0.01	0.26	–	1.88	m^2	**2.14**
more than 10° to horizontal	0.02	0.39	–	1.88	m^2	**2.27**
Organic mulching of timber mulch to a depth of 75 mm in planting areas to surfaces						
at 10° or less to the horizontal	0.02	0.39	–	3.75	m^2	**4.14**
more than 10° to horizontal	0.02	0.53	–	3.75	m^2	**4.28**
WEED CONTROL						
Weed and handfork planted areas including removing and dumping weed and debris on site.	0.07	1.84	–	–	m^2	**1.84**
Supply and apply selective weed killer						
35 g/m^2	0.01	0.13	–	0.13	m^2	**0.26**
70 g/m^2	0.01	0.18	–	0.26	m^2	**0.44**
100 g/m^2	0.01	0.26	–	0.37	m^2	**0.63**
MAINTENANCE OF ESTABLISHED TREES AND SHRUBS						
Initial cut back to shrubs and hedge plants, clear away all cuttings	0.04	1.05	–	–	m	**1.05**
Protect planted areas with windbreak fencing fixed to stakes 1.5 m × 50 mm diameter at 2.0 m centres and clear away on completion						
1.0 m high	0.06	1.58	–	5.34	m	**6.92**
Cut out dead and diseased wood, prune, trim and cut to shape; treat wounds with sealant; clear away cuttings						
shrubs ne 1.0 m high	0.20	5.26	–	–	each	**5.26**
shrubs 1.0–2.0 m high	0.30	7.90	–	–	each	**7.90**
shrubs 2.0–3.0 m high	0.40	10.53	–	–	each	**10.53**

SERIES 3000: LANDSCAPING AND ECOLOGY

Item	Gang hours	Labour £	Plant £	Material £	Unit	Total rate £
Cut and trim ornamental hedge to specified profile; clear away cuttings						
ornamental hedge 2.0 m high	0.60	15.79	–	–	m	**15.79**
ornamental hedge 4.0 m high	0.90	23.69	–	–	m	**23.69**
Trim field hedge to specified height and shape; clear away cuttings						
using flail	0.05	1.32	–	–	m	**1.32**
using cutting bar	0.07	1.84	–	–	m	**1.84**
MAINTENANCE ESTABLISHED GRASSED AREAS						
Grass cutting at medium frequency						
on central reserves	–	–	–	–	m^2	**6.75**

SERIES 5000 MAINTENANCE PAINTING OF STEELWORK

Item	Gang hours	Labour £	Plant £	Material £	Unit	Total rate £
SURFACE PREPARATION						
Surface preparation to general surfaces by dry blast cleaning to DfT Clause 5003 to remove unsound paint down to sound paint	–	–	–	–	m²	**9.25**
Protective System						
Protective system Type I (M) to DfT Table 50/2 to general surfaces prepared down to sound paint	–	–	–	–	m²	**26.60**

PART 8

Daywork

ESSENTIAL READING FROM TAYLOR AND FRANCIS

Reinforced and Prestressed Concrete Design to EC2

2nd Edition

Eugene O'Brien et al

Fully revised to Eurocode 2, this second edition retains the original's emphasis on qualitative understanding of the overall behaviour of concrete structures. With a new chapter dedicated to case studies, worked examples, and exercise examples, it is an even more comprehensive guide to conceptual design, analysis, and detailed design of concrete structures, providing civil and structural engineering students with complete coverage of the analysis and design of reinforced and prestressed concrete structures.

January 2012: 246x189: 236pp
Hb: 978-0-415-57194-4: **£99.00**
Pb: 978-0-415-57195-1: **£34.99**

To Order: Tel: +44 (0) 1235 400524 **Fax:** +44 (0) 1235 400525
or Post: Taylor and Francis Customer Services,
Bookpoint Ltd, Unit T1, 200 Milton Park, Abingdon, Oxon, OX14 4TA UK
Email: book.orders@tandf.co.uk

For a complete listing of all our titles visit:
www.tandf.co.uk

Taylor & Francis
Taylor & Francis Group

INTRODUCTION

'Dayworks' relates to work which is carried out incidental to a contract but where no other rates have been agreed.

The most recent Schedule is dated July 2011 and is published by the Civil Engineering Contractors Association. Copies may be obtained from:

CIVIL ENGINEERING CONTRACTORS ASSOCIATION
55 Tufton Street
London
SW1P 3QL
Tel: 020 7227 4620
Fax: 020 7227 4621

These schedules identify a number of items that are excluded from the rates and percentages quoted, but should be recovered by the Contractor when valuing his Daywork account. Under normal circumstances it is impractical to accurately value these items for each Daywork event, although certain items can be allowed for by means of a percentage addition. Suggested methods of calculating such additions are included in this section.

In summary the Daywork schedule allows for an addition:

- of 148% to wages paid to workmen
- of 88% to labour only subcontractors and hired plant drivers
- of 12.5% to subsistence allowances and travel paid to workmen
- of 12.5% to materials used on dayworks
- of 12.5% to full cost of plant hired for dayworks
- of 12.5% to cost of operating welfare facilities
- of 12.5% to cost of additional insurance premiums for abnormal contract work or special conditions

Part 4 of this book (Resources – Plant) includes detailed references to the plant section of the 2011 Daywork Schedule.

The text of the document is as follows:

SCHEDULES OF DAYWORKS CARRIED OUT INCIDENTAL TO CONTRACT WORK
The General clauses in the Schedule are repeated in full as follows:

1 Labour

Add to the amount of wages paid to operatives – 148%

1   'Amount of wages' means:

   Actual wages and bonus paid, daily travelling allowances (fare and/or time), tool allowance and all prescribed payments including those in respect of time lost due to inclement weather paid to operatives at plain time rates and/or at overtime rates.

2   The percentage addition provides for all statutory charges at the date of publication and other charges including:

   - National Insurance and Surcharge.
   - Normal Contract Works, Third Party & Employer's Liability Insurances
   - Annual and Public Holidays with pay.
   - Statutory and industry sick pay.
   - Welfare benefits.

- Industrial Training Levy.
- Redundancy Payments.
- Employment Rights Act 1996.
- Employment Relations Act 1999.
- Employment Act 2002.
- Site Supervision and staff including foremen and walking gangers, but the time of the gangers or charge hands' working with their gangs is to be paid for as for operatives.
- Small tools – such as picks, shovels, barrows, trowels, hand saws, buckets, trestles, hammers, chisels and all items of a like nature.
- Protective clothing.
- Head Office charges and profit.

3. The time spent in training, mobilization, demobilization etc. for the Dayworks operation is chargeable.
4. All hired plant drivers and labour subcontractor's accounts to be charged in full (without deduction of any cash discounts not exceeding 2.5%) plus 88%.
5. Subsistence or lodging allowances and periodic travel allowances (fare and/or time) paid to or incurred on behalf of operatives are chargeable at cost plus 12.5%.

2 Materials

Add to the cost of materials – 12.5%

1 The percentage addition provides for Head office charges and Profit
2 The cost of materials means the invoiced price of materials including delivery to site without deduction of any cash discounts not exceeding 2.5%
3 Unloading of materials:

> The percentage added to the cost of materials excludes the cost of handling which shall be charged in addition. An allowance for unloading into site stock or storage including wastage should be added where materials are taken from existing stock.

3 Supplementary Charges

1 Transport provided by contractors for operatives to, from, in and around the site to be charged at the appropriate Schedule rates.
2 Any other charges incurred in respect of any Dayworks operation including, tipping charges, professional fees, subcontractor's accounts and the like shall be paid for in full plus 12.5% (without deduction of cash discounts not exceeding 2.5%). Labour subcontractors being dealt with in Section 1.
3 The cost of operating welfare facilities to be charged by the contractor at cost plus 12.5%.
4 The cost of additional insurance premiums for abnormal contract work or special site conditions to be charged at cost plus 12.5%.
5 The cost of watching and lighting specially necessitated by Dayworks is to be paid for separately at Schedule rates.

4 Plant

1 These rates apply only to plant already on site, exclusive of drivers and attendants, but inclusive of fuel and consumable stores unless stated to be charged in addition, repairs and maintenance, insurance of plant but excluding time spent on general servicing.
2 Where plant is hired specifically for Dayworks: plant hire (exclusive of drivers and attendants), fuel, oil and grease, insurance, transport etc., to be charged at full amount of invoice (without deduction of any cash discount not exceeding 2.5%) to which should be added consumables where supplied by the contractor, all plus 12.5%.
3 Fuel distribution, mobilization and demobilization are not included in the rates quoted which shall be an additional charge.
4 Metric capacities are adopted and these are not necessarily exact conversions from their imperial equivalents, but cater for the variations arising from comparison of plant manufacturing firms' ratings.
5 SAE rated capacities of plant means rated in accordance with the standards specified by the Society of Automotive Engineers.
6 Minimum hire charge will be for the period quoted.
7 Hire rates for plant not included below shall be settled at prices reasonably related to the rates quoted.
8 The rates provide for Head Office charges and Profit.

The Schedule then gives 21 pages of hire rates for a wide range of plant and equipment.

APPLICATION OF DAYWORKS

Generally

A check should be made on the accuracy of the recorded resources and times.

Tender documents generally allow the contractor to tender percentage variations to the figure calculated using the published percentage additions. These vary widely but a reasonable average indication can be along the lines of:

Labour   20% less
Materials   10% less
Plant   30% less

Labour

The Contractor should provide substantiation of the hourly rates he wishes to be paid for the various classes of labour and should demonstrate that the basic 'amount of wages' does not include any of the items actually covered by the percentage addition.

The wage bill is intended to reflect the cost to the Contractor. The time involved is not restricted to the duration of the task, but also includes mobilization and demobilization, together with any training needed – which would include induction courses required for Health & Safety requirements. The rate paid is the actual value of wages and bonuses paid (not simply the basic rate promulgated for the labour grade involved, and includes overtime rates if applicable, tool money, time lost due to inclement weather. In addition, daily travelling allowances are included, as are periodic travel allowances and also subsistence or lodging allowances.

Care should be taken that the matters deemed included in the percentage addition are not duplicated in the amount of wages. For example, it should be noted that foremen, gangers and other supervisory staff are covered by the percentage addition, unless they work in which case they are paid for at the correct rate for the task involved. The proportion of the time they spend working rather than supervising must be agreed. Refer to the amplification of labour categories in Part 4.

Hired or subcontracted labour is paid at invoiced cost, adjusted only where any cash discount exceeded 2½%, in which case the excess percentage is deducted.

Materials

The cost of materials delivered to the site is simply the invoiced price of the materials plus any delivery charges.

Should the cash discount exceed 2½%, the excess percentage is deducted from the amount to be paid.

The percentage addition simply covers the cost of Head Office charges and profit. It does not include for unloading or temporarily storing the materials nor for distributing them on site to the work place. Such cost can be charged, even in cases where the materials may already be in the site stock.

Material waste should be added direct to the cost of materials used for each particular Daywork item as an appropriate percentage.

Handling and offloading materials

Schedule 2.3 states that an allowance for handling materials, and an allowance for unloading or storage including wastage should be an additional charge.

Example

For a 12 month, £10.0 m Civil Engineering scheme where the total cost of materials that require handling (excluding Ready Mix concrete, imported fills, fuels and similar items) is £2,500,000.
The following gang is employed (part time) throughout the contract for offloading and handling of materials.

		Net cost (£)
2 labourers		27.18
Lorry (8 t) with Hiab lift		<u>23.94</u>
	Rate per hour (£)	51.12

Allow an average of 5 hours per week over 50 weeks

250 hours @ £51.12 / hour = £12,780

This cost as a percentage of the materials element of the contract.

(£12,780/£2,500,000) × 100 = 0.51%

Supplementary charges

1 Transport

Schedule 3.1 states that transport provided by contractors to take operatives to and from the site as well as in and around the site shall be charged at the appropriate Schedule Rate. This would entail the driver and vehicle being included with the labour and plant parts of the Daywork calculation.

2 Any other charges

This relates to any other charges incurred in respect of any Dayworks operation and includes tipping charges, professional fees, subcontractor's accounts and the like. Schedule 3.2 provides for full payment of such charges in full plus the addition of 12½% for Head Office charges and Profit.

Should the cash discount exceed 2½%, the excess percentage is deducted from the amount to be paid.

Welfare Facilities

Schedule 3.3 allows for the net cost of operating these facilities plus 12.5%.

Example

How the costs of operating welfare facilities may be charged to the Daywork account on a 12 month, £10.0 m Civil Engineering scheme, where the total labour element is £1,400,000.

Facility	Weekly Cost £
Toilet unit (4 nr)	160.00
Jack leg hutments 24' (2 nr)	120.00
Jack leg hutment 12' (1 nr)	42.00
Labour to clean, maintain and make tea, etc. = 1 man, 2 hours per day, 6 days × 24.88	149.28
Consumables (heat, light, soap, disinfectant, etc.) say	50.00
Rates, insurance, taxes, etc. (add 2%)	10.43
Total weekly cost	**531.71**

Multiply by 50 weeks (construction period) plus 6 weeks (maintenance period)

Total cost to contract = £ 531.71 × 56	29,775.76
Thus cost as a percentage of the Labour element of the contract	2.13%
= (£29,776/£1,400,000) × 100 = £1,400,000	
Add, as schedule 3.3 12.5%	0.27%
Percentage addition for facilities	2.40%

Insurances

Schedule 3.4 allows for the cost of additional insurance premiums for abnormal work or special site conditions to be charged at cost plus 12.5%.

Watching and lighting

Schedule 3.5 allows for all such costs necessitated by Dayworks to be paid for separately at Schedule rates.

Plant

The cost of the driver(s) and any required attendants such as banksmen should be covered in the labour section of the dayworks calculation.

The Schedule rates include fuel and consumable stores, repairs and maintenance (but not the time spent on general servicing) and insurance.

The Schedule rates only apply to machinery which is on site at the time of the work – where plant is specifically hired for the task, the Contractor is entitled to be paid the invoiced value. If the invoice excludes consumables used (fuel, oil and grease) then the Contractor is entitled to add the cost – together with insurance and any transport costs incurred in getting the equipment on site all subject to a 12½% addition for Head Office costs and profit.

Head Office charges and Profit allowances are included in the Schedule rates, 12½% being added to the charged value of hired plant.

1   **General servicing of plant**

Schedule 4.1 specifically excludes time spent on general servicing from the Hire Rates.

General servicing in this context can be assumed to mean:

Checking, replenishing (or changing, if applicable). i.e.:

- engine lubrication
- transmission lubrications
- general greasing
- coolants
- hydraulic oils and brake systems
- filters
- tyres

Inspecting special items, e.g.:

- buckets
- hoses/airlines
- shank protectors
- Cables/ropes/hawsers
- rippers
- blades, steels, etc.

Example

Assuming an 8 hour working day these operations could take a plant operator on average:
large machine   20 minutes per day   (equating to 1 hour for each 24 worked)
medium machine   10 minutes per day   (equating to 1 hour for each 48 worked)
small machine   5 minutes per day   (equating to 1 hour for each 96 worked)

The cost of the servicing labour would be as follows:
large machine   £26.54/hr / 24hrs   = £1.11
medium machine   £17.58/hr / 48 hrs   = £0.37
small machine   £14.68/hr / 96 hrs   = £0.15

These labour costs can be expressed as a percentage of the schedule hire rates:
D8 Dozer   £1.11 / £121.54/hr × 100 = 0.9%
JCB 3CX   £0.37 / £18.16/hr × 100   = 2.0%
2 tonne dumper   £0.15 / £6.62/hr × 100   = 2.3%

Taking into account the range of these sizes of plant which are normally deployed on site, the following would provide a reasonable average percentage addition:
                      0.9% × 2   = 1.8%
                      2.0% × 3   = 6.0%
                      2.3% × 6   = 13.8%
Total = 21.6% for 11 items of plant   i.e. average = 2.0%

2 Fuel distribution

Schedule 4.3 allows for charging for fuel distribution.

Assuming this is done with a towed fuel bowser behind a farm type tractor with driver/labourer in attendance, the operation cycle would involve visiting, service, and return or continue on to the next machine.

The attendance cost based on the hourly rate tractor/bowser/driver would be: = £15.96 + £1.38 + £13.59 = £30.93

Example

A heavy item of plant, for example a Cat D8R Tractor Bulldozer with a 212 kW engine
    Fuel consumption is 30.4 litres/hr (38 litres/hr × 80% site utilization factor)
    Fuel capacity is 200 litres
    Requires filling after 6.5 working hours operation (200 litres divided by 30.40 l/hr)
    Tractor/bowser service taking 30 minutes

The machine cost during this 6.5 hr period would be £121.54 × 6.5, i.e. £ 790.01
The attendance cost for the 30 min cycle would be £30.93 × 0.5, i.e. £15.47
The percentage addition for fuelling the machine would be: (£15.47/£790.01) × 100 = 2.0%

Example

A medium sized item of plant, for example a JCB 3CX
    Fuel consumption is 5.6 litres/hr (7.5 litres/hr × 75% site utilization factor)
    Fuel capacity is 90 litres
    Requires filling after 16 working hours operation (90 litres divided by 5.60 l/hr)
    Tractor/bowser service taking 10 minutes

The machine cost during this 16 hr period would be £18.16 × 16 hrs, i.e. £290.56
The attendance cost for the 10 min cycle would be £30.93 × 0.17 hrs, i.e. £5.26
The percentage addition for fuelling the machine would be: (£5.26/£290.56) × 100 = 1.8%

Example

A medium sized item of plant, for example a 2 tonne dumper
    Fuel consumption is 2.4 litres/hr (3 litres/hr × 80% site utilization factor)
    Fuel capacity is 35 litres
    Requires filling after 14.5 working hours operation (35 litres divided by 2.40 l/hr)
    Tractor/bowser service taking 5 minutes

The machine cost during this 14.5 hr period would be £6.62 × 14.5 hrs, i.e. £95.99
The attendance cost for the 5 min cycle would be £30.93 × 0.08 hrs, i.e. £2.47
The percentage addition for fuelling the machine would be: (£2.47/£95.99) × 100 = 2.6%

Considering the range of these categories of plant which are normally deployed on site, the following would provide a reasonable average percentage addition for the above:

    2.0% × 2   = 4.0%
    1.8% × 3   = 5.4%
    2.6% × 6   = 15.6%

    Total   = 25.0% for 11 items of plant – average = 2.3%

- **Mobilization**

Schedule 4.3 allows for charging for mobilization and demobilization.

PART 9

Professional Fees

This part of the book contains the following sections:

	page
Consulting engineers' fees	513
Quantity surveyors' fees	518

ESSENTIAL READING FROM TAYLOR AND FRANCIS

Self-Consolidating Concrete
Applying What we Know

Daczko

Self-consolidating, or self-compacting, concrete (SCC) uses mixture proportioning techniques and admixture technology to create a concrete mixture with increased fluidity, thereby avoiding the need for mechanical consolidation/vibration. SCC mixtures with various performance levels can be produced in the laboratory, but the know-how is sometimes lacking to do this on site. This book presents a foundation for consistently producing and applying SCC in a daily production environment. It links science with practical application and will be useful for both new practitioners and those already using SCC. The author uses stories and case studies throughout as examples of specific types of applications.

February 2012: 234x156: 304pp
Hb: 978-0-415-59064-8: **£50.00**

To Order: Tel: +44 (0) 1235 400524 **Fax:** +44 (0) 1235 400525
or Post: Taylor and Francis Customer Services,
Bookpoint Ltd, Unit T1, 200 Milton Park, Abingdon, Oxon, OX14 4TA UK
Email: book.orders@tandf.co.uk

For a complete listing of all our titles visit:
www.tandf.co.uk

Taylor & Francis
Taylor & Francis Group

CONSULTING ENGINEERS' CONDITIONS OF APPOINTMENT

Introduction

A scale of professional charges for consulting engineering services in connection with civil engineering works is published by the Association for Consultancy and Engineering (ACE)

Copies of the document can be obtained direct from:

>Association for Consultancy and Engineering
>Alliance House
>12 Caxton Street
>London SW1H OQL
>Tel 020 7222 6557
>Fax 020 7222 0750

Comparisons

Instead of the previous arrangement of having different agreements designed for each major discipline of engineering, the current agreements have been developed primarily to suit the different roles that Consulting Engineers may be required to perform, with variants of some of them for different disciplines. The agreements have been standardized as far as possible whilst retaining essential differences.

Greater attention is required than with previous agreements to ensure the documents are completed properly. This is because of the perceived need to allow for a wider choice of arrangements, particularly of methods of payment.

The agreements are not intended to be used as unsigned reference material with the details of an engagement being covered in an exchange of letters, although much of their content could be used as a basis for drafting such correspondence.

In 2009 the ACE published a new suite of Agreements with a broader set of services, these are listed below.

Forms of Agreement

>ACE Agreement 1: Design
>
>ACE Agreement 2: Advise and Report
>
>ACE Agreement 3: Design and Construct
>
>ACE Agreement 4: Sub-Consultancy
>
>ACE Agreement 5: Homeowner
>
>ACE Agreement 6: Expert Witness (Sole Practitioner)
>
>ACE Agreement 7: Expert Witness (Firm)
>
>ACE Agreement 8: Adjudicator

CONSULTING ENGINEERS' CONDITIONS OF APPOINTMENT

To a number of the the above ACE Agreements Schedules of Services are appended and currently these are:

For use with ACE Agreement 1: Design

> ACE Schedule of Services – Part G(a):
> Civil and Structural Engineer – single consultant or non-lead consultant
>
> ACE Schedule of Services – Part G(b):
> Mechanical and Electrical Engineering (detailed design in buildings)
>
> ACE Schedule of Services – Part G(c):
> Mechanical and Electrical Engineering (performance design in buildings)
>
> ACE Schedule of Services – Part G(d):
> Civil and Structural Engineer – Lead consultant
>
> ACE Schedule of Services – Part G(e):
> Mechanical and Electrical Engineering Design in buildings – Lead consultant

For use with ACE Agreement 3: Design and Construct

> ACE Schedule of Services – Part G(f):
> Civil and Structural Engineer
>
> > ACE Schedule of Services – Part G(g):
> > Mechanical and Electrical Engineering (detailed design in buildings)
> >
> > ACE Schedule of Services – Part G(h):
> > Mechanical and Electrical Engineering (performance design in buildings)

ACE Agreement 1: Design
Design for the appointment of a consultant by a client to undertake detailed design and/or specification of permanent works to be undertaken or installed by a contractor including any studies, appraisals, investigations, contract administration or construction monitoring leading to or resulting from such detailed design and/or specification.

ACE Agreement 2: Advise and Report
Advise and report for the appointment of a consultant by a client to provide any type of advisory, research, checking, reviewing, investigatory, monitoring, reporting or technical services in the built and natural environments where such services do not consist of detailed design or specification of permanent works to be constructed or installed by a contractor.

ACE Agreement 3: Design and Construct
For the appointment of a consultant by a contractor in circumstances where the contractor is to construct permanent works designed by the consultant.

ACE Agreement 4: Sub-consultancy
For the appointment of a sub-consultant by a consultant in circumstances where the consultant is appointed on the terms of an ACE Agreement by its client.

ACE Agreement 5: Homeowner
Model letter for the appointment of a consultant by a homeowner.

ACE Agreement 6: Expert Witness (Sole Practitioner)
For the appointment of an individual to act as an expert witness.

CONSULTING ENGINEERS' CONDITIONS OF APPOINTMENT

ACE Agreement 7: Expert Witness (Firm)
For the appointment of a firm to provide an expert witness.

ACE Agreement 8: Adjudicator
For the appointment of an adjudicator.

Collateral Warranties

The association is convinced that collateral warranties are generally unnecessary and should only be used in exceptional circumstances. The interests of clients, employers and others are better protected by taking out project or BUILD type latent defects insurance. Nevertheless, in response to observations raised when the pilot editions excluded any mention of warranties, references and arrangements have been included in the Memorandum and elsewhere by which Consulting Engineers may agree to enter into collateral warranty agreements; these should however only be given when the format and requirements thereof have been properly defined and recorded in advance of undertaking the commission.

Requirements for the provision of collateral warranties will be justified even less with commissions under Agreement D than with those under the other ACE agreements. Occasional calls may be made for them, such as when a client intends to dispose of property and needs evidence of a duty of care being owed to specific third parties, but these will be few and far between.

Remuneration

Guidance on appropriate levels of fees to be charged is given at the end of each agreement. Firms and their clients may use this or other sources, including their own records, to determine suitable fee arrangements.

Need for formal documentation

The Association for Consultancy and Engineering recommends that formal written documentation should be executed to record the details of each commission awarded to a Consulting Engineer. These Conditions are published as model forms of agreement suitable for the purpose. However, even if these particular Conditions are not used, it is strongly recommended that, whenever a Consulting Engineer is appointed, there should be at least an exchange of letters defining the duties to be performed and the terms of payment.

Appointments outside the United Kingdom

These conditions of Engagement are designed for use within the UK. For work overseas it is impracticable to give definite recommendations; circumstances differ too widely between countries. There are added complications in documentation relating to local legislation, import customs, conditions of payment, insurance, freight, etc. Furthermore, it is often necessary to arrange for visits to be made by principals and senior staff whose absence abroad during such periods represents a serious reduction of their earning power. The additional duties, responsibilities and non-recoverable costs involved, and the extra work on general co-ordination, should be reflected in the levels of fees. Special arrangements are also necessary to cover travelling and other out-of-pocket expenses in excess of those normally incurred on similar work in the UK, including such matters as local cost-of-living allowances and the cost of providing home-leave facilities for expatriate staff.

CONDITIONS OF ENGAGEMENT

Obligations of the Consulting Engineer

The responsibilities of the Consultant Engineer for the works are as set out in the actual agreement The various standard clauses in the Conditions relate to such matters as differentiating between Normal and Additional services, the duty to exercise skill and care, the need for Client's written consent to the assignment or transfer of any benefit or obligation of the agreement, the rendering of advice if requested on the appointment of other consultants and specialist sub-consultants, any recommendations for design of any part of the Works by Contractors or

CONSULTING ENGINEERS' CONDITIONS OF APPOINTMENT

Subcontractors (with the proviso that the Consulting Engineer is not responsible for detailed design of contractors or for defects or omissions in such design), the designation of a Project Leader, the need for timeliness in requests to the Client for information etc., freezing the design once it has been given Client approval and the specific exclusion of any duty to advise on the actual or possible presence of pollution or contamination or its consequences.

Obligations of the Client

The Consultant Engineer shall be supplied with all necessary data and information in good time. The Client shall designate a Representative authorized to make decisions on his behalf and ensure that all decisions, instructions, and approvals are given in time so as not to delay or disrupt the Consultant Engineer.

Site Staff

The Consulting Engineer may employ site staff he feels are required to perform the task, subject to the prior written agreement of the Client. The Client shall bear the cost of local office accommodation, equipment and running costs.

Commencement, Determination, Postponement, Disruption and Delay

The Consulting Engineer's appointment commences at the date of the execution of the Memorandum of Agreement or such earlier date when the Consulting Engineer first commenced the performance of the Services, subject to the right of the Client to determine or postpone all or any of the Services at any time by Notice.

The Client or the Consulting Engineer may determine the appointment in the event of a breach of the Agreement by the other party after two weeks notice. In addition, the Consulting Engineer may determine his appointment after two weeks notice in the event of the Client failing to make proper payment.

The Consulting Engineer may suspend the performance of all or any of the Services for up to twenty-six weeks if he is prevented or significantly impeded from performance by circumstances outside his control. The appointment may be determined by either party in the event of insolvency subject to the issue of notice of determination.

Payments

The Client shall pay fees for the performance of the agreed service(s) together with all fees and charges to the local or other authorities for seeking and obtaining statutory permissions, for all site staff on a time basis, together with additional payments for any variation or the disruption of the Consulting Engineer's work due to the Client varying the task list or brief or to delay caused by the Client, others or unforeseeable events.

If any part of any invoice submitted by the Consulting Engineer is contested, payment shall be made in full of all that is not contested.

Payments shall be made within 28 days of the date of the Consulting Engineer's invoice; interest shall be added to all amounts remaining unpaid thereafter.

Ownership of Documents and Copyright

The Consulting Engineer retains the copyright in all drawings, reports, specifications, calculations etc. prepared in connection with the Task; with the agreement of the Consulting Engineer and subject to certain conditions, the Client may have a licence to copy and use such intellectual property solely for his own purpose on the Task in hand, subject to reservations.

The Consulting Engineer must obtain the client's permission before he publishes any articles, photographs or other illustrations relating to the Task, nor shall he disclose to any person any information provided by the Client as private and confidential unless so authorized by the Client.

CONSULTING ENGINEERS' CONDITIONS OF APPOINTMENT

Liability, Insurance and Warranties

The liability of the Consulting Engineer is defined, together with the duty of the Client to indemnify the Consulting Engineer against all claims etc. in excess of the agreed liability limit.

The Consulting Engineer shall maintain Professional Indemnity Insurance for an agreed amount and period at commercially reasonable rates, together with Public Liability Insurance and shall produce the brokers' certificates for inspection to show that the required cover is being maintained as and when requested by the Client.

The Consulting Engineer shall enter into and provide collateral warranties for the benefit of other parties if so agreed.

Disputes and Differences

Provision is made for mediation to solve disputes, subject to a time limit of six weeks of the appointment of the mediator at which point it should be referred to an independent adjudicator. Further action could be by referring the dispute to an arbitrator.

QUANTITY SURVEYORS' CONDITIONS OF APPOINTMENT

QUANTITY SURVEYORS' CONDITIONS OF APPOINTMENT

Introduction

Authors' Note:
The Royal Institution of Chartered Surveyors formally abolished standard Quantity Surveyor's fee scales with effect from 31st December 1998. However, in the absence of any alternative guidance and for the benefit of readers, extracts from relevant fee scales have been reproduced in part with the permission of the Royal Institution of Chartered Surveyors, which owns the copyright.

Summary of Scale of Professional Charges

Scale No 38. issued by The Royal Institution of Chartered Surveyors provides an itemized scale of professional charges for Quantity Surveying Services for Civil Engineering Works which is summarized as follows:

1.1 Generally
1.2 The Scale of professional charges is applicable where the contract provides for the bills of quantities and final account to be based on measurements prepared in accordance with or based on the principles of the Standard Method of Measurement of Civil Engineering Quantities issued by the Institution of Civil Engineers.
1.3 The fees are in all cases exclusive of travelling and other expenses (for which the actual disbursement is recoverable unless there is some prior arrangement for such charges) and of the cost of reproduction of bills of quantities and other documents, which are chargeable in addition at net cost.
1.4 The fees are in all cases exclusive of services in connection with the allocation of the cost of the works for purposes of calculating value added tax for which there shall be an additional fee based on the time involved.
1.5 If any of the materials used in the works are supplied by the employer or charged at a preferential rate, then the actual or estimated market value thereof shall be included in the amounts upon which fees are to be calculated.
1.6 The fees are in all cases exclusive of preparing a specification of the materials to be used and the works to be done.
1.7 If the quantity surveyor incurs additional costs due to exceptional delays in construction operations or any other cause beyond the control of the quantity surveyor then the fees may be adjusted by agreement between the employer and the quantity surveyor.
1.8 If the works are substantially varied at any stage or if the quantity surveyor is involved in abortive work there shall be an additional fee based on the time involved.
1.9 The fees and charges are in all cases exclusive of value added tax which will be applied in accordance with legislation.
1.10 The scale is not intended to apply to works of a civil engineering nature which form a subsidiary part of a building contract or to buildings which are ancillary to a civil engineering contract. In these cases the fees to be charged for quantity surveying services shall be in accordance with the scales applicable to building works.
1.11 When works of both categories I* and II** are included in one contract the fee to be charged shall be calculated by taking the total value of the sections of work in each of the categories and applying the appropriate scale from the beginning in each case. General items such as preliminaries (and in the case of post contract fees contract price fluctuations) or sections of works which cannot be specifically allocated to either category shall be apportioned pro-rata to the values of the other sections of the works and added thereto in order to ascertain the total value of works in each category.
1.12 When a project is the subject of a number of contracts then, for the purpose of calculating fees, the value of such contracts shall not be aggregated but each contract shall be taken separately and the scale of charges applied as appropriate.
1.13 Roads, railways, earthworks and dredging which are ancillary only to any Category II** work shall be regarded as Category II** work. Works or sections of works of Category I* which incorporate piled construction shall be regarded as Category II** works.

QUANTITY SURVEYORS' CONDITIONS OF APPOINTMENT

1.14 No addition to the fees given hereunder shall be made in respect of works of alteration or repair where such works are incidental to the new works. If the work covered by a single contract is mainly one of alteration or repair then an additional fee shall be negotiated.

1.15 In the absence of agreement to the contrary, payments to the quantity surveyor shall be made by instalments by arrangement between the employer and the quantity surveyor.

1.16 Copyright in bills of quantities and other documents prepared by the quantity surveyor is reserved to the quantity surveyor.

* Category I   Works or sections of works such as monolithic walls for quays, jetties dams and reservoirs; caissons; tunnels; airport runways and tracks roads; railways; and earthworks and dredging

** Category II   Works or sections of works such as piled quay walls; suspended jetties and quays; bridges and their abutments; culverts; sewers; pipe-lines; electric mains; storage and treatment tanks; water cooling towers and structures for housing heavy industrial and public utility furnace houses and rolling mills to steel works; and boiler houses, plant, e.g. reactor blocks and turbine halls to electricity generating stations.

Scale of charges

For the full Scale of Fees for Professional Charges for Quantity Surveying Services together with a detailed description of the full service provided the appropriate RICS Fee Scale should be consulted.

ESSENTIAL READING FROM TAYLOR AND FRANCIS

Spon's Asia-Pacific Construction Costs Handbook

4th Edition

Edited by **Davis Langdon & Seah**

Spon's Asia Pacific Construction Costs Handbook includes construction cost data for 20 countries. This new edition has been extended to include Pakistan and Cambodia. Australia, UK and America are also included, to facilitate comparison with construction costs elsewhere. Information is presented for each country in the same way, as follows:

- key data on the main economic and construction indicators.

- an outline of the national construction industry, covering structure, tendering and contract procedures, materials cost data, regulations and standards

- labour and materials cost data

- Measured rates for a range of standard construction work items

- Approximate estimating costs per unit area for a range of building types

- price index data and exchange rate movements against £ sterling, $US and Japanese Yen

The book also includes a Comparative Data section to facilitate country-to-country comparisons. Figures from the national sections are grouped in tables according to national indicators, construction output, input costs and costs per square metre for factories, offices, warehouses, hospitals, schools, theatres, sports halls, hotels and housing.

This unique handbook will be an essential reference for all construction professionals involved in work outside their own country and for all developers or multinational companies assessing comparative development costs.

April 2010: 234x156: 480pp
Hb: 978-0-415-46565-6: **£120.00**

To Order: Tel: +44 (0) 1235 400524 **Fax:** +44 (0) 1235 400525
or Post: Taylor and Francis Customer Services,
Bookpoint Ltd, Unit T1, 200 Milton Park, Abingdon, Oxon, OX14 4TA UK
Email: book.orders@tandf.co.uk

For a complete listing of all our titles visit:
www.tandf.co.uk

Taylor & Francis
Taylor & Francis Group

PART 10

Outputs

This part lists a selection of OUTPUT CONSTANTS for use within various areas of Civil Engineering Work

This part of the book contains the following sections:

	page
Disposal of excavated materials	523
Breaking out obstructions by hand	524
Concrete work	525
Drainage	528

ESSENTIAL READING FROM TAYLOR AND FRANCIS

Manual for Detailing Reinforced Concrete Structures to EC2

J. Calavera

Detailing 210 structural elements, each with a double page spread, this book is an indispensible guide to designing for reinforced concrete. Drawing on Jose Calavera's many years of experience, each entry provides commentary and recommendations for the elementconcerned, as 0077ell as summarising the appropriate Eurocode legislation and referring to further standards and literature.

Includes a CD ROM with AutoCad files of all of the models, which you can adapt to suit your own purposes.

Section 1. General Rules for Bending, Placing and Lapping of Reinforcement
Section 2. Constructive Details

1. Foundations 2. Walls 3. Columns and Joints 4. Walls Subjected to Axial Loads. Shear Walls and Cores 5. Beams and Lintels 6. Slabs 7. Flat Slabs 8. Stairs 9. Supports 10. Brackets 11. Pavements and Galleries 12. Cylindrical Chimneys, Towers and Hollow Piles 13. Rectangular Silos, Caissons and Hollow Piles 14. Reservoirs, Tanks and Swimming Pools 15. Special Details for Seismic Areas

October 2011: 297 x210: 528pp
Hb: 978-0-415-66348-9: **£80.00**

To Order: Tel: +44 (0) 1235 400524 **Fax:** +44 (0) 1235 400525
or Post: Taylor and Francis Customer Services,
Bookpoint Ltd, Unit T1, 200 Milton Park, Abingdon, Oxon, OX14 4TA UK
Email: book.orders@tandf.co.uk

For a complete listing of all our titles visit:
www.tandf.co.uk

Taylor & Francis
Taylor & Francis Group

DISPOSAL OF EXCAVATED MATERIALS

Outputs Per Hundred Cubic Metres

Tipper capacity and length of haul	Driver (hours)	Attendant labour (hours)	Number of cycles	Average speed (km/hr)	Loading	Haul	Discharge	Return	Total
3 m³ tipper:									
1 km haul	8.8	0.0	43.3	10	0.70	6.0	0.7	5.5	12.2
5 km haul	26.3	0.0	43.3	16	0.70	18.8	0.7	17.0	36.5
9 km haul	41.9	0.0	43.3	18	0.70	30.0	0.7	27.3	58.0
12 km haul	52.7	0.0	43.3	19	0.70	37.9	0.7	34.4	73.0
15 km haul	65.8	0.0	43.3	19	0.70	47.4	0.7	43.1	91.2
8 m³ tipper:									
1 km haul	10.6	0.8	16.3	3	2.0	20.0	0.9	18.2	39.1
5 km haul	19.7	0.8	16.3	8	2.0	37.5	0.9	34.1	72.5
9 km haul	25.7	0.8	16.3	11	2.0	49.1	0.9	44.6	94.6
12 km haul	26.9	0.8	16.3	14	2.0	51.4	0.9	46.8	99.1
15 km haul	31.4	0.8	16.3	15	2.0	60.0	0.9	54.5	115.4
12 m³ tipper:									
1 km haul	7.1	0.7	10.8	3	2.90	20.0	1.1	18.2	39.3
5 km haul	13.1	0.7	10.8	8	2.90	37.5	1.1	34.1	72.7
9 km haul	17.1	0.7	10.8	11	2.90	49.1	1.1	44.6	94.8
12 km haul	17.9	0.7	10.8	14	2.90	51.4	1.1	46.8	99.3
15 km haul	20.8	0.7	10.8	15	2.90	60.0	1.1	54.5	115.6
15 m³ tipper:									
1 km haul	5.8	0.8	8.7	3	3.70	20.0	1.5	18.2	39.7
5 km haul	10.6	0.8	8.7	8	3.70	37.5	1.5	34.1	73.1
9 km haul	13.8	0.8	8.7	11	3.70	49.1	1.5	44.6	95.2
12 km haul	14.5	0.8	8.7	14	3.70	51.4	1.5	46.8	99.7
15 km haul	16.8	0.8	8.7	15	3.70	60.0	1.5	54.5	116.0

Man hours include round trip for tipper and driver together with attendant labour for positioning during loading and unloading.

The number of cycles are based on the stated heaped capacity of the tipper divided into the total volume of 100 m³ to be moved, being multiplied by a bulking factor of × 1.30 to the loose soil volume.

The average speeds are calculated assuming that the vehicles run on roads or on reasonably level firm surfaces and allow for acceleration and deceleration with the return empty journey being say 10% faster than the haul.

The cycle time shows in detail the time spent being loaded (calculated using a 1.5 m³ loader with a cycle time of 22 seconds), haul journey, discharge (turning/manoeuvring/tipping) and return journey.

BREAKING OUT OBSTRUCTIONS BY HAND

Breaking out pavements, brickwork, concrete and masonry by hand and pneumatic breaker

		By hand using picks, shovels & points	Using compressor				
			7 m³ Compressor (2 Tool)		10 m³ Compressor (3 Tool)	
Description	Unit	Labour	Compressor	Labour	Compressor	Labour
Break out bitmac surfaces on subbase or hardcore						
75 mm thick	m²/hr	1.00	25	13	50	17
100 mm thick	m²/hr	1.00	20	10	33	11
Break out asphalt roads on hardcore:						
150 mm thick	m²/hr	0.60	8	4	11	4
225 mm thick	m²/hr	0.50	6	3	8	3
300 mm thick	m²/hr	0.40	4	2	7	2
Remove existing set paving	m²/hr	0.80	10	5	17	6
Break out brickwork in cement mortar: 215 mm thick	m²/hr	0.20	3	2	5	2
Break out concrete in areas						
100 mm thick	m²/hr	0.90	9	5	14	5
150 mm thick	m²/hr	0.50	7	4	10	4
225 mm thick	m²/hr	0.30	4	2	6	2
300 mm thick	m²/hr	0.20	3	1	4	1
Break out reinforced concrete	m³/hr	0.02	0.40	0.20	0.60	0.20
Break out sandstone	m³/hr	0.03	0.50	0.30	0.80	0.30

Loading loose materials and items by hand

		Loading into vehicles	
Material	Unit	Tonne	m³
Bricks	hr	1.7	2.9
Concrete, batches	hr	1.4	1.2
Gulley grates and frames	hr	1.0	
Kerb	hr	1.1	
Paving slabs	hr	0.9	
Pipes, concrete and clayware	hr	0.9	
Precast concrete items	hr	0.9	
Soil	hr	1.4	2.0
Steel reinforcement	hr	0.8	
Steel sections, etc.	hr	1.0	
Stone and aggregates:			
bedding material	hr	1.5	1.8
filter/subbase	hr	1.3	1.4
rock fill (6" down)	hr	1.3	1.3
Trench planking and shoring	hr	0.9	2.9

CONCRETE WORK

Placing ready mixed concrete in the works

Description	Labour gang (m³ per hour)
MASS CONCRETE	
Blinding	
150 mm thick	5.50
150–300 mm thick	6.25
300–500 mm thick	7.00
Bases and oversite concrete	
not exceeding 150 mm thick	5.00
not exceeding 300 mm thick	5.75
not exceeding 500 mm thick	6.75
exceeding 500 mm thick	7.00
REINFORCED CONCRETE	
Bases	
not exceeding 300 mm thick	5.50
not exceeding 500 mm thick	6.25
exceeding 500 mm thick	6.75
Suspended slabs (not exceeding 3 m above pavement level)	
not exceeding 150 mm thick	3.75
not exceeding 300 mm thick	4.75
exceeding 300 mm thick	5.75
Walls and stems (not exceeding 3 m above pavement)	
not exceeding 150 mm thick	3.50
not exceeding 300 mm thick	4.50
exceeding 300 mm thick	5.00
Beams, columns and piers (not exceeding 3 m above pavement)	
sectional area not exceeding 0.03 m²	2.00
sectional area not exceeding 0.03–1.0 m²	2.50
sectional area exceeding 1.0 m²	3.50

Fixing bar reinforcement

All bars delivered to site cut and bent and marked, including craneage and hoisting (maximum height 5 m)

Description (fix only)	Unit	Up to 6 mm Steelfixer	Up to 6 mm Labourer	7 to 12 mm Steelfixer	7 to 12 mm Labourer	13 to 19 mm Steelfixer	13 to 19 mm Labourer	Over 19 mm Steelfixer	Over 19 mm Labourer
Straight round bars to beams, floors, roofs and walls	t/hr	0.03	0.03	0.04	0.04	0.06	0.06	0.08	0.08
to braces, columns, sloping roofs and battered walls	t/hr	0.01	0.01	0.02	0.02	0.03	0.03	0.05	0.05
Bent round bars to beams, floors, roofs and walls	t/hr	0.02	0.02	0.03	0.03	0.03	0.03	0.04	0.04
to braces, columns, sloping roofs and battered walls	t/hr	0.01	0.01	0.01	0.01	0.02	0.02	0.03	0.03
Straight, indented or square bars to beams, floors, roofs and walls	t/hr	0.02	0.02	0.04	0.04	0.05	0.05	0.07	0.07
to braces, columns, sloping roofs and battered walls	t/hr	0.01	0.01	0.02	0.02	0.02	0.02	0.04	0.04
Bent, indented or square bars to beams, floors, roofs and walls	t/hr	0.02	0.02	0.02	0.02	0.03	0.03	0.03	0.03
to braces, columns, sloping roofs and battered walls	t/hr	0.01	0.01	0.01	0.01	0.02	0.02	0.02	0.02
(Average based on Gang D)		0.125		0.148		0.163		0.225	

Erecting formwork to beams and walls

Erect and strike formwork	Unit	Joiner	Labourer
Walls – vertical face (first fix)			
up to 1.5 m	m²/hr	1.7	0.8
1.5 to 3.0 m	m²/hr	1.4	0.7
3.0 to 4.5 m	m²/hr	1.2	0.6
4.5 to 6.0 m	m²/hr	1.0	0.5

Erecting formwork to slabs

Erect and strike formwork	Unit	Joiner	Labourer
Horizontal flat formwork at heights (first fix)			
up to 3.0 m	m²/hr	1.11	1.11
3.0 to 3.6 m	m²/hr	1.05	1.05
3.6 to 4.2 m	m²/hr	1.00	1.00
4.2 to 4.8 m	m²/hr	0.95	0.95
4.8 to 5.4 m	m²/hr	0.90	0.90
5.4 to 6.0 m	m²/hr	0.83	0.83

Multipliers for formwork

Description	Multiplier
Walls built to batter	1.20
Walls built circular to large radius	1.70
Walls built circular to small radius	2.10
Formwork used once	1.00
Formwork used twice, per use	0.85
Formwork used three times, per use	0.75
Formwork used four times, per use	0.72
Formwork used five times, per use	0.68
Formwork used six times, per use	0.66
Formwork used seven or more times, per use	0.63
Formwork to slope not exceeding 45° from horizontal	1.25

DRAINAGE

Laying and jointing flexible-jointed clayware pipes

Diameter of pipe in mm	Drainage gang	In trench not exceeding 1.5 m	In trench not exceeding 3 m	In trench 3–5 m
Pipework				
100 mm	m/hr	10	8	7
150 mm	m/hr	7	6	5
225 mm	m/hr	5	4	3
300 mm	m/hr	4	3	2
375 mm	m/hr	3	2	2
450 mm	m/hr	2	2	1
Bends				
100 mm	nr/hr	20	17	14
150 mm	nr/hr	17	14	13
225 mm	nr/hr	13	10	8
300 mm	nr/hr	10	8	7
375 mm	nr/hr	7	6	5
450 mm	nr/hr	3	3	2
Single junctions				
100 mm	nr/hr	13	10	8
150 mm	nr/hr	7	6	5
225 mm	nr/hr	6	5	4
300 mm	nr/hr	4	3	3
375 mm	nr/hr	3	2	2
450 mm	nr/hr	2	2	1

Precast concrete manholes in sections

Description	Unit	Pipelayer	Labourer
Place 675 mm dia. shaft rings	m/hr	1.00	0.30
Place 900 mm manhole rings	m/hr	0.60	0.20
Place 1200 mm manhole rings	m/hr	0.40	0.10
Place 1500 mm manhole rings	m/hr	0.30	0.10
Place 900/675 mm tapers	nr/hr	1.00	0.30
Place 1200/675 mm tapers	nr/hr	0.60	0.20
Place 1500/675 mm tapers	nr/hr	0.50	0.20
Place cover slabs to 675 mm rings	nr/hr	2.50	0.80
Place cover slabs to 900 mm rings	nr/hr	2.00	0.70
Place cover slabs to 1200 mm rings	nr/hr	1.40	0.50
Place cover slabs to 1500 mm rings	nr/hr	0.90	0.30
Build in pipes and make good base:			
150 mm diameter	nr/hr	4.00	–
300 mm diameter	nr/hr	2.00	–
450 mm diameter	nr/hr	1.00	–
Benching 150 mm thick	m²/hr	1.20	1.20
Benching 300 mm thick	m²/hr	0.60	0.60
Render benching 25 mm thick	m²/hr	1.10	1.10
Fix manhole covers frame	nr/hr	1.30	1.30

Part 11

Useful Addresses for Further Information

ACOUSTICAL INVESTIGATION AND
RESEARCH ORGANISATION LTD
Duxon's Turn, Maylands Avenue, Hemel Hempstead
Herts HP2 4SB
Tel: 01442 247 146
Fax: 01442 256 749
Website: www.airo.co.uk

AGGREGATE CONCRETE BLOCK ASSOCIATION
See BRITISH PRECAST CONCRETE FEDERATION

ALUMINIUM FEDERATION LTD
National Metalforming Centre
47 Birmingham Road, West Bromwich
West Midlands B70 6PY
Tel: 0121 601 6363
Fax: 0870 138 9714
Website: www.alfed.org.uk

AMERICAN HARDWOOD EXPORT COUNCIL
3 St Michael's Alley
London EC3V 9DS
Tel: 0207 626 4111
Fax: 0207 626 4222
Website: www.ahec-europe.org

ARBORICULTURAL ADVISORY &
INFORMATION SERVICE
Alice Holt Lodge, Wrecclesham, Farnham
Surrey GU10 4LH
Tel: 090 651 61147

ARBORICULTURAL ASSOCIATION
The Moult House, Stround Green, Standish Stonehouse
Gloustershire GL10 3DL
Tel: 01242 522152
Fax: 01242 577766
Website: www.trees.org.uk

ARCHITECTURAL ASSOCIATION SCHOOL OF
ARCHITECTURE
34–36 Bedford Square
London WC1B 3ES
Tel: 020 7887 4000
Fax: 020 7414 0782
Website: www.aaschool.ac.uk

ASBESTOS INFORMATION CENTRE LTD
5a The Maltings, Station Road East, Stowmarket
Suffolk IP14 1RQ
Tel: 01449 676900
Fax: 01449 770028
Website: www.aic.org.uk

ASBESTOS REMOVAL CONTRACTORS
ASSOCIATION
ARCA House, 237 Branston Road,
Burton upon Trent
Staffordshire DE14 3BT
Tel: 01283 513 126
Fax: 01283 568 228
Website: www.arca.org.uk

ASSOCIATION FOR CONSULTANCY AND
ENGINEERING
Alliance House, 12 Caxton Street
London SW1H 0QL
Tel: 0207 222 6557
Fax: 0207 222 0750
Website: www.acenet.co.uk

ASSOCIATION OF LOADING AND ELEVATING
EQUIPMENT MANUFACTURERS
Airport House, Purley Way, Croydon
Surrey CR0 0XY
Tel: 020 8253 4501
Fax: 020 8253 4510
Website: www.alem.org.uk

ASSOCIATION OF PROJECT MANAGEMENT
IBIS House, Summerleys Road, Princes Risborough,
Bucks HP27 9LE
Tel: 0845 458 1944
Fax: 01494 528 937
Website: www.apm.org.uk

BRITISH AGGREGATE CONSTRUCTION
MATERIALS INDUSTRIES LTD
See QUARRY PRODUCTS ASSOCIATION

BRITISH AIRPORTS AUTHORITY PLC
The Compass Centre, Nelson Road,
London Heathrow Airport
Hounslow TW6 2GW
Tel: 0207 87459800
Website: www.baa.com

BRITISH ANODISING ASSOCIATION
See ALUMINIUM FEDERATION LTD

BRITISH ARCHITECTURAL LIBRARY
RIBA, 66 Portland Place
London W1B 1AD
Tel: 0207 580 5533
Fax: 0207 251 1541
Website: www.architecture.com

BRITISH ASSOCIATION OF LANDSCAPE
INDUSTRIES
Landscape House, Stoneleigh Park
Warwickshire CV8 2LG
Tel: 024 7669 0333
Fax: 024 7669 0777
Website: www.bali.org.uk

BRITISH BOARD OF AGRÉMENT
PO Box 195, Bucknalls Lane, Garston, Watford
Herts WA25 9BA
Tel: 01923 665 300
Fax: 01923 665 301
Website: www.bbacerts.co.uk

BRITISH CABLE ASSOCIATION
37a Walton Rd, East Molesey
Surrey KT8 9DW
Tel: 0208 941 4079
Fax: 0208 783 0104

BRITISH CERAMIC RESEARCH
Queens Road, Penkhull, Stoke-on-Trent
Staffs ST4 7LQ
Tel: 01782 764 444
Fax: 01782 412 331
Website: www.ceram.com

BRITISH CONSTRUCTIONAL STEELWORK
ASSOCIATION LTD
4 Whitehall Court, Westminster
London SW1A 2ES
Tel: 0207 839 8566
Fax: 0207 976 1634
Website: www.bcsa.org.uk

BRITISH ELECTROTECHNICAL AND ALLIED
MANUFACTURERS ASSOCIATION
Westminster Tower, 3 Albert Embankment
London SE1 7SL
Tel: 0207 793 3000
Fax: 0207 793 3003
Website: www.beama.org.uk

BRITISH FIRE PROTECTION SYSTEMS
ASSOCIATION LTD
55 Eden Street, Kingston-upon-Thames,
Surrey KT1 1BW
Tel: 0208 549 5855
Fax: 0208 547 1564
Website: www.fia.uk.com

BRITISH FIRE SERVICES ASSOCIATION
7 Clover House, Boston Road
Lincolnshire NG34 7HD
Tel: 01526 830255
Website: www.bfsa.org.uk

BRITISH FLUE & CHIMNEY MANUFACTURERS
ASSOCIATION
See FEDERATION OF ENVIRONMENTAL TRADE
ASSOCIATIONS

BRITISH GEOLOGICAL SURVEY
Environmental Science Centre
Nicker Hill, Keyworth
Nottingham NG12 5GG
Tel: 0115 936 3100
Fax: 0115 936 3200
Website: www.bgs.ac.uk

BRITISH LIBRARY LENDING DIVISION
Thorpe Arch, Boston Spa, Wetherby
West Yorks LS23 7BQ
Tel: 01937 546 000

BRITISH LIBRARY, SCIENCE REFERENCE AND
INFORMATION LIBRARY
The British Library, St Pancras, 96 Euston Road
London NW1 2DB
Tel: 0870 444 1500
Website: www.bl.uk

BRITISH NON-FERROUS METALS FEDERATION
5 Grovelands Business Centre, Boundary Way
Hemel Hempstead HP2 7TE
Tel: 01442 275705
Fax: 01442 275716

BRITISH PLASTICS FEDERATION
5–6 Bath Place, Rivington Street
London EC2A 3JE
Tel: 0207 457 5000
Fax: 0207 457 5045
Website: www.bpf.co.uk

BRITISH PRECAST CONCRETE FEDERATION
60 Charles St
Leicester LE1 1FB
Tel: 0116 253 6161
Fax: 0116 251 4568
Website: www.britishprecast.org

BRITISH REINFORCEMENT MANUFACTURERS
ASSOCIATION (BRMA)
See UK STEEL ASSOCIATION

BRITISH RUBBER & POLYURETHANE PRODUCTS
ASSOCIATION
5 Berwyk Hall Court, White Colne, Colchester
Essex
Tel: 01787 226 995
Fax: 0845 301 6853

BRITISH STAINLESS STEEL ASSOCIATION
Broomgrove, 59 Clarkhouse Road
Sheffield S10 2LE
Tel: 0114 267 1260
Fax: 0114 266 1252
Website: www.bssa.org.uk

BRITISH STANDARDS INSTITUTION
389 Chiswick High Road
Chiswick W4 4AL
Tel: 0208 996 9000
Fax: 0208 996 7001
Website: www.bsiglobal.com

BRITISH WATER
1 Queen Anne's Gate
London SW1 9BT
Tel: 0207 957 4554
Fax: 0207 957 4565
Website: www.britishwater.co.uk

BRITISH WOODWORKING FEDERATION
26 Store Street
London WC1E 7BT
Tel: 0844 209 2610
Fax: 0844 209 2611
Website: www.bwf.org.uk

THE BUILDING CENTRE GROUP
26 Store Street
London WC1E 7BT
Tel: 0207 692 4000
Fax: 0207 580 9641
Website: www.buildingcentre.co.uk

BUILDING MAINTENANCE INFORMATION (BMI)
3 Cadogan Gate
London SW1X 0AS
Tel: 0207 695 1500
Fax: 0207 695 1501

BUILDING RESEARCH ESTABLISHMENT (BRE)
Bucknalls Lane, Garston, Watford
Herts WD25 9XX
Tel: 01923 664 000
Fax: 01923 664 010
Website: www.bre.co.uk

BUILDING RESEARCH ESTABLISHMENT:
SCOTLAND (BRE)
Kelvin Rd, East Kilbride
Glasgow G75 0RZ
Tel: 01355 576 200
Fax: 01355 576 210

BUILDING SERVICES RESEARCH AND
INFORMATION ASSOCIATION
Old Bracknell Lane West, Bracknell
Berks RG12 7 AH
Tel: 01344 465 600
Fax: 01344 465 626
Website: www.bsria.co.uk

CASTINGS TECHNOLOGY INTERNATIONAL
7 East Bank Road
Sheffield S2 3PT
Tel: 0114 272 8647
Fax: 0114 273 0854
Website: www.castingsdev.com

CEMENT ADMIXTURES ASSOCIATION
38 Tilehouse, Green Lane, Knowle
West Midlands B93 9EY
Tel: + Fax: 01564 776 362
Website: www.admixtures.org.uk

CHARTERED INSTITUTE OF ARBITRATORS
12 Bloomsbury Square
London WC1A 2LP
Tel: 0207 421 7444
Fax: 0207 404 4023
Website: www.ciarb.org

CHARTERED INSTITUTE OF ARCHITECTURAL
TECHNOLOGISTS
397 City Road
London EC1V 1NH
Tel: 0207 278 2206
Fax: 0207 837 3194
Website: www.ciat.org.uk

CHARTERED INSTITUTE OF BUILDING (CIOB)
Englemere, Kings Ride, Ascot
Berks SL5 7TB
Tel: 01344 630 700
Fax: 01344 630 777
Website: www.ciob.org.uk

CHARTERED INSTITUTION OF WATER AND
ENVIRONMENTAL MANAGEMENT
15 John Street
London WC1N 2EB
Tel: 0207 831 3110
Fax: 0207 405 4967
Website: www.ciwem.org.uk

THE CHARTERD QUALITY INSTITUTE
10 Furnival Street
London EC4A 1AB
Tel: 0207 245 6722
Fax: 0207 245 6788
Website: www.thecqi.org

CIVIL ENGINEERING CONTRACTORS
ASSOCIATION
1 Birdcage Walk
London SW1H 9JJ
Tel: 0207 340 0450
Website: www.ceca.co.uk

CLAY PIPE DEVELOPMENT ASSOCIATION
Copsham House, 53 Broad Street, Chesham
Bucks HP5 3EA
Tel: 01494 791 456
Fax: 01494 792 378
Website: www.cpda.co.uk

COLD ROLLED SECTIONS ASSOCIATIONS
National Metal Forming Centre, 47 Birmingham Road
West Bromwich B70 6PY
Tel: 0121 601 6350
Fax: 0121 601 6373
Website: www.crsauk.com

THE CONCRETE CENTRE
Riverside House, 4 Meadow Business Park,
Stanchion Approach
Camberley GU17 9AB
Tel: 01276 606 800
Website: www.concretecentre.com

CONCRETE PIPELINE SYSTEMS ASSOCIATION
60 Charles St
Leicester LE1 1FB
Tel: 0116 253 6161
Fax: 0116 251 4568
Website: www.concretepipes.co.uk

CONFEDERATION OF BRITISH INDUSTRY
Centre Point, 103 New Oxford St
London WC1A 1DU
Tel: 0207 379 7400
Fax: 0207 240 1578
Website: www.cbi.org.uk

CONSTRUCTION CONFEDERATION
55 Tufton Street, Westminster
London SW1P 3QL
Tel: 0870 8989090
Fax: 0207 8989095
Website: www.thecc.org.uk

CONSTRUCTION EMPLOYERS FEDERATION
143 Malone Rd
Belfast BT9 6SU
Tel: 02890 877 143
Fax: 02890 877 155
Website: www.cefni.co.uk

CONSTRUCTION INDUSTRY RESEARCH & INFORMATION ASSOCIATION (CIRIA)
Classic House, 174–180 Old Street
London EC1V 9BP
Tel: 020 7549 3000
Fax: 020 7253 0523
Website: www.ciria.org

DEPARTMENT OF TRADE AND INDUSTRY
1 Victoria Street
London SW1H 0ET
Tel: 0207 215 5000
Fax: 0207 828 3258

DEPARTMENT FOR TRANSPORT
Great Minister House, 33 Horseferry Road
London SW1 P 4DR
Tel: 0300 330 3000
Website: www.dft.gov.uk

DOORS & HARDWARE FEDERATION
42 Heath Street, Tamworth
Staffs B79 7JH
Tel: 01827 52337
Fax: 01827 310 827
Website: www.abhm.org.uk

ELECTRICAL CONTRACTORS ASSOCIATION (ECA)
ESCA House, 34 Palace Court, Bayswater
London W2 4HY
Tel: 0207 313 4800
Fax: 0207 221 7344
Website: www.eca.co.uk

ELECTRICAL CONTRACTORS ASSOCIATION OF SCOTLAND (select)
The Walled Garden, Bush Estate
Midlothian EH26 0SB
Tel: 0131 445 5577
Fax: 0131 445 5548
Website: www.select.org.uk

FEDERATION OF ENVIRONMENTAL TRADE ASSOCIATIONS
2 Waltham Court, Milley Lane, Hare Hatch
Reading RG10 9TH
Tel: 0118 940 3416
Fax: 0118 940 6258
Website: www.feta.co.uk

FEDERATION OF MASTER BUILDERS
Gordon Fisher House, 14–15 Gt James St
London WC1N 3DP
Tel: 0207 242 7583
Fax: 0207 404 0296
Website: www.fmb.org.uk

FEDERATION OF PILING SPECIALISTS
Forum Court, 83 Copers Cope Road, Beckenham
Kent BR3 1NR
Tel: 0208 663 0947
Fax: 0208 663 0949
Website: www.fps.org.uk

FEDERATION OF PLASTERING AND DRYWALL CONTRACTORS
61 Cheapside
London EC2V 6AX
Tel: 020 7634 9480
Fax: 020 7248 3685
Website: www.fpdc.org.uk

FENCING CONTRACTORS ASSOCIATION
Warren Road, Trelleck
Monmouthshire NP25 4PQ
Tel: 07000 560 722
Fax: 01600 860 614
Website: www.fencingcontractors.org

HEALTH & SAFETY LABORATORY
Business Development Unit Health & Safety Laboratory, Harpur Hill, Buxton
Derbyshire SK17 9JN
Tel: 01298 218 000
Fax: 01298 218 822
Website: www.hsl.gov.uk

HEATING AND VENTILATORS CONTRACTORS
ASSOCIATION
ESCA House, 34 Palace Court
London W2 4JG
Tel: 0207 313 4900
Fax: 0207 727 9268
Website: www.hvca.org.uk

HM LAND REGISTRY (HQ)
32 Lincolns Inn Fields
London WC2A 3PH
Tel: 0207 917 8888
Fax: 0207 955 0110
Website: www.landreg.gov.uk

ICOM ENERGY ASSOCIATION
Camden House, Warwick Road, Kenilworth
Warwickshire CV8 1TH
Tel: 01926 513748
Fax: 01926 21 855017
Website: www.icome.org.uk

INSTITUTE OF CONCRETE TECHNOLOGY
4 Meadows Business Park, Station Approach,
Blackwater
Camberley GU17 9 AB
Tel: 01276 37831
Website: www.ict.concrete.org.uk

INSTITUTE OF MATERIALS, MINERALS & MINING
1 Carlton House Terrace
London SW1Y 5DB
Tel: 0207 451 7300
Fax: 0207 839 1702
Website: www.materials.org.uk

INSTITUTION OF CIVIL ENGINEERS
1 Great George St
London SW1P 3AA
Tel: 0207 222 7722
Fax: 0207 222 7500
Website: www.ice.org.uk

THE INSTITUTION OF ENGINEERING &
TECHNOLOGY
Michael Faraday House, Six Hills Way Stevenage
Herts SG1 2 AY
Tel: 01438 313 3111
Fax: 01438 765 526
Website: www.theiet.org

INSTITUTION OF MECHANICAL ENGINEERS
1 Birdcage Walk
London SW1H 9JJ
Tel: 0207 222 7899
Fax: 0207 222 4557
Website: www.imeche.org.uk

INSTITUTION OF STRUCTURAL ENGINEERS
11 Upper Belgrave Street
London SW1X 8BH
Tel: 0207 235 4535
Fax: 0207 235 4294
Website: www.istructe.org.uk

INSTITUTION OF WASTES MANAGEMENT
9 Saxon Court, St Peter's Gardens, Marefair
Northampton NN1 1SX
Tel: 01604 620 426
Fax: 01604 621 339
Website: www.ciwm.co.uk

INSTITUTION OF WATER AND ENVIRONMENTAL
MANAGEMENT
See CHARTERED INSTITUTION OF WATER AND
ENVIRONMENTAL MANAGEMENT

INTERNATIONAL CONCRETE BRICK ASSOCIATION
See BRITISH PRECAST CONCRETE FEDERATION

INTERNATIONAL LEAD ASSOCIATION
17 A Welbeck Way
London W1G 9YJ
Tel: 0207 499 8422
Fax: 0207 493 1555
Website: www.ldaint.org

THE JOINT CONTRACTS TRIBUNAL
9 Cavendish Place
London W1G 0QD
Tel: 0207 637 8650
Fax: 0207 637 8670
Website: www.jctltd.co.uk

LANDSCAPE INSTITUTE
33 Great Portland Street
London W1W 8QG
Tel: 0207 299 4500
Fax: 0207 299 4501
Website: www.l-i.org.uk

MASTIC ASPHALT COUNCIL LTD
PO Box 77, Hastings
Kent TN35 4WL
Tel: 01424 814 400
Fax: 01424 814 446
Website: www.masticasphaltcouncil.co.uk

MET OFFICE
Fitzroy Road, Exeter
Devon EX1 3PB
Tel: 0870 900 0100
Fax: 0870 900 5050
Website: www.met-office.gov.uk

NATIONAL ASSOCIATION OF STEEL
STOCKHOLDERS
The Citadel, 190 Corporation Street
Birmingham B4 6QD
Tel: 0121 200 2288
Fax: 0121 236 7444
Website: www.nass.org.uk

NATIONAL FEDERATION OF ROOFING
CONTRACTORS
Roofing House, 31 Worship Street
London EC2A 2DY
Tel: 020 7638 7663
Fax: 020 7256 2125
Website: www.nfrc.co.uk

ORDNANCE SURVEY
Romsey Road, Maybush
Southampton SO16 4GU
Tel: 0845 605 0505
Fax: 02380 792 615
Website: www.ordsvy.gov.uk

PIPELINE INDUSTRIES GUILD
14/15 Belgrave Square
London SW1X 8PS
Tel: 0207 235 7938
Fax: 0207 235 0074
Website: www.pipeguild.co.uk

PUBLIC RECORDS OFFICE
Ruskin Avenue, Q Richmond
Surrey TW9 4DU
Tel: 0208 876 3444
Fax: 0208 392 5286
Website: www.nationalarchives.gov.uk

QUARRY PRODUCTS ASSOCIATION
Gillingham House, 38–44 Gillingham Street
London SW1V 1HU
Tel: 0207 963 8000
Fax: 0207 963 8001
Website: www.qpa.org

THE RESIN FLOORING FEDERATION
16 Edward Road, Farnham
Surrey GU9 8NP
Tel: 01252 714 250
Website: www.ferfa.org.uk

ROYAL INSTITUTE OF BRITISH ARCHITECTS
66 Portland Place
London W1B 1AD
Tel: 0207 580 5533
Fax: 0207 255 1541
Website: www.riba.org.uk

ROYAL INSTITUTION OF CHARTERED
SURVEYORS (RICS)
12 Great George Street, Parliament Square
London SW1P 3AD
Tel: 0870 333 1600
Fax: 0207 222 9430
Website: www.rics.org.uk

SCOTTISH BUILDING EMPLOYERS
FEDERATION
Crichton House, 4 Crichton Close
Edinburgh, EH8 8DT
Tel: 01315 568 866
Website: www.scottish-building.co.uk

SCOTTISH ENTERPRISE
5 Atlantic Quay, 150 Broomielaw
Glasgow G2 8LU
Tel: 0141 248 2700
Fax: 0141 221 3217
Website: www.scottish-enterprise.com

SOCIETY OF GLASS TECHNOLOGY
9 Churchill Way, Chapeltown
Sheffield S35 2PY
Tel: 0114 263 4455
Fax: 0871 875 4085
Website: www.societyofglasstechnology.org

SPECIALISED ACCESS ENGINEERING AND
MAINTENANCE ASSOCIATION
19 Joseph Fletcher Drive, Wingerworth
Chesterfield S42 6TZ
Tel: 0124 622 4175
Website: www.saema.org

STEEL CONSTRUCTION INSTITUTE
Silwood Park, Ascot
Berks SL5 7QN
Tel: 01344 623 345
Fax: 01344 622 944
Website: www.steel-sci.org

SWIMMING POOL AND ALLIED TRADES
ASSOCIATION (SPATA)
4 Eastgate House, East Street, Andover
Hampshire SP10 1EP
Tel: 01264 356 210
Website: www.spata.co.uk

TATA STEEL
30 Millbank
London SW1P 4WY
Tel: 0207 717 444
Website: www.tatasteeleurope.com

THERMAL INSULATION CONTRACTORS
ASSOCIATION
TICA House, Allington Way, Yarm Road Business
Park, Darlington
County Durham DL1 4QB
Tel: 01325 466 704
Fax: 01325 487 691
Website: www.tica-acad.co.uk

THERMAL INSULATION MANUFACTURERS
AND SUPPLIERS ASSOCIATION
Kingsley House, Ganders Business Park,
Kingsley, Bordon
Hampshire GU35 LU
Tel: 01420 471 624
Fax: 01420 471 611
Website: www.timsa.org.uk

TIMBER TRADE FEDERATION LTD
The Building Centre, 26 Store Street
London WC1E 7BT
Tel: 0203 205 0067
Fax: 0207 291 5379
Website: www.ttf.co.uk

TOWN AND COUNTRY PLANNING
ASSOCIATION
17 Carlton House Terrace
London SW1Y 5AS
Tel: 0207 930 8903
Fax: 0207 930 3280
Website: www.tcpa.org.uk

TRADA TECHNOLOGY LTD
Stocking Lane, Hughenden Valley, High Wycombe
Bucks HP14 4ND
Tel: 01494 569 600
Fax: 01494 565 487
Website: www.trada.co.uk

TWI – THE WELDING INSTITUTE
Granta Park, Great Abington
Cambridge CB1 6AL
Tel: 01123 899 000
Fax: 01223 892 588
Website: www.twi.co.uk

UK STEEL ASSOCIATION
Broadway House, Tothill Street
London SW1H 9NQ
Tel: 0207 7654 1518
Fax: 0207 222 2782
Website: www.uksteel.org.uk

WOOD PROTECTION ASSOCIATION
5C Fleming Court, Castlefield
West Yorkshire WF10 5HW
Tel: 01977 558274
Website: www.wood-protection.org

PART 12

Tables and Memoranda

This part of the book contains the following sections:

	page
Conversion Tables	539
Geometry	543
Formulae	547
Typical thermal conductivity of building materials	548
Earthwork	549
Concrete Work	555
Reinforcement	560
Formwork	562
Masonry	563
Timber	566
Roofing	571
Glazing	575
Metal	576
Kerbs, Paving, etc.	594
Seeding/Turfing and Planting	601
Fencing and Gates	604
Drainage	609
Electrical Supply/Power/Lighting Systems	616
Rail Tracks	617
Fractions, Decimals and Millimetre Equivalents	620
Imperial Standard Wire Gauge (SWG)	621
Pipes, Water, Storage, Insulation	622

ESSENTIAL READING FROM TAYLOR AND FRANCIS

Construction Contracts Questions and Answers

2nd Edition

David Chappell

What they said about the first edition: "A fascinating concept, full of knowledgeable gems put in the most frank of styles... A book to sample when the time is right and to come back to when another time is right, maybe again and again."
– David A Simmonds, Building Engineer magazine

- Is there a difference between inspecting and supervizing?
- What does 'time-barred' mean?
- Is the contractor entitled to take possession of a section of the work even though it is the contractor's fault that possession is not practicable?

Construction law can be a minefield. Professionals need answers which are pithy and straightforward, as well as legally rigorous. The two hundred questions in the book are real questions, picked from the thousands of telephone enquiries David Chappell has received as a Specialist Adviser to the Royal Institute of British Architects. Although the enquiries were originally from architects, the answers to most of them are of interest to project managers, contractors, QSs, employers and others involved in construction.

The material is considerably updated from the first edition – weeded, extended and almost doubled in coverage. The questions range in content from extensions of time, liquidated damages and loss and/or expense to issues of warranties, bonds, novation, practical completion, defects, valuation, certificates and payment, architects' instructions, adjudication and fees. Brief footnotes and a table of cases will be retained for those who may wish to investigate further.

August 2010: 216x138: 344pp
Pb: 978-0-415-56650-6: **£34.99**

To Order: Tel: +44 (0) 1235 400524 **Fax:** +44 (0) 1235 400525
or Post: Taylor and Francis Customer Services,
Bookpoint Ltd, Unit T1, 200 Milton Park, Abingdon, Oxon, OX14 4TA UK
Email: book.orders@tandf.co.uk

For a complete listing of all our titles visit:
www.tandf.co.uk

Taylor & Francis
Taylor & Francis Group

CONVERSION TABLES

CONVERSION TABLES

Length	Unit	Conversion factors			
Millimetre	mm	1 in	= 25.4 mm	1 mm	= 0.0394 in
Centimetre	cm	1 in	= 2.54 cm	1 cm	= 0.3937 in
Metre	m	1 ft	= 0.3048 m	1 m	= 3.2808 ft
		1 yd	= 0.9144 m		= 1.0936 yd
Kilometre	km	1 mile	= 1.6093 km	1 km	= 0.6214 mile

Note:
- 1 cm = 10 mm
- 1 m = 1 000 mm
- 1 km = 1 000 m
- 1 ft = 12 in
- 1 yd = 3 ft
- 1 mile = 1 760 yd

Area	Unit	Conversion factors			
Square Millimetre	mm^2	$1\ in^2$	$= 645.2\ mm^2$	$1\ mm^2$	$= 0.0016\ in^2$
Square Centimetre	cm^2	$1\ in^2$	$= 6.4516\ cm^2$	$1\ cm^2$	$= 1.1550\ in^2$
Square Metre	m^2	$1\ ft^2$	$= 0.0929\ m^2$	$1\ m^2$	$= 10.764\ ft^2$
		$1\ yd^2$	$= 0.8361\ m^2$	$1\ m^2$	$= 1.1960\ yd^2$
Square Kilometre	km^2	$1\ mile^2$	$= 2.590\ km^2$	$1\ km^2$	$= 0.3861\ mile^2$

Note:
- $1\ cm^2 = 100\ mm^2$
- $1\ m^2 = 10\ 000\ cm^2$
- $1\ km^2 = 100$ hectares
- $1\ ft^2 = 144\ in^2$
- $1\ yd^2 = 9\ ft^2$
- 1 acre $= 4\ 840\ yd^2$
- $1\ mile^2 = 640$ acres

Volume	Unit	Conversion factors			
Cubic Centimetre	cm^3	$1\ cm^3$	$= 0.0610\ in^3$	$1\ in^3$	$= 16.387\ cm^3$
Cubic Decimetre	dm^3	$1\ dm^3$	$= 0.0353\ ft^3$	$1\ ft^3$	$= 28.329\ dm^3$
Cubic Metre	m^3	$1\ m^3$	$= 35.3147\ ft^3$	$1\ ft^3$	$= 0.0283\ m^3$
		$1\ m^3$	$= 1.3080\ yd^3$	$1\ yd^3$	$= 0.7646\ m^3$
Litre	l	1 l	= 1.76 pint	1 pint	= 0.5683 l
			= 2.113 US pt		= 0.4733 US l

Note:
- $1\ dm^3 = 1\ 000\ cm^3$
- $1\ m^3 = 1\ 000\ dm^3$
- 1 l $= 1\ dm^3$
- $1\ ft^3 = 1\ 728\ in^3$
- $1\ yd^3 = 27\ ft^3$
- 1 pint = 20 fl oz
- 1 gal = 8 pints

Neither the Centimetre nor Decimetre are SI units, and as such their use, particularly that of the Decimetre, is not widespread outside educational circles.

Mass	Unit	Conversion factors			
Milligram	mg	1 mg	= 0.0154 grain	1 grain	= 64.935 mg
Gram	g	1 g	= 0.0353 oz	1 oz	= 28.35 g
Kilogram	kg	1 kg	= 2.2046 lb	1 lb	= 0.4536 kg
Tonne	t	1 t	= 0.9842 ton	1 ton	= 1.016 t

Note:
- 1 g = 1000 mg
- 1 kg = 1000 g
- 1 t = 1000 kg
- 1 oz = 437.5 grains
- 1 lb = 16 oz
- 1 stone = 14 lb
- 1 cwt = 112 lb
- 1 ton = 20 cwt

CONVERSION TABLES

Force	Unit	Conversion factors			
Newton	N	1 lbf	= 4.448 N	1 kgf	= 9.807 N
Kilonewton	kN	1 lbf	= 0.004448 kN	1 ton f	= 9.964 kN
Meganewton	MN	100 tonf	= 0.9964 MN		

Pressure and stress	Unit	Conversion factors	
Kilonewton per square metre	kN/m^2	1 lbf/in^2	= 6.895 kN/m^2
		1 bar	= 100 kN/m^2
Meganewton per square metre	MN/m^2	1 $tonf/ft^2$	= 107.3 kN/m^2 = 0.1073 MN/m^2
		1 kgf/cm^2	= 98.07 kN/m^2
		1 lbf/ft^2	= 0.04788 kN/m^2

Coefficient of consolidation (Cv) or swelling	Unit	Conversion factors	
Square metre per year	m^2/year	1 cm^2/s	= 3 154 m^2/year
		1 ft^2/year	= 0.0929 m^2/year

Coefficient of permeability	Unit	Conversion factors	
Metre per second	m/s	1 cm/s	= 0.01 m/s
Metre per year	m/year	1 ft/year	= 0.3048 m/year
			= 0.9651 × $(10)^8$ m/s

Temperature	Unit	Conversion factors	
Degree Celsius	°C	°C = 5/9 × (°F − 32)	°F = (9 × °C)/ 5 + 32

CONVERSION TABLES

SPEED CONVERSION

km/h	m/min	mph	fpm
1	16.7	0.6	54.7
2	33.3	1.2	109.4
3	50.0	1.9	164.0
4	66.7	2.5	218.7
5	83.3	3.1	273.4
6	100.0	3.7	328.1
7	116.7	4.3	382.8
8	133.3	5.0	437.4
9	150.0	5.6	492.1
10	166.7	6.2	546.8
11	183.3	6.8	601.5
12	200.0	7.5	656.2
13	216.7	8.1	710.8
14	233.3	8.7	765.5
15	250.0	9.3	820.2
16	266.7	9.9	874.9
17	283.3	10.6	929.6
18	300.0	11.2	984.3
19	316.7	11.8	1038.9
20	333.3	12.4	1093.6
21	350.0	13.0	1148.3
22	366.7	13.7	1203.0
23	383.3	14.3	1257.7
24	400.0	14.9	1312.3
25	416.7	15.5	1367.0
26	433.3	16.2	1421.7
27	450.0	16.8	1476.4
28	466.7	17.4	1531.1
29	483.3	18.0	1585.7
30	500.0	18.6	1640.4
31	516.7	19.3	1695.1
32	533.3	19.9	1749.8
33	550.0	20.5	1804.5
34	566.7	21.1	1859.1
35	583.3	21.7	1913.8
36	600.0	22.4	1968.5
37	616.7	23.0	2023.2
38	633.3	23.6	2077.9
39	650.0	24.2	2132.5
40	666.7	24.9	2187.2

CONVERSION TABLES

km/h	m/min	mph	fpm
41	683.3	25.5	2241.9
42	700.0	26.1	2296.6
43	716.7	26.7	2351.3
44	733.3	27.3	2405.9
45	750.0	28.0	2460.6
46	766.7	28.6	2515.3
47	783.3	29.2	2570.0
48	800.0	29.8	2624.7
49	816.7	30.4	2679.4
50	833.3	31.1	2734.0

GEOMETRY

Two dimensional figures

Figure	Diagram of figure	Surface area	Perimeter
Square		a^2	$4a$
Rectangle		ab	$2(a+b)$
Triangle		$\frac{1}{2}ch$	$a+b+c$
Circle		πr^2 $\frac{1}{4}\pi d^2$ where $2r = d$	$2\pi r$ πd
Parallelogram		ah	$2(a+b)$
Trapezium		$\frac{1}{2}h(a+b)$	$a+b+c+d$
Ellipse		Approximately πab	$\pi(a+b)$
Hexagon		$2.6 \times a^2$	

GEOMETRY

Figure	Diagram of figure	Surface area	Perimeter
Octagon		$4.83 \times a^2$	$6a$
Sector of a circle		$\frac{1}{2}rb$ or $\frac{q}{360}\pi r^2$ note b = angle $\frac{q}{360} \times \pi 2r$	
Segment of a circle		$S - T$ where S = area of sector, T = area of triangle	
Bellmouth		$\frac{3}{14} \times r^2$	

GEOMETRY

Three dimensional figures

Figure	Diagram of figure	Surface area	Volume
Cube		$6a^2$	a^3
Cuboid/ rectangular block		$2(ab + ac + bc)$	abc
Prism/ triangular block		$bd + hc + dc + ad$	$\frac{1}{2}hcd$
Cylinder		$2\pi r^2 + 2\pi h$	$\pi r^2 h$ $\frac{1}{4}\pi d^2 h$
Sphere		$4\pi r^2$	$\frac{4}{3}\pi r^3$
Segment of sphere		$2\pi Rh$	$\frac{1}{6}\pi h(3r^2 + h^2)$ $\frac{1}{3}\pi h^2(3R - H)$
Pyramid		$(a + b)l + ab$	$\frac{1}{3}abh$

GEOMETRY

Figure	Diagram of figure	Surface area	Volume
Frustum of a pyramid		$l(a+b+c+d) + \sqrt{(ab+cd)}$ [rectangular figure only]	$\frac{h}{3}(ab + cd + \sqrt{abcd})$
Cone		πrl (excluding base) $\pi rl + \pi r^2$ (including base)	$\frac{1}{3}\pi r^2 h$ $\frac{1}{12}\pi d^2 h$
Frustum of a cone		$\pi r^2 + \pi R^2 + \pi l(R+r)$	$\frac{1}{3}\pi(R^2 + Rr + r^2)$

Formulae

FORMULAE

Formula	Description
Pythagoras Theorem	$A^2 = B^2 + C^2$ where A is the hypotenuse of a right-angled triangle and B and C are the two adjacent sides
Simpsons Rule	The Area is divided into an even number of strips of equal width, and therefore has an odd number of ordinates at the division points $$\text{area} = \frac{S(A + 2B + 4C)}{3}$$ where S = common interval (strip width) A = sum of first and last ordinates B = sum of remaining odd ordinates C = sum of the even ordinates The Volume can be calculated by the same formula, but by substituting the area of each coordinate rather than its length
Trapezoidal Rule	A given trench is divided into two equal sections, giving three ordinates, the first, the middle and the last $$\text{volume} = \frac{S \times (A + B + 2C)}{2}$$ where S = width of the strips A = area of the first section B = area of the last section C = area of the rest of the sections
Prismoidal Rule	A given trench is divided into two equal sections, giving three ordinates, the first, the middle and the last $$\text{volume} = \frac{L \times (A + 4B + C)}{6}$$ where L = total length of trench A = area of the first section B = area of the middle section C = area of the last section

TYPICAL THERMAL CONDUCTIVITY OF BUILDING MATERIALS

(Always check manufacturer's details – variation will occur depending on product and nature of materials)

	Thermal conductivity (W/mK)		Thermal conductivity (W/mK)
Acoustic plasterboard	0.25	Oriented strand board	0.13
Aerated concrete slab (500 kg/m^3)	0.16	Outer leaf brick	0.77
Aluminium	237	Plasterboard	0.22
Asphalt (1700 kg/m^3)	0.5	Plaster dense (1300 kg/m^3)	0.5
Bitumen-impregnated fibreboard	0.05	Plaster lightweight (600 kg/m^3)	0.16
Blocks (standard grade 600 kg/m^3)	0.15	Plywood (950 kg/m^3)	0.16
Blocks (solar grade 460 kg/m^3)	0.11	Prefabricated timber wall panels (check manufacturer)	0.12
Brickwork (outer leaf 1700 kg/m^3)	0.84	Screed (1200 kg/m^3)	0.41
Brickwork (inner leaf 1700 kg/m^3)	0.62	Stone chippings (1800 kg/m^3)	0.96
Dense aggregate concrete block 1800 kg/m^3 (exposed)	1.21	Tile hanging (1900 kg/m^3)	0.84
Dense aggregate concrete block 1800 kg/m^3 (protected)	1.13	Timber (650 kg/m^3)	0.14
Calcium silicate board (600 kg/m^3)	0.17	Timber flooring (650 kg/m^3)	0.14
Concrete general	1.28	Timber rafters	0.13
Concrete (heavyweight 2300 kg/m^3)	1.63	Timber roof or floor joists	0.13
Concrete (dense 2100 kg/m^3 typical floor)	1.4	Roof tile (1900 kg/m^3)	0.84
Concrete (dense 2000 kg/m^3 typical floor)	1.13	Timber blocks (650 kg/m^3)	0.14
Concrete (medium 1400 kg/m^3)	0.51	Cellular glass	0.045
Concrete (lightweight 1200 kg/m^3)	0.38	Expanded polystyrene	0.034
Concrete (lightweight 600 kg/m^3)	0.19	Expanded polystyrene slab (25 kg/m^3)	0.035
Concrete slab (aerated 500 kg/m^3)	0.16	Extruded polystyrene	0.035
Copper	390	Glass mineral wool	0.04
External render sand/cement finish	1	Mineral quilt (12 kg/m^3)	0.04
External render (1300 kg/m^3)	0.5	Mineral wool slab (25 kg/m^3)	0.035
Felt – Bitumen layers (1700 kg/m^3)	0.5	Phenolic foam	0.022
Fibreboard (300 kg/m^3)	0.06	Polyisocyanurate	0.025
Glass	0.93	Polyurethane	0.025
Marble	3	Rigid polyurethane	0.025
Metal tray used in wriggly tin concrete floors (7800 kg/m^3)	50	Rock mineral wool	0.038
Mortar (1750 kg/m^3)	0.8		

EARTHWORK

Weights of Typical Materials Handled by Excavators

The weight of the material is that of the state in its natural bed and includes moisture Adjustments should be made to allow for loose or compacted states

Material	Mass (kg/m^3)	Mass (lb/cu yd)
Ashes, dry	610	1028
Ashes, wet	810	1365
Basalt, broken	1954	3293
Basalt, solid	2933	4943
Bauxite, crushed	1281	2159
Borax, fine	849	1431
Caliche	1440	2427
Cement, clinker	1415	2385
Chalk, fine	1221	2058
Chalk, solid	2406	4055
Cinders, coal, ash	641	1080
Cinders, furnace	913	1538
Clay, compacted	1746	2942
Clay, dry	1073	1808
Clay, wet	1602	2700
Coal, anthracite, solid	1506	2538
Coal, bituminous	1351	2277
Coke	610	1028
Dolomite, lumpy	1522	2565
Dolomite, solid	2886	4864
Earth, dense	2002	3374
Earth, dry, loam	1249	2105
Earth, Fullers, raw	673	1134
Earth, moist	1442	2430
Earth, wet	1602	2700
Felsite	2495	4205
Fieldspar, solid	2613	4404
Fluorite	3093	5213
Gabbro	3093	5213
Gneiss	2696	4544
Granite	2690	4534
Gravel, dry 1/4 to 2 inch	1682	2835
Gravel, dry, loose	1522	2565
Gravel, wet 1/4 to 2 inch	2002	3374
Gypsum, broken	1450	2444
Gypsum, solid	2787	4697
Hardcore (consolidated)	1928	3249
Lignite, dry	801	1350
Limestone, broken	1554	2619
Limestone, solid	2596	4375
Magnesite, magnesium ore	2993	5044
Marble	2679	4515
Marl, wet	2216	3735
Mica, broken	1602	2700
Mica, solid	2883	4859
Peat, dry	400	674
Peat, moist	700	1179

EARTHWORK

Material	Mass (kg/m^3)	Mass (lb/cu yd)
Peat, wet	1121	1889
Potash	1281	2159
Pumice, stone	640	1078
Quarry waste	1438	2423
Quartz sand	1201	2024
Quartz, solid	2584	4355
Rhyolite	2400	4045
Sand and gravel, dry	1650	2781
Sand and gravel, wet	2020	3404
Sand, dry	1602	2700
Sand, wet	1831	3086
Sandstone, solid	2412	4065
Shale, solid	2637	4444
Slag, broken	2114	3563
Slag, furnace granulated	961	1619
Slate, broken	1370	2309
Slate, solid	2667	4495
Snow, compacted	481	810
Snow, freshly fallen	160	269
Taconite	2803	4724
Trachyte	2400	4045
Trap rock, solid	2791	4704
Turf	400	674
Water	1000	1685

Transport Capacities

Type of vehicle	Capacity of vehicle	
	Payload	Heaped capacity
Wheelbarrow	150	0.10
1 tonne dumper	1250	1.00
2.5 tonne dumper	4000	2.50
Articulated dump truck (Volvo A20 6 × 4)	18500	11.00
Articulated dump truck (Volvo A35 6 × 6)	32000	19.00
Large capacity rear dumper (Euclid R35)	35000	22.00
Large capacity rear dumper (Euclid R85)	85000	50.00

EARTHWORK

Machine Volumes for Excavating and Filling

Machine type	Cycles per minute	Volume per minute (m³)
1.5 tonne excavator	1	0.04
	2	0.08
	3	0.12
3 tonne excavator	1	0.13
	2	0.26
	3	0.39
5 tonne excavator	1	0.28
	2	0.56
	3	0.84
7 tonne excavator	1	0.28
	2	0.56
	3	0.84
21 tonne excavator	1	1.21
	2	2.42
	3	3.63
Backhoe loader JCB3CX excavator Rear bucket capacity 0.28 m³	1	0.28
	2	0.56
	3	0.84
Backhoe loader JCB3CX loading Front bucket capacity 1.00 m³	1	1.00
	2	2.00

Machine Volumes for Excavating and Filling

Machine type	Loads per hour	Volume per hour (m³)
1 tonne high tip skip loader Volume 0.485 m³	5	2.43
	7	3.40
	10	4.85
3 tonne dumper Max volume 2.40 m³ Available volume 1.9 m³	4	7.60
	5	9.50
	7	13.30
	10	19.00
6 tonne dumper Max volume 3.40 m³ Available volume 3.77 m³	4	15.08
	5	18.85
	7	26.39
	10	37.70

EARTHWORK

Bulkage of Soils (after excavation)

Type of soil	Approximate bulking of 1 m³ after excavation
Vegetable soil and loam	25–30%
Soft clay	30–40%
Stiff clay	10–20%
Gravel	20–25%
Sand	40–50%
Chalk	40–50%
Rock, weathered	30–40%
Rock, unweathered	50–60%

Shrinkage of Materials (on being deposited)

Type of soil	Approximate bulking of 1 m³ after excavation
Clay	10%
Gravel	8%
Gravel and sand	9%
Loam and light sandy soils	12%
Loose vegetable soils	15%

Voids in Material Used as Subbases or Beddings

Material	m³ of voids/m³
Alluvium	0.37
River grit	0.29
Quarry sand	0.24
Shingle	0.37
Gravel	0.39
Broken stone	0.45
Broken bricks	0.42

Angles of Repose

Type of soil		Degrees
Clay	– dry	30
	– damp, well drained	45
	– wet	15–20
Earth	– dry	30
	– damp	45
Gravel	– moist	48
Sand	– dry or moist	35
	– wet	25
Loam		40

EARTHWORK

Slopes and Angles

Ratio of base to height	Angle in degrees
5:1	11
4:1	14
3:1	18
2:1	27
1½:1	34
1:1	45
1:1½	56
1:2	63
1:3	72
1:4	76
1:5	79

Grades (in Degrees and Percents)

Degrees	Percent	Degrees	Percent
1	1.8	24	44.5
2	3.5	25	46.6
3	5.2	26	48.8
4	7.0	27	51.0
5	8.8	28	53.2
6	10.5	29	55.4
7	12.3	30	57.7
8	14.0	31	60.0
9	15.8	32	62.5
10	17.6	33	64.9
11	19.4	34	67.4
12	21.3	35	70.0
13	23.1	36	72.7
14	24.9	37	75.4
15	26.8	38	78.1
16	28.7	39	81.0
17	30.6	40	83.9
18	32.5	41	86.9
19	34.4	42	90.0
20	36.4	43	93.3
21	38.4	44	96.6
22	40.4	45	100.0

EARTHWORK

Bearing Powers

Ground conditions		Bearing power		
		kg/m^2	lb/in^2	Metric t/m^2
Rock,	broken	483	70	50
	solid	2415	350	240
Clay,	dry or hard	380	55	40
	medium dry	190	27	20
	soft or wet	100	14	10
Gravel,	cemented	760	110	80
Sand,	compacted	380	55	40
	clean dry	190	27	20
Swamp and alluvial soils		48	7	5

Earthwork Support

Maximum depth of excavation in various soils without the use of earthwork support

Ground conditions	Feet (ft)	Metres (m)
Compact soil	12	3.66
Drained loam	6	1.83
Dry sand	1	0.3
Gravelly earth	2	0.61
Ordinary earth	3	0.91
Stiff clay	10	3.05

It is important to note that the above table should only be used as a guide. Each case must be taken on its merits and, as the limited distances given above are approached, careful watch must be kept for the slightest signs of caving in

CONCRETE WORK

Weights of Concrete and Concrete Elements

Type of material		kg/m³	lb/cu ft
Ordinary concrete (dense aggregates)			
Non-reinforced plain or mass concrete			
Nominal weight		2305	144
Aggregate	– limestone	2162 to 2407	135 to 150
	– gravel	2244 to 2407	140 to 150
	– broken brick	2000 (av)	125 (av)
	– other crushed stone	2326 to 2489	145 to 155
Reinforced concrete			
Nominal weight		2407	150
Reinforcement	– 1%	2305 to 2468	144 to 154
	– 2%	2356 to 2519	147 to 157
	– 4%	2448 to 2703	153 to 163
Special concretes			
Heavy concrete			
Aggregates	– barytes, magnetite	3210 (min)	200 (min)
	– steel shot, punchings	5280	330
Lean mixes			
Dry-lean (gravel aggregate)		2244	140
Soil-cement (normal mix)		1601	100

CONCRETE WORK

Type of material		kg/m² per mm thick	lb/sq ft per inch thick
Ordinary concrete (dense aggregates)			
Solid slabs (floors, walls etc.) Thickness:	75 mm or 3 in	184	37.5
	100 mm or 4 in	245	50
	150 mm or 6 in	378	75
	250 mm or 10 in	612	125
	300 mm or 12 in	734	150
Ribbed slabs Thickness:	125 mm or 5 in	204	42
	150 mm or 6 in	219	45
	225 mm or 9 in	281	57
	300 mm or 12 in	342	70
Special concretes			
Finishes etc.	Rendering, screed etc. Granolithic, terrazzo	1928 to 2401	10 to 12.5
	Glass-block (hollow) concrete	1734 (approx)	9 (approx)
Prestressed concrete		Weights as for reinforced concrete (upper limits)	
Air-entrained concrete		Weights as for plain or reinforced concrete	

CONCRETE WORK

Average Weight of Aggregates

Materials	Voids %	Weight kg/m³
Sand	39	1660
Gravel 10–20 mm	45	1440
Gravel 35–75 mm	42	1555
Crushed stone	50	1330
Crushed granite (over 15 mm)	50	1345
(n.e. 15 mm)	47	1440
'All-in' ballast	32	1800–2000

Material	kg/m³	lb/cu yd
Vermiculite (aggregate)	64–80	108–135
All-in aggregate	1999	125

Applications and Mix Design

Site mixed concrete

Recommended mix	Class of work suitable for	Cement (kg)	Sand (kg)	Coarse aggregate (kg)	Nr 25 kg bags cement per m³ of combined aggregate
1:3:6	Roughest type of mass concrete such as footings, road haunching over 300 mm thick	208	905	1509	8.30
1:2.5:5	Mass concrete of better class than 1:3:6 such as bases for machinery, walls below ground etc..	249	881	1474	10.00
1:2:4	Most ordinary uses of concrete, such as mass walls above ground, road slabs etc. and general reinforced concrete work	304	889	1431	12.20
1:1.5:3	Watertight floors, pavements and walls, tanks, pits, steps, paths, surface of 2 course roads, reinforced concrete where extra strength is required	371	801	1336	14.90
1:1:2	Works of thin section such as fence posts and small precast work	511	720	1206	20.40

CONCRETE WORK

Ready mixed concrete

Application	Designated concrete	Standardized prescribed concrete	Recommended consistence (nominal slump class)
Foundations			
Mass concrete fill or blinding	GEN 1	ST2	S3
Strip footings	GEN 1	ST2	S3
Mass concrete foundations			
Single storey buildings	GEN 1	ST2	S3
Double storey buildings	GEN 3	ST4	S3
Trench fill foundations			
Single storey buildings	GEN 1	ST2	S4
Double storey buildings	GEN 3	ST4	S4
General applications			
Kerb bedding and haunching	GEN 0	ST1	S1
Drainage works – immediate support	GEN 1	ST2	S1
Other drainage works	GEN 1	ST2	S3
Oversite below suspended slabs	GEN 1	ST2	S3
Floors			
Garage and house floors with no embedded steel	GEN 3	ST4	S2
Wearing surface: Light foot and trolley traffic	RC30	ST4	S2
Wearing surface: General industrial	RC40	N/A	S2
Wearing surface: Heavy industrial	RC50	N/A	S2
Paving			
House drives, domestic parking and external parking	PAV 1	N/A	S2
Heavy-duty external paving	PAV 2	N/A	S2

CONCRETE WORK

Prescribed Mixes for Ordinary Structural Concrete

Weights of cement and total dry aggregates in kg to produce approximately one cubic metre of fully compacted concrete together with the percentages by weight of fine aggregate in total dry aggregates

Conc. grade	Nominal max size of aggregate (mm)	40		20		14		10	
	Workability	Med.	High	Med.	High	Med.	High	Med.	High
	Limits to slump that may be expected (mm)	50–100	100–150	25–75	75–125	10–50	50–100	10–25	25–50
7	Cement (kg)	180	200	210	230	–	–	–	–
	Total aggregate (kg)	1950	1850	1900	1800	–	–	–	–
	Fine aggregate (%)	30–45	30–45	35–50	35–50	–	–	–	–
10	Cement (kg)	210	230	240	260	–	–	–	–
	Total aggregate (kg)	1900	1850	1850	1800	–	–	–	–
	Fine aggregate (%)	30–45	30–45	35–50	35–50	–	–	–	–
15	Cement (kg)	250	270	280	310	–	–	–	–
	Total aggregate (kg)	1850	1800	1800	1750	–	–	–	–
	Fine aggregate (%)	30–45	30–45	35–50	35–50	–	–	–	–
20	Cement (kg)	300	320	320	350	340	380	360	410
	Total aggregate (kg)	1850	1750	1800	1750	1750	1700	1750	1650
	Sand								
	Zone 1 (%)	35	40	40	45	45	50	50	55
	Zone 2 (%)	30	35	35	40	40	45	45	50
	Zone 3 (%)	30	30	30	35	35	40	40	45
25	Cement (kg)	340	360	360	390	380	420	400	450
	Total aggregate (kg)	1800	1750	1750	1700	1700	1650	1700	1600
	Sand								
	Zone 1 (%)	35	40	40	45	45	50	50	55
	Zone 2 (%)	30	35	35	40	40	45	45	50
	Zone 3 (%)	30	30	30	35	35	40	40	45
30	Cement (kg)	370	390	400	430	430	470	460	510
	Total aggregate (kg)	1750	1700	1700	1650	1700	1600	1650	1550
	Sand								
	Zone 1 (%)	35	40	40	45	45	50	50	55
	Zone 2 (%)	30	35	35	40	40	45	45	50
	Zone 3 (%)	30	30	30	35	35	40	40	45

REINFORCEMENT

Weights of Bar Reinforcement

Nominal sizes (mm)	Cross-sectional area (mm²)	Mass (kg/m)	Length of bar (m/tonne)
6	28.27	0.222	4505
8	50.27	0.395	2534
10	78.54	0.617	1622
12	113.10	0.888	1126
16	201.06	1.578	634
20	314.16	2.466	405
25	490.87	3.853	260
32	804.25	6.313	158
40	1265.64	9.865	101
50	1963.50	15.413	65

Weights of Bars (at specific spacings)

Weights of metric bars in kilogrammes per square metre

Size (mm)	\multicolumn{10}{c}{Spacing of bars in millimetres}
	75	100	125	150	175	200	225	250	275	300
6	2.96	2.220	1.776	1.480	1.27	1.110	0.99	0.89	0.81	0.74
8	5.26	3.95	3.16	2.63	2.26	1.97	1.75	1.58	1.44	1.32
10	8.22	6.17	4.93	4.11	3.52	3.08	2.74	2.47	2.24	2.06
12	11.84	8.88	7.10	5.92	5.07	4.44	3.95	3.55	3.23	2.96
16	21.04	15.78	12.63	10.52	9.02	7.89	7.02	6.31	5.74	5.26
20	32.88	24.66	19.73	16.44	14.09	12.33	10.96	9.87	8.97	8.22
25	51.38	38.53	30.83	25.69	22.02	19.27	17.13	15.41	14.01	12.84
32	84.18	63.13	50.51	42.09	36.08	31.57	28.06	25.25	22.96	21.04
40	131.53	98.65	78.92	65.76	56.37	49.32	43.84	39.46	35.87	32.88
50	205.51	154.13	123.31	102.76	88.08	77.07	68.50	61.65	56.05	51.38

Basic weight of steelwork taken as 7850 kg/m³
Basic weight of bar reinforcement per metre run = 0.00785 kg/mm²
The value of π has been taken as 3.141592654

REINFORCEMENT

Fabric Reinforcement

Preferred range of designated fabric types and stock sheet sizes

Fabric reference	Longitudinal wires			Cross wires			
	Nominal wire size (mm)	Pitch (mm)	Area (mm²/m)	Nominal wire size (mm)	Pitch (mm)	Area (mm²/m)	Mass (kg/m²)
Square mesh							
A393	10	200	393	10	200	393	6.16
A252	8	200	252	8	200	252	3.95
A193	7	200	193	7	200	193	3.02
A142	6	200	142	6	200	142	2.22
A98	5	200	98	5	200	98	1.54
Structural mesh							
B1131	12	100	1131	8	200	252	10.90
B785	10	100	785	8	200	252	8.14
B503	8	100	503	8	200	252	5.93
B385	7	100	385	7	200	193	4.53
B283	6	100	283	7	200	193	3.73
B196	5	100	196	7	200	193	3.05
Long mesh							
C785	10	100	785	6	400	70.8	6.72
C636	9	100	636	6	400	70.8	5.55
C503	8	100	503	5	400	49.0	4.34
C385	7	100	385	5	400	49.0	3.41
C283	6	100	283	5	400	49.0	2.61
Wrapping mesh							
D98	5	200	98	5	200	98	1.54
D49	2.5	100	49	2.5	100	49	0.77

Stock sheet size 4.8 m × 2.4 m, Area 11.52 m²

Average weight kg/m³ of steelwork reinforcement in concrete for various building elements

Substructure	kg/m³ concrete	Substructure	kg/m³ concrete
Pile caps	110–150	Plate slab	150–220
Tie beams	130–170	Cant slab	145–210
Ground beams	230–330	Ribbed floors	130–200
Bases	125–180	Topping to block floor	30–40
Footings	100–150	Columns	210–310
Retaining walls	150–210	Beams	250–350
Raft	60–70	Stairs	130–170
Slabs – one way	120–200	Walls – normal	40–100
Slabs – two way	110–220	Walls – wind	70–125

Note: For exposed elements add the following %:
Walls 50%, Beams 100%, Columns 15%

FORMWORK

Formwork Stripping Times – Normal Curing Periods

Conditions under which concrete is maturing	Minimum periods of protection for different types of cement					
	Number of days (where the average surface temperature of the concrete exceeds 10°C during the whole period)			Equivalent maturity (degree hours) calculated as the age of the concrete in hours multiplied by the number of degrees Celsius by which the average surface temperature of the concrete exceeds 10°C		
	Other	SRPC	OPC or RHPC	Other	SRPC	OPC or RHPC
1. Hot weather or drying winds	7	4	3	3500	2000	1500
2. Conditions not covered by 1	4	3	2	2000	1500	1000

KEY
OPC – Ordinary Portland Cement
RHPC – Rapid-hardening Portland Cement
SRPC – Sulphate-resisting Portland Cement

Minimum Period before Striking Formwork

	Minimum period before striking		
	Surface temperature of concrete		
	16°C	17°C	t°C (0–25)
Vertical formwork to columns, walls and large beams	12 hours	18 hours	300 hours t+10
Soffit formwork to slabs	4 days	6 days	100 days t+10
Props to slabs	10 days	15 days	250 days t+10
Soffit formwork to beams	9 days	14 days	230 days t+10
Props to beams	14 days	21 days	360 days t+10

MASONRY

Number of Bricks required for Various Types of Work per m² of Walling

Description	Brick size	
	215 × 102.5 × 50 mm	215 × 102.5 × 65 mm
Half brick thick		
Stretcher bond	74	59
English bond	108	86
English garden wall bond	90	72
Flemish bond	96	79
Flemish garden wall bond	83	66
One brick thick and cavity wall of two half brick skins		
Stretcher bond	148	119

Quantities of Bricks and Mortar required per m² of Walling

	Unit	No of bricks required	Mortar required (cubic metres)		
			No frogs	Single frogs	Double frogs
Standard bricks					
Brick size 215 × 102.5 × 50 mm					
half brick wall (103 mm)	m²	72	0.022	0.027	0.032
2 × half brick cavity wall (270 mm)	m²	144	0.044	0.054	0.064
one brick wall (215 mm)	m²	144	0.052	0.064	0.076
one and a half brick wall (322 mm)	m²	216	0.073	0.091	0.108
Mass brickwork	m³	576	0.347	0.413	0.480
Brick size 215 × 102.5 × 65 mm					
half brick wall (103 mm)	m²	58	0.019	0.022	0.026
2 × half brick cavity wall (270 mm)	m²	116	0.038	0.045	0.055
one brick wall (215 mm)	m²	116	0.046	0.055	0.064
one and a half brick wall (322 mm)	m²	174	0.063	0.074	0.088
Mass brickwork	m³	464	0.307	0.360	0.413
Metric modular bricks			Perforated		
Brick size 200 × 100 × 75 mm					
90 mm thick	m²	67	0.016	0.019	
190 mm thick	m²	133	0.042	0.048	
290 mm thick	m²	200	0.068	0.078	
Brick size 200 × 100 × 100 mm					
90 mm thick	m²	50	0.013	0.016	
190 mm thick	m²	100	0.036	0.041	
290 mm thick	m²	150	0.059	0.067	
Brick size 300 × 100 × 75 mm					
90 mm thick	m²	33	–	0.015	
Brick size 300 × 100 × 100 mm					
90 mm thick	m²	44	0.015	0.018	

Note: Assuming 10 mm thick joints

MASONRY

Mortar required per m² Blockwork (9.88 blocks/m²)

Wall thickness	75	90	100	125	140	190	215
Mortar m³/m²	0.005	0.006	0.007	0.008	0.009	0.013	0.014

Mortar Group	Cement: lime: sand	Masonry cement: sand	Cement: sand with plasticizer
1	1:0–0.25:3		
2	1:0.5:4–4.5	1:2.5-3.5	1:3–4
3	1:1:5–6	1:4–5	1:5–6
4	1:2:8–9	1:5.5–6.5	1:7–8
5	1:3:10–12	1:6.5–7	1:8

Group 1: strong inflexible mortar
Group 5: weak but flexible

All mixes within a group are of approximately similar strength
Frost resistance increases with the use of plasticizers
Cement: lime: sand mixes give the strongest bond and greatest resistance to rain penetration
Masonry cement equals ordinary Portland cement plus a fine neutral mineral filler and an air entraining agent

Calcium Silicate Bricks

Type	Strength	Location
Class 2 crushing strength	14.0 N/mm²	not suitable for walls
Class 3	20.5 N/mm²	walls above dpc
Class 4	27.5 N/mm²	cappings and copings
Class 5	34.5 N/mm²	retaining walls
Class 6	41.5 N/mm²	walls below ground
Class 7	48.5 N/mm²	walls below ground

The Class 7 calcium silicate bricks are therefore equal in strength to Class B bricks
Calcium silicate bricks are not suitable for DPCs

Durability of Bricks	
FL	Frost resistant with low salt content
FN	Frost resistant with normal salt content
ML	Moderately frost resistant with low salt content
MN	Moderately frost resistant with normal salt content

MASONRY

Brickwork Dimensions

No. of horizontal bricks	Dimensions (mm)	No. of vertical courses	Height of vertical courses (mm)
½	112.5	1	75
1	225.0	2	150
1½	337.5	3	225
2	450.0	4	300
2½	562.5	5	375
3	675.0	6	450
3½	787.5	7	525
4	900.0	8	600
4½	1012.5	9	675
5	1125.0	10	750
5½	1237.5	11	825
6	1350.0	12	900
6½	1462.5	13	975
7	1575.0	14	1050
7½	1687.5	15	1125
8	1800.0	16	1200
8½	1912.5	17	1275
9	2025.0	18	1350
9½	2137.5	19	1425
10	2250.0	20	1500
20	4500.0	24	1575
40	9000.0	28	2100
50	11250.0	32	2400
60	13500.0	36	2700
75	16875.0	40	3000

TIMBER

Weights of Timber

Material	kg/m³	lb/cu ft
General	806 (avg)	50 (avg)
Douglas fir	479	30
Yellow pine, spruce	479	30
Pitch pine	673	42
Larch, elm	561	35
Oak (English)	724 to 959	45 to 60
Teak	643 to 877	40 to 55
Jarrah	959	60
Greenheart	1040 to 1204	65 to 75
Quebracho	1285	80

Material	kg/m² per mm thickness	lb/sq ft per inch thickness
Wooden boarding and blocks		
Softwood	0.48	2.5
Hardwood	0.76	4
Hardboard	1.06	5.5
Chipboard	0.76	4
Plywood	0.62	3.25
Blockboard	0.48	2.5
Fibreboard	0.29	1.5
Wood-wool	0.58	3
Plasterboard	0.96	5
Weather boarding	0.35	1.8

TIMBER

Conversion Tables (for timber only)

Inches	Millimetres	Feet	Metres
1	25	1	0.300
2	50	2	0.600
3	75	3	0.900
4	100	4	1.200
5	125	5	1.500
6	150	6	1.800
7	175	7	2.100
8	200	8	2.400
9	225	9	2.700
10	250	10	3.000
11	275	11	3.300
12	300	12	3.600
13	325	13	3.900
14	350	14	4.200
15	375	15	4.500
16	400	16	4.800
17	425	17	5.100
18	450	18	5.400
19	475	19	5.700
20	500	20	6.000
21	525	21	6.300
22	550	22	6.600
23	575	23	6.900
24	600	24	7.200

Planed Softwood

The finished end section size of planed timber is usually 3/16" less than the original size from which it is produced. This however varies slightly depending upon availability of material and origin of the species used.

Standards (timber) to cubic metres and cubic metres to standards (timber)

Cubic metres	Cubic metres standards	Standards
4.672	1	0.214
9.344	2	0.428
14.017	3	0.642
18.689	4	0.856
23.361	5	1.070
28.033	6	1.284
32.706	7	1.498
37.378	8	1.712
42.050	9	1.926
46.722	10	2.140
93.445	20	4.281
140.167	30	6.421
186.890	40	8.561
233.612	50	10.702
280.335	60	12.842
327.057	70	14.982
373.779	80	17.122

TIMBER

1 cu metre = 35.3148 cu ft = 0.21403 std

1 cu ft = 0.028317 cu metres

1 std = 4.67227 cu metres

Basic sizes of sawn softwood available (cross-sectional areas)

Thickness (mm)	Width (mm)								
	75	100	125	150	175	200	225	250	300
16	X	X	X	X					
19	X	X	X	X					
22	X	X	X	X					
25	X	X	X	X	X	X	X	X	X
32	X	X	X	X	X	X	X	X	X
36	X	X	X	X					
38	X	X	X	X	X	X	X		
44	X	X	X	X	X	X	X	X	X
47*	X	X	X	X	X	X	X	X	X
50	X	X	X	X	X	X	X	X	X
63	X	X	X	X	X	X	X		
75	X	X	X	X	X	X	X	X	
100		X		X		X		X	X
150				X		X			X
200						X			
250								X	
300									X

* This range of widths for 47 mm thickness will usually be found to be available in construction quality only

Note: The smaller sizes below 100 mm thick and 250 mm width are normally but not exclusively of European origin. Sizes beyond this are usually of North and South American origin

Basic lengths of sawn softwood available (metres)

1.80	2.10	3.00	4.20	5.10	6.00	7.20
	2.40	3.30	4.50	5.40	6.30	
	2.70	3.60	4.80	5.70	6.60	
		3.90			6.90	

Note: Lengths of 6.00 m and over will generally only be available from North American species and may have to be recut from larger sizes

TIMBER

Reductions from basic size to finished size by planning of two opposed faces

Purpose	Reductions from basic sizes for timber			
	15–35 mm	36–100 mm	101–150 mm	over 150 mm
a) Constructional timber	3 mm	3 mm	5 mm	6 mm
b) Matching interlocking boards	4 mm	4 mm	6 mm	6 mm
c) Wood trim not specified in BS 584	5 mm	7 mm	7 mm	9 mm
d) Joinery and cabinet work	7 mm	9 mm	11 mm	13 mm

Note: The reduction of width or depth is overall the extreme size and is exclusive of any reduction of the face by the machining of a tongue or lap joints

Maximum Spans for Various Roof Trusses

Maximum permissible spans for rafters for Fink trussed rafters

Basic size (mm)	Actual size (mm)	Pitch (degrees)								
		15 (m)	17.5 (m)	20 (m)	22.5 (m)	25 (m)	27.5 (m)	30 (m)	32.5 (m)	35 (m)
38 × 75	35 × 72	6.03	6.16	6.29	6.41	6.51	6.60	6.70	6.80	6.90
38 × 100	35 × 97	7.48	7.67	7.83	7.97	8.10	8.22	8.34	8.47	8.61
38 × 125	35 × 120	8.80	9.00	9.20	9.37	9.54	9.68	9.82	9.98	10.16
44 × 75	41 × 72	6.45	6.59	6.71	6.83	6.93	7.03	7.14	7.24	7.35
44 × 100	41 × 97	8.05	8.23	8.40	8.55	8.68	8.81	8.93	9.09	9.22
44 × 125	41 × 120	9.38	9.60	9.81	9.99	10.15	10.31	10.45	10.64	10.81
50 × 75	47 × 72	6.87	7.01	7.13	7.25	7.35	7.45	7.53	7.67	7.78
50 × 100	47 × 97	8.62	8.80	8.97	9.12	9.25	9.38	9.50	9.66	9.80
50 × 125	47 × 120	10.01	10.24	10.44	10.62	10.77	10.94	11.00	11.00	11.00

TIMBER

Sizes of Internal and External Doorsets

Description	Internal size (mm)	Permissible deviation	External size (mm)	Permissible deviation
Coordinating dimension: height of door leaf height sets	2100		2100	
Coordinating dimension: height of ceiling height set	2300 2350 2400 2700 3000		2300 2350 2400 2700 3000	
Coordinating dimension: width of all doorsets S = Single leaf set D = Double leaf set	600 S 700 S 800 S&D 900 S&D 1000 S&D 1200 D 1500 D 1800 D 2100 D		900 S 1000 S 1200 D 1800 D 2100 D	
Work size: height of door leaf height set	2090	± 2.0	2095	± 2.0
Work size: height of ceiling height set	2285 2335 2385 2685 2985	± 2.0	2295 2345 2395 2695 2995	± 2.0
Work size: width of all doorsets S = Single leaf set D = Double leaf set	590 S 690 S 790 S&D 890 S&D 990 S&D 1190 D 1490 D 1790 D 2090 D	± 2.0	895 S 995 S 1195 D 1495 D 1795 D 2095 D	± 2.0
Width of door leaf in single leaf sets F = Flush leaf P = Panel leaf	526 F 626 F 726 F&P 826 F&P 926 F&P	± 1.5	806 F&P 906 F&P	± 1.5
Width of door leaf in double leaf sets F = Flush leaf P = Panel leaf	362 F 412 F 426 F 562 F&P 712 F&P 826 F&P 1012 F&P	± 1.5	552 F&P 702 F&P 852 F&P 1002 F&P	± 1.5
Door leaf height for all doorsets	2040	± 1.5	1994	± 1.5

ROOFING

Total Roof Loadings for Various Types of Tiles/Slates

	Roof load (slope) kg/m²		
	Slate/Tile	Roofing underlay and battens²	Total dead load kg/m
Asbestos cement slate (600 × 300)	21.50	3.14	24.64
Clay tile interlocking	67.00	5.50	72.50
plain	43.50	2.87	46.37
Concrete tile interlocking	47.20	2.69	49.89
plain	78.20	5.50	83.70
Natural slate (18" × 10")	35.40	3.40	38.80
	Roof load (plan) kg/m²		
Asbestos cement slate (600 × 300)	28.45	76.50	104.95
Clay tile interlocking	53.54	76.50	130.04
plain	83.71	76.50	60.21
Concrete tile interlocking	57.60	76.50	134.10
plain	96.64	76.50	173.14

ROOFING

Tiling Data

Product		Lap (mm)	Gauge of battens	No. slates per m^2	Battens (m/m^2)	Weight as laid (kg/m^2)
CEMENT SLATES						
Eternit slates (Duracem)	600 × 300 mm	100	250	13.4	4.00	19.50
		90	255	13.1	3.92	19.20
		80	260	12.9	3.85	19.00
		70	265	12.7	3.77	18.60
	600 × 350 mm	100	250	11.5	4.00	19.50
		90	255	11.2	3.92	19.20
	500 × 250 mm	100	200	20.0	5.00	20.00
		90	205	19.5	4.88	19.50
		80	210	19.1	4.76	19.00
		70	215	18.6	4.65	18.60
	400 × 200 mm	90	155	32.3	6.45	20.80
		80	160	31.3	6.25	20.20
		70	165	30.3	6.06	19.60
CONCRETE TILES/SLATES						
Redland Roofing						
Stonewold slate	430 × 380 mm	75	355	8.2	2.82	51.20
Double Roman tile	418 × 330 mm	75	355	8.2	2.91	45.50
Grovebury pantile	418 × 332 mm	75	343	9.7	2.91	47.90
Norfolk pantile	381 × 227 mm	75	306	16.3	3.26	44.01
		100	281	17.8	3.56	48.06
Renown interlocking tile	418 × 330 mm	75	343	9.7	2.91	46.40
'49' tile	381 × 227 mm	75	306	16.3	3.26	44.80
		100	281	17.8	3.56	48.95
Plain, vertical tiling	265 × 165 mm	35	115	52.7	8.70	62.20
Marley Roofing						
Bold roll tile	420 × 330 mm	75	344	9.7	2.90	47.00
		100	–	10.5	3.20	51.00
Modern roof tile	420 × 330 mm	75	338	10.2	3.00	54.00
		100	–	11.0	3.20	58.00
Ludlow major	420 × 330 mm	75	338	10.2	3.00	45.00
		100	–	11.0	3.20	49.00
Ludlow plus	387 × 229 mm	75	305	16.1	3.30	47.00
		100	–	17.5	3.60	51.00
Mendip tile	420 × 330 mm	75	338	10.2	3.00	47.00
		100	–	11.0	3.20	51.00
Wessex	413 × 330 mm	75	338	10.2	3.00	54.00
		100	–	11.0	3.20	58.00
Plain tile	267 × 165 mm	65	100	60.0	10.00	76.00
		75	95	64.0	10.50	81.00
		85	90	68.0	11.30	86.00
Plain vertical tiles (feature)	267 × 165 mm	35	110	53.0	8.70	67.00
		34	115	56.0	9.10	71.00

ROOFING

Slate Nails, Quantity per Kilogram

Length	Type			
	Plain wire	Galvanized wire	Copper nail	Zinc nail
28.5 mm	325	305	325	415
34.4 mm	286	256	254	292
50.8 mm	242	224	194	200

Metal Sheet Coverings

Thicknesses and weights of sheet metal coverings									
Lead to BS 1178									
BS Code No	3	4	5	6	7	8			
Colour code	Green	Blue	Red	Black	White	Orange			
Thickness (mm)	1.25	1.80	2.24	2.50	3.15	3.55			
Density kg/m^2	14.18	20.41	25.40	30.05	35.72	40.26			
Copper to BS 2870									
Thickness (mm)		0.60	0.70						
Bay width									
Roll (mm)		500	650						
Seam (mm)		525	600						
Standard width to form bay	600	750							
Normal length of sheet	1.80	1.80							
Zinc to BS 849									
Zinc Gauge (Nr)	9	10	11	12	13	14	15	16	
Thickness (mm)	0.43	0.48	0.56	0.64	0.71	0.79	0.91	1.04	
Density (kg/m^2)	3.1	3.2	3.8	4.3	4.8	5.3	6.2	7.0	
Aluminium to BS 4868									
Thickness (mm)	0.5	0.6	0.7	0.8	0.9	1.0	1.2		
Density (kg/m^2)	12.8	15.4	17.9	20.5	23.0	25.6	30.7		

ROOFING

Type of felt	Nominal mass per unit area (kg/10 m)	Nominal mass per unit area of fibre base (g/m^2)	Nominal length of roll (m)
Class 1			
1B fine granule	14	220	10 or 20
surfaced bitumen	18	330	10 or 20
	25	470	10
1E mineral surfaced bitumen	38	470	10
1F reinforced bitumen	15	160 (fibre) 110 (hessian)	15
1F reinforced bitumen, aluminium faced	13	160 (fibre) 110 (hessian)	15
Class 2			
2B fine granule surfaced bitumen asbestos	18	500	10 or 20
2E mineral surfaced bitumen asbestos	38	600	10
Class 3			
3B fine granule surfaced bitumen glass fibre	18	60	20
3E mineral surfaced bitumen glass fibre	28	60	10
3E venting base layer bitumen glass fibre	32	60*	10
3H venting base layer bitumen glass fibre	17	60*	20

* Excluding effect of perforations

GLAZING

GLAZING

Nominal thickness (mm)	Tolerance on thickness (mm)	Approximate weight (kg/m²)	Normal maximum size (mm)
Float and polished plate glass			
3	+ 0.2	7.50	2140 × 1220
4	+ 0.2	10.00	2760 × 1220
5	+ 0.2	12.50	3180 × 2100
6	+ 0.2	15.00	4600 × 3180
10	+ 0.3	25.00)	6000 × 3300
12	+ 0.3	30.00)	
15	+ 0.5	37.50	3050 × 3000
19	+ 1.0	47.50)	3000 × 2900
25	+ 1.0	63.50)	
Clear sheet glass			
2 *	+ 0.2	5.00	1920 × 1220
3	+ 0.3	7.50	2130 × 1320
4	+ 0.3	10.00	2760 × 1220
5 *	+ 0.3	12.50)	2130 × 2400
6 *	+ 0.3	15.00)	
Cast glass			
3	+ 0.4 / − 0.2	6.00)	2140 × 1280
4	+ 0.5	7.50)	
5	+ 0.5	9.50	2140 × 1320
6	+ 0.5	11.50)	3700 × 1280
10	+ 0.8	21.50)	
Wired glass (Cast wired glass)			
6	+ 0.3 / − 0.7	−)	3700 × 1840
7	+ 0.7	−)	
(Polished wire glass)			
6	+ 1.0	−	330 × 1830

* The 5 mm and 6 mm thickness are known as *thick drawn sheet*. Although 2 mm sheet glass is available it is not recommended for general glazing purposes

METAL

METAL

Weights of Metals

Material	kg/m³	lb/cu ft
Metals, steel construction, etc.		
Iron		
– cast	7207	450
– wrought	7687	480
– ore – general	2407	150
– (crushed) Swedish	3682	230
Steel	7854	490
Copper		
– cast	8731	545
– wrought	8945	558
Brass	8497	530
Bronze	8945	558
Aluminium	2774	173
Lead	11322	707
Zinc (rolled)	7140	446
	g/mm² per metre	lb/sq ft per foot
Steel bars	7.85	3.4
Structural steelwork	Net weight of member @ 7854 kg/m³	
riveted	+ 10% for cleats, rivets, bolts, etc.	
welded	+ 1.25% to 2.5% for welds, etc.	
Rolled sections		
beams	+ 2.5%	
stanchions	+ 5% (extra for caps and bases)	
Plate		
web girders	+ 10% for rivets or welds, stiffeners, etc.	
	kg/m	lb/ft
Steel stairs: industrial type		
1 m or 3 ft wide	84	56
Steel tubes		
50 mm or 2 in bore	5 to 6	3 to 4
Gas piping		
20 mm or ¾ in	2	1¼

Universal Beams BS 4: Part 1: 2005

Designation	Mass (kg/m)	Depth of section (mm)	Width of section (mm)	Thickness Web (mm)	Thickness Flange (mm)	Surface area (m²/m)
1016 × 305 × 487	487.0	1036.1	308.5	30.0	54.1	3.20
1016 × 305 × 438	438.0	1025.9	305.4	26.9	49.0	3.17
1016 × 305 × 393	393.0	1016.0	303.0	24.4	43.9	3.15
1016 × 305 × 349	349.0	1008.1	302.0	21.1	40.0	3.13
1016 × 305 × 314	314.0	1000.0	300.0	19.1	35.9	3.11
1016 × 305 × 272	272.0	990.1	300.0	16.5	31.0	3.10
1016 × 305 × 249	249.0	980.2	300.0	16.5	26.0	3.08
1016 × 305 × 222	222.0	970.3	300.0	16.0	21.1	3.06
914 × 419 × 388	388.0	921.0	420.5	21.4	36.6	3.44
914 × 419 × 343	343.3	911.8	418.5	19.4	32.0	3.42
914 × 305 × 289	289.1	926.6	307.7	19.5	32.0	3.01
914 × 305 × 253	253.4	918.4	305.5	17.3	27.9	2.99
914 × 305 × 224	224.2	910.4	304.1	15.9	23.9	2.97
914 × 305 × 201	200.9	903.0	303.3	15.1	20.2	2.96
838 × 292 × 226	226.5	850.9	293.8	16.1	26.8	2.81
838 × 292 × 194	193.8	840.7	292.4	14.7	21.7	2.79
838 × 292 × 176	175.9	834.9	291.7	14.0	18.8	2.78
762 × 267 × 197	196.8	769.8	268.0	15.6	25.4	2.55
762 × 267 × 173	173.0	762.2	266.7	14.3	21.6	2.53
762 × 267 × 147	146.9	754.0	265.2	12.8	17.5	2.51
762 × 267 × 134	133.9	750.0	264.4	12.0	15.5	2.51
686 × 254 × 170	170.2	692.9	255.8	14.5	23.7	2.35
686 × 254 × 152	152.4	687.5	254.5	13.2	21.0	2.34
686 × 254 × 140	140.1	383.5	253.7	12.4	19.0	2.33
686 × 254 × 125	125.2	677.9	253.0	11.7	16.2	2.32
610 × 305 × 238	238.1	635.8	311.4	18.4	31.4	2.45
610 × 305 × 179	179.0	620.2	307.1	14.1	23.6	2.41
610 × 305 × 149	149.1	612.4	304.8	11.8	19.7	2.39
610 × 229 × 140	139.9	617.2	230.2	13.1	22.1	2.11
610 × 229 × 125	125.1	612.2	229.0	11.9	19.6	2.09
610 × 229 × 113	113.0	607.6	228.2	11.1	17.3	2.08
610 × 229 × 101	101.2	602.6	227.6	10.5	14.8	2.07
533 × 210 × 122	122.0	544.5	211.9	12.7	21.3	1.89
533 × 210 × 109	109.0	539.5	210.8	11.6	18.8	1.88
533 × 210 × 101	101.0	536.7	210.0	10.8	17.4	1.87
533 × 210 × 92	92.1	533.1	209.3	10.1	15.6	1.86
533 × 210 × 82	82.2	528.3	208.8	9.6	13.2	1.85
457 × 191 × 98	98.3	467.2	192.8	11.4	19.6	1.67
457 × 191 × 89	89.3	463.4	191.9	10.5	17.7	1.66
457 × 191 × 82	82.0	460.0	191.3	9.9	16.0	1.65
457 × 191 × 74	74.3	457.0	190.4	9.0	14.5	1.64
457 × 191 × 67	67.1	453.4	189.9	8.5	12.7	1.63
457 × 152 × 82	82.1	465.8	155.3	10.5	18.9	1.51
457 × 152 × 74	74.2	462.0	154.4	9.6	17.0	1.50
457 × 152 × 67	67.2	458.0	153.8	9.0	15.0	1.50
457 × 152 × 60	59.8	454.6	152.9	8.1	13.3	1.50
457 × 152 × 52	52.3	449.8	152.4	7.6	10.9	1.48
406 × 178 × 74	74.2	412.8	179.5	9.5	16.0	1.51
406 × 178 × 67	67.1	409.4	178.8	8.8	14.3	1.50
406 × 178 × 60	60.1	406.4	177.9	7.9	12.8	1.49

METAL

Designation	Mass (kg/m)	Depth of section (mm)	Width of section (mm)	Thickness Web (mm)	Thickness Flange (mm)	Surface area (m²/m)
406 × 178 × 50	54.1	402.6	177.7	7.7	10.9	1.48
406 × 140 × 46	46.0	403.2	142.2	6.8	11.2	1.34
406 × 140 × 39	39.0	398.0	141.8	6.4	8.6	1.33
356 × 171 × 67	67.1	363.4	173.2	9.1	15.7	1.38
356 × 171 × 57	57.0	358.0	172.2	8.1	13.0	1.37
356 × 171 × 51	51.0	355.0	171.5	7.4	11.5	1.36
356 × 171 × 45	45.0	351.4	171.1	7.0	9.7	1.36
356 × 127 × 39	39.1	353.4	126.0	6.6	10.7	1.18
356 × 127 × 33	33.1	349.0	125.4	6.0	8.5	1.17
305 × 165 × 54	54.0	310.4	166.9	7.9	13.7	1.26
305 × 165 × 46	46.1	306.6	165.7	6.7	11.8	1.25
305 × 165 × 40	40.3	303.4	165.0	6.0	10.2	1.24
305 × 127 × 48	48.1	311.0	125.3	9.0	14.0	1.09
305 × 127 × 42	41.9	307.2	124.3	8.0	12.1	1.08
305 × 127 × 37	37.0	304.4	123.3	7.1	10.7	1.07
305 × 102 × 33	32.8	312.7	102.4	6.6	10.8	1.01
305 × 102 × 28	28.2	308.7	101.8	6.0	8.8	1.00
305 × 102 × 25	24.8	305.1	101.6	5.8	7.0	0.992
254 × 146 × 43	43.0	259.6	147.3	7.2	12.7	1.08
254 × 146 × 37	37.0	256.0	146.4	6.3	10.9	1.07
254 × 146 × 31	31.1	251.4	146.1	6.0	8.6	1.06
254 × 102 × 28	28.3	260.4	102.2	6.3	10.0	0.904
254 × 102 × 25	25.2	257.2	101.9	6.0	8.4	0.897
254 × 102 × 22	22.0	254.0	101.6	5.7	6.8	0.890
203 × 133 × 30	30.0	206.8	133.9	6.4	9.6	0.923
203 × 133 × 25	25.1	203.2	133.2	5.7	7.8	0.915
203 × 102 × 23	23.1	203.2	101.8	5.4	9.3	0.790
178 × 102 × 19	19.0	177.8	101.2	4.8	7.9	0.738
152 × 89 × 16	16.0	152.4	88.7	4.5	7.7	0.638
127 × 76 × 13	13.0	127.0	76.0	4.0	7.6	0.537

METAL

Universal Columns BS 4: Part 1: 2005

Designation	Mass (kg/m)	Depth of section (mm)	Width of section (mm)	Thickness Web (mm)	Thickness Flange (mm)	Surface area (m²/m)
356 × 406 × 634	633.9	474.7	424.0	47.6	77.0	2.52
356 × 406 × 551	551.0	455.6	418.5	42.1	67.5	2.47
356 × 406 × 467	467.0	436.6	412.2	35.8	58.0	2.42
356 × 406 × 393	393.0	419.0	407.0	30.6	49.2	2.38
356 × 406 × 340	339.9	406.4	403.0	26.6	42.9	2.35
356 × 406 × 287	287.1	393.6	399.0	22.6	36.5	2.31
356 × 406 × 235	235.1	381.0	384.8	18.4	30.2	2.28
356 × 368 × 202	201.9	374.6	374.7	16.5	27.0	2.19
356 × 368 × 177	177.0	368.2	372.6	14.4	23.8	2.17
356 × 368 × 153	152.9	362.0	370.5	12.3	20.7	2.16
356 × 368 × 129	129.0	355.6	368.6	10.4	17.5	2.14
305 × 305 × 283	282.9	365.3	322.2	26.8	44.1	1.94
305 × 305 × 240	240.0	352.5	318.4	23.0	37.7	1.91
305 × 305 × 198	198.1	339.9	314.5	19.1	31.4	1.87
305 × 305 × 158	158.1	327.1	311.2	15.8	25.0	1.84
305 × 305 × 137	136.9	320.5	309.2	13.8	21.7	1.82
305 × 305 × 118	117.9	314.5	307.4	12.0	18.7	1.81
305 × 305 × 97	96.9	307.9	305.3	9.9	15.4	1.79
254 × 254 × 167	167.1	289.1	265.2	19.2	31.7	1.58
254 × 254 × 132	132.0	276.3	261.3	15.3	25.3	1.55
254 × 254 × 107	107.1	266.7	258.8	12.8	20.5	1.52
254 × 254 × 89	88.9	260.3	256.3	10.3	17.3	1.50
254 × 254 × 73	73.1	254.1	254.6	8.6	14.2	1.49
203 × 203 × 86	86.1	222.2	209.1	12.7	20.5	1.24
203 × 203 × 71	71.0	215.8	206.4	10.0	17.3	1.22
203 × 203 × 60	60.0	209.6	205.8	9.4	14.2	1.21
203 × 203 × 52	52.0	206.2	204.3	7.9	12.5	1.20
203 × 203 × 46	46.1	203.2	203.6	7.2	11.0	1.19
152 × 152 × 37	37.0	161.8	154.4	8.0	11.5	0.912
152 × 152 × 30	30.0	157.6	152.9	6.5	9.4	0.901
152 × 152 × 23	23.0	152.4	152.2	5.8	6.8	0.889

METAL

Joists BS 4: Part 1: 2005 (retained for reference, Corus have ceased manufacture in UK)

Designation	Mass (kg/m)	Depth of section (mm)	Width of section (mm)	Thickness Web (mm)	Thickness Flange (mm)	Surface area (m²/m)
254 × 203 × 82	82.0	254.0	203.2	10.2	19.9	1.210
203 × 152 × 52	52.3	203.2	152.4	8.9	16.5	0.932
152 × 127 × 37	37.3	152.4	127.0	10.4	13.2	0.737
127 × 114 × 29	29.3	127.0	114.3	10.2	11.5	0.646
127 × 114 × 27	26.9	127.0	114.3	7.4	11.4	0.650
102 × 102 × 23	23.0	101.6	101.6	9.5	10.3	0.549
102 × 44 × 7	7.5	101.6	44.5	4.3	6.1	0.350
89 × 89 × 19	19.5	88.9	88.9	9.5	9.9	0.476
76 × 76 × 13	12.8	76.2	76.2	5.1	8.4	0.411

Parallel Flange Channels

Designation	Mass (kg/m)	Depth of section (mm)	Width of section (mm)	Thickness Web (mm)	Thickness Flange (mm)	Surface area (m²/m)
430 × 100 × 64	64.4	430	100	11.0	19.0	1.23
380 × 100 × 54	54.0	380	100	9.5	17.5	1.13
300 × 100 × 46	45.5	300	100	9.0	16.5	0.969
300 × 90 × 41	41.4	300	90	9.0	15.5	0.932
260 × 90 × 35	34.8	260	90	8.0	14.0	0.854
260 × 75 × 28	27.6	260	75	7.0	12.0	0.79
230 × 90 × 32	32.2	230	90	7.5	14.0	0.795
230 × 75 × 26	25.7	230	75	6.5	12.5	0.737
200 × 90 × 30	29.7	200	90	7.0	14.0	0.736
200 × 75 × 23	23.4	200	75	6.0	12.5	0.678
180 × 90 × 26	26.1	180	90	6.5	12.5	0.697
180 × 75 × 20	20.3	180	75	6.0	10.5	0.638
150 × 90 × 24	23.9	150	90	6.5	12.0	0.637
150 × 75 × 18	17.9	150	75	5.5	10.0	0.579
125 × 65 × 15	14.8	125	65	5.5	9.5	0.489
100 × 50 × 10	10.2	100	50	5.0	8.5	0.382

METAL

Equal Angles BS EN 10056-1

Designation	Mass (kg/m)	Surface area (m²/m)
200 × 200 × 24	71.1	0.790
200 × 200 × 20	59.9	0.790
200 × 200 × 18	54.2	0.790
200 × 200 × 16	48.5	0.790
150 × 150 × 18	40.1	0.59
150 × 150 × 15	33.8	0.59
150 × 150 × 12	27.3	0.59
150 × 150 × 10	23.0	0.59
120 × 120 × 15	26.6	0.47
120 × 120 × 12	21.6	0.47
120 × 120 × 10	18.2	0.47
120 × 120 × 8	14.7	0.47
100 × 100 × 15	21.9	0.39
100 × 100 × 12	17.8	0.39
100 × 100 × 10	15.0	0.39
100 × 100 × 8	12.2	0.39
90 × 90 × 12	15.9	0.35
90 × 90 × 10	13.4	0.35
90 × 90 × 8	10.9	0.35
90 × 90 × 7	9.61	0.35
90 × 90 × 6	8.30	0.35

Unequal Angles BS EN 10056-1

Designation	Mass (kg/m)	Surface area (m²/m)
200 × 150 × 18	47.1	0.69
200 × 150 × 15	39.6	0.69
200 × 150 × 12	32.0	0.69
200 × 100 × 15	33.7	0.59
200 × 100 × 12	27.3	0.59
200 × 100 × 10	23.0	0.59
150 × 90 × 15	26.6	0.47
150 × 90 × 12	21.6	0.47
150 × 90 × 10	18.2	0.47
150 × 75 × 15	24.8	0.44
150 × 75 × 12	20.2	0.44
150 × 75 × 10	17.0	0.44
125 × 75 × 12	17.8	0.40
125 × 75 × 10	15.0	0.40
125 × 75 × 8	12.2	0.40
100 × 75 × 12	15.4	0.34
100 × 75 × 10	13.0	0.34
100 × 75 × 8	10.6	0.34
100 × 65 × 10	12.3	0.32
100 × 65 × 8	9.94	0.32
100 × 65 × 7	8.77	0.32

METAL

Structural Tees Split from Universal Beams BS 4: Part 1: 2005

Designation	Mass (kg/m)	Surface area (m²/m)
305 × 305 × 90	89.5	1.22
305 × 305 × 75	74.6	1.22
254 × 343 × 63	62.6	1.19
229 × 305 × 70	69.9	1.07
229 × 305 × 63	62.5	1.07
229 × 305 × 57	56.5	1.07
229 × 305 × 51	50.6	1.07
210 × 267 × 61	61.0	0.95
210 × 267 × 55	54.5	0.95
210 × 267 × 51	50.5	0.95
210 × 267 × 46	46.1	0.95
210 × 267 × 41	41.1	0.95
191 × 229 × 49	49.2	0.84
191 × 229 × 45	44.6	0.84
191 × 229 × 41	41.0	0.84
191 × 229 × 37	37.1	0.84
191 × 229 × 34	33.6	0.84
152 × 229 × 41	41.0	0.76
152 × 229 × 37	37.1	0.76
152 × 229 × 34	33.6	0.76
152 × 229 × 30	29.9	0.76
152 × 229 × 26	26.2	0.76

Universal Bearing Piles BS 4: Part 1: 2005

Designation	Mass (kg/m)	Depth of Section (mm)	Width of Section (mm)	Thickness Web (mm)	Thickness Flange (mm)
356 × 368 × 174	173.9	361.4	378.5	20.3	20.4
356 × 368 × 152	152.0	356.4	376.0	17.8	17.9
356 × 368 × 133	133.0	352.0	373.8	15.6	15.7
356 × 368 × 109	108.9	346.4	371.0	12.8	12.9
305 × 305 × 223	222.9	337.9	325.7	30.3	30.4
305 × 305 × 186	186.0	328.3	320.9	25.5	25.6
305 × 305 × 149	149.1	318.5	316.0	20.6	20.7
305 × 305 × 126	126.1	312.3	312.9	17.5	17.6
305 × 305 × 110	110.0	307.9	310.7	15.3	15.4
305 × 305 × 95	94.9	303.7	308.7	13.3	13.3
305 × 305 × 88	88.0	301.7	307.8	12.4	12.3
305 × 305 × 79	78.9	299.3	306.4	11.0	11.1
254 × 254 × 85	85.1	254.3	260.4	14.4	14.3
254 × 254 × 71	71.0	249.7	258.0	12.0	12.0
254 × 254 × 63	63.0	247.1	256.6	10.6	10.7
203 × 203 × 54	53.9	204.0	207.7	11.3	11.4
203 × 203 × 45	44.9	200.2	205.9	9.5	9.5

METAL

Hot Formed Square Hollow Sections EN 10210 S275J2H & S355J2H

Size (mm)	Wall thickness (mm)	Mass (kg/m)	Superficial area (m²/m)
40 × 40	2.5	2.89	0.154
	3.0	3.41	0.152
	3.2	3.61	0.152
	3.6	4.01	0.151
	4.0	4.39	0.150
	5.0	5.28	0.147
50 × 50	2.5	3.68	0.194
	3.0	4.35	0.192
	3.2	4.62	0.192
	3.6	5.14	0.191
	4.0	5.64	0.190
	5.0	6.85	0.187
	6.0	7.99	0.185
	6.3	8.31	0.184
60 × 60	3.0	5.29	0.232
	3.2	5.62	0.232
	3.6	6.27	0.231
	4.0	6.90	0.230
	5.0	8.42	0.227
	6.0	9.87	0.225
	6.3	10.30	0.224
	8.0	12.50	0.219
70 × 70	3.0	6.24	0.272
	3.2	6.63	0.272
	3.6	7.40	0.271
	4.0	8.15	0.270
	5.0	9.99	0.267
	6.0	11.80	0.265
	6.3	12.30	0.264
	8.0	15.00	0.259
80 × 80	3.2	7.63	0.312
	3.6	8.53	0.311
	4.0	9.41	0.310
	5.0	11.60	0.307
	6.0	13.60	0.305
	6.3	14.20	0.304
	8.0	17.50	0.299
90 × 90	3.6	9.66	0.351
	4.0	10.70	0.350
	5.0	13.10	0.347
	6.0	15.50	0.345
	6.3	16.20	0.344
	8.0	20.10	0.339
100 × 100	3.6	10.80	0.391
	4.0	11.90	0.390
	5.0	14.70	0.387
	6.0	17.40	0.385
	6.3	18.20	0.384
	8.0	22.60	0.379
	10.0	27.40	0.374
120 × 120	4.0	14.40	0.470
	5.0	17.80	0.467

METAL

Size (mm)	Wall thickness (mm)	Mass (kg/m)	Superficial area (m²/m)
	6.0	21.20	0.465
	6.3	22.20	0.464
	8.0	27.60	0.459
	10.0	33.70	0.454
	12.0	39.50	0.449
	12.5	40.90	0.448
140 × 140	5.0	21.00	0.547
	6.0	24.90	0.545
	6.3	26.10	0.544
	8.0	32.60	0.539
	10.0	40.00	0.534
	12.0	47.00	0.529
	12.5	48.70	0.528
150 × 150	5.0	22.60	0.587
	6.0	26.80	0.585
	6.3	28.10	0.584
	8.0	35.10	0.579
	10.0	43.10	0.574
	12.0	50.80	0.569
	12.5	52.70	0.568
Hot formed from seamless hollow	16.0	65.2	0.559
160 × 160	5.0	24.10	0.627
	6.0	28.70	0.625
	6.3	30.10	0.624
	8.0	37.60	0.619
	10.0	46.30	0.614
	12.0	54.60	0.609
	12.5	56.60	0.608
	16.0	70.20	0.599
180 × 180	5.0	27.30	0.707
	6.0	32.50	0.705
	6.3	34.00	0.704
	8.0	42.70	0.699
	10.0	52.50	0.694
	12.0	62.10	0.689
	12.5	64.40	0.688
	16.0	80.20	0.679
200 × 200	5.0	30.40	0.787
	6.0	36.20	0.785
	6.3	38.00	0.784
	8.0	47.70	0.779
	10.0	58.80	0.774
	12.0	69.60	0.769
	12.5	72.30	0.768
	16.0	90.30	0.759
250 × 250	5.0	38.30	0.987
	6.0	45.70	0.985
	6.3	47.90	0.984
	8.0	60.30	0.979
	10.0	74.50	0.974
	12.0	88.50	0.969
	12.5	91.90	0.968

METAL

Size (mm)	Wall thickness (mm)	Mass (kg/m)	Superficial area (m²/m)
	16.0	115.00	0.959
300 × 300	6.0	55.10	1.18
	6.3	57.80	1.18
	8.0	72.80	1.18
	10.0	90.20	1.17
	12.0	107.00	1.17
	12.5	112.00	1.17
	16.0	141.00	1.16
350 × 350	8.0	85.40	1.38
	10.0	106.00	1.37
	12.0	126.00	1.37
	12.5	131.00	1.37
	16.0	166.00	1.36
400 × 400	8.0	97.90	1.58
	10.0	122.00	1.57
	12.0	145.00	1.57
	12.5	151.00	1.57
	16.0	191.00	1.56
(Grade S355J2H only)	20.00*	235.00	1.55

Note: * SAW process

METAL

Hot Formed Square Hollow Sections JUMBO RHS: JIS G3136

Size (mm)	Wall thickness (mm)	Mass (kg/m)	Superficial area (m²/m)
350 × 350	19.0	190.00	1.33
	22.0	217.00	1.32
	25.0	242.00	1.31
400 × 400	22.0	251.00	1.52
	25.0	282.00	1.51
450 × 450	12.0	162.00	1.76
	16.0	213.00	1.75
	19.0	250.00	1.73
	22.0	286.00	1.72
	25.0	321.00	1.71
	28.0 *	355.00	1.70
	32.0 *	399.00	1.69
500 × 500	12.0	181.00	1.96
	16.0	238.00	1.95
	19.0	280.00	1.93
	22.0	320.00	1.92
	25.0	360.00	1.91
	28.0 *	399.00	1.90
	32.0 *	450.00	1.89
	36.0 *	498.00	1.88
550 × 550	16.0	263.00	2.15
	19.0	309.00	2.13
	22.0	355.00	2.12
	25.0	399.00	2.11
	28.0 *	443.00	2.10
	32.0 *	500.00	2.09
	36.0 *	555.00	2.08
	40.0 *	608.00	2.06
600 × 600	25.0 *	439.00	2.31
	28.0 *	487.00	2.30
	32.0 *	550.00	2.29
	36.0 *	611.00	2.28
	40.0 *	671.00	2.26
700 × 700	25.0 *	517.00	2.71
	28.0 *	575.00	2.70
	32.0 *	651.00	2.69
	36.0 *	724.00	2.68
	40.0 *	797.00	2.68

Note: * SAW process

METAL

Hot Formed Rectangular Hollow Sections: EN10210 S275J2h & S355J2H

Size (mm)	Wall thickness (mm)	Mass (kg/m)	Superficial area (m²/m)
50 × 30	2.5	2.89	0.154
	3.0	3.41	0.152
	3.2	3.61	0.152
	3.6	4.01	0.151
	4.0	4.39	0.150
	5.0	5.28	0.147
60 × 40	2.5	3.68	0.194
	3.0	4.35	0.192
	3.2	4.62	0.192
	3.6	5.14	0.191
	4.0	5.64	0.190
	5.0	6.85	0.187
	6.0	7.99	0.185
	6.3	8.31	0.184
80 × 40	3.0	5.29	0.232
	3.2	5.62	0.232
	3.6	6.27	0.231
	4.0	6.90	0.230
	5.0	8.42	0.227
	6.0	9.87	0.225
	6.3	10.30	0.224
	8.0	12.50	0.219
76.2 × 50.8	3.0	5.62	0.246
	3.2	5.97	0.246
	3.6	6.66	0.245
	4.0	7.34	0.244
	5.0	8.97	0.241
	6.0	10.50	0.239
	6.3	11.00	0.238
	8.0	13.40	0.233
90 × 50	3.0	6.24	0.272
	3.2	6.63	0.272
	3.6	7.40	0.271
	4.0	8.15	0.270
	5.0	9.99	0.267
	6.0	11.80	0.265
	6.3	12.30	0.264
	8.0	15.00	0.259
100 × 50	3.0	6.71	0.292
	3.2	7.13	0.292
	3.6	7.96	0.291
	4.0	8.78	0.290
	5.0	10.80	0.287
	6.0	12.70	0.285
	6.3	13.30	0.284
	8.0	16.30	0.279

METAL

Size (mm)	Wall thickness (mm)	Mass (kg/m)	Superficial area (m²/m)
100 × 60	3.0	7.18	0.312
	3.2	7.63	0.312
	3.6	8.53	0.311
	4.0	9.41	0.310
	5.0	11.60	0.307
	6.0	13.60	0.305
	6.3	14.20	0.304
	8.0	17.50	0.299
120 × 60	3.6	9.70	0.351
	4.0	10.70	0.350
	5.0	13.10	0.347
	6.0	15.50	0.345
	6.3	16.20	0.344
	8.0	20.10	0.339
120 × 80	3.6	10.80	0.391
	4.0	11.90	0.390
	5.0	14.70	0.387
	6.0	17.40	0.385
	6.3	18.20	0.384
	8.0	22.60	0.379
	10.0	27.40	0.374
150 × 100	4.0	15.10	0.490
	5.0	18.60	0.487
	6.0	22.10	0.485
	6.3	23.10	0.484
	8.0	28.90	0.479
	10.0	35.30	0.474
	12.0	41.40	0.469
	12.5	42.80	0.468
160 × 80	4.0	14.40	0.470
	5.0	17.80	0.467
	6.0	21.20	0.465
	6.3	22.20	0.464
	8.0	27.60	0.459
	10.0	33.70	0.454
	12.0	39.50	0.449
	12.5	40.90	0.448
200 × 100	5.0	22.60	0.587
	6.0	26.80	0.585
	6.3	28.10	0.584
	8.0	35.10	0.579
	10.0	43.10	0.574
	12.0	50.80	0.569
	12.5	52.70	0.568
	16.0	65.20	0.559
250 × 150	5.0	30.40	0.787
	6.0	36.20	0.785
	6.3	38.00	0.784
	8.0	47.70	0.779
	10.0	58.80	0.774
	12.0	69.60	0.769
	12.5	72.30	0.768
	16.0	90.30	0.759

METAL

Size (mm)	Wall thickness (mm)	Mass (kg/m)	Superficial area (m^2/m)
300 × 200	5.0	38.30	0.987
	6.0	45.70	0.985
	6.3	47.90	0.984
	8.0	60.30	0.979
	10.0	74.50	0.974
	12.0	88.50	0.969
	12.5	91.90	0.968
	16.0	115.00	0.959
400 × 200	6.0	55.10	1.18
	6.3	57.80	1.18
	8.0	72.80	1.18
	10.0	90.20	1.17
	12.0	107.00	1.17
	12.5	112.00	1.17
	16.0	141.00	1.16
450 × 250	8.0	85.40	1.38
	10.0	106.00	1.37
	12.0	126.00	1.37
	12.5	131.00	1.37
	16.0	166.00	1.36
500 × 300	8.0	98.00	1.58
	10.0	122.00	1.57
	12.0	145.00	1.57
	12.5	151.00	1.57
	16.0	191.00	1.56
	20.0	235.00	1.55

METAL

Hot Formed Circular Hollow Sections EN 10210 S275J2H & S355J2H

Outside diameter (mm)	Wall thickness (mm)	Mass (kg/m)	Superficial area (m²/m)
21.3	3.2	1.43	0.067
26.9	3.2	1.87	0.085
33.7	3.0	2.27	0.106
	3.2	2.41	0.106
	3.6	2.67	0.106
	4.0	2.93	0.106
42.4	3.0	2.91	0.133
	3.2	3.09	0.133
	3.6	3.44	0.133
	4.0	3.79	0.133
48.3	2.5	2.82	0.152
	3.0	3.35	0.152
	3.2	3.56	0.152
	3.6	3.97	0.152
	4.0	4.37	0.152
	5.0	5.34	0.152
60.3	2.5	3.56	0.189
	3.0	4.24	0.189
	3.2	4.51	0.189
	3.6	5.03	0.189
	4.0	5.55	0.189
	5.0	6.82	0.189
76.1	2.5	4.54	0.239
	3.0	5.41	0.239
	3.2	5.75	0.239
	3.6	6.44	0.239
	4.0	7.11	0.239
	5.0	8.77	0.239
	6.0	10.40	0.239
	6.3	10.80	0.239
88.9	2.5	5.33	0.279
	3.0	6.36	0.279
	3.2	6.76	0.27
	3.6	7.57	0.279
88.9	4.0	8.38	0.279
	5.0	10.30	0.279
	6.0	12.30	0.279
	6.3	12.80	0.279
114.3	3.0	8.23	0.359
	3.2	8.77	0.359
	3.6	9.83	0.359
	4.0	10.09	0.359
	5.0	13.50	0.359
	6.0	16.00	0.359
	6.3	16.80	0.359

METAL

Outside diameter (mm)	Wall thickness (mm)	Mass (kg/m)	Superficial area (m²/m)
139.7	3.2	10.80	0.439
	3.6	12.10	0.439
	4.0	13.40	0.439
	5.0	16.60	0.439
	6.0	19.80	0.439
	6.3	20.70	0.439
	8.0	26.00	0.439
	10.0	32.00	0.439
168.3	3.2	13.00	0.529
	3.6	14.60	0.529
	4.0	16.20	0.529
	5.0	20.10	0.529
	6.0	24.00	0.529
	6.3	25.20	0.529
	8.0	31.60	0.529
	10.0	39.00	0.529
	12.0	46.30	0.529
	12.5	48.00	0.529
193.7	5.0	23.30	0.609
	6.0	27.80	0.609
	6.3	29.10	0.609
	8.0	36.60	0.609
	10.0	45.30	0.609
193.7	12.0	53.80	0.609
	12.5	55.90	0.609
219.1	5.0	26.40	0.688
	6.0	31.50	0.688
	6.3	33.10	0.688
	8.0	41.60	0.688
	10.0	51.60	0.688
	12.0	61.30	0.688
	12.5	63.70	0.688
	16.0	80.10	0.688
244.5	5.0	29.50	0.768
	6.0	35.30	0.768
	6.3	37.00	0.768
	8.0	46.70	0.768
	10.0	57.80	0.768
	12.0	68.80	0.768
	12.5	71.50	0.768
	16.0	90.20	0.768
273.0	5.0	33.00	0.858
	6.0	39.50	0.858
	6.3	41.40	0.858
	8.0	52.30	0.858
	10.0	64.90	0.858
	12.0	77.20	0.858
	12.5	80.30	0.858
	16.0	101.00	0.858

METAL

Outside diameter (mm)	Wall thickness (mm)	Mass (kg/m)	Superficial area (m²/m)
323.9	5.0	39.30	1.02
	6.0	47.00	1.02
	6.3	49.30	1.02
	8.0	62.30	1.02
	10.0	77.40	1.02
	12.0	92.30	1.02
	12.5	96.00	1.02
	16.0	121.00	1.02
355.6	6.3	54.30	1.12
	8.0	68.60	1.12
	10.0	85.30	1.12
	12.0	102.00	1.12
	12.5	106.00	1.12
	16.0	134.00	1.12
406.4	6.3	62.20	1.28
	8.0	79.60	1.28
	10.0	97.80	1.28
	12.0	117.00	1.28
	12.5	121.00	1.28
	16.0	154.00	1.28
457.0	6.3	70.00	1.44
	8.0	88.60	1.44
	10.0	110.00	1.44
	12.0	132.00	1.44
	12.5	137.00	1.44
	16.0	174.00	1.44
508.0	6.3	77.90	1.60
	8.0	98.60	1.60
	10.0	123.00	1.60
	12.0	147.00	1.60
	12.5	153.00	1.60
	16.0	194.00	1.60

METAL

Spacing of Holes in Angles

Nominal leg length (mm)	Spacing of holes						Maximum diameter of bolt or rivet		
	A	B	C	D	E	F	A	B and C	D, E and F
200		75	75	55	55	55		30	20
150		55	55					20	
125		45	60					20	
120									
100	55						24		
90	50						24		
80	45						20		
75	45						20		
70	40						20		
65	35						20		
60	35						16		
50	28						12		
45	25								
40	23								
30	20								
25	15								

KERBS, PAVING, ETC.

KERBS/EDGINGS/CHANNELS

Precast Concrete Kerbs to BS 7263
Straight kerb units: length from 450 to 915 mm

150 mm high × 125 mm thick		
bullnosed	type BN	
half battered	type HB3	
255 mm high × 125 mm thick		
45° splayed	type SP	
half battered	type HB2	
305 mm high × 150 mm thick		
half battered	type HB1	
Quadrant kerb units		
150 mm high × 305 and 455 mm radius to match	type BN	type QBN
150 mm high × 305 and 455 mm radius to match	type HB2, HB3	type QHB
150 mm high × 305 and 455 mm radius to match	type SP	type QSP
255 mm high × 305 and 455 mm radius to match	type BN	type QBN
255 mm high × 305 and 455 mm radius to match	type HB2, HB3	type QHB
225 mm high × 305 and 455 mm radius to match	type SP	type QSP
Angle kerb units		
305 × 305 × 225 mm high × 125 mm thick		
bullnosed external angle	type XA	
splayed external angle to match type SP	type XA	
bullnosed internal angle	type IA	
splayed internal angle to match type SP	type IA	
Channels		
255 mm wide × 125 mm high flat	type CS1	
150 mm wide × 125 mm high flat type	CS2	
255 mm wide × 125 mm high dished	type CD	

KERBS, PAVING, ETC.

Transition kerb units		
from kerb type SP to HB	left handed	type TL
	right handed	type TR
from kerb type BN to HB	left handed	type DL1
	right handed	type DR1
from kerb type BN to SP	left handed	type DL2
	right handed	type DR2

Number of kerbs required per quarter circle (780 mm kerb lengths)

Radius (m)	Number in quarter circle
12	24
10	20
8	16
6	12
5	10
4	8
3	6
2	4
1	2

Precast Concrete Edgings

Round top type ER	Flat top type EF	Bullnosed top type EBN
150 × 50 mm	150 × 50 mm	150 × 50 mm
200 × 50 mm	200 × 50 mm	200 × 50 mm
250 × 50 mm	250 × 50 mm	250 × 50 mm

KERBS, PAVING, ETC.

BASES

Cement Bound Material for Bases and Subbases

CBM1:	very carefully graded aggregate from 37.5–75 ym, with a 7-day strength of 4.5 N/mm^2
CBM2:	same range of aggregate as CBM1 but with more tolerance in each size of aggregate with a 7-day strength of 7.0 N/mm^2
CBM3:	crushed natural aggregate or blast furnace slag, graded from 37.5 mm – 150 ym for 40 mm aggregate, and from 20–75 ym for 20 mm aggregate, with a 7-day strength of 10 N/mm^2
CBM4:	crushed natural aggregate or blast furnace slag, graded from 37.5 mm – 150 ym for 40 mm aggregate, and from 20–75 ym for 20 mm aggregate, with a 7-day strength of 15 N/mm^2

INTERLOCKING BRICK/BLOCK ROADS/PAVINGS

Sizes of Precast Concrete Paving Blocks

Type R blocks
200 × 100 × 60 mm
200 × 100 × 65 mm
200 × 100 × 80 mm
200 × 100 × 100 mm

Type S
Any shape within a 295 mm space

Sizes of clay brick pavers
200 × 100 × 50 mm
200 × 100 × 65 mm
210 × 105 × 50 mm
210 × 105 × 65 mm
215 × 102.5 × 50 mm
215 × 102.5 × 65 mm

Type PA: 3 kN
Footpaths and pedestrian areas, private driveways, car parks, light vehicle traffic and over-run

Type PB: 7 kN
Residential roads, lorry parks, factory yards, docks, petrol station forecourts, hardstandings, bus stations

KERBS, PAVING, ETC.

PAVING AND SURFACING

Weights and Sizes of Paving and Surfacing

Description of item	Size	Quantity per tonne
Paving 50 mm thick	900 × 600 mm	15
Paving 50 mm thick	750 × 600 mm	18
Paving 50 mm thick	600 × 600 mm	23
Paving 50 mm thick	450 × 600 mm	30
Paving 38 mm thick	600 × 600 mm	30
Path edging	914 × 50 × 150 mm	60
Kerb (including radius and tapers)	125 × 254 × 914 mm	15
Kerb (including radius and tapers)	125 × 150 × 914 mm	25
Square channel	125 × 254 × 914 mm	15
Dished channel	125 × 254 × 914 mm	15
Quadrants	300 × 300 × 254 mm	19
Quadrants	450 × 450 × 254 mm	12
Quadrants	300 × 300 × 150 mm	30
Internal angles	300 × 300 × 254 mm	30
Fluted pavement channel	255 × 75 × 914 mm	25
Corner stones	300 × 300 mm	80
Corner stones	360 × 360 mm	60
Cable covers	914 × 175 mm	55
Gulley kerbs	220 × 220 × 150 mm	60
Gulley kerbs	220 × 200 × 75 mm	120

KERBS, PAVING, ETC.

Weights and Sizes of Paving and Surfacing

Material	kg/m³	lb/cu yd
Tarmacadam	2306	3891
Macadam (waterbound)	2563	4325
Vermiculite (aggregate)	64–80	108–135
Terracotta	2114	3568
Cork – compressed	388	24
	kg/m²	lb/sq ft
Clay floor tiles, 12.7 mm	27.3	5.6
Pavement lights	122	25
Damp-proof course	5	1
	kg/m² per mm thickness	lb/sq ft per inch thickness
Paving Slabs (stone)	2.3	12
Granite setts	2.88	15
Asphalt	2.30	12
Rubber flooring	1.68	9
Polyvinyl chloride	1.94 (avg)	10 (avg)

Coverage (m²) Per Cubic Metre of Materials Used as Subbases or Capping Layers

Consolidated thickness laid in (mm)	Square metre coverage		
	Gravel	Sand	Hardcore
50	15.80	16.50	–
75	10.50	11.00	–
100	7.92	8.20	7.42
125	6.34	6.60	5.90
150	5.28	5.50	4.95
175	–	–	4.23
200	–	–	3.71
225	–	–	3.30
300	–	–	2.47

KERBS, PAVING, ETC.

Approximate Rate of Spreads

Average thickness of course (mm)	Description	Approximate rate of spread			
		Open Textured		Dense, Medium & Fine Textured	
		(kg/m^2)	(m^2/t)	(kg/m^2)	(m^2/t)
35	14 mm open textured or dense wearing course	60–75	13–17	70–85	12–14
40	20 mm open textured or dense base course	70–85	12–14	80–100	10–12
45	20 mm open textured or dense base course	80–100	10–12	95–100	9–10
50	20 mm open textured or dense, or 28 mm dense base course	85–110	9–12	110–120	8–9
60	28 mm dense base course, 40 mm open textured of dense base course or 40 mm single course as base course		8–10	130–150	7–8
65	28 mm dense base course, 40 mm open textured or dense base course or 40 mm single course	100–135	7–10	140–160	6–7
75	40 mm single course, 40 mm open textured or dense base course, 40 mm dense roadbase	120–150	7–8	165–185	5–6
100	40 mm dense base course or roadbase	–	–	220–240	4–4.5

KERBS, PAVING, ETC.

Surface Dressing Roads: Coverage (m²) per Tonne of Material

Size in mm	Sand	Granite chips	Gravel	Limestone chips
Sand	168	–	–	–
3	–	148	152	165
6	–	130	133	144
9	–	111	114	123
13	–	85	87	95
19	–	68	71	78

Sizes of Flags

Reference	Nominal size (mm)	Thickness (mm)
A	600 × 450	50 and 63
B	600 × 600	50 and 63
C	600 × 750	50 and 63
D	600 × 900	50 and 63
E	450 × 450	50 and 70 chamfered top surface
F	400 × 400	50 and 65 chamfered top surface
G	300 × 300	50 and 60 chamfered top surface

Sizes of Natural Stone Setts

Width (mm)		Length (mm)		Depth (mm)
100	×	100	×	100
75	×	150 to 250	×	125
75	×	150 to 250	×	150
100	×	150 to 250	×	100
100	×	150 to 250	×	150

SEEDING/TURFING AND PLANTING

Topsoil Quality

Topsoil grade	Properties
Premium	Natural topsoil, high fertility, loamy texture, good soil structure, suitable for intensive cultivation.
General purpose	Natural or manufactured topsoil of lesser quality than Premium, suitable for agriculture or amenity landscape, may need fertilizer or soil structure improvement.
Economy	Selected subsoil, natural mineral deposit such as river silt or greensand. The grade comprises two subgrades; 'Low clay' and 'High clay' which is more liable to compaction in handling. This grade is suitable for low-production agricultural land and amenity woodland or conservation planting areas.

Forms of Trees

Standards:	Shall be clear with substantially straight stems. Grafted and budded trees shall have no more than a slight bend at the union. Standards shall be designated as Half, Extra light, Light, Standard, Selected standard, Heavy, and Extra heavy.
Sizes of Standards	
Heavy standard	12–14 cm girth × 3.50 to 5.00 m high
Extra Heavy standard	14–16 cm girth × 4.25 to 5.00 m high
Extra Heavy standard	16–18 cm girth × 4.25 to 6.00 m high
Extra Heavy standard	18–20 cm girth × 5.00 to 6.00 m high
Semi-mature trees:	Between 6.0 m and 12.0 m tall with a girth of 20 to 75 cm at 1.0 m above ground.
Feathered trees:	Shall have a defined upright central leader, with stem furnished with evenly spread and balanced lateral shoots down to or near the ground.
Whips:	Shall be without significant feather growth as determined by visual inspection.
Multi-stemmed trees:	Shall have two or more main stems at, near, above or below ground.

Seedlings grown from seed and not transplanted shall be specified when ordered for sale as:
1+0 one year old seedling
2+0 two year old seedling
1+1 one year seed bed, one year transplanted = two year old seedling
1+2 one year seed bed, two years transplanted = three year old seedling
2+1 two year seed bed, one year transplanted = three year old seedling
1u1 two years seed bed, undercut after 1 year = two year old seedling
2u2 four years seed bed, undercut after 2 years = four year old seedling

SEEDING/TURFING AND PLANTING

Cuttings

The age of cuttings (plants grown from shoots, stems, or roots of the mother plant) shall be specified when ordered for sale. The height of transplants and undercut seedlings/cuttings (which have been transplanted or undercut at least once) shall be stated in centimetres. The number of growing seasons before and after transplanting or undercutting shall be stated.

0 + 1	one year cutting
0 + 2	two year cutting
0 + 1 + 1	one year cutting bed, one year transplanted = two year old seedling
0 + 1 + 2	one year cutting bed, two years transplanted = three year old seedling

Grass Cutting Capacities in m² per hour

Speed mph	\multicolumn{12}{c}{Width of cut in metres}
	0.5	0.7	1.0	1.2	1.5	1.7	2.0	2.0	2.1	2.5	2.8	3.0	3.4
1.0	724	1127	1529	1931	2334	2736	3138	3219	3380	4023	4506	4828	5472
1.5	1086	1690	2293	2897	3500	4104	4707	4828	5069	6035	6759	7242	8208
2.0	1448	2253	3058	3862	4667	5472	6276	6437	6759	8047	9012	9656	10944
2.5	1811	2816	3822	4828	5834	6840	7846	8047	8449	10058	11265	12070	13679
3.0	2173	3380	4587	5794	7001	8208	9415	9656	10139	12070	13518	14484	16415
3.5	2535	3943	5351	6759	8167	9576	10984	11265	11829	14082	15772	16898	19151
4.0	2897	4506	6115	7725	9334	10944	12553	12875	13518	16093	18025	19312	21887
4.5	3259	5069	6880	8690	10501	12311	14122	14484	15208	18105	20278	21726	24623
5.0	3621	5633	7644	9656	11668	13679	15691	16093	16898	20117	22531	24140	27359
5.5	3983	6196	8409	10622	12834	15047	17260	17703	18588	22128	24784	26554	30095
6.0	4345	6759	9173	11587	14001	16415	18829	19312	20278	24140	27037	28968	32831
6.5	4707	7322	9938	12553	15168	17783	20398	20921	21967	26152	29290	31382	35566
7.0	5069	7886	10702	13518	16335	19151	21967	22531	23657	28163	31543	33796	38302

Number of Plants per m²: For Plants Planted on an Evenly Spaced Grid

Planting distances

mm	0.10	0.15	0.20	0.25	0.35	0.40	0.45	0.50	0.60	0.75	0.90	1.00	1.20	1.50
0.10	100.00	66.67	50.00	40.00	28.57	25.00	22.22	20.00	16.67	13.33	11.11	10.00	8.33	6.67
0.15	66.67	44.44	33.33	26.67	19.05	16.67	14.81	13.33	11.11	8.89	7.41	6.67	5.56	4.44
0.20	50.00	33.33	25.00	20.00	14.29	12.50	11.11	10.00	8.33	6.67	5.56	5.00	4.17	3.33
0.25	40.00	26.67	20.00	16.00	11.43	10.00	8.89	8.00	6.67	5.33	4.44	4.00	3.33	2.67
0.35	28.57	19.05	14.29	11.43	8.16	7.14	6.35	5.71	4.76	3.81	3.17	2.86	2.38	1.90
0.40	25.00	16.67	12.50	10.00	7.14	6.25	5.56	5.00	4.17	3.33	2.78	2.50	2.08	1.67
0.45	22.22	14.81	11.11	8.89	6.35	5.56	4.94	4.44	3.70	2.96	2.47	2.22	1.85	1.48
0.50	20.00	13.33	10.00	8.00	5.71	5.00	4.44	4.00	3.33	2.67	2.22	2.00	1.67	1.33
0.60	16.67	11.11	8.33	6.67	4.76	4.17	3.70	3.33	2.78	2.22	1.85	1.67	1.39	1.11
0.75	13.33	8.89	6.67	5.33	3.81	3.33	2.96	2.67	2.22	1.78	1.48	1.33	1.11	0.89
0.90	11.11	7.41	5.56	4.44	3.17	2.78	2.47	2.22	1.85	1.48	1.23	1.11	0.93	0.74
1.00	10.00	6.67	5.00	4.00	2.86	2.50	2.22	2.00	1.67	1.33	1.11	1.00	0.83	0.67
1.20	8.33	5.56	4.17	3.33	2.38	2.08	1.85	1.67	1.39	1.11	0.93	0.83	0.69	0.56
1.50	6.67	4.44	3.33	2.67	1.90	1.67	1.48	1.33	1.11	0.89	0.74	0.67	0.56	0.44

SEEDING/TURFING AND PLANTING

Grass Clippings Wet: Based on 3.5 m³/tonne

Annual kg/100 m²	Average 20 cuts kg/100 m²	m²/tonne	m²/m³
32.0	1.6	61162.1	214067.3

Nr of cuts	22	20	18	16	12	4
kg/cut	1.45	1.60	1.78	2.00	2.67	8.00
	Area capacity of 3 tonne vehicle per load					
m²	206250	187500	168750	150000	112500	37500
Load m³	100 m² units/m³ of vehicle space					
1	196.4	178.6	160.7	142.9	107.1	35.7
2	392.9	357.1	321.4	285.7	214.3	71.4
3	589.3	535.7	482.1	428.6	321.4	107.1
4	785.7	714.3	642.9	571.4	428.6	142.9
5	982.1	892.9	803.6	714.3	535.7	178.6

Transportation of Trees

To unload large trees a machine with the necessary lifting strength is required. The weight of the trees must therefore be known in advance. The following table gives a rough overview. The additional columns with root ball dimensions and the number of plants per trailer provide additional information for example about preparing planting holes and calculating unloading times.

Girth in cm	Rootball diameter in cm	Ball height in cm	Weight in kg	Numbers of trees per trailer
16–18	50–60	40	150	100–120
18–20	60–70	40–50	200	80–100
20–25	60–70	40–50	270	50–70
25–30	80	50–60	350	50
30–35	90–100	60–70	500	12–18
35–40	100–110	60–70	650	10–15
40–45	110–120	60–70	850	8–12
45–50	110–120	60–70	1100	5–7
50–60	130–140	60–70	1600	1–3
60–70	150–160	60–70	2500	1
70–80	180–200	70	4000	1
80–90	200–220	70–80	5500	1
90–100	230–250	80–90	7500	1
100–120	250–270	80–90	9500	1

Data supplied by Lorenz von Ehren GmbH
The information in the table is approximate; deviations depend on soil type, genus and weather

FENCING AND GATES

Types of Preservative

Creosote (tar oil) can be 'factory' applied	by pressure to BS 144: pts 1&2 by immersion to BS 144: pt 1 by hot and cold open tank to BS 144: pts 1&2
Copper/chromium/arsenic (CCA)	by full cell process to BS 4072 pts 1&2
Organic solvent (OS)	by double vacuum (vacvac) to BS 5707 pts 1&3 by immersion to BS 5057 pts 1&3
Pentachlorophenol (PCP)	by heavy oil double vacuum to BS 5705 pts 2&3
Boron diffusion process (treated with disodium octaborate to BWPA Manual 1986)	

Note: Boron is used on green timber at source and the timber is supplied dry

Cleft Chestnut Pale Fences

Pales	Pale spacing	Wire lines	
900 mm	75 mm	2	temporary protection
1050 mm	75 or 100 mm	2	light protective fences
1200 mm	75 mm	3	perimeter fences
1350 mm	75 mm	3	perimeter fences
1500 mm	50 mm	3	narrow perimeter fences
1800 mm	50 mm	3	light security fences

Close-Boarded Fences

Close-boarded fences 1.05 to 1.8 m high
Type BCR (recessed) or BCM (morticed) with concrete posts 140 × 115 mm tapered and Type BW with timber posts

Palisade Fences

Wooden palisade fences
Type WPC with concrete posts 140 × 115 mm tapered and Type WPW with timber posts

For both types of fence:
Height of fence 1050 mm: two rails
Height of fence 1200 mm: two rails
Height of fence 1500 mm: three rails
Height of fence 1650 mm: three rails
Height of fence 1800 mm: three rails

Tables and Memoranda

FENCING AND GATES

Post and Rail Fences

Wooden post and rail fences Type MPR 11/3 morticed rails and Type SPR 11/3 nailed rails Height to top of rail 1100 mm Rails: three rails 87 mm, 38 mm Type MPR 11/4 morticed rails and Type SPR 11/4 nailed rails Height to top of rail 1100 mm Rails: four rails 87 mm, 38 mm Type MPR 13/4 morticed rails and Type SPR 13/4 nailed rails Height to top of rail 1300 mm Rail spacing 250 mm, 250 mm, and 225 mm from top Rails: four rails 87 mm, 38 mm

Steel Posts

Rolled steel angle iron posts for chain link fencing

Posts	Fence height	Strut	Straining post
1500 × 40 × 40 × 5 mm	900 mm	1500 × 40 × 40 × 5 mm	1500 × 50 × 50 × 6 mm
1800 × 40 × 40 × 5 mm	1200 mm	1800 × 40 × 40 × 5 mm	1800 × 50 × 50 × 6 mm
2000 × 45 × 45 × 5 mm	1400 mm	2000 × 45 × 45 × 5 mm	2000 × 60 × 60 × 6 mm
2600 × 45 × 45 × 5 mm	1800 mm	2600 × 45 × 45 × 5 mm	2600 × 60 × 60 × 6 mm
3000 × 50 × 50 × 6 mm with arms	1800 mm	2600 × 45 × 45 × 5 mm	3000 × 60 × 60 × 6 mm

Concrete Posts

Concrete posts for chain link fencing

Posts and straining posts	Fence height	Strut
1570 mm 100 × 100 mm	900 mm	1500 mm × 75 × 75 mm
1870 mm 125 × 125 mm	1200 mm	1830 mm × 100 × 75 mm
2070 mm 125 × 125 mm	1400 mm	1980 mm × 100 × 75 mm
2620 mm 125 × 125 mm	1800 mm	2590 mm × 100 × 85 mm
3040 mm 125 × 125 mm	1800 mm	2590 mm × 100 × 85 mm (with arms)

FENCING AND GATES

Rolled Steel Angle Posts

Rolled steel angle posts for rectangular wire mesh (field) fencing

Posts	Fence height	Strut	Straining post
1200 × 40 × 40 × 5 mm	600 mm	1200 × 75 × 75 mm	1350 × 100 × 100 mm
1400 × 40 × 40 × 5 mm	800 mm	1400 × 75 × 75 mm	1550 × 100 × 100 mm
1500 × 40 × 40 × 5 mm	900 mm	1500 × 75 × 75 mm	1650 × 100 × 100 mm
1600 × 40 × 40 × 5 mm	1000 mm	1600 × 75 × 75 mm	1750 × 100 × 100 mm
1750 × 40 × 40 × 5 mm	1150 mm	1750 × 75 × 100 mm	1900 × 125 × 125 mm

Concrete Posts

Concrete posts for rectangular wire mesh (field) fencing

Posts	Fence height	Strut	Straining post
1270 × 100 × 100 mm	600 mm	1200 × 75 × 75 mm	1420 × 100 × 100 mm
1470 × 100 × 100 mm	800 mm	1350 × 75 × 75 mm	1620 × 100 × 100 mm
1570 × 100 × 100 mm	900 mm	1500 × 75 × 75 mm	1720 × 100 × 100 mm
1670 × 100 × 100 mm	600 mm	1650 × 75 × 75 mm	1820 × 100 × 100 mm
1820 × 125 × 125 mm	1150 mm	1830 × 75 × 100 mm	1970 × 125 × 125 mm

Cleft Chestnut Pale Fences

Timber Posts

Timber posts for wire mesh and hexagonal wire netting fences

Round timber for general fences

Posts	Fence height	Strut	Straining post
1300 × 65 mm dia.	600 mm	1200 × 80 mm dia.	1450 × 100 mm dia.
1500 × 65 mm dia.	800 mm	1400 × 80 mm dia.	1650 × 100 mm dia.
1600 × 65 mm dia.	900 mm	1500 × 80 mm dia.	1750 × 100 mm dia.
1700 × 65 mm dia.	1050 mm	1600 × 80 mm dia.	1850 × 100 mm dia.
1800 × 65 mm dia.	1150 mm	1750 × 80 mm dia.	2000 × 120 mm dia.

Squared timber for general fences

Posts	Fence height	Strut	Straining post
1300 × 75 × 75 mm	600 mm	1200 × 75 × 75 mm	1450 × 100 × 100 mm
1500 × 75 × 75 mm	800 mm	1400 × 75 × 75 mm	1650 × 100 × 100 mm
1600 × 75 × 75 mm	900 mm	1500 × 75 × 75 mm	1750 × 100 × 100 mm
1700 × 75 × 75 mm	1050 mm	1600 × 75 × 75 mm	1850 × 100 × 100 mm
1800 × 75 × 75 mm	1150 mm	1750 × 75 × 75 mm	2000 × 125 × 100 mm

FENCING AND GATES

Steel Fences to BS 1722: Part 9: 1992

	Fence height	Top/bottom rails and flat posts	Vertical bars
Light	1000 mm	40 × 10 mm 450 mm in ground	12 mm dia. at 115 mm cs
	1200 mm	40 × 10 mm 550 mm in ground	12 mm dia. at 115 mm cs
	1400 mm	40 × 10 mm 550 mm in ground	12 mm dia. at 115 mm cs
Light	1000 mm	40 × 10 mm 450 mm in ground	16 mm dia. at 120 mm cs
	1200 mm	40 × 10 mm 550 mm in ground	16 mm dia. at 120 mm cs
	1400 mm	40 × 10 mm 550 mm in ground	16 mm dia. at 120 mm cs
Medium	1200 mm	50 × 10 mm 550 mm in ground	20 mm dia. at 125 mm cs
	1400 mm	50 × 10 mm 550 mm in ground	20 mm dia. at 125 mm cs
	1600 mm	50 × 10 mm 600 mm in ground	22 mm dia. at 145 mm cs
	1800 mm	50 × 10 mm 600 mm in ground	22 mm dia. at 145 mm cs
Heavy	1600 mm	50 × 10 mm 600 mm in ground	22 mm dia. at 145 mm cs
	1800 mm	50 × 10 mm 600 mm in ground	22 mm dia. at 145 mm cs
	2000 mm	50 × 10 mm 600 mm in ground	22 mm dia. at 145 mm cs
	2200 mm	50 × 10 mm 600 mm in ground	22 mm dia. at 145 mm cs

Notes: Mild steel fences: round or square verticals; flat standards and horizontals. Tops of vertical bars may be bow-top, blunt, or pointed. Round or square bar railings

Timber Field Gates to BS 3470: 1975

Gates made to this standard are designed to open one way only
All timber gates are 1100 mm high
Width over stiles 2400, 2700, 3000, 3300, 3600, and 4200 mm
Gates over 4200 mm should be made in two leaves

Steel Field Gates to BS 3470: 1975

All steel gates are 1100 mm high
Heavy duty: width over stiles 2400, 3000, 3600 and 4500 mm
Light duty: width over stiles 2400, 3000, and 3600 mm

FENCING AND GATES

Domestic Front Entrance Gates to BS 4092: Part 1: 1966

Metal gates:	Single gates are 900 mm high minimum, 900 mm, 1000 mm and 1100 mm wide

Domestic Front Entrance Gates to BS 4092: Part 2: 1966

Wooden gates:	All rails shall be tenoned into the stiles Single gates are 840 mm high minimum, 801 mm and 1020 mm wide Double gates are 840 mm high minimum, 2130, 2340 and 2640 mm wide

Timber Bridle Gates to BS 5709:1979 (Horse or Hunting Gates)

Gates open one way only Minimum width between posts Minimum height	1525 mm 1100 mm

Timber Kissing Gates to BS 5709:1979

Minimum width	700 mm
Minimum height	1000 mm
Minimum distance between shutting posts	600 mm
Minimum clearance at mid-point	600 mm

Metal Kissing Gates to BS 5709:1979

Sizes are the same as those for timber kissing gates
Maximum gaps between rails 120 mm

Categories of Pedestrian Guard Rail to BS 3049:1976

Class A for normal use
Class B where vandalism is expected
Class C where crowd pressure is likely

DRAINAGE

Width required for Trenches for Various Diameters of Pipes

Pipe diameter (mm)	Trench n.e. 1.50 m deep	Trench over 1.50 m deep
n.e. 100 mm	450 mm	600 mm
100–150 mm	500 mm	650 mm
150–225 mm	600 mm	750 mm
225–300 mm	650 mm	800 mm
300–400 mm	750 mm	900 mm
400–450 mm	900 mm	1050 mm
450–600 mm	1100 mm	1300 mm

Weights and Dimensions – Vitrified Clay Pipes

Product	Nominal diameter (mm)	Effective length (mm)	BS 65 limits of tolerance min (mm)	BS 65 limits of tolerance max (mm)	Crushing strength (kN/m)	Weight (kg/pipe)	Weight (kg/m)
Supersleve	100	1600	96	105	35.00	14.71	9.19
	150	1750	146	158	35.00	29.24	16.71
Hepsleve	225	1850	221	236	28.00	84.03	45.42
	300	2500	295	313	34.00	193.05	77.22
	150	1500	146	158	22.00	37.04	24.69
Hepseal	225	1750	221	236	28.00	85.47	48.84
	300	2500	295	313	34.00	204.08	81.63
	400	2500	394	414	44.00	357.14	142.86
	450	2500	444	464	44.00	454.55	181.63
	500	2500	494	514	48.00	555.56	222.22
	600	2500	591	615	57.00	796.23	307.69
	700	3000	689	719	67.00	1111.11	370.45
	800	3000	788	822	72.00	1351.35	450.45
Hepline	100	1600	95	107	22.00	14.71	9.19
	150	1750	145	160	22.00	29.24	16.71
	225	1850	219	239	28.00	84.03	45.42
	300	1850	292	317	34.00	142.86	77.22
Hepduct (conduit)	90	1500	–	–	28.00	12.05	8.03
	100	1600	–	–	28.00	14.71	9.19
	125	1750	–	–	28.00	20.73	11.84
	150	1750	–	–	28.00	29.24	16.71
	225	1850	–	–	28.00	84.03	45.42
	300	1850	–	–	34.00	142.86	77.22

DRAINAGE

Weights and Dimensions – Vitrified Clay Pipes

Nominal internal diameter (mm)	Nominal wall thickness (mm)	Approximate weight (kg/m)
150	25	45
225	29	71
300	32	122
375	35	162
450	38	191
600	48	317
750	54	454
900	60	616
1200	76	912
1500	89	1458
1800	102	1884
2100	127	2619

Wall thickness, weights and pipe lengths vary, depending on type of pipe required
The particulars shown above represent a selection of available diameters and are applicable to strength class 1 pipes with flexible rubber ring joints
Tubes with Ogee joints are also available

DRAINAGE

Weights and Dimensions – PVC-u Pipes

	Nominal size	Mean outside diameter (mm) min	Mean outside diameter (mm) max	Wall thickness (mm)	Weight (kg/m)
Standard pipes	82.4	82.4	82.7	3.2	1.2
	110.0	110.0	110.4	3.2	1.6
	160.0	160.0	160.6	4.1	3.0
	200.0	200.0	200.6	4.9	4.6
	250.0	250.0	250.7	6.1	7.2
Perforated pipes heavy grade	As above	As above	As above	As above	As above
thin wall	82.4	82.4	82.7	1.7	–
	110.0	110.0	110.4	2.2	–
	160.0	160.0	160.6	3.2	–

Width of Trenches Required for Various Diameters of Pipes

Pipe diameter (mm)	Trench n.e. 1.5 m deep (mm)	Trench over 1.5 m deep (mm)
n.e. 100	450	600
100–150	500	650
150–225	600	750
225–300	650	800
300–400	750	900
400–450	900	1050
450–600	1100	1300

DRAINAGE

DRAINAGE BELOW GROUND AND LAND DRAINAGE

Flow of Water Which Can Be Carried by Various Sizes of Pipe

Clay or concrete pipes

Pipe size	Gradient of pipeline							
	1:10	1:20	1:30	1:40	1:50	1:60	1:80	1:100
	Flow in litres per second							
DN 100 15.0	8.5	6.8	5.8	5.2	4.7	4.0	3.5	
DN 150 28.0	19.0	16.0	14.0	12.0	11.0	9.1	8.0	
DN 225 140.0	95.0	76.0	66.0	58.0	53.0	46.0	40.0	

Plastic pipes

Pipe size	Gradient of pipeline							
	1:10	1:20	1:30	1:40	1:50	1:60	1:80	1:100
	Flow in litres per second							
82.4 mm i/dia.	12.0	8.5	6.8	5.8	5.2	4.7	4.0	3.5
110 mm i/dia.	28.0	19.0	16.0	14.0	12.0	11.0	9.1	8.0
160 mm i/dia.	76.0	53.0	43.0	37.0	33.0	29.0	25.0	22.0
200 mm i/dia.	140.0	95.0	76.0	66.0	58.0	53.0	46.0	40.0

Vitrified (Perforated) Clay Pipes and Fittings to BS En 295-5 1994

Length not specified		
75 mm bore	250 mm bore	600 mm bore
100	300	700
125	350	800
150	400	1000
200	450	1200
225	500	

Precast Concrete Pipes: Prestressed Non-pressure Pipes and Fittings: Flexible Joints to BS 5911: Pt. 103: 1994

Rationalized metric nominal sizes: 450, 500	
Length:	500–1000 by 100 increments
	1000–2200 by 200 increments
	2200–2800 by 300 increments
Angles: length:	450–600 angles 45, 22.5, 11.25°
	600 or more angles 22.5, 11.25°

Tables and Memoranda

DRAINAGE

Precast Concrete Pipes: Un-reinforced and Circular Manholes and Soakaways to BS 5911: Pt. 200: 1994

Nominal sizes:	
Shafts:	675, 900 mm
Chambers:	900, 1050, 1200, 1350, 1500, 1800, 2100, 2400, 2700, 3000 mm
Large chambers:	To have either tapered reducing rings or a flat reducing slab in order to accept the standard cover
Ring depths:	1. 300–1200 mm by 300 mm increments except for bottom slab and rings below cover slab, these are by 150 mm increments
	2. 250–1000 mm by 250 mm increments except for bottom slab and rings below cover slab, these are by 125 mm increments
Access hole:	750 × 750 mm for DN 1050 chamber
	1200 × 675 mm for DN 1350 chamber

Calculation of Soakaway Depth

The following formula determines the depth of concrete ring soakaway that would be required for draining given amounts of water.

$$h = \frac{4ar}{3\pi D^2}$$

h = depth of the chamber below the invert pipe
a = the area to be drained
r = the hourly rate of rainfall (50 mm per hour)
π = pi
D = internal diameter of the soakaway

This table shows the depth of chambers in each ring size which would be required to contain the volume of water specified. These allow a recommended storage capacity of ⅓ (one third of the hourly rainfall figure).

Table Showing Required Depth of Concrete Ring Chambers in Metres

Area m²	50	100	150	200	300	400	500
Ring size							
0.9	1.31	2.62	3.93	5.24	7.86	10.48	13.10
1.1	0.96	1.92	2.89	3.85	5.77	7.70	9.62
1.2	0.74	1.47	2.21	2.95	4.42	5.89	7.37
1.4	0.58	1.16	1.75	2.33	3.49	4.66	5.82
1.5	0.47	0.94	1.41	1.89	2.83	3.77	4.72
1.8	0.33	0.65	0.98	1.31	1.96	2.62	3.27
2.1	0.24	0.48	0.72	0.96	1.44	1.92	2.41
2.4	0.18	0.37	0.55	0.74	1.11	1.47	1.84
2.7	0.15	0.29	0.44	0.58	0.87	1.16	1.46
3.0	0.12	0.24	0.35	0.47	0.71	0.94	1.18

DRAINAGE

Precast Concrete Inspection Chambers and Gullies to BS 5911: Part 230: 1994

Nominal sizes:	375 diameter, 750, 900 mm deep
	450 diameter, 750, 900, 1050, 1200 mm deep
Depths:	from the top for trapped or un-trapped units:
	centre of outlet 300 mm
	invert (bottom) of the outlet pipe 400 mm
Depth of water seal for trapped gullies:	
	85 mm, rodding eye int. dia. 100 mm
Cover slab:	65 mm min

Bedding Flexible Pipes: PVC-u Or Ductile Iron

Type 1 =	100 mm fill below pipe, 300 mm above pipe: single size material
Type 2 =	100 mm fill below pipe, 300 mm above pipe: single size or graded material
Type 3 =	100 mm fill below pipe, 75 mm above pipe with concrete protective slab over
Type 4 =	100 mm fill below pipe, fill laid level with top of pipe
Type 5 =	200 mm fill below pipe, fill laid level with top of pipe
Concrete =	25 mm sand blinding to bottom of trench, pipe supported on chocks, 100 mm concrete under the pipe, 150 mm concrete over the pipe

DRAINAGE

Bedding Rigid Pipes: Clay or Concrete
(for vitrified clay pipes the manufacturer should be consulted)

Class D:	Pipe laid on natural ground with cut-outs for joints, soil screened to remove stones over 40 mm and returned over pipe to 150 m min depth. Suitable for firm ground with trenches trimmed by hand.
Class N:	Pipe laid on 50 mm granular material of graded aggregate to Table 4 of BS 882, or 10 mm aggregate to Table 6 of BS 882, or as dug light soil (not clay) screened to remove stones over 10 mm. Suitable for machine dug trenches.
Class B:	As Class N, but with granular bedding extending half way up the pipe diameter.
Class F:	Pipe laid on 100 mm granular fill to BS 882 below pipe, minimum 150 mm granular fill above pipe: single size material. Suitable for machine dug trenches.
Class A:	Concrete 100 mm thick under the pipe extending half way up the pipe, backfilled with the appropriate class of fill. Used where there is only a very shallow fall to the drain. Class A bedding allows the pipes to be laid to an exact gradient.
Concrete surround:	25 mm sand blinding to bottom of trench, pipe supported on chocks, 100 mm concrete under the pipe, 150 mm concrete over the pipe. It is preferable to bed pipes under slabs or wall in granular material.

PIPED SUPPLY SYSTEMS

Identification of Service Tubes From Utility to Dwellings

Utility	Colour	Size	Depth
British Telecom	grey	54 mm od	450 mm
Electricity	black	38 mm od	450 mm
Gas	yellow	42 mm od rigid 60 mm od convoluted	450 mm
Water	may be blue	(normally untubed)	750 mm

ELECTRICAL SUPPLY/POWER/LIGHTING SYSTEMS

Electrical Insulation Class En 60.598 BS 4533

Class 1:	luminaires comply with class 1 (I) earthed electrical requirements
Class 2:	luminaires comply with class 2 (II) double insulated electrical requirements
Class 3:	luminaires comply with class 3 (III) electrical requirements

Protection to Light Fittings

BS EN 60529:1992 Classification for degrees of protection provided by enclosures.
(IP Code – International or ingress Protection)

1st characteristic: against ingress of solid foreign objects		
The figure	2	indicates that fingers cannot enter
	3	that a 2.5 mm diameter probe cannot enter
	4	that a 1.0 mm diameter probe cannot enter
	5	the fitting is dust proof (no dust around live parts)
	6	the fitting is dust tight (no dust entry)
2nd characteristic: ingress of water with harmful effects		
The figure	0	indicates unprotected
	1	vertically dripping water cannot enter
	2	water dripping 15° (tilt) cannot enter
	3	spraying water cannot enter
	4	splashing water cannot enter
	5	jetting water cannot enter
	6	powerful jetting water cannot enter
	7	proof against temporary immersion
	8	proof against continuous immersion
Optional additional codes:		A–D protects against access to hazardous parts
	H	High voltage apparatus
	M	fitting was in motion during water test
	S	fitting was static during water test
	W	protects against weather
Marking code arrangement:		(example) IPX5S = IP (International or Ingress Protection)
		X (denotes omission of first characteristic)
		5 = jetting
		S = static during water test

RAIL TRACKS

	kg/m of track	lb/ft of track
Standard gauge Bull-head rails, chairs, transverse timber (softwood) sleepers etc.	245	165
Main lines Flat-bottom rails, transverse prestressed concrete sleepers, etc. Add for electric third rail Add for crushed stone ballast	418 51 2600	280 35 1750
	kg/m²	lb/sq ft
Overall average weight – rails connections, sleepers, ballast, etc.	733	150
	kg/m of track	lb/ft of track
Bridge rails, longitudinal timber sleepers, etc.	112	75

RAIL TRACKS

Heavy Rails

British Standard Section No.	Rail height (mm)	Foot width (mm)	Head width (mm)	Min web thickness (mm)	Section weight (kg/m)
Flat Bottom Rails					
60 A	114.30	109.54	57.15	11.11	30.62
70 A	123.82	111.12	60.32	12.30	34.81
75 A	128.59	114.30	61.91	12.70	37.45
80 A	133.35	117.47	63.50	13.10	39.76
90 A	142.88	127.00	66.67	13.89	45.10
95 A	147.64	130.17	69.85	14.68	47.31
100 A	152.40	133.35	69.85	15.08	50.18
110 A	158.75	139.70	69.85	15.87	54.52
113 A	158.75	139.70	69.85	20.00	56.22
50 'O'	100.01	100.01	52.39	10.32	24.82
80 'O'	127.00	127.00	63.50	13.89	39.74
60R	114.30	109.54	57.15	11.11	29.85
75R	128.59	122.24	61.91	13.10	37.09
80R	133.35	127.00	63.50	13.49	39.72
90R	142.88	136.53	66.67	13.89	44.58
95R	147.64	141.29	68.26	14.29	47.21
100R	152.40	146.05	69.85	14.29	49.60
95N	147.64	139.70	69.85	13.89	47.27
Bull Head Rails					
95R BH	145.26	69.85	69.85	19.05	47.07

Light Rails

British Standard Section No.	Rail height (mm)	Foot width (mm)	Head width (mm)	Min web thickness (mm)	Section weight (kg/m)
Flat Bottom Rails					
20M	65.09	55.56	30.96	6.75	9.88
30M	75.41	69.85	38.10	9.13	14.79
35M	80.96	76.20	42.86	9.13	17.39
35R	85.73	82.55	44.45	8.33	17.40
40	88.11	80.57	45.64	12.3	19.89
Bridge Rails					
13	48.00	92	36.00	18.0	13.31
16	54.00	108	44.50	16.0	16.06
20	55.50	127	50.00	20.5	19.86
28	67.00	152	50.00	31.0	28.62
35	76.00	160	58.00	34.5	35.38
50	76.00	165	58.50	–	50.18
Crane Rails					
A65	75.00	175.00	65.00	38.0	43.10
A75	85.00	200.00	75.00	45.0	56.20
A100	95.00	200.00	100.00	60.0	74.30
A120	105.00	220.00	120.00	72.0	100.00
175CR	152.40	152.40	107.95	38.1	86.92

RAIL TRACKS

Fish Plates

British Standard Section No.	Overall plate length		Hole diameter	Finished weight per pair	
	4 Hole (mm)	6 Hole (mm)	(mm)	4 Hole (kg/pair)	6 Hole (kg/pair)
For British Standard Heavy Rails: Flat Bottom Rails					
60 A	406.40	609.60	20.64	9.87	14.76
70 A	406.40	609.60	22.22	11.15	16.65
75 A	406.40	–	23.81	11.82	17.73
80 A	406.40	609.60	23.81	13.15	19.72
90 A	457.20	685.80	25.40	17.49	26.23
100 A	508.00	–	pear	25.02	–
110 A (shallow)	507.00	–	27.00	30.11	54.64
113 A (heavy)	507.00	–	27.00	30.11	54.64
50 'O' (shallow)	406.40	–	–	6.68	10.14
80 'O' (shallow)	495.30	–	23.81	14.72	22.69
60R (shallow)	406.40	609.60	20.64	8.76	13.13
60R (angled)	406.40	609.60	20.64	11.27	16.90
75R (shallow)	406.40	–	23.81	10.94	16.42
75R (angled)	406.40	–	23.81	13.67	–
80R (shallow)	406.40	609.60	23.81	11.93	17.89
80R (angled)	406.40	609.60	23.81	14.90	22.33
For British Standard Heavy Rails: Bull head rails					
95R BH (shallow)	–	457.20	27.00	14.59	14.61
For British Standard Light Rails: Flat Bottom Rails					
30M	355.6	–	–	–	2.72
35M	355.6	–	–	–	2.83
40	355.6	–	–	3.76	–

FRACTIONS, DECIMALS AND MILLIMETRE EQUIVALENTS

Fractions	Decimals	(mm)	Fractions	Decimals	(mm)
1/64	0.015625	0.396875	33/64	0.515625	13.096875
1/32	0.03125	0.79375	17/32	0.53125	13.49375
3/64	0.046875	1.190625	35/64	0.546875	13.890625
1/16	0.0625	1.5875	9/16	0.5625	14.2875
5/64	0.078125	1.984375	37/64	0.578125	14.684375
3/32	0.09375	2.38125	19/32	0.59375	15.08125
7/64	0.109375	2.778125	39/64	0.609375	15.478125
1/8	0.125	3.175	5/8	0.625	15.875
9/64	0.140625	3.571875	41/64	0.640625	16.271875
5/32	0.15625	3.96875	21/32	0.65625	16.66875
11/64	0.171875	4.365625	43/64	0.671875	17.065625
3/16	0.1875	4.7625	11/16	0.6875	17.4625
13/64	0.203125	5.159375	45/64	0.703125	17.859375
7/32	0.21875	5.55625	23/32	0.71875	18.25625
15/64	0.234375	5.953125	47/64	0.734375	18.653125
1/4	0.25	6.35	3/4	0.75	19.05
17/64	0.265625	6.746875	49/64	0.765625	19.446875
9/32	0.28125	7.14375	25/32	0.78125	19.84375
19/64	0.296875	7.540625	51/64	0.796875	20.240625
5/16	0.3125	7.9375	13/16	0.8125	20.6375
21/64	0.328125	8.334375	53/64	0.828125	21.034375
11/32	0.34375	8.73125	27/32	0.84375	21.43125
23/64	0.359375	9.128125	55/64	0.859375	21.828125
3/8	0.375	9.525	7/8	0.875	22.225
25/64	0.390625	9.921875	57/64	0.890625	22.621875
13/32	0.40625	10.31875	29/32	0.90625	23.01875
27/64	0.421875	10.71563	59/64	0.921875	23.415625
7/16	0.4375	11.1125	15/16	0.9375	23.8125
29/64	0.453125	11.50938	61/64	0.953125	24.209375
15/32	0.46875	11.90625	31/32	0.96875	24.60625
31/64	0.484375	12.30313	63/64	0.984375	25.003125
1/2	0.5	12.7	1.0	1	25.4

IMPERIAL STANDARD WIRE GAUGE (SWG)

SWG No.	Diameter (inches)	Diameter (mm)	SWG No.	Diameter (inches)	Diameter (mm)
7/0	0.5	12.7	23	0.024	0.61
6/0	0.464	11.79	24	0.022	0.559
5/0	0.432	10.97	25	0.02	0.508
4/0	0.4	10.16	26	0.018	0.457
3/0	0.372	9.45	27	0.0164	0.417
2/0	0.348	8.84	28	0.0148	0.376
1/0	0.324	8.23	29	0.0136	0.345
1	0.3	7.62	30	0.0124	0.315
2	0.276	7.01	31	0.0116	0.295
3	0.252	6.4	32	0.0108	0.274
4	0.232	5.89	33	0.01	0.254
5	0.212	5.38	34	0.009	0.234
6	0.192	4.88	35	0.008	0.213
7	0.176	4.47	36	0.008	0.193
8	0.16	4.06	37	0.007	0.173
9	0.144	3.66	38	0.006	0.152
10	0.128	3.25	39	0.005	0.132
11	0.116	2.95	40	0.005	0.122
12	0.104	2.64	41	0.004	0.112
13	0.092	2.34	42	0.004	0.102
14	0.08	2.03	43	0.004	0.091
15	0.072	1.83	44	0.003	0.081
16	0.064	1.63	45	0.003	0.071
17	0.056	1.42	46	0.002	0.061
18	0.048	1.22	47	0.002	0.051
19	0.04	1.016	48	0.002	0.041
20	0.036	0.914	49	0.001	0.031
21	0.032	0.813	50	0.001	0.025
22	0.028	0.711			

PIPES, WATER, STORAGE, INSULATION

WATER PRESSURE DUE TO HEIGHT

Imperial

Head (Feet)	Pressure (lb/in^2)	Head (Feet)	Pressure (lb/in^2)
1	0.43	70	30.35
5	2.17	75	32.51
10	4.34	80	34.68
15	6.5	85	36.85
20	8.67	90	39.02
25	10.84	95	41.18
30	13.01	100	43.35
35	15.17	105	45.52
40	17.34	110	47.69
45	19.51	120	52.02
50	21.68	130	56.36
55	23.84	140	60.69
60	26.01	150	65.03
65	28.18		

Metric

Head (m)	Pressure (bar)	Head (m)	Pressure (bar)
0.5	0.049	18.0	1.766
1.0	0.098	19.0	1.864
1.5	0.147	20.0	1.962
2.0	0.196	21.0	2.06
3.0	0.294	22.0	2.158
4.0	0.392	23.0	2.256
5.0	0.491	24.0	2.354
6.0	0.589	25.0	2.453
7.0	0.687	26.0	2.551
8.0	0.785	27.0	2.649
9.0	0.883	28.0	2.747
10.0	0.981	29.0	2.845
11.0	1.079	30.0	2.943
12.0	1.177	32.5	3.188
13.0	1.275	35.0	3.434
14.0	1.373	37.5	3.679
15.0	1.472	40.0	3.924
16.0	1.57	42.5	4.169
17.0	1.668	45.0	4.415

1 bar	=	14.5038 lbf/in^2
1 lbf/in^2	=	0.06895 bar
1 metre	=	3.2808 ft or 39.3701 in
1 foot	=	0.3048 metres
1 in wg	=	2.5 mbar (249.1 N/m^2)

PIPES, WATER, STORAGE, INSULATION

Dimensions and Weights of Copper Pipes to BSEN 1057, BSEN 12499, BSEN 14251

Outside Diameter (mm)	Internal Diameter (mm)	Weight per Metre (kg)	Internal Diameter (mm)	Weight per Metre (kg)	Internal Diameter (mm)	Weight per Metre (kg)
	Formerly Table X		Formerly Table Y		Formerly Table Z	
6	4.80	0.0911	4.40	0.1170	5.00	0.0774
8	6.80	0.1246	6.40	0.1617	7.00	0.1054
10	8.80	0.1580	8.40	0.2064	9.00	0.1334
12	10.80	0.1914	10.40	0.2511	11.00	0.1612
15	13.60	0.2796	13.00	0.3923	14.00	0.2031
18	16.40	0.3852	16.00	0.4760	16.80	0.2918
22	20.22	0.5308	19.62	0.6974	20.82	0.3589
28	26.22	0.6814	25.62	0.8985	26.82	0.4594
35	32.63	1.1334	32.03	1.4085	33.63	0.6701
42	39.63	1.3675	39.03	1.6996	40.43	0.9216
54	51.63	1.7691	50.03	2.9052	52.23	1.3343
76.1	73.22	3.1287	72.22	4.1437	73.82	2.5131
108	105.12	4.4666	103.12	7.3745	105.72	3.5834
133	130.38	5.5151	–	–	130.38	5.5151
159	155.38	8.7795	–	–	156.38	6.6056

Dimensions of Stainless Steel Pipes to BS 4127

Outside Diameter (mm)	Maximum Outside Diameter (mm)	Minimum Outside Diameter (mm)	Wall Thickness (mm)	Working Pressure (bar)
6	6.045	5.940	0.6	330
8	8.045	7.940	0.6	260
10	10.045	9.940	0.6	210
12	12.045	11.940	0.6	170
15	15.045	14.940	0.6	140
18	18.045	17.940	0.7	135
22	22.055	21.950	0.7	110
28	28.055	27.950	0.8	121
35	35.070	34.965	1.0	100
42	42.070	41.965	1.1	91
54	54.090	53.940	1.2	77

PIPES, WATER, STORAGE, INSULATION

Dimensions of Steel Pipes to BS 1387

Nominal Size (mm)	Approx. Outside Diameter (mm)	Outside Diameter Light Max (mm)	Outside Diameter Light Min (mm)	Outside Diameter Medium & Heavy Max (mm)	Outside Diameter Medium & Heavy Min (mm)	Thickness Light (mm)	Thickness Medium (mm)	Thickness Heavy (mm)
6	10.20	10.10	9.70	10.40	9.80	1.80	2.00	2.65
8	13.50	13.60	13.20	13.90	13.30	1.80	2.35	2.90
10	17.20	17.10	16.70	17.40	16.80	1.80	2.35	2.90
15	21.30	21.40	21.00	21.70	21.10	2.00	2.65	3.25
20	26.90	26.90	26.40	27.20	26.60	2.35	2.65	3.25
25	33.70	33.80	33.20	34.20	33.40	2.65	3.25	4.05
32	42.40	42.50	41.90	42.90	42.10	2.65	3.25	4.05
40	48.30	48.40	47.80	48.80	48.00	2.90	3.25	4.05
50	60.30	60.20	59.60	60.80	59.80	2.90	3.65	4.50
65	76.10	76.00	75.20	76.60	75.40	3.25	3.65	4.50
80	88.90	88.70	87.90	89.50	88.10	3.25	4.05	4.85
100	114.30	113.90	113.00	114.90	113.30	3.65	4.50	5.40
125	139.70	–	–	140.60	138.70	–	4.85	5.40
150	165.1*	–	–	166.10	164.10	–	4.85	5.40

* 165.1 mm (6.5in) outside diameter is not generally recommended except where screwing to BS 21 is necessary
All dimensions are in accordance with ISO R65 except approximate outside diameters which are in accordance with ISO R64
Light quality is equivalent to ISO R65 Light Series II

Approximate Metres Per Tonne of Tubes to BS 1387

Nom. Size (mm)	BLACK Plain/screwed ends L (m)	M (m)	H (m)	BLACK Screwed & socketed L (m)	M (m)	H (m)	GALVANIZED Plain/screwed ends L (m)	M (m)	H (m)	GALVANIZED Screwed & socketed L (m)	M (m)	H (m)
6	2765	2461	2030	2743	2443	2018	2604	2333	1948	2584	2317	1937
8	1936	1538	1300	1920	1527	1292	1826	1467	1254	1811	1458	1247
10	1483	1173	979	1471	1165	974	1400	1120	944	1386	1113	939
15	1050	817	688	1040	811	684	996	785	665	987	779	661
20	712	634	529	704	628	525	679	609	512	673	603	508
25	498	410	336	494	407	334	478	396	327	474	394	325
32	388	319	260	384	316	259	373	308	254	369	305	252
40	307	277	226	303	273	223	296	268	220	292	264	217
50	244	196	162	239	194	160	235	191	158	231	188	157
65	172	153	127	169	151	125	167	149	124	163	146	122
80	147	118	99	143	116	98	142	115	97	139	113	96
100	101	82	69	98	81	68	98	81	68	95	79	67
125	–	62	56	–	60	55	–	60	55	–	59	54
150	–	52	47	–	50	46	–	51	46	–	49	45

The figures for 'plain or screwed ends' apply also to tubes to BS 1775 of equivalent size and thickness
Key:
L – Light
M – Medium
H – Heavy

PIPES, WATER, STORAGE, INSULATION

Flange Dimension Chart to BS 4504 & BS 10

Normal Pressure Rating (PN 6) 6 Bar

Nom. Size	Flange Outside Dia.	Table 6/2 Forged Welding Neck	Table 6/3 Plate Slip on	Table 6/4 Forged Bossed Screwed	Table 6/5 Forged Bossed Slip on	Table 6/8 Plate Blank	Raised Face Dia.	Raised Face T'ness	Nr. Bolt Hole	Size of Bolt
15	80	12	12	12	12	12	40	2	4	M10 × 40
20	90	14	14	14	14	14	50	2	4	M10 × 45
25	100	14	14	14	14	14	60	2	4	M10 × 45
32	120	14	16	14	14	14	70	2	4	M12 × 45
40	130	14	16	14	14	14	80	3	4	M12 × 45
50	140	14	16	14	14	14	90	3	4	M12 × 45
65	160	14	16	14	14	14	110	3	4	M12 × 45
80	190	16	18	16	16	16	128	3	4	M16 × 55
100	210	16	18	16	16	16	148	3	4	M16 × 55
125	240	18	20	18	18	18	178	3	8	M16 × 60
150	265	18	20	18	18	18	202	3	8	M16 × 60
200	320	20	22	–	20	20	258	3	8	M16 × 60
250	375	22	24	–	22	22	312	3	12	M16 × 65
300	440	22	24	–	22	22	365	4	12	M20 × 70

Normal Pressure Rating (PN 16) 16 Bar

Nom. Size	Flange Outside Dia.	Table 6/2 Forged Welding Neck	Table 6/3 Plate Slip on	Table 6/4 Forged Bossed Screwed	Table 6/5 Forged Bossed Slip on	Table 6/8 Plate Blank	Raised Face Dia.	Raised Face T'ness	Nr. Bolt Hole	Size of Bolt
15	95	14	14	14	14	14	45	2	4	M12 × 45
20	105	16	16	16	16	16	58	2	4	M12 × 50
25	115	16	16	16	16	16	68	2	4	M12 × 50
32	140	16	16	16	16	16	78	2	4	M16 × 55
40	150	16	16	16	16	16	88	3	4	M16 × 55
50	165	18	18	18	18	18	102	3	4	M16 × 60
65	185	18	18	18	18	18	122	3	4	M16 × 60
80	200	20	20	20	20	20	138	3	8	M16 × 60
100	220	20	20	20	20	20	158	3	8	M16 × 65
125	250	22	22	22	22	22	188	3	8	M16 × 70
150	285	22	22	22	22	22	212	3	8	M20 × 70
200	340	24	24	–	24	24	268	3	12	M20 × 75
250	405	26	26	–	26	26	320	3	12	M24 × 90
300	460	28	28	–	28	28	378	4	12	M24 × 90

PIPES, WATER, STORAGE, INSULATION

Minimum Distances Between Supports/Fixings

Material	BS Nominal Pipe Size		Pipes – Vertical	Pipes – Horizontal on to low gradients
	(inch)	(mm)	Support distance in metres	Support distance in metres
Copper	0.50	15.00	1.90	1.30
	0.75	22.00	2.50	1.90
	1.00	28.00	2.50	1.90
	1.25	35.00	2.80	2.50
	1.50	42.00	2.80	2.50
	2.00	54.00	3.90	2.50
	2.50	67.00	3.90	2.80
	3.00	76.10	3.90	2.80
	4.00	108.00	3.90	2.80
	5.00	133.00	3.90	2.80
	6.00	159.00	3.90	2.80
muPVC	1.25	32.00	1.20	0.50
	1.50	40.00	1.20	0.50
	2.00	50.00	1.20	0.60
Polypropylene	1.25	32.00	1.20	0.50
	1.50	40.00	1.20	0.50
uPVC	–	82.40	1.20	0.50
	–	110.00	1.80	0.90
	–	160.00	1.80	1.20
Steel	0.50	15.00	2.40	1.80
	0.75	20.00	3.00	2.40
	1.00	25.00	3.00	2.40
	1.25	32.00	3.00	2.40
	1.50	40.00	3.70	2.40
	2.00	50.00	3.70	2.40
	2.50	65.00	4.60	3.00
	3.00	80.40	4.60	3.00
	4.00	100.00	4.60	3.00
	5.00	125.00	5.50	3.70
	6.00	150.00	5.50	4.50
	8.00	200.00	8.50	6.00
	10.00	250.00	9.00	6.50
	12.00	300.00	10.00	7.00
	16.00	400.00	10.00	8.25

PIPES, WATER, STORAGE, INSULATION

Litres of Water Storage Required Per Person Per Building Type

Type of Building	Storage (litres)
Houses and flats (up to 4 bedrooms	120/bedroom
Houses and flats (more than 4 bedrooms)	100/bedroom
Hostels	90/bed
Hotels	200/bed
Nurses homes and medical quarters	120/bed
Offices with canteen	45/person
Offices without canteen	40/person
Restaurants	7/meal
Boarding schools	90/person
Day schools – Primary	15/person
Day schools – Secondary	20/person

Recommended Air Conditioning Design Loads

Building Type	Design Loading
Computer rooms	500 W/m² of floor area
Restaurants	150 W/m² of floor area
Banks (main area)	100 W/m² of floor area
Supermarkets	25 W/m² of floor area
Large office block (exterior zone)	100 W/m² of floor area
Large office block (interior zone)	80 W/m² of floor area
Small office block (interior zone)	80 W/m² of floor area

PIPES, WATER, STORAGE, INSULATION

Capacity and Dimensions of Galvanized Mild Steel Cisterns – BS 417

Capacity (litres)	BS type (SCM)	Length (mm)	Dimensions Width (mm)	Depth (mm)
18	45	457	305	305
36	70	610	305	371
54	90	610	406	371
68	110	610	432	432
86	135	610	457	482
114	180	686	508	508
159	230	736	559	559
191	270	762	584	610
227	320	914	610	584
264	360	914	660	610
327	450/1	1220	610	610
336	450/2	965	686	686
423	570	965	762	787
491	680	1090	864	736
709	910	1070	889	889

Capacity of Cold Water Polypropylene Storage Cisterns – BS 4213

Capacity (litres)	BS type (PC)	Maximum height (mm)
18	4	310
36	8	380
68	15	430
91	20	510
114	25	530
182	40	610
227	50	660
273	60	660
318	70	660
455	100	760

PIPES, WATER, STORAGE, INSULATION

Minimum Insulation Thickness to Protect Against Freezing for Domestic Cold Water Systems (8 Hour Evaluation Period)

Pipe size (mm)	Condition 1 $\lambda = 0.020$	Condition 1 $\lambda = 0.030$	Condition 1 $\lambda = 0.040$	Condition 2 $\lambda = 0.020$	Condition 2 $\lambda = 0.030$	Condition 2 $\lambda = 0.040$
Copper pipes						
15	11	20	34	12	23	41
22	6	9	13	6	10	15
28	4	6	9	4	7	10
35	3	5	7	4	5	7
42	3	4	5	8	4	6
54	2	3	4	2	3	4
76	2	2	3	2	2	3
Steel pipes						
15	9	15	24	10	18	29
20	6	9	13	6	10	15
25	4	7	9	5	7	10
32	3	5	6	3	5	7
40	3	4	5	3	4	6
50	2	3	4	2	3	4
65	2	2	3	2	3	3

Condition 1: water temperature 7°C; ambient temperature −6°C; evaluation period 8 h; permitted ice formation 50%; normal installation, i.e. inside the building and inside the envelope of the structural insulation
Condition 2: water temperature 2°C; ambient temperature −6°C; evaluation period 8 h; permitted ice formation 50%; extreme installation, i.e. inside the building but outside the envelope of the structural insulation
λ = thermal conductivity [W/(mK)]

Insulation Thickness for Chilled And Cold Water Supplies to Prevent Condensation

On a Low Emissivity Outer Surface (0.05, i.e. Bright Reinforced Aluminium Foil) with an Ambient Temperature of +25°C and a Relative Humidity of 80%

Steel pipe size (mm)	$t = +10$ $\lambda = 0.030$	$t = +10$ $\lambda = 0.040$	$t = +10$ $\lambda = 0.050$	$t = +5$ $\lambda = 0.030$	$t = +5$ $\lambda = 0.040$	$t = +5$ $\lambda = 0.050$	$t = 0$ $\lambda = 0.030$	$t = 0$ $\lambda = 0.040$	$t = 0$ $\lambda = 0.050$
15	16	20	25	22	28	34	28	36	43
25	18	24	29	25	32	39	32	41	50
50	22	28	34	30	39	47	38	49	60
100	26	34	41	36	47	57	46	60	73
150	29	38	46	40	52	64	51	67	82
250	33	43	53	46	60	74	59	77	94
Flat surfaces	39	52	65	56	75	93	73	97	122

t = temperature of contents (°C)
λ = thermal conductivity at mean temperature of insulation [W/(mK)]

PIPES, WATER, STORAGE, INSULATION

Insulation Thickness for Non-domestic Heating Installations to Control Heat Loss

Steel pipe size (mm)	$t = 75$ Insulation thickness (mm)			$t = 100$ Insulation thickness (mm)			$t = 150$ Insulation thickness (mm)		
	$\lambda = 0.030$	$\lambda = 0.040$	$\lambda = 0.050$	$\lambda = 0.030$	$\lambda = 0.040$	$\lambda = 0.050$	$\lambda = 0.030$	$\lambda = 0.040$	$\lambda = 0.050$
10	18	32	55	20	36	62	23	44	77
15	19	34	56	21	38	64	26	47	80
20	21	36	57	23	40	65	28	50	83
25	23	38	58	26	43	68	31	53	85
32	24	39	59	28	45	69	33	55	87
40	25	40	60	29	47	70	35	57	88
50	27	42	61	31	49	72	37	59	90
65	29	43	62	33	51	74	40	63	92
80	30	44	62	35	52	75	42	65	94
100	31	46	63	37	54	76	45	68	96
150	33	48	64	40	57	77	50	73	100
200	35	49	65	42	59	79	53	76	103
250	36	50	66	43	61	80	55	78	105

t = hot face temperature (°C)
λ = thermal conductivity at mean temperature of insulation [W/(mK)]

ESSENTIAL READING FROM TAYLOR AND FRANCIS

Failures in Concrete Structures: Case Studies in Reinforced and Prestressed Concrete

R. Whittle

Some lessons are only learned from mistakes. But, it's much cheaper to learn from someone else's mistakes than to have to do so from your own. Drawing on over 50 years of working with concrete structures, Robin Whittle examines the problems which he has seen occur and shows how they could have been avoided.

The first and largest part of the book tells the stories of a number of cases where things have gone wrong with concrete structures. Each case is analysed to identify its cause and how it might have been prevented. It then looks at how failures in structural modelling can lead to big problems if they are not identified before construction is undertaken. Beyond this it examines how contract arrangements can encourage or prevent problems in the designing and building processes. It concludes with an examination of the role research and development in preventing failures.

By identifying the differences between shoddy economisations and genuine efficiency savings, this book offers savings in the short term which won't be at the expense of a structure's long-term performance. Invaluable reading if you're designing or building concrete structures and want to avoid problems which could be expensive or embarrassing further down the line.

July 2012:: 234x156: 154pp
Hb: 978-0-415-56701-5: **£ 60.00**

To Order: Tel: +44 (0) 1235 400524 **Fax:** +44 (0) 1235 400525
or Post: Taylor and Francis Customer Services,
Bookpoint Ltd, Unit T1, 200 Milton Park, Abingdon, Oxon, OX14 4TA UK
Email: book.orders@tandf.co.uk

**For a complete listing of all our titles visit:
www.tandf.co.uk**

Taylor & Francis
Taylor & Francis Group

ESSENTIAL READING FROM TAYLOR AND FRANCIS

Performance Based Fire Engineering of Structures

Yong Wang et al

This authoritative book summarizes the latest knowledge on performance-based approaches to the fire safety engineering of structures. Going beyond codes of practice, it addresses issues that have become more important in recent years, including the integrity of fire-resistant compartmentation, the progressive collapse of structures in fire and provisions for connection robustness, and accurate information on material properties at high temperatures. Written by core members of the COST C26 working group on fire resistance, it explains how to use advanced fire engineering design to improve structural safety and reduce construction costs.

June 2012: 234x156: 394pp
Hb: 978-0-415-55733-7: **£80.00**

To Order: Tel: +44 (0) 1235 400524 **Fax:** +44 (0) 1235 400525
or Post: Taylor and Francis Customer Services,
Bookpoint Ltd, Unit T1, 200 Milton Park, Abingdon, Oxon, OX14 4TA UK
Email: book.orders@tandf.co.uk

For a complete listing of all our titles visit:
www.tandf.co.uk

Taylor & Francis
Taylor & Francis Group

Index

abandoned mine workings, Ground stabilization 154
Abbreviations xi
Access covers and frames, manholes 224
Access roads, provision and placing of concrete 425
Access shaft
 excavation 300
 linings – concrete and formwork 301
Acknowledgements xiii
Aftercare and maintenance, planting 498
Aggregates levy 13
Air conditioning 130
Alteration work
 dense aggregate blockwork 490
 lightweight aggregate blockwork 490
Aluminium
 cladding 119, 246
 external doors 121
 glazed internal walls 122
 glazed partitions 122
 patent glazing 111
 roof covering 115
 roof decking 111
 windows 121
Anchorages
 ground anchors 417
 ground stabilization 150
Ancillary building works, Earth Retention ansd Stabilization 95
Anti-condensation paint 321
Antiskid, pavements 426
Approximate Estimating, building prices per square metre 91
Ashlar masonry, Portland whitbed 315, 491
Asphalt
 extruded channels and kerbs 285
 floor finishes 125
 roof covering 111, 115
 tanking 328

bailey bridges 102
Balustrading to stairs and walkways 116

Bar reinforcement
 high yield steel 192, 282, 475
 mild steel 191, 282, 475
 stainless steel 192, 474
Bases, footings, pile caps and ground beams
 placing of mass concrete 469
 placing of reinforced concrete 183
Beams
 formwork (outputs) 527
 in situ concrete 108
 precast concrete 198, 471
 prestressed concrete 184
 reinforced concrete 183
 steelwork 242
Beany block kerb/drainage block 284, 430
Bench seats
 precast concrete 441
 steel 441
 street furniture 441
 timber 441
Benching to manholes, provision and placing of concrete 224, 528
Bill of quantities, initial appraisal 7
Bitumen
 macadam 279, 424
 spray applied 115
Bituminous screeds, roof coverings 113
Bituminous spray
 flexible pavements 281
 road construction 427
 tack coat 427
Blast cleaning, steelwork 243
Blasting boulders, rock removal 177
Blockwork 489
 dense aggregate 313, 490
 external walls 117
 internal walls 122
 lightweight aggregate 312, 489
 partitions 122
 wall ties 317
Bollards
 precast concrete 441
 steel 441

Bolts, woodwork 252
Bored in situ reinforced concrete piling 256, 450
Box girder main members, steelwork 477
Bracings, steelwork 242
Bradstone, reconstituted stone 314
Breaking out
  brickwork 524
  concrete 524
  masonry 524
  pavement 418, 524
  sandstone 524
Brickwork 486
  chambers 384, 447, 449
  class A engineering bricks 310, 487
  class B engineering bricks 310, 488
  common bricks 305, 486
  Dritherm cavity insulation 317
  Expamet reinforcement 317
  external walls 118
  facing bricks 308, 488
  fair facing work 308
  Hyload damp-proof course 317
  internal walls 122
  joint filler 317
  manholes 224
  movement joints 317
  partitions 122
Bridge Bearings 244, 248
  laminated elastomeric rubber 482
  mechanical guide 482
  plain rubber 482
  pot, fixed and sliding 482
  testing 483
Bridge Works
  footbridges 96
  road bridges 96
Build in pipes and ducts
  common brickwork 308
  dense aggregate blockwork 312
  engineering brickwork 310–311
  facing brickwork 309
  lightweight aggregate blockwork 312
Building paper 112
Building Price per Square Metre
  airport facilities 93
  car parks 93
  garages 93
  Industrial facilities 94
  stations 93
  telephone exchanges 93
  Water treatment facilities 94
Built-up girders, steelwork 242, 477
Bulbs, planting 172, 498
Bush hammering, external walls 119

Cabling
  communication 448
  feeder pillars 447
  joints and terminations 446
  lighting 445
  tiles 445
  trench 445, 448
Canals, dredging 159
Capital allowances 19
car parking 133
Carbon steel pipes BS 10126, drainage below ground 215
Carpentry and joinery 338
Casings to metal sections, placing of reinforced concrete 184
Cast iron
  rainwater gutters 334
  rainwater gutters and pipes 114
  rainwater pipes 334
  Timesaver 212
Catseye road studs, road markings 440
Ceilings
  acoustic plaster 128
  artex 128
  Gyproc 128
  integrated 128
  plaster 128
  plasterboard 128
  softwood tongued and grooved 128
  Supalux 128
  suspended 128, 350
  Tyrolean finish 128
  woodwool slabs 128
Cement and sand screed, floor finishes 125
Cement paint to external walls 119
Ceramic tile floor and wall finishes 347
Chambers
  brick 384
  concrete 387
Channels
  extruded asphalt 284
  precast concrete 283, 429
Chestnut pale, temporary fencing 367

Chipboard
   floor finishes 125
   wall finishes 124
Clay
   cable ducts 381
   cable tiles 445, 448
   channels – manholes 224
   grease traps/interceptors 234, 394
   laying and jointing pipes (outputs) 528
   perforated drains 381
   pipes BS EN 295 206
   tile roof covering 114
Closeboarded permanent fencing 368
Coated chippings
   flexible pavements 280
   road construction 426
Coated corrugated steel culverts, drainage below ground 236
Cobble paving 287, 435
coffer dams 102
Cold asphalt
   flexible pavements 280
   road construction 425
Cold milling (planing), pavements 427
Cold water installations 129
columns
   bases – foundations 105
   common brickwork 307
   concrete encased steelwork 108
   engineering brickwork 311
   facing brickwork 309
   in situ concrete 108
   lightweight aggregate blockwork 312
   provision and placing of concrete 183, 470
   steelwork 242
Communications – motorway 448
Compaction, plant costs 88
Concrete
   ancillaries 185
   cable tiles 445, 448
   chambers 387
   concrete floors 108
   designed mix 180
   dowels 195
   drain pipe 239–241
   encased columns 108
   floor finishes 125
   formwork 187, 473
   grout base plates 197
   in situ beams 108
   in situ work 179, 469
   inserts 197
   internal walls 122
   joints in concrete 194, 282, 425
   labour gang rate allowances 468
   pavements 282
   piles 269
   piling 256, 450
   pipe support stools and thrust blocks 241
   pipes to BS 5911 209, 380
   placing of reinforced concrete 183
   porous pipe 382
   post-tensioned prestressing 195
   precast 198, 471
   prescribed mix 180
   provision and placing 469
   ready mixed 525
   reinforced 183
   reinforcement 474
   resources 185
   retaining walls 95
   roads and pavings 282, 425
   roofs 111
   safety barriers 372
   sprayed or brushed waterproofing 330
   stairs 116
   stools and thrust blocks 241
   structural 468
   sulphate-resisting 181
   tile roof covering 113
   walls 118, 122, 469
   waterstops 194
Conditions of contract, initial appraisal 7
Construction joints in concrete 283, 426
Contraction joints in concrete 283, 426
Contractor's Site Oncosts
   Non Time Related 56
   Time Related 54
Contractors, plant costs 85
Contractual Requirements
   Insurances of works 141
   performance bond 141
   Third Party insurance 141
Contents v
Copings
   common brickwork 307, 487
   engineering brickwork 311, 487
   facing brickwork 309, 489
   precast concrete 314
Copper lead conductors, lightning protection 447

Corner cube road studs, road markings 292
Corrosion protection of steelwork 478
Corrugated steel, drainage below ground 381
Cost and tender price indices 53
   Constructed Civil Engineering Cost Index 73
   Introduction 71
   Price Adjustment Formula Indices 71
   Road Construction Price Indes 74
Cost estimate, outline of the tendering and estimating process 10
Costs of equipment
   ancillary equipment 177
   earth moving 174
   earth moving equipment 175
   excavation equipment 174
   transportation equipment 175
Crash cushion 374
Crib walling 95
   precast concrete 416
   timber 416
Cultivation additives
   compost 495
   fertilizer 495
   granular treatments 495
   herbicide 495
   landscaping 171
   manure 495
   weed killer 495
Culverts, cattle creeps, subway units
   coated corrugated steel 236
   precast concrete 471
Curtain walling, external walls 119
Curved formwork 473
Cutting, pavements 281, 426

Damp-Proofing 327
Damp-proofing
   Bituthene 328, 480
   waterproof render 329
Daywork
   application of dayworks 506
   insurances 508
   introduction 503
   mobilization 510
   schedules of daywork 503
   servicing of plant 509
   watching and lighting 508
   welfare facilities 508

Decking
   prestressed concrete 198
   timber 251
Demolition and Clearance
   Buildings 156
   General 155
   Pipelines 156
   redundant drains or sewers 156
   wall structures 156
   work to redundant drains 156
Demolitions, buildings 364
Demountable partitions, internal partitions 122
Dense aggregate blockwork
   build in pipes etc. 313
   walls 313, 490
Dense bitumen macadam
   flexible pavements 280
   pavements 424
Dense tar surfacing
   flexible pavements 279
   road construction 424
Dense tarmacadam
   flexible pavements 279
   pavements 424
Design and construct variations, outline of the tendering and estimating process 12
Design of temporary works, etc., outline of the tendering and estimating process 9
Detector, feeder cables 447, 449
Dewatering, earthworks 178
Diamond crossings 297
diaphragm walling 150
   ground stabilization 467
Disposal
   basis of rates 157
   double handling 165
   earthworks 160
   excavated earth 160
   excavated material 402
   hazardous waste 42
   installations 129
   non-hazardous waste 42
   rock or other hard material 164
Distribution of fuel, daywork 510
Ditches, land drainage 234
dock levellers 131
Doors
   external 342
   frames 343
   internal 123, 343

Internal 123
special partitioning and doors 122
stainless steel entrance screens and doors 121
Double handling
　disposal 165
　excavated topsoil 165
　rock or other hard material 165
Dowels
　concrete 195
　movement joints 485
Drainage 151
　above ground 334
　below ground 203
　carbon steel pipes BS 10126 215
　Cast iron pipes 202
　cast iron pipes BS 437 212
　clay pipes BS 1196 201
　coated corrugated steel culverts 236
　concrete bedding 239
　concrete haunchings 239
　Concrete pipes 201
　concrete pipes BS 5911 209, 380
　concrete porous pipes 382
　concrete stools and thrust blocks 241
　corrugated steel 381
　dewatering system 178
　drainage pipework 134, 201
　ductile spun iron pipes BS 4772 214
　excavation in mass concrete 237
　excavation in reinforced concrete 237
　excavation in rock 237
　Field Drains 203
　gullies 135, 234, 394
　land drainage 135, 234
　large bore concrete pipes 209
　manholes 135, 224
　non-return tidal flap valves 222
　penstocks 222
　pipe bedding 238–239
　pipe connections 234
　pipe haunching – mass concrete 239
　pipe jacking 238
　PVC-u rainwater pipes 335
　redundant drains or sewers 156, 336
　renovation of sewers 337
　Safeticurb channels 284, 429
　sewer renovation 337
　soft spots 237, 394
　Ultrarib PVC-u 220

valves and penstocks 222
vitrified clay BS 65 378
vitrified clay BS EN295 206
work to redundant drains 156, 396
Drainage and Service Ducts 377
Drawings, initial appraisal 8
Dredgers
　floating craft 159
　trailing suction hopper dredger 159
Dredging
　canals 160
　dock systems 160
　general bed lowering 160
　harbour beds 160
　industrial lagoons 160
　lakes 160
　maintaining shipping channels 160
　new channels in tidal location 160
　rivers 160
　silted dock systems 161
Drilling and blasting
　in open cut 177
　in quarry 177
　in rock 177
　in trenches 177
　rock removal 177
Drilling for grout holes 148
Duct covers, Stelduct 245
Ductile spun iron, pipes BS 4772 214
Durbar flooring, steelwork 245
Dynamic compaction, Ground stabilization 153, 415

Earth moving
　earth moving equipment 175
　excavation by hand 177
　excavation by pneumatic equipment 177
　excavation equipment 174
　load material 177
　transportation equipment 175
Earth retaining walls 95
Earth Retention ansd Stabilization
　crib walls 95
　gabion walls 95
　retaining walls 95
Earth rods 447
Earth support
　earthworks 165
　manholes 224
earthwork support 103

Earthworks
  crib walling 416
  dewatering 178
  disposal 164
  dredging 159
  drilling and blasting in rock 177
  earth moving 174
  earthwork supports 165
  excavation 157
  excavation for foundations 163
  excavation rates 396
  excavation to general areas 163
  gabions 416
  general excavation 163
  geotextiles 168
  ground anchors 417
  labour gang rate allowances 158
  landscaping 171, 495
  membranes 168
  treatment of surfaces 164
  trial pits 418
Edgings, precast concrete 283, 429
Electrical Installation, fittings/rates per each point 130
Electrical work road lighting and signs 445
Ellis Trief safety kerb, precast concrete 284, 430
Emergency lighting 130
Employer's and Consultants Site Requirements
  Non Time Related 59
  Time Related 58
Enamelled steel
  internal walls 122
  partitions 122
Entrance screens 121
EPDM membrane, roof coverings 116
Epoxy, painting of steelwork 243
Erection of steelwork, permanent, trial 242
Erosion control mat, geotextiles 170
Erosion protection, geotextiles 170
Eurogrid, flooring 247
Excavated material, disposal outputs 523
Excavated rock, filling 167
Excavated surfaces, treatment of surfaces 164
Excavated topsoil, double handling 165
Excavation
  access shafts 300
  backacter 162, 174
  basis of disposal rates 157
  basis of rates generally 157
  bulldozer 164
  cable trench 445, 448
  deposition 406
  disposal 406
  dragline 174
  Earthworks 157, 163
  Excavation by Dredging 159
  excavation rates 401
  face shovel 174
  foundations 163
  general areas 164
  Ground conditions 157
  hand 177
  hard material 396, 405
  highway works 404
  Labour Resources 158
  lining watercourses 414
  manholes 224
  multi-purpose loader/back hoe 175
  plant costs 88
  Plant Resources 158
  pneumatic equipment 177
  preparation of filled surfaces 168
  road lighting 445
  tractor loader 165, 175
  tunnels 300
Excavation gang, labour gang rate allowances 158
Expamet reinforcement 317
Expansion joints and sealing of gaps to bridge deck 484
Expansion joints in concrete 283, 426
Exposed aggregate external walls 120
External doors and windows 120
External Fixtures
  benches 134
  bus stops 134
  cycle stand 134
  flag pole 134
  lighting to pedestrian areas 134
  lighting to roads 134
  litter bins 134
  traffic signs 134
External Services
  electric services 135
  gas services 135
  telephone services 135
  water services 135
External Walls
  brick/block walling 117
  curtain/glazed walling 119

external finishes 119
reinforced concrete walling 118
wall claddings 118
External works
  brick paviors 132
  cobblestones 132
  concrete paviors 132
  concrete slabs 132
  granite setts 132
  York stone paving 132
Externally illuminated signs, traffic signs on gantries 438
Extruded asphalt
  channels 285
  kerbs 285

Fabric reinforcement, high yield steel 193, 476
Facing brickwork 118, 308, 488
Fair faced
  formwork 189, 473
  wall finishes 124
Fees
  consulting engineers' conditions of appointment 513
  quantity surveyors' conditions of appointment 518
Felt roof coverings 115
Fencing 366
  chain link fencing 133
  closeboarded 332
  environmental barriers 366
  gates and stiles 333, 367
  guard rails 134
  parking bollards 134
  pedestrian guard rails 333, 376
  post and wire 332
  remove, store, re-erect 369
  safety 333, 372
  timber 331
  timber fencing 133
  vehicle control barrier 134
Fertilizer, cultivation additives 495
Field drains BS 1196, clay pipes 203
Field gate 368
Filled surfaces, treatment of surfaces 168
Filling
  compaction 410
  excavated earth 166
  excavated rock 167
  French and rubble drains 235

highways 408
imported topsoil 166
Filter drains, clay perforated pipes 204, 381
Fin drains 383
Finance costs, contract investment oncosts and profit 49
Finishing surfaces, concrete 196
Finishings
  floors 125
  wall 124
Fire
  doors 123
  Fire fighting/protective installations 131
  ladder 117
  lightning protection 131
  proof partitions 123
  resistant ceiling finishes 128
  shutters 121
  sprinkler installations 131
Fittings
  kitchen fittings 129
  office furniture 129
  reception area furniture 129
  signs 129
  wrought softwood shelving 128
Fix only equipment, motorway communications 448
Fixing reinforcement, outputs 526
Flexible construction, roads and paving 279
Flexible pavements
  bitumen macadam 279
  bituminous spray 281
  coated chippings 281
  cold asphalt 280
  dense bitumen macadam 280
  dense tar surfacing 280
  dense tarmacadam 279
  rolled asphalt 281
  slurry sealing 280
Floor finishes
  carpet 349
  flexible sheet 348
  rubber 348
  vinyl 348
Flooring
  open grid 247
  plate 245
Floors
  access floors 127
  acrylic polymer 126

Altro safety 127
carpet tiles/ underlay 127
cement and sand 126
ceramic tile 126
chipboard 125
composition block 126
concrete finishes 125
entrance matting 127
epoxy 126
floor coverings, common 127
granolithic 126
latex cement screeds 125
mastic asphalt 126
polyester resin 126
quarry tile 126
rubber tiles 127
skirtings 127
Synthanite 126
vinyl sheet 127
vinyl tile 126
wrought softwood 125
Fog warning – motorway communication 449
Foot bridges 96
Footways 432
Formwork
  class F1 finish 473
  class F2 finish 474
  class F3 finish 474
  curved 188–189, 191, 473
  definitions 473
  extra for smooth finish 108
  extra smooth finish 190
  fair finish 189, 473
  for voids 188, 190
  labour gang rate allowances 468
  multipliers 527
  outputs 527
  rough finish 187, 473
  tunnel access shaft lining 301
  tunnel lining 301
Foundations
  column bases 105
  foundations other/alternate 108
  ground slabs 107
  pile caps 106
  piled foundations 107
  raft foundations 107
  strip footings 104
  strip or base foundations 106
  trench fill 104

Frame and Upper Floors
  beams 108
  columns 108
  precast concretefloors; reinforced concrete frame 110
  precast concretefloors; steel frame 110
  reinforced concrete floor and frame 109
  reinforced concrete floor and steel frame 110
  reinforced concrete floor; no frame 109
  softwood jointed floor; no frame 109
Frames, gullies and gratings, steelwork 234, 394
French and rubble drains, filling 235
Fuel consumption, plant costs 89
Fuel distribution, daywork 510

Gabions 416
  rock 95, 335, 416
Galvanized steel roof decking 111
Galvanizing steelwork 243, 478
garages 133
Gates and stiles 333
Geogrid, highways 410
Geotechnical and other specialist processes
  Anchorages 150
  Ground Consolidation 152
  sand, band and wick drains 151
Geotextiles, applications costs 168
Geoxtiles, highways 410
Glazed walling, external walls 119
Glazing 344
goods hoist 131
goods lift 131
Granite chippings, roof coverings 115
Granite setts 287, 434
Granular material, subbases 279, 423
Granular treatments, cultivation additives 169, 495
Grass sowing
  hydraulic mulch grass seed 496
  landscaping 495
  laying imported turf 171, 496
  planting 172
Grease interceptors, clay 234, 394
Greenheart 262
  hardwood 249
Grit blasting, steelwork 478
Ground anchorages, earthworks 417
Ground Consolidation
  Dynamic compaction 153
  Vibro Replacement 152

Ground investigation
  instrumental observations 146
  Light cable percussion boreholes 145
  Professional services 147
  Resources 144
  Rotary drilled boreholes 145
ground stabilization 151
  dynamic compaction 415
  grout holes 154
Ground water lowering 417
Grout
  base plates 197
  holes – ground stabilization 154
  redundant drains 396
grout holes 149
grout materials and injection 149
Guard rails 333, 376
Gullies
  clay 234, 394
  precast concrete 234
  sealing redundant 396
Gutters 114

Handrails 116
  pedestrian guard railing 376
  stairs 117
  steelwork 245
Harbour bed control, dredging locations 160
Hardcore, road subbases 279
Hardwood
  decking 251
  external doors 121
  Greenheart 249
  internal door frames 123
  stain 324
  windows 121
hazardous waste, disposal 42
Head Office Overheads, Oncosts and profit 49
Heating
  gas fired radiator heating 130
  oil fired radiator heating 130
Hedges, planting 173, 366, 496
Herbaceous and ground cover plants,
  planting 497
Herbicide, cultivation additives 495
Highway Works 357
  all purpose roads 98
  disused sewers, drains 413
  excavation 404
  geogrid 410

  geoxtiles 410
  motorways 97
  road crossings 100
  roads other 99
  rural all purpose roads 98
  rural motorways 98
  scour and erosion protection 412
  topsoil 414
  underpasses 100
  urban all purpose roads 99
  Urban motorways 98
  watercourse, lining 414
  Zebra crossing 100
Holding down bolts, steelwork 242
Holorib, permanent formwork 110–111
Hydraulic mulch grass seed, grass sowing 172,
  496

Imported, granular material 166–167
Initial Appraisal 6
  Bill of quantities 7
  Conditions of contract 7
  Drawings 8
  Specification 7
Insulation
  boards 112, 341
  quilts 112, 341
Insurances and Bons 61
Integrated ceilings 128
Interceptors, clay 234
Interlocking Steel Piles 273
Internal Doors
  doors perimeter treatment 124
  doors/door linings/frames 123
Internally illuminated signs, traffic signs on
  gantries 288, 438
Ironmongery 343

Jointing clayware pipes, outputs 528
Joints
  concrete 194, 282, 425
  construction 282, 426
  contraction 426
  expansion 282, 426, 484
  longitudinal 282, 425
  movement joint sealers 484

Kerbs
  extruded asphalt 430
  extruded concrete 431

precast concrete 283, 429
quadrants 283, 429
remove from store and relay 430

Laboratory tests, Ground investigation 146
Labour
  basis 77
  labour categories 82
  labour costs 80
Ladders 117
lakes, Landscaping 101
Lamps, lanterns and luminaires, road lighting 443
Land drainage
  culverts 236
  ditches 234
  filter drains 381
  fin drains 383
  French and rubble drains 235
  narrow filter drains 383
  PVC-u perforated pipes 217
Land reclamation, dredging 159
Land remediation
  Conditions 45
  Cost Considerations 41
  Definition 44
  Effect on Disposal Costs 39
  Land Remediation Techniques 40
  Statutory Framework 37
  The EU Landfill Directive 38
  Types of Hazardous wastes 39
Landfill tax 35
Landscaping 495
  cultivation additives 171, 495
  earthworks 495
  grassing 495
  land drains 234
  mulching 498
  parklands 101
  plant support netting 172, 498
  planted areas 133
  planters 497
  planting 495–496, 498–499
  preparation of surfaces 495
  seeded areas 133
  shrub stakes 498
  sports pitches 100
  tree guards 496
  tree stakes 496
  tree supports 496
  trees 133
  turfed areas 133
  weed control 495, 498
Lifting traffic barrier, steel 441
Light duty pavements
  asphalt/ bitumen macadam 285
  brick paviors 287, 434
  cobble paving 287, 435
  concrete base 285
  Ennstone Johnston Golden gravel 285, 433
  granite setts 287, 434
  gravel 285, 432
  gravel base 285
  hardcore base 285
  precast concrete paving 286, 433
  sand base 285
Lighting
  precast concrete bracket arms 443
  precast concrete columns 443
  road lighting cable 445
  road lighting cable jointing 446
  road lighting cable terminations 447
Lightning protection
  copper lead conductors 447
  earth rods 447
lights, roof 112
Lightweight aggregate blocks 312, 489
Linings to tunnels, precast concrete 301
Loading loose material by hand, outputs 524
Locating buried cables
  communications 448
  road lighting and traffic signs 445

Mains electrical work 130
Maintaining shipping channels, dredging
  locations 160
Management appraisal, outline of the tendering
  and estimating process 12
Manholes 396
  abandonment, alteration 337
  backdrops 233
  brick 224
  civil engineering works – unit costs 224
  concrete 397
  precast concrete 227
Marker posts, traffic signs 441
Masonry
  ashlar masonry 315, 491
  civil engineering works – unit costs 304
  painting 322
  rubble walling 315, 491

Mechanical services installations 130
Membranes
  geotextiles 168
  roof coverings 115
Metal
  ladders, stairs 117
  roof decking 111
  stud partitions and doors 121
Metalwork
  miscellaneous 477
  structural 242
Method statement and programme, outline of the tendering and estimating process 9
Mild steel, bar reinforcement 191, 282, 475
mine workings, abandoned mine workings 154
Motorway communications 448
Movement joints
  dowels 484
  expansion joints 484
  joint sealers 484
  waterstops 484
Multipliers for formwork, outputs 527

Narrow filter drains, land drainage 383
New channels in tidal location, dredging locations 160
Noise barriers
  noise reflective barriers 367
  sound absorptive barriers 367
non-hazardous waste, disposal 42
Non-return tidal flap valves, drainage below ground 222

Obtaining prices, outline of the tendering and estimating process 8
Offloading materials, daywork 507
Oncosts and profit 49
Outline of estimating and tender process
  Cost estimate 10
  Design and construct variations 12
  Submission of tender 12
Outline of the Tendering and Estimating Process 5
  Design of temporary works, etc. 9
  Management Appraisal 12
  Method statement and programme 9
  Pricing the Tender 11
  Quantities check 10
  Submission of Tender 12

Outputs
  beam formwork 527
  breaking out by hand 524
  clayware pipes – laying and jointing 528
  concrete – placing 525
  disposal of excavated material 523
  fixing reinforcement 526
  formwork 527
  manholes – precast concrete 528
  material loading by hand 524
  placing concrete 525
  precast concrete manholes 528
  ready mix concrete – placing 525
  reinforcement – fixing 526
  slab formwork 527
  wall formwork 527

Painting 318
  anti-condensation 321
  Artex plastic compound 125
  bituminous 325
  emulsion 125, 321
  flame retardant paint 125
  gloss 125, 319
  hardwood stain 324
  masonry 322
  polyurethane varnish 323
  preservative (wood) 325
  primer 318
  sprayed 125
  steelwork protection 243, 478
Parapets
  aluminium 374
  steel 373
parklands, playgrounds 101
Partitions, brick/blockwork 122
Patent glazing 112
Paved areas 133
Pavements 421
  antiskid 426
  breaking up 418, 524
  cutting 426
  flexible 279–280, 282, 424
  highway works 423
  kerbs 283, 429
  perforation 418
  rigid 282
  scarify surface 281, 427
  surface planing 426

Paving 285
  highways 418
Pavings
  labour gang rate 276
  plant rate 278
Paviors, brickwork 286, 434
Pedestrian crossings, guard rails 376
Penstocks, drainage below ground 222
Perforated drains 217
  clay pipes 205
Perforating pavements 418
Permanent fencing
  closeboarded 368
  field gate 368
  plastic coated chain mesh 368
  post and rail 368
  steel gate 368
  timber stile 368
Permeable membrane, geotextiles 168
Piers
  common brickwork 307
  engineering brickwork 311
  facing brickwork 309
  lightweight aggregate blockwork 312
Pile tests 274
Piling 256, 450
  bored in situ reinforced concrete 256, 450
  concrete injected piles (flight augered) 451
  driven cast-in-place reinforced concrete 258, 451
  driven timber piles 261
  interlocking steel sheet 265, 463
  pile test 274
  piling ancillaries 269
  plant 450
  precast concrete 450
  preformed concrete 260
  proof loading 462–463
  segmental casing method 259, 451
  steel tubular piles 459
  steel universal bearing piles 263, 451
  temporary steel casing 259
  U section 452, 465
  Z section 451, 464
Pipe
  beds 238–239
  carbon steel BS 10126 215
  cast iron BS 437 212
  clay pipes 205
  connections 234

  ductile spun iron BS 4772 214
  jacking 238
  porous precast concrete BS 5911 209
  precast concrete BS 5911 209
  PVC-u 217
  surrounds 240–241
Piped building services
  equipment 351
  pipework 351
  sanitary appliances 351
Pipework, drainage 200
Placing concrete
  mass concrete 183
  outputs 525
  prestressed concrete 184
  reinforced concrete 183–184
Planing surface, pavements 426
Plant, fuel costs 89
Plant costs
  contractor owned plant 85
  fuel consumption 89
  owning costs 86
  utilization factors 87
Plaster
  ceiling finish 128
  lightweight Thistle 346
  wall finish 124
Plasterboard
  ceiling finish 128
  wall finish 124
Plasterboard faced
  internal wall finish 124
  partitions 121
Plastic coated chain mesh, permanent fencing 368
Plate flooring, steelwork 245
Plywood
  floor finishes 341
  soffit 341
  wall finishes 124, 340–341
Polyethylene sheet, waterproof and protective layers 283, 425
Polymer, spray applied 115
Polypropylene, geotextile sheet 169
Polythene sheet, waterproof and protective layers 327
Polyurethane
  spray – roof covering 115
  varnish – painting 323
Portal frames, steelwork 242

Portland whitbed
  ashlar masonry 315, 491
  copings 315, 491
  corbels 315, 491
  keystones 315, 491
Post and rail fencing, permanent fencing 368
Post-tensioned prestressing 195
Power
  floated floor finishes 125
  installation, electrical 130
Precast concrete
  beams 198, 471
  beanie block kerb/drainage block 430
  Beany block kerb/drainage block 284
  bench seats 441
  bollards 441
  cable tiles 445
  channels 283, 429
  copings 199
  crib walling 416
  culverts, cattle creeps, subway units 472
  deckings 199
  drop kerbs 283, 429
  edgings 283, 429
  Ellis Trief safety kerb 284, 430
  kerb quadrants 283, 429
  kerbs 283, 429
  large bore pipes 209
  lighting bracket arms 443
  lighting columns 443
  linings to tunnels 301
  manholes 528
  paving blocks 286, 434
  paving flags 286, 436
  pipes 209
  porous pipes BS 5911 209
  prestressed pretensioned beams 198
  road gulley 234
  road lighting columns 443
  Safeticurb channels 284, 429
  slabs 199, 436
  sleepers 295
Preface ix
Preformed segmental linings, tunnels 302
Preliminaries
  General Purpose Plant 60
  Other Charges and Fees 59
  Worked Example 63
Preliminaries and general items 361
  Check List of Items 53

  General Requirementsand Method Related Items 53
Preparation of surfaces, prior to landscaping 171, 495
Pressure grouting 303
Prestressed concrete
  decking 199
  post-tensioned 195
  pretensioned beams 198, 471
Pricing the tender, outline of the tendering and estimating process 11
Profit and oncosts 49
Protection
  coatings 478
  steelwork 478
Provision and placing of concrete
  access shaft linings 301
  bases 469
  blinding 469
  columns, piers and beams 470
  slabs 282
  suspended slabs 469
  tunnel linings 301
  walls 469
Purlins, steelwork 242
Purpose and content of book 3
PVC-u
  drainage above ground 335
  drainage below ground 217
  gutters 114
  marker tape 445
  perforated pipes 217
  pipes 217
  rainwater pipes 114
  Ultrarib pipes 220

Quantities check, outline of the tendering and estimating process 10
Quarry products, drilling and blasting rock 177

Rail track 293
  Decauville track 299
  foundations 294
  lifting, packing, slewing 294
  permanent way 293
  possessioncosts 293
  standard gauge track 295, 297
  sundries 297
  taking up 294

Rainwater goods
  gutters 114, 334
  pipes 114, 334
Raise level of lid, water stop tap boxes, covers and frames 395
Rebates and chases
  common brickwork 307
  engineering brickwork 311
  facing brickwork 309
Reconstituted stone walls 314, 490
Red tinted sand asphalt, waterproofing structures 480
Redundant drains, grout drains and sewers 396
Reflectorised, traffic signs 287
Reinforced concrete, internal walls 122
Reinforcement
  fabric 107, 193, 282, 425, 476
  fixing (outputs) 526
  high yield steel bars 192, 270, 475
  mild steel bars 107, 191, 270, 475
  slope reinforcement 411
  stainless steel bars 192, 474
Reinstatement of pavements 396
Removal
  pipework and services 365
  road markings 291
Render, wall finishes 120, 124
Resources, plant 85
Risk/Opportunity, Oncosts and profit 49
Road markings, arrows, letters, lines 290, 439
Road studs 292
roads 97
  bridgeworks 96
  communications 448
  construction 279–282, 424, 426
  labour gang rate 276
  Lay-bys 99
  lighting 443, 445, 447
  motorways 97
  plant rate 278
  road markings 439
Rock
  bolts 303
  gabions 95, 335
  imported fill material 167
  other than hard 164, 167
  preparation of filled surfaces 168
  preparation of surfaces 168
  removal 177
  treatment of filled surfaces 168

Rolled asphalt, flexible pavements 280, 424
Roller shutter doors 120
Roof
  Cladding 244
  cladding finishes 115
  concrete flat roof 111
  covering 112
  felt 112, 116
  flat roof decking and finishes 111
  roof claddings 112
  roof over/under lays 112
  rooflights/patent glazing and glazed work 112
  softwood flat roof 110
  softwood trussed pitch roof 110
  steel trussed pitch roof 111
  tiling and slating 113
  waterproof finishes 115
Rotary drilled boreholes, Ground investigation 145
Rubble walling, masonry 315, 491
Runways, concrete 425

Safeticurb channels, precast concrete 284, 429
Safety barriers
  remove from store and re-erect 374
  tensioned beam safety fence 371
  untensioned beam safety fence 371
  wire rope safety fence 372
Sand
  blasting of steelwork 478
  cable bed and cover 445
Sanitary Fittings
  cold water installations 129
  hot water installations 129
  sanitary installation 129
  Soil and waste pipework 129
scaffolding, access scaffolding 101
Scarify surface, pavements 281, 427
Scour and erosion protection, highways 412
Screed, sand and cement 125, 330, 345
Seal chambers 396
Sealing of gaps 484
security, Closed circuit TV 132
Segmental arches, facing brickwork 309, 489
Selected fill, imported fill material 168
Service ducts
  clay cable ducts 381
  reinstatement of pavements 396
Servicing of plant, daywork 509

Sewer renovation
  preparation 337
  renovation 337
  stabilization 336
Sheet steel piling 265, 463
shop fronts 121
Shot blasting
  external walls 120
  steelwork 479
shrubs, planting 172
Shutters and grilles 121
Signs, traffic 436
Site clearance 363
site tests, Ground investigation 146
Site visit, outline of the tendering and estimating
    process 8
Slab
  formwork (outputs) 527
  provision and placing of concrete 282, 469
Slate roof coverings 114
Sleepers
  precast concrete 295
  timber 295
Slurry sealing
  flexible pavements 280
  road construction 426
Soft spots
  drainage below ground 237
  highways 413
  manholes and voids 394
Softwood
  decking 251
  flat and pitched roofs 339
  stress graded 250
  structural and carcassing 338
  wrought, tanalized 340
Soil and waste pipework 129
Sound
  absorptive barriers 367
  reduction quilt under slab 108
Soundproof, partitions 123
Spacer blocks, cast iron for rail 296
Specification, initial appraisal 7
sports pitches
  cricket 101
  football 100
  golf 101
  tennis 101
Stainless Steel
  bar reinforcement 192, 474

entrance screens 121
straps 254
Stairs
  finishes/balustrading 117
  handrails 117
  Metal construction 117
  reinforced concrete construction 116
  steelwork 245
  timber construction 116
  treads and riser finishings 117
Standard signs
  internally illuminated 437
  reflectorised 437
Steel
  access stairs 117
  bearing piles 451
  bench seats 441
  bollards 441
  bracket arms for road lighting 443
  cladding 119, 246
  coach screws 255
  coated corrugated culverts 236
  columns for road lighting 443
  connectors 255
  external doors 120
  flitch plates 254
  gates 368
  internal walls 123
  lifting traffic barrier 441
  parapets 373
  roof decking 111
  shear plate connections 254
  spikes 252
  straps 253
  support arches 303
  timber connections 254
  trussed pitched roofs 111
  tubular piles 459
  windows 120
steel Bearing Piles 271
Steelwork
  anchorages 242
  beams 242
  blast cleaning 243
  box girder main members 477
  bracings 242
  built-up girders 477
  cladding rails 242
  columns 242
  concrete encased beams 108

concrete encased columns 108
Durbar flooring 245
epoxy painting 243
erection 242
  for structures 477
frames 477
galvanizing 243, 479
grit blasting 479
handrails 245
holding down bolts 242
landings 245
miscellaneous framing 245
open grid flooring 247
plate flooring 245
plate girder main members 477
portal frames 242
priming 243
protective system 479, 500
purlins 242
rail track 294
rolled section main members 477
sand/shot blasting 479
site bolts 242
stairs 245
surface preparation 500
towers 242
trestles 242
trial erection 242
trusses 242
Stelduct 245
Stone, roof covering 114
Stonework
  ashlar, Portland Whitbed 315
  cobble paving 435
  dry rubble 558
  granite setts 287, 434
  reconstituted stone 314, 490
  rubble walling 315, 491
Street furniture
  bench seats 441
  bollards 441
  lifting traffic barrier 441
Stress graded softwood components 250
strip, foundations 104
Structural concrete 179, 468
  formwork 188
  post-tensioned 196
  precast concrete 198
  prestressing 196
  reinforcement 191

Subbases
  granular material 279
  hardcore 279
  pavements 285, 423
  roads and paving 279
  wet mix macadam 280
Submission of Tender, Outline of the Tendering and Estimating Process 12
Substructure
  column bases 105
  strip foundations 104
  trench fill foundations 104
Supalux, ceilings 128
Support and stabilization of tunnels 303
Surface treatment, pavements 426
Suspended slabs
  placing of prestressed concrete 184
  placing of reinforced concrete 184
  provision and placing of concrete 470

Tables and memoranda 537
Tack coat, bituminous spray 427
Tanking, asphalt 328
Tanks, covered and uncovered 245
Temporary
  accommodation 361
  fencing 367
Temporary Works
  bailey bridges 102
  coffer dams 102
  earthwork support 103
Tender summary, oncosts and profit 51
Tensar 411
Tensioned beam safety barriers 371
Terminal blocks, motorway communications 449
Terram 411
Thermoplastic road markings 290
Thermoplastic screed, road markings 289
tile
  ceramic wall tiles 125
  roofing 114
Timber
  bench seats 441
  boarding roof covering 113
  chipboard flooring 109
  crib walling 416
  external doors 120
  fencing 368
  field gate 368

floor finishes 125
gates and stiles 368
hardwood 249
roofs 110
sleepers 295
stairs 116
stud partitions 121
windows 120
Timber piles 261, 271
Timesaver cast iron pipe 212
Topsoil
  highways 414
  preparation of filled surfaces 168
  soil 168
  trimming of filled surfaces 168
Traffic signal installations 292
Traffic signals, installation 440
Traffic signs and road markings 436
  marker posts 441
  standard reflectorised signs 437
  traffic signals 436
Treatment
  earthworks 168
  excavated surfaces 164
Trees
  guards 173, 496
  planting 173, 496
  removal 155
  stakes 173, 496
  stump removal 155
  supports 173, 496
  Vegetation Clearance 155
trench fill, foundations 104
Trenches and cabling, road lighting 445
Trestles, steelwork 242
Trial holes, pits and trenches, Ground investigation 144
Trimming
  earth 168
  edges of pavements 281, 426
  rock 168
  topsoil 168
Trussed roofs
  steel 111
  timber 111
Trusses, steelwork 242
Tubular guard railing, pedestrian guard railing 376
tunnelling field costs 97
Tunnels 300

excavation 300
in situ concrete linings 301
preformed segmental linings 302
support and stabilization 303
Turf, laying imported turf 171, 496
Turnouts 296
  taking up 294
Tyrolean, ceiling finish 128

Untensioned beam safety barrier 333, 371
Useful addresses for further information 529
Utilization factors, plant costs 88

Value Added Tax, oncosts and profit 51
Value Added Tax 27
Valves and penstocks 222
Vapour barrier 112–113, 341
Varnish, polyurethane 323
Vegetation Clearance
  Stumps 155
  trees 155
Ventilation, systems 130
Verynyl strip, road markings 291
Vibro Replacement, Ground stabilization 152

Wallpaper 125
Walls
  dry lining/sheet panelling 124
  dry plasterboard lining 124
  reinforced concrete walls 122
  timber or metal stud partitions and doors 121
  walls other finishes 124
Watercourse, lining, highways 414
Waterproofing 326, 480
  pea gravel protective layer 330
  polythene sheet 327
  protection board 330
  sand protective layer 330
  screed 330
  sprayed/ brushed 330
Waterstops
  concrete 194
  movement joints 484
WC/changing cubicles 123
Weed killer 495
Welfare facilities, daywork 508
Wet mix macadam
  pavements subbases 423
  road subbases 280

Windows
　metal 342
　PVC-u 342
　shutters and grills 121
　timber 342
Windows and external doors 342
　softwood windows and external doors 120
　steel roller shutters 120
　steel windows and external doors 120
Wire rope safety fence safety barriers 332
Wire rope safety fences 372
Wooden post and wire temporary fencing 367
Woodwool roof decking, ceiling finish 128

ESSENTIAL READING FROM TAYLOR AND FRANCIS

Damage to Concrete Structures

De Schutter

Unlike the more specialist books which deal with diagnosis techniques or non-destructive testing, with repair and strengthening of concrete structures or with specific degradation mechanisms such as sulfate attack or alkali-silica reaction, this book describes and explains the different types of damage to concrete structures comprehensively. It is written as a textbook for undergraduate and masters students, and is also helpful to practitioners such as design engineers, architects or consultants.

November 2012: 234x156: 204pp
Pb: 978-0-415-60388-1: **£35.00**

To Order: Tel: +44 (0) 1235 400524 **Fax:** +44 (0) 1235 400525
or Post: Taylor and Francis Customer Services,
Bookpoint Ltd, Unit T1, 200 Milton Park, Abingdon, Oxon, OX14 4TA UK
Email: book.orders@tandf.co.uk

For a complete listing of all our titles visit:
www.tandf.co.uk

Taylor & Francis
Taylor & Francis Group

ESSENTIAL READING FROM TAYLOR AND FRANCIS

Reinforced and Prestressed Concrete Design to EC2

2nd Edition

Eugene O'Brien et al

Fully revised to Eurocode 2, this second edition retains the original's emphasis on qualitative understanding of the overall behaviour of concrete structures. With a new chapter dedicated to case studies, worked examples, and exercise examples, it is an even more comprehensive guide to conceptual design, analysis, and detailed design of concrete structures, providing civil and structural engineering students with complete coverage of the analysis and design of reinforced and prestressed concrete structures.

January 2012: 246x189: 236pp
Hb: 978-0-415-57194-4: **£99.00**
Pb: 978-0-415-57195-1: **£34.99**

To Order: Tel: +44 (0) 1235 400524 **Fax:** +44 (0) 1235 400525
or Post: Taylor and Francis Customer Services,
Bookpoint Ltd, Unit T1, 200 Milton Park, Abingdon, Oxon, OX14 4TA UK
Email: book.orders@tandf.co.uk

For a complete listing of all our titles visit:
www.tandf.co.uk

Taylor & Francis
Taylor & Francis Group

ESSENTIAL READING FROM TAYLOR AND FRANCIS

Self-Consolidating Concrete
Applying What we Know

Daczko

Self-consolidating, or self-compacting, concrete (SCC) uses mixture proportioning techniques and admixture technology to create a concrete mixture with increased fluidity, thereby avoiding the need for mechanical consolidation/vibration. SCC mixtures with various performance levels can be produced in the laboratory, but the know-how is sometimes lacking to do this on site. This book presents a foundation for consistently producing and applying SCC in a daily production environment. It links science with practical application and will be useful for both new practitioners and those already using SCC. The author uses stories and case studies throughout as examples of specific types of applications.

February 2012: 234x156: 304pp
Hb: 978-0-415-59064-8: **£50.00**

To Order: Tel: +44 (0) 1235 400524 **Fax:** +44 (0) 1235 400525
or Post: Taylor and Francis Customer Services,
Bookpoint Ltd, Unit T1, 200 Milton Park, Abingdon, Oxon, OX14 4TA UK
Email: book.orders@tandf.co.uk

For a complete listing of all our titles visit:
www.tandf.co.uk

Taylor & Francis
Taylor & Francis Group

SPON'S PRICEBOOKS 2013

Spon's Architects' and Builders' Price Book 2013
DAVIS LANGDON

The most detailed, professionally relevant source of UK construction price information currently available anywhere.

What's new in this year's Spon's A&B Price Book? There's a new cost model for an office fit out. And new measured works: Concrete canvas, Dreadnought Handformed roof tiles, UPVC external doors, Thermalite paint grade concrete blocks, EPS/PIR hybrid roof insulation, Omnidec composite flooring systems, sheep wool insulation, Kingspan TEK cladding panels, 750 & 900 dia CFA piling, Lignacite architectural blocks. The book also provides more on cork flooring. Plus lots else has been changed throughout.

Hbk & Spon's Online
800pp approx.: 978-0-415-69077-5: **£150**

ebook & Spon's Online
978-0-203-07708-5: **£150**
(inc. sales tax where appropriate)

Spon's Mechanical and Electrical Services Price Book 2013
DAVIS LANGDON ENGINEERING SERVICES

Our M&E price book continues to be the most comprehensive and best annual services engineering price book currently available, providing detailed pricing information across the full range of mechanical and electrical services, together with higher-level costs for a diverse range of systems and different building applications. This year sees very slight increases in material and labour rates. The book keeps its focus on CO2 control and renewables. Feed-In Tariffs have been brought up to date with the current rates and processes.

Hbk & Spon's Online
864pp approx.: 978-0-415-69080-5: **£150**

ebook & Spon's Online
978-0-203-07705-4: **£150**
(inc. sales tax where appropriate)

Spon's External Works and Landscape Price Book 2013
DAVIS LANGDON

Now in its 32nd edition, the External Works and Landscape Price Book provides a special focus on water related issues. It remains the indispensable source of UK cost information for detailed external works and landscaping. This year brings rainwater harvesting tanks; Cellweb load support systems and erosion control; outdoor rooms – with several design options of home-offices and garden studios; green roofs – updated systems of green roof treatments and installations; and permeable pavings – permeable macadams, bricks and more.

Hbk & Spon's Online
720pp approx.: 978-0-415-69079-9: **£125**

ebook & Spon's Online
978-0-203-07706-1: **£125**
(inc. sales tax where appropriate)

Spon's Civil Engineering and Highway Works Price Book 2013
DAVIS LANGDON

With pressure through the National Infrastructure Plan to deliver "value for money", accurate pricing counts. This new edition of the Civils and Highways Price Book reflects the moderate overall price rises of the last year, and the higher than average price increases of anything affected by crude oil costs. This is more than just a price book. It provides a comprehensive work manual for the UK's civil engineering, surveying and construction business.

Hbk & Spon's Online
752pp approx.: 978-0-415-69078-2: **£160**

ebook & Spon's Online
978-0-203-07707-8: **£160**
(inc. sales tax where appropriate)

Receive *Spon's Online* free when you order any hard copy or ebook Spon 2013 Price Book. Visit **www.pricebooks.co.uk**

To Order: Tel: +44 (0) 1235 400524 **Fax:** +44 (0) 1235 400525
or Post: Taylor and Francis Customer Services, Bookpoint Ltd, Unit T1, 200 Milton Park, Abingdon, Oxon, OX14 4SB, UK
Email: book.orders@tandf.co.uk
A complete listing of all our books is on: **www.sponpress.com**

Spon Press
an imprint of Taylor & Francis

Electronic Version Single-User Licence Agreement

We welcome you as a user of this Spon Price Book Software and hope that you find it a useful and valuable tool. Please read this document carefully. **This is a legal agreement** between you (hereinafter referred to as the "Licensee") and Taylor and Francis Books Ltd. (the "Publisher"), which defines the terms under which you may use the Product. **By accessing and retrieving the access code on the label inside the back cover of this book you agree to these terms and conditions outlined herein. If you do not agree to these terms you must return the Product to your supplier intact, with the seal on the label unbroken and with the access code not accessed.**

1. **Definition of the Product**
   The product which is the subject of this Agreement, (the "Product") consists of internet access for searching and viewing the price book data from *Spon's Civil Engineering and Highway Works Price Book 2013* online.

2. **Commencement and licence**
   2.1 This Agreement commences upon the breaking open of the document containing the access code by the Licensee (the "Commencement Date").
   2.2 This is a licence agreement (the "Agreement") for the use of the Product by the Licensee, and not an agreement for sale.
   2.3 The Publisher licenses the Licensee on a non-exclusive and non-transferable basis to use the Product on condition that the Licensee complies with this Agreement. The Licensee acknowledges that it is only permitted to use the Product in accordance with this Agreement.

3. **Multiple use**
   For more than one user or for a wide area network or consortium, use of the Product is only permissible with the purchase from the Publisher of a multiple-user licence and adherence to the terms and conditions of that licence.

4. **Installation and Use**
   4.1 The Licensee may provide access to the Product for individual study in the following manner: The Licensee may install the Product on a secure local area network on a single site for use by one user.
   4.2 The Licensee shall be responsible for installing the Product and for the effectiveness of such installation.
   4.3 Text from the Product may be incorporated in a coursepack. Such use is only permissible with the express permission of the Publisher in writing and requires the payment of the appropriate fee as specified by the Publisher and signature of a separate licence agreement.
   4.4 The Product is a free addition to the book and the Publisher is under no obligation to provide any technical support.

5. **Permitted Activities**
   5.1 The Licensee shall be entitled to use the Product for its own internal purposes;
   5.2 The Licensee acknowledges that its rights to use the Product are strictly set out in this Agreement, and all other uses (whether expressly mentioned in Clause 6 below or not) are prohibited.

6. **Prohibited Activities**
   The following are prohibited without the express permission of the Publisher:
   6.1 The commercial exploitation of any part of the Product.
   6.2 The rental, loan, (free or for money or money's worth) or hire purchase of this product, save with the express consent of the Publisher.
   6.3 Any activity which raises the reasonable prospect of impeding the Publisher's ability or opportunities to market the Product.
   6.4 Any networking, physical or electronic distribution or dissemination of the product save as expressly permitted by this Agreement.
   6.5 Any reverse engineering, decompilation, disassembly or other alteration of the Product save in accordance with applicable national laws.
   6.6 The right to create any derivative product or service from the Product save as expressly provided for in this Agreement.
   6.7 Any alteration, amendment, modification or deletion from the Product, whether for the purposes of error correction or otherwise.

7. **General Responsibilities of the License**
   7.1 The Licensee will take all reasonable steps to ensure that the Product is used in accordance with the terms and conditions of this Agreement.
   7.2 The Licensee acknowledges that damages may not be a sufficient remedy for the Publisher in the event of breach of this Agreement by the Licensee, and that an injunction may be appropriate.
   7.3 The Licensee undertakes to keep the Product safe and to use its best endeavours to ensure that the product does not fall into the hands of third parties, whether as a result of theft or otherwise.
   7.4 Where information of a confidential nature relating to the product of the business affairs of the Publisher comes into the possession of the Licensee pursuant to this Agreement (or otherwise), the Licensee agrees to use such information solely for the purposes of this Agreement, and under no circumstances to disclose any element of the information to any third party save strictly as permitted under this Agreement. For the avoidance of doubt, the Licensee's obligations under this sub-clause 7.4 shall survive the termination of this Agreement.

8. **Warrant and Liability**
   8.1 The Publisher warrants that it has the authority to enter into this agreement and that it has secured all rights and permissions necessary to enable the Licensee to use the Product in accordance with this Agreement.
   8.2 The Publisher warrants that the Product as supplied on the Commencement Date shall be free of defects in materials and workmanship, and undertakes to replace any defective Product within 28 days of notice of such defect being received provided such notice is received within 30 days of such supply. As an alternative to replacement, the Publisher agrees fully to refund the Licensee in such circumstances, if the Licensee so requests, provided that the Licensee returns this copy of *Spon's Civil Engineering and Highway Works Price Book 2013* to the Publisher. The provisions of this sub-clause 8.2 do not apply where the defect results from an accident or from misuse of the product by the Licensee.
   8.3 Sub-clause 8.2 sets out the sole and exclusive remedy of the Licensee in relation to defects in the Product.
   8.4 The Publisher and the Licensee acknowledge that the Publisher supplies the Product on an "as is" basis. The Publisher gives no warranties:
      8.4.1 that the Product satisfies the individual requirements of the Licensee; or
      8.4.2 that the Product is otherwise fit for the Licensee's purpose; or
      8.4.3 that the Product is compatible with the Licensee's hardware equipment and software operating environment.
   8.5 The Publisher hereby disclaims all warranties and conditions, express or implied, which are not stated above.
   8.6 Nothing in this Clause 8 limits the Publisher's liability to the Licensee in the event of death or personal injury resulting from the Publisher's negligence.
   8.7 The Publisher hereby excludes liability for loss of revenue, reputation, business, profits, or for indirect or consequential losses, irrespective of whether the Publisher was advised by the Licensee of the potential of such losses.
   8.8 The Licensee acknowledges the merit of independently verifying the price book data prior to taking any decisions of material significance (commercial or otherwise) based on such data. It is agreed that the Publisher shall not be liable for any losses which result from the Licensee placing reliance on the data under any circumstances.
   8.9 Subject to sub-clause 8.6 above, the Publisher's liability under this Agreement shall be limited to the purchase price.

9. **Intellectual Property Rights**
   9.1 Nothing in this Agreement affects the ownership of copyright or other intellectual property rights in the Product.
   9.2 The Licensee agrees to display the Publishers' copyright notice in the manner described in the Product.
   9.3 The Licensee hereby agrees to abide by copyright and similar notice requirements required by the Publisher, details of which are as follows:
   "© 2013 Taylor & Francis. All rights reserved. All materials in *Spon's Civil Engineering and Highway Works Price Book 2013* are copyright protected. All rights reserved. No such materials may be used, displayed, modified, adapted, distributed, transmitted, transferred, published or otherwise reproduced in any form or by any means now or hereafter developed other than strictly in accordance with the terms of the licence agreement enclosed with *Spon's Civil Engineering and Highway Works Price Book 2013*. However, text and images may be printed and copied for research and private study within the preset program limitations. Please note the copyright notice above, and that any text or images printed or copied must credit the source."

9.4 This Product contains material proprietary to and copyedited by the Publisher and others. Except for the licence granted herein, all rights, title and interest in the Product, in all languages, formats and media throughout the world, including copyrights therein, are and remain the property of the Publisher or other copyright holders identified in the Product.

10. **Non-assignment**
This Agreement and the licence contained within it may not be assigned to any other person or entity without the written consent of the Publisher.

11. **Termination and Consequences of Termination.**
    11.1 The Publisher shall have the right to terminate this Agreement if:
    11.1.1 the Licensee is in material breach of this Agreement and fails to remedy such breach (where capable of remedy) within 14 days of a written notice from the Publisher requiring it to do so; or
    11.1.2 the Licensee becomes insolvent, becomes subject to receivership, liquidation or similar external administration; or
    11.1.3 the Licensee ceases to operate in business.
    11.2 The Licensee shall have the right to terminate this Agreement for any reason upon two month's written notice. The Licensee shall not be entitled to any refund for payments made under this Agreement prior to termination under this sub-clause 11.2.
    11.3 Termination by either of the parties is without prejudice to any other rights or remedies under the general law to which they may be entitled, or which survive such termination (including rights of the Publisher under sub-clause 7.4 above).
    11.4 Upon termination of this Agreement, or expiry of its terms, the Licensee must destroy all copies and any back up copies of the product or part thereof.

12. **General**
    12.1 *Compliance with export provisions*
    The Publisher hereby agrees to comply fully with all relevant export laws and regulations of the United Kingdom to ensure that the Product is not exported, directly or indirectly, in violation of English law.
    12.2 *Force majeure*
    The parties accept no responsibility for breaches of this Agreement occurring as a result of circumstances beyond their control.
    12.3 *No waiver*
    Any failure or delay by either party to exercise or enforce any right conferred by this Agreement shall not be deemed to be a waiver of such right.
    12.4 *Entire agreement*
    This Agreement represents the entire agreement between the Publisher and the Licensee concerning the Product. The terms of this Agreement supersede all prior purchase orders, written terms and conditions, written or verbal representations, advertising or statements relating in any way to the Product.
    12.5 *Severability*
    If any provision of this Agreement is found to be invalid or unenforceable by a court of law of competent jurisdiction, such a finding shall not affect the other provisions of this Agreement and all provisions of this Agreement unaffected by such a finding shall remain in full force and effect.
    12.6 *Variations*
    This agreement may only be varied in writing by means of variation signed in writing by both parties.
    12.7 *Notices*
    All notices to be delivered to: Spon's Price Books, Taylor & Francis Books Ltd., 3 Park Square, Milton Park, Abingdon, Oxfordshire, OX14 4RN, UK.
    12.8 *Governing law*
    This Agreement is governed by English law and the parties hereby agree that any dispute arising under this Agreement shall be subject to the jurisdiction of the English courts.

If you have any queries about the terms of this licence, please contact:

Spon's Price Books
Taylor & Francis Books Ltd.
3 Park Square, Milton Park, Abingdon, Oxfordshire, OX14 4RN
Tel: +44 (0) 20 7017 6000
www.sponpress.com

Spon Press
Taylor & Francis Group

Multiple-user use of the Spon Press Software and eBook

To buy a licence to install *Spon's Online* on a secure local area network or a wide area network, and for the supply of network key files, for an agreed number of users please contact:

Spon's Price Books
Taylor & Francis Books Ltd.
2 Park Square, Milton Park, Abingdon, Oxfordshire, OX14 4RN
Tel: +44 (0) 207 017 6000
www.pricebooks.co.uk

Number of users	Licence cost
2–5	£450
6–10	£915
11–20	£1400
21–30	£2150
31–50	£4200
51–75	£5900
76–100	£7100
Over 100	Please contact Spon for details

Spon's Online Installation and Use Instructions

General

Spon's Online is available via the web using the given access code as outlined below.

No CD-ROM is supplied. For set up and use you will need internet access.

Setting up

Open your internet browser, type in www.pricebooks.co.uk/downloads and follow the instructions on screen.

Retrieve your unique access code by scratching off the security-protected label inside the back cover of this book.

Type in your access code as directed.

When your access code is validated successfully, please give your email address and choose a memorable password to register yourself.

You will be sent an activation email instantly and can begin to use *Spon's Online*.

Please note: you will only be allowed one use of the access code and one opportunity to register, although you will be able to change your password later if you wish.

Use

The set-up process will give you access via www.pricebooks.co.uk/SponsOnline through your regular internet browser.

Technical Support

Spon's Online is a free addition to the book.

For help with setting up please visit
www.pricebooks.co.uk/esupport

Guidance on use is provided on
www.pricebooks.co.uk/SponsOnline

All materials in *Spon's Online*, which is supplied with *Spon's Civil Engineering and Highway Works Price Book 2013*, are copyright protected. No such materials may be used, displayed, modified, adapted, distributed, transmitted, transferred, published or otherwise reproduced in any form or by any means now or hereafter developed other than strictly in accordance with the terms of the attached Electronic Version Single-User Licence Agreement.

Spon's Online is furnished under the attached Electronic Version Single-User Licence Agreement. The software may be used only in accordance with the terms of this Agreement, unless the additional multi-user licence agreement is purchased from Electronic Sales, Spon Press, 2 Park Square, Milton Park, Abingdon, Oxon OX14 4RN Tel: +44 (0) 20 7017 6000

© COPYRIGHT ALL RIGHTS RESERVED

Free Updates

with three easy steps...

1. Register today on www.pricebooks.co.uk/updates

2. We'll alert you by email when new updates are posted on our website

3. Then go to www.pricebooks.co.uk/updates
   and download the update.

All four Spon Price Books – *Architects' and Builders'*, *Civil Engineering and Highway Works*, *External Works and Landscape* and *Mechanical and Electrical Services* – are supported by an updating service. Two or three updates are loaded on our website during the year, typically in November, February and May. Each gives details of changes in prices of materials, wage rates and other significant items, with regional price level adjustments for Northern Ireland, Scotland and Wales and regions of England. The updates terminate with the publication of the next annual edition.

As a purchaser of a Spon Price Book you are entitled to this updating service for this 2013 edition – free of charge. Simply register via the website www.pricebooks.co.uk/updates and we will send you an email when each update becomes available.

If you haven't got internet access or if you've some questions about the updatees please write to us at Spon Price Book Updates, Spon Press Marketing Department, 3 Park Square, Milton Park, Abingdon, Oxfordshire, OX14 4RN.

Find out more about Spon books
Visit www.sponpress.com for more details.